Structure and Concentration of Point Defects in Selected Spinels and Simple Oxides

Structure and Concentration of Point Defects in Selected Spinels and Simple Oxides

Andrzej Stokłosa and Stefan S. Kurek

CRC Press
Taylor & Francis Group
Boca Raton London New York

CRC Press is an imprint of the
Taylor & Francis Group, an **informa** business

First edition published 2021
by CRC Press
6000 Broken Sound Parkway NW, Suite 300, Boca Raton, FL 33487-2742

and by CRC Press
2 Park Square, Milton Park, Abingdon, Oxon, OX14 4RN

Library of Congress Cataloging-in-Publication Data
Names: Stokłosa, Andrzej, author. | Kurek, Stefan S., author.
Title: Structure and concentration of point defects in selected spinels and simple oxides / Andrzej Stokłosa and Stefan S. Kurek.
Description: First edition. | Boca Raton : CRC Press, 2021. | Includes bibliographical references and index.
Identifiers: LCCN 2020046258 (print) | LCCN 2020046259 (ebook) | ISBN 9780367617127 (hardback) | ISBN 9781003106166 (ebook)
Subjects: LCSH: Spinel group. | Point defects. | Crystallography.
Classification: LCC QD947 .S77 2021 (print) | LCC QD947 (ebook) | DDC 549/.526--dc23
LC record available at https://lccn.loc.gov/2020046258
LC ebook record available at https://lccn.loc.gov/2020046259

ISBN: 978-0-367-61712-7 (hbk)
ISBN: 978-0-367-61715-8 (pbk)
ISBN: 978-1-003-10616-6 (ebk)

Typeset in Times
by SPi Global, India

Contents

PART IV Diagrams of Concentrations of Point Defects in Oxide Solid Solutions $(Fe_{1-x}M_x)_{1-\delta}O$ (where M = Mn and Co)

Preface

Transport properties, as well as electrical, magnetic and optical properties of metal oxides, subject of this monograph, depend on their chemical composition and can be modified by doping with other ions; however, they may be governed by the type and concentration of point defects. The theory of point defects, developed in 1930s by Frenkel, Schottky and Wagner, was intensively developed in the second half of the 20th century. The problem of structure and concentration of point defects in non-stoichiometric compounds, especially in metal oxides, was the object of numerous experimental and theoretical studies. They have shown that the technology of preparation affects the concentration of point defects; thus, it may determine material properties. Despite the development of the theory of point defects, the interpretation of the results of the studies of the deviation from the stoichiometry, resulting from the concentration of ionic point defects, was very difficult without the method for the determination of diagrams of defects' concentrations, showing their relative concentrations depending on the pressure of the oxidant. For some systems, the use of a simplified method developed by Brouwer, and further by Kröger and Vink, enabled plotting incomplete diagrams of point defects, based on the theoretically calculated equilibrium constants of reactions of formation of point defects. The author proposed a new method for the determination of diagrams of point defects (Stokłosa 2015), employing interdependence of defect reactions. It consists of sequential adjustments of standard Gibbs energies of formation of defects for the considered model of defects and calculations of their concentration. The match between the calculated dependence of the deviation from the stoichiometry on the oxygen pressure and the experimental values of the deviation permitted determination of the concentration of dominating point defects present in cation and anion sublattices, as well as the maximum concentration of minority defects. The work (Stokłosa 2015) contains the results of calculations of the diagrams of concentrations of point defects for oxides: $Ni_{1-\delta}O$, $Co_{1-\delta}O$, $Mn_{1-\delta}O$, $Fe_{1-\delta}O$, $Cu_{2-\delta}O$ and $TiO_{2-\delta}$, pure and doped with ions of other metals, using the results of the studies of the deviation from the stoichiometry obtained by different authors between 1950 and 2010. The performed calculations have shown that the method of calculation is correct, and they allowed for a complete interpretation of the results of the studies of the deviation from the stoichiometry and electrical conductivity. Above all, they fully confirmed the assumptions of the defects' theory, stating that the defects can be considered as quasi-particles following the laws of chemical thermodynamics. They permitted also to demonstrate, quantitatively, the effect of doping ions with the oxidation state different from that of the parent ion on the relative concentrations of point defects.

The monograph presents the problem of the structure and concentration of ionic and electronic defects in magnetite and hausmannite, pure and doped with M^{3+} cations (Ti^{3+} (Ti^{4+}), Cr^{3+}, Al^{3+}) and M^{2+} cations (Co^{2+}, Mn^{2+}, Co^{2+}/Mn^{2+}, $Co^{2+}/M^{2+}/Zn^{2+}$), and in spinels exhibiting magnetic properties and high electric conductance. Complete equations of formation of ionic and electronic defects, and defect equilibria in spinels, pure and doped with various ions. Based on the developed method and using the results of the studies on the deviation from stoichiometry in the analysed spinels, the standard Gibbs energies of formation of point defects were determined, which allowed the determination of joint defect concentrations (diagrams of Vince point defect concentrations). Employing the obtained concentrations of ionic and electronic defects, and self-diffusion coefficients and electric conductance from numerous literature sources, ion mobilities via defects and electronic defect mobilities in analysed spinels and oxides were determined.

Part I presents the results of calculations of diagrams of defect concentrations for magnetite ($Fe_{3\pm\delta}O_4$) and hausmannite ($Mn_{3\pm\delta}O_4$). The defect mobilities were determined by using their computed concentrations and self-diffusion coefficients of tracers. Part II presents the effect of doping M^{3+} and M^{2+} ions on the concentration of point defects in magnetite, and Part III, the same in hausmannite. Ion mobilities via cation vacancies and interstitial cations (diffusion coefficients of defects) in doped spinels were also determined with the use of tracer self-diffusion coefficients. Part IV presents the calculations of point defect diagrams in iron oxide (wüstite) doped with manganese and cobalt $(Fe_{1-x}Mn_x)_{1-\delta}O$ and $(Fe_{1-x}Co_x)_{1-\delta}O$, and the determination of ion mobilities based on tracer self-diffusion coefficients.

The calculation method is explained in detail in Chapter 1, where also structures of magnetite and wüstite are discussed.

We hope that the determined diagrams of concentrations of defects will be helpful in the interpretation of properties of analysed oxide systems and that they will encourage young scientists to study non-stoichiometric compounds with a complete interpretation.

Spinels and non-stoichiometric oxides have not lost their importance. On the contrary, a brief search in the Web of Science yields dozens of publications that in just two years have gained more than 100 citations, some exceeding 150. Now, most of them deal with the application of these materials, also in nanoparticle or nanosheet forms. As these materials are often characterised by the presence of vacancies, it is no surprise that many of them show superb catalytic properties. Materials with oxygen vacancies have proven to be excellent catalysts of oxygen evolution reaction (Kim et al. 2018, Liu et al. 2019), which also means they are applicable for the anode in electrolytic water splitting (Kim et al. 2018, Peng et al. 2018), but also for hydrogen production (Zeng et al. 2018), and photocatalytic water oxidation in combination with a light transducer (Zhang et al. 2018). They proved also superior in the catalysis of the opposite reaction, oxygen reduction, demonstrating to be outstanding as cathodes materials for Li batteries (Nayak et al. 2018, Yan et al. 2018, Zhan et al. 2018), and also for Zn-air (Hao et al. 2018). Interestingly, they also catalyse electrolytic reduction of N_2 to NH_3 (Liu et al. 2018, Zhang et al. 2019), and NOx reduction with ammonia (Han et al. 2019, Meng et al. 2018). Spinels, among them in the form of nanoparticles, have been used in supercapacitors (Li et al. 2019, Repp et al. 2018), their magneto-optical properties have also found application (Abraham et al. 2018, Slimani et al. 2018), as well as they are employed for electromagnetic absorption (Lan et al. 2020, Qin et al. 2020). Of great importance in these applications is electrical conductivity, concentrations of defects, including vacations, but also the equilibrium between dioxygen in the gas phase and oxide anions in oxides, the data that could be found in this monography, collected from the classical, old literature on the subject.

The authors are grateful to the Faculty of Chemical Engineering and Technology of Cracow University of Technology for financial support of this monography. Special thanks are due to Dr Katarzyna Pirowska for her help in linguistic editing of the text.

REFERENCES

Abraham, A. G., A. Manikandan, E. Manikandan, S. Vadivel, S. K. Jaganathan, A. Baykal, and P. S. Renganathan. 2018. Enhanced Magneto-Optical and Photo-Catalytic Properties of Transition Metal Cobalt (Co^{2+} Ions) Doped Spinel $MgFe_2O_4$ Ferrite Nanocomposites. *J. Magn. Magn. Mater.* 452:380–388.

Han, L. P., M. Gao, C. Feng, L. Y. Shi, and D. S. Zhang. 2019. Fe_2O_3-CeO_2-Al_2O_3 Nanoarrays on Al-Mesh as SO_2-Tolerant Monolith Catalysts for NO, Reduction by NH_3. *Environ. Sci. Technol.* 53:5946–5956.

Hao, J. W., J. Mou, J. W. Zhang, L. B. Dong, W. B. Liu, C. J. Xu and F. Y. Kang. 2018. Electrochemically Induced Spinel-Layered Phase Transition of Mn_3O_4 in High Performance Neutral Aqueous Rechargeable Zinc Battery. *Electrochim. Acta.* 259:170–178.

Kim, J. S., B. Kim, H. Kim, and K. Kang. 2018. Recent Progress on Multimetal Oxide Catalysts for the Oxygen Evolution Reaction. *Adv. Energy Mater.* 8: 1702774.

Lan, D., M. Qin, J. L. Liu, G. L. Wu, Y. Zhang, and H. J. Wu. 2020. Novel Binary Cobalt Nickel Oxide Hollowed-Out Spheres for Electromagnetic Absorption Applications. *Chem. Eng. J.* 382: 122797.

Lee, B. Y., C. T. Chu, M. Krajewski, M. Michalska, and J. Y. Lin. 2020. Temperature-Controlled Synthesis of Spinel Lithium Nickel Manganese Oxide Cathode Materials for Lithium-Ion Batteries. *Ceram. Int.* 46: 20856–20864.

Li, Y. M., X. Han, T. F. Yi, Y. B. He, and X. F. Li. 2019. Review and Prospect of $NiCo_2O_4$-Based Composite Materials for Supercapacitor Electrodes. *J. Energy Chem.* 31:54–78.

Liu, Q., X. X. Zhang, B. Zhang, Y. L. Luo, G. W. Cui, F. Y. Xie, and X. P. Sun. 2018. Ambient N_2 Fixation to NH_3 Electrocatalyzed by a Spinel Fe_3O_4 Nanorod. *Nanoscale.* 10:14386–14389.

Liu, Y., Y. R. Ying, L. F. Fei, Y. Liu, Q. Z. Hu, G. G. Zhang, S. Y. Pang et al. 2019. Valence Engineering via Selective Atomic Substitution on Tetrahedral Sites in Spinel Oxide for Highly Enhanced Oxygen Evolution Catalysis. *J. Am. Chem. Soc.* 141:8136–8145.

Meng, D. M., Q. Xu, Y. L. Jiao, Y. Guo, Y. L. Guo, L. Wang, G. Z. Lu, and W. C. Zhan. 2018. Spinel structured $Co_a Mn_b O_x$ Mixed Oxide Catalyst for the Selective Catalytic Reduction of NO_x with NH_3. *Appl. Catal. B-Environ.* 221:652–663.

Nayak, P. K., E. M. Erickson, F. Schipper, T. R. Penki, N. Munichandraiah, P. Adelhelm, H. Sclar et al. 2018. Review on Challenges and Recent Advances in the Electrochemical Performance of High Capacity Li- and Mn-Rich Cathode Materials for Li-Ion Batteries. *Adv. Energy Mater.* 8:1702397.

Peng, S. J., F. Gong, L. L. Li, D. S. Yu, D. X. Ji, T. R. Zhang, Z. Hu et al. 2018. Necklace-like Multishelled Hollow Spinel Oxides with Oxygen Vacancies for Efficient Water Electrolysis. *J. Am. Chem. Soc.* 140:13644–13653.

Qin, M., D. Lan, G. L. Wu, X. G. Qiao, and H. J. Wu. 2020. Sodium Citrate Assisted Hydrothermal Synthesis of Nickel Cobaltate Absorbers with Tunable Morphology and Complex Dielectric Parameters Toward Efficient Electromagnetic Wave Absorption. *Appl. Surf. Sci.* 504:144480,

Repp, S., E. Harputlu, S. Gurgen, M. Castellano, N. Kremer, N, Pompe, J. Worner. 2018. Synergetic Effects of Fe^{3+} Doped Spinel $Li_4Ti_5O_{12}$ Nanoparticles on Reduced Graphene Oxide for High Surface Electrode Hybrid Supercapacitors. *Nanoscale.* 10:1877–1884.

Slimani, Y., H. Gungunes, M. Nawaz, A. Manikandan, H. S. El Sayed, M. A. Almessiere, H. Sozeri et al. 2018. Magneto-Optical and Microstructural Properties of Spinel Cubic Copper Ferrites with Li-Al co-Substitution. *Ceram. Int.* 44:14242–14250.

Stokłosa, A. 2015. *Non-Stoichiometric Oxides of 3d Metals. (Mat. Sci. Foundations 79)*, Pfaffikon: Trans. Tech. Pub

Yan, P. F., J. M. Zheng, J. Liu, B. Q. Wang, X. P. Cheng, Y. F. Zhang, X. L. Sun, C. M. Wang, and J. G. Zhang. 2018. Tailoring Grain Boundary Structures and Chemistry of Ni-Rich Layered Cathodes for Enhanced Cycle Stability of Lithium-Ion Batteries. *Nat. Energy.* 3:600–605.

Zeng, D. W., Y. Qiu, S. Peng, C. Chen, J. M. Zeng, S. Zhang, and R. Xiao. 2018. Enhanced Hydrogen Production Performance through Controllable Redox Exsolution within $CoFeAlO_x$ Spinel Oxygen Carrier Materials. *J. Mater. Chem. A.* 6:11306–11316.

Zhan, C., T. P. Wu, J. Lu, and K. Amine. 2018. Dissolution, Migration, and Deposition of Transition Metal Ions in Li-Ion Batteries Exemplified by Mn-Based Cathodes – A Critical Review. *Energy Environ. Sci.* 11:243–257.

Zhang, L. Z., C. Yang, Z. L. Xi, and X. C. Wang. 2018. Cobalt Manganese Spinel as an Effective Cocatalyst for Photocatalytic Water Oxidation. *Appl. Catal. B-Environ.* 224:886–894.

Zhang, L., X. Y. Xie, H. B. Wang, L. Ji, Y. Zhang, H. Y. Chen, T. S. Li et al. 2019. Boosting Electrocatalytic N_2 Reduction by MnO_2 with Oxygen Vacancies. *Chem. Commun.* 55:4627–4630.

Authors

Andrzej Stokłosa is Professor Emeritus at Cracow University of Technology, Faculty of Chemical Engineering and Technology in Krakow, Poland. After receiving MSc degree in chemistry, at the Jagiellonian University in Krakow, he was employed at the AGH University of Mining and Metallurgy in Krakow in the Department of Solid State Chemistry in the group of Professor Stanisław Mrowec. He received his PhD degree in 1972, DSc (habilitation) degree in 1984. He moved to Cracow University of Technology in 1989 and became Head of Physical Chemistry Department as Associate Professor, and was promoted to full professor position in 1992. He is an author and co-author of more than 150 scientific papers, the author of books: Non-Stoichiometric Oxides of 3d Metals (Trans. Tech. Pub. Pfaffikon, Switzerland, 2015 (Mat. Sci. Foundations 79 (2015)). "Basic Phenomenological and Statistical Thermodynamics", "Introduction to Physical Chemistry" and several other standard textbooks for students (in Polish). He has published in the area of solid-state materials associated with crystallochemistry; structure and thermodynamics of point defects in metal oxides and sulfides; the defects and electronic structure modification for the use in electrochemical intercalation; kinetics and mechanism of catalytic and electrocatalytic reactions.

Stefan S. Kurek is Assistant Professor at Cracow University of Technology in Krakow, Poland. He graduated from this University in 1974, and got his PhD in chemical technology there in 1982. He worked as postdoc in the UK, at the Universities of Birmingham and Bristol, on coordination compounds, including electron transfer from photoexcited porphyrins. Now, his research interests focus on molecular electrochemistry.

Notations

Kröger - Vink notation of defects in M_aO_b oxides

S_X^q	structural elements; S denotes the chemical unit (element) or point defect in sublattice (X). The electric charge q is referred to the perfect crystal (excess quantity (•) - positive, (′) - negative, (x) -neutral)
M_M^x	M^{n+} metal ion in its lattice (M) site (neutral)
O_O^x	O^{2-} oxygen ion in oxygen lattice site
$V_M^{n'}$	cation (M) vacancy with n effective charge
V_M^x	neutral metal (M) vacancy
$V_O^{\bullet\bullet}$	oxygen vacancy with double positive effective charge in relation to the lattice
$V_O^{\bullet}$	oxygen vacancy with positive effective charge in relation to the lattice
V_O^x	neutral oxygen vacancy
V_i	interstitial void in the lattice
$M_i^{n\bullet}$	interstitial metal (M) ion with n positive effective charge
M_i^x	neutral interstitial metal atom (M)
O_i''	interstitial oxygen ion with double negative charge
O_i'	interstitial oxygen ion with negative charge
O_i^x	neutral interstitial oxygen atom
$h^{\bullet}$	quasi-free electron hole in a crystal with positive effective charge (in the valence band or $M_M^{\bullet}$ metal ion with positive charge in its lattice (M) site or $O_O^{\bullet}$ single-valent oxygen ions (O^-) in its lattice site, frequently designated by p)
e'	quasi-free electron in a crystal (in the conduction band or M_M' metal ion with negative charge in its lattice (M) site, frequently designated by n)
A_M'	A^{n+} foreign metal ion with negative effective charge in relation to the lattice in M lattice site
$D_M^{\bullet}$	D^{n+} foreign metal ion with positive effective charge in a M lattice site

SYMBOLS

b_D	ratio of the coefficient of diffusion of ions via interstitial ions to the coefficient of diffusion of ions via vacancies $\left(D^o_{I(M)}/D^o_{V(M)}\right)$
$b_{e/h}$	ratio of the mobility of electrons to the mobility of electron holes (μ_e/μ_h)
C_2'	defect complexes $(2{:}1)'-\left\{\left(V_M''\right)_2 M_i^{3\bullet}\right\}'$ with negative effective charge
$C_4^{5'}$	defect complexes $(4{:}1)^{5'}-\left\{\left(V_M''\right)_4 M_i^{3\bullet}\right\}^{5'}$ with fire negative effective charge
$C_6^{6'}$	defect complexes $(6{:}2)^{6'}-\left\{\left(V_M''\right)_6 \left(M_i^{3\bullet}\right)_2\right\}^{6'}$ with six negative effective charge
D_M^*	coefficient of diffusion of tracer M*
$D^o_{V(M)}$	effective coefficients of diffusion (mobility) of metal ions M via cation vacancies
$D^o_{I(M)}$	effective coefficients of diffusion (mobility) of metal ions M via interstitial ions
F	Faraday constant
ΔG_i^o	standard Gibbs energy change of reaction
ΔG_i^o	standard Gibbs energy change of formation of ionic defects i
ΔG_{eh}^o	standard Gibbs energy change of formation of electronic defects
$\Delta G^{o(\delta)}_{[def]}$	"resultant" standard Gibbs energy change of the defects' formation per one mole of oxygen atom at deviation from stoichiometry δ (Eq. 1.32).
K_i	equilibrium constants of the formation of defects i
p_{O2}	oxygen partial pressure – oxygen activity (atm)
$p^{(s)}_{O_2}$	partial pressure of oxygen where the oxide reaches the stoichiometric composition
R	molar gas constant (8.314 ($J{\cdot}K^{-1}mol^{-1}$))
T	absolute temperature (K)
V_{MO}	molar volume of the oxide MO
$y^o_{A'}$	concentration of dopant A_M' (mol/mol oxide)
$y_{D^{\bullet}}$	concentration of dopant $D_M^{\bullet}$ (mol/mol oxide)
δ	deviation from stoichiometric composition in oxide (mol/mol)
δ_c	deviation from the stoichiometry in the cationic sublattice (mol/mol)

δ_a	deviation from the stoichiometry in the oxygen (anionic) sublattice
[n]	square brackets around the symbol of defect *n* denote the concentration in mol/mol oxide
η_i	electrochemical potential of defects *i*
μ_i	chemical potential of component *i*
μ_e	mobility of electrons ($cm^2\ V^{-1}\ s^{-1}$)
μ_h	mobility of electron holes ($cm^2\ V^{-1}\ s^{-1}$)
σ	electrical conductivity ($\Sigma\sigma_i$ (Scm^{-1})

Part I

Diagrams of the Concentrations of Point Defects

1 Magnetite $Fe_{3\pm\delta}O_4$

1.1 INTRODUCTION

Iron oxide, $Fe_{3\pm\delta}O_4$ (magnetite) belongs to a large group of ferrimagnetic compounds. Magnetic properties of magnetite result from the presence of iron ions with the oxidation state Fe^{2+} and Fe^{3+}, which form an ordered, cubic spinel structure (point group $Fd\bar{3}m$, Z = 8), specifically, inverse spinel structure. In the structure of magnetite, oxygen atoms form a close-packed cubic sublattice, where a part of tetrahedral voids is occupied by Fe^{3+} iron ions, while a part of octahedral voids is occupied in half with Fe^{2+} ions and in half with Fe^{3+} ions, hence, the formula of the oxide, $Fe^{3+}(Fe^{2+}Fe^{3+})O_4$; it is a mixed-valence oxide. Magnetite is one of the three iron oxides ($Fe_{1-\delta}O$, $Fe_{3\pm\delta}O_4$, Fe_2O_3) that differ significantly by structure and concentration of point defects. The concentration of ionic point defects in $Fe_{3\pm\delta}O_4$ is much lower than in $Fe_{1-\delta}O$ iron oxide and higher than in Fe_2O_3. The range where magnetite exists is much wider than the range for $Fe_{1-\delta}O$ (Bryant and Smeltzer 1969, Giddings and Gordon 1973). Analysis and discussion of multiple studies concerning the range of existence of iron oxides (Fe-O system) can be found in Sundman (1991) and Wriedt (1991).

First studies on the deviation from the stoichiometry were conducted using the method of isothermal decomposition of Fe_2O_3. These studies determined the composition (Fe/O ratio) and the oxygen pressure below which magnetite exists, as well as the value of the deviation from the stoichiometry at higher oxygen pressures (White et al. 1935, Schmahl 1941, Richards and White 1954, Meyer 1959, Schmahl et al. 1969). Several authors analysed the composition of magnetite samples with frozen structure, which were annealed at high temperatures at a defined oxygen pressure and then quenched (Greig et al. 1935, Darken and Gurry 1945, 1946, Smilten 1957, Brynestad and Flood 1958, Katsura and Muan 1964). The obtained results of the studies indicated that dominating defects are iron vacancies and it was assumed that magnetite reaches stoichiometric composition near the phase boundary with wüstite, $Fe_{1-\delta}O/Fe_3O_4$.

The studies on the dependence of diffusion of iron, cobalt and chromium (radioisotope tracers) in magnetite on the oxygen pressure (Dieckmann and Schmalzried 1977, Dieckmann et al. 1978) showed that at low oxygen pressures the transport of ions occurs via interstitial iron ions, while at higher oxygen pressures it occurs via iron vacancies. The above diffusion mechanism was confirmed by diffusion studies by other authors (Halloran and Bowen 1980, Peterson et al. 1980, Lewis et al. 1985, Hallström et al. 2011) and Mössbauer spectroscopy (Becker et al. 1993). The above model of defects' structure and the fact of reaching the stoichiometric composition in the range of its existence of magnetite were confirmed by the studies on the deviation from the stoichiometry in dependence on the oxygen pressure (Dieckmann 1982).

The first studies on the deviation from the stoichiometry in dependence of the oxygen pressure (at equilibrium state) were performed by White (White 1938) and, later, using the method of coulometric titration at 1473 K (Sockel and Schmalzried 1968), and the thermogravimetry at high temperatures (Dieckmann 1982, Nakamura et al. 1978, Yamauchi et al. 1983). The problem of defects' concentration and iron diffusion in magnetite was also discussed in works by Dieckmann (1983), Dieckmann and Schmalzried (1986), and Dieckmann et al. (1987).

In turn, other studies (Millot and Niu 1997) showed that, in magnetite, the self-diffusion coefficients of oxygen are by nearly an order of magnitude lower than coefficients of iron ion diffusion, which was confirmed by computer simulations (Lewis et al. 1985). This indicates that the concentration of point defects in the oxygen sublattice should be significantly lower than that in the cation sublattice.

Studies of electrical properties have shown that magnetite is an *n*-type semiconductor (negative values of the thermo-electric power and Hall constant, (Lavine 1959, Haubenreisser 1961, Kuipers and Brabers 1976)), with a band gap of 0.609 eV. At 120 K, magnetite shows a phase transition (so-called Verwey transition), where the structure changes from ortho-rhombic to spinel cubic (Verwey and Haayman 1941, Verwey et al. 1947).

Studies of electrical properties of magnetite, mainly at low temperatures, were the object of many works (Lavine 1959, Haubenreisser 1961, Kuipers and Brabers 1976, Verwey and Haayman 1941, Verwey et al. 1947, Domenciali 1950, Smith 1956, Miles et al. 1957, Calhoun 1954, Griffiths et al. 1970, Pan and Evans 1976, Parker and Tinsley 1976, Klinger and Samokhralov 1977, Srinavasan and Srivastava 1981, Shepherd et al. 1985, Shepherd et al. 1985, Rasmussen et al. 1987, Kąkol and Honig 1989, Shepherd et al. 1991, Honig and Spałek 1989, Honig and Spałek 1992). It was found that below 120 K, magnetite shows large resistance, from $0.1 \cdot 10^7$ Ω·cm. The activation energy of electrical conductivity is about 0.1 eV (Calhoun 1954, Griffiths et al. 1970). At the transition temperature, electrical conductivity increases by two orders of magnitude. However, the transition temperature depends on the deviation from the stoichiometry of magnetite samples with frozen structure (Verwey and Haayman 1941, Verwey et al. 1947) and on impurities and dopants. This problem was studied in detail by Honig et al. (Shepherd et al. 1985, Shepherd et al. 1985, Rasmussen et al. 1987, Kąkol and Honig 1989, Shepherd et al. 1991, Honig and Spałek 1989, Honig and Spałek 1992), who performed studies on specific heat capacity, electrical conductivity, thermoelectric power and magnetisation. A theoretical model was proposed to explain the changes of magnetite properties near the Verwey transition, with the use of elements of order-disorder theory of strongly correlated electrons (Honig and Spałek 1989, Honig and Spałek 1992).

On the other hand, studies on electrical conductivity in the temperature range from 120 K to about 300 K have shown its increase up to about 250 $\Omega^{-1}cm^{-1}$ (Verwey et al. 1947, Domenciali 1950, Smith 1956, Miles et al. 1957, Calhoun 1954, Griffiths et al. 1970, Pan and Evans 1976, Parker and Tinsley 1976, Klinger and Samokhralov 1977, Srinavasan and Srivastava 1981). Further temperature increase, up to the Curie temperature (T_C = 858 K) causes a small decrease in electrical conductivity, to 190 $\Omega^{-1}cm^{-1}$, and then, up to about 1273 K, the conductivity increases to about 198–210 $\Omega^{-1}cm^{-1}$ (Domenciali 1950, Smith 1956, Miles et al. 1957, Calhoun 1954, Griffiths et al. 1970, Pan and Evans 1976, Parker and Tinsley 1976, Klinger and Samokhralov 1977, Srinavasan and Srivastava 1981). When the temperature increases, the activation energy of electrical conductivity changes within the range of 0.1–0.7 eV. In the temperature range 1273–1573, electrical conductivity slightly decreases, however, as shown in (Dieckmann et al. 1983), the process of charge transport is activated; they obtained an activation energy of 0.7 eV for the model of small polarons, and 0.6 eV for the band model.

In the high temperature range, electrical conductivity depends only slightly on the oxygen pressure (concentration of ionic defects) (Dieckmann et al. 1983, Wagner and Koch 1936, Tannhauser 1962). Therefore, magnetite conductivity in high temperatures is high and similar to the conductivity of $Fe_{1-\delta}O$ at maximum concentration of defects (Tannhauser 1962). Negative values of thermoelectric power at higher temperatures (Mason and Bowen 1981, Wu and Mason 1981, Wu et al. 1981, Erickson and Mason 1985) indicate that electron transport occurs according to the hopping mechanism of small polarons, via electron jumping between ions Fe^{2+} and Fe^{3+} in a sublattice with octahedral configuration (Haubenreisser 1961, Parker and Tinsley 1976, Tannhauser 1962, Mason and Bowen 1981, Wu and Mason 1981, Wu et al. 1981, Erickson and Mason 1985). The hopping mechanism has been also confirmed by Mössbauer spectroscopy (Kündig and Hargrove 1969, Daniels and Rosencwaig 1969).

Therefore, the increase in electrical conductivity above 120 K is related to a change in the oxidation state of iron atoms located in lattice nodes with octahedral and tetrahedral configuration, with random distribution. The distribution of iron ions with different charge was determined based on thermodynamic calculations (Navrotsky and Kleppa 1967), crystal field theory (Dunitz and Orgel 1957, McClure and Miller 1957, Miller 1959), and based on studies of thermoelectric power (Mason and Bowen 1981). The above studies have shown that, in magnetite, the concentration of iron ions Fe^{2+} and Fe^{3+} with octa- and tetrahedral configuration changes when the temperature increases. At a constant temperature, an equilibrium is set between the concentrations of iron ions Fe^{2+} and Fe^{3+} in both sublattices, similarly as in heterocation spinels (see Chapters 3–8).

1.2 MODEL OF POINT DEFECTS BY SCHMALZRIED, WAGNER AND DIECKMANN

1.2.1 Model of Schmalzried and Wagner

As already mentioned, the first studies on the deviation from the stoichiometry indicated that dominating defects are iron vacancies. It was assumed that iron ions Fe^{3+} and Fe^{2+} are statistically distributed over the whole lattice and that the formation of iron vacancies occurs according to the reaction (Schmalzried and Wagner 1962, Schmalzried and Tretiakov 1966):

$$8Fe_{Fe}^{2+} + 2O_2 = 8Fe_{Fe}^{3+} + 3V_{Fe} + 4O_O^{2-} \tag{1.1}$$

where V_{Fe} denotes iron vacancies, which, according to the form of the equation, should be electroneutral. The equilibrium constant for the reaction (1.1) is expressed by the equation:

$$K_V = \frac{\left[Fe_{Fe}^{3+}\right]^8 \left[V_{Fe}\right]^3}{\left[Fe_{Fe}^{2+}\right]^8 p_{O_2}^2} \tag{1.2a}$$

where the concentration of iron vacancies is 1/3 of the deviation from the stoichiometry ($[V_{Fe}] = \delta/3$). Assuming that ions Fe^{2+} and Fe^{3+} are randomly distributed, their concentrations will be:

$$\left[Fe_{Fe}^{2+}\right] = \frac{1-3\delta}{3} \quad \text{and} \quad \left[Fe_{Fe}^{3+}\right] = \frac{2(1+\delta)}{3}$$

Therefore, the expression for the equilibrium constant assumes the form of:

$$\frac{\delta\left[2(1+\delta)\right]^{8/3}}{3(1-3\delta)^{8/3}} = K_V^{1/3} p_{O_2}^{2/3} \tag{1.2b}$$

If the deviation from the stoichiometry is small, the concentration of iron vacancies is proportional to the oxygen pressure with the exponent of 2/3, which is approximatively consistent with the experimental results (Darken and Gurry 1945, Smilten 1957):

$$\delta \cong K'_V p_{O_2}^{2/3} \tag{1.2c}$$

1.2.2 Model of Schmalzried and Dieckmann

As it was already mentioned, studies on diffusion have shown that interstitial iron ions and iron vacancies are present in magnetite. Concentrations of both types of ionic defects were correlated with the expressions describing the equilibrium constants (Dieckmann and Schmalzried 1977, Dieckmann 1982). The formation of iron vacancies occurs according to the equation:

$$3Fe_{Fe}^{2+} + 2/3O_2 = 2Fe_{Fe}^{3+} + V_{Fe} + 1/3Fe_3O_4 \tag{1.3}$$

The expression describing the equilibrium constant gives the following relation between the concentration of iron vacancies and oxygen pressure:

$$\left[V_{Fe}\right] = \frac{\left[Fe_{Fe}^{2+}\right]^3}{\left[Fe_{Fe}^{3+}\right]^2} a_{Fe_3O_4}^{-1/3} K_V p_{O_2}^{2/3} = \frac{(1-3\delta)^3}{4(1+\delta)^3} K_V p_{O_2}^{2/3} \tag{1.4}$$

which for small concentrations of defects assumes a simple form:

$$\left[V_{Fe}\right] \cong \frac{K_V}{4} p_{O_2}^{2/3} \cong K'_V p_{O_2}^{2/3} \tag{1.5}$$

For higher concentrations of defects, the experimental results follow the following equation (Dieckmann 1982):

$$\left[V_{Fe}\right] \cong \frac{K_V}{4} \frac{1}{1+2K_V p_{O_2}^{2/3}} p_{O_2}^{2/3} \tag{1.6}$$

In turn, using the equilibrium between iron vacancies and interstitial iron ions (Frenkel defects) is described by the equation (Dieckmann 1982), determining the dependence of the concentration of interstitial ions on the oxygen pressure:

$$\left[Fe_I^{n+}\right] = \frac{(2+2\delta)^2}{(1-3\delta)^3} \frac{4K_F}{K_V} p_{O_2}^{-2/3} \cong 4K_I p_{O_2}^{-2/3} \cong K'_I p_{O_2}^{-2/3} \tag{1.7}$$

According to Equation 1.5 and 1.7, the deviation from the stoichiometry δ is the difference between the concentration of iron vacancies and interstitial iron ions and it is the following function of oxygen pressure:

$$\delta = \left[V_{Fe}\right] - \left[Fe_I^{n+}\right] \cong K'_V p_{O_2}^{2/3} + K'_I p_{O_2}^{-2/3} \tag{1.8}$$

It was shown that the dependence of the deviation from the stoichiometry on the oxygen pressure in $Fe_{3\pm\delta}O_4$ is fully consistent with Equation 1.6–1.8 (Dieckmann 1982) and the values of equilibrium constants of reactions of formation of cation vacancies K_V and interstitial cations K_I were determined (Dieckmann 1982, Dieckmann and Schmalzried 1986).

The model of point defects and description of defect equilibria (Dieckmann and Schmalzried 1977, Dieckmann 1982, Dieckmann and Schmalzried 1986) is not fully consistent with the assumed notation of Kröger–Vink. The concentration of iron vacancies and interstitial ions has been averaged; moreover, these point defects can have different ionisation degrees (the equations do not include the effective charge of point defects in relation to the lattice). The formation of ionic defects is accompanied by electronic defects which are not taken into account in defect equations. Besides, at the stoichiometric composition, apart from intrinsic ionic defects of Frenkel type, in magnetite, there should occur a significant concentration of intrinsic electronic defects, which is indicated by high values of electrical conductivity. The change in concentration of electronic defects is affected by the change in concentration of ionic defects due to the change in oxygen pressure. This is due to the fact that during the formation of cation vacancies, oxygen atoms are incorporated into the surface; they accept electrons and cause an increase in the concentration of electron holes. The formation of interstitial cations is also associated with the release of oxygen atoms that leave electrons, causing an increase in their concentration. Therefore, there occurs a perturbation of the equilibrium between the electronic defects.

1.3 COMPARISON OF MAGNETITE AND WÜSTITE STRUCTURES

The structure of magnetite is to a certain degree similar to the cubic NaCl-type structure of wüstite ($Fe_{1-\delta}O$). In wüstite, oxygen ions form a close-packed sublattice, where all octahedral voids are occupied by iron ions. Thus, all Fe^{2+} iron ions have octahedral coordination. The deviation from the stoichiometry results from the presence of lattice nodes that are not occupied by iron ions. Therefore, it is associated with the presence of iron vacancies $\left(V_{Fe}''\right)$ and electron holes, which balance their charge. A high concentration of iron vacancies causes the formation of defect complexes in wüstite. They are formed due to a transition of an iron ion into a tetrahedral void and the creation of an interstitial ion ($Fe_i^{3\bullet}$), which interacts with two $\left\{\left(V_{Fe}''\right)_2 Fe_i^{3\bullet}\right\}$ or four $\left\{\left(V_{Fe}''\right)_4 Fe_i^{3\bullet}\right\}$ iron vacancies. At higher concentrations of iron vacancies, more complex defect clusters are formed. A wider discussion on the thermodynamics of defect complexes and their concentration of wüstite can be found in (Stokłosa 2015).

Magnetite, which is formed as a result of oxidation of $Fe_{1-\delta}O$, has similarly a close-packed oxygen ion sublattice. The lattice parameter is practically twice as high as that for wüstite ($Fe_{1-\delta}O$). As it was already mentioned, magnetite has spinel cubic structure ($Fd\bar{3}m$ point group). A unit cell contains 8 "molecules" of Fe_3O_4 (Z = 8). In contrast to $Fe_{1-\delta}O$, in magnetite only a part of octahedral voids is occupied by iron ions. In particular, in a unit cell of stoichiometric magnetite (Fe_3O_4) containing 32 oxygen ions, only 16 octahedral voids are occupied by 8 Fe^{2+} ions and 8 Fe^{3+} ions. On the other hand, the remaining 16 octahedral voids (potential cation sites) are unoccupied. In turn, a part of 64 tetrahedral voids is occupied by 8 Fe^{3+} ions.

Figure 1.1 shows the structure of magnetite (Verwey et al. 1947). Differently from a unit cell of cubic system of NaCl-type, and from the spinel system presented in crystallography, Verwey assumed not an oxygen ion, but an iron ion in a tetrahedral void as the coordinate system origin.

As a result, as can be seen in Figure 1.1, in a unit cell we can distinguish two different cubic blocks, A and B, which are placed alternatively (4A and 4B). As can be seen in Figure 1.1b, A and B block differ by atom layout inside. Block A contains an iron ion Fe^{3+} surrounded by four oxygen atoms that form a tetrahedron (the iron ion has tetrahedral coordination). The placement of these atoms can be visualised with a cube with a side equal 1/4 of the lattice parameter *a*, where four iron ions are missing (if all octahedral atoms were occupied, as in the NaCl-type structure). This layout is identical to the symmetry of complexes $\left\{\left(V_{Fe}''\right)_4 Fe_i^{3\bullet}\right\}$ formed by ions $Fe_i^{3\bullet}$ with four vacancies V_{Fe}'' in $Fe_{1-\delta}O$ oxide. Therefore, in an ideal structure of magnetite, a lattice of ordered "defect complexes" was formed. Iron ions Fe^{3+} with tetrahedral coordination with four iron "vacancies" (unoccupied octahedral voids) became elements of the lattice, and not defect complexes, as in the case of wüstite. Due to the ordering of defect complexes in magnetite structure they lost their properties of quasi-free

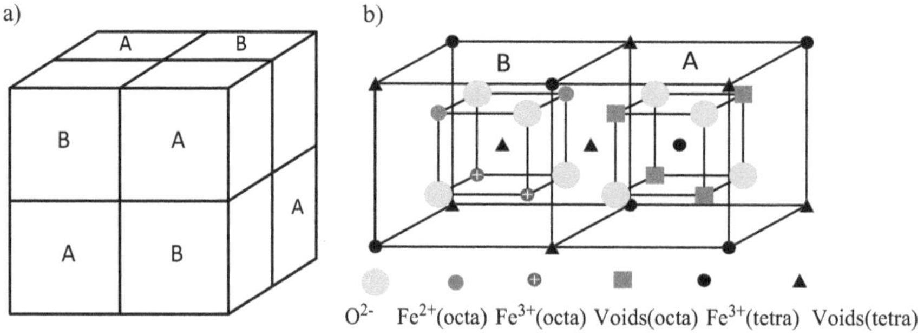

FIGURE 1.1 An unit cell of magnetite (spinel), (a) pattern of alternating A and B blocks, (b) distribution of ions in blocks A and B: oxygen O^{2-} (■), iron ions Fe^{2+} (◯) and Fe^{3} (⊕) with octahedral coordination, and Fe^{3+} (●) with tetrahedral coordination and unoccupied voids with octahedral coordination (●) and tetrahedral configuration (▲).

defects. Considering the fact that unoccupied octahedral voids are common for adjacent "cubes", their ratio to iron ions in tetrahedral voids is 2:1.

As can be seen in Figure 1.1b, apart from block A, containing Fe^{3+} ion with tetrahedral coordination, there is block B, containing a cubic block that also contains four oxygen ions and two Fe^{2+} ions and two Fe^{3+} ions with octahedral coordination. Inside the cube there is a vacant tetrahedral void that could be occupied by iron ions. However, due to the presence of two Fe^{3+} ions, there will be a stronger repulsive interaction in comparison to an analogous block of a cubic structure, where there are only Fe^{2+} ions, as for example in $Fe_{1-\delta}O$. Between the discussed A and B block there is a third "cube block" containing a tetrahedral void; there are two unoccupied octahedral voids at its corners. Such a void could potentially be occupied by iron ions, as there is a larger free space and smaller repulsive interaction. The different charges of Fe^{2+} and Fe^{3+} in octahedral and tetrahedral coordination cause that these ions interact with oxygen ions with different energy. This leads to the formation of polarised, ionic-covalent bonds of variable nature (Goodenough and Loeb 1955). An important role here is played also by the energy of stabilisation of ions in a crystalline field (Dunitz and Orgel 1957).

Thus, it can be assumed, that in magnetite there is a maximum (limit) concentration of complexes of type $\left\{\left(V''_{Fe}\right)_4 Fe_i^{3\bullet}\right\}$, similar as in wüstite; their ordering caused the formation of a new, stable structure with lower energy. Comparing to the structure of FeO oxide, the structure of magnetite ($Fe_{0.75}O$) is looser, as there are 25% less iron ions and there is a significant number of unoccupied voids with octahedral coordination. In the structure of magnetite, "iron vacancies" and "interstitial ions" lost their properties of quasi-free ionic defects. Despite the loose structure and the presence of a high number of unoccupied octahedral voids, the diffusion coefficient of iron in magnetite is by one order of magnitude lower than that in $Fe_{1-\delta}O$ oxide (Dieckmann and Schmalzried 1977, Franke and Dieckmann 1989, Lu et al. 1993, Töpfer et al. 1995, Aggarwal and Dieckmann 2002). Also, at lower temperatures, below the Verwey transition temperature (120 K), despite the presence of Fe^{2+} and Fe^{3+} ions in magnetite, the electrical conductivity is very low.

At temperatures above the phase transition, there occurs a change in the distribution of Fe^{2+} and Fe^{3+} ions in spinel sublattices. The studies of thermoelectric power of magnetite (Wu and Mason 1981) indicate that the ordered structure of iron ions that is present in low temperatures gets disturbed when the temperature increases. Above the Curie temperature (585 °C), in the case of stoichiometric oxide, the mole fraction of Fe^{3+} ions in octahedral voids increases and the concentration of Fe^{2+} ions with octahedral coordination decreases. Simultaneously, an equivalent number of Fe^{2+} ions appears in tetrahedral voids; their concentration increases when the temperature increases and it reaches 0.32 mol at 1773 K. At a constant temperature, an equilibrium state is set between concentrations of Fe^{2+} and Fe^{3+} with different coordination. Such a large change in mole fraction of iron ions allows assuming random distribution of Fe^{2+} and Fe^{3+} ions. A consequence of changes in the oxidation state of iron atoms in magnetite with increasing temperature is an increase in electrical conductivity. The change in oxidation state of iron atoms with octahedral and tetrahedral coordination will also be related to the concentration of electronic defects, electrons and electron holes. The problem of charge carriers in high temperatures is not fully explained and it will be further discussed when analysing electronic defects (see Section 1.4.5).

The "loose" structure of magnetite and the existence of a high number of octahedral and tetrahedral voids favours the formation, at high temperatures, of quasi-free ionic defects; these are iron vacancies, but also interstitial iron ions, with the concentration depending on the oxygen pressure. Different ordering of iron ions Fe^{2+} and Fe^{3+} in magnetite causes different interaction with surrounding ions. Owing to that, the energy of formation of individual types of point defects is different and it is closely related to the crystallographic structure. From the studies on the deviation from the stoichiometry and tracer ions diffusion in magnetite it results that in the range of its existence, there is an oxygen pressure at which the stoichiometric composition is reached. Therefore, the range of existence of magnetite will be divided into two subranges, where iron vacancies and interstitial iron ions will dominate.

1.4 REACTIONS OF FORMATION OF POINT DEFECTS

1.4.1 Ionic Defects with the Highest Ionisation Degree

As already mentioned, Fe^{2+} and Fe^{3+} ions with octahedral coordination, due to different charge, have different interaction energies; therefore, energies of formation of iron vacancies $V''_{Fe(o)}$ and $V'''_{Fe(o)}$, as well as interstitial iron ions $Fe_i^{\bullet\bullet}$ and $Fe_i^{3\bullet}$ will be different. Similarly, energy of formation of vacancies in the sublattice of ions with tetrahedral coordination, $V'''_{Fe(t)}$, will be different (standard Gibbs energies of reactions of their formation will be different). As already mentioned, the problem of interactions in the spinel structure was broadly discussed (Goodenough and Loeb 1955).

Thus, in the magnetite, we can formally distinguish three sublattices of iron ions. Namely: sublattice of Fe^{2+} ions with octahedral coordination, Fe^{3+} ions with octahedral coordination and Fe^{3+} ions with tetrahedral coordination. At higher temperatures, quasi-free ionic and electronic defects will form in these sublattices; at a constant temperature and at a specific oxygen pressure, a thermodynamic equilibrium will be set between these defects.

In order to correctly write the reaction of formation of point defects, it is necessary to formally assign an appropriate number of oxygen ions (oxygen nodes) to iron ions with different coordination. From Figure 1.1 it results that there are 4 oxygen atoms per 2 Fe^{2+} ions and two Fe^{3+} with octahedral coordination. Similarly, there are also four oxygen ions per one Fe^{3+} ion with tetrahedral coordination. If we assume that the charge of oxygen ion is 2−, then in a unit cell (also in a volume per one "molecule" of Fe_3O_4 in a crystal), the electroneutrality condition for the lattice is fulfilled. On the other hand, the electroneutrality condition is not fulfilled for individual sublattices. In order for the ratio of cation to anion sites and the electroneutrality condition to be fulfilled in sublattices, we must formally assign an appropriate number of oxygen atoms (oxygen sites) per one iron ion (mole of ions) with different charge and coordination. That is why the molecular formula of magnetite, for the purpose of writing defect reactions, should have the form of:

$$\left(\mathrm{Fe}_{(o)}^{(2+)}\mathrm{O}\right)\left(\mathrm{Fe}_{(o)}^{(3+)}\mathrm{O}_{3/2}\right)\left(\mathrm{Fe}_{(t)}^{(3+)}\mathrm{O}_{3/2}\right)$$

According to the notation of Kröger–Vink, the reaction of formation of iron vacancies in the sublattice of iron ions with octahedral coordination can be presented with defect equations:

$$1/2\mathrm{O}_2 = \mathrm{O}_\mathrm{O}^\mathrm{x} + \mathrm{V}_{\mathrm{Fe(o)}}'' + 2\mathrm{h}^\bullet \qquad \Delta G^\mathrm{o}_{\mathrm{V}_\mathrm{Fe}''} \tag{1.9a}$$

$$3/4\mathrm{O}_2 = 3/2\mathrm{O}_\mathrm{O}^\mathrm{x} + \mathrm{V}_{\mathrm{Fe(o)}}''' + 3\mathrm{h}^\bullet \qquad \Delta G^\mathrm{o}_{\mathrm{V}_{\mathrm{Fe(o)}}'''} \tag{1.9b}$$

A similar equation describes the formation of vacancies in the sublattice of iron ions with tetrahedral coordination:

$$3/4\mathrm{O}_2 = 3/2\mathrm{O}_\mathrm{O}^\mathrm{x} + \mathrm{V}_{\mathrm{Fe(t)}}''' + 3\mathrm{h}^\bullet \qquad \Delta G^\mathrm{o}_{\mathrm{V}_{\mathrm{Fe(t)}}'''} \tag{1.9c}$$

Therefore, the formation of iron vacancies is associated with the incorporation of oxygen atoms, which, in order to become ions must accept electrons, causing an increase in the concentration of electron holes. Expressions for the equilibrium constant of reactions (1.9a)–(1.9c) assume the form of:

$$K_{\mathrm{V}_\mathrm{Fe}''} = \frac{\left[\mathrm{V}_\mathrm{Fe}''\right]\left[\mathrm{h}^\bullet\right]^2}{p_{\mathrm{O}_2}^{1/2}} \tag{1.10a}$$

$$K_{\mathrm{V}_{\mathrm{Fe(o)}}'''} = \frac{\left[\mathrm{V}_{\mathrm{Fe(o)}}'''\right]\left[\mathrm{h}^\bullet\right]^3}{p_{\mathrm{O}_2}^{3/4}} \tag{1.10b}$$

$$K_{\mathrm{V}_{\mathrm{Fe(t)}}'''} = \frac{\left[\mathrm{V}_{\mathrm{Fe(t)}}'''\right]\left[\mathrm{h}^\bullet\right]^3}{p_{\mathrm{O}_2}^{3/4}} \tag{1.10c}$$

The concentration of iron vacancies depends on the oxygen pressure, and, above all, on the value of equilibrium constant K_i for the reaction of formation of individual vacancies. The K_i constant depends on the standard Gibbs energy of the formation of defects and on the temperature, according to the equation:

$$\ln K_i = \frac{-\Delta G_i^0}{RT} = \frac{-\Delta H_i^0 + T\Delta S_i^0}{RT} \tag{1.11}$$

From Equation 1.10a it results that if dominating defects are iron vacancies V_Fe'', then, with a simplified electroneutrality condition $2\left[\mathrm{V}_\mathrm{Fe}''\right] = \left[\mathrm{h}^\bullet\right]$, their concentration will depend on the oxygen pressure with the exponent of 1/6. On the other hand, with $\left[\mathrm{V}_\mathrm{Fe}'''\right]$ as dominating defects, the exponent would be 3/16. The real exponent will depend on the contribution of the individual iron vacancies and it will have an intermediate value, much different from the experimental value, 2/3 (Dieckmann 1982).

If the concentration of electronic defects is high, which is indicated by high values of the electrical conductivity, it can be assumed that the concentration of iron vacancies is independent of the concentration of holes (electronic defects). According to Equation 1.10a, 1.10b, 1.10c, the concentration of vacancies $\left[\mathrm{V}_\mathrm{Fe}''\right]$ will depend on the oxygen pressure with an exponent of 1/2, and vacancies $\left[\mathrm{V}_\mathrm{Fe}'''\right]$ with an exponent of 3/4. The experimental exponent 2/3 is therefore a resultant value derived from an approximate expression containing the sum of concentrations of iron vacancies:

$$\left[\mathrm{V}_\mathrm{Fe}\right] = \alpha\left[\mathrm{V}_\mathrm{Fe}''\right] + \beta\left[\mathrm{V}_{\mathrm{Fe(o)}}'''\right] + \gamma\left[\mathrm{V}_{\mathrm{Fe(t)}}'''\right] \cong K_\mathrm{V} p_{\mathrm{O}_2}^{2/3} \tag{1.12}$$

We can write in a similar way the reaction of formation of quasi-free interstitial iron ions. They occupy tetrahedral voids, but they are quasi-free ionic defects, differently from ordered Fe^{3+} ions located in lattice nodes with tetrahedral coordination, which are adjacent to four octahedral voids. Thus, these ions can occupy the remaining tetrahedral voids, for example between adjacent A and B blocks or between internal "cubes" (see Figure 1.1b). The formation of an interstitial iron cation is associated with the release of oxygen, which leaves two electrons. They perturb the equilibrium of electronic defects and cause an increase in the concentration of electrons, causing their reaction with holes (partial annihilation occurs, causing a decrease in the concentration of holes). As a result, an equilibrium is attained between electronic defects:

$$e' + h^{\bullet} = nil \qquad -\Delta G^{o}_{eh} \qquad (1.13a)$$

where ΔG^{o}_{eh} is the standard Gibbs energy of the process of formation of electronic defects. Therefore, the reactions of formation of interstitial ions, involving electron holes, assume the form of:

$$Fe^{x(2+)}_{Fe(o)} + O^{x}_{O} + 2h^{\bullet} + V_i = Fe_i^{\bullet\bullet} + 1/2O_2 \qquad \Delta G^{o}_{Fe_i^{\bullet\bullet}} \qquad (1.13b)$$

$$Fe^{x(3+)}_{Fe(o)} + 3/2O^{x}_{O} + 3h^{\bullet} + V_i = Fe_i^{3\bullet} + 3/4O_2 \qquad \Delta G^{o}_{Fe_i^{3\bullet}} \qquad (1.13c)$$

$$Fe^{x(3+)}_{Fe(t)} + 3/2O^{x}_{O} + 3h^{\bullet} + V_i = Fe_i^{3\bullet} + 3/4O_2 \qquad \Delta G^{o}_{Fe_i^{3\bullet}} \qquad (1.13d)$$

As can be seen from the reactions (1.13c) and (1.13d), the concentration of $\left[Fe_i^{3\bullet}\right]$ ions results from achieving an equilibrium between Fe^{3+} ions located in octahedral sublattice and tetrahedral sublattice and they will be in equilibrium with the concentration of vacancies $\left[V'''_{Fe(o)}\right]$ and $\left[V'''_{Fe(t)}\right]$. Expressions for the equilibrium constants of reactions (1.13b, 1.13c, 1.13d) assume the form of:

$$K_{Fe_i^{\bullet\bullet}} = \frac{\left[Fe_i^{\bullet\bullet}\right] p_{O_2}^{1/2}}{\left[h^{\bullet}\right]^2} \qquad (1.14a)$$

$$K_{Fe_i^{3\bullet}} = \frac{\left[Fe_i^{3\bullet}\right] p_{O_2}^{3/4}}{\left[h^{\bullet}\right]^3} \qquad (1.14b)$$

$$K'_{Fe_i^{3\bullet}} = \frac{\left[Fe_i^{3\bullet}\right] p_{O_2}^{3/4}}{\left[h^{\bullet}\right]^3} \qquad (1.14c)$$

Despite the fact that we analyse the formation of point defects in individual sublattices, their concentrations are mutually related, as they depend on the oxygen pressure and on the concentration of point defects (defect Equations 1.9a–1.9c and 1.13a–1.13d are mutually related – linear relation).

Similarly, as in the case of iron vacancies, if the concentration of electronic defects is high, it can be assumed that the concentration of interstitial iron ions is independent of the concentration of electrons. According to Equation 1.14a to 1.14c, the concentration of interstitial cations $\left[Fe_i^{\bullet\bullet}\right]$ will depend on the oxygen pressure with an exponent of −1/2, and interstitial ions $\left[Fe_i^{3\bullet}\right]$ with an exponent of −3/4. The experimental exponent −2/3 is therefore a resultant value derived from an approximate expression containing the sum of concentrations of interstitial iron ions:

$$\left[Fe_i\right] = \gamma\left[Fe_i^{\bullet\bullet}\right] + \varepsilon\left[Fe_i^{3\bullet}\right] \cong K_I p_{O_2}^{-2/3} \qquad (1.15)$$

1.4.2 Defects with Lower Ionisation Degrees

In magnetite, near the stoichiometric composition, the concentration of ionic defects is at its minimum, and defects with the highest ionisation degree should dominate. At higher concentration of ionic defects, as a result of their interaction with electronic defects, also defects with lower ionisation degrees should be present in magnetite. This is additionally favoured by a high concentration of electronic defects in magnetite. Ionic defects with lower ionisation degrees are defect complexes formed as a result of interaction of ionic defects with electronic defects. Therefore, due to an interaction of iron vacancies

$V''_{Fe(o)}$ in the sublattice of ions with octahedral coordination with electron holes, defect with lower ionisation degrees (defect complexes) are formed, according to the reactions:

$$V''_{Fe(o)} + h^{\bullet} = \left\{V''_{Fe(o)}, h^{\bullet}\right\}' \equiv V'_{Fe(o)} \quad (1.16a)$$

$$V'_{Fe(o)} + h^{\bullet} = \left\{V''_{Fe(o)}, 2h^{\bullet}\right\}^{x} \equiv V^{x}_{Fe(o)} \quad (1.16b)$$

Thus, the formation of a complex $\left\{V''_{Fe(o)}, h^{\bullet}\right\}$ will occur as a result of an interaction of an iron vacancy with a hole localised on an adjacent nodal iron ion $\left\{V''_{Fe(o)}, Fe^{\bullet}_{Fe}\right\}$. As can be seen, the formation of vacancies with a lower charge does not cause a change in the concentration of ions Fe^{3+} $\left(Fe^{\bullet}_{Fe}\right)$, but the concentration of quasi-free electron holes decreases.

Similar complexes will be formed in the case of an interaction of holes with vacancies $V'''_{Fe(o)}$; they will be different from complexes with vacancies $V''_{Fe(o)}$, because in the vicinity of vacancies $V'''_{Fe(o)}$, Fe^{3+} ions with octahedral coordination must be present, which increases repulsive interactions between ions:

$$V'''_{Fe(o)} + h^{\bullet} = \left\{V'''_{Fe(o)}, h^{\bullet}\right\}'' \quad (1.17a)$$

$$\left\{V'''_{Fe(o)}, h^{\bullet}\right\}'' + h^{\bullet} = \left\{V'''_{Fe(o)}, 2h^{\bullet}\right\}' \quad (1.17b)$$

$$\left\{V'''_{Fe(o)}, 2h^{\bullet}\right\}' + h^{\bullet} = \left\{V'''_{Fe(o)}, 3h^{\bullet}\right\}^{x} \quad (1.17c)$$

Similarly, there will occur an interaction of holes with vacancies $V'''_{Fe(t)}$ in the sublattice with tetrahedral coordination:

$$V'''_{Fe(t)} + h^{\bullet} = \left\{V'''_{Fe(t)}, h^{\bullet}\right\}'' \quad (1.18a)$$

$$\left\{V'''_{Fe(t)}, h^{\bullet}\right\}'' + h^{\bullet} = \left\{V'''_{Fe(t)}, 2h^{\bullet}\right\}' \quad (1.18b)$$

$$\left\{V'''_{Fe(t)}, 2h^{\bullet}\right\}' + h^{\bullet} = \left\{V'''_{Fe(t)}, 3h^{\bullet}\right\}^{x} \quad (1.18c)$$

The formation of a complex of an iron vacancy with an electron hole ($Fe^{\bullet}_{Fe}$ ion) causes an increase in the number of Fe^{3+} around an unoccupied node, which will affect mutual repulsive interactions, especially in the case of Fe^{3+} ion with tetrahedral coordination. Therefore, the concentration of vacancies with lower ionisation degrees, despite the high concentration of electronic defects, should be small.

Conversely, quasi-free interstitial iron ions present in tetrahedral voids will interact with electrons, which can be presented by the reactions:

$$Fe^{3\bullet}_{i} + e' = \left\{Fe^{3\bullet}_{i}, e'\right\}^{\bullet\bullet} \equiv Fe^{\bullet\bullet}_{i} \quad (1.19a)$$

$$Fe^{\bullet\bullet}_{i} + e' = \left\{Fe^{\bullet\bullet}_{i}, e'\right\}^{\bullet} \equiv Fe^{\bullet}_{i} \quad (1.19b)$$

$$Fe^{\bullet}_{i} + e' = \left\{Fe^{\bullet\bullet}_{i}, 2e'\right\}^{x} \equiv Fe^{x}_{i} \quad (1.19c)$$

The reaction between interstitial iron ions and electron leads to the decrease in their effective charge; thus, the number of Fe^{2+} $\left(Fe'_{Fe}\right)$ ion. Also, complexes with electrons localised on nodal iron ions with negative charge in relation to the lattice $\left\{Fe^{\bullet\bullet}_{i}, Fe'_{Fe}\right\}^{\bullet}$ can be formed; in such a case the number of Fe^{2+} will not change, but the concentration of quasi-free electrons will decrease.

The expression for the equilibrium constant for the reaction of formation of defect complexes, for example for the reaction (1.18a) has the form of:

$$K_{V'_{Fe(o)}} = \frac{\left[\left\{V''_{Fe(o)}, h^{\bullet}\right\}\right]}{V''_{Fe(o)}\left[h^{\bullet}\right]} \quad (1.20)$$

Similar expressions for equilibrium constants can be obtained for the formation of the remaining defects with lower ionisation degrees.

The high concentration of electronic defects in magnetite causes, beside the equilibria of ionic defects with the highest ionisation degrees, that the defects with lower ionisation degrees will also be in equilibrium. Their concentration will depend on the equilibrium constant of the individual reactions (ΔG_i^o of formation of the individual complexes – ionic defect and electronic defect) and on the concentration of electronic defects and the concentration of ionic defects, necessary for their formation.

1.4.3 Defects in the Oxygen Sublattice

Minority defects will be also defects in the oxygen sublattice; the oxygen diffusion proceeds via these defects (Millot and Niu 1997). These defects will be oxygen vacancies $V_O^{\bullet\bullet}$ at lower oxygen pressures and interstitial oxygen ions O_i'' at higher pressures. The formation of oxygen vacancies occurs according to the reaction:

$$O_O^x = 1/2O_2 + V_O^{\bullet\bullet} + 2e' \tag{1.21a}$$

The formation of interstitial oxygen ions occurs according to the reaction:

$$1/2O_2 + 2e' + V_i = O_i'' \tag{1.21b}$$

Expressions for the equilibrium constants of reactions (1.21a and 1.21b) assume the form of:

$$K_{V_O^{\bullet\bullet}} = \left[V_O^{\bullet\bullet}\right]\left[e'\right]^2 p_{O_2}^{1/2} \tag{1.22a}$$

$$K_{O_i''} = \frac{\left[O_i''\right]}{p_{O_2}^{1/2}\left[e'\right]^2} \tag{1.22b}$$

1.4.4 Deviation from the Stoichiometry

At constant temperature, at a specific oxygen pressure, the existence of a specific concentration of ionic defects causes a perturbation in the stoichiometric ratio of iron ions to oxygen. The value of the deviation from the Fe/O ratio equal 3/4 is called the deviation from the stoichiometry, which is expressed by the relation:

$$\frac{[Fe]}{[O]} = \frac{3-\left(\Sigma[V_{Fe}]-\Sigma[Fe_i]\right)}{4-\left(\Sigma[V_O]-\Sigma[O_i]\right)} = \frac{3-y_c}{4-x_a} = \frac{3-\delta}{4} \tag{1.23a}$$

where y_c and x_a are deviations from the stoichiometry in cation sublattice and anion sublattice, respectively. If we assume, that in magnetite the concentration of oxygen defects and the concentration of cation defects with lower ionisation degrees is small and can be neglected, then the deviation from the stoichiometry will be the difference between the concentrations of iron vacancies and interstitial iron ions:

$$\delta \cong y_c = \left[V_{Fe(o)}'''\right] + \left[V_{Fe(t)}'''\right] + \left[V_{Fe(o)}''\right] - \left[Fe_i^{3\bullet}\right] - \left[Fe_i^{\bullet\bullet}\right] \tag{1.23b}$$

If the concentration of iron vacancies in the sublattice with tetrahedral coordination at the stoichiometric composition is small, the following relation will be true:

$$\delta = \left[V_{Fe(o)}'''\right] + \left[V_{Fe(o)}''\right] - \left[Fe_i^{3\bullet}\right] - \left[Fe_i^{\bullet\bullet}\right] = 0 \tag{1.23c}$$

Due to a difference in interaction of ions Fe^{2+} and Fe^{3+} with their immediate surroundings, we can treat sublattices of those ions independently. Due to that, at the stoichiometric composition, the following relations between the concentration of iron vacancies and interstitial ions can be approximately assumed:

$$\left[V_{Fe(o)}'''\right] = \left[Fe_i^{3\bullet}\right] \text{and} \left[V_{Fe(o)}''\right] = \left[Fe_i^{\bullet\bullet}\right] \tag{1.23d}$$

1.4.5 Electronic Defects

In discussing the reactions of formation of point defects we considered defect equilibria including quasi-free electrons and electron holes. Such an approach is correct in the case of a small concentration of electronic defects, thus, in wide-band semiconductors. For example, for an oxide of MO type, the formation of electronic intrinsic defects is associated with a transition of an electron into the conductivity band and a formation of a quasi-free hole and an electron. The formation of ionic defects as a result of a change in oxygen pressure causes that an oxygen atom accepts or releases electrons, which changes the concentration of electronic defects. In the band gap, acceptor and donor levels are created, while the existence of electronic defects causes, in the lattice of, for example, MO-type oxide, the formation of M^{3+} and M^{+} ions, with a charge different from parent ions M^{2+}, which should have been treated as ions (point defects) that are positive ($M_M^{\bullet}$) and negative ($M_M^{\prime}$), respectively, in relation to the lattice. An equivalent equation describing the formation of electronic defects can therefore have a form of:

$$2M_M^x = M_M^{\bullet} + M_M^{\prime} \equiv h^{\bullet} + e^{\prime} \qquad \Delta G_{eh}^{o} \qquad (1.24a)$$

The expression for the equilibrium constant of the reaction (1.24a) assumes the form of:

$$K_{eh} = \frac{\left[M_M^{\bullet}\right]\left[M_M^{\prime}\right]}{\left[M_M^x\right]^2} \cong \left[M_M^{\bullet}\right]\left[M_M^{\prime}\right] \equiv \left[h^{\bullet}\right]\left[e^{\prime}\right] \qquad (1.24b)$$

Therefore, the change in the concentration of electronic defects is associated with the change in the concentration of ions with an ionisation degree different from that for parent ions in a given sublattice (they are positive or negative relative to electroneutral ions of the sublattice). When oxygen pressure changes and ionic defects are formed, electronic defects take part in the reaction of their formation, thus, relative concentrations of holes and electrons change.

In magnetite, in high temperatures, the problem of electronic defects is not as obvious as in the case of MO-type oxides. As already mentioned, magnetite, in high temperatures, shows a high electrical conductivity (about 200 $\Omega^{-1}cm^{-1}$) (Dieckmann et al. 1983), this, also the concentration of charge carriers, must be high. From the studies of thermoelectric power, it results that the carriers are electrons, and their transport occurs by jumping of electrons between iron ions Fe^{2+} and Fe^{3+} in the sublattice with octahedral coordination, according to the hopping mechanism of small polarons (Wu and Mason 1981).

The studies on the deviation from the stoichiometry (Dieckmann 1982) showed that magnetite in high temperatures, in the range of its existence, reaches the stoichiometric composition, thus, there is a small equilibrium concentration of intrinsic ionic defects of Frenkel-type. In these conditions, also the concentrations of holes and electrons (intrinsic electronic defects) should be equal. This conclusion is consistent with the results of the studies on the electrical conductivity (Dieckmann et al. 1983), which showed that the electrical conductivity, in a wide range of oxygen pressures, is independent of p_{O_2}. On the other hand, at higher oxygen pressures, where concentrations of iron vacancies and electron holes increase, the electrical conductivity decreases. This indicates that the concentration of electronic defects is affected by the concentration of ionic defects. In the range of higher oxygen pressures, electron holes should dominate and the electrical conductivity should increase. The decrease in the conductivity indicates that the mobility of electrons is much higher than the mobility of holes and the decrease in the concentration of electrons causes a decrease in the electrical conductivity (this problem will be widely discussed in Section 1.6.3).

Therefore, there is a problem about how to relate the oxidation state of iron ions in magnetite with the presence of electronic defects; how to distinguish them from ions that do not participate in charge transport, and how high is their concentration?

Everybody agrees that in higher temperatures, in magnetite with the stoichiometric composition, when the temperature increases, the concentration of Fe^{3+} iron ions in the sublattice of iron ions with octahedral coordination increases, and simultaneously, the concentration of Fe^{2+} ions in the sublattice with tetrahedral coordination does too, while the changes are mutually related and large. From the calculations of the distribution of iron ions in magnetite with the stoichiometric composition, performed based on the results of the studies on the thermoelectric power (Wu and Mason 1981) it results that their concentration reaches about 0.3 mol/mol (at 1770 K). Thus, there occurs an exchange of electrons between iron ions in the individual cation sublattices. Fe^{2+} ions with octahedral coordination release electrons that are accepted by Fe^{3+} ions with tetrahedral coordination, which can be represented with the reactions:

$$Fe_{(o)}^{2+} = Fe_{(o)}^{3+} + e^{-} \qquad (1.25a)$$

$$Fe_{(t)}^{3+} + e^{\prime} = Fe_{(t)}^{2+} \qquad (1.25b)$$

The change in oxidation state of ions Fe^{2+} and Fe^{3+} in spinel sublattice with the temperature increase can result from an optimisation of interactions between ions, similarly as in the case of formation of heterocation spinels, where ions with a permanent oxidation state occupy both cation sublattices in a specific ratio.

At high temperatures, a thermodynamic equilibrium state is set between iron ions Fe^{2+} and Fe^{3+} in both sublattices. Therefore, by adding the Equation 1.25a and 1.25b we get an equation describing the equilibrium state:

$$Fe^{2+}_{(o)} + Fe^{3+}_{(t)} \leftrightarrows Fe^{3+}_{(o)} + Fe^{2+}_{(t)} \tag{1.26a}$$

Equilibrium constant for the above reactions assumes the form of:

$$K\prod_i \gamma_i = K' = \frac{\left[Fe^{3+}_{(o)}\right]\left[Fe^{2+}_{(t)}\right]}{\left[Fe^{2+}_{(o)}\right]\left[Fe^{3+}_{(t)}\right]} \tag{1.26b}$$

where γ_i denotes the ion activity coefficient. Expressions for the equilibrium constant, resulting from the equity of chemical potentials of components, are determined by the relation between their activities. In the case of such a high concentration of ions (system components), as all the cations are included, the activity and the activity coefficient are hard to define and determine. Equation 1.26b describes the equilibrium state, but the constant K' is an effective equilibrium constant, because it is determined by the concentrations of ions. As in high temperatures the changes of concentrations of ions with different changes in the individual sublattices, due to the change in concentrations of ionic defects, are small, the constant K' is approximatively unchanged, thus Equation 1.26b determines the equilibrium state.

Thus, the change in the oxidation state of iron ions can result from the optimisation of interactions between ions, but it should be also associated with the change in the concentration of electronic defects that are localised on iron ions and they cause an increase in electrical conductivity when the temperature increases.

Let us assume that the change in the oxidation state (charge) of iron ions is associated with the formation of electronic defects, which seems to be the simplest model. In such a case, according to the nomenclature of point defects by Kröger–Vink, iron ions Fe^{2+} in cation nodes with tetrahedral coordination will be negative ions in relation to the lattice $\left(Fe'_{Fe(t)}\right)$ and it should be assumed that their concentration should be equal to the concentration of electrons $\left[Fe'_{Fe(t)}\right] = \left[e'\right]$. A decrease in the concentration of Fe^{2+} ions and an increase in the concentration of Fe^{3+} in the cation sublattice with octahedral coordination should be considered as an appearance of ions that are positive in relation to the lattice $\left[Fe^{\bullet}_{Fe(o)}\right]$; their concentration would be equal to the concentration of electron holes $\left[Fe^{\bullet}_{Fe(o)}\right]$ = [h$^{\bullet}$]. As can be seen, the ions, on which electronic defects are localised, $\left(Fe'_{Fe}\right)$, $\left(Fe^{\bullet}_{Fe}\right)$, are ions with the same charge as parent ions, "not participating" in charge transport (as remaining node ions, but they are in another sublattice). Therefore, the notation for their symbols should be different.

Thus, question arise: how to treat holes or electrons, which should not be quasi-free electronic defects due their high concentration, and how their concentration is related to the distribution of ions with different oxidation states? There is also a problem, how, in such a case, properly write the equations describing the defect equilibria, according to the assumptions and the notation of Kröger–Vink.

Therefore, if we assume that in magnetite with the stoichiometric composition, the change in oxidation state of iron atoms located in the sublattice with octahedral coordination is associated with the formation of electron holes, then the nodal ion releasing an electron becomes an ion that is positive in relation to the lattice, and, according to the notation of Kröger–Vink, its formation is presented by the reaction:

$$Fe^{x}_{Fe(o)} \leftrightarrows Fe^{\bullet}_{Fe(o)} + e' \tag{1.27a}$$

The concentration of these ions equals the concentration of electron holes $\left[Fe^{\bullet}_{Fe}\right]$ = [h$^{\bullet}$]. Thus, the movement of holes will be associated with electron hopping from a nodal ion Fe^{2+} to ion $Fe^{\bullet}_{Fe(o)}$.

If an iron ion, for example in the sublattice with tetrahedral coordination, accepts an electron (it will be located on it), it will become negative in relation to the lattice:

$$Fe^{x}_{Fe(t)} + e' \leftrightarrows Fe'_{Fe(t)} \tag{1.27b}$$

The concentration of these ions will equal the concentration of electrons and electron transport will consist in their hopping from an Fe^{2+} ion to Fe^{3+} nodal ions. Adding Equation 1.27a and 1.27b we get:

$$\mathrm{Fe}^{\mathrm{x}}_{\mathrm{Fe(o)}} + \mathrm{Fe}^{\mathrm{x}}_{\mathrm{Fe(t)}} \leftrightarrows \mathrm{Fe}^{\bullet}_{\mathrm{Fe(o)}} + \mathrm{Fe}'_{\mathrm{Fe(t)}} \tag{1.27c}$$

Equilibrium constant for the above reactions assumes the form of:

$$K\prod_i \gamma_i = K'_{\mathrm{ef}} = \frac{\left[\mathrm{Fe}^{\bullet}_{\mathrm{Fe}}\right]\left[\mathrm{Fe}'_{\mathrm{Fe)}}\right]}{\left[\mathrm{Fe}^{\mathrm{x}}_{\mathrm{Fe(o)}}\right]\left[\mathrm{Fe}^{\mathrm{x}}_{\mathrm{Fe(t)}}\right]} \tag{1.28a}$$

Equation 1.27c describes an equilibrium state between the concentration of node ions that are electroneutral in relation to the lattice and ions that have an effective charge; it is similar to Equation 1.26b. The equilibrium constant K'_{ef} is an effective constant, because it is determined by concentrations of ions and it is a product of the equilibrium constant and activity coefficients ($\prod_i \gamma_i$). If changes in concentrations of ions are small, we can assume that activity coefficients are approximately constant. Thus, the equilibrium constant will be a product of their concentration and activity coefficients. Approximatively, Equation 1.28a assumes the form of:

$$K'_{\mathrm{ef}}\left[\mathrm{Fe}^{\mathrm{x}}_{\mathrm{Fe(o)}}\right]\left[\mathrm{Fe}^{\mathrm{x}}_{\mathrm{Fe(t)}}\right] \cong K_{\mathrm{eh}} = \left[\mathrm{Fe}^{\bullet}_{\mathrm{Fe}}\right]\left[\mathrm{Fe}'_{\mathrm{Fe}}\right] \equiv \left[\mathrm{h}^{\bullet}\right]\left[\mathrm{e}'\right]\Delta G^{\mathrm{o}}_{\mathrm{eh}} \tag{1.28b}$$

Therefore, the effective equilibrium constant K_{eh} depends on the "effective" standard Gibbs energy of the formation of ions $\left(\mathrm{Fe}^{\bullet}_{\mathrm{Fe}}\right)$ and $\left(\mathrm{Fe}'_{\mathrm{Fe}}\right)$, i.e. electronic defects $\left(\Delta G^{\mathrm{o}}_{\mathrm{eh}}\right)$, as Equation is equivalent to Equation 1.24b describing the equilibrium between quasi-free electronic defects. The model of electronic defects assumed above seems quite obvious and simple, as holes and electrons are localised in two different sublattices and their transport occurs as a result of electrons hopping between the ions Fe^{2+} and Fe^{3+} in the whole crystal.

On the other hand, if we assume that electronic defects are localised only in the octahedral sublattice, as many authors state, this would mean that electronic defects are localised on part of Fe^{2+} and Fe^{3+} ions. These ions will have an effective charge in relation to the lattice ions, while their oxidation state will be the same as that of parent ions in the other sublattice. As can be seen in Figure 1.1b, in the magnetite structure the presence of an electron causes that three Fe^{2+} ions, and, in the case of a hole, three Fe^{3+} ions would be present in the internal "cube" of block B. Such a state is significantly different from the ideal layout (cubes in Figure 1.1b) where there are two Fe^{2+} ions and two Fe^{3+} ions, and such a magnetite has very low electrical conductivity. Therefore, in magnetite with the stoichiometric composition, there can exist a determined concentration of ions [Fe^{2+}] and [Fe^{3+}], which are charged in relation to ions of a given sublattice ($\left[\mathrm{Fe}'_{\mathrm{Fe}}\right]$ and $\left[\mathrm{Fe}^{\bullet}_{\mathrm{Fe}}\right]$) and their concentration is equal to the concentration of electronic defects. Thus, the distribution of iron ions with different charges (Equation 1.26b) is also an effect of the presence of electronic defects.

A similar problem for the notation of electronic defects is present in the case of the reaction of ionic defects formation. From the reaction, e.g. of formation of iron vacancies (Equation 1.9a to 1.9c), both in octahedral and tetrahedral lattice, it results that as an effect of incorporation of an oxygen atom and formation of O^{2-} ions, two electrons are taken from Fe^{2+} iron ions. This causes an increase in the concentration of Fe^{3+} ions and these ions should be considered quasi-free defects $\left(\mathrm{Fe}^{\bullet}_{\mathrm{Fe}}\right)$. Their concentration is equivalent of the concentration of the electrons taken and it determines the increase in the concentration of electron holes. In the defect equation, instead of the symbol of quasi-free electron holes, it would be then necessary to use the symbol of iron ions positive in relation to the lattice ($\mathrm{Fe}^{\bullet}_{\mathrm{Fe}}$) and Equation 1.9a would assume the form of:

$$1/2\mathrm{O}_2 + 2\mathrm{Fe}^{\mathrm{x}}_{\mathrm{Fe(2+)}} = \mathrm{O}^{\mathrm{x}}_{\mathrm{O}} + \mathrm{V}''_{\mathrm{Fe(o)}} + 2\mathrm{Fe}^{\bullet}_{\mathrm{Fe(2+)}} \tag{1.29a}$$

Ions $\mathrm{Fe}^{\bullet}_{\mathrm{Fe}}$ are identical with iron node ions Fe^{3+} and their concentration increases. In the case of formation of vacancies $\mathrm{V}'''_{\mathrm{Fe(o)}}$, ions Fe^{4+} with octahedral coordination could be formed, and the reaction of their formation would assume the form of:

$$3/4\mathrm{O}_2 + 3\mathrm{Fe}^{\mathrm{x}}_{\mathrm{Fe(3+)}} = 3/2\mathrm{O}^{\mathrm{x}}_{\mathrm{O}} + \mathrm{V}'''_{\mathrm{Fe(o)}} + 3\mathrm{Fe}^{\bullet}_{\mathrm{Fe(o)(3+)}} \tag{1.29b}$$

As the transition $Fe^{2+} \rightarrow Fe^{3+} + e'$ is more favourable energetically, the formation of Fe^{4+} ions seems unlikely, and, form magnetite, their presence was not found experimentally.

In turn, in the case of formation of interstitial iron ions (Equation 1.13a to 1.13d), the freed oxygen releases the electrons and the oxidation state of an iron ion decreases. Accepting an electron by an Fe^{3+} ion with tetrahedral coordination causes an increase in the concentration of Fe^{2+} ions, negative in relation to the lattice ($Fe'_{Fe(t)}$), which corresponds to an increase in the concentration of electrons. Thus, the reaction of formation of interstitial ions $Fe_i^{3\bullet}$ assumes the form of:

$$4Fe^{x}_{Fe(t)} + 3/2O^{x}_{O} = Fe_i^{3\bullet} + 3Fe'_{Fe(t)} + 3/4O_2 \qquad (1.29c)$$

By analogy, in the case of a transition of Fe^{3+} ions with octahedral coordination into an interstitial void, electrons left by the removed oxygen cause the formation of a Fe^{2+} ion, which is negative in relation to the lattice $\left(Fe'_{Fe(o)}\right)$, and the reaction (1.13c) would have the form of:

$$4Fe^{x}_{Fe(o)(3+)} + 3/2O^{x}_{O} = Fe_i^{3\bullet} + 3Fe'_{Fe(o)(2+)} + 3/4O_2 \qquad (1.29d)$$

The $Fe'_{Fe(o)}$ ion is formally identical with the Fe^{2+} node ion, but it should be treated as a point defect that participates in electron transport. On this ion, an electron could be localised; or an increase in the concentration of these ions is equivalent to an increase in the concentration of quasi-free electrons.

Therefore, in magnetite at the stoichiometric composition, at a constant temperature, in a given sublattice there would exist a part of cations that would be quasi-free defects (charged relative to the lattice). Their concentration, even if they were not localised electronic defects, would be equivalent to their concentration. In turn, as a result of an increase in the oxygen pressure that causes an increase in the concentration of iron vacancies, electrons accepted by oxygen increase the concentration of electron holes and the concentration of ions $\left[Fe^{\bullet}_{Fe}\right]$ increases. On the other hand, electrons released by oxygen, due to an increase inf the concentration of interstitial iron ions when the oxygen pressure decreases, cause an increase in the concentration of ions $\left[Fe'_{Fe}\right]$. The introduced notation of quasi-free ions $Fe^{\bullet}_{Fe}$ and Fe'_{Fe} does not require their accurate localisation in tetrahedral or octahedral sublattice. The studies of the charge transport mechanism could indicate their localisation in a given sublattice. In the crystal, an equilibrium is set between electronic defects and ionic defects, thus, there is a determined distribution of Fe^{2+} and Fe^{3+} in individual sublattices. Apart from ions with different oxidation states, related to the presence of electronic defects, the distribution of ions with different charges will result from optimisation of energy of interaction between ions.

Thus, in magnetite there will be an equilibrium between $Fe^{\bullet}_{Fe}$ and Fe'_{Fe} ions resulting from the existence of electrons and holes, as presented by the reaction. Therefore, instead of Equation 1.24b describing the equilibrium between quasi-free electronic defects, Equation 1.28b should be used; it describes the equilibrium between $Fe^{\bullet}_{Fe}$ and Fe'_{Fe} ions, with a charge in relation to the lattice. They are closely related to electronic defects and Equation 1.28b is one of the elementary reactions of formation of point defects. Moreover, the condition of lattice electroneutrality de facto describes the sum of the concentration of ionic defects and ions $Fe^{\bullet}_{Fe}$ and Fe'_{Fe}, with a charge in relation to the lattice. These ions have a charge in relation to ions of the sublattice, despite the fact that for magnetite they have the same oxidation state as parent ions of the adjacent sublattice. Node ions of the lattice are treated, according to the notation of Kröger–Vink, as electroneutral in relation to the lattice. The value $\Delta G^{o}_{eh}s$ would be related to the formation of ions Fe'_{Fe} and $Fe^{\bullet}_{Fe}$; their concentration is equivalent to the concentration of electronic defects. The character of electronic defects and mechanism of their transport should not affect the reaction of formation of ionic defects and setting the equilibrium between iron vacancies and interstitial ions and ions Fe'_{Fe} and $Fe^{\bullet}_{Fe}$. Due to a high concentration of ions $\left[Fe'_{Fe}\right]$ and $\left[Fe^{\bullet}_{Fe}\right]$ in magnetite, their activity coefficients are different from one, but, as the changes of the concentration of ionic defects are small, changes in the concentrations of these ions are small. Thus, effective (apparent) equilibrium constants for the reaction of formation of "electronic defects" and the value ΔG^{o}_{eh} should not depend on the oxygen pressure.

Due to a long tradition and simplicity of the notation of electronic defects and their differentiation from ionic defects, symbols of holes ($h^{\bullet}$) and electrons (e') should be allowed also in the case where they cannot be treated as quasi-free electronic defects, because the concentration of electronic defects is always related to the concentration of ions with a charge in relation to the lattice (their concentration changes when the concentration of electronic defects changes).

The consistency of the calculated values of concentrations of electronic defects and ionic defects (deviation from the stoichiometry) with the experimental results for many oxides indicates that treating point defects as quasi-free components of the crystal is correct. In the crystal, they form an approximately ideal solution and their description is possible with equations resulting from chemical thermodynamics (thermodynamics of point defects). Changes of concentration of point defects in the range of oxides' existence are small and the equilibrium constants, even at high concentrations of defects, practically do not change when the oxygen pressure changes (see for example oxides: $Fe_{1-\delta}O$, $TiO_{2-\delta}$ (Stokłosa 2015)).

1.4.6 Resultant Standard Gibbs Energy of Formation of Defects

In the case of a simple structure of point defects, when, in a non-stoichiometric compound, in a specific range of oxygen pressure, one type of ionic defects dominates, it is possible to determine the equilibrium constant and standard Gibbs energy of their formation from the dependence of the deviation from the stoichiometry on the oxygen pressure. The problem becomes complex when there are several types of ionic defects, as in the case of magnetite, and their relative concentrations change when the oxygen pressure changes (see Equations 1.10a to 1.10c and 1.14a to 1.14c).

As it was shown (Stokłosa 2015), when considering the change in the concentration of defects in relation to the stoichiometric composition, it is possible to write a general equation of formation of ionic defects and determine a resultant standard Gibbs energy of formation of defects; its values change with the oxygen pressure. In the case of defects with the highest ionisation degrees, due to an increase in oxygen pressure, the concentration of cation vacancies increases, and, simultaneously, the concentration of interstitial ions decreases. Changes in concentrations in relation to the concentration at the stoichiometric composition are:

$$\Delta y_i = y_i - y_i^o \quad \text{or} \quad \Delta y_j = y_j^o - y_j$$

where y_i^o and y_j^o denote the concentrations of the individual defects (cation vacancies and interstitial ions) at the stoichiometric composition. Multiplying the reactions (1.9a, 1.9b and 1.13a) by appropriate changes in concentrations of defects, we get reactions of formation of individual defects:

$$\frac{\Delta y_{V''_{Fe}}}{2}O_2 = \Delta y_{V''_{Fe}}O_O^x + \Delta y_{V''_{Fe}}V''_{Fe(o)} + 2\Delta y_{V''_{Fe}}h^\bullet \qquad \Delta y_{V''_{Fe}}\Delta G^o_{V''_{Fe}} \qquad (1.30a)$$

$$\frac{3\Delta y_{V'''_{Fe(o)}}}{4}O_2 = \frac{3\Delta y_{V'''_{Fe(o)}}}{2}O_O^x + \Delta y_{V'''_{Fe(o)}}V'''_{Fe(okta)} + 3\Delta y_{V'''_{Fe(o)}}h^\bullet \qquad \Delta y_{V'''_{Fe(o)}}\Delta G^o_{V'''_{Fe(o)}} \qquad (1.30b)$$

$$\frac{3\Delta y_{V'''_{Fe(t)}}}{4}O_2 = \frac{3\Delta y_{V'''_{Fe(t)}}}{2}O_O^x + \Delta y_{V'''_{Fe(t)}}V'''_{Fe(tetra)} + 3\Delta y_{V'''_{Fe(t)}}h^\bullet \qquad \Delta y_{V'''_{Fe(t)}}\Delta G^o_{V'''_{Fe(t)}} \qquad (1.30c)$$

On the other hand, the concentration of interstitial iron ions will decrease when the oxygen pressure decreases, thus, the reactions in the opposite direction will occur; their concentration will decrease (Equation 1.13a to 1.13d describe their formation):

$$\Delta y_{Fe_i^{\bullet\bullet}}Fe_i^{\bullet\bullet} + \frac{\Delta y_{Fe_i^{\bullet\bullet}}}{2}O_2 = \Delta y_{Fe_i^{\bullet\bullet}}Fe_{Fe(o)}^{x(2+)} + \Delta y_{Fe_i^{\bullet\bullet}}O_O^x + 2\Delta y_{Fe_i^{\bullet\bullet}}h^\bullet \qquad -\Delta y_{Fe_i^{\bullet\bullet}}\Delta G^o_{Fe_i^{\bullet\bullet}} \qquad (1.30d)$$

$$\Delta y_{Fe_i^{3\bullet}}Fe_i^{3\bullet} + \frac{3\Delta y_{Fe_i^{3\bullet}}}{4}O_2 = \Delta y_{Fe_i^{3\bullet}}Fe_{Fe(o)}^{x(3+)} + \frac{3\Delta y_{Fe_i^{3\bullet}}}{2}O_O^x + 3\Delta y_{Fe_i^{3\bullet}}h^\bullet \qquad -\Delta y_{Fe_i^{3\bullet}}\Delta G^o_{Fe_i^{3\bullet}} \qquad (1.30e)$$

$$\Delta y_{Fe_i^{3\bullet}}Fe_i^{3\bullet} + \frac{3\Delta y_{Fe_i^{3\bullet}}}{4}O_2 = \Delta y_{Fe_i^{3\bullet}}Fe_{Fe(t)}^{x(3+)} + \frac{3\Delta y_{Fe_i^{3\bullet}}}{2}O_O^x + 3\Delta y_{Fe_i^{3\bullet}}h^\bullet \qquad -\Delta y_{Fe_i^{3\bullet}}\Delta G^o_{Fe_i^{3\bullet}} \qquad (1.30f)$$

Adding the Equation 1.30a to 1.30f we get an equation describing the formation of cation vacancies and the decrease in the concentration of interstitial iron ions:

$$\frac{2\Delta y_{V''_{Fe}} + 3\Delta y_{V'''_{Fe(o)}} + 3\Delta y_{V'''_{Fe(t)}} + 2\Delta y_{Fe_i^{\bullet\bullet}} + 6\Delta y_{Fe_i^{3\bullet}}}{4}O_2 + \Delta y_{Fe_i^{\bullet\bullet}}Fe_i^{\bullet\bullet} + 2\Delta y_{Fe_i^{3\bullet}}Fe_i^{3\bullet}$$

$$= \frac{2\Delta y_{V''_{Fe}} + 3\Delta y_{V'''_{Fe(o)}} + 3\Delta y_{V'''_{Fe(t)}} + 2\Delta y_{Fe_i^{\bullet\bullet}} + 6\Delta y_{Fe_i^{3\bullet}}}{4}O_O^X + \Delta y_{Fe_i^{\bullet\bullet}}Fe_{Fe}^X + 2\Delta y_{Fe_i^{3\bullet}}Fe_{Fe}^X + \Delta y_{V''_{Fe}}V''_{Fe(o)}$$

$$+\Delta y_{V'''_{Fe(o)}}V'''_{Fe(okta)} + \Delta y_{V'''_{Fe(t)}}V'''_{Fe(tetra)} + (2\Delta y_{V''_{Fe}} + 3\Delta y_{V'''_{Fe(o)}} + 3\Delta y_{V'''_{Fe(t)}} + 2\Delta y_{Fe_i^{\bullet\bullet}} + 6\Delta y_{Fe_i^{3\bullet}})h^\bullet \qquad \Delta G^{o(\delta)}_{[def]} \qquad (1.31a)$$

We can divide Equation 1.31a by the amount of the incorporated oxygen, that is:

$$2\Delta y_{V''_{Fe}} + 3\Delta y_{V'''_{Fe(o)}} + 3\Delta y_{V'''_{Fe(t)}} + 2\Delta y_{Fe_i^{\bullet\bullet}} + 6\Delta y_{Fe_i^{3\bullet}} = \Delta y_O \qquad (1.31b)$$

As a result, we get an equation of formation of defects with the highest ionisation degree in magnetite, due to the incorporation of 1/2 mol of oxygen atoms (1/2 O_O^x):

$$\frac{1}{4}O_2 + \frac{\Delta y_{Fe_i^{\bullet\bullet}}}{\Delta y_O}Fe_i^{\bullet\bullet} + \frac{2\Delta y_{Fe_i^{3\bullet}}}{\Delta y_O}Fe_i^{3\bullet} = \frac{1}{2}O_O^x + \left(\frac{\Delta y_{Fe_i^{\bullet\bullet}}}{\Delta y_O} + \frac{2\Delta y_{Fe_i^{3\bullet}}}{\Delta y_O}\right)Fe_{Fe}^x$$

$$+\frac{\Delta y_{V_{Fe}''}}{\Delta y_O}V_{Fe(o)}'' + \frac{\Delta y_{V_{Fe(o)}'''}}{\Delta y_O}V_{Fe(o)}''' + \frac{\Delta y_{V_{Fe(t)}'''}}{\Delta y_O}V_{Fe(t)}''' + h^\bullet \qquad \Delta G^{o(\delta)}_{[def]} \qquad (1.32a)$$

The standard Gibbs energy of the formation of defects $\Delta G^o_{[def]}$, according to the reaction 1.32a, will be also a sum ΔG_i^o of reactions 1.30a-f multiplied by appropriate ratios of the change in defects' concentrations:

$$\Delta G^{o(\delta)}_{[def]} = \frac{\Delta y_{V_{Fe}''}}{\Delta y_O}\Delta G^o_{V_{Fe}''} + \frac{\Delta y_{V_{Fe(o)}'''}}{\Delta y_O}\Delta G^o_{V_{Fe(o)}'''} + \frac{\Delta y_{V_{Fe(t)}'''}}{\Delta y_O}\Delta G^o_{V_{Fe(t)}'''} - \frac{\Delta y_{Fe_i^{\bullet\bullet}}}{\Delta y_O}\Delta G^o_{Fe_i^{\bullet\bullet}} - \frac{2\Delta y_{Fe_i^{3\bullet}}}{\Delta y_O}\Delta G^o_{Fe_i^{3\bullet}} \qquad (1.32b)$$

Therefore, knowing the values of ΔG_i^o of formation of the individual defects, we can determine the value of $\Delta G^o_{[def]}$ at a given oxygen pressure. It will change with the change in relative concentrations of defects (with the change in oxygen pressure). If, for example, iron vacancies $V_{Fe(o)}'''$ dominate at the highest oxygen pressures, it will be close to the value of $1/3\Delta G^o_{V_{Fe(o)}'''}$, as the value $\Delta G^o_{V_{Fe(o)}'''}$ is associated with the incorporation of 3/2 moles of oxygen. At the lowest oxygen pressures, where ions $Fe_i^{3\bullet}$ dominate, it will be close to $1/3\Delta G^o_{Fe_i^{3\bullet}}$. From the temperature dependence of $\Delta G^o_{[def]}$ we can determine the resultant Gibbs energy of their formation at a given oxygen pressure. It should be close to the value of the Gibbs energy of formation of defects, that is determined from the dependence of the deviation from the stoichiometry on the inverse of temperature in specified ranges of oxygen pressure.

According to the thermodynamics of the point defects, at equilibrium state, at a given oxygen pressure, there is an equality of the virtual chemical potentials (μ) and electrochemical potentials (η) of the point defects present in the spinel. The equilibrium condition for the reaction 1.32a assumes the form of:

$$\frac{1}{4}\mu_{O_2} + \frac{\Delta y_{Fe_i^{\bullet\bullet}}}{\Delta y_O}\eta_{Fe_i^{\bullet\bullet}} + \frac{2\Delta y_{Fe_i^{3\bullet}}}{\Delta y_O}\eta_{Fe_i^{3\bullet}} = \frac{1}{2}\mu_{O_O^x} + \left(\frac{\Delta y_{Fe_i^{\bullet\bullet}}}{\Delta y_O} + \frac{2\Delta y_{Fe_i^{3\bullet}}}{\Delta y_O}\right)\mu_{Fe_{Fe}^x} + \frac{\Delta y_{V_{Fe}''}}{\Delta y_O}\eta_{V_{Fe}''} + \frac{\Delta y_{V_{Fe(o)}'''}}{\Delta y_O}\eta_{V_{Fe(o)}'''} + \frac{\Delta y_{V_{Fe(t)}'''}}{\Delta y_O}\eta_{V_{Fe(t)}'''} + \eta_{h^\bullet} \qquad (1.33a)$$

The chemical potential of a component i can be presented as a sum of the standard potential and the component concentration (activity) function, ($\mu_i = \mu_i^o + RT\ln a_i$). The standard Gibbs energy of the individual processes (Equation 1.30a to 1.30f) is the difference of the standard chemical potentials of the products and reactants of these reactions. The sum of the standard Gibbs energies of these reactions will be equal to the sum of the standard chemical potentials of the defects appearing in Equation 1.33a and finally, it will assume the form of:

$$\frac{\Delta y_{V_{Fe}''}}{\Delta y_O}\ln\left[V_M''\right] + \frac{\Delta y_{V_{Fe(o)}'''}}{\Delta y_O}\ln\left[V_{Fe(o)}'''\right] + \frac{\Delta y_{V_{Fe(t)}'''}}{\Delta y_O}\ln\left[V_{Fe(t)}'''\right] + \ln\left[h^\bullet\right] - \frac{\Delta y_{Fe_i^{\bullet\bullet}}}{\Delta y_O}\ln\left[Fe_i^{\bullet\bullet}\right] - \frac{2\Delta y_{Fe_i^{3\bullet}}}{\Delta y_O}\ln\left[Fe_i^{3\bullet}\right] - \ln p_{O_2}^{1/4} = \frac{\Delta G^{o(\delta)}_{[def]}}{RT} \qquad (1.33b)$$

Equation 1.33b describes the dependence of concentration of all ionic defects on the oxygen pressure, at the equilibrium state (for defects assumed in the model). The derivative of the log of deviation from the stoichiometry with respect to the log of oxygen pressure (d(log δ)/d(log p_{O_2}), determines the values of the resultant "exponent" in the relation of concentration of defects that changes when the oxygen pressure changes. In the range of oxygen pressure, where one type of ionic defects dominates, this exponent will have a constant value, close to the value resulting from the expression describing the equilibrium constant of the reaction of formation of this defect (Equation 1.10a, 1.10b, 1.10c and 1.14a, 1.14b, 1.14c).

1.5 METHODS FOR CALCULATING THE DIAGRAM OF THE CONCENTRATIONS OF POINT DEFECTS

For the calculations of diagrams of concentrations of defects, it was assumed that point defects in a crystal can be treated as quasi-free components which form a solid solution that is approximately ideal. For their description it is possible to use chemical thermodynamics (thermodynamics of point defects). In the case of magnetite, the above assumption is valid, as the concentration of ionic defects (deviation from the stoichiometry) is small. In order to calculate the diagram of concentrations of defects it is necessary to know the equilibrium constants of reactions of formation of defects (Equation 1.10a, 1.10b, 1.10c and 1.14a, 1.14b, 1.14c), which are lacking and we can only attempt to calculate them theoretically. The remaining solution is then the method consisting in adjusting such values of the equilibrium constants that the calculated sum of concentrations of defects is close to the experimental values of the deviation from the stoichiometry at the individual oxygen pressures. From the equations presented in Section 1.4.1, describing equilibria between defects, it results that, in the case of magnetite, this would mean adjusting at least six equilibrium constants for defects with the highest ionisation degree and the equilibrium constant of the reaction of formation of electronic defects. Adjusting simultaneously seven constant parameters causes leads to the ambiguity of the solution. The problem can be solved by performing the adjustment of equilibrium constants in several sequential steps. Instead of adjusting equilibrium constants, with values that can vary in a wide range, it is more comfortable to track the influence of sequentially adjusted values of standard Gibbs energies (ΔG_i^{o}) of formation of individual defects, that are used to calculate the equilibrium constants. As the concentrations of the individual defects are mutually related (they are dependent on the oxygen pressure and on the concentration of electron holes), during their calculation we can use the relations resulting from the expression for equilibrium constants (see Equation 1.10a, 1.10b, 1.10c and 1.14a, 1.14b, 1.14c). The relative concentrations of defects can be determined for example in relation to the concentration of iron vacancies $\left[V_{Fe(o)}^{'''}\right]$ or $\left[V_{Fe(o)}^{''}\right]$.

Thus, we adjust for example the values $\Delta G_{V_{Fe(o)}^{'''}}^{o}$ and calculate equilibrium constants according to Equation 1.11 and the concentration of vacancies $\left[V_{Fe(o)}^{'''}\right]$. The concentration of interstitial iron ions is associated with the change in the concentration of vacancies $\left[V_{Fe(o)}^{'''}\right]$,. At the same oxygen pressure, from Equation 1.10b and 1.14b, a relation between the concentration of vacancies $\left[V_{Fe(o)}^{'''}\right]$ and the concentration of interstitial ions $\left[Fe_i^{3\bullet}\right]$ is derived:

$$\frac{\left[V_{Fe(o)}^{'''}\right]\left[h^{\bullet}\right]}{K_{V_{Fe(o)}^{'''}}} = \frac{K_{Fe_i^{3\bullet}}^{2}\left[h^{\bullet}\right]}{\left[Fe_i^{3\bullet}\right]} \tag{1.34a}$$

Transforming Equation 1.34a we obtain:

$$\left[V_{Fe(o)}^{'''}\right]\left[Fe_i^{3\bullet}\right] = K_{V_{Fe(o)}^{'''}} K_{Fe_i^{3\bullet}}^{2} = K_{F_1} \tag{1.34b}$$

Adjusting the value of $\Delta G_{F_1}^{o}$ of formation of Frenkel-type defects, we calculate the value of K_{F^1} and determine the concentration of ions $\left[Fe_i^{3\bullet}\right]$ depending on the concentration $\left[V_{Fe(o)}^{'''}\right]$.

The standard Gibbs energy of the formation of interstitial $\left[Fe_i^{3\bullet}\right]$ ions according to Equation 1.34b and 1.11 will be:

$$\Delta G_{Fe_i^{3\bullet}}^{o} = \Delta G_{F_1}^{o} - \Delta G_{V_{Fe(o)}^{'''}}^{o} \tag{1.34c}$$

In turn, from Equation 1.10a and 1.10b it results that at the same oxygen pressure there is a relation between the concentration of vacancies $\left[V_{Fe(o)}^{'''}\right]$ and $\left[V_{Fe(o)}^{''}\right]$. The ratio of concentration of vacancies $\left[V_{Fe(o)}^{'''}\right]/\left[V_{Fe(o)}^{''}\right]$ depends on the equilibrium constants $K_{V_{Fe(o)}^{'''}}$ and $K_{V_{Fe}^{''}}$ according to the relation:

$$\frac{\left[V_{Fe(o)}^{'''}\right]^{3/4}\left[h^{\bullet}\right]^{4}}{K_{V_{Fe(o)}^{'''}}^{3/4}} = \frac{\left[V_{Fe(o)}^{''}\right]^{2}\left[h^{\bullet}\right]^{4}}{K_{V_{Fe(o)}^{''}}^{2}} \tag{1.35a}$$

After rearrangement we get:

$$\frac{\left[V_{Fe}^{''}\right]^{2}}{\left[V_{Fe(o)}^{'''}\right]^{4/3}} = \frac{K_{V_{Fe}^{''}}^{2}}{K_{V_{Fe(o)}^{'''}}^{4/3}} = K_{V_{Fe}^{''}/V_{Fe(o)}^{'''}} = K_1 \tag{1.35b}$$

The concentration of vacancies $\left[V''_{Fe(o)}\right]$ at the stoichiometric composition should not be greater than that of vacancies $\left[V'''_{Fe(o)}\right]$. Therefore, choosing the value of ΔG_1^o we calculated the constant K_1 and, depending on the concentration $\left[V'''_{Fe(o)}\right]$ we calculate the concentration of vacancies $\left[V''_{Fe}\right]$, which, at the stoichiometric composition, should be close to the concentration of vacancies $\left[V'''_{Fe(o)}\right] \cong \left[V''_{Fe(o)}\right]$:

$$\left[V''\right] = K_1^{3/4}\left[V'''_{Fe(o)}\right]^{3/2} \tag{1.35c}$$

The standard Gibbs energy of the formation of vacancies V''_{Fe}, according to Equation 1.11 and 1.35b, is:

$$\Delta G^o_{V''_{Fe}} = 1/2\Delta G_1^o + 2/3\Delta G^o_{V'''_{Fe(o)}} \tag{1.35d}$$

With the change in the concentration of vacancies $\left[V''_{Fe}\right]$, the concentration of interstitial iron ions $\left[Fe_i^{\bullet\bullet}\right]$ is also associated. At the same oxygen pressure, from Equation 1.10a and 1.14a, the following relation results:

$$\frac{\left[V''_{Fe(o)}\right]^2\left[h^\bullet\right]^4}{K^2_{V''_{Fe(o)}}} = \frac{K^2_{Fe_i^{\bullet\bullet}}\left[h^\bullet\right]^4}{\left[Fe_i^{\bullet\bullet}\right]^2} \tag{1.36a}$$

Transforming Equation 1.36a we obtain:

$$\left[V''_{Fe(o)}\right]\left[Fe_i^{\bullet\bullet}\right] = K_{V''_{Fe(o)}}K^2_{Fe_i^{\bullet\bullet}} = K_{F_2} \tag{1.36b}$$

Thus, adjusting the value of the standard Gibbs energy of the formation of Frenkel defects $\Delta G^o_{F_2}$ we calculate the equilibrium constant K_{F2} and determine the concentration of interstitial ions $\left[Fe_i^{\bullet\bullet}\right]$ dependence on the concentration of vacancies $\left[V''_{Fe(o)}\right]$.

The standard Gibbs energy of the formation of interstitial $\left[Fe_i^{\bullet\bullet}\right]$ ions according to Equation 1.36b is:

$$\Delta G^o_{Fe_i^{\bullet\bullet}} = \Delta G^o_{F_2} - \Delta G^o_{V''_{Fe(o)}} \tag{1.36c}$$

The concentration of interstitial iron ions $\left[Fe_i^{3\bullet}\right]$ is in equilibrium with the concentration of iron vacancies in the sublattice with tetrahedral coordination $\left[V'''_{Fe(t)}\right]$. Considering the mutual interactions between ions in octahedral and tetrahedral sublattice we can expect that the energy of formation of iron vacancies $V'''_{Fe(o)}$ should be less than $V'''_{Fe(t)}$, thus the concentration of vacancies in the tetrahedral sublattice should be much less than that in the octahedral sublattice $\left[V'''_{Fe(t)}\right] \ll \left[V'''_{Fe(o)}\right]$

In turn, at the same oxygen pressure, from Equation 1.10c and 1.14c, a relation between the concentration of interstitial ions $\left[Fe_i^{3\bullet}\right]$ and iron vacancies in the tetrahedral sublattice $\left[V'''_{Fe(t)}\right]$ results:

$$\frac{K^2_{Fe_i^{3\bullet}}\left[h^\bullet\right]^3}{\left[Fe_i^{3\bullet}\right]} = \frac{\left[V'''_{Fe(t)}\right]\left[h^\bullet\right]^3}{K_{V'''_{Fe(t)}}} \tag{1.37a}$$

Transforming Equation 1.37a we obtain:

$$\left[V'''_{Fe(t)}\right]\left[Fe_i^{3\bullet}\right] = K_{V'''_{Fe(t)}}K^2_{Fe_i^{3\bullet}} = K_{F_3} \tag{1.37b}$$

The standard Gibbs energy of the formation of vacancies $V'''_{Fe(t)}$, according to Equation 1.36b is:

$$\Delta G^o_{V'''_{Fe(t)}} = \Delta G^o_{F_3} - \Delta G^o_{Fe_i^{3\bullet}} \tag{1.37c}$$

Therefore, in order to determine the relative concentrations of ionic defects, we must choose such values of $\Delta G^{o}_{V'''_{Fe}}$ s, $\Delta G^{o}_{F_1}$, $\Delta G^{o}_{F_2}$, $\Delta G^{o}_{F_3}$ and ΔG^{o}_{I}, (which will serve to calculate the appropriate equilibrium constants) that the calculated deviation from the stoichiometry, being the difference between the concentration of iron vacancies and interstitial ions:

$$\delta = \left[V''_{Fe(o)}\right] + \left[V'''_{Fe(o)}\right] + \left[V'''_{Fe(t)}\right] - \left[Fe_i^{\bullet\bullet}\right] - \left[Fe_i^{3\bullet}\right] \quad (1.38)$$

is consistent with the experimental results δ. Simultaneously, at the stoichiometric composition, for simplification, we can assume the following conditions that are fulfilled by the concentrations of defects:

$$\left[V'''_{Fe(o)}\right] + \left[V'''_{Fe(t)}\right] \cong \left[Fe_i^{3\bullet}\right]$$
$$\text{or, when } \left[V'''_{Fe(o)}\right] \gg \left[V'''_{Fe(t)}\right] \quad (1.39a)$$
$$\left[V'''_{Fe(o)}\right] \cong \left[Fe_i^{3\bullet}\right]$$

$$\text{and } \left[V''_{Fe}\right] \cong \left[Fe_i^{\bullet\bullet}\right] \quad (1.39b)$$

The concentrations of all ionic and electronic defects are bound by the electroneutrality condition, which assumes the form of:

$$\left[e'\right] + 2\left[V''_{Fe(o)}\right] + 3\left[V'''_{Fe(o)}\right] + 3\left[V'''_{Fe(t)}\right] = \left[h^{\bullet}\right] + 2\left[Fe_i^{\bullet\bullet}\right] + 3\left[Fe_i^{3\bullet}\right] \quad (1.40)$$

Thus, the concentration of ionic defects is dependent on the concentration of electronic defects. Between electronic defects there exists also an equilibrium, which was discussed in Section 1.4.5 (see Equation 1.24a, 1.24b or 1.28b) Therefore, in order to determine the concentration of electronic defects it is necessary to know the value of the standard Gibbs energy of formation of electronic defects ΔG^{o}_{eh}. Calculating from Equation 1.40 e.g. electron concentration and inserting into Equation 1.24b, we get the dependence of the concentration of electron holes on the concentration of ionic defects:

$$K_{eh} = \left[h^{\bullet}\right]\left[e'\right] = \left[h^{\bullet}\right]\left(\left[h^{\bullet}\right] + 2\left[Fe_i^{\bullet\bullet}\right] + 3\left[Fe_i^{3\bullet}\right] - 2\left[V''_{Fe(o)}\right] - 3\left[V'''_{Fe(o)}\right] - 3\left[V'''_{Fe(t)}\right]\right) = \left[h^{\bullet}\right]\left(\left[h^{\bullet}\right] + \Sigma\left[def\right]\right) \quad (1.41)$$

Solving the quadratic Equation 1.41, knowing the equilibrium constant K_{eh} which is derived from the value of ΔG^{o}_{eh}, we determine the dependence of the concentration of electron holes on the concentration of ionic defects and, next, the concentration of quasi-free electrons.

Therefore, we calculate the relative concentrations of ionic defects from the dependence on the concentration of, for example, vacancies $\left[V'''_{Fe(o)}\right]$, according to Equation 1.34a, 1.34b, 1.34c to 1.37a, 1.37b, 1.37c and Equation (1.41). The oxygen pressure for the equilibrium concentrations of point defects can be calculated using the expression for the equilibrium constant of formation of, for example, vacancies $V'''_{Fe(o)}$ (Equation 1.10b), which after transformation has the form of:

$$p_{O_2} = \frac{\left[V'''_{Fe(o)}\right]^{4/3}\left[h^{\bullet}\right]^{4}}{K^{4/3}_{V'''_{Fe(o)}}} \quad (1.42)$$

Thus, as a first approximation we assume that in $Fe_{3\pm\delta}O_4$, defects with the highest ionisation degrees dominate and we adjust the values $\Delta G^{o}_{V'''_{Fe}}$, $\Delta G^{o}_{F_1}$, $\Delta G^{o}_{F_2}$, $\Delta G^{o}_{F_3}$ and ΔG^{o}_{I}, which are not entirely arbitrary, because at the stoichiometric composition they fulfil particular conditions (1.39a, b). The process of adjusting the values ΔG^{o}_{i} of formation of individual defects is conducted until a consistency is reached between the dependence of the deviation from the stoichiometry on the oxygen pressure and the experimental values of deviation *δ*. At each change in the value of ΔG^{o}_{i}, the concentration of these defects changes, which causes a perturbation of the electroneutrality condition (Equation 1.41). Consequently, the calculations are performed until this condition is fulfilled. For the calculations, it is possible to use an Excel sheet with one macro that allows, as a result of a few dozen of sequential steps, reaching the electroneutrality condition.

In the first stage of calculations, as small as possible values of $\Delta G^{o}_{F_1} \cong \Delta G^{o}_{F_2}$ and ΔG^{o}_{I} are chosen in such a way that the concentrations of iron vacancies are close, $\left[V'''_{Fe(o)}\right] \cong \left[V''_{Fe}\right]$ (according to Equation 1.35a). This determines the maximum concentration of vacancies $\left[V''_{Fe}\right]$, which does not deteriorate the fit of the calculated dependence of the deviation from the stoichiometry to the experimental results *δ*. Similarly, such smallest values of $\Delta G^{o}_{F_3}$ are chosen that the concentration of vacancies $\left[V_{Fe(t)}'''\right]$ does not deteriorate the quality of match between the dependence of the deviation from the stoichiometry on p_{O_2} and the experimental results δ.

The initial calculations performed have shown that the assumed value of ΔG_{eh}^{o} very significantly affects the adjusted value of $\Delta G_{V_{Fe}'''}^{o}$. The lack of experimental values of ΔG_{eh}^{o}, and effective masses of holes and electrons at high temperatures, does not permit to determine the equilibrium constant K_{eh}. A low value of the band gap and high electrical conductivity showed by magnetite indicates that the concentration of electronic defects should be high, thus, the value of ΔG_{eh}^{o} should be as small as possible. The performed calculations have shown that for the values ΔG_{eh}^{o} up to 50 kJ/mol it is possible to adjust the values of $\Delta G_{V_{Fe(o)}'''}^{o}$ (without changing the value of $\Delta G_{F_1}^{o} \cong \Delta G_{F_2}^{o}$), for which a very good match is obtained between the dependence of the deviation from the stoichiometry on the oxygen pressure and the experimental value of δ. Figure 1.2 shows changes in the value of $\Delta G_{V_{Fe(o)}'''}^{o}$ depending on ΔG_{eh}^{o} at temperatures 1173–1673 K, for which the same degree of match was obtained between the dependence of δ on p_{O_2} and the experimental value of δ.

As can be seen in Figure 1.2, at individual temperatures the dependences are linear and, within the error limit, have similar slope. Table 1.1 presents the parameters of the obtained linear equations.

From the performed calculations it results that the concentration of holes and electrons is slightly lower than the concentration of ions Fe^{2+} in the sublattice with tetrahedral coordination, which was determined based on the studies on the thermoelectric power (Wu and Mason 1981). Due to that, finally, such values of ΔG_{eh}^{o} were adopted, at which the concentration of electronic defects at the stoichiometric composition was equal to the concentration of ions $[Fe^{2+}]$ in the sublattice with tetrahedral coordination. At higher values of $\Delta G_{eh}^{o} > 50$ kJ/mol, in order to obtain a match between the dependence of δ on p_{O_2} and the experimental values, it was necessary to take into account the concentration of ionic defects with lower ionisation degrees (Equations 1.16a, 1.16b to 1.19a, 1.19b, 1.19c).

Thus, at the next stage of calculations it is possible to adjust the values of ΔG_i^{o} of formation of individual defects with lower ionisation degrees, reactions (1.16–1.19), calculate the equilibrium constants and concentrations of defects with lower ionisation degree, taking into account the concentration of defects necessary for their formation (according to Equation 1.20). The concentrations of these defects must be also taken into account in the electroneutrality condition.

When calculating the concentration of defects in the oxygen sublattice, we proceed in the same way as for cation defects. From Equation 1.10b and 1.22b it results that at the same oxygen pressure there is a relation between the concentration of iron vacancies $\left[V_{Fe(o)}'''\right]$ and interstitial oxygen ions $\left[O_i''\right]$:

$$\frac{\left[V_{Fe(o)}'''\right]^{4/3}\left[h^{\cdot}\right]^{4}}{K_{V_{Fe(o)}'''}^{4/3}} = \frac{\left[O_i''\right]^{2}}{K_{O_i''}^{2}\left[e'\right]^{4}} \tag{1.43a}$$

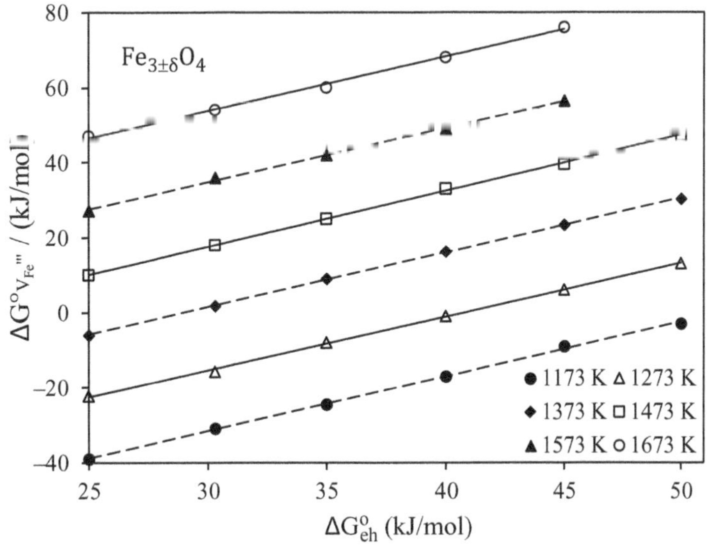

FIGURE 1.2 Dependence of the standard Gibbs energy $\Delta G_{V_{Fe}'''}^{o}$ of formation of iron vacancies V_{Fe}''' in magnetite on the assumed values of ΔG_{eh}^{o} of formation of electronic defects at 1173–1673 K, at which for the defects with the highest ionisation degree there is no deterioration of the match between the dependence of the deviation from the stoichiometry on the oxygen pressure and the experimental results δ (Dieckmann 1982)

TABLE 1.1
Parameters of the dependence of standard Gibbs energy $\Delta G^o_{V_{Fe}'''}$ of formation of iron vacancies V_{Fe}''' in magnetite on the assumed standard Gibbs energy ΔG^o_{eh} of formation of electronic defects ($\Delta G^o_{V_{Fe}'''} = a\Delta G^o_{eh} + b$) in the range of ΔG^o_{eh} = 25–50 kJ/mol at 1173–1673 K determined based on the dependences shown in Figure 1.2.

	$\Delta G^o_{V_{Fe}'''} = a\Delta G^o_{eh} + b$	
T/(K)	a	b
1173	1.45 ± 0.02	−75.2 ± 0.9
1273	1.43 ± 0.02	−58.3 ± 0.8
1373	1.46 ± 0.01	−42.2 ± 0.5
1473	1.48 ± 0.02	−27.0 ± 0.6
1573	1.44 ± 0.03	−8.4 ± 1.2
1673	1.45 ± 0.05	10.2 ± 1.6

After rearrangement we get:

$$\frac{[O_i'']^2}{[V'''_{Fe(o)}]^{4/3}} = \frac{K^2_{O_i''}[e']^4[h^\bullet]^4}{K^{4/3}_{V'''_{Fe(o)}}} = \frac{K^2_{O_i''}K^4_{eh}}{K^{4/3}_{V'''_{Fe(o)}}} = K_{O_i''/V_{Fe}'''} = K_3 \quad (1.43b)$$

Adjusting the value ΔG_3^o, which allows the calculation of the constant K_3, we calculate the concentrations of interstitial oxygen ions $[O_i'']$, depending on the concentration of iron vacancies $[V'''_{Fe(o)}]$.

The concentration of oxygen vacancies $[V_O^{\bullet\bullet}]$ is associated with the change in the concentration of interstitial oxygen ions $[O_i'']$. At the same oxygen pressure, from Equation 1.22a and 1.22b, the following relation results:

$$\frac{K^2_{V_O^{\bullet\bullet}}}{[V_O^{\bullet\bullet}]^2[e']^4} = \frac{[O_i'']^2}{K^2_{O_i''}[e']^4} \quad (1.44a)$$

Transforming Equation 1.44a we obtain:

$$[V_O^{\bullet\bullet}][O_i''] = K_{V_O^{\bullet\bullet}}K^2_{O_i''} = K_{F_O} \quad (1.44b)$$

Thus, adjusting the value of the standard Gibbs energy of the formation of defects of anti-Frenkel type $\Delta G^o_{F_O}$ depending on the concentration of interstitial oxygen ions $[O_i'']$, we calculated the concentration of oxygen vacancies $[V_O^{\bullet\bullet}]$. It is also possible to adjust such a value of $\Delta G^o_{V_O^{\bullet\bullet}}$ (and calculate the equilibrium constant $K_{V_O^{\bullet\bullet}}$), that the concentration of oxygen vacancies is equal to the concentration of interstitial oxygen ions. The deviation from the stoichiometry will be related to the deviation in the cation sublattice and oxygen sublattice, according to Equation 1.23a. We must take into account the concentration of oxygen defects in the electroneutrality condition. The presented method of the calculations of the diagrams of the concentrations of defects and its theoretical foundations were more widely discussed in (Stokłosa 2015).

1.6 RESULTS OF CALCULATION OF THE DIAGRAMS OF THE CONCENTRATIONS OF POINTS DEFECTS AND DISCUSSION

1.6.1 Ionic Defects

Figures 1.3a–c and 1.4a–d present the calculation results of concentrations of ionic and electronic defects (diagrams of the defect concentrations (solid lines)), with the method for the calculation of the defect concentrations diagrams described

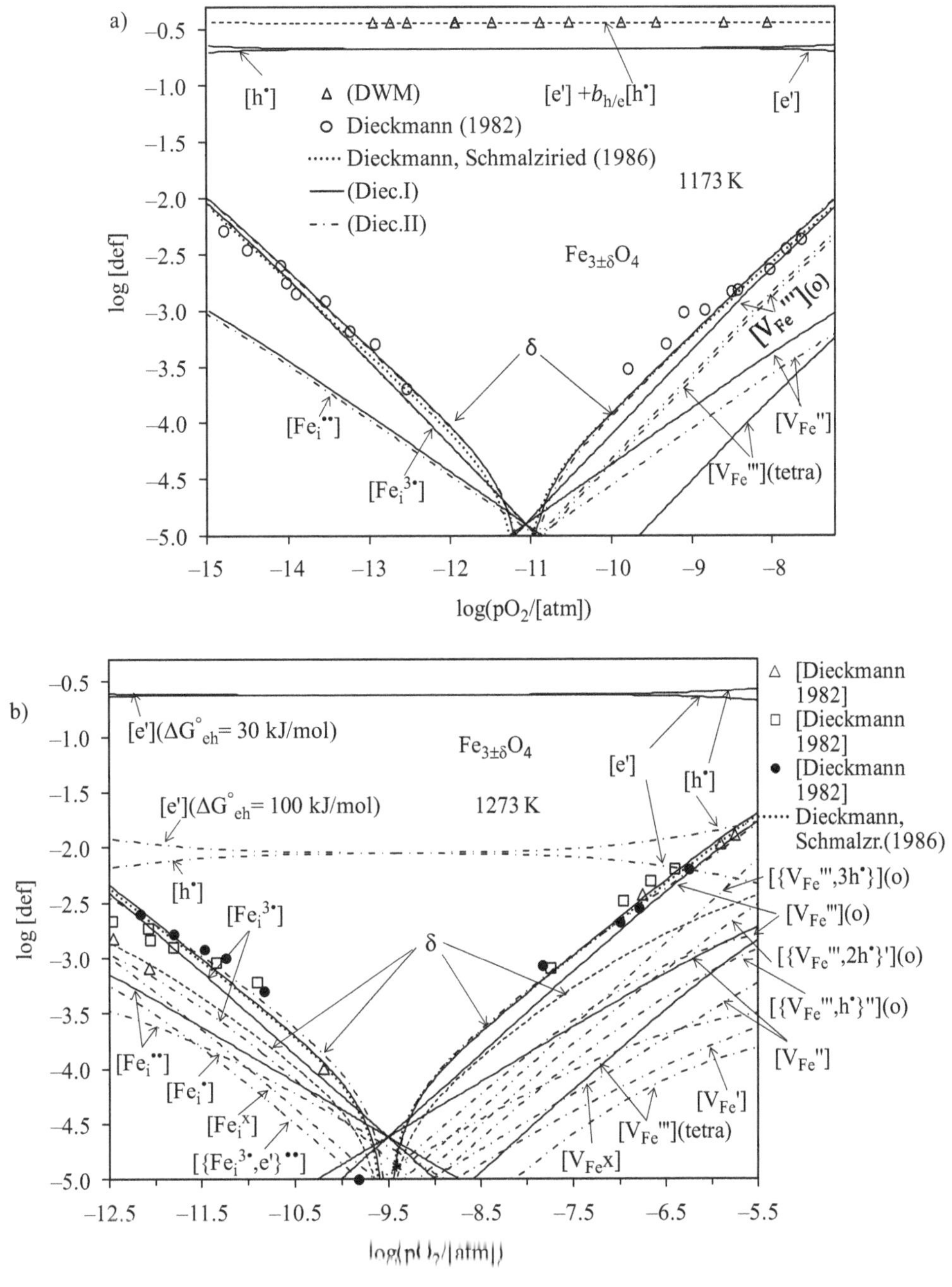

FIGURE 1.3 (a) Diagram of concentrations of point defects for magnetite at 1173 K (solid lines), obtained using the results of the studies of the deviation from the stoichiometry (Dieckmann 1982) ((○) points). The (···) line shows the dependence of the deviation from the stoichiometry on the oxygen pressure, calculated using the "equilibrium constants" of reactions of formation of cation vacancies $\boldsymbol{K}_V$ and interstitial ions $\boldsymbol{K}_I$, (Dieckmann 1982, Dieckmann and Schmalzried 1986). Lines (-··) show the concentration of ionic defects when the concentrations of iron vacancies are similar $\left[V'''_{Fe(o)}\right] \cong \left[V'''_{Fe(t)}\right]$. The points (△)(DWM) represent the values of the sum of concentrations of electronic defects ($[e'] + \boldsymbol{b}_{e/h}[h^{\bullet}]$) calculated according to Equation 1.49, for the fitted value of the mobility of electrons and using the values of the electrical conductivity obtained by Dieckmann et al. (Dieckmann et al. 1983). Line (---) shows the sum of concentrations of electronic defects ($[e'] + \boldsymbol{b}_{e/h}[h^{\bullet}]$) at the fitted value of $\boldsymbol{b}_{e/h}$. (b) at 1273 K (solid lines). Lines (-··) show the concentrations of defects for the standard Gibbs energy of formation of electronic defects equal ΔG^{o}_{eh} = 100 kJ/mol. Line (---) represents the dependence of the deviation from the stoichiometry with ionic defects with lower ionisation degrees not taken into account. (c) at 1373 K (solid lines). Lines (-··) represent the maximum concentration of oxygen vacancies and interstitial oxygen ions that do not deteriorate the degree of match between the dependence of the deviation from the stoichiometry on the oxygen pressure and the experimental values of δ. (d) at 1473 K (solid lines). Lines (-···) denote the maximum concentration of defects with lower ionisation degrees, which does not deteriorate the match between the dependence of the deviation from the stoichiometry on the oxygen pressure and the experimental values of δ. The points represent the results of the studies of the deviation δ (Dieckmann 1982) – (○,●) points and (Sockel and Schmalzried 1968) – (□) points after correction (Dieckmann 1982).

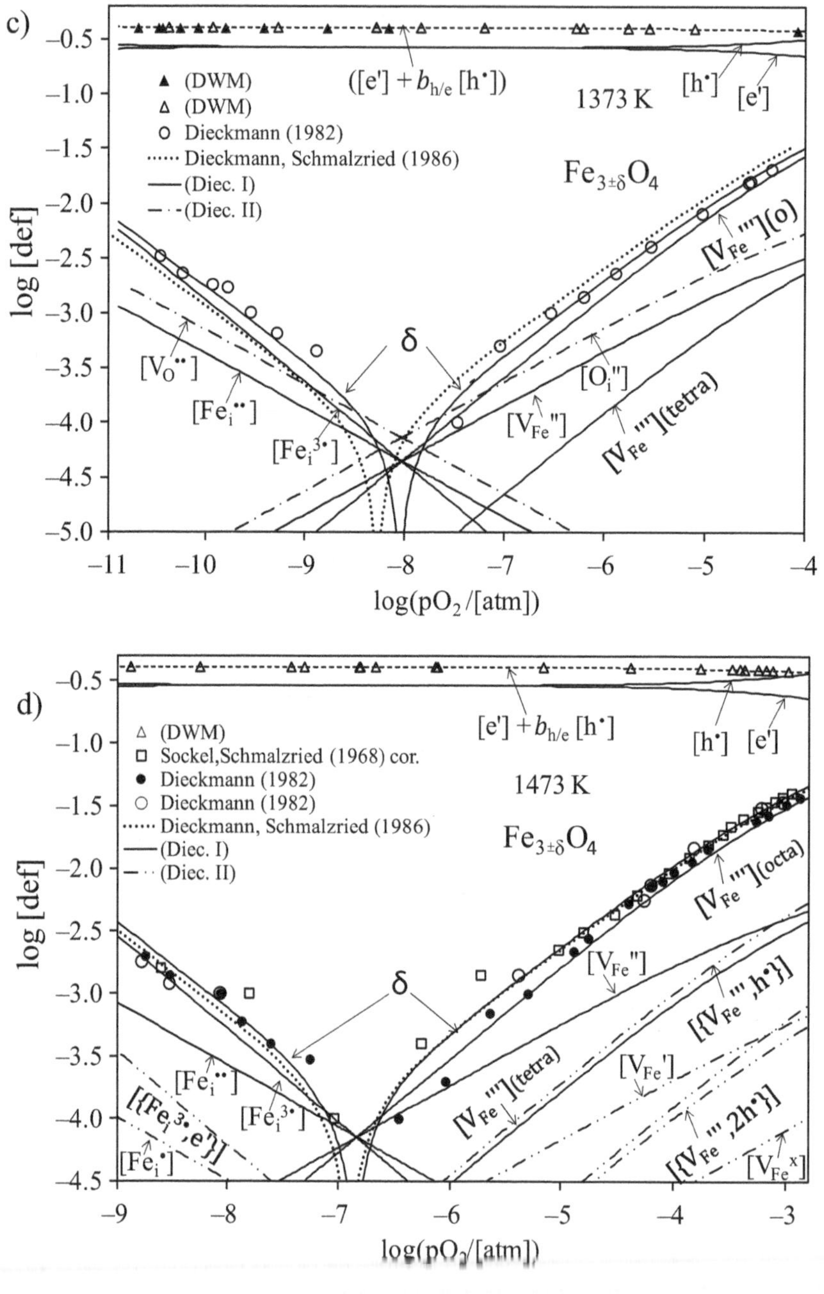

FIGURE 1.3 (Continued)

above, using the results of the deviation from the stoichiometry (Dieckmann 1982) ((○) points) in the temperature range of 1273–1673 K in the whole range of existence of $Fe_{3\pm\delta}O_4$ and (Nakamura et al. 1978, Yamauchi et al. 1983) ((◇,◆) points) in the temperature range of 1573–1723 K at higher oxygen pressures ((---) line). Figures 1.3a–c and 1.4a–d show the dependence of δ on p_{O_2} obtained when using "equilibrium constants" of reactions of formation of cation vacancies K_V and interstitial ions K_I (dotted line), which were calculated from their dependences on temperature, presented in (Dieckmann and Schmalzried 1986).

As can be seen in Figures 1.3a–c and 1.4a–d a very good match was obtained between the dependence of the deviation from the stoichiometry on p_{O_2} and the experimental results δ (Dieckmann 1982); also, an agreement with the dependence determined using the values of "equilibrium constants" K_V and K_I (Dieckmann and Schmalzried 1986) was obtained. As can be seen in Figure 1.4, there are slight differences between the results of studies of the deviation from the stoichiometry (Dieckmann 1982) and (Nakamura et al. 1978), which can be an effect of a measurement error associated with the assumed reference value of the deviation from the stoichiometry, in relation to which the changes were determined.

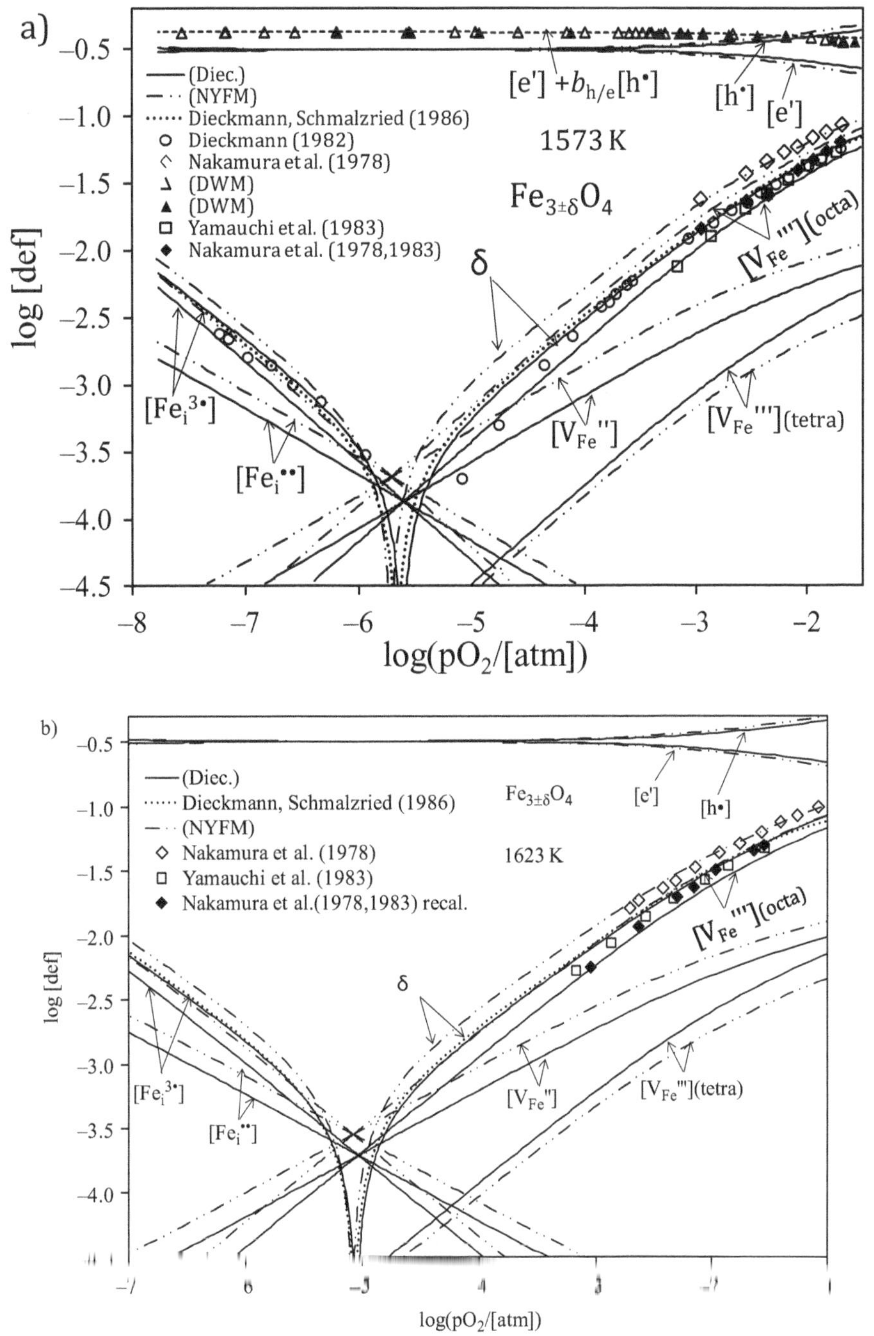

FIGURE 1.4 (a) Diagram of concentrations of point defects for $Fe_{3\pm\delta}O_4$ at 1573 K (solid lines), obtained using the results of the studies of the deviation from the stoichiometry (Dieckmann 1982) and (Nakamura et al. 1978) ((-··-) line). The points represent the results of the studies of deviation from the stoichiometry δ (Dieckmann 1982) ((○) points) and (Nakamura et al. 1978) ((◇,◆) points), and (Yamauchi et al. 1983) ((□) points). The (···) line shows the dependence of δ on p_{O_2} calculated using "equilibrium constants" of reactions of formation of cation vacancies K_V and interstitial ions K_I (Dieckmann 1982, Dieckmann and Schmalzried 1986). The points (△,▲) (DWM) represent the values of the sum of concentrations of electronic defects ([e′] + $b_{e/h}$[h$^{\bullet}$]) calculated according to Equation 1.49 for the fitted value of the mobility of electrons and using the values of the electrical conductivity (Dieckmann et al. 1983), and line (---) denotes the dependence of the sum of concentrations of electronic defects ([e′] + $b_{e/h}$[h$^{\bullet}$]) on p_{O_2} at the fitted value of $b_{e/h}$. (b) at 1623 K (solid lines). (c) at 1673 K (solid lines). (d) at 1723 K (solid lines).

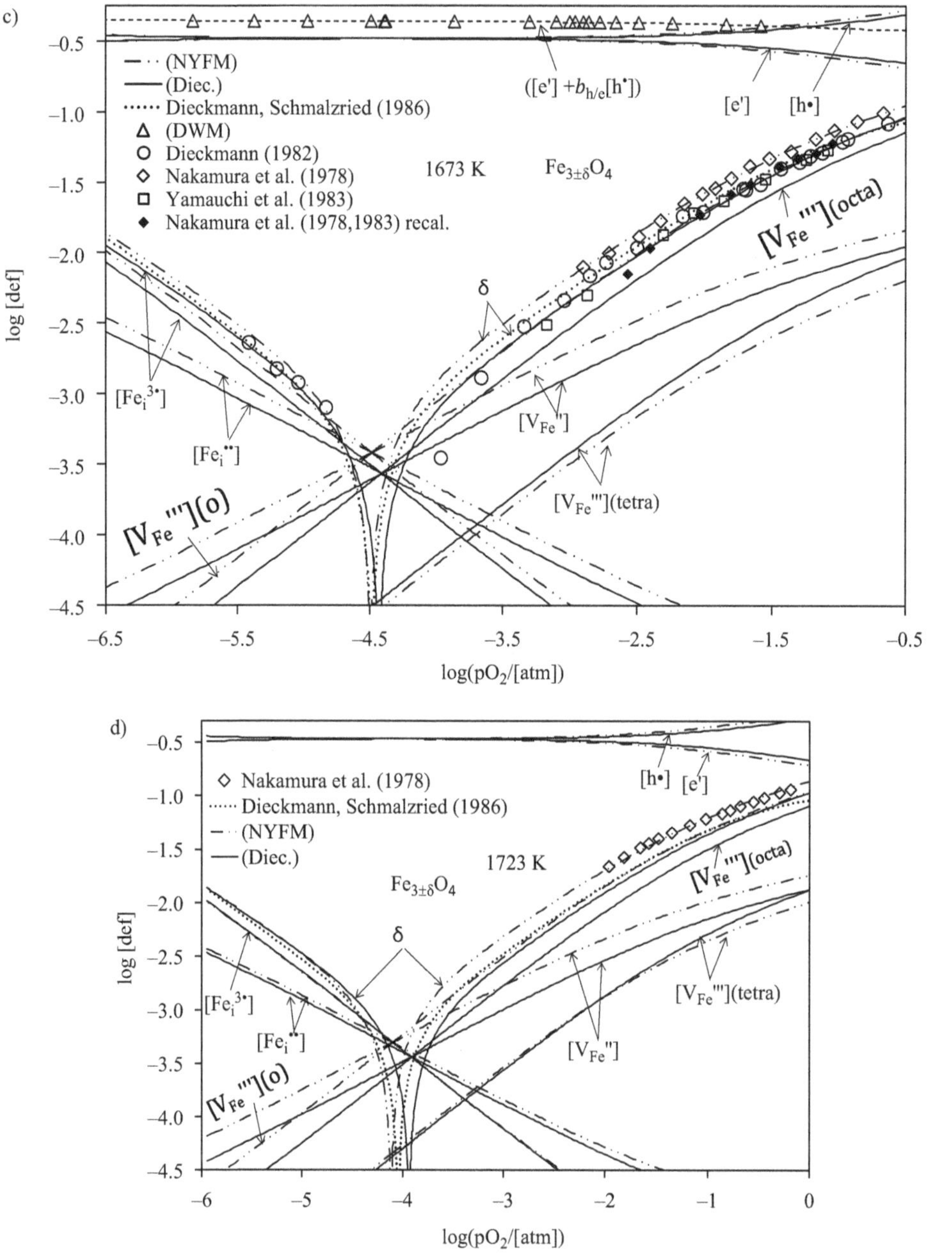

FIGURE 1.4 (Continued)

From the diagrams of concentrations of defects drawn in Figures 1.3a–c and 1.4a–d it results that the dominating defects are iron vacancies in the sublattice with octahedral coordination $\left(V_{Fe(o)}'''\right)$ and interstitial iron ions $\left(Fe_i^{3\bullet}\right)$, while the maximum concentration of vacancies $\left[V_{Fe(o)}''\right]$ and $\left[Fe_i^{\bullet\bullet}\right]$ at the stoichiometric composition might be similar, but at the highest oxygen pressures and the lowest oxygen pressures they are lower by one order of magnitude.

Figure 1.3a presents also the results of calculations for such a value of $\Delta G^o_{F_3}$ of formation of Frenkel defects in the tetrahedral sublattice for which the concentrations of iron vacancies in the octahedral sublattice and the tetrahedral sublattice are similar $\left[V_{Fe(o)}'''\right] \cong \left[V_{Fe(t)}'''\right]$ ((–··–) lines). In order to obtain a good match between the dependence of the deviation from the stoichiometry and the experimental results δ, it was necessary to increase the values $\Delta G^o_{F_1} \cong \Delta G^o_{F_2} \cong \Delta G^o_{F_3}$ to 225 kJ/mol and increase the value $\Delta G^o_{V'''Fe(o)} = -26$ kJ/mol. As can be seen in Figure 1.3a, concentrations of vacancies $\left[V_{Fe}'''\right]$ in octahedral and tetrahedral sublattices have the same character of the function and the curves practically overlap (a difference between $\Delta G^o_{F_1}$ and $\Delta G^o_{F_3}$ equal 1 kJ/mol was assumed). The sum of their concentrations determines the values of the deviation from the stoichiometry and it does not affect the degree of match of the dependence of δ on p_{O_2} and the experimental values δ.

As the character of the dependence of the concentration of iron vacancies is the same (Equation 1.10b and 1.10c), the ratio of concentrations of vacancies $\left[V_{Fe(o)}'''\right]/\left[V_{Fe(t)}'''\right]$ does not affect the accuracy of match between the dependence of δ on p_{O_2} and the experimental values δ. On the other hand, a higher concentration $\left[V_{Fe(t)}'''\right]$ must cause a decrease in the concentration $\left[V_{Fe(o)}'''\right]$. Therefore, based on the results of the deviation from the stoichiometry, it is not possible to draw conclusions concerning the ratio of concentrations of vacancies $\left[V_{Fe(o)}'''\right]/\left[V_{Fe(t)}'''\right]$.

As presented in Figure 1.2, in the range of values ΔG_{eh}^{o} = 25–45 kJ/mol for defects with the highest ionisation degree, a good match was obtained between the dependence of δ on p_{O_2} and the experimental values δ. At higher values of $\Delta G_{eh}^{o} >$ 50 kJ/mol it is also possible to obtain a good match, however, it is necessary to consider defects with lower ionisation degrees. Figure 1.3b presents, for the purpose of comparison, the results of the calculations of defects' concentrations ((-·-) line) for thrice the value of ΔG_{eh}^{o} = 100 kJ/mol. The match of the dependence of δ on p_{O_2} was obtained at the following values of standard Gibbs energies:

$$\Delta G_{V_{Fe(o)}'''}^{o} = 97 \text{ kJ/mol}, \Delta G_{1}^{o} = -80, \Delta G_{F_1}^{o} = 238, \Delta G_{F_2}^{o} = 263, \Delta G_{F_3}^{o} = 235$$

$$\Delta G_{V_{Fe}^{x}}^{o} s = \Delta G_{\left\{V_{Fe(o)}''',2h^{\bullet}\right\}'}^{o} = 53 \text{ and } \Delta G_{V_{Fe}'}^{o} = \Delta G_{\left\{V_{Fe(o)}''',h^{\bullet}\right\}''}^{o} = \Delta G_{\left\{V_{Fe(o)}''',2h^{\bullet}\right\}'}^{o} = 40.$$

As can be seen in Figure 1.3b, the concentration of electronic defects is by over an order of magnitude lower than at the value ΔG_{eh}^{o} = 30 kJ/mol. When the defects with lower ionisation degrees are not considered, at the same values of $\Delta G_{V_{Fe(o)}'''}^{o}$, $\Delta G_{F_1}^{o}$, $\Delta G_{F_2}^{o}$, the deviation from the stoichiometry on p_{O_2} ((---) line) significantly deviates from the experimental values δ. As can be seen in Figure 1.3b, in order to obtain a consistency with the experimental values of δ, there must be present a high concentration of complexes $\left\{V_{Fe(o)}''',3h^{\bullet}\right\}$ and interstitial one-positive ions ($Fe_i^{\bullet}$) (or complexes $\left\{Fe_i^{3\bullet},2e'\right\}$ and electroneutral $\left(Fe_i^{x}\right)$. From the above example it results that at higher values of ΔG_{eh}^{o} it is also possible to obtain a good match between the dependence of δ on p_{O_2} and the experimental values δ, but there must be present a significant concentration of defects with lower ionisation degrees. Higher values of ΔG_{eh}^{o} cause that the concentration of electronic defects in magnetite is lower. From the performed calculations it results that when an accurate value of ΔG_{eh}^{o} of formation of electronic defects in magnetite is missing, it is not possible to determine the concentration of defects with lower ionisation degree based on the results of the studies on the deviation from the stoichiometry.

Figure 1.3d presents the results of the calculations for the diagram of the defects' concentrations, where apart from defects with the highest ionisation degree (solid lines), the maximum concentration of defects with lower ionisation degree was taken into account ((-···-) line), that do not deteriorate the match between the calculated dependence of δ on p_{O_2} and the experimental results δ. As can be seen in Figure 1.3d, the concentration of ionic defects with lower ionisation degree (complexes of ionic defects with holes or electrons) is by one order of magnitude lower than the concentration of defects with the highest ionisation degree.

In turn, Figure 1.3c shows the diagram of defects that takes into account the maximum concentration of defects in the oxygen sublattice. According to Equation 1.43b and 1.44b, the values of ΔG_{3}^{o} and $\Delta G_{F_O}^{o}$ were chosen in order not to deteriorate the match between the dependence of the deviation from the stoichiometry and the experimental results δ. As can be seen in Figure 1.3c, the maximum concentration of oxygen defects can be quite significant but lower than that of cation defects, which agrees with the difference between the coefficients of self-diffusion of iron and oxygen (Dieckmann et al. 1978, Millot and Niu 1997). The relatively high concentration of oxygen defects, which does not deteriorate the match between the dependence δ on p_{O_2} and the experimental results δ is due to their effect on the values of the deviation (there is an effect of "compensation" of the deviation in the cation sublattice and oxygen sublattice). Also, the effect of decreasing interactions between ions if these defects are present should be noted. In particular, interstitial oxygen ions are present in the range where iron vacancies dominate. On the other hand, where there is a high concentration of interstitial iron ions, oxygen vacancies are present. Thus, the presence of oxygen vacancies should decrease the repulsive interactions between cations in interstitial positions and nodal positions. Similarly, a high concentration of iron vacancies causes an increase in free space that allows for the existence of interstitial oxygen ions, and, therefore, the repulsive interaction between ions decreases.

Figure 1.5 compares the dependences of the deviation from the stoichiometry on the oxygen pressure in magnetite, at temperatures 1173–1673 K, which were presented in Figures 1.3 and 1.4.

As can be seen in Figure 1.5, the character of the dependence of the deviation from the stoichiometry at different temperatures is very similar. Magnetite-wüstite transition occurs at a similar concentration of defects of about 0.01 mol/mol. On the other hand, magnetite-hematite transition, when the temperature increases, occurs at increasingly higher concentration of defects 0.01–0.1 mol/mol at 1673 K. The ranges of existence of magnetite only slightly decrease when the temperature increases, but the limit pressures and the oxygen pressure at which magnetite reaches the stoichiometric composition, shift towards higher oxygen pressures.

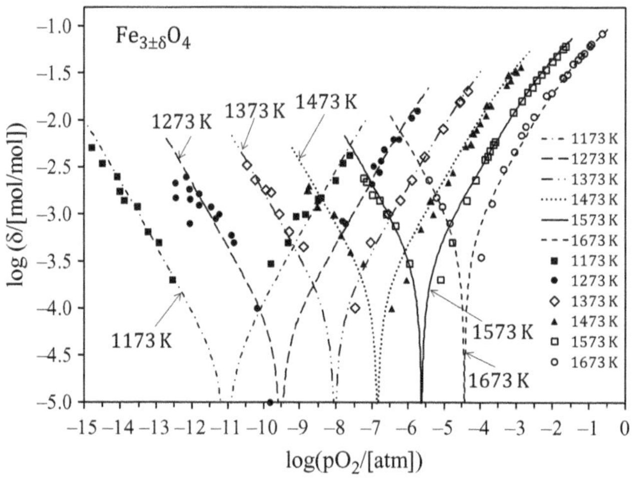

FIGURE 1.5 Dependences of the deviation from the stoichiometry in magnetite on the oxygen pressure at temperatures 1173–1673 K (presented in Figures 1.3 and 1.4).

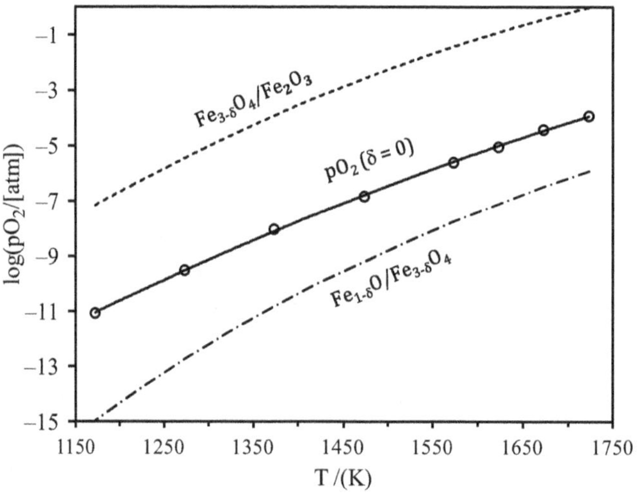

FIGURE 1.6 The temperature dependence of the oxygen pressure at which magnetite reaches the stoichiometric composition ((O) points), obtained using the results of the studies on the deviation from the stoichiometry (Dieckmann 1982). Dashed lines denote the equilibrium oxygen pressures at the phase boundary $Fe_{1-\delta}O/Fe_{3+\delta}O_4$ and $Fe_{3-\delta}O_4/Fe_2O_3$. Sources (Bryant and Smeltzer 1969, Giddings and Gordon 1973).

Figure 1.6 shows the range of existence of magnetite (Bryant and Smeltzer 1969, Giddings and Gordon 1973) and the values of the oxygen pressure at which magnetite reaches stoichiometric composition at a given temperature.

As can be seen in Figure 1.6, the range of oxygen pressures where magnetite exists decreases only slightly when the temperature increases, but limit oxygen pressures increase. Moreover, when the temperature increases, also the oxygen pressure at which magnetite reaches stoichiometric composition increases.

In turn, Figure 1.7 shows the dependence of the concentration of Frenkel defects (at the stoichiometric composition) on the temperature. When comparing the concentration of defects with that present in MO-type oxides, the values should be divided by four.

In turn, Figure 1.8 shows concentrations of electrons and holes at the stoichiometric composition, resulting from the diagrams of defects' concentrations at given temperatures ((O) points). For comparison, concentrations of iron ions Fe^{2+} in the sublattice with tetrahedral coordination are also shown ((□) points), along with the increment above one of the concentration of ions Fe^{3+} in the octahedral sublattice ((▲) points); this concentration, for magnetite with the stoichiometric composition was determined (Wu and Mason 1981) based on the studies of thermoelectric power.

As can be seen in Figure 1.8, concentrations of electronic defects at the stoichiometric composition are by three orders of magnitude higher than concentrations of ionic defects. As was shown in Section 1.4.5, the dependence of the concentration of point defects on the oxygen pressure is expressed by a quite complex equation (Equation 1.33b). In the case where

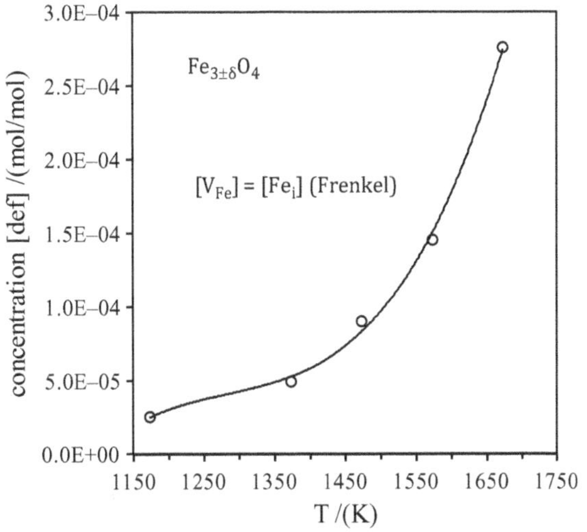

FIGURE 1.7 Temperature dependence of the concentration of Frenkel defects, obtained using the results of the studies of the deviation from the stoichiometry (Dieckmann 1982).

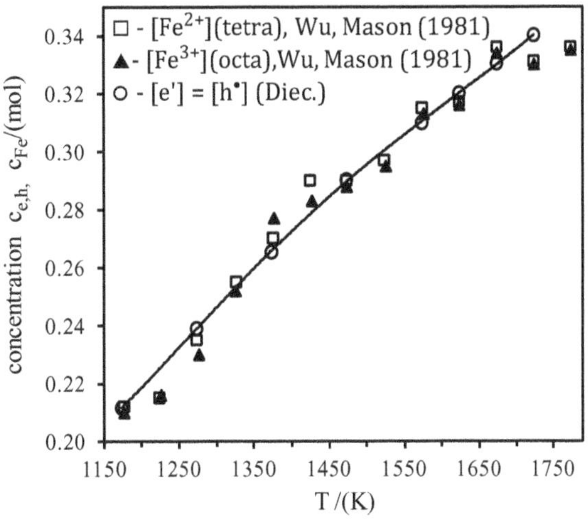

FIGURE 1.8 Dependence of the concentration of electrons and electron holes at the stoichiometric composition in magnetite on the temperature ((○) points), resulting from the calculated diagrams of concentrations of point defects presented in Figures 1.3a–c and 1.4a–d. Points (□) – denote the concentration of Fe^{2+} in the sublattice with tetrahedral coordination, while points (▲) denote an increase in the concentration (above 1) of ions Fe^{3+} in the sublattice with octahedral coordination determined using the values of the thermoelectric power (Wu and Mason 1981).

one type of defects dominates in a certain range, the dependence of the concentration of defects of a given type is an exponential function of oxygen pressure, with an exponent of $1/n$ (see Equation 1.10a, 1.10b, 1.10c, 1.14a, 1.14b, 1.14c). The exponent, or the derivative (d log δ/d log p_{O_2}) = $1/n_\delta$, in such a case indicates the type of dominating defects. Figure 1.9a shows the dependence of the derivative (log p_{O_2}/logδ) = n_δ on the oxygen pressure at temperatures 1173–1723 K, determined for the dependences of the deviation δ on p_{O_2} shown in Figures 1.3 and 1.4.

As can be seen in Figure 1.9a, at the lowest oxygen pressures the parameter n reaches the value of −1.5 ($1/n$ = −2/3), while at the highest pressures, in a certain narrow range it reaches the value of n = 1.5, but when the oxygen pressure increases further, the values are higher. In turn, Figure 1.9b, c presents the dependence of the derivative (d log p_{O_2}/d log [def]) = n on the oxygen pressure at temperatures 1173–1723 K, determined from the dependence of concentration of iron vacancies $\left[V_{Fe}'''\right]$ and $\left[V_{Fe}''\right]$ as well as $\left[Fe_i^{3\bullet}\right]$ and $\left[Fe_i^{\bullet\bullet}\right]$ on p_{O_2} shown in Figures 1.3 and 1.4. As can be seen in Figure 1.9b, in the case of vacancies $\left[V_{Fe}'''\right]$ the exponent n in a wide range of oxygen pressures is 1.3, but at the lowest oxygen pressures it increases up to 1.6, and at the highest oxygen pressures, when the temperature increases, it increases up to 1.6–2.8. In the case of vacancies $\left[V_{Fe}''\right]$ in a wide range of oxygen pressures n = 2 and at the lowest oxygen pressures it increases up to 2.6;

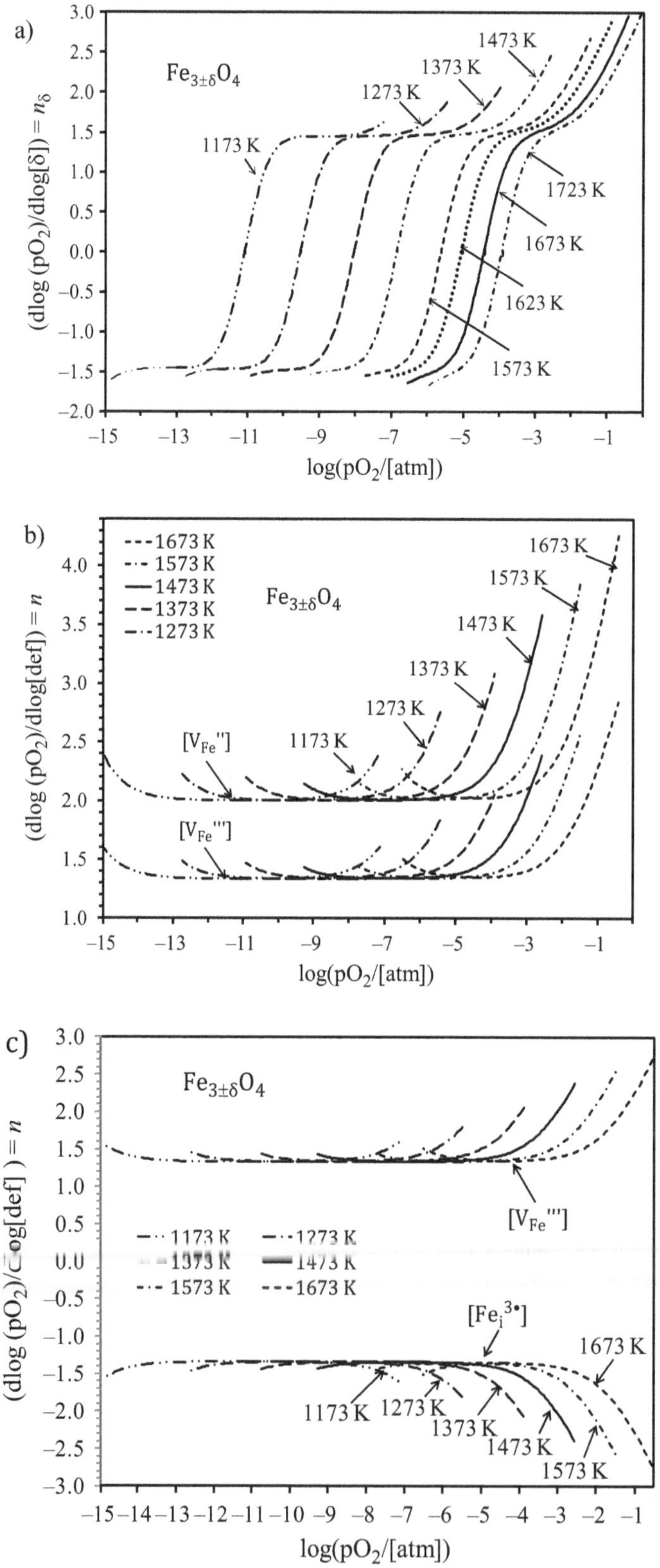

FIGURE 1.9 (a) Dependence of the derivative: $n_\delta = \mathrm{dlog}(p_{O_2})/\mathrm{dlog}\delta$ on the oxygen pressure at 1173–1723 K, obtained from the dependence of the deviation from the stoichiometry on p_{O_2} in magnetite (presented in Figures 1.3a–c and 1.4a–d). (b) Dependence of the derivative: $n_\delta = \mathrm{dlog}(p_{O_2})/\mathrm{dlog[def]}$ on the oxygen pressure at 1173–1673 K, obtained from the dependence of the concentration of cation vacancies $[V_{Fe}''']$ and $[V_{Fe}'']$ on the oxygen pressure p_{O_2} in magnetite (presented in Figures 1.3a–c and 1.4a–d). (c) Dependence of the derivative: $n = \mathrm{dlog}(p_{O_2})/\mathrm{dlog[def]}$ on the oxygen pressure at 1173–1673 K, obtained from the dependence of the concentration of interstitial iron ions $[Fe_i^{3\bullet}]$ and $[Fe_i^{\bullet\bullet}]$ on the oxygen pressure p_{O_2} in magnetite (presented in Figures 1.3a–c and 1.4a–d).

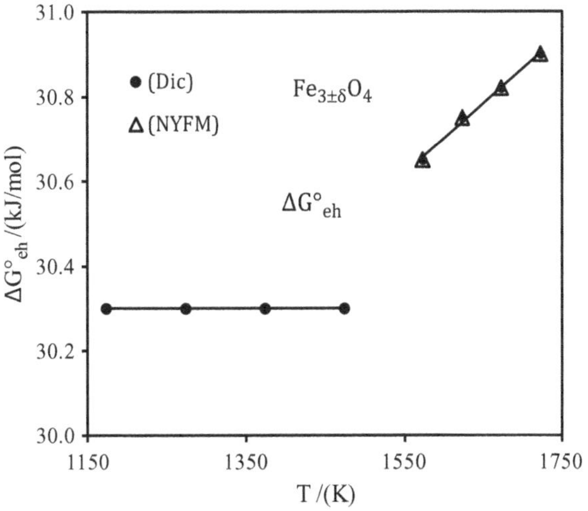

FIGURE 1.10 Temperature dependence of the standard Gibbs energy ΔG^o_{eh} of formation of electronic defects in magnetite obtained when assumed that the concentration of electronic defects at the stoichiometric composition equals the concentration of ions $[Fe^{2+}]$ in the sublattice with tetrahedral coordination, obtained when using the values of thermoelectric power (Wu and Mason 1981).

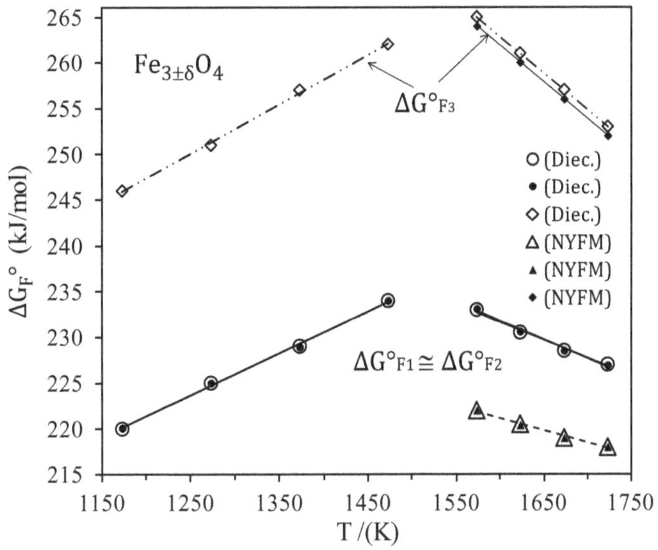

FIGURE 1.11 Temperature dependence of standard Gibbs energy $\Delta G^o_{F_1}$ (empty points) and $\Delta G^o_{F_2}$ (solid points) and $\Delta G^o_{F_3}$ ((◇,♦) points) of formation of Frenkel defects with charge two and three in magnetite, obtained using the results of the studies on the deviation from the stoichiometry (Dieckmann 1982) ((○,●,◇)(Dic) points) and (Nakamura et al. 1978) ((△,▲,◆)(NYFM) points).

at the highest oxygen pressures, when the temperature increases, it increases up to 2–4.3. In turn, as can be seen in Figure 1.9c, in the case of the dependence of the concentration of interstitial iron ions $\left[Fe_i^{3\bullet}\right]$ and $\left[Fe_i^{\bullet\bullet}\right]$ on p_{O_2}, the dependences of the analogue derivative in respect to the oxygen pressure in given temperatures are practically symmetrical, but with negative values. Changes in the above derivative (exponent) of the dependence of the concentration of defects in respect to the oxygen pressure are due to a slight change in the concentration of electronic defects that are in equilibrium with ionic defects. Thus, the exponent $1/n = 2/3$ and $-2/3$, assumed by (Dieckmann 1982, Dieckmann and Schmalzried 1986) is close to the derivative $(d\log\delta/d\log p_{O_2}) = 1/n_\delta$ but it is not constant, as it depends on the concentration of ionic defects and electronic defects.

Figures 1.10–1.13 present the temperature dependences of adjusted values of standard Gibbs energies ΔG^o_{eh}, $\Delta G^o_{F_1}$, $\Delta G^o_{F_2}$ $\Delta G^o_{F_3}$, $\Delta G^o_{V''_{Fe(o)}}$ and ΔG^o_I. On the other hand, Figures 1.12 and 1.14 show the temperature dependences of the standard Gibbs

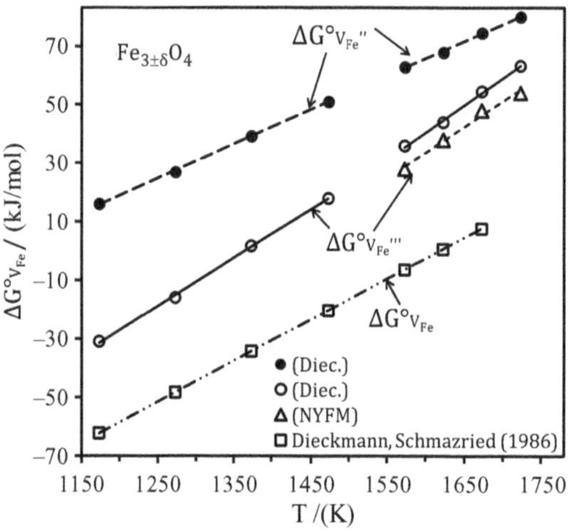

FIGURE 1.12 The temperature dependence of $\Delta G^o_{V'''_{Fe}}$ of formation of vacancies $V'''_{Fe(o)}$ and $\Delta G^o_{V''_{Fe}}$ of formation of vacancies $V''_{Fe(o)}$ (calculated according to Equation 1.35c) in magnetite, obtained using the results of the studies of the deviation from the stoichiometry (Dieckmann 1982) ((●,○)(Dic) points) and (Nakamura et al. 1978) ((△)(NYFM) points). Points (□) denote the values of standard Gibbs energy of formation of iron vacancies calculated using the "equilibrium constants" of reactions of formation of iron vacancies K_V (Dieckmann 1982, Dieckmann and Schmalzried 1986).

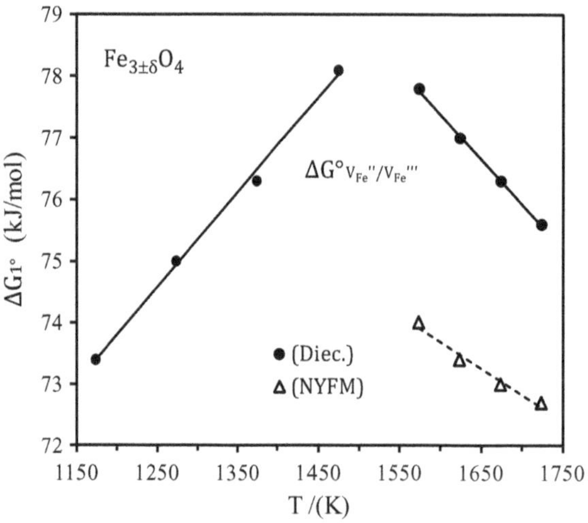

FIGURE 1.13 Temperature dependence of ΔG^o_1 in magnetite, permitting to determine the ratio of concentrations of iron vacancies $\left[V'''_{Fe(o)}\right]/\left[V''_{Fe(o)}\right]$ (according to Equation 1.36b), obtained using the results of the studies of the deviation from the stoichiometry (Dieckmann 1982) – (●)(Dic) points and (Nakamura et al. 1978) – (Δ)(NYFM) points.

energy ($\Delta G^o_{V''_{Fe}}$) of formation of vacancies $V''_{Fe(o)}$ (calculated according to Equation 1.35d) and $\Delta G^o_{Fe_i^{3\bullet}}$ and ($\Delta G^o_{Fe_i^{\bullet\bullet}}$) of formation of interstitial iron ions $Fe_i^{3\bullet}$ (according to Equation 1.36c) and $Fe_i^{\bullet\bullet}$ (according to Equation 1.34c).

As can be seen in the above figures, the obtained dependences are linear functions, although above 1473 K their slope changes. Therefore, it can be expected that magnetite will have different properties in the above temperature range. Table 1.2 compares the values of enthalpies and entropies of the formation of individual defects calculated from the relation ($\Delta G^o_i = \Delta H^o_i - T\Delta S^o_i$). As it was mentioned, such values of ΔG^o_{eh} of the formation of electronic defects were adjusted, at which the concentration of electronic defects at the stoichiometric composition is equal to the concentration of ions [Fe^{2+}] in the sublattice with tetrahedral coordination (Wu and Mason 1981); it was assumed that they are equal to the concentration of electronic defects. As can be seen in Figure 10.7, in the temperature range 1173–1473 K the values of ΔG^o_{eh} are practically constant, and in higher temperatures they increase only slightly. The real values of ΔG^o_{eh} can be different, but the changes with temperature increase should be small.

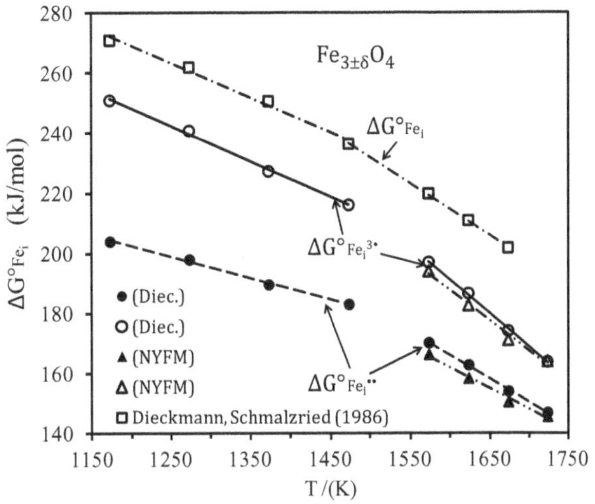

FIGURE 1.14 The temperature dependence of $\Delta G^o_{Fe_i^{3\bullet}}$ and $\Delta G^o_{Fe_i^{\bullet\bullet}}$ of formation of interstitial ions $Fe_i^{3\bullet}$ and $Fe_i^{\bullet\bullet}$ (calculated according to Equations 1.34c and 1.36c) in magnetite, obtained using the results of the studies on the deviation from the stoichiometry (Dieckmann 1982) ((●)(Dic) points) and (Nakamura et al. 1978) ((△)(NYFM) points). Points □ denote the values of standard Gibbs energy of formation of interstitial cations $\Delta G^o_{Fe_i}$, calculated using the "equilibrium constants" of reactions of formation of interstitial cations $\boldsymbol{K}_I$ (Dieckmann 1982, Dieckmann and Schmalzried 1986).

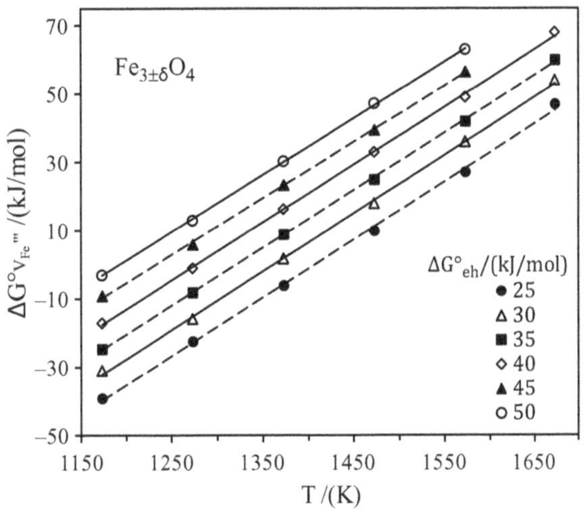

FIGURE 1.15 Temperature dependence of $\Delta G^o_{V_{Fe}}$ of formation of vacancies $V'''_{Fe(o)}$ in magnetite depending on the assumed values of ΔG^o_{eh} of formation of electronic defects.

As can be seen in Figure 1.12, a particularly large change in the slope occurs for the standard Gibbs energy ΔG^o_F of formation of Frenkel defects. In the temperature range of 1173–1473 K the values of ΔG^o_F increase with temperature, while in the range of 1473–1463 K they decrease and these changes are not associated with the value of ΔG^o_{eh}.

Figures 1.12 and 1.14 present the dependence of the standard Gibbs energy of formation of iron vacancies and interstitial iron ions, calculated using of "equilibrium constants" of reactions of formation of cation vacancies K_V and interstitial ions K_I (Dieckmann and Schmalzried 1986) (see Equation 1.5 and 1.7). As can be seen in Figures 1.12 and 1.14, the dependence of $\Delta G^o_{V_{Fe}}$ and $\Delta G^o_{Fe_i}$ on temperature has a similar character to that of the temperature dependence of $\Delta G^o_{V'''_{Fe(o)}}$ and $\Delta G^o_{Fe_i^{3\bullet}}$. The differences that occur are due to the fact that they refer to different defect reactions describing the equilibrium state in magnetite.

As can be seen in Figure 1.4, slight differences between the results of studies of the deviation from the stoichiometry (Dieckmann 1982) and (Nakamura et al. 1978) cause quite significant differences in the adjusted values of defects' formation (see Figures 1.11–1.14).

TABLE 1.2

Values of enthalpies (ΔH_i^o) and entropies $\left(\Delta S_i^o\right)$ of formation of intrinsic electronic defects, Frenkel defects, iron vacancies V_{Fe}'' and V_{Fe}''' and interstitial iron ions $Fe_i^{\bullet\bullet}$ and $Fe_i^{3\bullet}$ in magnetite, based on the dependences shown in Figures 1.10–1.14 (at ΔG_{eh}^o = 30.3 kJ/mol of formation of electronic defects). Data for the calculations were taken from (Dieckmann 1982) and (Nakamura et al. 1978) denoted (Dieck) and (NYFM) in the Data source column.

Equation	ΔH_i^o (kJ/mol)	ΔS_i^o (J/(mol·K))	ΔT (K)	Data Source
$K_{eh} = [h^\bullet][e']$				
Equation 1.28b	28.1 ± 0.1	−1.6 ± 0.1	1573–1673	(Dieck.)
$\left[V_{Fe(o)}'''\right]\left[Fe_i^{3\bullet}\right] = K_{F_1}$	166 ± 2	−46 ± 1	1173–1473	(Dieck.)
Equation 1.34b	296 ± 5	40 ± 3	1573–1673	(Dieck.)
-”-	264 ± 3	27 ± 2	1573–1723	(NYFM)
$\left[V_{Fe(o)}''\right]\left[Fe_i^{\bullet\bullet}\right] = K_{F_2}$	183 ± 2	−54 ± 2	1173–1473	(Dieck.)
Equation 1.36b	391	80	1573–1673	(Dieck.)
-”-	390	80	1573–1723	(NYFM)
$V_{Fe(o)}'''$	−224 ± 4	−164 ± 3	1173–1473	(Dieck.)
Equation 1.9b	−254 ± 11	−184 ± 6	1573–1673	(Dieck.)
-”-	−248 ± 23	−176 ± 14	1573–1723	(NYFM)
$K_{V_{Fe}^x/V_{Fe(o)}^x} = K_1$	55 ± 1	−15 ± 1	1173–1473	(Dieck.)
Equation 1.35b	101 ± 1	15 ± 1	1573–1673	(Dieck.)
-”-	87 ± 2	9 ± 1	1573–1723	(NYFM)
$V_{Fe(o)}''$	−122 ± 3	−117 ± 2	1173–1473	(Dieck.)
Equation 1.9a	−119 ± 7	−116 ± 4	1573–1673	(Dieck.)
-”-	−122 ± 15	−113 ± 9	1573–1723	(NYFM)
$Fe_i^{\bullet\bullet}$	288 ± 4	72 ± 3	1173–1473	(Dieck.)
Equation 1.14a	417 ± 7	157 ± 4	1573–1673	(Dieck.)
-”-	386 ± 18	140 ± 11	1573–1723	(NYFM)
$Fe_i^{3\bullet}$	390 ± 6	118 ± 5	1173–1473	(Dieck)
Equation 1.14b	550 ± 10	224 ± 6	1573–1673	(Dieck.)
-”-	512 ± 26	203 ± 16	1573–1723	(NYFM)
$\Delta G_{[def]}^{o(\delta=0)}$	−204 ± 4	−93 ± 3	1173–1473	(Dieck)
Equation 1.33b	−261 ± 7	−131 ± 4	1573–1673	(Dieck.)
V_{Fe}	−226 ± 7	−131 ± 1	1173–1673	(DS)
Fe_i	407 ± 11	113 ± 8	1173–1423	(DS)
Fe_i	491 ± 4	173 ± 3	1473–1673	(DS)

The (xx) symbols stand for the results of the deviation from the stoichiometry (used in the calculations noted in the present work) taken from (Dieckmann 1982) (Dieck.) and (Nakamura et al. 1978)(NYFN). (DS) (Dieckmann and Schmalzried 1986).

It should be noted that temperature dependences of ΔG_i^o of formation of individual defects, shown in Figures 1.11–1.14, depend on the value of ΔG_{eh}^o assumed in the calculations. Figure 1.15 shows, as an example, dependence of the standard Gibbs energy of the formation of vacancies $V_{Fe(o)}'''$ ($\Delta G_{V_{Fe}'''}^o$) on temperature, fitted for a series of ΔG_{eh}^o values. As can be seen in Figure 1.15, the relations are linear, with a high correlation coefficient ($R^2 = 0.999$), with practically the same slope. Table 1.3 compares the values of enthalpies and entropies of formation of iron vacancies and interstitial iron ions depending on the assumed value of ΔG_{eh}^o. As can be seen, depending on the assumed value of ΔG_{eh}^o, changes in entropy are small, of the order of confidence interval, while enthalpies of formation of defects differ significantly. Thus, it can be expected that the real values of enthalpy ΔH_i^o of formation of defects will be in the range of changes of these values.

Due to a high concentration of electronic defects in magnetite there should be also a significant concentration of defects at lower ionisation degrees. Also, there is no data about the concentration of vacancies $\left[V_{Fe(t)}'''\right]$ in the sublattice of ions with tetrahedral coordination.

The parameter, on which the concentration of point defect at a given oxygen pressure is dependent, is the resultant standard Gibbs energy of formation of defects ($\Delta G_{[def]}^{o(\delta)}$) which can be calculated according to Equation 1.33b. Figure 1.16

TABLE 1.3
Values of enthalpies (ΔH_i^o) and entropies (ΔS_i^o) of formation of iron vacancies V_{Fe}''' (see Figure 1.15) and V_{Fe}'' and interstitial iron ions $Fe_i^{3\bullet}$ and $Fe_i^{\bullet\bullet}$ in magnetite on the value, assumed in calculations, of standard Gibbs energy ΔG_{eh}^o of formation of electronic defects ($\Delta G_i^o = \Delta H_i^o - T\Delta S_i^o$).

ΔG_{eh}^o	$\Delta H_{V_{Fe}'''}^o$ (kJ/mol)	$\Delta S_{V_{Fe}'''}^o$ (J/(mol·K))	T (K)
25	−238.9 ± 4.4	−169.9 ± 3.1	1173–1673
30.3	−231.8 ± 3.4	−170.3 ± 2.4	1173–1673
35	−222.0 ± 1.9	−168.1 ± 1.3	1173–1673
40	−215.9 ± 2.8	−169.1 ± 1.9	1173–1673
45	−202.6 ± 2.8	−164.5 ± 2.0	1173–1673
50	−198.1 ± 2.0	−166.2 ± 1.5	1173–1673
	$\Delta H_{V_{Fe}''}^o$	$\Delta S_{V_{Fe}''}^o$	
25	−126.2 ± 1.0	−116.8 ± 0.7	1173–1673
30.3	−121.5 ± 1.2	−117.1 ± 0.9	1173–1673
35	−115.0 ± 1.2	−115.6 ± 0.8	1173–1673
40	−110.8 ± 1.6	−116.2 ± 1.1	1173–1673
45	−105.2 ± 1.7	−115.6 ± 1.2	1173–1673
50	−102.2 ± 2.5	−116.8 ± 1.8	1173–1673
	$\Delta H_{Fe_i^{3\bullet}}^o$	$\Delta S_{Fe_i^{3\bullet}}^o$	
25	397 ± 2	117 ± 2	1173–1473
25	538 ± 28	213 ± 18	1473–1673
30.3	390 ± 6	118 ± 5	1173–1473
30.3	522 ± 16	208 ± 10	1473–1673
35	385 ± 3	120 ± 3	1173–1473
35	508 ± 20	203 ± 13	1473–1673
40	380 ± 4	121 ± 3	1173–1473
40	500 ± 30	203 ± 19	1473–1673
45	367 ± 6	117 ± 4	1173–1473
45	505 ± 27	210 ± 17	1473–1673
50	366 ± 4	122 ± 3	1173–1473
50	434	168	1473–1573
	$\Delta H_{Fe_i^{\bullet\bullet}}^o$	$\Delta S_{Fe_i^{\bullet\bullet}}^o$	
25	293 ± 2	71 ± 2	1173–1473
25	405 ± 22	146 ± 14	1473–1673
30.3	[illegible]	72 ± 3	1173–1473
30.3	394 ± 14	143 ± 9	1473–1673
35	285 ± 3	72 ± 2	1173–1473
35	385 ± 17	140 ± 11	1473–1673
40	281 ± 3	74 ± 2	1173–1473
40	380 ± 23	140 ± 15	1473–1673
45	273 ± 4	71 ± 3	1173–1473
45	383 ± 22	145 ± 14	1473–1673
50	273 ± 3	74 ± 2	1173–1473
50	330	113	1473–1573

shows the dependence of the resultant standard Gibbs energy $\Delta G_{[def]}^{o(\delta)}$ of defects formation on the oxygen pressure (expressed in kJ/mole of oxygen atoms).

As can be seen in Figure 1.16, for the individual temperatures, the values tend to a constant value which determines the standard Gibbs energy of formation of iron vacancies and interstitial ions. Depending on the temperature, they are, for iron

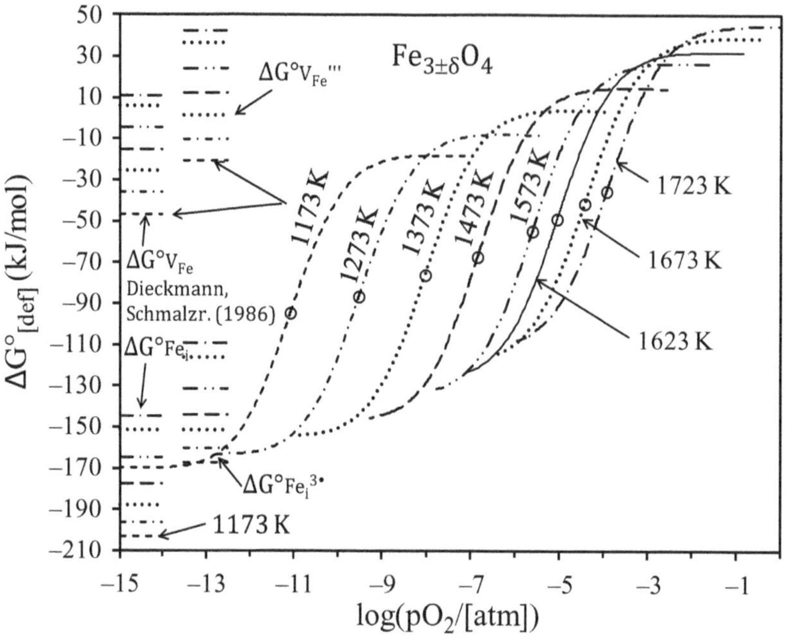

FIGURE 1.16 Dependence of the resultant standard Gibbs energy of formation of defects $\Delta G^{o(\delta)}_{[def]}$ in magnetite on the oxygen pressure, determined according to Equation 1.33b (expressed in moles per one mole of oxygen atoms). Points (O) denote the oxygen pressures at which the magnetite reaches the stoichiometric composition. Short lines denote the values of $\Delta G^{o}_{V_{Fe}}$ and $\left(-\Delta G^{o}_{Fe_i}\right)$ calculated using the "equilibrium constants" $\boldsymbol{K}_V$ and $\boldsymbol{K}_I$ (Dieckmann and Schmalzried 1986) and the values of $\Delta G^{o}_{V'''_{Fe}}$ and $\left(-\Delta G^{o}_{Fe_i^{3\bullet}}\right)$ presented in Figures 1.12 and 1.14 (per 1 mol of oxygen atoms).

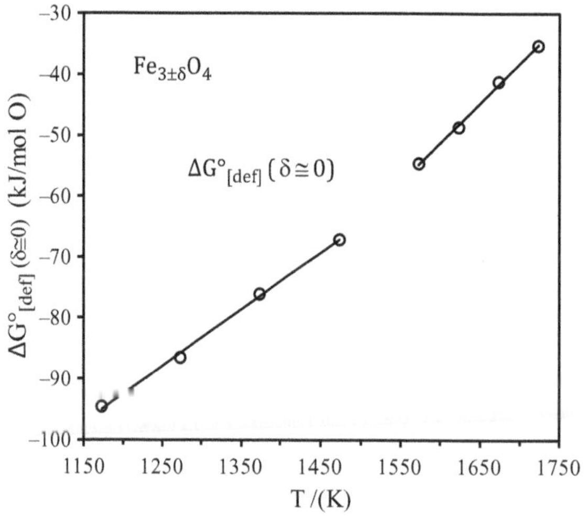

FIGURE 1.17 Temperature dependence of the resultant standard Gibbs energy of formation of defects $\Delta G^{o(\delta=0)}_{[def]}$ in magnetite, at the stoichiometric composition (see Equation 1.33b).

vacancies, from $\Delta G^{o(\delta)}_{V} = -20$ to 45 kJ/mole of oxygen and $\Delta G^{o(\delta)}_{Fe_i} = -165$ to -105 kJ/mole of oxygen. The values of $\Delta G^{o}_{Fe_i}$ are negative, because they refer to the process of decrease in the concentration of interstitial ions.

Figure 1.16 shows, for comparison, the values of $\Delta G^{o}_{V'''_{Fe}}$ and $\Delta G^{o}_{Fe_i^{3\bullet}}$, calculated per 1 mole of oxygen atoms, and the values of $\Delta G^{o}_{V_{Fe}}$ and $\Delta G^{o}_{Fe_i}$ calculated using the values of "equilibrium constants" (Dieckmann and Schmalzried 1986), also calculated per 1 mole of oxygen atoms. As can be seen in Figure 1.16, the values of $\Delta G^{o}_{V'''_{Fe}}$ and $\Delta G^{o}_{Fe_i^{3\bullet}}$ are close to the limit values of $\Delta G^{o(\delta)}_{V}$ and $\Delta G^{o(\delta)}_{Fe_i}$ at given temperatures.

The characteristic value is the value of $\Delta G^{o(\delta=0)}_{[def]}$ when the oxide reaches the stoichiometric composition. This value is associated to the ΔG^{o}_{O} of incorporation of an oxygen atoms and to the difference of $\left(\Delta G^{o}_{V_{Fe}} - \Delta G^{o}_{Fe_i}\right)$ of formation of iron vacancies and interstitial iron ions (see discussion and Equation 15.1 in (Stokłosa 2015)). Figure 1.17 presents the dependences of $\Delta G^{o(\delta=0)}_{[\mathrm{def}]}$ on the temperature, and Table 1.2 shows the values of enthalpy and entropy of the process, mainly, of the incorporation of an oxygen atom. Therefore, it will be the standard Gibbs energy of the process of adsorption of oxygen, its dissociation and electron affinity of the adsorbed oxygen atom.

1.6.2 Diffusion of Ions via Defects – Their Mobility

The diffusion of ions in a crystal depends on their mobility and on the concentration of defects through which they move. The relation between the coefficient of self-diffusion of metal ions and the coefficients of diffusion of defects participating in their transport and the concentration of defects is given by:

$$D_{\mathrm{M}} c_{\mathrm{M}} = \sum_i D_i c_i \tag{1.45a}$$

where D_{M} denotes the coefficient of self-diffusion of metal ions, c_{M} – the concentration of metal, c_i – the concentration of a given type of defects i. The coefficient of diffusion D_i is called the coefficient of diffusion of defects, because it is independent of their concentration, which suggests that it is a rate of movement of defects. More accurately, it is the rate of movement of metal ions (i) via defects. Thus, the coefficient of diffusion D_i is the mobility of ion i, similarly to the definition of mobility of electrons in solids or ions in a solution. Parent ions or other atoms (ions), being the object of the studies of diffusion, can move through defects.

The coefficients of self-diffusion of ions are usually determined by a radioisotope method, called *tracer method*, where atoms of radioactive elements are used. Due to a "correlation effect", the coefficients of diffusion of tracers (D^*_{M}) differ from the coefficients of self-diffusion of ions D_{M}:

$$D^*_{\mathrm{M}} = D_{\mathrm{M}} f^*_{\mathrm{M}} \tag{1.45b}$$

where D^*_{M} denotes the coefficient of diffusion of tracer M*, and f^*_{M} is the correlation coefficient depending on the diffusion mechanism.

The presence of different types of defects in magnetite causes that the coefficients of diffusion via these defects will be different. Therefore, transport of metal ions in magnetite will occur by different types of iron vacancies and interstitial cations. According to a general Equation 1.45a, the resultant coefficient of self-diffusion of metal D_{M} is a sum of products of coefficients of diffusion via defects and concentrations of individual defects. The differences between the coefficients of diffusion of ions via vacancies V'''_{Fe} and V''_{Fe} in magnetite should not be large. As defects with the highest ionisation degree dominate in magnetite, it can be approximately assumed that the resultant coefficient of diffusion of metal in the range of higher oxygen pressures will be dependent on the sum of the concentrations of iron vacancies $[V_{\mathrm{Fe}}]$, and in the range of low oxygen pressures on the sum of the concentrations of interstitial cations $[Fe_{\mathrm{i}}]$. Equation 1.45a assumes then an approximated form of:

$$D^*_{\mathrm{M}} = D_{\mathrm{V(M)}} f^*_{\mathrm{M(V)}} \frac{c_V}{c_M} + D_{\mathrm{I(M)}} f^*_{\mathrm{M(I)}} \frac{c_I}{c_M} = D^o_{\mathrm{V(M)}}\left[V_{\mathrm{Fe}}\right] + D^o_{\mathrm{I(M)}}\left[Fe_{\mathrm{i}}\right] \tag{1.46a}$$

where $D_{\mathrm{V(M)}}$ and $f^*_{\mathrm{M(V)}}$ denote the coefficient of diffusion of metal ions M via cation vacancies and the correlation coefficient for this diffusion mechanism, $D_{\mathrm{I(M)}}$ and $f^*_{\mathrm{M(I)}}$ - denote the coefficient of diffusion via interstitial cations and the correlation coefficient for the above diffusion mechanism. In turn, $D^o_{\mathrm{V(M)}}$ and $D^o_{\mathrm{I(M)}}$ are effective coefficients of diffusion (mobility) of metal ions M via cation vacancies and interstitial ions. $[V_{\mathrm{Fe}}]$ and $[Fe_{\mathrm{i}}]$ denote the sum of the concentration of iron vacancies and interstitial ions in magnetite, that are dependent on the oxygen pressure.

Thus, using the concentration of ionic defects resulting from the diagram of concentrations of defects and the values of coefficients of self-diffusion of tracers, we can determine the coefficients of diffusion of defects, and, more precisely, the mobility of a given type of metal ions via cation vacancies and interstitial cations. Equation 1.46a can be written as:

$$D^*_{\mathrm{M}} \cong D^o_{\mathrm{V(M)}}\left(\left[V_{\mathrm{Fe}}\right] + \frac{D^o_{\mathrm{I(M)}}}{D^o_{\mathrm{V(M)}}}\left[Fe_{\mathrm{i}}\right]\right) = D^o_{\mathrm{V(M)}}\left(\left[V_{\mathrm{Fe}}\right] + b_{\mathrm{D}}\left[Fe_{\mathrm{i}}\right]\right) \tag{1.46b}$$

where b_{D} denotes the ratio of the coefficient of diffusion of ions via interstitial ions to the coefficient of diffusion of ions via vacancies $D^o_{\mathrm{I(M)}}/D^o_{\mathrm{V(M)}}$. Transforming Equation 1.46b we get:

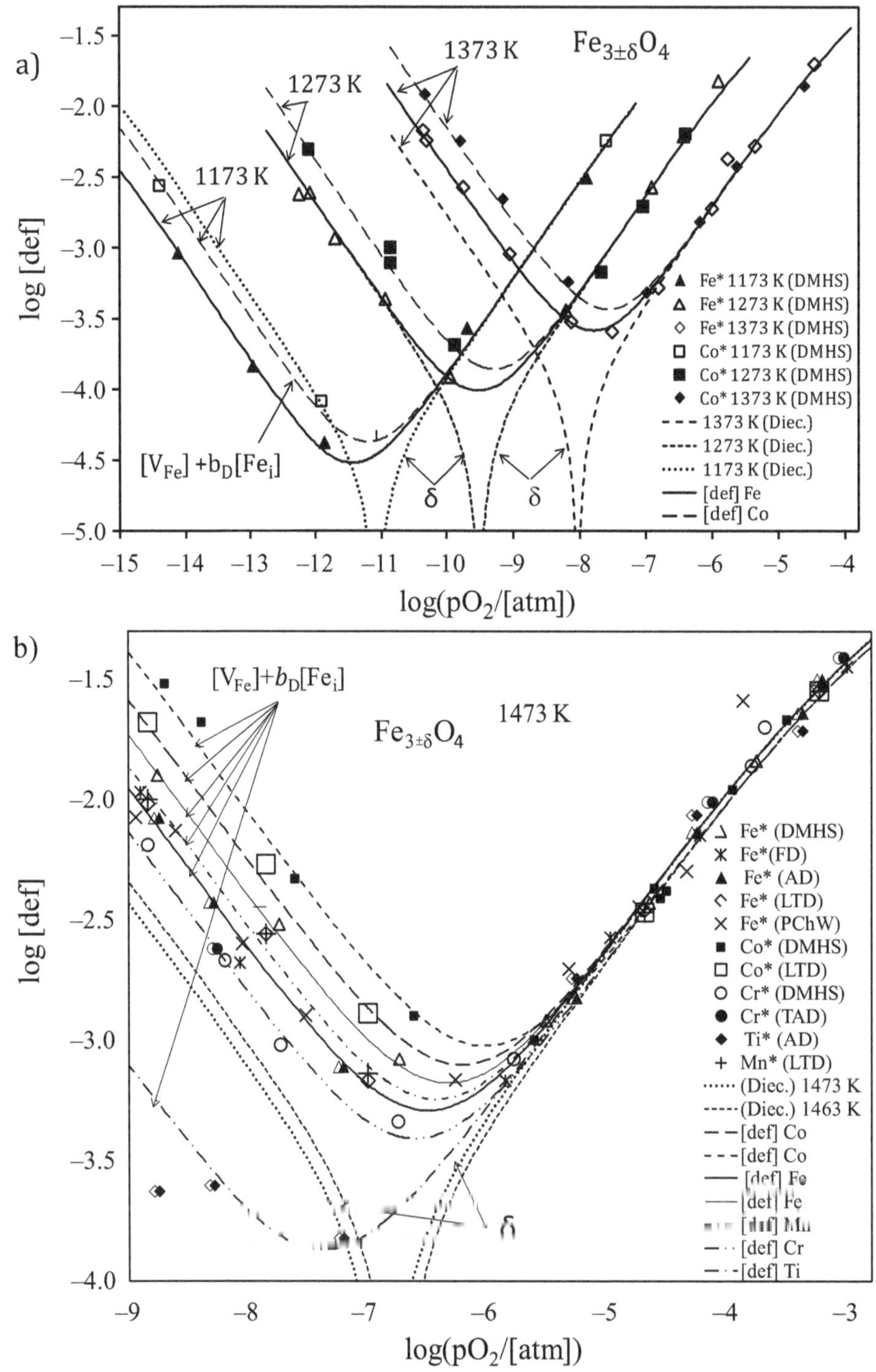

FIGURE 1.18 (a) Dependences of the deviation from the stoichiometry on the oxygen pressure in magnetite, at temperatures 1173–1373 K (fine dashed lines) presented in Figure 1.3. The points represent the values of the sum of concentrations of ionic defects ($[V_{Fe}] + \boldsymbol{b}_D[M_i]$) calculated according to Equation 1.47a for fitted values of $D^{o}_{V(M)}$ for the highest oxygen pressures and when using the values of coefficients of self-diffusion of tracers: Fe* -, ((△, ▲,◆)(DHMS) points), Co* - (□,■,◆)(DHMS) (Dieckmann et al. 1978). Solid line and dashed line denote, respectively, the dependence of the sum of concentrations of ionic defects $[V_{Fe}] + \boldsymbol{b}_D[M_i]$) on p_{O_2} obtained for the fitted value of $\boldsymbol{b}_D$. (b) at 1473 K ((···) line) and 1463 K (---). The points represent the values of the sum of concentrations of ionic defects ($[V_{Fe}] + \boldsymbol{b}_D[M_i]$) calculated according to Equation 1.47a for fitted values of $D^{o}_{V(M)}$ for the highest oxygen pressures and when using the values of coefficients of self-diffusion of tracers: Fe* - (△,x,✳, ▲,◆) points (Dieckmann et al. 1978, Peterson et al. 1980, Franke and Dieckmann 1989, Lu et al. 1993, Töpfer et al. 1995, Aggarwal and Dieckmann 2002), Co* - (□,■) (Dieckmann et al. 1978, Lu et al. 1993), Cr* - (●,○) (Dieckmann et al. 1978, Töpfer et al. 1995) Mn* - (+) (Lu et al. 1993), Ti - (◆) (Aggarwal and Dieckmann 2002). Solid line and dashed lines denote, respectively, the dependence of the sum of concentrations of ionic defects $[V_{Fe}] + \boldsymbol{b}_D[M_i]$) on p_{O_2} obtained for the fitted value of $\boldsymbol{b}_D$.

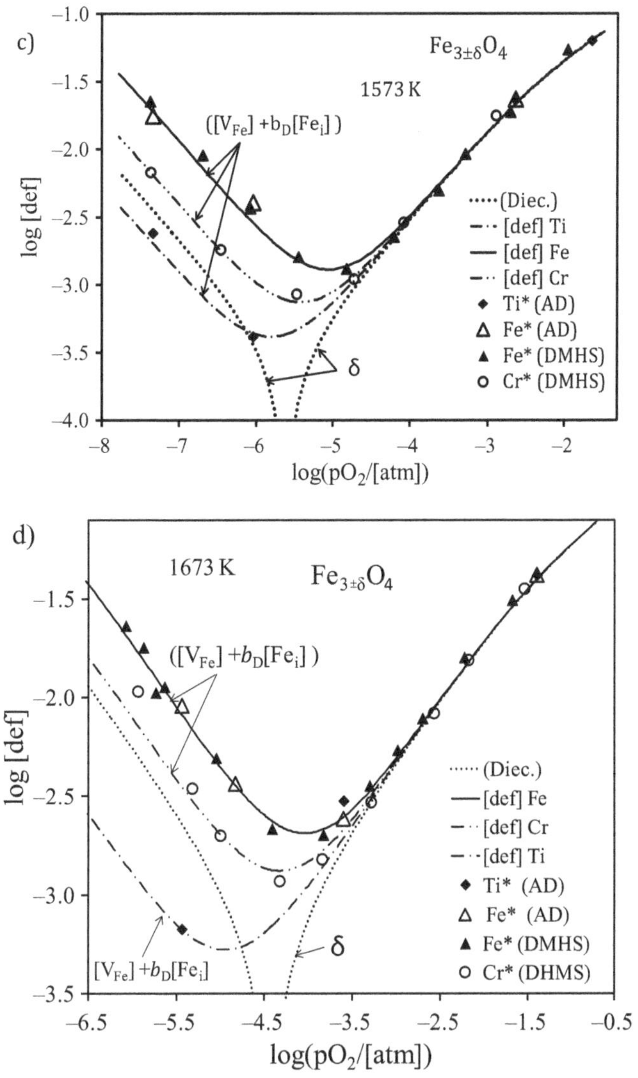

FIGURE 1.18 (Continued) (c) Dependence of the deviation from the stoichiometry on the oxygen pressure in magnetite at 1573 K ((···) line) (presented in Figure 1.4a). The points represent the values of the sum of concentrations of ionic defects ($[V_{Fe}] + \boldsymbol{b}_D[M_i]$), calculated according to Equation 1.47a for fitted values of $D^o_{V(M)}$ for the highest oxygen pressures and when using the values of coefficients of self-diffusion of tracers: Fe* - (△, ▲)(DHMS)(AD) points (Dieckmann et al. 1978, Aggarwal and Dieckmann 2002), Cr* - (○)(DHMS) (Dieckmann et al. 1978), Ti - (◆)(AD) (Aggarwal and Dieckmann 2002). Solid line and dashed lines denote the dependence of the sum of concentrations of ionic defects $[V_{Fe}] + \boldsymbol{b}_D[M_i]$) on p_{O_2} obtained for the fitted value of $\boldsymbol{b}_D$. (d) at 1673 K ((···) line) (presented in Figure 1.4c). The points represent the values of the sum of concentrations of ionic defects ($[V_{Fe}] + \boldsymbol{b}_D[M_i]$), calculated according to Equation 1.47a for fitted values of $D^o_{V(M)}$ for the highest oxygen pressures and when using the values of coefficients of self-diffusion of tracers: Fe* - ((△, ▲)(AD)(DMHS) points) (Dieckmann et al. 1978, Aggarwal and Dieckmann 2002), Cr* - (○)(DHMS) (Dieckmann et al. 1978), Ti - (◆)(AD) (Aggarwal and Dieckmann 2002). Solid line and dashed lines denote the dependence of the sum of concentrations of ionic defects $[V_{Fe}] + \boldsymbol{b}_D[M_i]$) on p_{O_2} obtained for the fitted value of $\boldsymbol{b}_D$.

$$\left[V_{Fe}\right] + b_D\left[Fe_i\right] = \frac{D^*_M}{D^o_{V(M)}} \tag{1.47a}$$

From Equation 1.47a and 1.47b it results that if we assume that at the highest oxygen pressures the diffusion of metal ions occurs via cation vacancies, we can fit such a value of b_D that the dependence of the sum of concentrations of ionic defects ($[V_{Fe}] + b_D[Fe_i]$) on p_{O_2} at the adjusted value of the parameter b_D, were consistent with the values of this sum

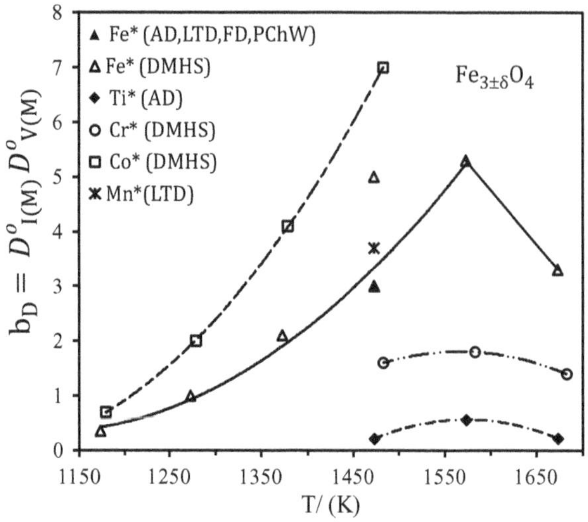

FIGURE 1.19 Temperature dependence of the ratio of the coefficient of diffusion of tracer ions via interstitial cations to the coefficient of diffusion of cations via vacancies $\left(b_D = D^o_{I(M)}/D^o_{V(M)}\right)$ in magnetite, obtained using the coefficients of self-diffusion of tracers: Fe* ((▲,△) points) (Dieckmann et al. 1978, Franke and Dieckmann 1989, Lu et al. 1993, Töpfer et al. 1995, Aggarwal and Dieckmann 2002), Co* (□) (Dieckmann et al. 1978), Cr* (○)(GMHS) (Töpfer et al. 1995), Ti* (◆)(AD) (Aggarwal and Dieckmann 2002), Mn* (✱)(LTD) (Lu et al. 1993) and concentrations of ionic defects.

($[V_{Fe}] + b_D[Fe_i]$) calculated using the values of the coefficients of self-diffusion of tracer M* (D^*_M) and the adjusted value of the coefficient of diffusion via vacancies $D^o_{V(M)}$.

Figure 1.18a presents the dependence of the deviation from the stoichiometry on the oxygen pressure (short-dashed lines) at temperatures from 1173 K to 1273 K and 1373 K, which were shown in Figures 1.3a–c (obtained using the results of the studies (Dieckmann 1982)). The figure shows also the dependence of the "sum" of the concentrations of ionic defects ($[V_{Fe}] + b_D[Fe_i]$) on p_{O_2} for the fitted value b_D (solid and dashed lines) and the values of this sum, calculated according to Equation 1.47a using the values of coefficients of self-diffusion of iron ^{59}Fe ((△,▲,◆) points) and cobalt ^{60}Co ((□,■,◆) points) (Dieckmann and Schmalzried 1977, Dieckmann et al. 1978). Thus, the values of the coefficient of diffusion $D^o_{V(M)}$ of iron an cobalt via iron vacancies at the highest oxygen pressure and of the parameter b_D, being the ratio of the coefficients of self-diffusion of ions Fe* and Co* via defects ($D^o_{I(M)}/D^o_{V(M)}$), were adjusted in such a way as to obtain a match between of the dependence ($[V_{Fe}] + b_D[Fe_i]$) resulting from the diagram of defects with the values of the sum ($[V_{Fe}] + b_D[Fe_i]$) calculated using the diffusion coefficients. In turn, Figure 1.18b–d show the results of analogous calculations for temperatures 1473 K, 1573 K and 1673 K, using the coefficients of self-diffusion of tracer ions: ^{59}Fe, ^{60}Co ^{53}Mn, ^{51}Cr and ^{44}Ti, reported in (Dieckmann and Schmalzried 1977, Dieckmann et al. 1978, Peterson et al. 1980, Franke and Dieckmann 1989, Lu et al. 1993, Töpfer et al. 1995, Aggarwal and Dieckmann 2002).

As can be seen in Figure 1.18, a very good agreement was obtained between the dependence of the sum of the concentrations of ionic defects ($[V_{Fe}] + b_D[Fe_i]$) on p_{O_2} and the values of this sum determined using the coefficients of self-diffusion of tracers. In the calculations of the coefficients of diffusion for cobalt and chromium isotopes, the concentrations of defects at the measurement temperature, which was by 10 degrees higher, were taken into account.

Figure 1.19 presents the temperature dependence of parameter b_D for the tracers, Fe*, Co*, Mn*, Cr*, Ti*, in magnetite.

As can be seen in Figure 1.19, the mobility of iron ions and cobalt ions via interstitial ions is higher than the mobility via vacancies. On the other hand, in the case of chromium diffusion, the mobilities are similar ($b_D > 1$), and in the case of titanium diffusion, the mobility via cation vacancies is larger than the mobility via interstitial cations and the ratio $D^o_{V(M)}/D^o_{I(M)} = b'_D$ is 1.7–4.7. Such significant differences indicate that the mobility of ions via defects depends on the effective charge of diffusing ion, its radius and its interaction with node ions. As can be seen in Figure 1.19, different values of the parameter b_D for the diffusion of individual metal ions cause that the minima of the obtained dependences of ($[V_{Fe}] + b_D[Fe_i]$) on the oxygen pressure are shifted in relation to the oxygen pressure at which there occurs the minimum concentration of ionic defects at the stoichiometric composition (see Figure 1.18). The shift in the above oxygen pressures results from different mobilities of ions $D^o_{V(M)}$ and $D^o_{I(M)}$ (rates of their diffusion via cation vacancies and interstitial ions).

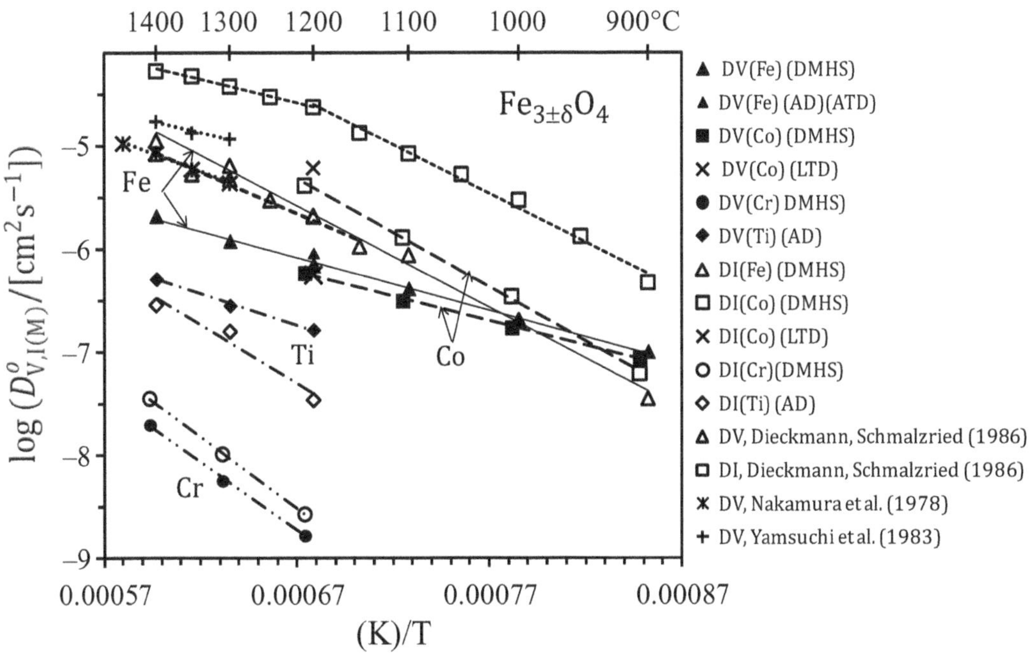

FIGURE 1.20 Temperature dependence of the coefficient of diffusion of defects via cation vacancies $D^o_{V(M)}$ (solid points) and via interstitial cations $D^o_{I(M)}$ (empty points), obtained when using the values of the coefficients of self-diffusion of tracers: Fe* ((▲,△) points) (Dieckmann et al. 1978, Lu et al. 1993, Töpfer et al. 1995, Aggarwal and Dieckmann 2002), Co* (■,□,×)(DMHS)(LTD) (Dieckmann et al. 1978, Smilten 1957), Cr* (●○)(DMHS) (Dieckmann et al. 1978), Ti* (◆,◇)(AD) (Aggarwal and Dieckmann 2002) and concentrations of ionic defects. Points (⊟,△,∗,+) denote the values of coefficients $D^o_{V(M)}$ and $D^o_{I(M)}$ determined using the coefficients of chemical diffusion taken from (Nakamura et al. 1978, Yamauchi et al. 1983, Dieckmann and Schmalzried 1986).

TABLE 1.4
Enthalpies ΔH^o_i of activation of mobility of ions Fe, Co, Cr, Ti via cation vacancies ($D^o_{V(M)}$) and via interstitial cations ($D^o_{I(M)}$) in magnetite. determined based on the dependences shown in Figure 1.20 $D^o_{V.I(M)} = D_o \exp\left(-\Delta H^o_{V,I(M)}/RT\right)$

$D^o_{V,I(M)}$	$\Delta H^o_{V,I(M)}$ (kJ/mol)	log D_o	ΔT (K)
$D^o_{V(Fe)}$ (DMHS)	98 ± 3	−2.65 ± 0.10	1173–1673
$D^o_{V(Co)}$ (DMHS)	90 ± 3	−3.07 ± 0.13	1173–1473
$D^o_{V(Cr)}$ (DMHS)	255 ± 11	0.21 ± 0.35	1473–1673
$D^o_{V(Ti)}$ (AD)	116 ± 7	−2.68 ± 0.24	1473–1673
$D^o_{V(Fe)}$ (DS)	155 ± 12	−0.24 ± 0.4	1323–1673
$D^o_{V(Fe)}$ (NYFN)	136 ± 8	−0.97 ± 0.25	1573–1723
$D^o_{V(Fe)}$ (YNSF)	87 ± 18	−2.06 ± 0.57	1573–1673
$D^o_{I(Fe)^S}$ (DMHS)	189 ± 7	1.02 ± 0.26	1173–1673
$D^o_{I(Co)}$ (DMHS)	200 ± 4	1.72 ± 0.15	1173–1473
$D^o_{I(Cr)}$ (DMHS)	268 ± 4	0.86 ± 0.14	1473–1673
$D^o_{I(Ti)}$ (AD)	218 ± 47	0.34 ± 1.57	1473–1673
$D^o_{I(Fe)}$ (DS)	180 ± 8	1.77 ± 0.32	1173–1473
$D^o_{I(Fe)}$ (DS)	85 ± 4	−1.60 ± 0.13	1473–1673

Initials in parentheses indicate the data source: (Dieckmann et al. 1978) – (DMHS), Aggarwal and Dieckmann 2002) – (AD), (Dieckmann and Schmalzried 1986) – (DS), (Nakamura et al. 1978) – (NYFN), and (Yamauchi et al. 2002) – (YNSF).

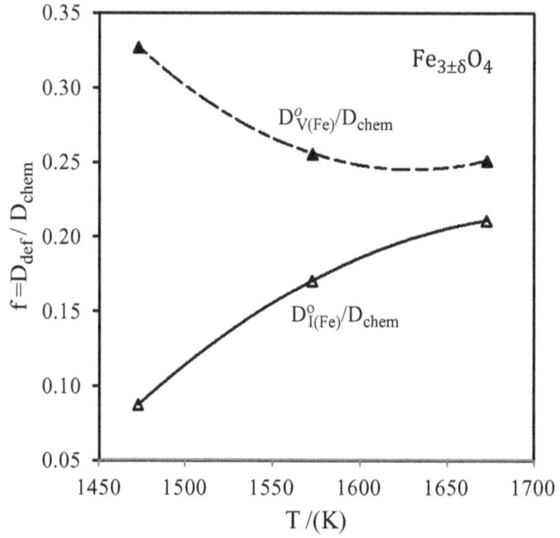

FIGURE 1.21 Temperature dependence of the ratio of the coefficient of iron diffusion via iron vacancies ((▲) points) and via interstitial cations ((△) points) to the coefficient of chemical diffusion in magnetite (Dieckmann and Schmalzried 1986).

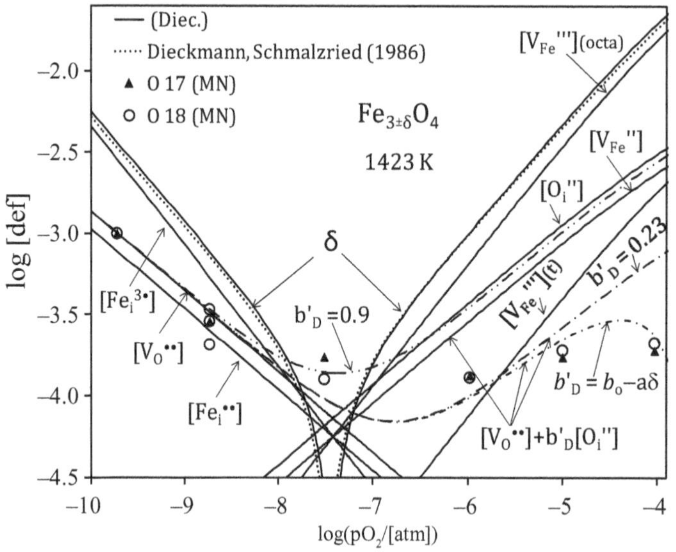

FIGURE 1.22 Diagram of concentrations of point defects for magnetite at 1423 K (solid lines). The (···) line shows the dependence of the deviation from the stoichiometry on the oxygen pressure, calculated using the "equilibrium constants" K_V and K_I (Dieckmann and Schmalzried 1986). The points (○,▲) represent the values of the sum of concentrations of ionic defects $\left(\left[V_O^{\bullet\bullet}\right]+b'_D\left[O_i''\right]\right)$ calculated according to Equation 1.47b for the fitted values of $D^{o}_{V(O)}$ for the lowest oxygen pressures, and using the published coefficients of self-diffusion of oxygen tracers (^{18}O, ^{17}O) (Millot and Niu 1997). The (-·····-) line denotes the calculated sum of the concentrations of ionic defects $\left(\left[V_O^{\bullet\bullet}\right]+b'_D\left[O_i''\right]\right)$ for the value of $b'_D = 0.9$, and the (-·-) line for the value of $b'_D = 0.23$ and when b'_D depends on the deviation from the stoichiometry ($b'_D = b_o - a\delta$, for the value $\boldsymbol{a} = 8$).

On the other hand, Figure 1.20 presents the temperature dependence of diffusion coefficients of ions via cation vacancies $D^{o}_{V(M)}$ and via interstitial ions $D^{o}_{I(M)}$.

As can be seen in Figure 1.20, in the above system of coordinates, the relations are approximately linear. Table 1.4 compares the enthalpies of activation of diffusion of $D^{o}_{V(M)}$ and $D^{o}_{I(M)}$ metal ions via defects. As can be seen in Figure 1.20, the coefficients of diffusion of Fe and Co ions via defects in magnetite differ only slightly. The mobility of titanium is lower by over an order of magnitude, and the mobility of chromium, via cation vacancies as well as via interstitial cations, is lower by over two orders of magnitude.

Figure 1.20 present the values of diffusion coefficients of iron via defects, obtained when using the published coefficients of chemical diffusion (Nakamura et al. 1978, Yamauchi et al. 1983, Dieckmann and Schmalzried 1986). As can be seen in Figure 1.20, they are higher than the mobilities determined using the coefficients of iron diffusion, by radioisotope method, but the temperature dependence is similar. Figure 1.21 presents the temperature dependence of the effective correlation coefficient, being the ratio of coefficients of diffusion via defects, determined in the present work, to the published coefficient of chemical diffusion ($f_{efe} = D^o_{V,I(M)}/D_{chem}$) (Nakamura et al. 1978, Dieckmann and Schmalzried 1986).

The obtained values deviate from the theoretical values of correlation coefficients, which is associated with the fact of participation of different defects and different diffusion mechanisms in magnetite, as well as with a complex dependence between the coefficients of the chemical diffusion and the calculated resultant coefficient of defect diffusion (diffusion of ions via defects).

In turn, Figure 1.22 presents the diagram of defects' concentration for 1423 K (solid lines), considering constant concentration of oxygen defects, which does not deteriorate the match between the dependence of δ on p_{O_2} and the dependence calculated using "equilibrium constants" of the reaction of formation of cation vacancies K_V and interstitial cations K_I (Dieckmann 1982, Dieckmann and Schmalzried 1986) ((···) line).

The figures show also the calculations of the dependence of the sum of the concentrations of oxygen defects $\left[V_O^{\bullet\bullet}\right]+b'_D\left[O_i''\right]$ on p_{O_2} and the values of this sum, calculated according to an equation analoguous to Equation 1.47a:

$$\left[V_O^{\bullet\bullet}\right]+b'_D\left[O_i''\right]=\frac{D_O^*}{D^o_{V(O)}} \tag{1.47b}$$

using the coefficients of self-diffusion of oxygen isotopes ^{17}O and ^{18}O determined at 1423 K (Millot and Niu 1997). For the lowest oxygen pressures, the coefficient of diffusion of oxygen via oxygen vacancies was adjusted, equal $D^o_{V(O)} = 1.1\cdot10^{-7}$ cm²/s. The estimated value of the diffusion coefficient $D^o_{V(O)}$ is the maximum value, as it was determined for the maximum concentration of oxygen defects, at which the match between the dependence of the deviation from the stoichiometry on the oxygen pressure and the experimental dependence of δ on p_{O_2} ((···) line) does not deteriorate. As can be seen in Figure 1.22, in the range of low pressures, a good agreement was obtained between the dependence of the sum of the concentrations of oxygen defects $\left[V_O^{\bullet\bullet}\right]+b'_D\left[O_i''\right]$ on p_{O_2} and the values of this sum determined using the coefficients of self-diffusion of oxygen. On the other hand, in the range of higher oxygen pressures, it was impossible to adjust a value of parameter b'_D allowing a match between the dependence of the sum $\left[V_O^{\bullet\bullet}\right]+b'_D\left[O_i''\right]$ with the values calculated using the coefficients of oxygen diffusion. As can be seen in Figure 1.22, for the value of $b'_D = 0.9$ and 0.23 no good match was obtained. The discrepancies could result from the fact that the coefficient $D^o_{I(O)}$ depends on the deviation from the stoichiometry. Due to that, an approximately linear dependence of the parameter b'_D on the deviation from the stoichiometry ($b'_D = b_o - a\delta$) was assumed. For the value of the parameter $a = 8$, quite a good match was obtained between the dependence of $\left(\left[V_O^{\bullet\bullet}\right]+b'_D\left[O_i''\right]\right)$ on p_{O_2} and the values of this sum calculated using the coefficients of oxygen diffusion.

1.6.2 Concentration of Electronic Defects and their Mobility

As can be seen in Figures 1.3 and 1.4, in magnetite, at a wide range of oxygen pressures the concentration of electrons and electron holes does not depend on p_{O_2}. This results from their high concentration and small influence of the concentration of ionic defects on the dependence of σ on p_{O_2}. The above character of the dependence of the concentration of electronic defects on the oxygen pressure is not typical of oxide semiconductors. Usually the concentration of electronic defects in oxides is comparable or lower than the concentration of ionic defects (it is of the same order of magnitude), as for example in MnO, TiO_2 oxides. If the electron and hole conductivities reach the same value at a specific oxygen pressure, there occurs the so-called *p/n*-type transition that is detectable through a clear minimum of the concentration of electronic defects and electrical conductivity. If the mobilities of electron holes and electrons are similar, *p/n* transition occurs at the stoichiometric composition. For magnetite, there is no change, characteristic of semiconductors, of the sign of the thermoelectric power, related to the *p/n* transition and the change in the character of the S-shaped dependence of Seebeck's coefficient on the oxygen pressure, as for example in $Mn_{1-\delta}O$ (Stokłosa 2015, Hed and Tannhauser 1967). The values, as well as the character of the dependence of thermoelectric power and electrical conductivity of magnetite with the stoichiometric composition allow assumption that the electron carriers in magnetite are mainly electrons (Dieckmann et al. 1983, Wu and Mason 1981). However, the decrease in the electrical conductivity at high oxygen pressures reported in (Dieckmann et al. 1983) and (Mason and Bowen 1981), indicates that the charge transfer is affected by the concentration of ionic defects.

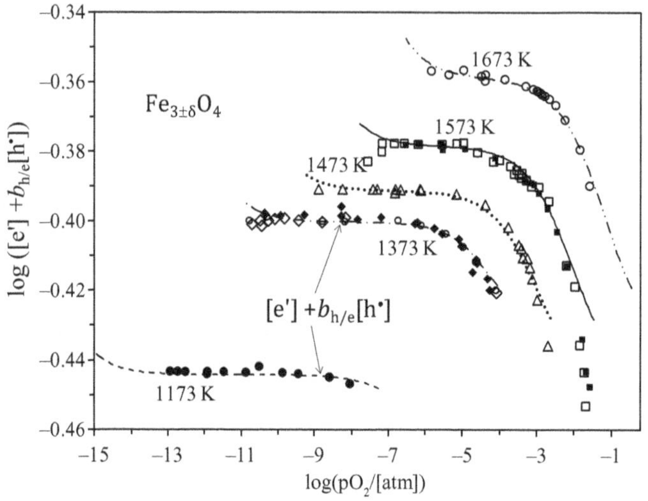

FIGURE 1.23 Dependences of the sum of concentrations of electrons and holes ($[e'] + b_{h/e}[h^\bullet]$) on the oxygen pressure for the fitted values of $b_{h/e}$ in magnetite, at 1173–1673 K, presented in Figures 1.3 and 1.4. The points represent the values of the sum of concentrations of electronic defects ($[e'] + b_{h/e}[h^\bullet]$) calculated according to Equation 1.49 using the values of the electrical conductivity (Dieckmann et al. 1983) and the fitted value of the mobility of electrons.

Thus, it can be assumed that in magnetite, in the temperature range of 1173–1673 K, there occurs a mixed electrical conductivity, described by the equation:

$$\sigma = \sigma_n + \sigma_p = \frac{F}{V_{Fe_3O_4}}\left(\mu_e\left[e'\right] + \mu_h\left[h^\bullet\right]\right) \tag{1.48a}$$

where σ_n and σ_p denote, respectively, electron conductivity and hole conductivity, μ_e and μ_h - mobility of electrons and holes, F - Faraday constant, V_{Fe3O4} - molar volume of magnetite, equal 44.53 cm^3. Factoring out the mobility of electrons we get:

$$\sigma = \frac{F}{V_{Fe_3O_4}}\mu_e\left(\left[e'\right] + \frac{\mu_h}{\mu_e}\left[h^\bullet\right]\right) = \frac{F}{V_{Fe_3O_4}}\mu_e\left(\left[e'\right] + b_{h/e}\left[h^\bullet\right]\right) \tag{1.48b}$$

where $b_{h/e}$ denotes the ratio of the mobility of electron holes to the mobility of electrons ($b_{h/e} = \mu_h/\mu_e$). Transforming Equation 1.48b we get:

$$\left[e'\right] + b_{h/e}\left[h^\bullet\right] = \frac{\sigma V_{Fe_3O_4}}{F\mu_e} \tag{1.49}$$

From Equation 1.49 it results that if we assume that at lower oxygen pressures in the range where σ does not depend on the oxygen pressure, electrical conductivity will depend only on the concentration of electrons, than we can adjust such a value of the parameter $b_{h/e}$ that the dependence of the sum of the concentrations of electronic defects ($[e'] + b_{h/e}[h^\bullet]$) resulting from the defects' diagram were consistent with the values of the sum ($[e'] + b_{h/e}[h^\bullet]$) calculated using the values of the electrical conductivity and the adjusted value of the mobility of electrons μ_e. The mobilities of electrons and electron holes are thus determined.

Figures 1.3 and 1.4, as well as Figure 1.23, show the dependence of the "sum" of the concentrations of electronic defects ($[e'] + b_{h/e}[h^\bullet]$) resulting from the diagram of concentrations of defects and the adjusted value of $b_{h/e}$ ((---) line) and the values of this sum, calculated according to Equation 1.49 using the value of the electrical conductivity (Dieckmann et al. 1983) and the adjusted value of the mobility of electrons μ_e.

Figure 1.24 presents the values of the adjusted parameter $b_{h/e}$, which changes from the value of 0.7 at 1173 K to about 0.3 at 1673 K.

As can be seen in Figures 1.3 and 1.4, the concentrations of electron holes and electrons are similar in a quite wide range of oxygen pressures near the oxygen pressure at which magnetite reaches the stoichiometric composition. Also, at low oxygen pressures, despite the small difference between the concentrations of electrons and electron holes, the "sum" of electronic defects ($[e'] + b_{h/e}[h^\bullet]$), as well as the electrical conductivity, practically does not depend on the oxygen pressure.

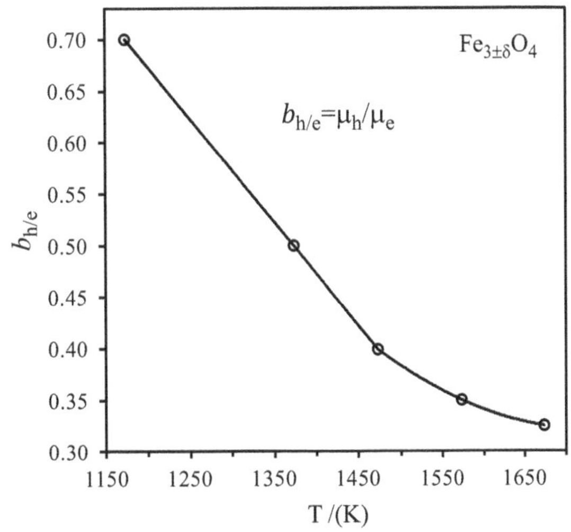

FIGURE 1.24 Temperature dependence of the ratio of the mobility of holes to the mobility of electrons ($\mu_h/\mu_e = b_{h/e}$) in magnetite, obtained using the values of electrical conductivity (Dieckmann et al. 1983) and using the concentration of electronic defects resulting from the diagrams.

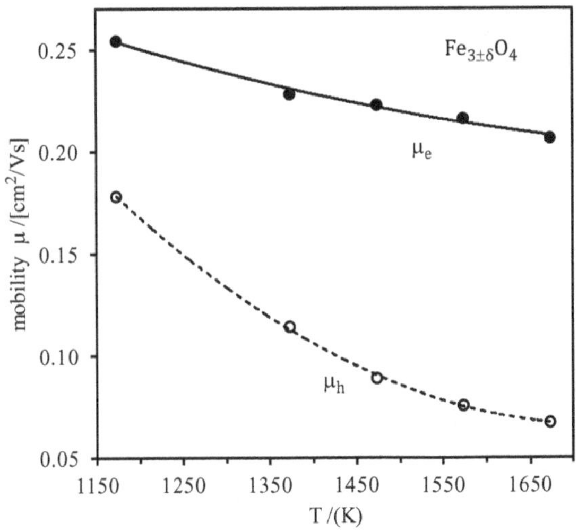

FIGURE 1.25 Temperature dependence of the mobility of electrons μ_e and electron holes μ_h in magnetite.

On the other hand, at higher oxygen pressures, in the range where the concentration of electron holes is higher than the concentration of electrons, the value of the "sum" of defects' concentration decreases (see Figures 1.23), as well as the value of electrical conductivity (Dieckmann et al. 1983, Mason and Bowen 1981). In accordance with the values of the parameter $b_{h/e}$, a decrease in the electrical conductivity when the oxygen pressure increases is due to a higher mobility of electrons than the mobility of electron holes.

Therefore, at higher oxygen pressures, the concentration of intrinsic electronic defects becomes perturbed by the increase in the concentration of electron holes resulting from an increase in the concentration of iron vacancies, and thus the concentration of electrons, and in consequence, also the electrical conductivity, decrease. At the lowest pressures, due to an increase in the concentration of interstitial iron ions, the concentration of electrons increases and a rise in the electrical conductivity should be observed. In a narrow range of oxygen pressures, where the concentration of electrons rises, the increase in the concentration of electrons is small enough for the rise in the electrical conductivity to be within the measurement error limits.

In turn, Figure 1.25 shows the dependence of the mobilities of electrons and electron holes on temperature; as can be seen, they decrease when the temperature increases.

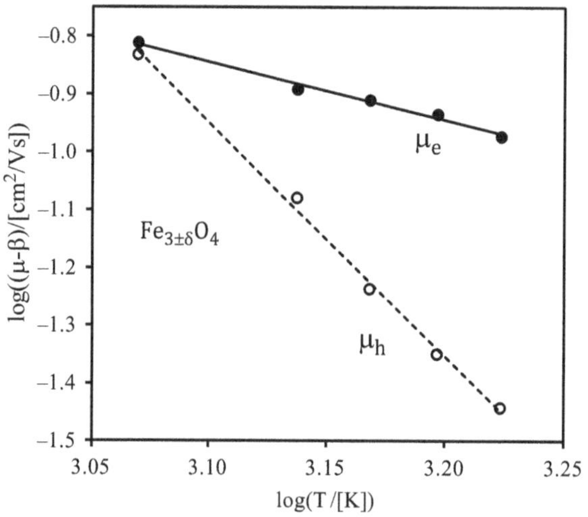

FIGURE 1.26 Temperature dependence of the mobility of electrons and electron holes in magnetite, in the temperature range 1473–1673 K, as a log–log plot.

TABLE 1.5
The exponent *n* and the parameter α of the dependence of the mobilities of electrons and electron holes on the temperature, in magnetite, according to the equation log(*μ* + *β*) = log *α* + *n* log *T* (Equation 1.50b) in the temperature range of 1173–1673 K, determined based on the dependences shown in Figure 1.26.

Equation	logα	*n*	β	R^2	ΔT/(K)
Electrons	2.3 ± 0.2	−1.00 ± 0.06	0.11	0.9885	1173–1673
Holes	11.5 ± 0.4	−4.03 ± 0.14	0.03	0.9966	1173–1673

The values of the determined mobility of electronic defects depend on the systematic error in the conductivity measurement, related mainly to an error in the measurement of the sample dimensions and inter-electrode distance. They depend also on the time of attaining a constant concentration of ionic defects (time of reaching the equilibrium state). The values of the determined mobilities of electron holes and electrons depend also on the concentration of charge carriers, which depend on the value ΔG^{o}_{eh} assumed in the calculations (see Figure 1.3b for the value of ΔG^{o}_{eh} three times as high). Taking into account a very small change in conductivity with the change in oxygen pressure and measurement error of conductivity at its relatively high values, the assumed value $\Delta G^{o}_{eh} = 30$ kJ/mol seems close to the real one. However, at a higher value of ΔG^{o}_{eh} (lower concentrations of electronic defects), the value of the parameter $b_{h/e}$ will be higher, and the determined value of electron mobility will be lower.

The decrease in the mobility of electronic defects when the temperature increases indicates that the effect of electron scattering on lattice ions dominates in their transport. The mobility of electronic defects, similarly as the electrical conductivity, should be therefore dependent on temperature, according to the relation:

$$\mu = \frac{\alpha}{T^n} + \beta \tag{1.50a}$$

Taking logarithms of both sides of Equation 1.50a we obtain:

$$\log(\mu - \beta) = \log\alpha + n\log T \tag{1.50b}$$

Figure 1.26 presents the dependence of the mobility of electronic defects on temperature as a log–log plot.

As can be seen, the points fit a straight line quite well. If the parameter β had a small value, it would be possible to determine the parameter n characterising the dependence of the mobility on the temperature associated with the scattering effect. The value of mobility depends on the measured values of the electrical conductivity, which, as was mentioned, are most often charged with a systematic error resulting for example from the measurement of the sample geometry. Thus, apart from the constant β, there is also the value of the systematic error ($\Delta\mu$) of the determined mobility. In calculations, the value of the sum $(\beta + \Delta\mu) = \beta'$ (which is subtracted from the value of mobility ($\log[\mu_e - (\beta + \Delta\mu_e)]$)) was chosen in such a way as to obtain the highest possible value of the correlation coefficient R^2 (close to 1). The determined values of the exponent n are given in Table 1.5. In the case of the mobility of electrons, the value of the exponent is about $n \cong 1$, while for the mobility of electron holes, the obtained value was about $n \cong 4$.

The model of mixed electrical conductivity, assumed in this calculation, explains the character of the dependence of electrical conductivity on the oxygen pressure. However, as it was already mentioned, the problem is in the assumed value of ΔG^{o}_{eh} of formation of electronic defects. Considering high values of the electrical conductivity and its small changes, the measurement error could be significant, which affects the determined values of mobilities of electronic defects and their temperature dependences. Additionally, it was assumed that the standard Gibbs energy of the formation of electronic defects (ΔG^{o}_{eh}) practically does not depend on the temperature, which causes that the calculated concentration of electronic could differ from the actual one.

As already mentioned, at the values of $\Delta G^{o}_{eh} = 30$ kJ/mol of formation of electronic defects assumed in the calculations, a good match was obtained between the dependence of the deviation from the stoichiometry on p_{O_2} and the experimental values δ for defects with the highest ionisation degree (as dominating defects). A good agreement of the dependence of δ on p_{O_2} with the values of the deviation from the stoichiometry δ can be also obtained at higher values of ΔG^{o}_{eh}, but defects with lower ionisation degrees, which are also likely to be present, must be considered as well. Higher values of ΔG^{o}_{eh} cause that the concentration of electronic defects is lower than the concentration determined in the present calculation. The presented calculation of ionic defects and electronic defects allow for a quite wide interpretation of the point defects structure; however, many issues are still open and require further studies on the electrical conductivity and the thermoelectric power in dependence on the oxygen pressure at high temperatures, in thermodynamic equilibrium conditions.

1.7 CONCLUSIONS

Diagrams of the concentrations of point defects in magnetite was determined using the results of deviation from the stoichiometry studies in the temperature range of 1173–1723 K (Dieckmann 1982) and (Nakamura et al. 1978),. From the performed calculations it results that the dominating defects should be defects with the highest ionisation degree, iron vacancies $V'''_{Fe(o)}$ in the sublattice of ions with octahedral coordination and interstitial iron ions $Fe_i^{3\bullet}$. It was assumed that the concentration of iron vacancies $\left[V''_{Fe}\right]$ and interstitial ions $\left[Fe_i^{\bullet\bullet}\right]$ at the stoichiometric composition is close to the concentration of defects with a charge of 3. The determined concentrations $\left[V''_{Fe}\right]$ and $\left[Fe_i^{\bullet\bullet}\right]$, when the oxygen pressure changes, are lower than the concentrations $\left[V'''_{Fe(o)}\right]$ and $\left[Fe_i^{3\bullet}\right]$ and they do not affect the degree of match between the dependence of the deviation from the stoichiometry on the oxygen pressure and the experimental δ. Therefore, these are the maximum concentrations of $\left[V''_{Fe}\right]$ and $\left[Fe_i^{\bullet\bullet}\right]$ that can be expected in magnetite.

Similarly, it was assumed that the concentration of iron vacancies $\left[V'''_{Fe(t)}\right]$ in the sublattice of ions with tetrahedral coordination is significantly smaller than the concentration of vacancies $\left[V'''_{Fe(o)}\right]$ in the octahedral sublattice. The determined concentration $\left[V'''_{Fe(t)}\right]$ is the maximum concentration which does not deteriorate the match between the dependence of the deviation from the stoichiometry on p_{O_2} and the experimental δ.

The problem is in the value ΔG^{o}_{eh} of formation of electronic defects. It was found that for the values of ΔG^{o}_{eh} smaller than 45 kJ/mol, a good match is obtained between the dependence of the deviation from the stoichiometry on the oxygen pressure and the experimental δ, considering only defects with the highest ionisation degree. Finally, it was assumed that the concentrations of electronic defects at the stoichiometric composition are equal to the concentration of Fe^{2+} ions in the tetrahedral sublattice and an increase in the concentration of Fe^{3+} ions in the octahedral sublattice. The concentrations of these ions were determined based on the studies of thermoelectric power (Wu and Mason 1981). The calculated equilibrium constant of formation of electronic defects corresponds to the value of $\Delta G^{o}_{eh} \cong 30$ kJ/mol, which practically does not depend on temperature.

Depending on the assumed value of ΔG^{o}_{eh} the values of $\Delta G^{o}_{V'''_{Fe(o)}}$, $\Delta G^{o}_{V''_{Fe(o)}}$ and $\Delta G^{o}_{Fe_i^{3\bullet}}$ and $\Delta G^{o}_{Fe_i^{\bullet\bullet}}$ change, respectively, while the changes are linear.

The determined values $\Delta G^{o}_{F_1}$, $\Delta G^{o}_{F_2}$, $\Delta G^{o}_{F_3}$ of formation of Frenkel defects in the individual sublattices practically do not depend on the value of ΔG^{o}_{eh} and unambiguously determine the concentration of defects in magnetite. It should be noted that the assumption of the same concentrations of vacancies $\left[V''_{Fe}\right] \cong \left[V'''_{Fe(o)}\right]$ causes an equity of the values of $\Delta G^{o}_{F_1} \cong \Delta G^{o}_{F_2}$. The value of $\Delta G^{o}_{F_3}$ of formation of defects in the sublattice of ions with tetrahedral coordination is derived from the maximum value of concentration of vacancies $\left[V'''_{Fe(t)}\right]$ in the tetrahedral sublattice at which the degree of match between the calculated dependence of the deviation from the stoichiometry p_{O_2} and the experimental δ is not deteriorated.

At the assumed values of $\Delta G^{o}_{eh} < 45$ kJ/mol, the maximum concentration of defects with lower ionisation degrees is significantly smaller than the concentration of defects with the highest ionisation degrees. At higher values of $\Delta G^{o}_{eh} > 50$ kJ/mol, it is possible to obtain a good match between the calculated dependence of deviation from the stoichiometry on p_{O_2} and the experimental δ, but it is necessary to consider significant concentrations of defects with lower ionisation degrees.

Reaching the stoichiometric composition in the middle of the range of existence of magnetite is not typical for non-stoichiometric oxides, for example MO ($Ni_{1-\delta}O$, $Co_{1-\delta}O$ etc.) with cubic structure, also for other oxides. The presence of strong repulsive interactions between ions in high temperatures causes that in the range of existence of these oxides, cation vacancies dominate. From the calculations presented in the work (Stokłosa 2015) it results that the stoichiometric composition of oxides of $M_{1-\delta}O$ type occurs at oxygen pressures a few orders of magnitude lower than the decomposition pressure. In turn, for oxides of MO_2 type, which have significantly lower space-packing density of structure, for example TiO_2, the stoichiometric composition is reached above the oxygen pressure of 1 atm.

In the case of magnetite, its structure is significantly looser ($M_{0.75}O$) in comparison to a cubic structure of MO oxides, which causes that in the range of existence of magnetite there is an oxygen pressure at which the stoichiometric composition is reached. A decrease in oxygen pressure ($p_{O_2} < p_{O_2}^{(\delta=0)}$) causes oxygen release and, in magnetite, increases the concentration of interstitial cations, which occupy empty tetrahedral voids. In turn, an increase in the oxygen pressure ($p_{O_2}^{(\delta=0)} < p_{O_2}$) forces the formation of cation vacancies, thus, further loosening the cubic spinel structure.

Magnetite-wüstite transformation occurs regardless of temperature at a similar concentration of interstitial iron ions, below 0.02 mol/mol, at oxygen pressures that increase when the temperature increases, from 10^{-15} atm at 1173 K to 10^{-8} atm at 1673 K. Magnetite-hematite transformation occurs at concentration of iron vacancies from 0.02 mol/mol at 1173 K to 0.1 mol/mol at 1673 K and at increasingly higher oxygen pressures, from 10^{-8} atm at 1173 K to about 0.5 atm at 1673 K.

The coefficients of self-diffusion of metal ions (isotope tracers ^{59}Fe, ^{60}Co, ^{51}Cr and ^{44}Ti) (Dieckmann et al. 1978, Dieckmann et al. 1987) and calculated from the diagram of concentrations of ionic defects, permitted to determine the diffusion coefficients of defects, or the mobility of tracer ions via iron vacancies ($D^o_{V(M)}$) and interstitial iron ions ($D^o_{I(M)}$). It was shown that the coefficients of diffusion via interstitial ions are higher than the coefficients of diffusion via iron vacancies $D^o_{I(M)} > D^o_{V(M)}$. It was also found that the mobilities of tracer ions differ despite the fact that they diffuse via the same defects. The coefficients $D^o_{I(M)}$ and $D^o_{V(M)}$ differ slightly for diffusion of iron and cobalt, while the mobility of titanium ions is smaller by almost an order of magnitude, and the mobility of chromium is even smaller. The differences are present in the values of $D^o_{I(M)}$ and $D^o_{V(M)}$ and in the activation energies of mobility of ions (diffusion of metal ions via defects).

The concentration of electronic defects, determined from the diagram of defects, using the results of studies of the electrical conductivity (Dieckmann et al. 1983) allowed the calculation of mobilities of electrons and electron holes. It was found that in the temperature range of 1173–1673 K the mobility of electrons is higher than the mobility of electron holes. The ratio of the mobility of electron holes to the mobility of electrons $b_{h/e} = \mu_h/\mu_e$ decreases when the temperature increases, from the value of $b_{h/e}$ = 0.7–0.3. A high concentration of electronic defects and the difference between their mobilities cause in a wide range of oxygen pressures that electrical conductivity of magnetite is independent of the oxygen pressure. At higher oxygen pressures, where electron holes dominate, electrical conductivity of magnetite decreases.

It was shown that in the temperature range of 1173–1673 K mobilities of electrons and electron holes in magnetite decrease, when the temperature increases, which indicates a strong effect of scattering of ions. The mobility of electrons is inversely proportional to the temperature, and the mobility of electron holes is inversely proportional to the temperature raised to the power of four (Equation 1.50a and 1.50b). The charge transport mechanism in magnetite at high temperatures requires verification and it is necessary to perform studies on electrical conductivity and thermoelectric power in dependence on the oxygen pressure in order to unambiguously decide on the structure of point defects in magnetite.

REFERENCES

Aggarwal, S., and R. Dieckmann, 2002, Point Defects and Cation Tracer Diffusion in $(Ti_xFe_{1-x})_{3-\delta}O_4$. II. Cation Tracer Diffusion. *J. Phys. Chem. Miner.* 29:707–718.

Becker, K. D., V. von Wurmb, and F. J. Litterst. 1993. Disorder and Diffusional Motion of Cations in Magnetite, $Fe_{3-\delta}O_4$, Studied by Mössbauer Spectroscopy at High Temperatures. *J. Phys. Chem. Solids* 54:923–935.

Bryant, P. E. C., and W. W. Smeltzer. 1969. The Dissociation Pressure of Hematite. *J. Electrochem. Soc.* 116:1409–1409.

Brynestad, J., and H. Flood. 1958. The Redox Equilibrium in Wüstite and Solid Solutions of Wüstite and Magnesium Oxide. *Z. Elektrochem.* 62: 953–958.

Calhoun, B. A. 1954. Magnetic and Electric Properties of Magnetite at Low Temperatures. *Phys. Rev.* 94: 1577–1585.

Daniels, J. M., and A. Rosencwaig. 1969. Mössbauer Spectroscopy of Stoichiometric and Non–stoichiometric Magnetite. *J. Phys. Chem. Solids* 30: 1561–1571.

Darken, L. S., and R. W. Gurry. 1945. The System Iron–Oxygen. I. The Wüstite Field and Related Equilibria. *J. Am. Chem. Soc.* 67: 1398–1412.

Darken, L. S., and R. W. Gurry. 1946. The System Iron—Oxygen. II. Equilibrium and Thermodynamics of Liquid Oxide and Other Phases. *J. Am. Chem. Soc.* 68: 798–816.

Dieckmann, R. 1982. Defects and Cation Diffusion in Magnetite (IV): Nonstoichiometry and Point Defect Structure of Magnetite ($Fe_{3-\delta}O_4$). *Ber. Bunsenges. Phys. Chem.* 86:112-118.

Dieckmann, R. 1983. Punktfehlordnung, Nichtstöchiometrie und Transporteigenschaften von Oxiden der Übergangsmetalle Kobalt, Eisen und Nickel (Point Defect Structure, Nonstoichiometry and Transport Properties of Oxides of the Transition Metals Cobalt, Iron and Nickel). Habilitation Thesis, University of Hannover, Germany.

Dieckmann, R., M. R. Hilton, and T. O. Mason. 1987. Defects and Cation Diffusion in Magnetite (VIII): Migration Enthalpies for Iron and Impurity Cations. *Ber. Bunsenges. Phys. Chem.* 91: 59–66.

Dieckmann, R., T. O. Mason, J. D. Hodge, and H. Schmalzried. 1978. Defects and Cation Diffusion in Magnetite (III.) Tracer Diffusion of Foreign Tracer Cations as a Function of Temperature and Oxygen Potential. *Ber. Bunsenges. Phys. Chem.* 82: 778–783.

Dieckmann, R., and H. Schmalzried. 1977. Defects and Cation Diffusion in Magnetite (I). *Ber. Bunsenges. Phys. Chem.* 81: 344–347

Dieckmann, R., and H. Schmalzried. 1986. Defects and Cation Diffusion in Magnetite (VI): Point Defect Relaxation and Correlation in Cation Tracer Diffusion. *Ber. Bunsenges. Phys. Chem.* 90: 564–575.

Dieckmann, R., C. A. Witt, and T. O. Mason. 1983. Defects and Cation Diffusion in Magnetite (V): Electrical Conduction, Cation Distribution and Point Defects in $Fe_{3-\delta}O_4$. *Ber. Bunsenges. Phys. Chem.* 87: 495–503.

Domenciali, C. A. 1950. Magnetic and Electric Properties of Natural and Synthetic Single Crystals of Magnetite. *Phys. Rev.* 78: 458–467.

Dunitz, J. D., and L. E. Orgel. 1957. Electronic Properties of Transition-metal Oxides-II: Cation Distribution amongst Octahedral and Tetrahedral Sites. *J. Phys. Chem. Solids* 3: 318–323.

Erickson, D. S., and T. O. Mason. 1985. Nonstoichiometry, Cation Distribution, and Electrical Properties in Fe_3O_4, $CoFe_2O_4$ at High Temperature. *J. Solid State Chem.* 59: 42–53.

Franke, P., and R. Dieckmann. 1989. Defect Structure and Transport Properties of Mixed Iron-manganese Oxides. *Solid State Ionics* 32–33: 817–823.

Giddings, R. A., and R. S. Gordon. 1973. Review of Oxygen Activities and Phase Boundaries in Wüstite as Determined by Electromotive-Force and Gravimetric Methods. *J. Am. Ceram. Soc.* 56:111–116.

Goodenough, J. B., and A. L. Loeb. 1955. Theory of Ionic Ordering, Crystal Distortion, and Magnetic Exchange Due to Covalent Forces in Spinels. *Phys. Rev.* 98: 391–408.

Greig, J. W., E. Posnjak, H. E. Merwin, and R. B. Sosman. 1935. Equilibrium Relationships of Fe_3O_4, Fe_2O_3, and Oxygen. *Am. J. Sci.* 30: 239–316.

Griffiths, B. A., D. Elwell, and R. Parker. 1970. The Thermoelectric Power of the System $NiFe_20_4$-Fe_30_4. *Phil. Mag. Ser.* 822:163–174.

Halloran, J. W., and H. K. Bowen. 1980. Iron Diffusion in Iron-Aluminate Spinels. *J. Am. Ceram. Soc.* 63: 58–65.

Hallström, S., L. Höglund, and J. Ågren, 2011. Modeling of Iron Diffusion in the Iron Oxides Magnetite and Hematite with Variable Stoichiometry. *Acta Materialia* 59: 53–60.

Haubenreisser, W. 1961. Zur Theorie der Elektronenleitung im stöchiometrischen Fe_3O_4–Kristall (Magnetit) oberhalb des Übergangspunktes $T_ü$ = 119,4 °K. *Phys. Status Solidi* 1: 619–635.

Hed, A. Z., and D. S. Tannhauser. 1967. High-Temperature Electrical Properties of Manganese Monoxide. *J. Chem. Phys.* 47:2090–2103.

Honig, J. M., and J. Spałek. 1989. Relation of the Verwey Transition in Magnetite to an Order-disorder Transition of Strongly Correlated Electrons. *J. Less–Common Met.* 156:423–438.

Honig, J. M., and J. Spałek. 1992. Elementary Formulation of the Verwey Transition in Magnetite via the Order-disorder Formalism. *J. Solid State Chem.* 96:115–122.

Kąkol, Z., and J. M. Honig. 1989. Influence of Deviations from Ideal Stoichiometry on the Anisotropy Parameters of Magnetite $Fe_{3(1-\delta)}O_4$. *Phys. Rev. B* 40:9090–9097.

Katsura, T., and H. Muan. 1964. Experimental Study of Equilibria in the System FeO-Fe_2O_3– Cr_2O_3 at 1300°C. *Trans. Met. Soc. AIME* 230:77–84.

Klinger, M. J., and A. A. Samokhralov. 1977. Electron Conduction in Magnetite and Ferrites. *Phys. Status Solidi (b)* 79:9–48.

Kuipers, A. J. M., and V. A. M. Brabers. 1976. Thermoelectric Properties of Magnetite at the Verwey Transition. *Phys. Rev. B* 14:1401–1405.

Kündig, W., and R. S. Hargrove. 1969. Electron Hopping in Magnetite. *Solid State Comm.* 7:223–227.

Lavine, J. M. 1959. Ordinary Hall Effect in Fe_3O_4 and $(NiO)_{0.75}(FeO)_{0.25}$ (Fe_2O_3) at Room Temperature. *Phys. Rev.* 114:482–488.

Lewis, G. V., C. R. A. Catlow, and A. N. Cormack. 1985. Defect Structure and Migration in Fe_3O_4. *J. Phys. Chem. Solids* 46:1227–1235.

Lu, F. H., S. Tinkler, and R. Dieckmann. 1993. Point Defects and Cation Tracer Diffusion in $(Co_xFe_{1-x})_{3-\delta}O_4$ Spinels. *Solid State Ionics* 62: 39–52.

Mason, T. O., and H. K. Bowen. 1981. Electronic Conduction and Thermopower of Magnetite and Iron-Aluminate Spinels. *J. Am. Ceram. Soc.* 64:237–242.

McClure, D. S., and A. Miller. 1957. The Distribution of Transition Metal Cations in Spinels. *J. Phys. Chem. Solids* 3:311–317.

Meyer, G. 1959. Phasenverhältnisse im System Eisen-Titan-Sauerstoff und Schmelzpunkte im System CaO-Al_2O_3-TiO_2. PhD Thesis, RWTH Aachen University, Germany.

Miles, P. A., W. B. Westphal, and A. von Hippel. 1957. Dielectric Spectroscopy of Ferromagnetic Semiconductors. *Rev. Mod. Phys.* 29:279–307.

Miller, A. 1959. Distribution of Cations in Spinels. *J. Appl. Phys.* 30:S24–S25.

Millot, F., and Y. Niu. 1997. Diffusion of O^{18} in Fe_3O_4: An Experimental Approach to Study the Behavior of Minority Defects in Oxides. *J. Phys. Chem. Solids* 58:63–72.

Nakamura, A., S. Yamauchi, K. Fueki, and T. Mukaibo. 1978. Vacancy Diffusion in Magnetite Containing Iron Oxides *J. Phys. Chem. Solids* 39:1203–1206.

Navrotsky, A., and O. J. Kleppa. 1967. The Thermodynamics of Cation Distributions in Simple Spinels. *J. Inorg. Nucl. Chem.* 29:2701–2714.

Pan, L. S., and B. J. Evans. 1976. High Temperature Electronic Structure of Fe_3O_4. *AIP Conf. Proc.* 34:181.

Parker, R., and C. J. Tinsley. 1976. Electrical Conduction in Magnetite. *Phys. Status Solidi (a)* 33:189–194.

Peterson, N. L., W. K. Chen, and D. Wolf. 1980. Correlation and Isotope Effects for Cation Diffusion in Magnetite. *J. Phys. Chem. Solids* 41:709–719.

Rasmussen, R. J., R. Aragón, and J. M. Honig. 1987. Influence of Magnetic Field Cooling on Electrical Properties of Magnetite. *J. Appl. Phys.* 61:4395–4396.

Richards, R. G., and J. White. 1954. Phase Relationships of Iron-Oxide-Containing Spinels : I, Relationships in the System Fe-Al-O. *Trans. Brit. Ceram. Soc.* 53:233–270.

Schmahl, N. G. 1941. The Relations among Oxygen, Pressure, Temperature and Composition in the System Fe_2O_3-Fe_3O_4. Z. *Elektrochem.* 47:821–843.

Schmahl, N. G., D. Henning, and C. Rübelt. 1969. Zersetzungsgleichgewichte im System Fe_2O_3—Fe_3O_4. *Arch. Eisenhüttenwes.* 40:375–379.

Schmalzried, H., and J. D. Tretjakov. 1966. Fehlordnung in Ferriten. *Ber. Bunsenges. Phys. Chem.* 70:180–189.

Schmalzried, H., and C. Wagner. 1962. Fehlordnung in ternären Ionenkristallen. *Z. Phys. Chem. N.F.* 31:198–221.

Shepherd, J. P., R. Aragón, and J. W. Koenitzer. 1985. Changes in the Nature of the Verwey Transition in Nonstoichiometric Magnetite (Fe_3O_4). *Phys. Rev. B* 32:1818–1819.

Shepherd, J. P., J. W. Koenitzer, R. Aragón, C. J. Sandberg, and J. M. Honig. 1985. Heat Capacity Studies on Single Crystal Annealed Fe_3O_4. *Phys. Rev. B* 31:1107–1113.

Shepherd, J. P., J. M. Koenitzer, R. Aragón, J. Spałek, and J. M. Honig. 1991. Heat Capacity and Entropy of Nonstoichiometric Magnetite $Fe_{3(1-\delta)}O_4$: The Thermodynamic Nature of the Verwey Transition. *Phys. Rev. B Condens. Matter* 43:8461–8471.

Smilten, J. 1957. Investigation of the Ferrite Region of the Phase Diagram Fe-Co-O. *J. Am. Chem. Soc.* 79:4881–4884. Smilten, J. 1957. The Standard Free Energy of Oxidation of Magnetite to Hematite at Temperatures above 1000°. *J. Am. Chem. Soc.* 79:4877–4880.

Smith, D. O. 1956. Magnetization of a Magnetite Single Crystal Near the Curie Point. *Phys. Rev.* 102:959–963.

Sockel, H. G., and H. Schmalzried. 1968. Coulometrische Titration an Übergangsmetalloxiden. *Ber. Bunsenges. Phys. Chem.* 72:745–754.

Srinavasan, C., and C. M. Srivastava. 1981. Electrical Conductivity Mechanism in Zinc and Copper Substituted Magnetite. *Phys. Status Solidi (b)* 103:665–671.

Stokłosa, A. 2015. *Non-Stoichiometric Oxides of 3d Metals*. pp. 313–376. Pfäffikon: Trans Tech Publications Ltd.

Sundman, B. 1991. An Assessment of the Fe-O System. *J. Phase Equilibria.* 12:127–140.

Tannhauser, D. S. 1962. Conductivity in Iron Oxides. *J. Phys. Chem. Solids* 23:25–34.

Töpfer, J., S. Aggarwal, and R. Dieckmann. 1995. Point Defects and Cation Tracer Diffusion in $(Cr_xFe_{1-x})_{3-\delta}O_4$ Spinels. *Solid State Ionics* 81:251–266.

Verwey, E. J. W., and P. W. Haayman. 1941. Electronic Conductivity and Transition Point of Magnetite ("Fe_3O_4"). *Physica* 8:979–987.

Verwey, E. J. W., P. W. Haayman, and F. C. Romeijn. 1947. Physical Properties and Cation Arrangement of Oxides with Spinel Structures II. Electronic Conductivity. *J. Chem. Phys.* 15:181–187.

Wagner, C., and E. Koch. 1936. Die elektrische Leitfähigkeit der Oxyde des Kobalts und Eisens (Mit einem Anhang über Rekristallisation von Zinkoxyd). *Z. Phys. Chem. B* 32:439–446.

White, J. 1938. Equilibrium at High Temperatures in Systems Containing Iron Oxides. *Iron Steel Inst. (London), Carnegie Schol. Mem.* 27:1–75.

White, J., R. Graham, R. Hay. 1935. An Investigation into the Oxidizing Power of Basic Slags. *J. Iron Steel Inst. London* 131:91–111.

Wriedt, H. A. 1991. The Fe-O (Iron-Oxygen) System. *J. Phase Equilibria.* 12:170–200.

Wu, C. C., S. Kumarakrishnan, and T. O. Mason. 1981. Thermopower Composition Dependence in Ferrospinels. *J. Solid State Chem.* 37:144–150.

Wu, C. C., and T. O. Mason. 1981. Thermopower Measurement of Cation Distribution in Magnetite. *J. Am. Ceram. Soc.* 64:520–522.

Yamauchi, S., A. Nakamura, T. Shimizu, and K. Fueki. 1983. Vacancy Diffusion in Magnetite-hercynite Solid Solution. *J. Solid State Chem.* 50:20–32.

2 Hausmannite $Mn_{3\pm\delta}O_4$

2.1 INTRODUCTION

Hausmannite, $Mn_{3\pm\delta}O_4$, is significantly different from magnetite. At low oxygen pressures it is in equilibrium with $Mn_{1-\delta}O$, while the oxygen equilibrium pressures are significantly higher than those for magnetite. In the temperature range of 1173–1673 K, they are equal to 10^{-6}–10^{-3} atm, while for magnetite they range from 10^{-12} to 10^{-7} atm. Also, the oxygen pressure over the $Mn_{3-\delta}O_4/Mn_2O_3$ phase boundary is higher, close to 1 atm, and above 1123 K it even reaches 100 atm. The range of existence of hausmannite was the subject of many studies (Hahn Jr. and Muan 1960, Kim et al. 1966, Keller and Dieckmann 1985a, Keller and Dieckmann 1985b, Schmahl and Hennings 1969a, Schmahl and Hennings 1969b, Terayama et al. 1983), and their critical review (Grundy et al. 2003, Kjellqvist and Selleby 2010) permitted determination of the characteristic thermodynamic values and a full phase diagram.

Hausmannite has two polymorphic variants, α-$Mn_{3\pm\delta}O_4$ with tetragonal structure (defected spinel structure – Jahn–Teller distortion), where the ratio of $c/a = 1.16$ (Mason 1947, McMurdie and Golovato 1948, Petzold 1971) and β-$Mn_{3\pm\delta}O_4$, with cubic spinel structure (McMurdie and Golovato 1948, McMurdie et al. 1950, Petzold 1971). Differently from magnetite, the transition temperature exceeds 1373 K and it is dependent on the oxygen pressure (see literature review by Grundy et al. (2003)). Systematic studies (Keller and Dieckmann 1985b) showed that at the equilibrium pressure, at the $Mn_{1-\delta}O/Mn_{3+\delta}O_4$ phase boundary, the transition occurs at 1448 K, at the stoichiometric composition at 1408 K, and at the pressure of 1 atm at 1428 K. From experimental results and thermodynamic calculations, it results that enthalpy of the phase transition is 18–22.8 kJ/mol (Grundy et al. (2003), Kjellqvist and Selleby 2010).

In contrast to magnetite, hausmannite has a structure of normal spinel, so that Mn^{2+} ions are in cation sites with tetrahedral coordination, and Mn^{3+} are in sites with octahedral coordination, which is consistent with the electron structure of ions and the energy of stability in the crystal field (Dunitz and Orgel 1957, Goodenough and Loeb 1955). Such distribution of ions in α-$Mn_{3\pm\delta}O_4$, especially at room temperature, has been supported by several authors (Irani et al. 1960, Kasper 1959, Larson et al. 1962, McClure 1957, Nogués and Poix 1974, Verney and de Boer 1936). At higher temperatures, a change in ion distribution occurs, and, similarly to magnetite, an equilibrium is set between the concentration of manganese ions, Mn^{2+} and Mn^{3+}, in both sublattices. In earlier studies it was assumed that in the octahedral sublattice, Mn^{2+} and Mn^{4+} ions are present (Driessens 1967, Fine and Chiou 1957, Fyfe 1949, Romeijn 1953, Rosenberg et al. 1963, Verney and de Boer 1936). In β-$Mn_{3\pm\delta}O_4$ it is postulated that the sites in the cation sublattice with tetrahedral coordination are occupied by Mn^{2+} ions. In turn, in the octahedral sublattice, a process of disproportionation of Mn^{3+} ions occurs, and Mn^{2+} and Mn^{4+} are created (an equilibrium $Mn^{3+}_{(o)} \rightleftarrows Mn^{2+}_{(o)} + Mn^{4+}_{(o)}$ is set). From the results of the studies of the thermoelectric power, the distribution of manganese ions in both cation sublattices were calculated (Dorris and Mason 1988). It resulted that in the temperature range of 1173–1400 K, the concentration of Mn^{3+} ions decreases from the value of 1.9–0.9 mol, while the concentration of Mn^{2+} and Mn^{4+} ions increase from 0.1–0.7 mol/mol. In the temperature range of 1450–1773 K, concentrations of Mn^{2+} and Mn^{4+} ions are similar and vary from 0.68 to 0.78 mol/mol. Therefore, the formula of hausmannite could be:

$$\left(Mn^{2+}_{1-x}Mn^{3+}_{x}\right)\left(Mn^{3+}_{2-x}Mn^{2+}_{x}\right)O_4 \quad \text{or} \quad Mn^{2+}_{(t)}\left(Mn^{3+}_{2-2y}Mn^{4+}_{y}Mn^{2+}_{y}\right)O_4$$

However, there are no unambiguous results of the studies concerning the distribution of manganese ions with a different charge in the spinel structure of hausmannite (in α- and β- Mn_3O_4 variations).

At high temperatures, α- and β-$Mn_{3\pm\delta}O_4$ variants show a deviation from the stoichiometry, smaller than that in magnetite. Numerous works dealt with the studies of the deviation from the stoichiometry (Bergstein and Vintera 1957, Hahn Jr. and Muan 1960, Keller and Dieckmann 1985b, Kim et al. 1966, LeBlanc and Wehner 1934, Millar 1928, Moore et al. 1950, Schmahl and Hennings 1969a, Schmahl and Hennings 1969b, Southard and Moore 1942, Terayama et al. 1983), however, they covered a narrow range of oxygen pressures and absolute values of deviation δ were determined with a large error. Extensive studies (Keller and Dieckmann 1985b) showed that hausmannite, similarly to magnetite, reaches the stoichiometric composition in the range of its existence. At low oxygen pressures, interstitial manganese ions dominate, and at higher pressures, manganese vacancies dominate. The above model of ionic defects is consistent with the results of the studies on diffusion of radioisotope elements ^{54}Mn, ^{59}Fe and ^{60}Co performed at 1473 K (Lu and Dieckmann 1993).

The studies of electrical conductivity showed that it is significantly lower than that of magnetite (Logothetis and Park 1975, Metselaar et al. 1981, Romeijn 1953, Rosenberg et al. 1963, Rosenberg et al. 1966). α-$Mn_{3\pm\delta}O_4$, at room temperature, shows large resistance, of the order of 10^8 Ω·cm, which decreases when the temperature increases. As has been shown (Dorris and Mason 1988, Logothetis and Park 1975), the activation energy is 1.3–1.67 eV in the temperature range of 973–1373 K. The discrepancies between the results of different authors are related to conducting the studies in the air; at this oxygen pressure a slow transformation of Mn_3O_4 into Mn_2O_3 occurs.

At the temperature of polymorphic $\alpha \rightarrow \beta$ transformation of hausmannite, about 1403 K, there occurs a jump in the value of electrical conductivity, of approximately a half of an order of magnitude, and then there occurs a hysteresis, in the range of 25 K (Metselaar et al. 1981, Romeijn 1953, Rosenberg et al. 1963). In the range of β-$Mn_{3\pm\delta}O_4$ variation, electrical conductivity also increases with temperature, while the activation energy is half as high (0.65 eV).

The results of studies of the oxygen pressure dependence on electrical conductivity of hausmannite under thermodynamic equilibrium conditions (Metselaar et al. 1981) showed that in a wide range of oxygen pressure, it does not depend on p_{O_2}. Inversely as for magnetite, electrical conductivity decreases only at low oxygen pressures, near the $Mn_{1-\delta}O$/$Mn_{3+\delta}O_4$ phase boundary. It could indicate that despite a higher concentration of electrons in this range, the mobility of electron holes is significantly higher than that of electrons.

In turn, studies of thermoelectric power (Dorris and Mason 1988, Metselaar et al. 1981) showed that the Seebeck coefficient has a positive value. Therefore, the α- variant, as well as β-Mn_3O_4, is a *p-type* semiconductor, which is confirmed by the fact that the mobility of electron holes is significantly higher than that of electrons (inversely as for magnetite). In the range of existence of α-Mn_3O_4, the inverse temperature dependence of Seebeck's coefficient is linear, with a slope corresponding to 1.0–1.4 eV and these values are close to the activation energy of electrical conductivity. However, in the range of existence of β-$Mn_{3\pm\delta}O_4$, the slopes correspond to the energy of 0.1–0.3 eV, thus, these values are significantly lower than the activation energy of electrical conductivity.

2.2 MODEL OF IONIC DEFECTS IN HAUSMANNITE AND METHODS OF CALCULATION OF DIAGRAMS OF THEIR CONCENTRATION

As already mentioned in the literature review section, extensive studies on the deviation from the stoichiometry were conducted by Keller and Dieckmann (Keller and Dieckmann 1985b). They found the concentration of defects by one order of magnitude lower than that for magnetite. They determined also the character of the oxygen pressure dependence of the deviation from the stoichiometry, analogously to the case of magnetite, with an exponent equal approximately 2/3 and −2/3. This indicates that in the range of higher oxygen pressures, manganese vacancies $\left(V_{Mn}'''\right)$ should dominate, and at low oxygen pressures, interstitial manganese ions $\left(Mn_i^{3\bullet}\right)$ should dominate. Therefore, for hausmannite, it is possible to assume a model of ionic defects analogous as in the case of magnetite (see Section 1.4). According to other studies it was assumed that hausmannite has a structure of normal spinel, which means that cation sites with tetrahedral coordination are occupied by Mn^{2+} ions, while Mn^{3+} ions occupy sites with octahedral coordination. Thus, hausmannite with the stoichiometric composition will have the formal formula of $\left(Mn_{(t)}^{(2+)}\right)\left(Mn_{(o)}^{(3+)}\right)_2 O_4$. For the purpose of writing reactions, we can assume the following formula: $\left(Mn_{(t)}^{(2+)}O\right)\left(Mn_{(o)}^{(3+)}O_{3/2}\right)_2$. Thus, formally, there is one mole of oxygen per 1 mole of Mn^{2+} ions, while there is 3/2 mole of oxygen per one mole of Mn^{3+} ions. Therefore, the reaction of formation of ionic defects have a form analogous to that of magnetite. As a result of the incorporation of oxygen atoms into the sublattice with tetrahedral coordination, manganese vacancies $\left(V_{Mn(t)}''\right)$ are formed, according to the reaction:

$$1/2O_2 = O_O^x + V_{Mn(t)}'' + 2h^{\circ} \qquad \Delta G^o_{V_{Mn(t)}''} \qquad (2.1a)$$

Creation of a manganese vacancy in the sublattice with octahedral coordination is associated with the incorporation of 3/2 mol of oxygen atoms:

$$3/4O_2 = 3/2O_O^x + V_{Mn(o)}''' + 3h^{\circ} \qquad \Delta G^o_{V_{Mn(o)}'''} \qquad (2.1b)$$

By analogy, the formation of interstitial manganese ions due to partial decomposition of the oxide will occur according to the reactions:

$$Mn_{Mn(t)}^{x(2+)} + O_O^x + 2h^{\bullet} + V_i = Mn_i^{\bullet\bullet} + 1/2O_2 \qquad \Delta G^o_{Mn_i^{\bullet\bullet}} \qquad (2.2a)$$

$$Mn^{x(3+)}_{Mn(o)} + 3/2O^{x}_{O} + 3h^{\bullet} + V_i = Mn^{3\bullet}_i + 3/4O_2 \qquad \Delta G^{o}_{Mn^{3\bullet}_i} \qquad (2.2b)$$

Expressions describing the "equilibrium constant" of reactions of defect formation have a form analogous to that of magnetite (see Equation 1.10a, 1.10b, 1.10c, 1.14a, 1.14b, 1.14c). Concentrations of vacancies $\left[V'''_{Mn}\right]$ and $\left[V''_{Mn}\right]$ at the same oxygen pressure are related by Equation 1.35b

$$\frac{\left[V''_{Mn}\right]^2}{\left[V'''_{Mn}\right]^{4/3}} = \frac{K^2_{V''_{Mn}}}{K^{4/3}_{V'''_{Mn}}} = K_{V''_{Mn}/V'''_{Mn}} = K_1 \qquad (2.3)$$

Concentrations of manganese vacancies and interstitial manganese ions at a constant oxygen pressure are related by equation describing the formation of Frenkel-type defects (see Equation 1.34a, 1.34b, 1.34c, 1.36a, 1.36b, 1.36c):

$$\left[V'''_{Mn(o)}\right]\left[Mn^{3\bullet}_i\right] = K_{V'''_{Mn(o)}} K_{Mn^{3\bullet}_i} = K_{F_1} \qquad (2.4a)$$

$$\left[V''_{Mn(t)}\right]\left[Mn^{\bullet\bullet}_i\right] = K_{V''_{Mn(t)}} K_{Mn^{\bullet\bullet}_i} = K_{F_2} \qquad (2.4b)$$

Apart from the defects with the highest ionisation degree, defects with lower charge should be expected, namely, complexes of ionic and electronic defects, which will be formed according to the reactions of type (1.16a, 1.16b–1.19a, 1.19b, 1.19c).

Also, a possibility of formation of oxygen defects in hausmannite should be considered; they will be formed according to the reactions 1.21a, 1.21b.

As it was previously shown, in hausmannite, if the dependence of the deviation from the stoichiometry on the oxygen pressure has an exponent of 2/3 and −2/3, there should occur quite high concentration of electronic defects, independent of the oxygen pressure. A high concentration of electronic defects in hausmannite is indicated by high values of electrical conductivity (Dorris and Mason 1988, Logothetis and Park 1975, Metselaar et al. 1981, Rosenberg et al. 1966).

The concentration of electronic defects in hausmannite will be related, similarly as in magnetite, with a change in the oxidation state of manganese ions. It is assumed that at room temperature and higher temperatures, the distribution of Mn^{2+} and Mn^{3+} in octa- and tetra-hedral lattices is random, and it is created as a result of electron transfer between manganese ions. Two possibilities can be considered. Mn^{2+} ions with tetrahedral coordination release electrons that are accepted by Mn^{3+} ions with octahedral coordination, according to the reactions:

$$Mn^{2+}_{(t)} = Mn^{3+}_{(t)} + e^- \qquad (2.5a)$$

$$Mn^{3+}_{(o)} + e' = Mn^{2+}_{(o)} \qquad (2.5b)$$

As a result, an equilibrium is set:

$$Mn^{2+}_{(t)} + Mn^{3+}_{(o)} \rightleftarrows Mn^{3+}_{(t)} + Mn^{2+}_{(o)} \qquad (2.5c)$$

The change in the oxidation state of ions can result from optimisation of energy of interaction between ions or it can be related to the formation of electronic defects. In this case, in cation sites with tetrahedral coordination, manganese ions $Mn^{3+}_{(t)}$ will be ions which are positive in relation to the lattice $\left[Mn^{\bullet}_{Mn(t)}\right]$, with the concentration equal to the concentration of electron holes $\left[Mn^{\bullet}_{Mn(t)}\right] = [h^{\bullet}]$. In turn, a decrease in the concentration of Mn^{3+} in the cation sublattice with octahedral coordination and an increase in the concentration of Mn^{2+} should be treated as an appearance of ions that are negative in relation to the lattice $\left(Mn'_{Mn(o)}\right)$, with the concentration equal to the concentration of electrons $\left[Mn'_{Mn(o)}\right] = \left[e'\right]$. Analogously as in the case of magnetite, the equilibrium constant K will be approximately a product of the concentrations of ions $\left[Mn^{\bullet}_{Mn(t)}\right]$ and $\left[Mn'_{Mn(o)}\right]$ (see Equation 1.28b):

$$K_{eh} = \left[Mn^{\bullet}_{Mn(o)}\right]\left[Mn'_{Mn(t)}\right] \equiv \left[h^{\bullet}\right]\left[e'\right] \qquad \Delta G^{o}_{eh} \qquad (2.6)$$

It was assumed that at high temperatures, exchange of electrons occurs only in the sublattice with octahedral coordination (Dorris and Mason 1988). Thus, production of an electron hole is associated with the release of an electron by Mn^{3+}

ion and the formation of Mn^{4+} ion ($Mn^{\bullet}_{Mn(o)}$). Production of an electron is associated with the acceptance of an electron by an Mn^{3+} ion and the formation of an Mn^{2+} ion, negative in relation to the lattice ($Mn'_{Mn(o)}$). Therefore, the following processes occur:

$$Mn^{3+}_{(o)} = Mn^{4+}_{(o)} + e^- \tag{2.7a}$$

$$Mn^{3+}_{(o)} + e' = Mn^{2+}_{(o)} \tag{2.7b}$$

As a result, an equilibrium is set:

$$2Mn^{3+}_{(o)} \rightleftarrows Mn^{4+}_{(o)} + Mn^{2+}_{(o)} \tag{2.7c}$$

The reaction (2.7c) is called disproportionation reaction and it can describe the formation of electronic defects. According to the notation of Kröger–Vink, Equation (2.7c) it assumes the form of:

$$2Mn^{x}_{Mn(o)} \leftrightarrows Mn^{\bullet}_{Mn(o)} + Mn'_{Mn(o)} \equiv h^{\bullet} + e' \tag{2.7d}$$

Therefore, the equilibrium constant for the reaction 2.7d has the form analogous to Equation 2.6

In turn, the formation of ionic defects in high temperatures causes also the formation of electronic defects. The formation of manganese vacancies $V'''_{Mn(o)}$, when the oxygen pressure increases, according to reaction 2.1, is accompanied by withdrawing of electrons from the lattice; therefore, in the sublattice of manganese ions with octahedral coordination, ions ($Mn^{4+}_{(o)}$) are created, or in the sublattice with tetrahedral coordination, ions ($Mn^{3+}_{(t)}$) are formed, positive in relation to the lattice ($Mn^{\bullet}_{Mn}$), and their concentration will increase the concentration of electron holes.

Analogously, when interstitial manganese ions $\left(Mn_i^{3\bullet}\right)$ are created, the electrons released by the oxygen will be accepted by Mn^{3+} ions in the sublattice with octahedral coordination, becoming ions that are negative in relation to the lattice ($Mn'_{Mn(o)}$). The concentrations of these ions will be equivalent to the concentration of electrons. Therefore, in the lattice of hausmannite, apart of node ions with the charge ($Mn^{2+}_{(t)}$) and ($Mn^{3+}_{(o)}$), there will be ions with formally the same charge, but these ions ($Mn^{\bullet}_{Mn(o)}$ and $Mn'_{Mn(o)}$), will be positive or negative in relation to the lattice and they have the properties of point defects. The concentration of these ions is equivalent to the concentrations of electronic defects.

At a constant temperature and under defined oxygen pressure, in hausmannite an equilibrium will set between ionic defects ($V^{z'}_{Mn}$), ($Mn_i^{z\bullet}$) and electronic defects (ions ($Mn^{\bullet}_{Mn}$) and (Mn'_{Mn}). For the purpose of writing the reactions of point defects, the localisation of electronic defects in cation sublattices has no significant meaning, while the energy of formation will depend on their localisation on cations in octahedral or tetrahedral sublattice.

2.3 METHODS AND RESULTS OF CALCULATION OF THE DEFECTS DIAGRAMS, AND DISCUSSION

As mentioned, the concentrations of ionic defects and the character of the dependence of the deviation from the stoichiometry on the oxygen pressure in hausmannite is analogous to that in magnetite, which allows using the same methods of calculation of the concentration of ionic defects (see Section 1.5). It was assumed that the dominating defects are manganese vacancies V'''_{Mn} and interstitial manganese ions $Mn_i^{3\bullet}$. In order to calculate concentrations of individual defects and the deviation from the stoichiometry depending on the oxygen pressure, it is necessary to choose the values of $\Delta G^o_{F_1}$ of formation of Frenkel defects, $\Delta G^o_{V'''_{Mn(o)}}$ of formation of manganese vacancies and it is necessary to adjust the value ΔG^o_1, which permits calculation of the constant $K_1 = K_{V''_{Mn(t)}/V'''_{Mn(o)}}$, relating the concentration of manganese vacancies $\left[V'''_{Mn(o)}\right]$ and $\left[V''_{Mn(t)}\right]$ in both cation sublattices (see Equation 2.3). From the initial calculations performed, it results that for the value $\Delta G^o_{F_1} \cong \Delta G^o_{F_2}$ the concentration of vacancies $\left[V''_{Mn}\right]$ in the sublattice with tetrahedral coordination does not affect the character of the dependence of the deviation from the stoichiometry on the oxygen pressure. Due to that, in the calculations it was assumed that at the stoichiometric composition, the concentration of vacancies $\left[V''_{Mn(t)}\right]$ is close to the concentration $\left[V'''_{Mn(o)}\right]$. Therefore, the maximum concentration $\left[V''_{Mn(t)}\right]$ and $\left[Mn_i^{\bullet\bullet}\right]$ was determined; actually, it may be lower. Accurate determination of the value of $\Delta G^o_{F_2}$ is ambiguous, because concentrations $\left[V''_{Mn(t)}\right]$ and $\left[Mn_i^{\bullet\bullet}\right]$ cause changes in the deviation from the stoichiometry at oxygen pressures near the stoichiometric composition, i.e. in the range where the error of determination of absolute values of δ is the highest. On the other hand, due to symmetrical dependence of the concentration of manganese vacancies $\left[V'''_{Mn(o)}\right]$ and interstitial manganese ions $\left[Mn_i^{3\bullet}\right]$ on the oxygen pressure and relatively small concentration of ionic defects, the adjusted values of $\Delta G^o_{F_I}$ of formation of Frenkel defects (Equation 2.4b) are determined in an unambiguous way.

Relating the concentration of ionic defects and electronic defects requires the knowledge of the standard Gibbs energy of their formation ΔG^o_{eh}. As there are no values of ΔG^o_{eh} available, the problem is – which values are possible? Studies of the oxygen pressure dependence of electrical conductivity (Dorris and Mason 1988, Metselaar et al. 1981) showed that it is only slightly lower than that for magnetite and similarly, in a wide range of oxygen pressures, it does not depend on the concentration of ionic defects. Electrical conductivity increases when the temperature increases (i.e., it shows activation energy), therefore, also the concentration and mobility of electronic defects should increase. Thus, also a decrease in the value ΔG^o_{eh} should be expected. In turn, from the studies of thermoelectric power it results that the transport occurs via electron holes. As hausmannite reaches the stoichiometric composition in the range of its existence, also a significant concentration of electrons should be expected, resulting from intrinsic electronic defects. A decrease in the electrical conductivity described in (Metselaar et al. 1981), at low oxygen pressures, where electrons should dominate indicates that the mobility of electron holes is much higher than that of electrons. These facts indicate that in hausmannite there is a significant concentration of intrinsic electronic defects, which are only slightly affected by the concentration of ionic defects. From the calculations of the distribution of ions Mn^{2+}, Mn^{3+}, Mn^{4+}, done using the values of the Seebeck coefficients (Dorris and Mason 1988), it results that at 1173–1753 K, the concentration of ions Mn^{2+} and Mn^{4+} in the range of existence of α-Mn_3O_4 changes from 0.1–0.45 mol/mol, and in the range of existence of β-$Mn_{3\pm\delta}O_4$ phase from 0.6–0.7 mol/mol. Changes in concentrations of these ions should be also associated with the concentration of electronic defects. However, these concentrations are twice as high as these in magnetite, thus, it is difficult to conclude about the value of ΔG^o_{eh} of formation of electronic defects based on the concentrations of Mn^{2+} and Mn^{4+} ions.

Initially performed calculations of diagrams of the concentrations of point defects showed, similarly as in the case of magnetite, that when adjusting several pairs of values ΔG^o_{eh} and $\Delta G^o_{V'''_{Mn(o)}}$ it is possible to obtain a very good match between the dependence of δ on p_{O_2} to experimental results δ. Figure 2.1 shows the dependence of the value of $\Delta G^o_{V'''_{Mn(o)}}$ of formation of cation vacancies on the values of ΔG^o_{eh} assumed in the calculations for several temperatures, for α- $Mn_{3\pm\delta}O_4$ (solid points) and β-$Mn_{3\pm\delta}O_4$ (empty points), for which a good match between the dependence of δ on p_{O_2} and the experimental results δ is obtained for defects with the highest ionisation degree.

As can be seen in Figure 2.1, the relationships are linear, with the correlation coefficient close to one ($R^2 = 0.9999$). Also, the slopes are similar; they change more for the α-$Mn_{3\pm\delta}O_4$ (solid points) and less for β-Mn_3O_4 (empty points). However, the free component in a-d the linear equation only slightly varies with the temperature.

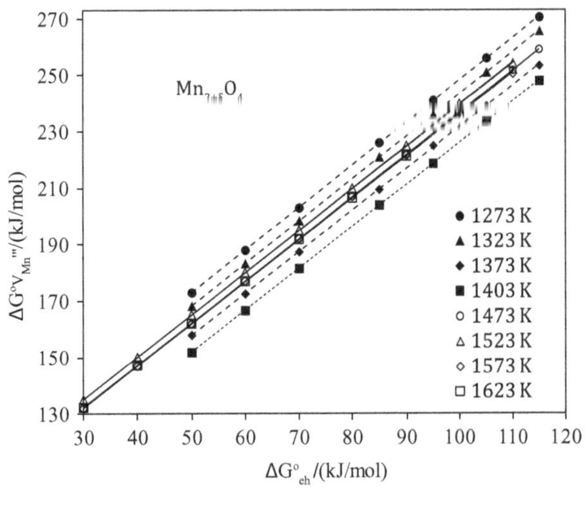

FIGURE 2.1 Dependence of the standard Gibbs energy $\Delta G^o_{V'''_{Fe}}$ of formation of vacancies $\left(V'''_{Fe(o)}\right)$, in $Mn_{3\pm\delta}O_4$ on the assumed value of ΔG^o_{eh} of formation of electronic defects at 1273–1623 K, at which, for the defects with the highest ionisation degree, there is no deterioration of the degree of match between the oxygen pressure dependence of the deviation from the stoichiometry and the experimental values of δ (Keller and Dieckmann 1985b).

Due to that, in the calculations, a constant value of ΔG^{o}_{eh} of formation of electronic defects was assumed, independent of temperature. In the range of existence of α-$Mn_{3\pm\delta}O_4$, a value of ΔG^{o}_{eh} = 85 kJ/mol was assumed, and for β-$Mn_{3\pm\delta}O_4$, ΔG^{o}_{eh} = 60 kJ/mol.

Figure 2.2 present the results of the calculations of the concentrations of point defects (diagrams of defects' concentrations) for α-$Mn_{3\pm\delta}O_4$ at 1273–1403 K (solid lines), obtained when using the results taken from (Keller and Dieckmann 1985b) ((O) points).

As can be seen in Figure 2.2, a very good match was obtained between the dependence of the deviation from the stoichiometry on p_{O_2} and the experimental results δ. In Figure 2.2, also the dependence of δ on p_{O_2} is shown (···), calculated using the values of "equilibrium constants" of reactions of formation of cation vacancies K_V and interstitial ions K_I, taken from Keller and Dieckmann (1985b); see equations of type 1.12 and 1.15.

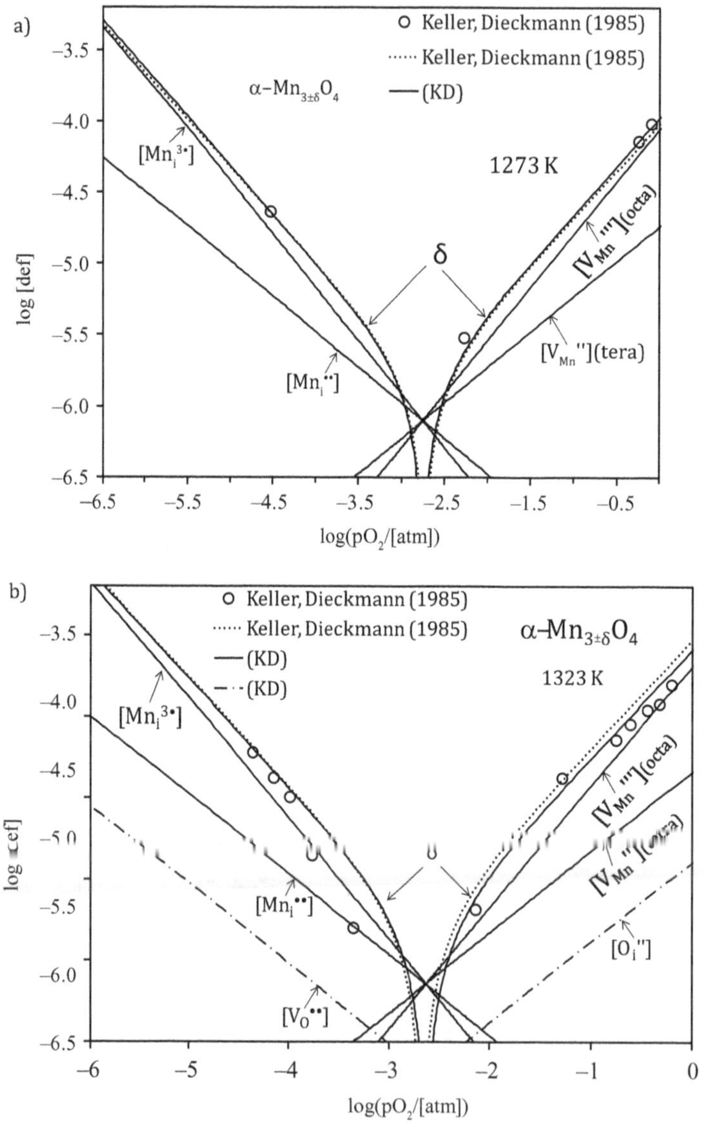

FIGURE 2.2 (a) Diagram of concentrations of point defects for α-Mn3±δ04 at 1273 K (solid lines), obtained using the results of the studies of the deviation from the stoichiometry taken from (Keller and Dieckmann 1985b) ((O) points). The (···) line shows the oxygen pressure dependence of the deviation from the stoichiometry, calculated using the "equilibrium constants" of reactions of formation of cation vacancies K_V and interstitial ions K_I, taken from (Keller and Dieckmann 1985b). (b) at 1323 K (solid lines). Lines (- · · -) represent the maximum concentration of oxygen vacancies and interstitial oxygen ions that do not deteriorate the degree of match between the oxygen pressure dependence of the deviation from the stoichiometry and the experimental results δ. (c) at 1373 K (solid lines). (d) at 1403 K (solid lines). Lines (- ·· -) represent the maximum concentrations of defects with lower ionisation degrees, which do not deteriorate the degree of match between the oxygen pressure dependence of the deviation from the stoichiometry and the experimental results.

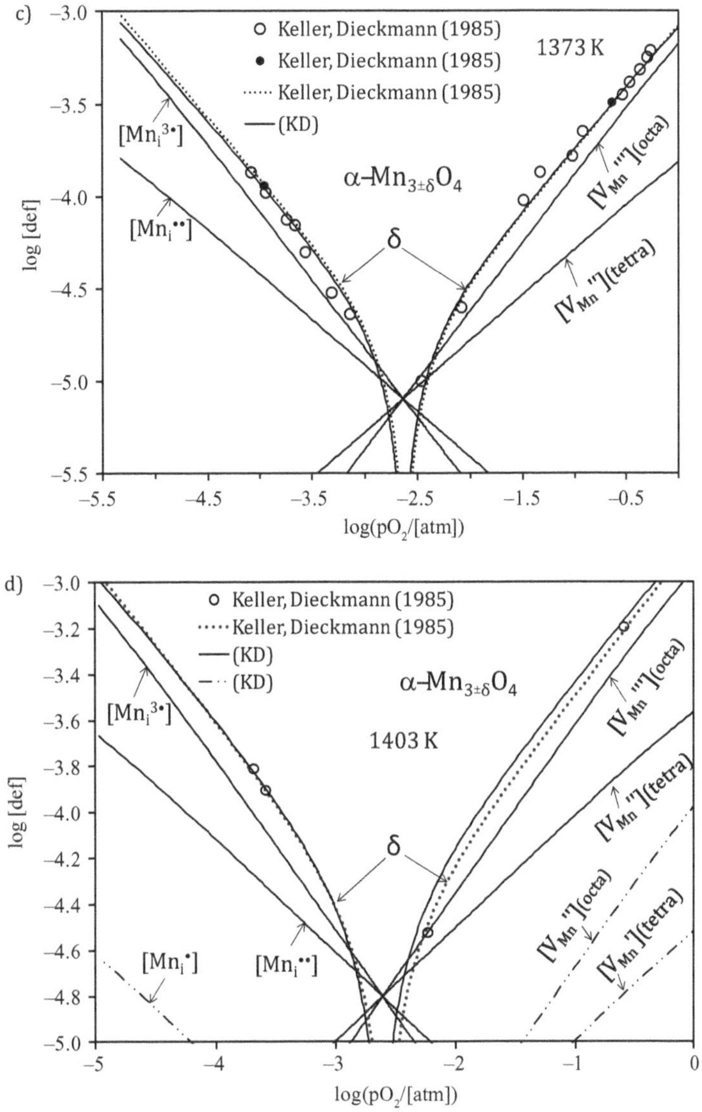

FIGURE 2.2 (Continued)

As can be seen in Figure 2.2, at higher oxygen pressures, manganese vacancies V_{Mn}''' dominate, while at low oxygen pressures, interstitial manganese ions $Mn_i^{3\bullet}$ dominate. In the middle of the range of its existence hausmannite reaches the stoichiometric composition. As can be seen in Figure 2.2, concentration of manganese vacancies $\left[V_{Mn(t)}''\right]$ and interstitial ions $\left[Mn_i^{\bullet\bullet}\right]$ is significantly smaller than that of the dominating defects (V_{Mn}''' and $Mn_i^{3\bullet}$), thus, it can be assumed that this is the maximum concentration of these defects.

Figure 2.2b shows, as an example, the results of calculations of concentrations of defects, considering the maximum concentration of oxygen defects, which does not deteriorate the match between the dependence of δ on p_{O_2} and the experimental values of δ. The calculations were performed in the same way as for magnetite (see Section 1.5). As can be seen in Figure 2.2b, the concentration of oxygen defects can be quite significant, but lower than the concentration of cation defects (actually it can be lower than that). Thus, the studies on the deviation from the stoichiometry do not allow unambiguous determination of the concentration of oxygen defects. The adjusted values of ΔG_i^o of formation of defects are $\Delta G^o_{O_i''} = 123.35$ kJ/mol and $\Delta G^o_{V_O^{\bullet\bullet}} = 314$ kJ/mol.

As it was mentioned, oxygen vacancies "loosen" the spinel structure in the range where interstitial manganese ions are present. However, the presence of a high concentration of manganese vacancies causes that at a higher oxygen pressure, a significant concentration of interstitial oxygen ions can be present. Quite high calculated values concentration of oxygen defects is due also to the fact that they do not affect significantly the deviation from the stoichiometry (an effect of compensation of the changes in the concentrations of defects occurs, see Equation 1.23a).

In turn, Figure 2.2d presents a diagram of the concentrations of defects at 1403 K, assuming the maximum concentration of defects with lower ionisation degrees, which does not deteriorate the match between the dependence of δ on p_{O_2} and the experimental values of δ. The same values of standard Gibbs energies were assumed for the first lower ionisation degree of defects and higher values for the second lower ionisation degree. The calculated concentrations of defects with lower ionisation degrees were obtained with the standard Gibbs energies: $\Delta G^{o}_{V'_{Mn}} = \Delta G^{o}_{V''_{Mn(o)}} = \Delta G^{o}_{Mn_i^{\bullet}} s = -15$ kJ/mol, $\Delta G^{o}_{V^{x}_{Mn}} = \Delta G^{o}_{V'_{Mn(o)}} = \Delta G^{o}_{Mn_i^{x}} = -18$ kJ/mol. As can be seen in Figure 2.2d, the maximum concentration of defects with lower ionisation degree are by over an order of magnitude lower than that of dominating defects.

Figure 2.3 present the diagrams of the concentrations of defects for β-$Mn_{3\pm\delta}O_4$ at 1473–1623 K (solid lines), obtained using the published results (Keller and Dieckmann 1985b) ((O) points).

As can be seen in Figure 2.3, a good match was also obtained between the dependence of the deviation from the stoichiometry and the experimental values of δ. Figure 2.3a shows also the maximum concentrations of defects with lower ionisation degrees, which do not deteriorate the match between the dependence of δ on p_{O_2} and the experimental values of δ. The calculated concentrations of defects with lower ionisation degrees were obtained at the values of the standard Gibbs energies: $\Delta G^{o}_{V'_{Mn(t)}} = \Delta G^{o}_{V''_{Mn(o)}} = \Delta G^{o}_{Mn} = -10$ kJ/mol, $\Delta G^{o}_{V^{x}_{Mn(t)}} = \Delta G^{o}_{V'_{Mn(o)}} = \Delta G^{o}_{Mn_i^{x}} = -14$ kJ/mol. As can be seen, these concentrations are by over an order of magnitude lower than that of dominating defects.

Figure 2.4a compares the dependence of deviation from the stoichiometry on the oxygen pressure for temperatures 1273–1403 K, for α-$Mn_{3\pm\delta}O_4$, and Figure 2.4b – for temperatures 1473–1623 K for α-$Mn_{3\pm\delta}O_4$ (shown in Figures 2.2 and 2.3).

As can be seen in Figure 2.4a and b, the character of the dependence of the deviation from the stoichiometry at given temperatures for α-$Mn_{3\pm\delta}O_4$ and β-$Mn_{3\pm\delta}O_4$ differs quite significantly. These dependences are different in relation to the dependences of δ on p_{O_2} in magnetite, obtained at the temperatures 1373–1673 K (see Figure 1.5). For α-Mn_3O_4, at every temperature, the spinel reaches the stoichiometric composition at practically the same oxygen pressure. When the temperature increases, at a constant oxygen pressure, the deviation from the stoichiometry increases, thus, the concentration of ionic defects increases. As can be seen in Figure 2.4a, oxygen pressures, at which a transition of hausmannite into $Mn_{1-\delta}O$ manganese oxide occurs, increase when the temperature increases, while the limit concentration of ionic defects increases

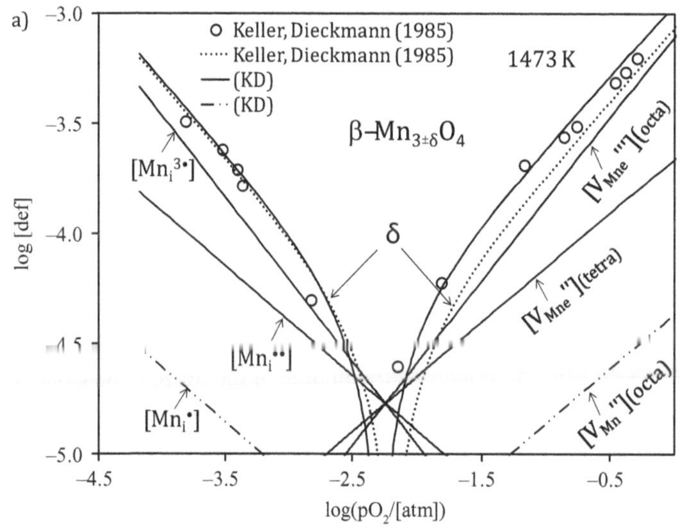

FIGURE 2.3 (a) Diagram of concentrations of point defects for β-$Mn_{3\pm\delta}O_4$ at 1473 K (solid lines), obtained using the results of the studies of the deviation from the stoichiometry (Keller and Dieckmann 1985b) ((O) points). The (···) line shows the dependence of the deviation from the stoichiometry on the oxygen pressure, calculated using the "equilibrium constants" of reactions of formation of cation vacancies K_V and interstitial ions K_I, taken from Keller and Dieckmann (1985b). Lines (-···) represent the maximum concentration of defects with lower ionisation degrees, which does not deteriorate the match between the curve of the deviation from the stoichiometry on the oxygen pressure and the experimental values of δ. (b) at 1523 K (solid lines). (c) at 1573 K (solid lines). (d) at 1623 K. Lines (---) denote the concentrations of electrons and electron holes and line (···) represents the sum of concentrations of electronic defects ($[h^{\bullet}] + b_{e/h}[e']$) at the value $\Delta G^{o}_{eh} s = 108$ kJ/mol, calculated according to Equation 2.4b for the value of $b_{e/h} = 0.01$. Points (□) denote the values of the sum of concentrations of electronic defects ($[h^{\bullet}] + b_{e/h}[e']$) obtained using the values of electrical conductivity taken from (Metselaar et al. 1981) at 1613 K and the fitted value of the mobility of electron holes.

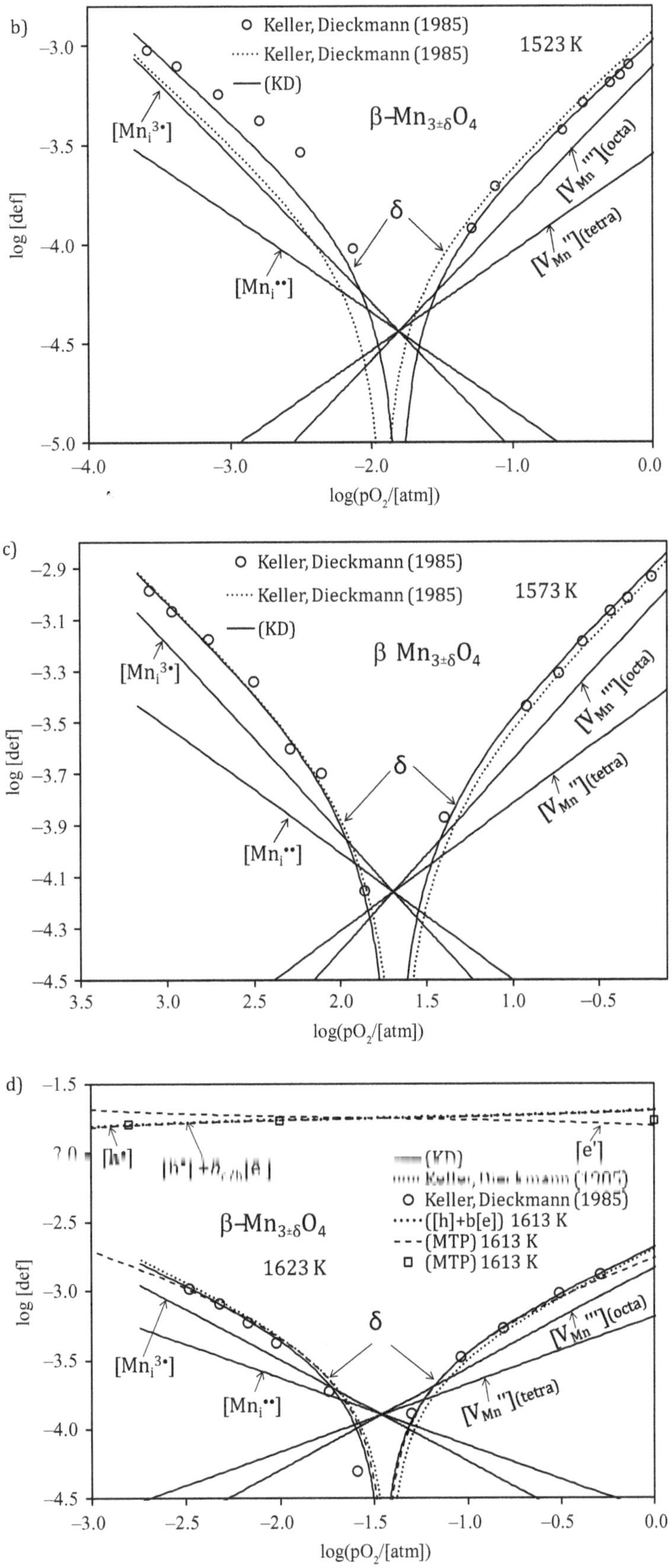

FIGURE 2.3 (Continued)

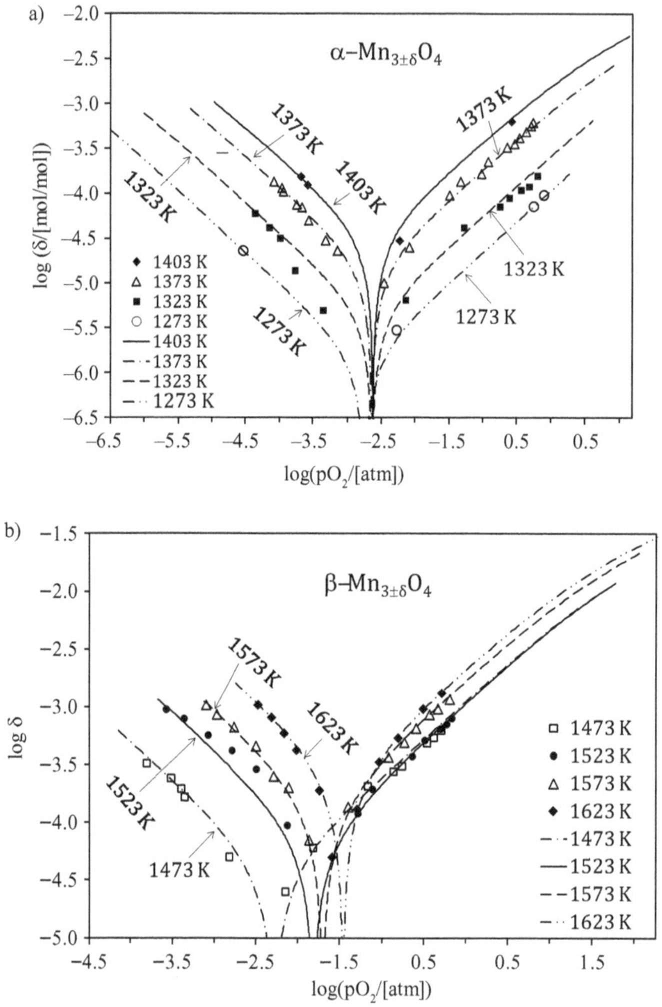

FIGURE 2.4 (a) Dependences of the deviation from the stoichiometry on the oxygen pressure in α-$Mn_{3\pm\delta}O_4$, at temperatures 1273–1403 K (presented in Figure 2.2). (b) Dependences of the deviation from the stoichiometry on the oxygen pressure in β-$Mn_{3\pm\delta}O_4$, at temperatures 1473–1623 K (presented in Figure 2.3).

slightly. Similarly, when the temperature increases, the oxygen pressure at which a transition of α-$Mn_{3\pm\delta}O_4$ into Mn_2O_3 oxide occurs, increases, while these pressures are in the range 2–10 atm. For the transition of α-$Mn_{3-\delta}O_4$ spinel into Mn_2O_3, the limit concentration of manganese vacancies increases significantly when the temperature increases. Figure 2.5 shows pressures over phase boundaries $Mn_{1-\delta}O/Mn_{3+\delta}O_4$ and $Mn_{3-\delta}O_4/Mn_2O_3$, determined in Keller and Dieckmann (1985a, 1985b), and the values of the oxygen pressure at which hausmannite reaches the stoichiometric composition at given temperatures.

In turn, as can be seen in Figure 2.4b, in the case of β-$Mn_{3\pm\delta}O_4$ variation, when the temperature increases, the stoichiometric composition is reached at increasingly higher oxygen pressures. Simultaneously, in the temperature range of 1473–1623 K, the extent of this shift in the pressure is within the limit of one order of magnitude.

Accurate values of the oxygen pressure at which β-$Mn_{3\pm\delta}O_4$ reaches the stoichiometric composition at given temperatures are shown in Figure 2.5.

In the range of low oxygen pressures, thus, in the range where interstitial manganese ions dominate, when the temperature increases, at a constant oxygen pressures, the deviation from the stoichiometry increases significantly. The concentration of manganese vacancies at the same oxygen pressure increases to a much smaller degree when the temperature increases. As can be seen in Figure 2.4b, when the temperature increases, the oxygen pressure at which the transformation of β-$Mn_{3\pm\delta}O_4$ into $Mn_{1-\delta}O$ occurs, increases by about one and a half order of magnitude and the limit concentration of defects increases slightly. In turn, the oxygen pressure at which the transformation of β-$Mn_{3\pm\delta}O_4$ into Mn_2O_3 oxide occurs, increases slightly when the temperature increases, while, as it results from the studies on the range of existence of hausmannite, these pressures are high, from 30 to 180 atm. Limit concentrations of manganese vacancies in β-$Mn_{3\pm\delta}O_4$, are also high, above 0.01 mol/mol; at these concentrations, spinel transformation occurs.

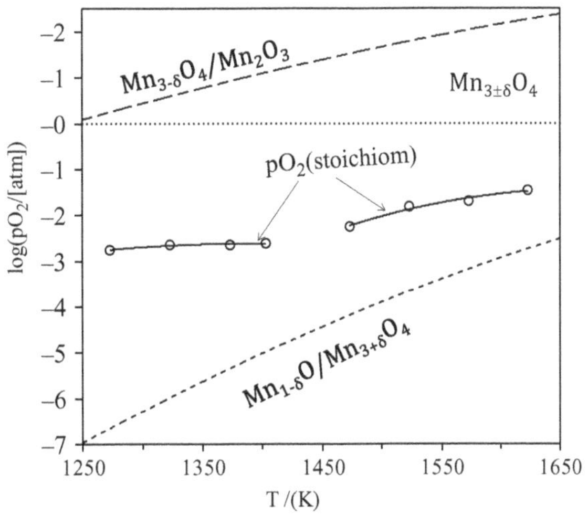

FIGURE 2.5 The temperature dependence of the oxygen pressure at which $Mn_{3\pm\delta}O_4$ reaches the stoichiometric composition ((O) points), obtained using the published results of the studies on the deviation from the stoichiometry (Keller and Dieckmann 1985b). Dashed lines denote the equilibrium oxygen pressures at the phase boundary $Mn_{1-\delta}O/Mn_{3+\delta}O_4$ (---) and $Mn_{3-\delta}O_4/Mn_2O_3$ (——) taken from Keller and Dieckmann (1985a, 1985b).

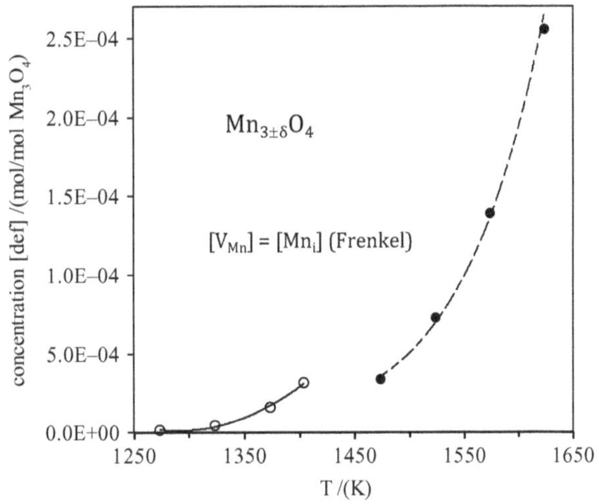

FIGURE 2.6 Temperature dependence of the concentration of Frenkel defects in hausmannite, resulting from concentration of ionic defects presented in Figures 2.2 and 2.3.

In turn, Figure 2.6 shows the dependence of the concentration of Frenkel defects in hausmannite on the temperature (at the stoichiometric composition). When comparing the concentration of defects with that present in MO-type oxides, the values should be divided by four.

As was shown in Section 1.4.6, the dependence of the concentration of defects on the oxygen pressure is expressed by a quite complex equation (Equation 1.33b. The nature of this function is determined by the derivative of the deviation δ with respect to the oxygen pressure $(d\log\delta/d\log p_{O_2}) = 1/n$. In the case, where in a certain range one type of defects dominates, the dependence of the concentration of defects of a given type is an exponential function of oxygen pressure, with an exponent of $1/n$ (see Equation 1.10a, 1.10b, 1.10c, 1.14a, 1.14b, 1.14c). The exponent, or the derivative $(d\log\delta/d\log p_{O_2}) = 1/n$, in such a case indicates the type of dominating defects. Figure 2.7 shows the oxygen pressure dependence of the derivative $(d\log p_{O_2}/\log\delta) = n$, for the dependences of the deviation δ on p_{O_2} shown in Figures 2.2 and 2.3.

As can be seen in Figure 2.7, at the lowest oxygen pressures the parameter n reaches the value of −1.5 ($1/n = -2/3$), while at the highest pressures, in a certain narrow range it reaches the value of $n = 1.5$. Therefore, the exponent $1/n = 2/3$ and −2/3, assumed in Keller and Dieckmann (1985b), is a good approximation.

Figures 2.8–2.10 present the temperature dependences of adjusted values of standard Gibbs energies of formation of defects ΔG^{o}_{F1}, ΔG^{o}_{F2}, $\Delta G^{o}_{V''_{Mn}}$, ΔG^{o}_{I}.

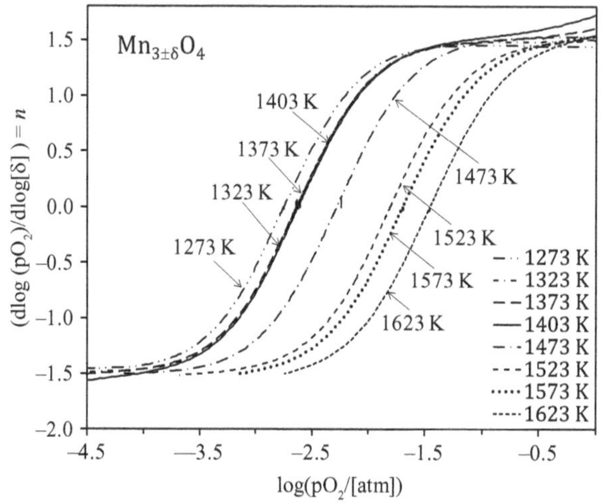

FIGURE 2.7 The oxygen pressure dependence of the derivative $n_\delta = d\log(p_{O_2})/d\log\delta$ at 1273–1623 K, obtained for dependences of the deviation from the stoichiometry on the oxygen pressure in hausmannite, presented in Figures 2.2 and 2.3.

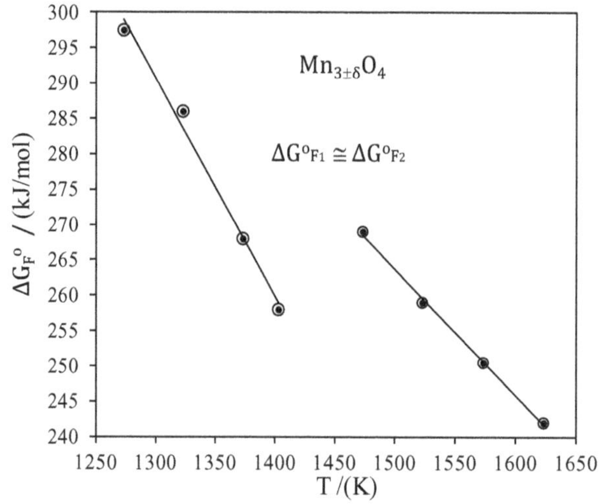

FIGURE 2.8 Temperature dependence of standard Gibbs energy of $\Delta G^o_{F_1}$ ((o) points) and $\Delta G^o_{F_2}$ ((●) points) of formation of Frenkel defects with charge two and three in hausmannite, obtained using the published results of the studies on the deviation from the stoichiometry (Keller and Dieckmann 1985b).

Figure 2.9 and 2.11 show the temperature dependence of standard Gibbs energy $\Delta G^o_{V''_{Mn}}$ of formation of vacancies V''_{Mn} (calculated according to Equation 1.35d) and $\Delta G^o_{Mn_i^{3\bullet}}$ of the formation of interstitial manganese ions $Mn_i^{3\bullet}$ (calculated according to Equation 1.34c) and $\Delta G^o_{Mn_i^{\bullet\bullet}}$ of formation of ions $Mn_i^{\bullet\bullet}$ (calculated according to Equation 1.36c; see Table 2.1).

As can be seen in the above figures, the obtained dependences for α- and β-$Mn_{3\pm\delta}O_4$ are linear functions with different slopes. Table 2.2 compares the values of ΔH_i^o and ΔS_i^o of formation of defects, resulting from the relations presented in Figures 2.8–2.11. The values of ΔG_i^o of formation of defects were determined for a constant value of the standard Gibbs energy of formation of electronic defects, equal, for α-$Mn_{3\pm\delta}O_4$, $\Delta G^o_{eh} = 85$ kJ/mol and, for α-$Mn_{3\pm\delta}O_4$, $\Delta G^o_{eh} = 60$ kJ/mol.

As it was mentioned, the adjusted standard Gibbs energies of formation of manganese vacancies and interstitial manganese ions depend on the assumed value of ΔG^o_{eh} (see Figure 2.1). Figure 2.12 shows, as an example, the dependence of $\Delta G^o_{V''_{Mn}}$ on the temperature for several assumed values of ΔG^o_{eh} for α-$Mn_{3\pm\delta}O_4$ in the range of values 50–115 kJ/mol and for β-$Mn_{3\pm\delta}O_4$ in the range of values 30–100 kJ/mol.

As can be seen, practically the same slope of the above dependences was obtained. Table 2.3 presents the parameters of the obtained linear equations ($\Delta G_i^o = \Delta H_i^o - T\Delta S_i^o$). As can be seen in Table 2.3, the differences between the values of ΔH_i^o in dependence on the value of ΔG^o_{eh} are significant. The real enthalpy and entropy of the reaction of formation of the individual defects should be within the range of the values obtained for the individual values of ΔG^o_{eh}. The fact of the dependence of ΔG^o_{eh} on temperature, not included in the calculations, should be also considered.

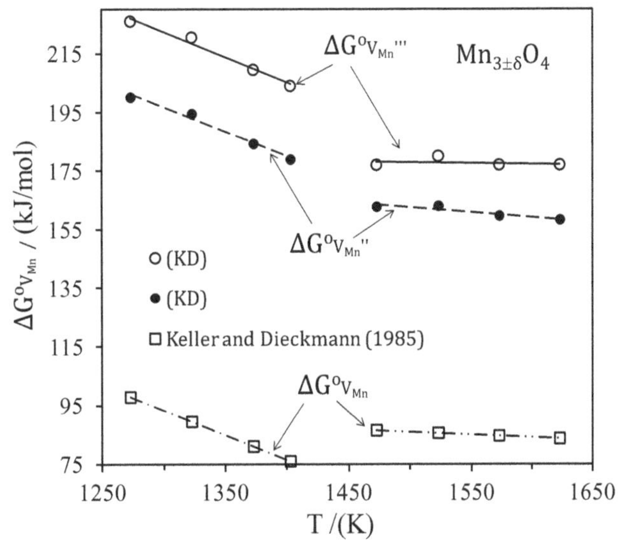

FIGURE 2.9 Temperature dependence of standard Gibbs energy $\Delta G^o_{V'''_{Mn}}$ and $\Delta G^o_{V''_{Mn}}$ of formation of vacancies $V'''_{Mn(o)}$ and V''_{Mn} (calculated according to Equation 1.35c) in hausmannite. Points (□) denote the values of standard Gibbs energy of formation of manganese vacancies calculated using the "equilibrium constants" of reactions of their formation K_V taken from (Keller and Dieckmann 1985b).

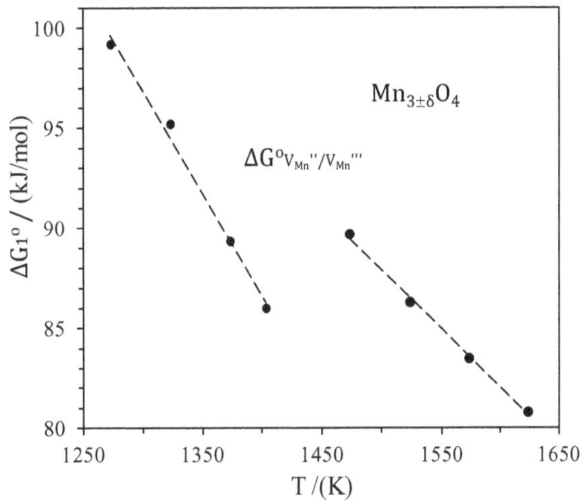

FIGURE 2.10 Temperature dependence of ΔG_I^o in hausmannite, permitting the determination of the ratio of concentrations of [illegible] vacancies $\left[V'''_{Mn(o)}\right]/\left[V''_{Mn(o)}\right]$ according to Equation 2.3.

Figures 2.9 and 2.11 show also the values of $\Delta G^o_{V_{Mn}}$ and $\Delta G^o_{Mn_i}$ calculated using the values of "equilibrium constants" of reaction of formation of manganese vacancies K_V and interstitial manganese ions K_I (Keller and Dieckmann 1985b). As can be seen in Figures 2.9 and 2.11, there are large differences between the values of $\Delta G^o_{V_{Mn}}$ and $\Delta G^o_{Mn_i}$ obtained in the present work with the use of other equations of formation of defects. The dependences have a similar slope to the values of ΔG_i^o obtained for a constant value of ΔG^o_{eh}, independent of temperature.

A parameter characterising the concentration of defect is the resultant standard Gibbs energy of formation of defects, which can be calculated according to Equation 1.33b. Figure 2.13 shows the dependence of $\Delta G^{o(\delta)}_{[def]}$ of formation of defects on the oxygen pressure.

As can be seen in Figure 2.13, for the given temperatures, the values tend to approach a constant value which determines the standard Gibbs energy of formation of manganese vacancies and interstitial ions, when they dominate. They are, for manganese vacancies, from $\Delta G^o_{V_{Mn}}$ = 150–170 kJ/mol oxygen atoms and $\Delta G^o_{Mn_i}$ = −70 to −50 kJ/mol O, and they are close to the values of $\Delta G^o_{V'''_{Mn(o)}}$ and $\Delta G^o_{Mn_i^{3\bullet}}$ per mole of incorporated oxygen atoms.

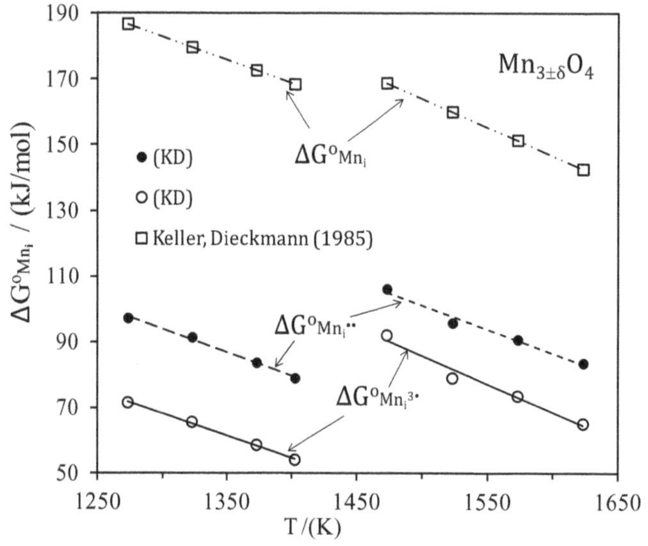

FIGURE 2.11 Temperature dependence of $\Delta G^o_{Mn_i^{3\bullet}}$ and $\Delta G^o_{Mn_i^{\bullet\bullet}}$ of formation of interstitial ions $Mn_i^{3\bullet}$ and $Mn_i^{\bullet\bullet}$ in hausmannite (calculated according to Equation 1.34c and 1.36c). Points (□) denote the values of standard Gibbs energy of formation of interstitial cations calculated using the "equilibrium constants" of reactions of their formation K_I taken from Keller and Dieckmann (1985b).

TABLE 2.1
Parameters of the dependence of the standard Gibbs energy $\Delta G^o_{V'''_{Mn}}$ of formation of manganese vacancies V'''_{Mn} on the value, assumed in the calculations, of the standard Gibbs energy of formation of electronic defects ΔG^o_{eh} ($\Delta G^o_{V'''_{Mn}} = a\Delta G^o_{eh} + b$) for the α-$Mn_{3\pm\delta}O_4$ variant in the range of ΔG^o_{eh} = 50–115 kJ/mol at 1273–1403 K and for β-$Mn_{3\pm\delta}O_4$ in the range of ΔG^o_{eh} = 30–100 kJ/mol at 1473–1623 K, determined based on the dependences shown in Figure 2.1.

	$\Delta G^o_{V'''_{Mn}} = a\Delta G^o_{eh} + b$	
T/(K)	b	a
α-$Mn_{3\pm\delta}O_4$		
1273	97.8 ± 0.4	1.505 ± 0.004
1323	93.4 ± 0.3	1.500 ± 0.003
1373	84.3 ± 0.5	1.473 ± 0.006
1403	77.9 ± 0.2	1.481 ± 0.003
β-$Mn_{3\pm\delta}O_4$		
1473	87.2 ± 0.1	1.498 ± 0.002
1523	90.5 ± 0.2	1.491 ± 0.003
1573	87.7 ± 0.3	1.486 ± 0.005
1623	88.0 ± 0.1	1.481 ± 0.002

A value characteristic of the process of defects' formation is the value of $\Delta G^{o(\delta=0)}_{[def]}$ at the oxygen pressure at which hausmannite reaches the stoichiometric composition. This value is associated to the ΔG^o_O of incorporation of atoms and to the difference of $\left(\Delta G^o_{V_{Mn}} - \Delta G^o_{Mn_i}\right)$ of formation of manganese vacancies and interstitial manganese ions (see discussion and Equation 15.1 in the work Keller and Dieckmann (1985b)). Figure 2.14 presents the dependence of $\Delta G^{o(\delta=0)}_{[def]}$ on temperature, and Table 2.2 shows the values of enthalpy and entropy of the process associated mainly to the incorporation of oxygen. Therefore, it will be the heat effect related to the process of adsorption of oxygen, its dissociation and electron affinity of adsorbed oxygen atoms.

TABLE 2.2
Values of enthalpies ΔH_i^o and entropies ΔS_i^o of formation of Frenkel defects, manganese vacancies V''_{Mn} and V'''_{Mn}, interstitial manganese ions $Mn_i^{\bullet\bullet}$ and $Mn_i^{3\bullet}$, in α-$Mn_{3\pm\delta}O_4$ in the temperature range of 1273–1403 K (at ΔG^o_{eh} = 85 kJ/mol) and in β-$Mn_{3\pm\delta}O_4$ in the temperature range of 1473–1623 K (at ΔG^o_{eh} = 60 kJ/mol), determined based on the dependences shown in Figures 2.8–2.14.

Equation	ΔH_i^o (kJ/mol)	ΔS_i^o (J/(mol·K))
α-$Mn_{3\pm\delta}O_4$		
$\left[V'''_{Mn(o)}\right]\left[Mn_i^{3\bullet}\right] = K_{F_1}$ (Equation 2.4a)	692 ± 29	309 ± 21
$V'''_{Mn(o)}$ (α) (Equation 2.1b)	449 ± 23	174 ± 17
$K_{V'_{Mn}/V''_{Mn}} = K_1(\alpha)$ (Equation 2.3)	231 ± 8	103 ± 6
$V''_{Mn(o)}$ (α) (Equation 2.1a)	414 ± 20	167 ± 15
$Mn_i^{3\bullet}$ (α) (Equation 2.2b))	243 ± 6	135 ± 4
$Mn_i^{\bullet\bullet}$ (α) (Equation 2.2a)	278 ± 9	141 ± 7
$\Delta G^o_{[def]}(\delta \cong 0)$ (α) (Equation 1.33b)	22 ± 5	−7 ± 4
V_{Mn} (α) (Keller and Dieckmann 1985b)	313 ± 1	169 ± 1
Mn_i (α) (Keller and Dieckmann 1985b)	365 ± 1	140 ± 1
β-$Mn_{3\pm\delta}O_4$		
$\left[V'''_{Mn(o)}\right]\left[Mn_i^{3\bullet}\right] = K_{F_1}$ (Equation 2.4a)	532 ± 8	179 ± 5
$V'''_{Mn(o)}$ (β (Equation 2.1b))	187 ± 25	6
$K_{V'_{Mn(o)}/V''_{Mn(o)}} = K_1(\beta)$ (Equation 2.3)	176 ± 4	59 ± 2
$V''_{Mn(o)}$ (β) (Equation 2.1a)	213 ± 15	33 ± 10
$Mn_i^{3\bullet}$ (β) (Equation 2.2b)	345 ± 32	173 ± 21
$Mn_i^{\bullet\bullet}$ (β) (Equation 2.2a)	319 ± 22	146 ± 14
$\Delta G^o_{[def]}(\delta \cong 0)$ (β) (Equation 1.32b)	−89 ± 18	−68 ± 12
V_{Mn} (β) (Keller and Dieckmann 1985b)	114 ± 1	19 ± 1
Mn_i (β) (Keller and Dieckmann 1985b)	427 ± 1	175 ± 1

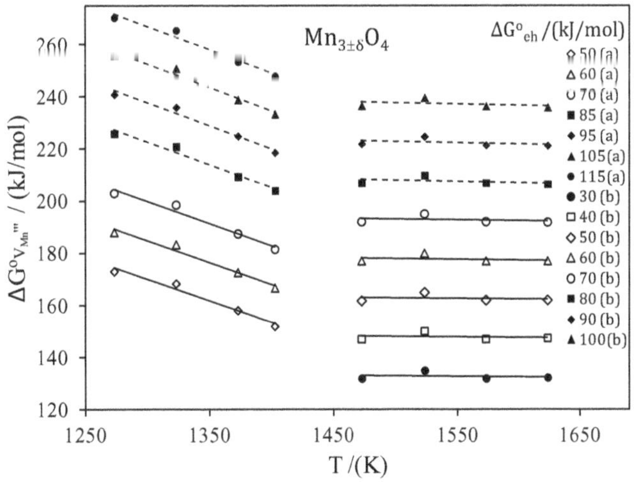

FIGURE 2.12 Temperature dependence of $\Delta G^o_{V'''_{Mn}}$ of formation of vacancies $\left(V'''_{Mn}\right)$ depending on the assumed values of ΔG^o_{eh} of formation of electronic defects in hausmannite.

TABLE 2.3
Values of enthalpies (ΔH_i^o) and entropies $\left(\Delta S_i^o\right)$ of formation of manganese vacancies V_{Mn}''' (see Figure 2.1) and V_{Mn}'' and interstitial manganese ions $Mn_i^{3\bullet}$. $Mn_i^{\bullet\bullet}$ on the value, assumed in calculations, of the standard Gibbs energy ΔG_{eh}^o of formation of electronic defects ($\Delta G_i^o = \Delta H_i^o - T\Delta S_i^o$) in α-$Mn_{3\pm\delta}O_4$ in the temperature range of 1273–1403 K and β-$Mn_{3\pm\delta}O_4$ in the temperature range of 1473–1623 K.

ΔG_{eh}^o	$\Delta H_{V_{Mn}'''}^o$ (kJ/mol)	$\Delta S_{V_{Mn}'''}^o$ (J/(mol·K))
α-$Mn_{3\pm\delta}O_4$		
50	385 ± 27	166 ± 20
60	403 ± 27	168 ± 20
70	421 ± 29	170 ± 22
85	450 ± 27	175 ± 20
95	466 ± 28	175 ± 21
105	486 ± 29	180 ± 22
115	501 ± 30	180 ± 22
β-$Mn_{3\pm\delta}O_4$		
30	139 ± 24	4.2
40	154 ± 24	4.2
50	169 ± 26	4.2
60	187 ± 25	6.0
70	204 ± 25	7.1
80	221 ± 24	8.8
90	239 ± 25	10.8
100	254 ± 26	10.6
α-$Mn_{3\pm\delta}O_4$	$\Delta H_{V_{Mn}''}^o$	$\Delta S_{V_{Mn}''}^o$
50	372 ± 22	162 ± 6
60	384 ± 22	163 ± 17
70	396 ± 24	165 ± 18
85	415 ± 22	168 ± 17
95	426 ± 23	168 ± 17
105	439 ± 23	171 ± 17
115	449 ± 24	171 ± 18
β-$Mn_{3\pm\delta}O_4$		
30	183 ± 14	33 ± 9
40	193 ± 14	33 ± 9
50	202 ± 15	33 ± 10
60	214 ± 14	34 ± 9
70	225 ± 14	35 ± 11
80	237 ± 14	36 ± 9
90	249 ± 15	38 ± 9
100	259 ± 15	37 ± 10
α-$Mn_{3\pm\delta}O_4$	$\Delta H_{Fe_i^{3\bullet}}^o$	$\Delta S_{Fe_i^{3\bullet}}^o$
50	307 ± 4	143 ± 3
60	289 ± 2	141 ± 1
70	271 ± 1	138 ± 1
85	242 ± 4	134 ± 3
95	226 ± 3	133 ± 2
105	206 ± 4	129 ± 3
115	191 ± 5	129 ± 4
β-$Mn_{3\pm\delta}O_4$		
30	393 ± 31	175 ± 20
40	378 ± 31	175 ± 20
50	363 ± 33	175 ± 22
60	345 ± 32	173 ± 21

(Continued)

TABLE 2.3
(Continued)

ΔG^o_{eh}	$\Delta H^o_{V_{Mn}'''}$ (kJ/mol)	$\Delta S^o_{V_{Mn}'''}$ (J/(mol·K))
70	329 ± 32	172 ± 21
80	311 ± 32	170 ± 21
90	293 ± 32	168 ± 21
100	279 ± 32	168 ± 22
α-$Mn_{3\pm\delta}O_4$	$\Delta H^o_{Fe_i^{\bullet\bullet}}$	$\Delta S^o_{Fe_i^{\bullet\bullet}}$
50	320 ± 7	147 ± 5
60	308 ± 7	145 ± 5
70	296 ± 5	144 ± 4
85	277 ± 7	141 ± 5
95	266 ± 6	140 ± 5
105	253 ± 6	138 ± 5
115	242 ± 6	137 ± 4
β-$Mn_{3\pm\delta}O_4$		
30	350 ± 21	146 ± 14
40	340 ± 21	146 ± 14
50	330 ± 22	146 ± 14
60	318 ± 21	145 ± 14
70	307 ± 22	144 ± 14
80	295 ± 22	143 ± 14
90	283 ± 22	141 ± 14
100	273 ± 22	142 ± 14

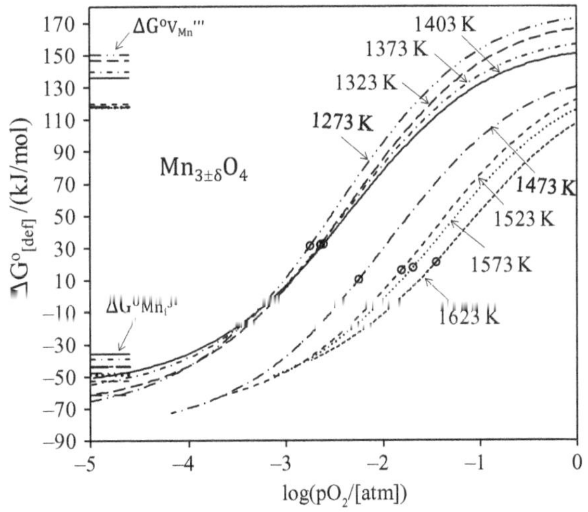

FIGURE 2.13 Oxygen pressure dependence of the resultant standard Gibbs energy of formation of defects $\Delta G^{o(\delta)}_{[def]}$ in hausmannite, calculated according to Equation 1.33b (per mole of oxygen atoms). Points (o) denote values of oxygen pressure at which Mn_3O_4 reaches the stoichiometric composition. Short lines denote the values of $\Delta G^o_{V_{Mn}'''}$ and $\left(-\Delta G^o_{Mn_i^{3\bullet}}\right)$ presented in Figures 2.9 and 2.11 (per mole of oxygen atoms).

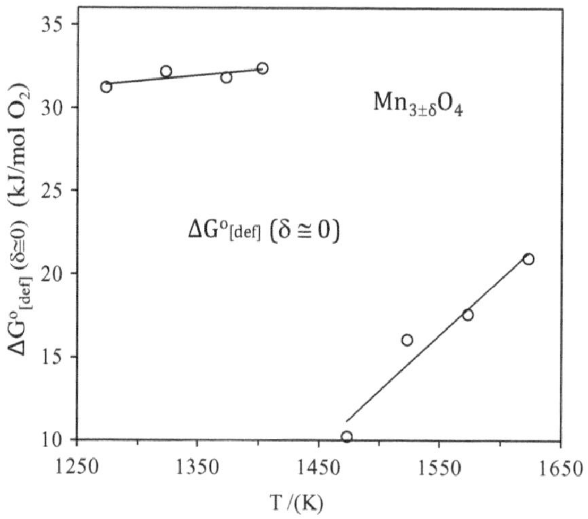

FIGURE 2.14 Temperature dependence of the resultant standard Gibbs energy of formation of defects $\Delta G^{o(\delta=0)}_{[def]}$ at the stoichiometric composition of Mn_3O_4 (according to Equation 1.33b).

2.4 CONCENTRATION OF ELECTRONIC DEFECTS AND ELECTRICAL CONDUCTIVITY

As mentioned in Introduction, the values of the electrical conductivity in hausmannite, are quite high (Dorris and Mason 1988, Metselaar et al. 1981). In the temperature range of 1273–1400 K (α-$Mn_{3\pm\delta}O_4$), electrical conductivity is 2–7 Ω^{-1} cm^{-1}, while in the range of 1473–1650 K (β-$Mn_{3\pm\delta}O_4$) it is 20–60 Ω^{-1} cm^{-1}. As hausmannite reaches the stoichiometric composition in the range of its existence, it should also have a high concentration of intrinsic electronic defects, however, lower than that in magnetite. The oxygen pressure dependence of electrical conductivity (Metselaar et al. 1981) indicates that it is in a wide range independent of the oxygen pressure. Only at oxygen pressures near the boundary with $Mn_{1-\delta}O$ it slightly decreases (inversely as in magnetite), which indicates that the concentration of electronic defects depends on the concentration of interstitial manganese ions. A decrease in electrical conductivity in the range where electrons should dominate indicates that the mobility of electron holes is significantly higher than that of electrons, which is consistent with positive values of thermoelectric power (Dorris and Mason 1988, Logothetis and Park 1975, Metselaar et al. 1981, Rosenberg et al. 1966). The extent of the difference in concentrations of electronic defects at limit concentrations of ionic defects depends on the value of ΔG^{o}_{eh} assumed in the calculations. At the values of ΔG^{o}_{eh} = 60 and 85 kJ/mol, which were assumed in the calculations practically in the whole range of existence of hausmannite, the concentration of electronic defects practically does not depend on the oxygen pressure (on the concentration of ionic defects), which is inconsistent with the oxygen pressure dependence of the electrical conductivity. The performed calculations showed that larger differences in the concentrations of electron holes and electrons at limit oxygen pressures occur at values of ΔG^{o}_{eh} significantly higher than 100 kJ/mol. Thus, it was decided to adjust such values of ΔG^{o}_{eh}, at which the dependence of the concentration of electronic defects on the oxygen pressure has the same character as the dependence of electrical conductivity given in Metselaar et al. (1981).

Let us assume, therefore, that in hausmannite, in the temperature range of 1273–1623 K, there occurs mixed electrical conductivity and it is a sum of the hole conductivity and electron conductivities, according to Equation 1.48a. Factoring out the mobility of electron holes we get:

$$\sigma = \frac{F}{V_{Mn_3O_4}}\mu_h\left(\left[h^{\bullet}\right]+\frac{\mu_e}{\mu_h}\left[e'\right]\right)=\frac{F}{V_{Mn_3O_4}}\mu_h\left(\left[h^{\bullet}\right]+b_{e/h}\left[e'\right]\right) \tag{2.8a}$$

where $b_{e/h}$ denotes the ratio of the mobility of electrons to the mobility of electron holes ($b_{e/h} = \mu_e/\mu_h$), F – Faraday constant, $V_{Mn_3O_4}$ – molar volume of hausmannite, which equals 47.276 cm^3. Transforming Equation 2.8a, we obtain:

$$\left[h^{\bullet}\right]+b_{e/h}\left[e'\right]=\frac{\sigma V_{Mn_3O_4}}{F\mu_h} \tag{2.8b}$$

From Equation 2.8b it results that if we assume that in the range where σ is independent of the oxygen pressure, electrical conductivity depends only on the concentration of electron holes, then we can adjust such a value of the parameter $b_{e/h}$, that the dependence of the sum of concentrations of electronic defects ($[h^\bullet] + b_{e/h}[e']$) on p_{O_2} resulting from the diagram of defects is consistent with the values of this sum ($[h^\bullet] + b_{e/h}[e']$) calculated according to Equation 2.8b, using the values of the electrical conductivity and the adjusted value of the mobility of electron holes μ_h.

The performed calculations showed that consistency of the dependence of ($[h^\bullet] + b_{e/h}[e']$) with the values of this sum based on the electrical conductivity is obtained only at the temperature of 1613 K for the value of ΔG^o_{eh} = 108 kJ/mol (see Figure 2.3d). At 1513 K, a match was obtained for the value of ΔG^o_{eh} = 143 kJ/mol, but for the dependence of the deviation from the stoichiometry on p_{O_2} to be consistent with the experimental values δ, it was necessary to consider defects with lower ionisation degrees. A full diagram of the concentrations of defects for the temperature of 1513 K is shown in Figure 2.15. At temperatures of 1288 K and 1388 K, when considering the concentration of defects with lower ionisation degrees, no satisfactory match was obtained between the dependence of δ on p_{O_2} and the experimental values δ (for similar temperatures, 1273 and 1373 K).

Figure 2.16 shows the dependence of the concentration of electron holes and electrons and the sums ($[h^\bullet] + b_{e/h}[e']$) on p_{O_2}, calculated at 1283 K, 1388 K, 1513 K and 1613 K for the values ΔG^o_{eh}, at which a match was obtained between the dependence of the sum ($[h^\bullet] + b_{e/h}[e']$) on p_{O_2} (dotted line) and the values of this sum ((♦,o,Δ,□) points) calculated using the electrical conductivity (Metselaar et al. 1981) and the adjusted value of the mobility of electron holes.

As can be seen in Figure 2.16, a good match of the dependence of the sum ($[h^\bullet] + b_{e/h}[e']$) and the values of this sum obtained using the electrical conductivity was obtained for the values of ΔG^o_{eh} equal, respectively: ΔG^o_{eh} = 199, 179, 143 and 108 kJ/mol and for the parameter $b_{e/h}$ = 0.01. Thus, at the mobility of electron holes practically one hundred times higher than the mobility of electrons. Figure 2.17 shows the temperature dependence of the determined mobility of electron holes; when the temperature increases, it decreases from 14 cm^2/Vs to a value about 2 cm^2/Vs. The value of the mobility of electron holes at 1613 K, due to a small value of the electrical conductivity, can be burdened with an error, because points corresponding to the lowest pressure can be outside the range of existence of hausmannite.

As can be seen in Figure 2.16, the calculated values of the sum of concentrations of electronic defects ($[h^\bullet] + b_{e/h}[e']$) at the oxygen pressure of 1 atm (points) are lower than the concentration of electron holes. If the values of the conductivity are determined correctly, it would mean that in the range of higher oxygen pressures, where manganese vacancies dominate, the mobility of electron holes decreases, while the mobility of electrons increases significantly and becomes higher than that of electron holes, which causes a decrease in electrical conductivity.

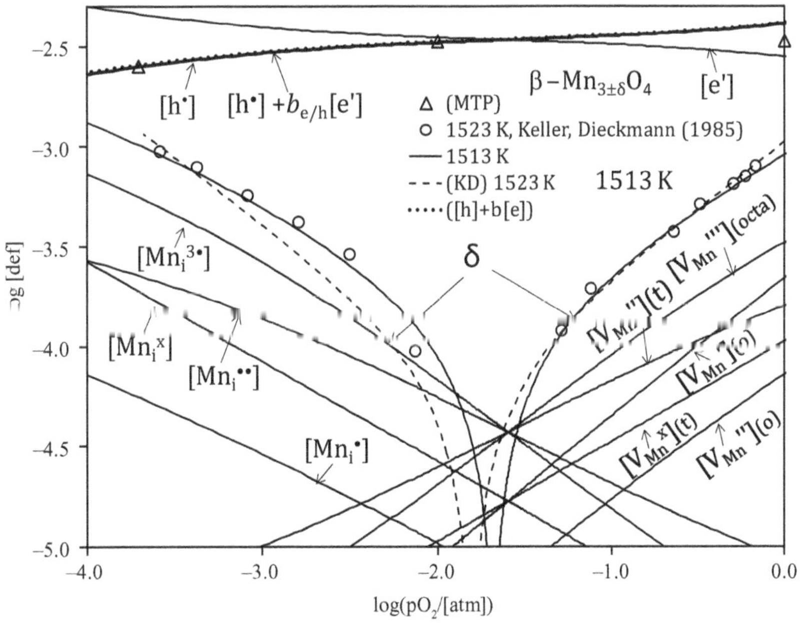

FIGURE 2.15 Diagram of concentrations of point defects for $Mn_{3\pm\delta}O_4$, at 1513 K, taking into account the concentration of defects with lower ionisation degree (solid lines). Points (O) and line (---) represent the oxygen pressure dependence of the deviation from the stoichiometry for the temperature of 1523 K, obtained using the published results of the studies on the deviation δ (Keller and Dieckmann 1985b). The points (△) (MTP) represent the values of the sum of concentrations of electronic defects ($[h^\bullet] + b_{e/h}[e']$), calculated according to Equation 2.8b, obtained using the results of the studies on the electrical conductivity given in Metselaar et al. (1981) and the fitted value of the mobility of electron holes. Line (···) represents the oxygen pressure dependence of the sum of the concentrations of electronic defects ($[h^\bullet] + b_{e/h}[e']$) at the value of ΔG^o_{eh} = 143 kJ/mol for the value $b_{e/h}$ = 0.01.

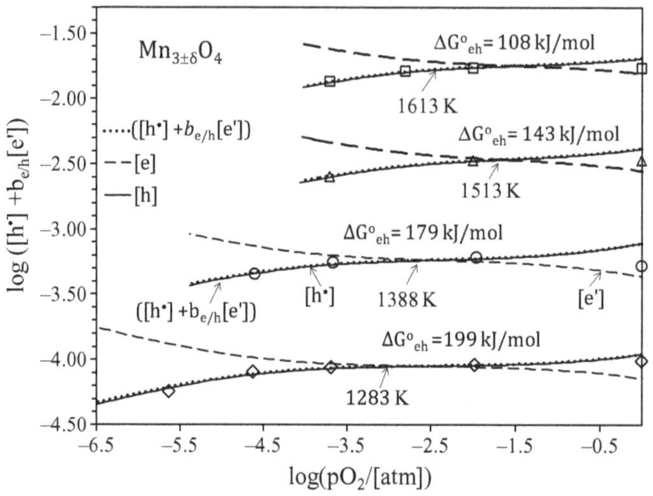

FIGURE 2.16 Dependence of concentrations of electrons and electron holes and the sum of concentrations of electronic defects ([h•] + $b_{e/h}$[e′]) on the oxygen pressure in hausmannite at 1283–1613 K, for the values of ΔG^o_{eh}, respectively ΔG^o_{eh} = 199, 179, 143 and 108 kJ/mol and for the value of the parameter $b_{e/h}$ = 0.01. Points denote the values of the sum ([h•] + $b_{e/h}$[e′]) calculated using the values of the electrical conductivity taken from Metselaar et al. (1981) and the fitted mobility of electron holes.

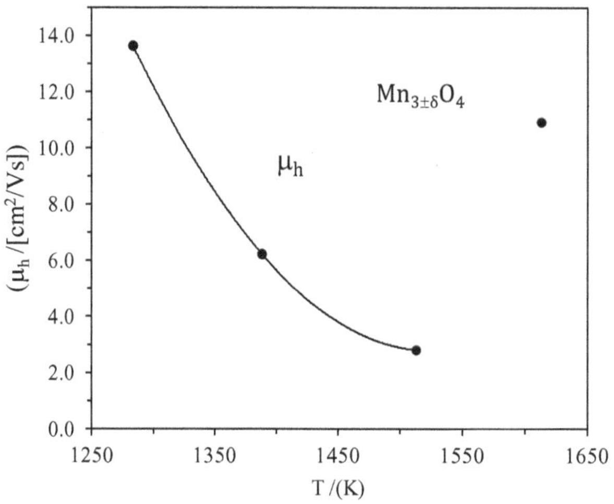

FIGURE 2.17 Temperature dependence of the mobility of electron holes μ_h in hausmannite, obtained using the values of electrical conductivity extracted from Metselaar et al. (1981) and the concentration of electronic defects resulting from the assumed values of ΔG^o_{eh} = 199, 179, 143 and 108 kJ/mol.

From the above calculations it results that the concentration of electronic defects assumed for the determination of the diagram of the concentrations of point defects is significantly higher than that resulting from the electrical conductivity. In turn, the obtained values of the mobility of electron holes are rather too high for an oxide material. Thus, in order to remove the existing discrepancies, wider studies, at high temperatures, of oxygen pressure dependence of electrical conductivity and thermoelectric power are necessary.

2.5 CONCLUSIONS

Using the results of deviation from the stoichiometry studies in the temperature range from 1273 K to 1623 K (Keller and Dieckmann 1985b), diagrams of the concentrations of point defects in hausmannite were determined, assuming normal spinel structure.

It was shown that similarly as in magnetite, the dominating ionic defects are defects with the highest ionisation degree, manganese vacancies $V'''_{Mn(o)}$ in the cation sublattice with octahedral coordination and interstitial manganese ions $Mn_i^{3\bullet}$. The adjusted values of ΔG^o_{F1} of their formation unambiguously determine the concentrations of Frenkel defects ($\left[V'''_{Mn}\right]$ and

$\left[Mn_i^{3\bullet}\right]$) in hausmannite. It was found that the adjusted values of $\Delta G_{F_2}^o$ of formation of Frenkel defects (V_{Mn}'' and $Mn_i^{\bullet\bullet}$), equal $\Delta G_{F_2}^o = \Delta G_{F_1}^o$, do not affect the degree of match between the dependence of δ on p_{O_2} and the experimental values of δ. The concentrations $\left[V_{Mn}''\right]$ and $\left[Mn_i^{\bullet\bullet}\right]$, resulting from the value of $\Delta G_{F_2}^o$ are thus the maximum concentrations of these defects that could be expected in hausmannite; in reality they could be lower.

A value that is determined in an ambiguous way is the value ΔG_{eh}^o of formation of electronic defects. It was found that for the model of defects with the highest ionisation degree, the values of ΔG_{eh}^o for α-$Mn_{3\pm\delta}O_4$ up to 110 kJ/mol, and for β-$Mn_{3\pm\delta}O_4$ up to ΔG_{eh}^o = 100 kJ/mol do not affect the degree of match between the oxygen pressure dependence of the deviation from the stoichiometry and the experimental values of δ. In the calculations of the diagram of concentrations of defects, the assumed values were: for α-$Mn_{3\pm\delta}O_4$, ΔG_{eh}^o = 85 kJ/mol and for β-$Mn_{3\pm\delta}O_4$, ΔG_{eh}^o = 60 kJ/mol (independent of temperature).

The performed attempts at the verification of the concentration determined based on the results of the electrical conductivity did not give satisfactory outcomes. The consistency of the sum of the concentrations of electronic defects ($[h^\bullet] + b_{e/h}[e']$) on p_{O_2} with the values of this sum determined based on the electrical conductivity was obtained at the values of ΔG_{eh}^o significantly higher than those assumed for the calculations of the diagram of defects' concentrations and at the parameter $b_{h/e}$ = 0.01, i.e. the mobility of electron holes a hundred times higher than the mobility of electrons. At temperatures: 1283 K, 1388 K, 1513 K and 1613 K, the values ΔG_{eh}^o would have to be, respectively: ΔG_{eh}^o = 199, 172, 143 and 108 kJ/mol. With so large values of ΔG_{eh}^o at 1283 K and 1388 K, no satisfactory match was obtained between the dependence of the deviation from the stoichiometry on p_{O_2} and the experimental values of δ, even after assuming a significant concentration of ionic defects with a lower ionisation degree. The determined values of the mobility of electron holes range from 2–14 cm^2/Vs; they are high and can be burdened with an error.

The character of the dependence of electrical conductivity on the oxygen pressure indicates that in the range of low oxygen pressures, where interstitial manganese ions dominate, the mobility of electron holes is much higher than that of electrons. In the range of higher oxygen pressures, where manganese vacancies dominate, the mobility of electrons is higher than that of electron holes. The above conclusions require experimental verification and wider studies, performed in high temperatures, of the oxygen pressure dependence of electrical conductivity and thermoelectric power.

From the comparison of the diagrams of concentrations of ionic defects for magnetite and hausmannite, it results that the concentration of ionic defects is by about one order of magnitude lower than in magnetite. The oxygen pressure dependences of the deviation from the stoichiometry, presented in Figure 1.5 and Figure 2.4, obtained at several temperatures for magnetite and hausmannite, show a similar character of the dependence of δ on p_{O_2} and the same type of dominating defects. In the middle of the range of existence of spinels, the stoichiometric composition is reached. However, the ranges of existence of magnetite and hausmannite and their changes with the temperature differ quite significantly. The range of existence of magnetite occurs at significantly lower oxygen pressures, and when the temperature increases, it symmetrically shifts towards higher pressures (by seven orders of magnitude). Also, the oxygen pressures at which spinels reach the stoichiometric composition, shifts by seven orders of magnitude (at 1273 K it is $p_{O_2} = 10^{-10}$ atm, and at 1623 K it is $p_{O_2} = 10^{-5}$ atm). However, the range of oxygen pressures where α-$Mn_{3\pm\delta}O_4$ variation exists, at temperatures 1273–1403 K is shifted by two orders of magnitude, while, independently of temperature, the stoichiometric composition is reached at the same oxygen pressure, $p_{O_2} = 2.3 \cdot 10^{-3}$ atm. Similarly, in the case of β-$Mn_{3\pm\delta}O_4$ variant, in the temperature range of 1473–1623 K, the range of existence is shifted by about two orders of magnitude, and this variant is stable up to the oxygen pressure of about 100 atm. The oxygen pressure, at which β-Mn_3O_4 reaches the stoichiometric composition shifts by about one order of magnitude (from $p_{O_2} = 5 \cdot 10^{-3} - 3.5 \times 10^{-2}$ atm). The differences that are present in the ranges of existence of magnetite and hausmannite and in their diagrams of defects' concentrations are a consequence of a difference in the energy of interaction of iron and manganese ions with oxygen ions. Similarly, high differences in the range of existence and in diagrams of defects' concentrations are present for simple oxides, $Fe_{1-\delta}O$ and $Mn_{1-\delta}O$ (see: analysis performed in the work (Stokłosa 2015)).

The fact that magnetite and hausmannite reach the stoichiometric composition and the presence of a range where cation vacancies and interstitial cations dominate is associated with relatively "loose" spinel structure. The M/O ratio magnetite and hausmannite is only 0.75 and in $Fe_{1-\delta}O$ and $Mn_{1-\delta}O$ it is close to one; however, they have the same cubic structure with a not much different lattice parameter ($4a_{FeO} = a_{Fe3O4}$). The lower number of cations in relation to oxygen ions is the reason for weaker repulsive interaction between ions in comparison to MO oxides. As a result, magnetite and hausmannite reach the stoichiometric composition in the range of their existence and there exists a range of oxygen pressures where interstitial cations and cation vacancies dominate. A consequence of mutual interactions of ions in magnetite and hausmannite is the narrow-band electron structure, and, at high temperatures, high concentration of electronic defects, significantly higher than that of ionic defects, which is evidenced by a high electrical conductivity of these spinels. On the other hand, $Fe_{1-\delta}O$ and $Mn_{1-\delta}O$ oxides have significantly lower concentration of electronic defects, comparable with the concentration of electronic defects (of the same order of magnitude). Wüstite $Fe_{1-\delta}O$ decays at the concentrations of iron vacancies 0.05 mol and at the oxygen pressure about 10^{-15} atm at 1273 K, which is by 20 orders of magnitude higher than the oxygen pressure at which it would reach the stoichiometric composition. $Mn_{1-\delta}O$ reaches the stoichiometric composition near the decay pressure, which, at 1273 K, is 10^{-24} atm.

REFERENCES

Bergstein, A., and J. Vintera. 1957. Die thermische Zersetzung von Mangancarbonat. *Collect. Czech. Chem. Commun.* 22:884–895.

Dorris, S. E., and T. O. Mason. 1988. Electrical Properties and Cation Valencies in Mn_3O_4. *J. Am. Ceram. Soc.* 71:379–385.

Driessens, F. C. M. 1967. Place and Valence of the Cations in Mn_3O_4 and Some Related Manganates. *Inorg. Chim. Acta* 1:193–201.

Dunitz, J. D., and L. E. Orgel. 1957. Electronic Properties of Transition-metal Oxides-II: Cation Distribution amongst Octahedral and Tetrahedral Sites. *J. Phys. Chem. Solids* 3:318–323.

Fine, M. E., and C. Chiou. 1957. Acoustic Relaxation Effect in Mn_3O_4. *Phys. Rev.* 105:121.

Fyfe, W. S. 1949. State of Manganese in Manganese Oxides. *Nature (London)* 164:790.

Goodenough, J. B., and A. L. Loeb. 1955. Theory of Ionic Ordering, Crystal Distortion, and Magnetic Exchange Due to Covalent Forces in Spinels. *Phys. Rev.* 98: 391–408.

Grundy, A. N. 2003. Assessment of the Mn-O system. *J. Phase Equilib.* 24: 21–39.

Hahn Jr., W. C. and A. Muan, 1960. Studies in the System Mn-O. The Mn_2O_3-Mn_3O_4 and Mn_3O_4-MnO Equilibria. *Am. J. Sci.* 258: 66–78.

Irani, K. S., A. P. B. Sinha, and A. B. Biswas. 1960. Crystal Distortion in Spinels Containing Mn^{3+} Ions. *J. Phys. Chem. Solids* 17: 101–111.

Kasper, J. S. 1959. Magnetic Structure of Hausmannite. Mn_3O_4. *Bull. Am. Phys. Soc.* II 4: 178.

Keller, M., and R. Dieckmann. 1985a. Defect Structure and Transport Properties of Manganese Oxides: (I) The Nonstoichiometry of Manganosite ($Mn_{1-\delta}O$). *Ber. Bunsenges. Phys. Chem.* 89: 883–893.

Keller, M., and R. Dieckmann. 1985b. Defect Structure and Transport Properties of Manganese Oxides: (II) The Nonstoichiometry of Hausmannite ($Mn_{3-\delta}O_4$). *Ber. Bunsenges. Phys. Chem.* 89: 1095–1104.

Kim, D. Q., Y. Wilbert, and F. Marion. 1966. Sur la détermination directe des équilibres des oxydes de manganèse à haute température. *C.R Acad. Sci. Sér. C: Sci. Chim.* 262: 756–758.

Kjellqvist, L. 2010. Thermodynamic Assessment of the Fe-Mn-O System. *J. Phase Equilib. Diff.* 31:113–134.

Larson, E. G., R. J. Arnott, and D. G. Wickham. 1962. Preparation, Semiconduction and Low-temperature Magnetization of the System $Ni_{1-x}Mn_{2+x}O_4$. *J. Phys. Chem. Solids* 23: 1771–1781.

LeBlanc, M., and G. Wehner. 1934. Contribution to the Knowledge of the Manganese Oxides. *Z. Phys. Chem. A* 168: 59–78.

Logothetis, E. M., and K. Park. 1975. The Electrical Conductivity of Mn_3O_4. *Solid State Commun.* 16: 909–912.

Lu, F. H., and R. Dieckmann, 1993. Point Defects and Cation Tracer Diffusion in $(Co_x\ Mn_{1-x})_{3-\delta}O_4$ Spinels. *Solid State Ionics* 67: 145–155.

Mason, B. 1947. Mineralogical Aspects of the System Fe_3O_4-Mn_3O_4-$ZnMn_2O_4$-$ZnFe_2O_4$. *Am. Mineral.* 32: 426–441.

McClure, D. S. 1957. The Distribution of Transition Metal Cations in Spinels. *J. Phys. Chem. Solids* 3:311–317.

McMurdie, H. F., and E. Golovato. 1948. Study of the Modifications of Manganese Dioxide. *J. Res. Nat. Bur. Stand. U.S.A.* 41: 589–600.

McMurdie, H. F., B. M. Sullivan, and F. A. Mauer. 1950. High-Temperature X-Ray Study of the System Fe_3O_4-Mn_3O_4. *J. Res. Nat. Bur. Stand. U.S.A.* 45: 35–41.

Metselaar, R., R. E. J. van Tol, and P. Piercy. 1981. The Electrical Conductivity and Thermoelectric Power of Mn_3O_4 at High Temperatures. *J. Solid State Chem.* 38: 335–341.

Millar, R. W. 1928. The Specific Heats at Low Temperatures of Manganous Oxide, Manganous-Manganic Oxide and Manganese Dioxide. *J. Am. Chem. Soc.* 50: 1875–1883.

Moore, T. E., N. Ellis, and P. W. Selwood. 1950. Solid Oxides and Hydroxides of Manganese. *J. Am. Chem. Soc.* 72:856–866.

Nogués, M., and P. Poix. 1974. Effet Jahn-Teller coopératif dans le systéme $Mn_3O_4 \times Mn_2SnO_4$. *J. Solid State Chem.* 9: 330–335.

Petzold, D. R. 1971. Die Temperaturabhängigkeit der Kationenverteilung bei inversen Spinellen. *Krist. all. Tech.* 6: K43–K45.

Romeijn, F. C. 1953. Physical and Crystallographical Properties of Some Spinels. *Philips Res. Rep.* 8: 304–320.

Rosenberg, R. M., P. Nicolau, and I. Bunget. 1966. Electrical Properties of the System $Fe_xMn_{3-x}O_4$. *Phys. Status Solidi* 14: K65–K69.

Rosenberg, M., P. Nicolau, R. Manaila, and P. Pausescu, 1963. Preparation, Electrical Conductivity and Tetragonal Distortion of Some Manganite–System. *J. Phys. Chem. Solids* 24: 1419–1434.

Schmahl, N. G., and D. F. K. Hennings, 1969a. The Phase Diagram of the Mn_3O_4-MnO System and Its Pressures of Dissociation. *Arch. Eisenhüttenwes* 40: 395–399.

Schmahl, N. G., and D. F. K. Hennings. 1969b. Die Nichtstöchiometrie des Mangan (II). *Z. Phys. Chem. N.F.* 63: 111–124.

Southard, J., and G. Moore. 1942. High-temperature Heat Content of Mn_3O_4, $MnSiO_3$ and Mn_3C. *J. Am. Chem. Soc.* 64: 1769–1770.

Stokłosa, A. 2015. *Non-Stoichiometric Oxides of 3d Metals.* Pfäffikon: Trans Tech Publications Ltd.

Terayama, K., M. Ikeda, and M. Taniguchi. 1983. Study on Thermal Decomposition of MnO_2 and Mn_2O_3 by Thermal Analysis. *Trans. Jap. Inst. Metals* 24:754–758.

Verney, E. J. W., and J. H. de Boer. 1936. Cation Arrangement in a Few Oxides with Crystal Structures of the Spinel Type. *Recl. Trav. Chim. Pays-Bas* 55: 531–540.

Part II

Magnetite Doped with M^{3+} and M^{2+} Ions

The features, mentioned in previous chapters, of the spinel structure of magnetite cause that iron ions Fe^{2+} and Fe^{3+} can be substituted with other cations of a stable oxidation state, which will form a solid solution in cation sublattices. The "loose" structure of magnetite causes the formation of solid solutions to be possible in practically the whole range of concentrations. The concentration limit of the dopant can reach 1 mol/mol of the sublattice and such spinels are treated as new heterocation spinels, where iron ions with tetrahedral coordination (AFe_2O_4) or with octahedral coordination $Fe(BFe)O_4$ are substituted. Analogous spinels are created by doped hausmannite, and by several other ions of metals of 2+ and 3+ oxidation states; such spinels are called 2–3 type spinels. Distinct from spinels of this type, there is a large group of heterocation spinels in which atoms have different stable oxidation states. Spinel properties were the object of many experimental and theoretical works (Romeijn 1953, Goodenough and Loeb 1955, Dunitz and Orgel 1957, McClure and Miller 1957, Miller 1959, Navrotsky and Kleppa 1967, Navrotsky and Kleppa 1968, Cullen and Cullen 1973, O'Neill and Navrotsky 1983, O'Neill and Navrotsky 1984, Snelling 1988, McCurrie 1994, Valenzuela 1994). Theoretical calculations took into account ionic radius, energy of stabilisation in the crystal field, electron structure and formation of specific types of bonds (interactions between metal ions and oxygen ions).

Due to the practical importance of ferrites, many studies concerned the properties of doped magnetite, where Fe^{2+} and Fe^{3+} iron ions are substituted with M^{2+} ions (e.g. Co, Ni, Zn etc.) and M^{3+} ions (e.g. Cr, Al, Ti etc.). The studies were conducted mainly on frozen samples; their magnetic properties and crystalline structure were determined using X-ray diffraction, neutron diffraction, spectroscopy, Mössbauer effects etc; these methods permit determination of the distribution of cations and their oxidation states. Most studies at high temperatures were limited to the determination of the range of existence of individual spinels. Only the presence of a small deviation from the stoichiometry was found or it was not present at all. Wide studies on the effect of dopants of foreign ions with different oxidation states on the concentration of ionic defects and their diffusion were performed by Dieckmann and co-workers (Franke and Dieckmann 1989, Franke and Dieckmann 1990, Lu and Dieckmann, 1992, Lu and Dieckmann 1993a, Lu and Dieckmann 1993b, Lu et al. 1993, Lu and Dieckmann 1995, Töpfer et al. 1995, Aggarwal et al. 1997, Aggarwal and Dieckmann 2002a, Aggarwal and Dieckmann 2002b, Töpfer et al. 2003, Töpfer and Dieckmann 2004). The following spinels were studied: $(Fe_{1-x}M_x)_{3-\delta}O_4$ doped with titanium (Töpfer and Dieckmann 2004), chromium (Aggarwal et al. 1997), manganese (Franke and Dieckmann 1989, Franke and Dieckmann 1990), cobalt (Lu et al. 1993), Mn and Co (Lu and Dieckmann, 1992) and with Mn and Zn (Töpfer et al. 2003, Töpfer and Dieckmann 2004) as well as $(Mn_{1-x}M_x)_{3-\delta}O_4$ doped with cobalt (Lu and Dieckmann 1993b), iron (Lu and Dieckmann 1995), Co and Fe (Lu and Dieckmann 1993a).

As the studies of the oxygen pressure dependence of the deviation from the stoichiometry were performed for the above doped spinels in the whole range of their existence, it was decided to use them in the determination of the diagrams of concentrations of point defects.

REFERENCES

Aggarwal, S., and R. Dieckmann. 2002a. Point Defects and Cation Tracer Diffusion in $(Ti_xFe_{1-x})_{3-\delta}O_4$ I. Non-stoichiometry and Point Defects. *J. Phys. Chem. Miner.* 29:695–706.

Aggarwal, S., and R. Dieckmann. 2002b. Point Defects and Cation Tracer Diffusion in $(Ti_xFe_{1-x})_{3-\delta}O_4$. II. Cation Tracer Diffusion. *J. Phys. Chem. Miner.* 29:707–718.

Aggarwal, S., J. Töpfer, T. L. Tsai, and R. Dieckmann. 1997. Point Defects and Transport in Binary and Ternary, Non-stoichiometric Oxides. *Solid State Ionics* 101–103:321–331.

Cullen, J. R., and E. Cullen. 1973. Multiple Ordering in Magnetite. *Phys. Rev. B* 7:397–402.

Dunitz, J. D., and L. E. Orgel. 1957. Electronic Properties of Transition-metal Oxides-II: Cation Distribution Amongst Octahedral and Tetrahedral Sites. *J. Phys. Chem. Solids* 3:318–323.

Franke, P., and R. Dieckmann. 1989. Defect Structure and Transport Properties of Mixed Iron-manganese Oxides. *Solid State Ionics* 32–33:817–823.

Franke, P., and R. Dieckmann. 1990. Thermodynamics of Iron Manganese Mixed Oxides at High Temperatures. *J. Phys. Chem. Solids* 51:49–57.

Goodenough, J. B., and A. L. Loeb. 1955. Theory of Ionic Ordering, Crystal Distortion, and Magnetic Exchange Due to Covalent Forces in Spinels. *Phys. Rev.* 98:391–408.

Lu, F. H., and R. Dieckmann. 1992. Point Defects and Cation Tracer Diffusion in $(Co,Fe,Mn)_{3-\delta}O_4$ Spinels: I. Mixed Spinels $(Co_xFe_{2y}Mn_y)_{3-\delta}O_4$. *Solid State Ionics* 53–56:290–302.

Lu, F. H., and R. Dieckmann. 1993a. Point Defects and Cation Tracer Diffusion in $(Co, Fe, Mn)_{3-\delta}O_4$ Spinels: II. Mixed Spinels $(Co_xFe_zMn_{2z})_{3-\delta}O_4$. *Solid State Ionics* 59:71–82.

Lu, F. H., and R. Dieckmann. 1993b. Point Defects and Cation Tracer Diffusion in $(Co_xMn_{1-x})_{3-\delta}O_4$ Spinels. *Solid State Ionics* 67:145–155.

Lu, F. H., and R. Dieckmann. 1995. Point Defects in Oxide Spinel Solid Solutions of the Type $(Co,Fe,Mn)_{3-\delta}O_4$ at 1200 °C. *J. Phys. Chem. Solids* 56:725–733.

Lu, F. H., S. Tinkler, and R. Dieckmann. 1993. Point Defects and Cation Tracer Diffusion in $(Co_xFe_{1-x})_{3-\delta}O_4$ Spinels. *Solid State Ionics* 62:39–52.

McClure, D. S., and A. Miller. 1957. The Distribution of Transition Metal Cations in Spinels. *J. Phys. Chem. Solids* 3:311–317.

McCurrie, R. A. 1994. *Ferromagnetic Materials: Structure and Properties*. Academic Press.

Miller, A. 1959. Distribution of Cations in Spinels. *J. Appl. Phys.* 30:S24–S25.

Navrotsky, A., and O. J. Kleppa. 1967. The Thermodynamics of Cation Distributions in Simple Spinels. *J. Inorg. Nucl. Chem.* 29:2701–2714.

Navrotsky, A., and O. J. Kleppa. 1968. Thermodynamics of Formation of Simple Spinels. *J. Inorg. Nucl. Chem.* 30:479–498.

O'Neill, H. S. C., and A. Navrotsky. 1983. Simple Spinels; Crystallographic Parameters, Cation Radii, Lattice Energies, and Cation Distribution, *Am. Mineral.* 68:181–194.

O'Neill, H. S. C., and A. Navrotsky. 1984. Cation Distributions and Thermodynamic Properties of Binary Spinel Solid Solutions. *Am. Mineral.* 69:733–753.

Romeijn, F. C. 1953. Physical and Crystallographical Properties of Some Spinels. *Philips Res. Rep.* 8:304–342.

Snelling, E. C. 1988. *Soft Ferrites Properties and Application*. Butterworth Co Ltd.

Töpfer, J., and R. Dieckmann. 2004. Point Defects and Deviation from Stoichiometry in $(Zn_{x-y/4}Mn_{1-x-3y/4}Fe_{2+y})_{1-\delta/3}O_4$. *J. Europ. Ceram. Soc.* 24:603–612.

Töpfer, J., L. Liu, and R. Dieckmann. 2003. Deviation from Stoichiometry and Point Defects in $(Zn_xMn_{1-x}Fe_2)_{1-\delta/3}O_4$. *Solid State Ionics* 159:397–404.

Töpfer, J., S. Aggarwal, and R. Dieckmann. 1995. Point Defects and Cation Tracer Diffusion in $(Cr_xFe_{1-x})_{3-\delta}O_4$ Spinels. *Solid State Ionics* 81:251–266.

Valenzuela, R. 1994. *Magnetic Ceramics*. Cambridge University Press.

3 Titanium-doped Magnetite – $(Fe_{1-x}Ti_x)_{3\pm\delta}O_4$

3.1 INTRODUCTION

Studies of several authors (Webster and Bright 1961, Buddington and Lindsley 1964, Taylor 1964, Hauptman 1974, Simons and Woermann 1978, Spencer and Lindsley 1981, Aragón and McCallister 1982, Senderov et al. 1993) showed that the range of existence of magnetite doped with titanium, $(Fe_{1-x}Ti_x)_{3\pm\delta}O_4$, is shifted towards lower oxygen pressures when the titanium content increases. However, most of the studies on the deviation from the stoichiometry are incomplete and they were performed on frozen samples, therefore, they are burdened with large errors. Systematic studies of magnetite doped with titanium were performed widely (Aggarwal and Dieckmann 2002a, Aggarwal and Dieckmann 2002b), particularly on the oxygen pressure dependence on the deviation from the stoichiometry in the spinel $(Fe_{1-x}Ti_x)_{3\pm\delta}O_4$ with a titanium content $x_{Ti} = 0.1$, 0.2 and 0.25 mol, in, at 1373–1673 K. Also, on the same spinels, the studies were performed on the self-diffusion of radioisotope tracers: ^{59}Fe, ^{60}Co, ^{54}Mn and ^{44}Ti, at the same temperatures and oxygen pressures as the studies on the deviation from the stoichiometry. It was shown that in the range of existence of spinel $(Fe_{1-x}Ti_x)_{3\pm\delta}O_4$, there are defects of the type similar to the ones in pure magnetite (Aggarwal and Dieckmann 2002a, Aggarwal and Dieckmann 2002b). At higher pressures, cation vacancies dominate, while at lower pressures, in a narrow range, interstitial cations dominate. The above model of point defects was confirmed by diffusion studies, which showed that the coefficients of self-diffusion depend on the concentration of ionic defects and on the type of diffused metal, and they differ depending on the range in which cation vacancies and interstitial cations dominate. The results of the studies on the diffusion of metal ions obtained by the Dieckmann group (Aggarwal and Dieckmann 2002a, Aggarwal and Dieckmann 2002b) are consistent with the studies of other authors (Creer et al. 1970, Petersen 1970, Ozima and Ozima 1972,Freer and Hauptman 1978, Price 1981, Aragón et al. 1984), which were performed at a constant oxygen pressure.

Many works treated the distribution of ions in sublattices of magnetite doped with titanium, which were conducted on frozen samples (Akimoto 1954, Akimoto et al. 1957, Banerjee et al. 1967, Bleil 1971, Bleil 1976, Chevallier et al. 1955, Fujino 1974, Ishikawa et al. 1971, Kąkol et al. 1991, Néel 1955, O'Donovan and O'Reilly 1980, O'Neill and Navrotsky 1984, O'Reilly and Banerjee 1965, Stephenson 1969, Wechsler et al. 1984), in order to determine their magnetic properties. It was found that titanium ions occupy lattice nodes with octahedral coordination, despite the fact that many other studies showed that they are present also in the sublattice with tetrahedral coordination (Foster and Hall 1965, Gorter 1957, Stout and Bayliss 1980). From most of the studies it results that the incorporated titanium ions have an oxidation state of four and they occupy cation nodes with octahedral coordination (Akimoto 1954, Akimoto et al. 1957, Banerjee et al. 1967, Bleil 1971, Bleil 1976, Chevallier et al. 1955, Fujino 1974, Ishikawa et al. 1971, Kąkol et al. 1991, Néel 1955, O'Donovan and O'Reilly 1980, O'Neill and Navrotsky 1984, O'Reilly and Banerjee 1965, Stephenson 1969, Trestman-Matts et al. 1983, Wechsler et al. 1984). Based on the studies on the variation of magnetisation depending on the concentration of titanium, several models of the distribution of ions with different charges were proposed. The model of Akimoto (Akimoto 1954, Akimoto et al. 1957) assumes that Ti^{4+} ions incorporate into the nodes of the sublattice of Fe^{3+} ions with octahedral coordination, which simultaneously causes an increase in the concentration of Fe^{2+} ions with the sublattice with octahedral or tetrahedral coordination.

$$Fe^{3+}_{1-x}Fe^{2+}_{x}\left(Fe^{2+}Fe^{3+}_{1-x}Ti^{4+}_{x}\right)O_4$$

The model of Néel and Chevallier (Chevallier et al. 1955, Néel 1955) assumes that for the titanium content of $0 \le x_{Ti} \le 0.5$ mol there occurs a decrease in the concentration of Fe^{2+} ions only in the cation sublattice with octahedral coordination. At the content of $x_{Ti} \ge 0.5$ mol, there are no more Fe^{3+} ions in the octahedral sublattice, and further increase in the concentration of titanium causes an increase in the concentration of Fe^{2+} ions with tetrahedral coordination, which can be expressed with the formula:

$$Fe^{3+}_{2-2x}Fe^{2+}_{2-x}\left(Fe^{2+}_{2-x}Ti^{4+}_{x}\right)O_4 \quad \text{for} \quad 0.5 \le x_{Ti} \le 1\,\text{mol/mol}$$

A slightly different model (O'Reilly and Banerjee 1965, Banerjee et al. 1967) was proposed for the titanium content of $0.2 \le x_{Ti} \le 0.8$ mol/mol:

$$Fe^{2+}_{x-0,2}Fe^{3+}_{1,2-x}\left(Fe^{2+}_{1,2}Fe^{3+}_{0,8-x}Ti^{4+}_{x}\right)O_4$$

Studies by other authors (Bleil 1971, Bleil 1976, Fujino 1974, Ishikawa et al. 1971, Kąkol et al. 1991, O'Donovan and O'Reilly 1980, O'Neill and Navrotsky 1984, Stephenson 1969, Wechsler et al. 1984), using various techniques, confirmed the above models. Mason et al. (Trestman-Matts et al. 1983), based on the studies of thermoelectric power in the temperature range of 873–1573 K, proposed the distribution of ions similar to that in the model of Akimoto (Akimoto 1954), however, depending on the titanium concentration, there are quite significant differences. Differences in the results concerning distribution of ions and magnetic properties of the obtained magnetite doped with titanium depend on the conditions of formation of the spinel, especially on the annealing time and the rate of freezing the samples. During the above process, also a change in the oxidation state of iron ions occurs. Therefore, there is no full consistency between the studies of different authors.

3.2 MECHANISM OF DOPING WITH TITANIUM AND METHODS OF CALCULATION OF THE DIAGRAMS OF DEFECTS' CONCENTRATIONS

Magnetite doped with titanium can be obtained for example in a reaction of FeO, Fe_2O_3 and Ti_2O_3 oxides, by annealing them at high temperatures:

$$Fe0 + x/2Ti_2O_3 + (2-x)/2Fe_2O_3 \rightarrow Fe^{3+}_{(t)}\left(Fe^{2+}_{(o)}Fe^{3+}_{1-x}Ti^{3+}_{x}\right)O_4 \quad (3.1a)$$

Titanium ions, Ti^{3+}, substitute Fe^{3+} ions with octahedral coordination, which should not change the structure of ionic defects significantly. The presence of titanium ions, however, will affect standard Gibbs energies of defects formation.

Titanium-doped magnetite is also obtained through a reaction of iron oxides with TiO_2. In this case the resultant reaction assumes the form of:

$$FeO + xTiO_2 + (2-x)/2Fe_2O_3 \rightarrow Fe^{3+}_{(t)}\left(Fe^{2+}_{(o)}Fe^{3+}_{1-x}Ti^{3+}_{x}\right)O_4 + x/4O_2 \quad (3.1b)$$

According to the reaction 3.1b, electrons left from the oxidation of oxygen reduce titanium ions, and a spinel is formed, analogously as in the case of reaction 3.1a. If titanium ions preserve their oxidation state, then the electrons released will reduce Fe^{3+} ions, and, in such a case, the concentration of Fe^{2+} ions increases. An equilibrium is set at high temperatures and titanium ions can be incorporated also in cation nodes of the sublattice with tetrahedral coordination. Setting of an equilibrium causes also setting of equilibrium concentrations of ionic defects and electronic defects. In the case of Ti^{3+} presence, creation of electron holes will be associated with releasing an electron by a titanium ion ($Ti^{3+} \rightarrow Ti^{4+} + e^-$), or by an Fe^{2+} ion. The release of electrons will be associated with binding an electron by an Fe^{3+} ion ($Fe^{3+} + e^- \rightarrow Fe^{2+}$) and their localisation, which will cause an increase in the concentration of Fe^{2+} ions and a decrease in the concentration of Fe^{3+} ions, or electrons will be taken up by Ti^{4+} ions.

Thus, incorporation of titanium ions into magnetite can be written with a general reaction:

$$Fe0 + xTiO_2 + (2-x)/2Fe_2O_3 \rightarrow Fe^{3+}_{a}Fe^{2+}_{b}\left(Fe^{2+}_{c}Fe^{3+}_{d}Ti^{4+}_{x}\right)O_4 + x/4O_2 \quad (3.1c)$$

where numbers of atoms in the spinel fulfil the condition (a + b = 1) and (c + d = 2 – x). In equilibrium conditions, a determined distribution of ions with different charges is set; it is a result of their optimal interactions. Distribution of ions depends also on the concentration of electronic defects, as they also cause a change in the oxidation state of an ion. The oxidation state of Ti^{4+} ions incorporated into the magnetite structure can be stable, and it probably occurs in titanium-doped magnetite. In that case, Ti^{4+} ions are the node ions (electroneutral in relation to the lattice), and the concentration of Fe^{2+} node ions must increase depending on their concentration. However, Ti^{4+} ions can participate in the charge transport, and a part of their concentration will be associated with the presence of electron holes. Their concentration will be also affected by the concentration of ionic defects which cause a change in the concentration of electronic defects.

The character of the oxygen pressure dependence of the deviation from the stoichiometry in $(Fe_{1-x}Ti_x)_{3\pm\delta}O_4$ spinel (Aggarwal and Dieckmann 2002a, Aggarwal and Dieckmann 2002b), is close to the dependence for pure magnetite (see Figure 1.3a, 1.3b, 1.3c, 1.3d). Therefore, it was assumed that $(Fe_{1-x}Ti_x)_{3\pm\delta}O_4$ spinel is formed according to the reaction 3.1c and incorporated titanium ions change interactions between ions. Due to that, the calculations of diagrams of the concentrations of point defects for spinels $(Fe_{1-x}Ti_x)_{3\pm\delta}O_4$ were conducted with a method analogous to the one used for pure magnetite (see Section 1.5), using the results of the deviation from the stoichiometry (Aggarwal and Dieckmann 2002a, Aggarwal and Dieckmann 2002b). It was assumed that titanium ions are incorporated into sites with octahedral coordination, they replace Fe^{3+} ions, and at first, they have a charge of 3+. A change in the oxidation state occurs as a result of internal oxidation, which does not cause nor increase the concentration of electronic defects. The distribution of ions in titanium-doped magnetite, proposed in (Trestman-Matts et al. 1983), due to a high concentration of possible electronic defects, resulting from the distribution of ions in sublattices, cannot be used for calculations of the value of ΔG^o_{eh} of formation of electronic defects. Due to that, in the calculations, a constant value of ΔG^o_{eh} equal 30.3 kJ/mol was assumed, the same as for pure magnetite. As shown by initial calculations, assuming the value of ΔG^o_{eh} in the range from 20–50 kJ/mol does not affect the degree of match between the calculated dependence of deviation from the stoichiometry on p_{O_2} and the experimental values of δ. Only the values of ΔG^o_i of formation of cation vacancies and interstitial cations change, analogously as for pure magnetite (see Table 1.3). Therefore, a change in the oxygen pressure will cause an increase/decrease in the concentration of point defects in doped magnetite (according to the reactions 1.14–1.17). Apart from iron vacancies and interstitial iron ions, also titanium vacancies $V^{4'}_{Ti(o)}$ and interstitial ions $Ti_i^{4\bullet}$ will be formed and their standard Gibbs energies of the formation should be different than those for iron point defects. In the case of a high concentration of Ti^{4+} ions it is difficult to evaluate the contribution of cation vacancies $\left[V^{4'}_{Ti(o)}\right]$ and interstitial cations $\left[Ti_i^{4\bullet}\right]$ to the total concentration of these defects. This fact was not taken into account in the calculations and in the notation of defects, and, for example, $V^{3'}_{Fe}$ denotes a vacancy in the cation sublattice, and $Fe_i^{3\bullet}$ or generally $M_i^{z\bullet}$ – an interstitial cation. However, the concentration of titanium in magnetite will affect the resultant standard Gibbs energy ΔG^o_i of formation of the individual point defects.

The created spinel structure, at even a small excess of TiO_2 oxide, according to the thermodynamics, should be doped with Ti^{4+} ions. Therefore, it is necessary to consider Ti^{4+} ions incorporated into the sublattice of Fe^{3+} ions as dopants with a higher charge than Fe^{3+} ions. This is analogous to the process of doping of $M_{1-\delta}O$ oxides with D^{3+} ions (D_2O_3). In magnetite, in comparison to MO-type oxides, the notation of the doping reaction is not so simple, because titanium ions, electroneutral relative to the lattice, are present in the lattice nodes. According to the principle of writing the doping reaction, the incorporation of Ti^{4+} ions in the form of TiO_2 oxide must lead to building-in of a molecule of the spinel crystal. It is necessary to preserve the number of newly created oxygen sites and cation sites. Thus, the incorporation of four oxygen sites must cause the formation of three cation sites. Therefore, the reaction of incorporation of titanium ions into the spinel structure will have the form of:

$$2TiO_2 \rightarrow 2Ti^{\bullet}_{Fe(3+)} + 4O^x_O + V''_{Fe(o)} \tag{3.2}$$

According to the reaction (3.2), two titanium ions Ti^{4+} ($2TiO_2$) occupy two Fe^{3+} nodes and, according to the Kröger-Vink notation, these ions are positive in relation to $Ti^{\bullet}_{Fe}$. They can be incorporated in the sublattice with octahedral and/or tetrahedral coordination and a cation vacancy $V''_{Fe(o)}$ must be formed. An equilibrium state is set through the diffusion process. Due to mutual interactions, substitution of ions between sublattices can be beneficial or titanium ions can occupy only sites with octahedral coordination. Thus, according to the reaction 3.2, incorporation of Ti^{4+} ions into the sublattice of Fe^{3+} ions (($Ti^{\bullet}_{Fe}$) ions) into the spinel with the stoichiometric composition will cause an increase in the concentration of iron vacancies $\left[V''_{Fe(o)}\right]$ and a decrease in the concentration of interstitial ions $\left[Fe_i^{\bullet\bullet}\right]$ as well as a change in concentrations of remaining point defects. The equilibrium pressure above the spinel doped in this way will significantly be shifted towards higher oxygen pressures. Annealing of the spinel at a lower oxygen pressure will cause a change in the concentrations of ionic defects and electronic defects. Due to an increase in the concentration of cation vacancies and a perturbation, by dopant ions, of electroneutrality of the crystal, the spinel will reach the stoichiometric composition at a lower oxygen pressure in comparison to the spinel not doped with $Ti^{\bullet}_{Fe}$ ions. The character of the dependence of δ on p_{O_2} changes in a specific way (symmetry of branches of the dependence of δ on p_{O_2} is perturbed). The above problems were widely discussed in (Stokłosa 2015), for $M_{1-\delta}O$ oxides ($Ni_{1-\delta}O$, $Co_{1-\delta}O$, $Mn_{1-\delta}O$) doped with D^{3+} ions (D_2O_3).

Therefore, according to the reaction 3.2, an introduction of titanium dopants in an amount of y_{Ti} into the spinel with the stoichiometric composition causes an increase in the concentration of iron vacancies $\left[V''_{Fe}\right]$ by $1/2y_{Ti}$:

$$\left[V''_{Fe}\right] = y^o_{V''_{Fe}} + 1/2y_{Ti},$$

where $y^o_{V''_{Fe}}$ denotes the concentration of vacancies $\left[V''_{Fe}\right]$ in the spinel with the stoichiometric composition. This fact causes a perturbation of the equilibrium between defects and leads, at a defined oxygen pressure, to a new equilibrium state. As the concentration of $\left[Ti^{\bullet}_{Fe}\right]$ ions is low, it can be assumed that it does not affect the values of standard Gibbs energies of formation of defects in the spinel. Due to that, it is possible to assume the same values of ΔG^o_i of defects' formation as for a spinel not doped with $Ti^{\bullet}_{Fe}$ ions. The calculations should take into account, under the electroneutrality condition, the concentration of ions $\left[Ti^{\bullet}_{Fe}\right] = y_{Ti^{\bullet}}$:

$$\left[e'\right] + 2\left[V''_{Fe(o)}\right] + 3\left[V'''_{Fe(o)}\right] + 3\left[V'''_{Fe(t)}\right] = \left[h^{\bullet}\right] + 2\left[Fe_i^{\bullet\bullet}\right] + 3\left[Fe_i^{3\bullet}\right] + y_{Ti^{\bullet}} \tag{3.3}$$

Also, when calculating the deviation from the stoichiometry, it is necessary to consider an increase in the number of cation sites and oxygen sites as a result of incorporation of ions:

$$\frac{\left[Fe\right] + \left[Ti\right]}{\left[O\right]} = \frac{3 + y_{Ti} - \Sigma\left[def\right]}{4 + 2y_{Ti}} \tag{3.4}$$

where,

$$\Sigma\left[def\right] = \left[V''_{Fe(o)}\right] + \left[V'''_{Fe(o)}\right] + \left[V'''_{Fe(t)}\right] - \left[Fe_i^{\bullet\bullet}\right] - \left[Fe_i^{3\bullet}\right]$$

If the concentration of titanium ions positive in relation to the lattice $\left[Ti^{\bullet}_{Fe}\right]$ is small, then their presence will affect the values of the deviation δ near the stoichiometric composition. Thus, when adjusting the values of ΔG_i^o of defects' formation, we should consider only possibly high values of the deviation from the stoichiometry, at low and high oxygen pressures (near the limit oxygen pressures). More details on the calculation of diagram of the concentrations of defects for doped oxides can be found in work by Stokłosa (2015).

Therefore, according to the methods of calculations presented in Section 1.5, at an assumed value of ΔG^o_{eh} of formation of electronic defects, the values of $\Delta G^o_{F_1}$ and $\Delta G^o_{F_2}$ of formation of Frenkel defects are fitted. As a first approximation it was assumed that $\Delta G^o_{F_1} = \Delta G^o_{F_2}$. Due to the possibility of doping the spinel according to the reaction 3.2, relative concentrations of defects were calculated as a function of the concentration of vacancies $\left[V''_{Fe}\right]$. Thus, the values of $\Delta G^o_{V''_{Fe(o)}}$ of formation of iron vacancies $V''_{Fe(o)}$ were fitted, and then the values of ΔG_I^o were chosen in a way to permit, according to Equation 1.35b, calculation of the concentration of vacancies $\left[V'''_{Fe(o)}\right]$. As in the calculations, the equity $\Delta G^o_{F_1} = \Delta G^o_{F_2}$ was assumed, the value of ΔG_I^o was fitted in such a way to obtain similar concentrations of iron vacancies at the stoichiometric composition $\left[V'''_{Fe(o)}\right] \cong \left[V''_{Fe}\right]$. Then, such values of $\Delta G^o_{F_3}$ are chosen that the concentration of vacancies $\left[V'''_{Fe(t)}\right]$ does not deteriorate the quality of match between the dependence of δ on p_{O_2} and the experimental values of δ. When taking into account the concentration of ions $Ti^{\bullet}_{Fe}$, simultaneously the value of $y_{Ti^{\bullet}}$ is adjusted in such a way that the dependence of the deviation from the stoichiometry on p_{O_2} is consistent with the experimental values of δ. The equilibrium oxygen pressure is calculated according to Equation 1.10a, using the value of $\Delta G^o_{V''_{Fe}}$, concentration of vacancies $\left[V''_{Fe}\right]$ and concentration of electron holes.

3.3 RESULTS OF CALCULATION OF THE DIAGRAMS OF THE CONCENTRATIONS OF DEFECTS, AND DISCUSSION

Figure 3.1a–f present the diagrams of the concentrations of ionic defects for the spinel $(Fe_{1-x}Ti_x)_{3\pm\delta}O_4$ (solid lines), for titanium content of $x_{Ti} = 0.1$ and 0.2 mol, at temperatures 1373–1573 K, calculated using the values of the deviation from the stoichiometry (Aggarwal and Dieckmann 2002a, Aggarwal and Dieckmann 2002b) ((○) points). The dotted line shows the dependence of δ on p_{O_2} obtained using "equilibrium constants" of reactions of formation of cation vacancies K_V and interstitial ions K_I, as determined in (Aggarwal and Dieckmann 2002a, Aggarwal and Dieckmann 2002b).

As can be seen in Figure 3.1, a very good match was obtained between the dependence of the deviation from the stoichiometry and the experimental values of δ. Dominating defects are cation vacancies $\left[V'''_{Fe(o)}\right]$ and interstitial cations $\left[Fe_i^{3\bullet}\right]$. Similarly, as in pure magnetite, the concentration of vacancies $\left[V''_{Fe}\right]$ and $\left[Fe_i^{\bullet\bullet}\right]$ is significantly lower, as well as the concentration of vacancies $\left[V''_{Fe(t)}\right]$ in the sublattice with tetrahedral coordination.

Figure 3.1a presents the oxygen pressure dependence of the deviation from the stoichiometry with pure magnetite doped with titanium according to the reaction 3.2, with the amount of $y_{Ti^{\bullet}} = 0.0046$ mol/mol ((-----) line). As can be seen in Figure 3.1a, for magnetite doped with $Ti^{\bullet}_{Fe}$ ions to reach the stoichiometric composition at the oxygen pressure which results from the studies on deviation from the stoichiometry (Aggarwal and Dieckmann 2002a, Aggarwal and Dieckmann 2002b), titanium concentration should be only $x_{Ti} = 0.046$ mol/mol. As can be seen, incorporation of $Ti^{\bullet}_{Fe}$ ions into the magnetite

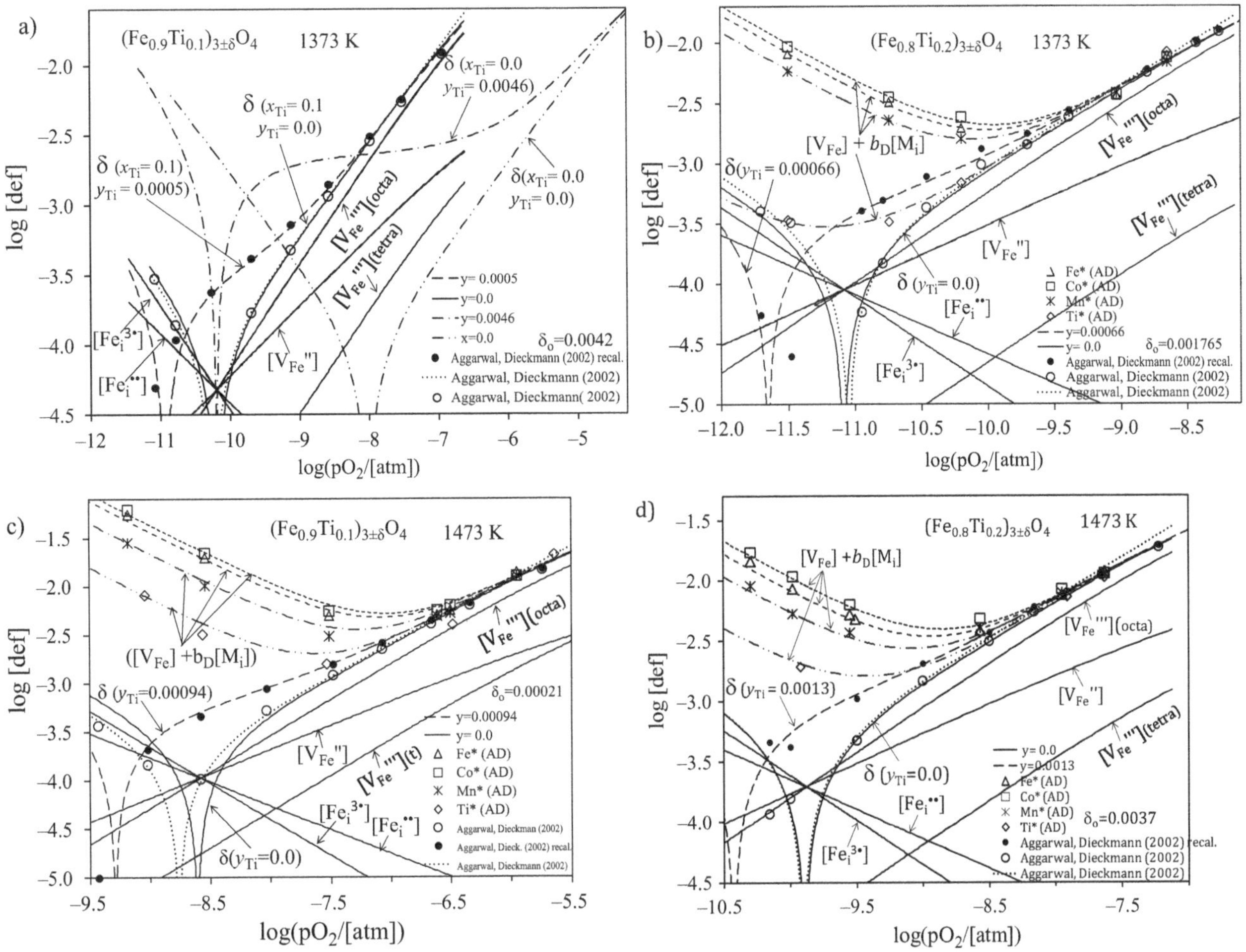

FIGURE 3.1 (a) Diagram of concentrations of point defects for the spinel $(Fe_{0.9}Ti_{0.1})_{3\pm\delta}O4$ at 1373K (solid lines), obtained using the results of the studies of the deviation from the stoichiometry taken from (Aggarwal and Dieckmann 2002a, Aggarwal and Dieckmann 2002b) ((○) points). The (····) line shows the oxygen pressure dependence of the deviation from the stoichiometry, determined using the values of "equilibrium constants" of reactions of formation of cation vacancies K_V and interstitial cations K_I, extracted from (Aggarwal and Dieckmann 2002a, Aggarwal and Dieckmann 2002b). Points (●) are the values recalculated relative to the value $\delta_o = 4.2 \cdot 10^{-3}$, and (---) line represents the dependence of δ on p_{O_2} for the spinel $(Fe_{0.9}Ti_{0.1})_{3-\delta}O_4$ doped with titanium ions $Ti^{•}_{Fe}$ in the amount of $y_{Ti}{}^{•}$ = $5 \cdot 10^{-4}$ mol/mol. Line (-···-) represents the dependence of the deviation from the stoichiometry for pure magnetite (Figure 1.3c), and line (-··-) – the p_{O_2} dependence of δ for magnetite doped with titanium ions $Ti^{•}_{Fe}$ in the amount of $y_{Ti}{}^{•} = 4.6 \cdot 10^{-3}$ mol/mol. (b) for the spinel $(Fe_{0.8}Ti_{0.2})_{3\pm\delta}O_4$ at 1373 K (solid lines). Points (●) are the recalculated values of the deviation from the stoichiometry relative to the value $\delta_o = 1{,}765 \cdot 10^{-3}$, and (- -) line represents the dependence of δ on p_{O_2} for concentration of titanium $y_{Ti}{}^{•} = 6.6 \cdot 10^{-4}$ mol/mol. The remaining points represent the values of the sum of concentrations of ionic defects ([V_{Fe}] + b_D[M_i]), calculated according to Equation 1.47a, 1.47b for fitted values of $D^{o}_{V(M)}$ for the highest oxygen pressures and when using coefficients of self-diffusion of tracers: Fe* – (△) points, Co* – (□), Mn* – (✳), Ti* – (◇), extracted from (Aggarwal and Dieckmann 2002a, Aggarwal and Dieckmann 2002b). Lines (---), (-·-·-), (-···-) denote, respectively, the oxygen pressure dependence of the sum of concentrations of ionic defects ([V_{Fe}] + b_D[M_i]) obtained for the fitted value of b_D. (c) for the spinel $(Fe_{0.9}Ti_{0.1})_{3\pm\delta}O_4$ at 1473 K (solid lines). Points (●) are the recalculated values of the deviation from the stoichiometry relative to the value $\delta_o = 2.1 \cdot 10^{-4}$, and (- -) line represents the dependence of δ on p_{O_2} for concentration of titanium $y_{Ti}{}^{•} = 9.4 \cdot 10^{-4}$ mol/mol. (d) for the spinel $(Fe_{0.8}Ti_{0.2})_{3\pm\delta}O_4$ at 1473 K (solid lines). Points (●) are the recalculated values of the deviation from the stoichiometry relative to the value $\delta_o = 3.7 \cdot 10^{-3}$, and (- -) line represents the dependence of δ on p_{O_2} for concentration of titanium $y_{Ti}{}^{•}$ = 1.3·10–3 mol/mol. (e) for the spinel $(Fe_{0.9}Ti_{0.1})_{3\pm\delta}O_4$ at 1573 K (solid lines). Points (●) are the recalculated values of the deviation from the stoichiometry relative to the value $\delta_o = 5.6 \cdot 10^{-3}$, and (- -) line represents the dependence of δ on p_{O_2} for concentration of titanium $y_{Ti}{}^{•} = 2.3 \cdot 10^{-3}$ mol/mol. (f) for the spinel $(Fe_{0.8}Ti_{0.2})_{3\pm\delta}O_4$ at 1573 K (solid lines). Points (●) are the recalculated values of the deviation from the stoichiometry relative to the value $\delta_o = 4.3 \cdot 10^{-3}$, and (- -) line represents the dependence of δ on p_{O_2} for concentration of titanium $y_{Ti}{}^{•} = 2.2 \cdot 10^{-3}$ mol/mol.

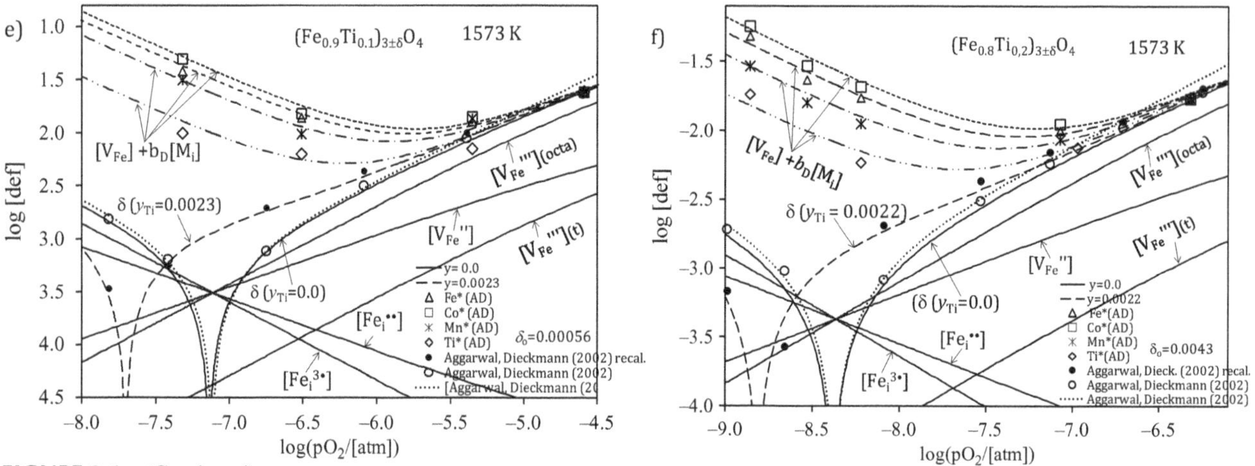

FIGURE 3.1 (Continued)

structure changes very significantly the character of the dependence of δ on p_{O_2}, which differs from experimental results (Aggarwal and Dieckmann 2002a, Aggarwal and Dieckmann 2002b). Changes in relative concentrations of point defects are small, but their differences cause a change in the deviation from the stoichiometry, which changes the shape of the dependence of δ on p_{O_2} in a log–log plot very significantly. As can be seen in Figure 3.1, symmetrical character of the dependence of δ on p_{O_2} in the spinel $(Fe_{1-x}Ti_x)_{3\pm\delta}O_4$ (Aggarwal and Dieckmann 2002a, Aggarwal and Dieckmann 2002b), indicates that titanium is incorporated into cation sites of Fe^{3+} ions and it forms a solid solution in the range of titanium content x_{Ti} = 0–0.333 mol, where titanium ions affect mainly the energy of interaction between ions.

As can be seen in Figure 3.1a–f, titanium-doped magnetite reaches the stoichiometric composition at the doped wüstite/spinel phase boundary. The fact that there is no range with higher concentrations of interstitial cations at low oxygen pressures causes that the fitting of the reference values δ_o can be ambiguous. The performed calculations showed that a small correction of the value of δ_o, the character of the dependence of δ on p_{O_2} changes and it becomes characteristic of spinel doped with a small amount of titanium ions in the form of $Ti_{Fe}^{\bullet}$. Figure 3.1a–f present the dependence of δ on p_{O_2} for the spinel $(Fe_{1-x}Ti_x)_{3\pm\delta}O_4$ with titanium content of x_{Ti} = 0.1 and 0.2 mol for the adjusted value of the concentration of ions $Ti_{Fe}^{\bullet}$ ($y_{Ti}^{\bullet}$) ((---) line). As can be seen in Figure 3.1a–f, for the spinel $(Fe_{1-x}Ti_x)_{3\pm\delta}O_4$ with titanium content of x_{Ti} = 0.1 and 0.2 mol, there is a very good match between the dependence of δ on p_{O_2} and the recalculated values of δ ((●) points) (values of δ_o, in relation to which the values of δ were calculated, shown in Figure 3.1a–f). For the titanium content of x_{Ti} = 0.25, correction of the value of δ_o changed the character of the dependence of δ on p_{O_2} only slightly (see Figure 3.3d). Figure 3.2 presents the fitted values of concentration of titanium ions $\left[Ti_{Fe}^{\bullet}\right]$ ($y_{Ti}^{\bullet}$), obtained for the spinel $(Fe_{1-x}Ti_x)_{3\pm\delta}O_4$ with titanium content of x_{Ti} = 0.1 and 0.2 mol at temperatures 1373–1573 K. The published values of the deviation from the stoichiometry

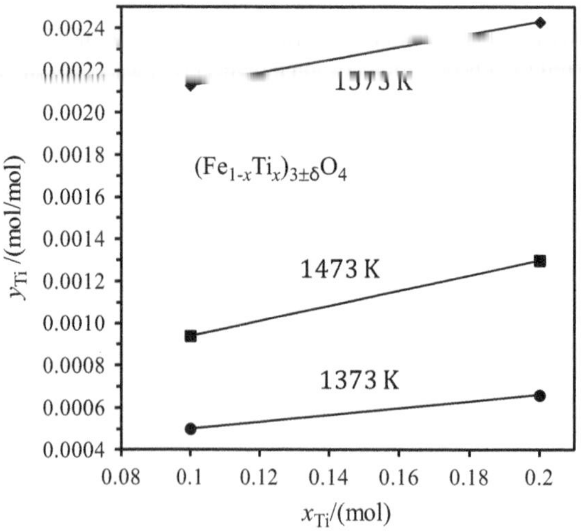

FIGURE 3.2 Dependence of the concentration of titanium ions $[Ti_{Fe}^{\bullet}] = 1/2y_{Ti}$ incorporated into the sublattice of ions Fe^{3+} on the titanium content x_{Ti} in the spinel $(Fe_{1-x}Ti_x)_{3\pm\delta}O_4$, at 1373–1573 K.

(Aggarwal and Dieckmann 2002a, Aggarwal and Dieckmann 2002b) do not exclude the possibility of doping $Ti^{\bullet}_{Fe}$ ions into the formed structure of the spinel $(Fe_{1-x}Ti_x)_{3\pm\delta}O_4$. Thus, Ti^{4+} titanium ions incorporate themselves into cation sites of the sublattice with octahedral coordination and they are electroneutral in relation to the lattice. However, a certain, small concentration of Ti^{4+} ions can be incorporated into the sublattice of Fe^{3+} ions with tetrahedral or octahedral coordination, which are positive in relation to the lattice $\left(Ti^{\bullet}_{Fe}\right)$.

Figure 3.3a compares the oxygen pressure dependences of deviation from the stoichiometry obtained at 1373–1573 K for pure magnetite and for $(Fe_{1-x}Ti_x)_{3\pm\delta}O_4$ spinel with titanium content of $x_{Ti} = 0.1$ mol, and Figure 3.3b,c – for the spinel $(Fe_{1-x}Ti_x)_{3\pm\delta}O_4$ with titanium content of $x_{Ti} = 0.2$ and 0.25 mol (curves fitted to the experimental values δ (Aggarwal and Dieckmann 2002a, Aggarwal and Dieckmann 2002b)) ((○) points).

In turn, Figure 3.3d presents the dependence of concentrations of defects for titanium content of = 0.3 mol, which were used for the interpretation of results of diffusion in doped magnetite. The values of formation of defects, for titanium content of = 0.3 mol, were extrapolated for the dependences presented in Figures. 3.10-3.13. In turn, Figure 3.4 compares the dependence of deviation from the stoichiometry on p_{O_2} obtained at 1473 K for $(Fe_{1-x}Ti_x)_{3\pm\delta}O_4$ spinel with titanium content of $x_{Ti} = 0.0$, 0.1, 0.2 and 0.25 mol.

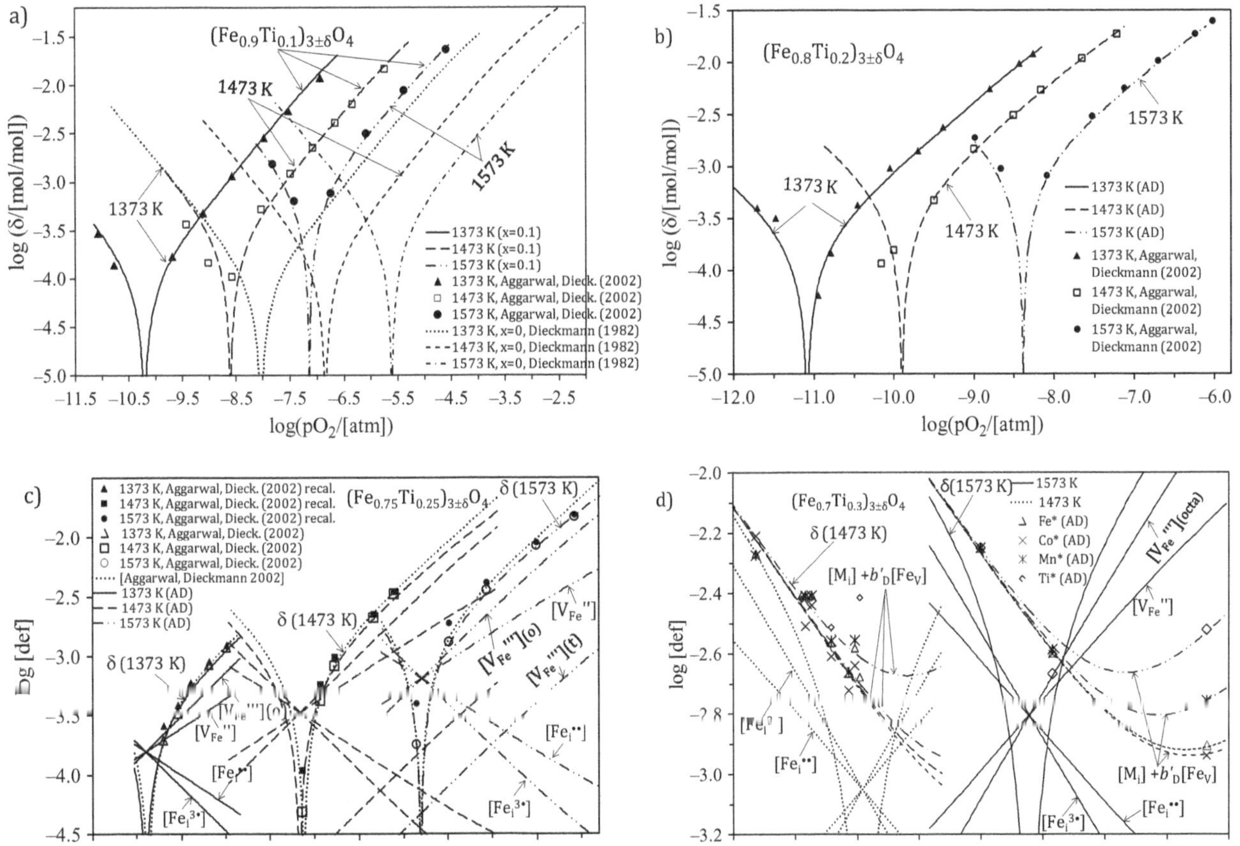

FIGURE 3.3 (a) Oxygen pressure dependence of the deviation from the stoichiometry in the spinel $(Fe_{0.9}Ti_{0.1})_{3\pm\delta}O_4$, at 1373–1573 K, obtained using the results of the studies of the deviation from the stoichiometry (Aggarwal and Dieckmann 2002a, Aggarwal and Dieckmann 2002b) ((▲,□,●) points) and for non-doped magnetite $Fe_{3-\delta}O_4$ ((···), (---),(-·-·-) line), obtained using the published results (Dieckmann 1982). (b) Dependence of the deviation from the stoichiometry for the spinel $(Fe_{0.8}Ti_{0.2})_{3\pm\delta}O_4$ on the oxygen pressure, at 1373–1573 K, obtained using the results of the studies of the deviation from the stoichiometry taken from (Aggarwal and Dieckmann 2002a, Aggarwal and Dieckmann 2002b) ((▲,□,●) points). (c) Diagram of concentrations of point defects for the spinel $(Fe_{0.75}Ti_{0.2})_{3\pm\delta}O_4$ at 1373–1573 K, obtained using the values of the deviation from the stoichiometry taken from (Aggarwal and Dieckmann 2002a, Aggarwal and Dieckmann 2002b) ((△,□,o) points). The (····) line shows the dependence of δ on p_{O_2} determined using "equilibrium constants" of reactions of formation of cation vacancies K_V and interstitial cations K_I, published in (Aggarwal and Dieckmann 2002a, Aggarwal and Dieckmann 2002b). Solid points – recalculated values of δ. (d) Diagrams of concentrations of point defects for the spinel $(Fe_{0.7}Ti_{0.3})_{3\pm\delta}O_4$, at 1473 K and 1573 K, obtained using extrapolated standard Gibbs energies of formation of defects, see Figures 3.7–3.9.

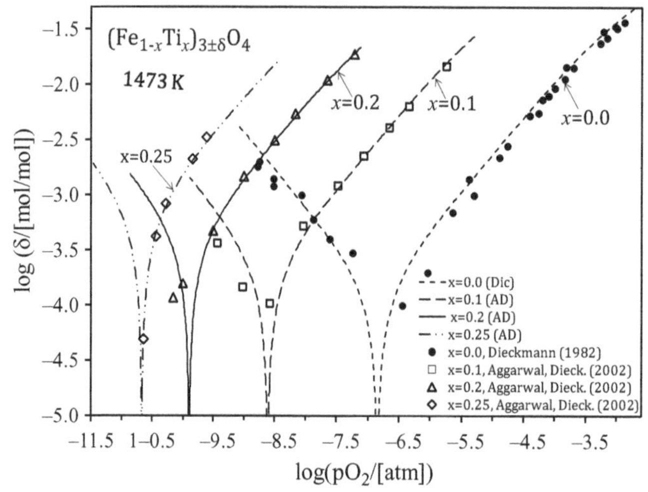

FIGURE 3.4 Oxygen pressure dependences of the deviation from the stoichiometry in the spinel $(Fe_{1-x}Ti_x)_{3\pm\delta}O_4$ with the titanium content of $x_{Ti} = 0.1$, 0.2 and 0.25 mol, at 1473 K, (presented in Figure 3.1) and for non-doped $Fe_{3\pm\delta}O_4$, obtained using the results of the studies of the deviation from the stoichiometry (Aggarwal and Dieckmann 2002a, Aggarwal and Dieckmann 2002b, Dieckmann 1982).

As can be seen in Figures 3.3a–d and 3.4, titanium doping of magnetite causes a very significant shift in the oxygen pressure at which the spinel reaches the stoichiometric composition towards lower pressures (in relation to pure magnetite). The dependence of δ on p_{O_2} is practically shifted in a parallel way. However, the range of oxygen pressures where the doped spinel exists, decreases; mainly, the range where interstitial cations dominate is shortened. The largest change in the range of existence can be observed by dopant content of $x_{Ti} = 0.25$ mol (see Figure 3.3c). When the temperature increases, the deviation from the stoichiometry (concentration of interstitial cation), at which there occurs a transformation of spinel into doped wüstite, increases.

As can be seen, titanium doping in wüstite, $Fe_{1-x}Ti_xO$, causes that iron oxide, also with a cubic structure, is transformed into spinel structure at a significantly lower pressure. Similarly, titanium-doped magnetite transforms into doped hematite at a lower pressure than pure magnetite. Therefore, titanium ions cause an increase in bonding energy, which is visible through, as mentioned above, lower limit oxygen pressures in comparison to pure magnetite. However, when titanium content increases in the spinel $(Fe_{1-x}Ti_x)_{3\pm\delta}O_4$, the lattice parameter a of the unit cell increases (Senderov et al. 1993,Wechsler et al. 1984), which indicates an increase in repulsive interactions between ions. As can be seen in Figure 3.4, at 1473 K, and similarly at other temperatures, when the content of titanium ions in the spinel $(Fe_{1-x}Ti_x)_{3\pm\delta}O_4$ increases, there occurs a decrease in the concentration of interstitial cations and shortening of the range of oxygen pressures in which these defects dominate. At a higher content of titanium, interstitial ions should be also titanium ions, $Ti_i^{4\bullet}$ or $Ti_i^{3\bullet}$, because their concentration in the octahedral sublattice is significant, and, additionally, a shortage of Fe^{3+} ions with octahedral coordination starts. Thus, titanium doping causes a decrease in the maximum concentration of interstitial ions, at which the transformation of spinel into doped wüstite occurs. The range of oxygen pressures where interstitial cations dominate is narrow and it shrinks when the dopant concentration increases and when the temperature decreases, which indicates an increase in repulsive interactions in titanium-doped magnetite.

As it was mentioned, when discussing defects in pure magnetite (Figure 1.3a), the studies of the deviation from the stoichiometry depending on the oxygen pressures are not suitable to decide on the concentration of cation vacancies in the sublattice with octahedral coordination. As can be seen in Figures 3.3a-d and 3.4, dependences of δ on p_{O_2} especially in the range where cation vacancies dominate, at given temperatures, independently of titanium content in the spinel $(Fe_{1-x}Ti_x)_{3\pm\delta}O_4$, are shifted practically in a parallel way in relation to pure magnetite. The maximum concentration of cation vacancies only slightly decreases when titanium content in the spinel increases, and it only slightly increases when the temperature increases. This indicates that titanium, being incorporated into the sublattice with octahedral coordination, takes the role of an Fe^{3+} ion, but its electrostatic interaction is much stronger, which is related to its charge (it has also a smaller ionic radius). It is also difficult to exclude incorporation of titanium ions into the sublattice with tetrahedral coordination (Gorter 1957, Foster and Hall 1965, Stout and Bayliss 1980). From other studies (Wechsler et al. 1984) it results that when the concentration of titanium in magnetite increases, the lattice parameter a increases, practically in a linear way, while the largest change occurs between oxygen ions and cation with tetrahedral coordination A (A–O) and oxygen, O–O, in the sublattice with tetrahedral coordination, which would indicate the presence of titanium ions (Senderov et al. 1993,Wechsler et al. 1984). The possibility of their presence in the sublattice with tetrahedral coordination is rejected based on the energy of interaction and the energy of stabilisation of ions in the crystal field (it is not known whether it is rejected rightly or not).

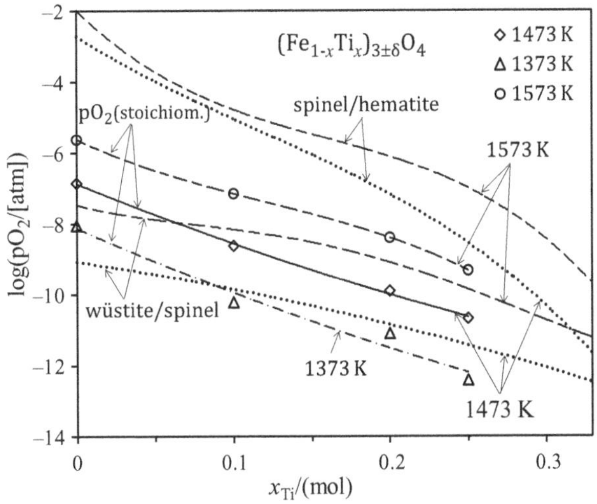

FIGURE 3.5 Dependence of the oxygen pressure at which the spinel $(Fe_{1-x}Ti_x)_3O_4$ reaches the stoichiometric concentration on the titanium content x_{Ti}, at 1373–1573 K, obtained using the published results of the studies on the deviation from the stoichiometry (Aggarwal and Dieckmann 2002a, Aggarwal and Dieckmann 2002b). Lines (---) and (····) denote equilibrium pressures at the phase boundary doped wüstite/spinel and spinel/hematite at 1473 K and 1573 K, taken from (Senderov et al. 1993, Webster and Bright 1961).

Figure 3.5 presents the dependence of oxygen pressure, at which spinels $(Fe_{1-x}Ti_x)_3O_4$ reach the stoichiometric composition, on the titanium content x_{Ti}.

As can be seen in Figure 3.5, these pressures decrease when the titanium content increases, and the dependences only slightly deviate from the linear relation. This is an effect of the presence of stronger repulsive interactions due to the incorporation of titanium into magnetite. In Figure 3.5, also the range of existence of doped magnetite at 1473–1573 K is marked, as given in (Senderov et al. 1993, Webster and Bright 1961). As can be seen in Figure 3.6, when the temperature increases, the oxygen pressure at which titanium-doped magnetite reaches the stoichiometric composition, increases linearly.

In turn, Figure 3.7 presents the dependence of the concentration of Frenkel defects at the stoichiometric composition depending on the titanium content x_{Ti} in $(Fe_{1-x}Ti_x)_{3\pm\delta}O_4$ spinel, at 1373–1573 K, and Figure 3.8 shows the temperature dependence.

As can be seen, when titanium content in the spinel increases, as well when the temperature increases, the concentration of Frenkel-type defects increases, which is a result of a decrease in the value of ΔG^{o}_{Fi} when temperature and titanium content increase (see Figure 3.10). The increase in the concentration of Frenkel defects when titanium content increases counteracts with the increase in bond energy (it reduces the repulsive interaction between ions).

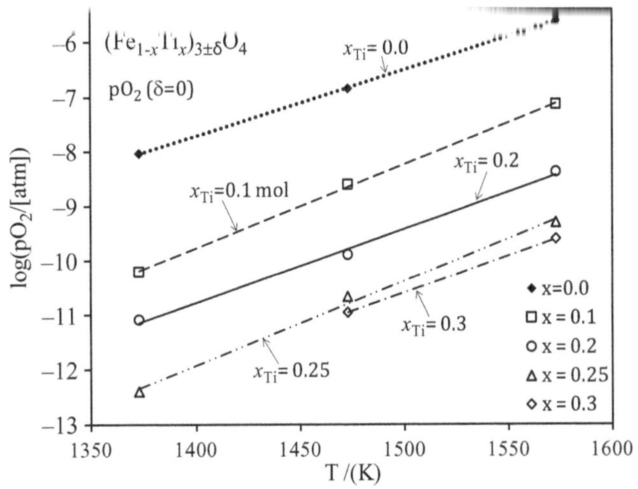

FIGURE 3.6 Temperature dependence of the oxygen pressure at which the spinel $(Fe_{1-x}Ti_x)_3O_4$ with the titanium content x_{Ti} = 0.0–0.25 mol reaches the stoichiometric composition.

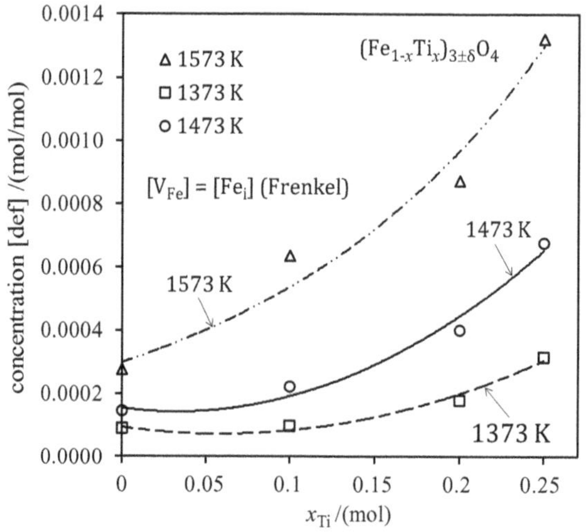

FIGURE 3.7 Dependence of the concentration of Frenkel defects in the spinel $(Fe_{1-x}Ti_x)_3O_4$ on the titanium content x_{Ti}, at 1373–1573 K.

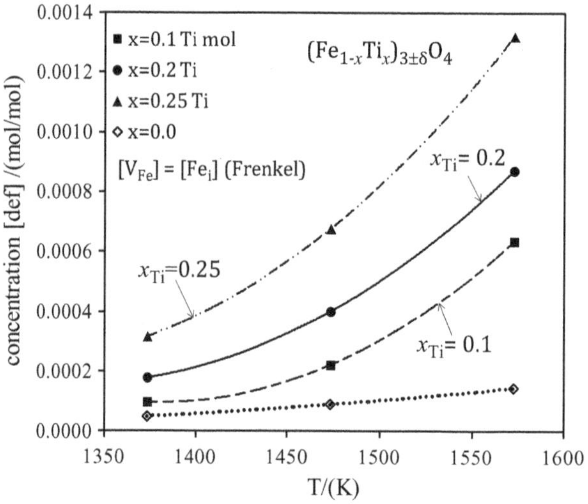

FIGURE 3.8 Temperature dependence of the concentration of Frenkel defects: in the spinel $(Fe_{1-x}Ti_x)_3O_4$ with the titanium content x_{Ti} = 0.1, 0.2, 0.25 mol.

In turn, Figure 3.9 shows the oxygen pressure dependence of the derivative $(d\log p_{O_2}/\log\delta) = n$, for the dependence of the deviation from the stoichiometry on p_{O_2} shown in Figure 3.1.

As can be seen in Figure 3.9, independently of titanium content, in a narrow range of higher oxygen pressures the parameter n reaches the value of 1.5 (1/n = 2/3); however, at the highest pressures it gets close to 2 (similarly as in pure magnetite). At low oxygen pressures n approaches −1.5, thus, these are the values similar to the ones adopted in the model of defects' structure (Aggarwal and Dieckmann 2002a, Aggarwal and Dieckmann 2002b).

Figures 3.10–3.12 present the dependence of adjusted standard Gibbs energies: $\Delta G^o_{F_1}$, $\Delta G^o_{F_2}$, $\Delta G^o_{F_3}$, $\Delta G^o_{V''_{Fe(o)}}$, ΔG^o_I on titanium content x_{Ti} in spinel $(Fe_{1-x}Ti_x)_{3\pm\delta}O_4$ at 1373–1573 K. Figures 3.13 and 3.14 show the dependence of standard Gibbs energy ($\Delta G^o_{V''_{Fe}}$) of formation of vacancies $V''_{Fe(o)}$ (calculated according to Equation (1.35d)), $\Delta G^o_{Fe_i^{3\bullet}}$ of the formation of interstitial iron ions $Fe_i^{3\bullet}$ (calculated according to Equation (1.34c)) and ($\Delta G^o_{Fe_i^{\infty}}$) of formation of ions $Fe_i^{\bullet\bullet}$ (calculated according to Equation (1.36c)) depending on titanium content x_{Ti} in spinel $(Fe_{1-x}Ti_x)_{3\pm\delta}O_4$.

As can be seen in Figure 3.10, standard Gibbs energy $\Delta G^o_{F_1}$ of formation of Frenkel defects decreases when the titanium content in spinel $(Fe_{1-x}Ti_x)_{3\pm\delta}O_4$ increases. Slight differences occur at 1373 K and 1473 K, a larger difference occurs at 1573 K at the content of x_{Ti} = 0.1 mol. In turn, as can be seen in Figures 3.11 and 3.13, when titanium content increases, the values ($\Delta G^o_{V''_{Fe}}$) and ($\Delta G^o_{V''_{Fe}}$) decrease, while the change is practically linear. As can be seen in Figure 3.14, the nature of the

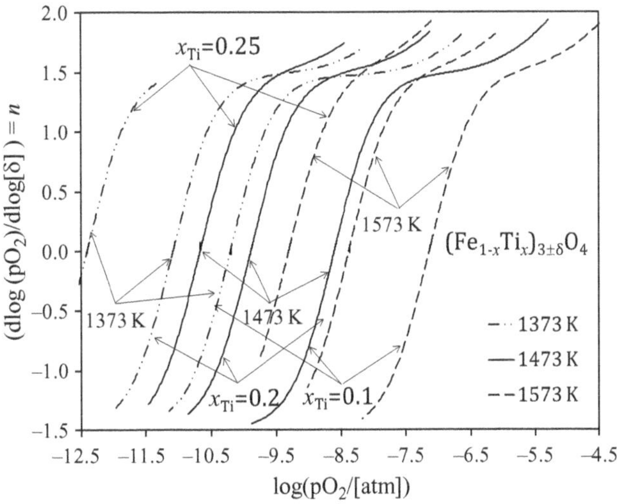

FIGURE 3.9 Oxygen pressure dependence of the derivative: $n_\delta = dlog(p_{O_2})/dlog(\delta)$, obtained for the dependence of the deviation from the stoichiometry on p_{O_2} in the spinel $(Fe_{1-x}Ti_x)_3O_4$ with the titanium content x_{Ti} = 0.1, 0.2, 0.25 mol, at 1373–1573 K (presented in Figure 3.1).

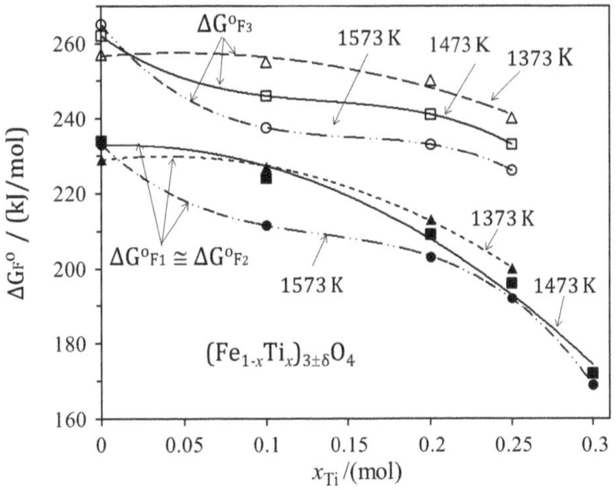

FIGURE 3.10 Dependence of the standard Gibbs energy $\Delta G^o_{F_1} \cong \Delta G^o_{F_2}$ (solid points) and $\Delta G^o_{F_3}$ (empty points) of formation of Frenkel defects with the charge of two and three on the titanium content x_{Ti} in the spinel $(Fe_{1-x}Ti_x)_{3\pm\delta}O_4$, at 1373–1573 K, obtained using the results of the studies of the deviation from the stoichiometry published in (Aggarwal and Dieckmann 2002a, Aggarwal and Dieckmann 2002b).

change in $\Delta G^o_{Fe_i^{3\bullet}}$ and $\Delta G^o_{Fe_i^{\bullet\bullet}}$ is inverse, they increase when titanium content in the spinel increases. Figures 3.11 and 3.14 show $\Delta G^o_{V_{Fe}}$ and $\Delta G^o_{Fe_i}$ calculated using the published values of "equilibrium constants" of reaction of formation of cation vacancies K_V and interstitial cations K_I (Aggarwal and Dieckmann 2002a, Aggarwal and Dieckmann 2002b). These values are different from those determined in the present work, as they describe another model of defects (another defect equation), but the character of the dependence on the titanium content x_{Ti} is similar.

The values of the above thermodynamic functions marked for a spinel with titanium content of x_{Ti} = 0.3 mol/mol, are extrapolated values which were used to determine the defect diagrams for the interpretation of the results of diffusion of metal ions in doped magnetite (see Figure 3.3d).

In turn, Figures 3.15–3.19 present temperature dependences of standard Gibbs energies $\Delta G^o_{F_1}$, $\Delta G^o_{F_2}$, $\Delta G^o_{F_3}$, $\Delta G^o_{V'''_{Fe(o)}}$, ΔG^o_{I}, $\Delta G^o_{V''_{Fe}}$, $\Delta G^o_{Fe_i^{3\bullet}}$, $\Delta G^o_{Fe_i^{\bullet\bullet}}$ of formation of individual defects in spinel $(Fe_{1-x}Ti_x)_{3\pm\delta}O_4$ with titanium content of x_{Ti} = 0.1, 0.2, 0.25 mol.

Table 3.1 compares the parameters of the obtained linear relations which indicate the enthalpy and entropy of formation of the defects. It should be noted that the above thermodynamic values depend on the assumed value of ΔG^o_{eh} and its

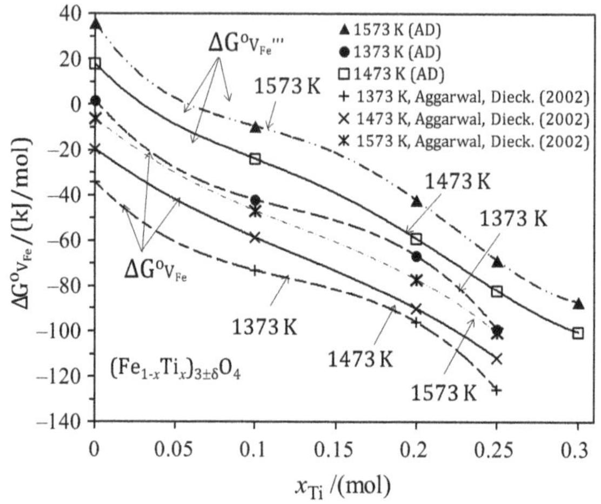

FIGURE 3.11 Dependence of standard Gibbs energy $\Delta G^{o}_{V_{Fe}'''}$ of formation of vacancies $\left(V'''_{Fe(o)}\right)$ on the titanium content x_{Ti} in the spinel $(Fe_{1-x}Ti_x)_{3\pm\delta}O_4$, at 1373–1573 K and the values $\Delta G^{o}_{V_{Fe}}$ calculated using the values of "equilibrium constants" of reaction of formation of vacancies K_V, taken from (Aggarwal and Dieckmann 2002a, Aggarwal and Dieckmann 2002b).

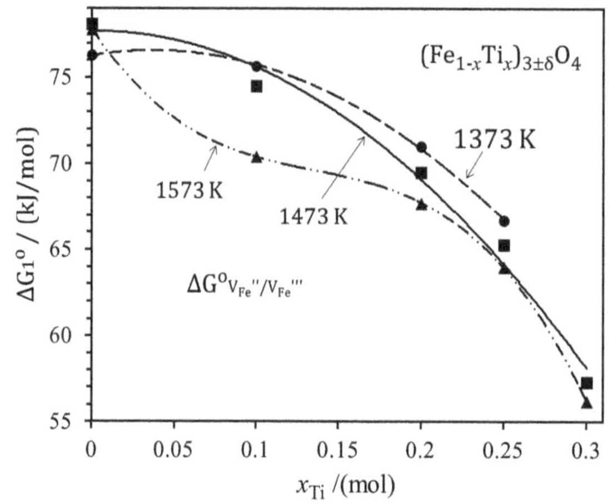

FIGURE 3.12 Dependence of standard Gibbs energy ΔG^{o}_{1} on the titanium content x_{Ti} in the spinel $(Fe_{1-x}Ti_x)_{3\pm\delta}O_4$, at 1373–1573 K, permitting the determination of the ratio of concentrations of iron vacancies $\left[V'''_{Fe(o)}\right]/\left[V''_{Fe(o)}\right]$.

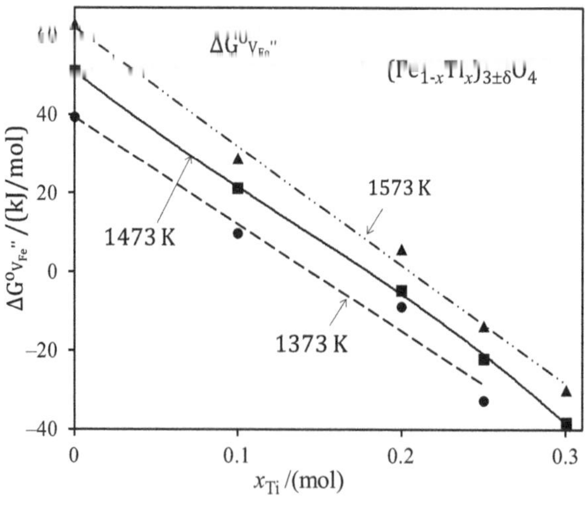

FIGURE 3.13 Dependence of standard Gibbs energy $\Delta G^{o}_{V''_{Fe}}$ of formation of vacancies $\left(V''_{Fe(o)}\right)$ on the titanium content x_{Ti} in the spinel $(Fe_{1-x}Ti_x)_{3\pm\delta}O_4$, at 1373–1573 K (calculated according to Equation 1.35d).

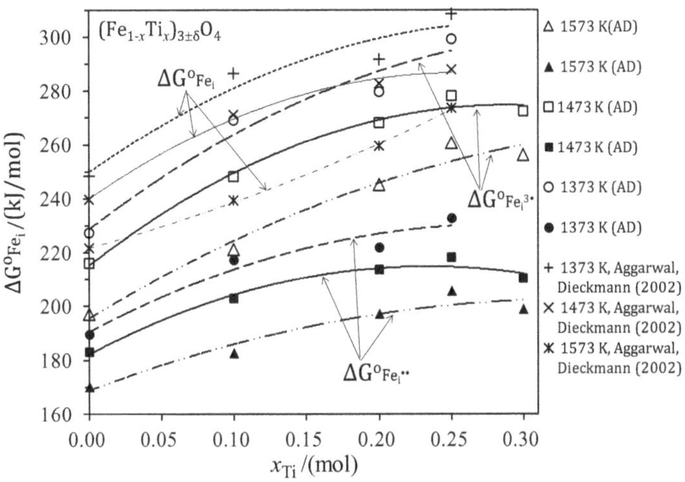

FIGURE 3.14 Dependence of the standard Gibbs energy $\Delta G^o_{Fe_i^{3\bullet}}$ of formation of interstitial ions $Fe_i^{3\bullet}$ (calculated according to Equation 1.34c) and $\Delta G^o_{Fe_i^{\bullet\bullet}}$ of formation of ions $Fe_i^{\bullet\bullet}$ (calculated according to Equation 1.36c) on the titanium content x_{Ti}, in the spinel $(Fe_{1-x}Ti_x)_{3\pm\delta}O_4$, at 1373–1573 K and the values $\Delta G^o_{Fe_i}$ calculated using the values of "equilibrium constants" of reaction of formation of interstitial iron ions K_I, taken from (Aggarwal and Dieckmann 2002a, Aggarwal and Dieckmann 2002b).

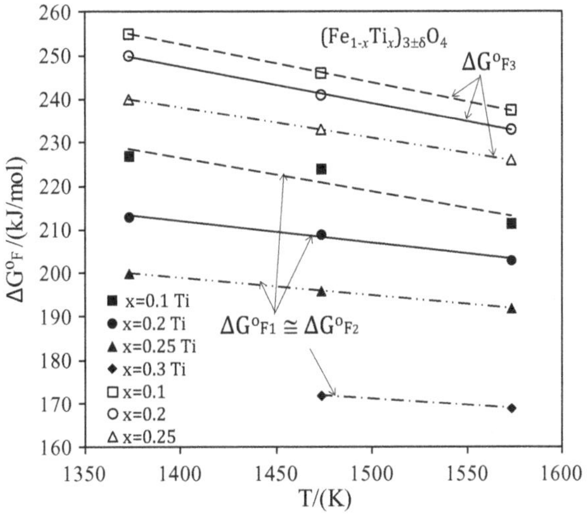

FIGURE 3.15 Temperature dependence of standard Gibbs energy $\Delta G^o_{F_1} = \Delta G^o_{F_2}$ (solid points) and $\Delta G^o_{F_3}$ (empty points) of formation of Frenkel defects in the spinel $(Fe_{1-x}Ti_x)_{3\pm\delta}O_4$ with the titanium content of $x_{Ti} = 0.1$, 0.2 and 0.25 mol.

temperature dependence, which was not taken into account in the present calculations (determined for the value of $\Delta G^o_{eh} = 30.3$ kJ/mol).

3.4 DIFFUSION OF METAL IONS IN TITANIUM-DOPED MAGNETITE

Similarly as for pure magnetite, using the concentration of ionic defects resulting from the diagram of defects' concentrations and the values of coefficients of self-diffusion of tracers Fe*, Co*, Mn* and Ti* in doped magnetite, coefficients of diffusion (mobility) of ions via cation vacancies and via interstitial cations were determined (see Section 1.6.2).

Figure 3.1 presents the dependence of the "sum" of concentrations of ionic defects ($[V_{Fe}] + b_D[M_i]$) on p_{O_2} for fitted values of b_D (dashed lines) and the values of this sum ($[V_{Fe}] + b_D[M_i]$) ((△,□,✱,◇) points, respectively) calculated according to Equation 1.47a using the values of coefficients of self-diffusion of iron ^{59}Fe, cobalt ^{60}Co, manganese ^{53}Mn and titanium ^{44}Ti (Aggarwal and Dieckmann 2002a, Aggarwal and Dieckmann 2002b). For the highest oxygen pressures, the values of the coefficient of diffusion of cations $D^o_{V(M)}$ via cation vacancies and the parameter b_D, being the ratio of coefficients of diffusion via defects ($D^o_{I(M)}/D^o_{V(M)}$) were adjusted in such a way as to obtain a match between the dependence of

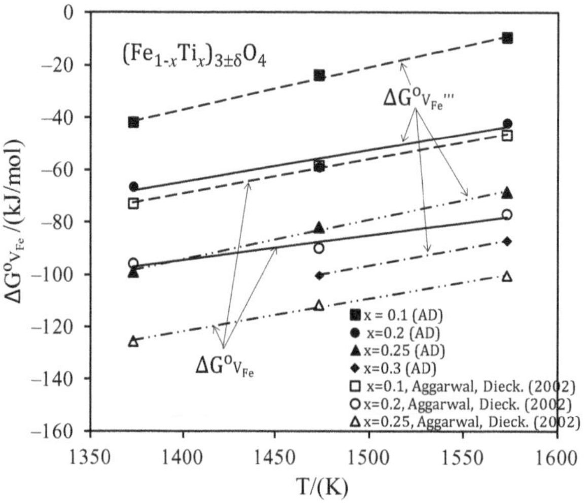

FIGURE 3.16 Temperature dependence of standard Gibbs energy $\Delta G^o_{V'''_{Fe}}$ of formation of vacancies $\left(V'''_{Fe(o)}\right)$ in the spinel $(Fe_{1-x}Ti_x)_{3\pm\delta}O_4$ with the titanium content of x_{Ti} = 0.1, 0.2 and 0.25 mol, and the values $\Delta G^o_{V_{Fe}}$ calculated using the values of "equilibrium constants" of reaction of formation of vacancies K_V, taken from (Aggarwal and Dieckmann 2002a, Aggarwal and Dieckmann 2002b).

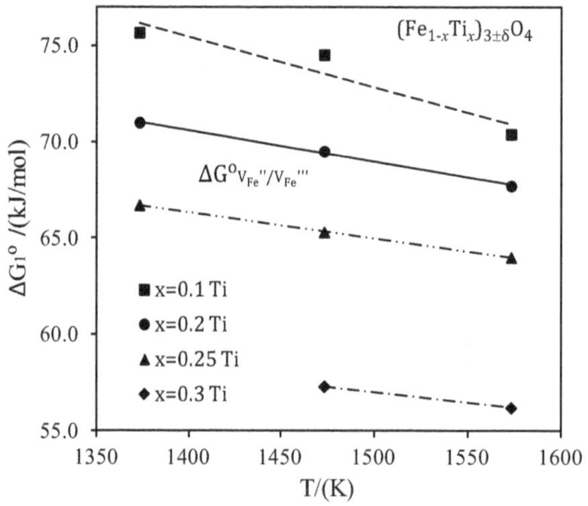

FIGURE 3.17 Temperature dependence of standard Gibbs energy ΔG_1^o in the spinel $(Fe_{1-x}Ti_x)_{3\pm\delta}O_4$ with the titanium content of x_{Ti} = 0.1, 0.2, 0.25 and 0.3 mol, permitting determination of the ratio of concentrations of iron vacancies $\left[V'''_{Fe(o)}\right]/\left[V''_{Fe(o)}\right]$ (according to Equation 1.35b).

the sum of the concentrations of defects ($[V_{Fe}] + b_D[M_i]$) on p_{O_2} with the values based on the coefficients of diffusion of radioactive tracers. As can be seen in Figure 3.1, a very good agreement was obtained between the dependence of the sum of the concentrations of ionic defects ($[V_{Fe}] + b_D[M_i]$) on p_{O_2} for the fitted value of b_D and the values of this sum determined using the coefficients of diffusion of tracers and the fitted value of $D^o_{V(M)}$. As can be seen in Figure 3.1, the minimum of the dependence of ($[V_{Fe}] + b_D[M_i]$) on p_{O_2} is shifted in relation to the oxygen pressure at which the spinel reaches the stoichiometric composition. The extent of this shift depends on the value of the ratio ($D^o_{I(M)}/D^o_{V(M)} = b_D$).

Figure 3.20 presents the dependence of the parameter b_D on titanium content in spinel $(Fe_{1-x}Ti_x)_{3\pm\delta}O_4$ for given tracers, at 1373–1573 K.

As can be seen in Figure 3.20, the ratio of the coefficients of diffusion via interstitial cations to the coefficients of diffusion via interstitial cations changes with titanium content and temperature. The largest increase and the largest differences in the parameter b_D occur at the titanium content of x_{Ti} = 0.1 mol (at the temperature of 1473 K; for cobalt b_D = 110; for iron, 95; for manganese, 55; and for titanium only 19), which significantly decrease at the content of x_{Ti} = 0.2 mol and for the content of 0.3 mol change only slightly. For the titanium content of x_{Ti} = 0.1 mol, the values of b_D are higher at 1473 K than at 1573 K, while at a higher titanium content the effect is opposite.

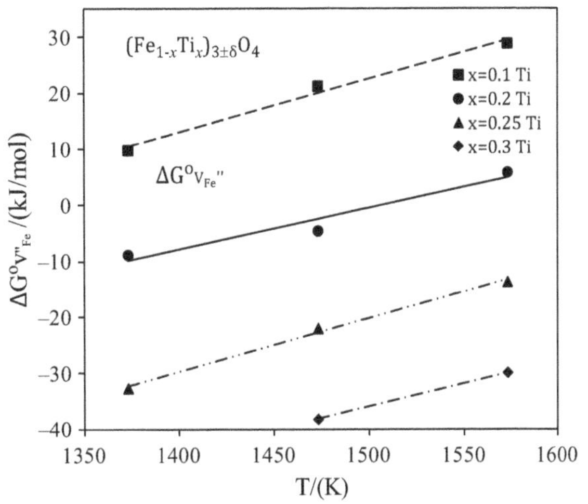

FIGURE 3.18 Temperature dependence of standard Gibbs energy $\Delta G^o_{V''_{Fe}}$ of formation of vacancies $\left(V''_{Fe(o)}\right)$ in the spinel $(Fe_{1-x}Ti_x)_{3\pm\delta}O_4$ with the titanium content x_{Ti} = 0.1, 0.2, 0.25 and 0.3 mol.

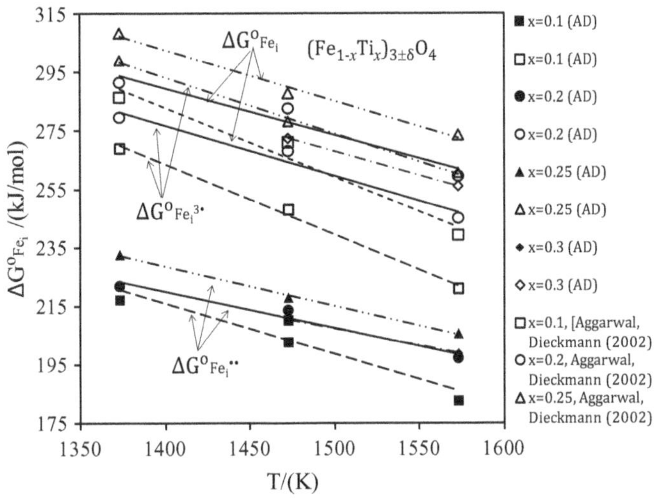

FIGURE 3.19 Temperature dependence of standard Gibbs energy $\Delta G^o_{Fe_i^{3\bullet}}$ and $\Delta G^o_{Fe_i^{\bullet\bullet}}$ of formation of interstitial ions $Fe_i^{3\bullet}$ and $Fe_i^{\bullet\bullet}$ in the spinel $(Fe_{1-x}Ti_x)_{3\pm\delta}O_4$ with the titanium content of x_{Ti} = 0.1, 0.2, 0.25 and 0.3 mol, and the values $\Delta G^o_{Fe_i}$ calculated using the values of "equilibrium constants" of reaction of formation of interstitial iron ions K_i, taken from (Aggarwal and Dieckmann 2002a, Aggarwal and Dieckmann 2002b).

In turn, Figure 3.21 presents the dependence of the coefficients of diffusion of tracer ions via cation vacancies $D^o_{V(M)}$ on the titanium content x_{Ti} in spinel $(Fe_{1-x}Ti_x)_{3\pm\delta}O_4$, which were determined for the highest value of the oxygen pressure.

As can be seen in Figure 3.21, in the case of diffusion of ions Fe*, Co* and Mn*, the coefficients of diffusion via cation vacancies at the same temperature differ only slightly. At the titanium content x_{Ti} = 0.1 mol they are lower than in pure magnetite, at the titanium content x_{Ti} = 0.2 mol they increase significantly, while at the titanium content x_{Ti} = 0.3 mol they are lower. Differences in the values of $D^o_{V(M)}$ can result from the accuracy of their determination (small number of measured values).

As can be seen in Figure 3.21, the mobility of titanium ions Ti* via cation vacancies $D^o_{V(Ti)}$ is significantly small, but the character of its change with the titanium content is similar

Figure 3.22 shows the dependence of coefficients of diffusion of tracer ions via interstitial cations $D^o_{I(M)}$ on titanium content x_{Ti} in spinel $(Fe_{1-x}Ti_x)_{3\pm\delta}O_4$, which, according to the values of the parameter b_D, are by over one order of magnitude higher than the coefficients of diffusion via vacancies.

Similarly, there are small differences between mobilities of ions Fe*, Co* and Mn*, however, they are a little larger than in the case of the diffusion of ions via vacancies. At the titanium content of x_{Ti} = 0.1 mol the mobility of ions is higher than

TABLE 3.1
Values of Enthalpies ΔH_i^o and Entropies ΔS_i^o of Formation of Intrinsic Frenkel Defects, Iron Vacancies V_{Fe}'' and V_{Fe}''', Interstitial iron ions $Fe_i^{\bullet\bullet}$ and $Fe_i^{3\bullet}$ in the Spinel $(Fe_{1-x}Ti_x)_{3\pm\delta}O_4$, Determined based on the Dependences shown in Figures 3.15–3.19.

Equation	x_{Ti} (mol)	ΔH_i^o (kJ/mol)	ΔS_i^o (J/(mol·K))	ΔT (K)
$\left[V_{Fe(o)}'''\right]\left[Fe_i^{3\bullet}\right] = K_{F_1}$	0.1	335 ± 40	78 ± 27	1373–1573
Equation 1.34b	0.2	282 ± 9	50 ± 6	1373–1573
-”-	0.25	255	40	1373–1573
-”-	0.3	216	30	1473–1573
$\left[V_{Fe(t)}''\right]\left[Fe_i^{\bullet\bullet}\right] = K_{F_2}$	0.1	375 ± 2	88 ± 1	1373–1573
Equation 1.36b	0.2	367 ± 4	85 ± 3	1373–1573
-”-	0.25	336	70	1373–1573
$V_{Fe(o)}'''$	0.1	−265 ± 15	−163 ± 10	1373–1573
Equation 1.9b	0.2	−236 ± 40	−123 ± 27	1373–1573
-”-	0.25	−308 ± 15	−153 ± 10	1373–1573
-”-	0.3	−296	−133	1473–1573
$V_{\mathbf{Fe}}$ (Aggarwal and Dieckmann 2002a, Aggarwal and Dieckmann 2002b)	0	−228 ± 3	−141 ± 2	1373–1573
-”-	0.1	−253 ± 12	−1318 ± 8	1373–1573
-”-	0.2	−227 ± 31	−95 ± 21	1373–1573
-”-	0.25	−298 ± 11	−126 ± 7	1373–1573
$K_{V_{Fe(o)}''/V_{Fe(o)}'''} = K_1 s$	0.1	113 ± 12	27 ± 8	1373–1573
Equation 1.35b	0.2	94 ± 1	17 ± 1	1373–1573
-”-	0.25	85	14	1373–1573
-”-	0.3	74	−11	1473–1573
$V_{Fe(o)}''$	0.1	−120 ± 16	−95 ± 11	1373–1573
Equation 1.9a	0.2	−111 ± 26	−73 ± 18	1373–1573
-”-	0.25	−163 ± 10	−95 ± 7	1373–1573
-”-	0.3	−161	−83	1473–1573
$Fe_i^{\bullet\bullet}$	0.1	455 ± 24	173 ± 16	1373–1573
Equation 1.14a	0.2	393 ± 35	123 ± 24	1373–1573
-”-	0.25	418 ± 10	135 ± 7	1373–1573
-”-	0.3	377	113	1473–1573
$Fe_i^{3\bullet}$	0.1	600 ± 26	240 ± 17	1373–1573
Equation 1.14b	0.2	610 ± 19	173 ± 33	1373–1573
-”-	0.25	562 ± 16	133 ± 10	1373–1573
-”-	0.3	512	163	1473–1573
$Fe_{\mathbf{i}}$ (Aggarwal and Dieckmann 2002a, Aggarwal and Dieckmann 2002b)	0	433 ± 40	133 ± 27	1373–1573
-”-	0.1	613 ± 70	236 ± 47	1373–1573
-”-	0.2	514 ± 61	160 ± 41	1373–1573
-”-	0.25	547 ± 27	174 ± 18	1373–1573

that in pure magnetite and it decreases only slightly when titanium content increases. Co and Fe ions diffuse with the highest speed, Mn diffuses slower.

The mobility of titanium ions (coefficients of diffusion of titanium $D_{I(Ti)}^o$) via cation vacancies is lower by over an order of magnitude, but the character of the dependence on the titanium content is analogous to other ions. This might be related to the radius of diffusing titanium ions and their effective charge (with electrostatic field interacting with adjacent ions during their movement).

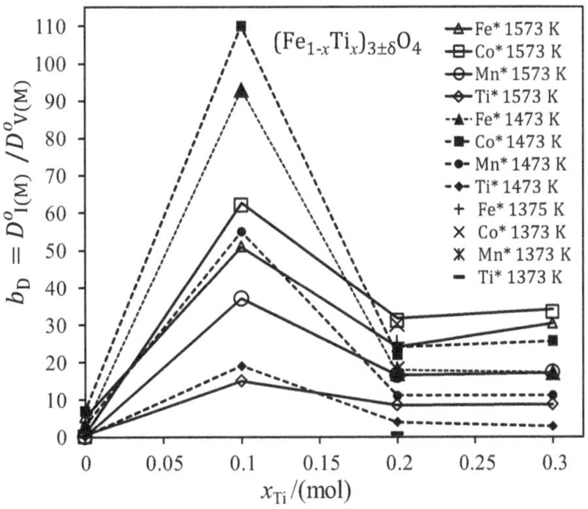

FIGURE 3.20 Dependence of the ratio of the coefficient of diffusion of tracer ions via interstitial cations to the coefficient of diffusion of cations via vacancies $D^o_{I(M)}/D^o_{V(M)}$ (parameter b_D) on the titanium content x_{Ti} in the spinel $(Fe_{1-x}Ti_x)_{3\pm\delta}O_4$ at 1373–1573 K, obtained using the values of coefficients of self-diffusion of tracers: Fe* ((+,▲,△) points), Co* (×,■,□), Mn* (✳,●,○), Ti* (△,◆,◇) published in (Aggarwal and Dieckmann 2002a, Aggarwal and Dieckmann 2002b, Dieckmann et al. 1978) and the concentration of ionic defects.

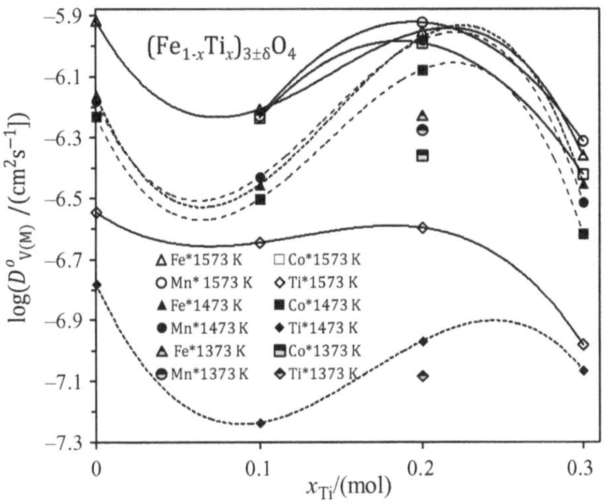

FIGURE 3.21 Dependence of the coefficient of diffusion of tracer ions via cation vacancies $D^o_{V(M)}$ on the titanium content x_{Ti} in the spinel $(Fe_{1-x}Ti_x)_{3\pm\delta}O_4$ at 1373–1573 K, using the values of coefficients of self-diffusion of tracers: Fe* ((▲, △, ⬒) points), Co* (■,□,◒), Mn* (●,○,◓), Ti* (◆,◇,△) taken from (Aggarwal and Dieckmann 2002a, Aggarwal and Dieckmann 2002b, Dieckmann et al. 1978) and the concentration of ionic defects.

In turn, Figures 3.23 and 3.24 present the temperature dependence of coefficients of diffusion $D^o_{V(M)}$ and $D^o_{I(M)}$ in spinel $(Fe_{1-x}Ti_x)_{3\pm\delta}O_4$ and it is compared with mobilities of iron and titanium in pure magnetite (shown in Figure 1.22).

As can be seen in Figures 3.23 and 3.24, quite a large deviation occurs of the values from the linear relation in the above system, which is associated with a measurement error of coefficients of self-diffusion of ions and the calculated concentration of defects and, above all, a small number of measured values. Table 3.2 compares the approximated values of the activation enthalpy of diffusion (mobility) of tracer metal ions via cation vacancies and interstitial cations, determined from the slopes of the obtained linear values. As can be seen in Table 3.2, for the diffusion via cation vacancies, the activation enthalpy of the mobility in pure magnetite is about 100 kJ/mol, while in the spinel $(Fe_{1-x}Ti_x)_{3\pm\delta}O_4$ with a titanium content of $x_{Ti} = 0.2$ it is lower by about 20 kJ/mol. For the diffusion via interstitial cations in pure magnetite it is twice as high, about 200 kJ/mol, and at titanium content of $x_{Ti} = 0.2$ mol it is close to the activation energy via cation vacancies. Activation enthalpy of the mobility of titanium differs, it increases up to 320 kJ/mol. Due to a small number of measured values for the titanium concentration of $x_{Ti} = 0.1$ and 0.3 mol in $(Fe_{1-x}Ti_x)_{3\pm\delta}O_4$, the values of activation energy of the mobility of ions are difficult to interpret.

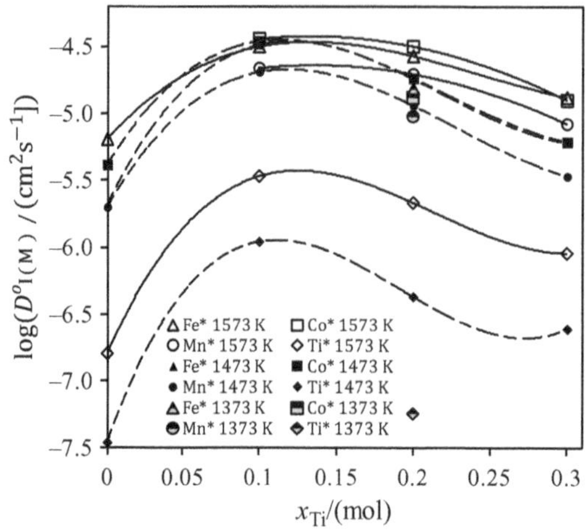

FIGURE 3.22 Dependence of the coefficient of diffusion of tracer ions via interstitial cations $D^o_{I(M)}$ on the titanium content x_{Ti} in the spinel $(Fe_{1-x}Ti_x)_{3\pm\delta}O_4$ at 1373–1573 K, using the values of coefficients of self-diffusion of tracers: Fe* ((▲,△,▬) points), Co* (■,□,◓), Mn* (●,○,⬘), Ti* (◆,◇,△) published in (Aggarwal and Dieckmann 2002a, Aggarwal and Dieckmann 2002b, Dieckmann et al. 1978) and the concentration of ionic defects resulting from diagrams.

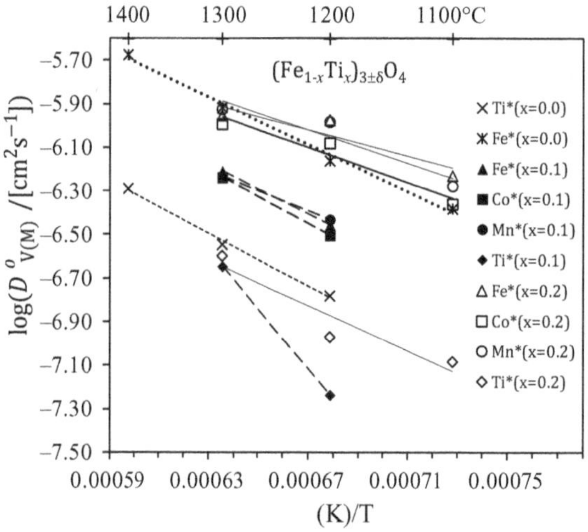

FIGURE 3.23 Temperature dependence of the coefficient of diffusion of tracer ions via cation vacancies $D^o_{V(M)}$ in the spinel $(Fe_{1-x}Ti_x)_{3\pm\delta}O_4$ with the titanium content of x_{Ti} = 0.0, 0.1, 0.2 and 0.3 mol, obtained using the values of coefficients of self-diffusion of tracers: Fe* ((✶,▲,△,▬) points), Co* (■,□,◓), Mn* (●,○,⬘), Ti* (×,◆,◇) taken from (Aggarwal and Dieckmann 2002a, Aggarwal and Dieckmann 2002b, Dieckmann et al. 1978) and the concentration of ionic defects resulting from diagrams.

3.5 CONCLUSIONS

Using the results of the studies of the deviation from the stoichiometry in spinel $(Fe_{1-x}Ti_x)_{3\pm\delta}O_4$ (Aggarwal and Dieckmann 2002a, Aggarwal and Dieckmann 2002b), diagrams of defects' concentrations were determined at 1373–1673 K, for the titanium content x_{Ti} = 0.1, 0.2 and 0.25 mol. It was found that titanium doping in magnetite only slightly affects the character of the dependence of the deviation from the stoichiometry on the oxygen pressure and these dependences are practically shifted in parallel towards lower oxygen pressures. Spinel $(Fe_{1-x}Ti_x)_{3\pm\delta}O_4$ reaches the stoichiometric composition in the range of its existence, while when titanium content increases, the oxygen pressure at which spinel reaches the stoichiometric composition is shifted towards lower values by about three orders of magnitude for the content of x_{Ti} = 0.25 mol. When titanium content increases, the range of oxygen pressures, in which the doped spinel exists, is reduced, while main reduction occurs in the range where interstitial cations dominate (in relation to pure magnetite). The largest change in the range of existence can be observed at the titanium content of x_{Ti} = 0.25 mol. When the temperature increases, the pressure at which spinels reach the stoichiometric composition is shifted towards higher pressures by a similar value as that for pure magnetite.

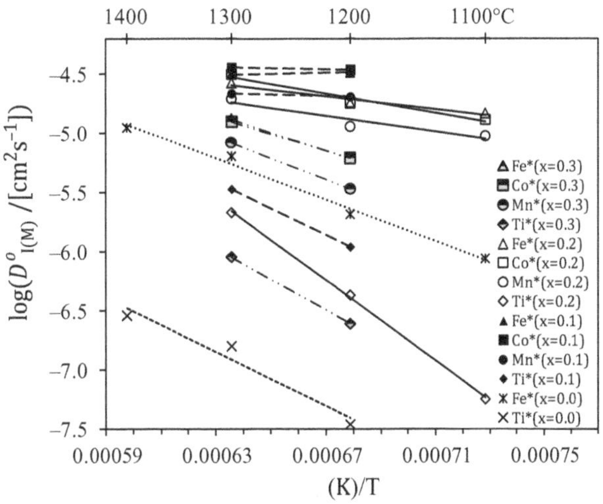

FIGURE 3.24 Temperature dependence of the coefficient of diffusion of tracer ions via interstitial cations $D^o_{I(M)}$ in the spinel $(Fe_{1-x}Ti_x)_{3\pm\delta}O_4$ with the titanium content of $x_{Ti} = 0.0, 0.1, 0.2$ and 0.3 mol, obtained using the values of coefficients of self-diffusion of tracers: Fe* ((✳,▲,△,◭) points), Co* (■,□,◧), Mn* (●,○,◒), Ti* (×,◆,◇,⬖) taken from (Aggarwal and Dieckmann 2002a, Aggarwal and Dieckmann 2002b, Dieckmann et al. 1978) and the concentration of ionic defects resulting from diagrams.

The structure of point defects practically does not change. In the range of higher oxygen pressures, cation vacancies $\left[V'''_{Fe(o)}\right]$ dominate, and, in a narrow range of low oxygen pressures, interstitial cations $\left[Fe_i^{3\bullet}\right]$. The concentration of vacancies $\left[V''_{Fe}\right]$ and $\left[Fe_i^{\bullet\bullet}\right]$ is significantly lower, as well as the concentration of vacancies $\left[V'''_{Fe(t)}\right]$ in the sublattice with tetrahedral coordination.

It was shown that it is possible to adjust a reference value δ_o which causes a change in the character of the oxygen pressure dependence of the deviation from the stoichiometry, indicating the occurrence of doping with Ti^{4+} ions in the sublattice of ions Fe^{3+} and/or Fe^{2+}. The result of this type of doping is the presence of ions $Ti^{\bullet}_{Fe}$ and $Ti^{\bullet\bullet}_{Fe}$ positive in relation to the lattice. Maximum concentration of these ions should be $y_{Ti}^{\bullet} = 1{\cdot}10^{-3}$ mol/mol at 1473 K and $y_{Ti}^{\bullet} = 2{\cdot}10^{-3}$ mol/mol at 1573 K.

It was found that titanium doping causes a decrease in the value of the standard Gibbs energy of formation of Frenkel defects, which causes an increase in the concentration of Frenkel defects (at the stoichiometric composition). The determined enthalpy of the formation of Frenkel defects in the spinel $(Fe_{1-x}Ti_x)_{3\pm\delta}O_4$ with titanium content of $x_{Ti} = 0.1$ mol is $\Delta H^o_{F1} = 335$ kJ/mol and it decreases to 255 kJ/mol at the content of $x_{Ti} = 0.25$ mol.

The standard Gibbs energies of formation of cation vacancies $\Delta G^o_{V''_{Fe}}$ and $\Delta G^o_{V'''_{Fe}}$ and interstitial cations $\Delta G^o_{Fe_i^{3\bullet}}$ and $\Delta G^o_{Fe_i^{\bullet\bullet}}$ in spinel $(Fe_{1-x}Ti_x)_{3\pm\delta}O_4$ depend on the assumed values of ΔG^o_{eh} of formation of electronic defects and they were determined for the value of $\Delta G^o_{eh} = 30$ kJ/mol, the same as the value assumed for pure magnetite. The enthalpy of formation of cation vacancies V'''_{Fe} in the spinel $(Fe_{1-x}Ti_x)_{3\pm\delta}O_4$ with titanium content of $x_{Ti} = 0.1$ mol is $\Delta H^o_{V'''_{Fe}} = -265$ kJ/mol and it decreases to -308 kJ/mol at the titanium content of $x_{Ti} = 0.25$ mol. The enthalpy of formation of interstitial cations decreases when titanium content increases, from $\Delta H^o_{Fe_i^{3\bullet}} = 600$ kJ/mol for $x_{Ti} = 0.1$ mol to 563 kJ/mol for $x_{Ti} = 0.25$ mol.

Using the determined concentration of ionic defects and the values of coefficients of self-diffusion of radioactive tracers M* (^{59}Fe, ^{60}Co, ^{53}Mn and ^{44}Ti) in spinel $(Fe_{1-x}Ti_x)_{3\pm\delta}O_4$, the ratio of coefficient of diffusion (mobility) of tracer ions via interstitial cations to the coefficient of diffusion via vacancies ($D^o_{I(M)}/D^o_{V(M)} = b_D$) was determined. It was found that in the spinel with titanium content of $x_{Ti} = 0.1$ mol, the mobility via interstitial cations is significantly higher than that via cation vacancies. At 1473 K, the value of the parameter b_D for Co and Fe ions reaches $b_D \approx 100$, and for manganese, $b_D \approx 50$; these values decrease when the temperature increases. The values of b_D for titanium ions are smaller and they are about $b_D \approx 15$ at titanium content of $x_{Ti} = 0.1$ mol.

Mobilities of ions in the spinel $(Fe_{1-x}Ti_x)_{3\pm\delta}O_4$ significantly depend on the titanium content. At titanium content of $x_{Ti} = 0.1$, there occurs a significant increase in mobility via interstitial cations, and during further increase in titanium content, the value of $D^o_{I(M)}$ decreases slightly. For the diffusion via cation vacancies, at the titanium content of $x_{Ti} = 0.1$ mol, the values of $D^o_{V(M)}$ are lower than in pure magnetite, at the content of $x_{Ti} = 0.2$ mol they increase, and then, at the content of 0.3 mol, they are significantly lower. Mobilities of ions Fe*, Co* and Mn* in spinel $(Fe_{1-x}Ti_x)_{3\pm\delta}O_4$ are similar, while the mobility of titanium is lower by almost one order of magnitude.

TABLE 3.2
Enthalpies ΔH_i^o and Entropies ΔS_i^o of Activation of Mobility (Diffusion) of Ions Fe, Co, Mn and Ti via Cation Vacancies ($D^o_{V(M)}$) and via Interstitial Cations ($D^o_{I(M)}$) in the Spinel $(Fe_{1-x}Ti_x)_{3\pm\delta}O_4$ with the Titanium Content x_{Ti} = 0.0–0.3 mol, Determined based on the Dependences shown in Figures 3.23 and 3.24 $D^o_{V,I(M)} = D_o \exp\left(-\Delta H^o_{V,I(M)}/RT\right)$.

$D^o_{V,I(M)}$ (cm²/s)	x Ti (mol/mol)	ΔH_i^o (kJ/mol)	logD$_o$	ΔT (K)
$D^o_{V(Fe)}$ (DMHS)	0.0	98 ± 3	−2.65 ± 0.10	1173–1673
$D^o_{V(Co)}$ (DMHS)	0.0	90 ± 3	−3.07 ± 0.13	1173–1473
$D^o_{V(Ti)}$(AD)	0.0	116 ± 7	−2.68 ± 0.24	1473–1673
$D^o_{V(Fe)}$(AD)	0.1	110	−6.6	1473–1573
$D^o_{V(Co)}$(AD)	0.1	118	−6.3	1473–1573
$D^o_{V(Mn)}$(AD)	0.1	122	−6.2	1473–1573
$D^o_{V(Ti)}$(AD)	0.1	262	−2.0	1473–1573
$D^o_{V(Fe)}$(AD)	0.2	58 ± 26	−4.0 ± 0.9	1373–1573
$D^o_{V(Co)}$(AD)	0.2	77 ± 20	−3.4 ± 0.7	1373–1573
$D^o_{V(Mn)}$(AD)	0.2	74 ± 25	−3.4 ± 0.9	1373–1573
$D^o_{V(Ti)}$(AD)	0.2	99 ± 35	−3.4 ± 1.2	1373–1573
$D^o_{V(Fe)}$(AD)	0.3	147	−5.4	1473–1573
$D^o_{V(Co)}$(AD)	0.3	140	−5.7	1473–1573
$D^o_{V(Mn)}$(AD)	0.3	122	−6.2	1473–1573
$D^o_{V(Ti)}$(AD)	0.3	65	−8.8	1473–1573
$D^o_{I(Fe)}$(DMHS)	0.0	189 ± 7	1.02 ± 0.26	1173–1673
$D^o_{I(Co)}$(DMHS)	0.0	200 ± 4	1.72 ± 0.15	1173–1473
$D^o_{I(Ti)}$(AD)	0.0	218 ± 47	0.34 ± 1.57	1473–1673
$D^o_{I(Fe)}(AD)$	0.1	−6	−8.7	1473–1573
$D^o_{I(Co)}(AD)$	0.1	22	−7.7	1473–1573
$D^o_{I(Mn)}(AD)$	0.1	12	−8.3	1473–1573
$D^o_{I(Ti)}(AD)$	0.1	216	−2.3	1473–1573
$D^o_{I(Fe)}(AD)$	0.2	53 ± 12	−2.8 ± 0.4	1373–1573
$D^o_{I(Co)}(AD)$	0.2	78 ± 14	−1.9 ± 0.5	1373–1573
$D^o_{I(Mn)}(AD)$	0.2	64 ± 21	−2.6 ± 0.7	1373–1573
$D^o_{I(Ti)}(AD)$	0.2	[illegible]	3.2 ± 0.3	1373–1573
$D^o_{I(Fe)}(AD)$	0.3	172	−3.2	1473–1573
$D^o_{I(Co)}(AD)$	0.3	128	−4.7	1473–1573
$D^o_{I(Mn)}(AD)$	0.3	197	−2.5	1473–1573
$D^o_{I(Ti)}(AD)$	0.3	256	−1.5	1473–1573

The (xx) symbols stand for the results of coefficients of self-diffusion of tracers: Fe*, Co*, Mn* and Ti* extracted from (Aggarwal and Dieckmann 2002a, Aggarwal and Dieckmann 2002b) (AD) and (Dieckmann et al. 1978) (DMHS).

REFERENCES

Aggarwal, S., and R. Dieckmann. 2002a. Point Defects and Cation Tracer Diffusion in $(Ti_xFe_{1-x})_{3-\delta}O_4$ I. Non-stoichiometry and Point Defects. *J. Phys. Chem. Miner.* 29:695–706.

Aggarwal, S., and R. Dieckmann. 2002b. Point Defects and Cation Tracer Diffusion in $(Ti_xFe_{1-x})_{3-\delta}O_4$. II. Cation Tracer Diffusion. *J. Phys. Chem. Miner.* 29:707–718.

Akimoto, S. 1954. Thermo-Magnetic Study of Ferromagnetic Minerals Contained in Igneous Rocks. *J. Geomagn. Geoelectr.* 6:1–14.

Akimoto, S., T. Katsura, and M. J. Yoshida. 1957. Magnetic Properties of $TiFe_2O_4$-Fe_3O_4 System and Their Change with Oxidation. *J. Geomagn. Geoelectr.* 9:165–178.

Aragón, R., and R. H. McCallister. 1982. Phase and Point Defect Equilibria in the Titanomagnetite Solid solution. *Phys. Chem. Miner.* 8:112–120.

Aragón, R., R. H. McCallister, and H. R. Harrison. 1984. Cation Diffusion in Titanomagnetites. *Contrib. Mineral. Petrol.* 85: 174–185.

Banerjee, S. K., W. O'Reilly, T. C. Gibb, and N. N. Greenwood. 1967. The Behaviour of Ferrous Ions in Iron-titanium Spinels. *J. Phys. Chem. Solids* 28:1323–1335.

Bleil, U. 1971. Cation Distribution in Titanomagnetites. *Z. Geophys.* 37: 305–319.

Bleil, U. 1976. An Experimental Study of the Titanomagnetite Solid Solution Series. *Pure Appl. Geophys.* 114: 165–175.

Buddington, A. F., and D. H. Lindsley. 1964. Iron-Titanium Oxide Minerals and Synthetic Equivalents. *J. Petrol.* 5: 310–357.

Chevallier, R., J. Bolfa, and S. Mathiell. 1955. Titanomagnetites et ilmenites ferromagnetiques. *Bull. Soc. Fr. Mineral. Cristall.* 78: 307–346.

Creer, K.M., J. Ibbetson, and W. Drew. 1970. Activation Energy of Cation Migration in Titanomagnetites. *Geophys. J. Royal Astron. Soc.* 19: 93–101.

Dieckmann, R. 1982. Defects and Cation Diffusion in Magnetite (IV): Nonstoichiometry and Point Defect Structure of Magnetite ($Fe_{3-\delta}O_4$). *Ber. Bunsenges. Phys. Chem.* 86:112–118.

Dieckmann, R., T. O. Mason, J. D. Hodge, and H. Schmalzried. 1978. Defects and Cation Diffusion in Magnetite (III.) Tracer Diffusion of Foreign Tracer Cations as a Function of Temperature and Oxygen Potential. *Ber. Bunsenges. Phys. Chem.* 82: 778–783.

Foster, R. H., and E. O. Hall. 1965. A neutron and X-ray diffraction study of ulvöspinel, Fe_2TiO_4. *Acta Crystall.* 18: 857–862.

Freer, R., and Z. Hauptman. 1978. An Experimental Study of Magnetite-titanomagnetite Interdiffusion. *Phys. Earth Planet Inter.* 16: 223–231.

Fujino, K. 1974. Cation Distribution and Local Variation of Site Symmetry in Solid Solution Series, Fe_3O_4-Fe_2TiO_4. *Mineralog. J.* 7: 472–488.

Gorter, E. W. 1957. Chemistry and Magnetic Properties of Some Ferrimagnetic Oxides Like Those Occurring in Nature. *Adv. Phys.* 6: 336–361.

Hauptman, Z. 1974. High Temperature Oxidation, Range of Non-stoichiometry and Curie Point Variation of Cation-deficient Titanomagnetite, $Fe_{2.4}Ti_{0.6}O_{4+\gamma}$. *Geophys. J. Royal Astron. Soc.* 38: 29–47.

Ishikawa, Y., S. Sato, and Y. Syono. 1971. Neutron and Magnetic Studies of a Single Crystal of Fe_2TiO_4. *J. Phys. Soc. Jpn.* 31: 452–460.

Kąkol, Z., J. Sabol, and J. M. Honig. 1991. Cation Distribution and Magnetic Properties of Titanomagnetites $Fe_{3-x}Ti_xO_4$ ($0 \leq x < 1$). *Phys. Rev. B* 43: 649–654.

Néel, L. 1955. Some Theoretical Aspects of Rock-magnetism. *Adv. Phys.* 4: 191–243.

O'Donovan, J. B., and W. O'Reilly. 1980. The Temperature Dependent Cation Distribution in Titanomagnetites. *Phys. Chem. Minerals* 5: 235–243.

O'Neill, H. S. C., and A. Navrotsky. 1984. Cation Distributions and Thermodynamic Properties of Binary Spinel Solid Solutions. *Am. Mineral.* 69:733–753.

O'Reilly, W., and S. K. Banerjee. 1965. Cation Distribution in Titanomagnetites $(1-x)Fe_3O_{4-x}Fe_2TiO_4$. *Phys. Lett.* 17: 237–238.

Ozima, M., and M. Ozima. 1972. Activation Energy of Unmixing of Titanomaghemite. *Phys. Earth Planet. Inter.* 5: 87–89.

Petersen, N, 1970, Calculation of Diffusion Coefficient and Activation Energy of Titanium in Titanomagnetite. *Phys. Earth Planet. Inter.* 2: 175–178

Price, G. D. 1981. Diffusion in the Titanomagnetite Solid Solution Series. *Miner. Mag.* 44: 195–200.

Senderov, E., A. U. Dogan, and A. Navrotsky, 1993. Nonstoichiometry of Magnetite-ulvöspinel Solid Solutions Quenched from 1300 °C. *Am. Mineral.* 78: 565–573.

Simons, B., and E. Woermann. 1978. Iron Titanium Oxides in Equilibrium with Metallic Iron. *Contrib. Mineral. Petrol.* 66: 81–89.

Spencer, K., and D. H. Lindsley. 1981. A Solution Model for Coexisting Iron–titanium Oxides. *Am. Mineral.* 66:1189–1201.

Stephenson, A. 1969. The Temperature Dependent Cation Distribution in Titanomagnetites. *Geophys. J. Royal Astron. Soc.* 18:199–210.

Stokłosa, A. 2015. *Non-Stoichiometric Oxides of 3d Metals* pp. 313-376. Pfäffikon: Trans Tech Publications Ltd.

Stout, M. Z., and P. Bayliss. 1980. Crystal Structure of Two Ferrian Ulvospinels from British Columbia. *Canad. Mineral.* 18: 339–341.

Taylor, R. W. 1964. Phase Equilibria in the System FeO-Fe_2O_3-TiO_2 at 1300° C. *Am. Mineral.* 49:1016–1030.

Trestman-Matts, A., S. E. Dorris, S. Kumarakrishnan, and T. O. Mason. 1983. Thermoelectric Determination of Cation Distribution in Fe_3O_4-Fe_2TiO_4. *J. Am. Ceram. Soc.* 66: 829–834.

Webster, A. H., and N. F. H. Bright. 1961. The System Iron-Titanium-Oxygen at 1200°C. and Oxygen Partial Pressures Between 1 Atm. and 2×10^{-14} Atm. *J. Am. Ceram. Soc.* 44: 110–116.

Wechsler, B. A., D. H. Lindsley, and C. T. Prewitt. 1984. Crystal Structure and Cation Distribution in Titanomagnetites ($Fe_{3-x}Ti_xO_4$). *Am. Mineral.* 69:754–770.

4 Chromium-Doped Magnetite – $(Fe_{1-x}Cr_x)_{3\pm\delta}O_4$

4.1 INTRODUCTION

An example of spinel, in which Fe^{3+} iron ions are substituted with M^{3+} dopant ions is chromium-doped magnetite $(Fe_{1-x}Cr_x)_{3\pm\delta}O_4$. As it is a product of oxidation of iron and chromium based alloys, mainly studies on the crystalline structure and composition at high temperatures were performed. The phase diagram at 1473 K was presented in (Birchenall 1971) based on the results of studies of Kinlod and it differs from the results of other authors (Snethlage and Klemm 1975Pelton et al. 1979). This system was also studied at 1573 K (Katsura and Muan 1964, Katsura et al. 1975). Extensive studies of the deviation from the stoichiometry depending on the oxygen pressures in the spinel $(Fe_{1-x}Cr_x)_{3\pm\delta}O_4$ with chromium content of $x_{Cr} = 0.1–0.5$ mol at 1473 K were also performed (Töpfer et al. 1995). The same group performed also studies on the diffusion of radioisotope elements: iron ^{59}Fe, cobalt ^{60}Co, manganese ^{53}Mn and chromium ^{51}Cr, in dependence on the oxygen pressure and on the chromium content in the spinel $(Fe_{1-x}Cr_x)_{3\pm\delta}O_4$. In chromium-doped magnetite, there are defects of the type similar to the ones in pure magnetite (Töpfer et al. 1995). At higher pressures, cation vacancies dominate, while at lower pressures, in a narrow range, interstitial cations dominate. Studies on the diffusion confirmed the structure of ionic defects and showed that the coefficients of self-diffusion of ions depend on the type of metal and they are different in the range in which cation vacancies and interstitial cations dominate.

4.2 MECHANISM OF INCORPORATION OF CHROMIUM IONS AND METHODS OF CALCULATION OF THE DIAGRAMS OF DEFECTS' CONCENTRATIONS

Chromium-doped magnetite can be obtained for example by annealing, at high temperatures, the oxides: FeO, Fe_2O_3 and Cr_2O_3. The process occurs as follows:

$$FeO + x/2Cr_2O_3 + (2-x)/2Fe_2O_3 \rightarrow Fe^{3+}_{(t)}\left(Fe^{2+}_{(o)}Fe^{3+}_{1-x}Cr^{3+}_{x}\right)O_4 \quad (4.1)$$

Chromium is incorporated into cation sites with octahedral coordination, reducing the concentration of Fe^{3+} ions, which does not perturb electroneutrality of the spinel and does not cause significant changes in the concentration of ionic defects. The presence of chromium ions affects the range of oxygen pressures, in which the spinel exists, and the standard Gibbs energies of formation of point defects. Due to the fact that the dependences of the deviation from the stoichiometry on the oxygen pressure in spinels $(Fe_{1-x}Cr_x)_{3\pm\delta}O_4$ (Töpfer et al. 1995) have a similar nature as in pure magnetite, the calculations of diagrams of concentrations of defects were performed with a method analogous to the one used for pure magnetite (see Section 1.5). It was assumed that Cr^{3+} chromium ions incorporate into nodes with octahedral or tetrahedral coordination and an equilibrium is set. Therefore, a change in the oxygen pressure causes an increase or reduction in the concentration of point defects according to the reactions 1.10–1.14. Due to the lack of the value of ΔG^{o}_{eh} of formation of electronic defects, it was assumed that it only slightly depends on the concentration of chromium. Due to that, in the calculations, a constant value of $\Delta G^{o}_{eh} = 30.3$ kJ/mol was assumed, the same as for pure magnetite. As shown by initial calculations, a constant value of ΔG^{o}_{eh} in the range from 20 to 50 kJ/mol does not affect the degree of match between the dependence of δ on p_{O_2} and the experimental values of δ, but it affects only the value of $\Delta G^{o}_{V'''_{Fe(o)}}$ of formation of cation vacancies and other defects.

Thus, according to the methods of calculation presented in Section 1.5, the values of $\Delta G^{o}_{F_1}$ of formation of Frenkel defects and $\Delta G^{o}_{V'''_{Fe(o)}}$ of formation of iron vacancies $V'''_{Fe(o)}$ were adjusted in such a way as to obtain a good match between the dependence of δ on p_{O_2} and the experimental values of δ. In the first stage of calculations it was assumed that at the stoichiometric composition, the concentrations of cation vacancies are equal $\left[V'''_{Fe}\right] = \left[V''_{Fe}\right]$ ($\Delta G^{o}_{F_1} \cong \Delta G^{o}_{F_2}$ was assumed). Thus, the values of ΔG^{o}_{1} were adjusted in order to obtain similar values of cation vacancies at the stoichiometric composition, $\left[V''_{Fe}\right] \cong \left[V'''_{Fe}\right]$ (according to Equation (1.35b). Then such values of $\Delta G^{o}_{F_3}$ were chosen that the concentration of vacancies $\left[V'''_{Fe(t)}\right]$ did not deteriorate the quality of match between the dependence of δ on p_{O_2} and the experimental values δ.

Analogously as for titanium-doped magnetite, the possibility of doping the formed structure of spinel $(Fe_{1-x}Cr_x)_{3\pm\delta}O_4$ with Cr^{3+} ions, into the sublattice of Fe^{2+} ions, that become positive in relation to the lattice $Cr^{\bullet}_{Fe}$, was considered.

Incorporation of chromium ions is associated with building-in of a "molecule" of spinel; due to that, the doping process can be represented with the following reaction:

$$4/3Cr_2O_3 \rightarrow Cr^{\bullet}_{Fe(2+)} + Cr^{x}_{Fe(3+)(t)} + 2/3Cr^{x}_{Fe(3+)(o)} + 4O^{x}_{O} + 1/3V'''_{Fe(o)} \quad (4.2a)$$

Therefore, incorporation of 4 oxygen ions will be associated with the incorporation of 1 chromium ion into the cation sublattice of iron, Fe^{2+}, and the formation of an ion positive in relation to the lattice $Cr^{\bullet}_{Fe}$, the remaining chromium ions will occupy nodes of Fe^{3+} ions in octahedral and tetrahedral sublattice. In order to preserve the ratio of oxygen sites to cation sites, the deficit of cation sites is replenished by cation vacancies $V'''_{Fe(o)}$.

By dividing Equation (4.2a) by 3/8, we get:

$$1/2Cr_2O_3 \rightarrow 3/8Cr^{\bullet}_{Fe(2+)} + 3/8Cr^{x}_{Fe(3+)(t)} + 3/4Cr^{x}_{Fe(3+)(o)} + 3/2O^{x}_{O} + 1/8V'''_{Fe(o)} \quad (4.2b)$$

According to the reaction 4.2b, incorporation of Cr^{3+} ions, in the amount of $y_{Cr^{\bullet}}$ mol/mol, into the structure of a spinel with stoichiometric composition will cause an increase in the concentration of cation vacancies $\left[V'''_{Fe}\right]$ by $1/8y_{Cr^{\bullet}}$:

$$\left[V'''_{Fe}\right] = y^{o}_{V'''_{Fe}} + 1/8y_{Cr^{\bullet}}$$

where $y^{o}_{V'''_{Fe}}$ is the concentration of vacancies at the stoichiometric composition. Increase in the concentration of vacancies causes perturbation of the equilibrium between defects and setting of a new equilibrium state at the oxygen pressure that is significantly higher than that over non-doped spinel. Decrease in the oxygen pressure causes a reduction in the concentration of vacancies and changes in concentration of the remaining ionic defects. Due to the increase in the concentration of cation vacancies, the stoichiometric composition of spinel doped with $Cr^{\bullet}_{Fe}$ ions will be reached at the oxygen pressure lower than that for non-doped spinel. According to Equation 4.2b, concentration of ions $\left[Cr^{\bullet}_{Fe}\right] = 3/8y_{Cr^{\bullet}}$ must be taken into account in the electroneutrality condition, which assumes the form of:

$$\left[e'\right] + 2\left[V''_{Fe(o)}\right] + 3\left[V'''_{Fe(o)}\right] + 3\left[V'''_{Fe(t)}\right] = \left[h^{\bullet}\right] + 2\left[Fe^{\bullet\bullet}_{i}\right] + 3\left[Fe^{3\bullet}_{i}\right] + 3/8y_{Cr^{\bullet}} \quad (4.3)$$

Also, when calculating the deviation from the stoichiometry, it is necessary to consider an increase in the number of sites due to incorporation of ions of Cr_2O_3 oxide:

$$\frac{\left[Fe\right] + \left[Cr\right]}{\left[O\right]} = \frac{3 + y_{Cr^{\bullet}} - \Sigma\left[def\right]}{4 + 3/2y_{Cr}} \quad (4.4)$$

where $\Sigma\left[def\right] = \left[V''_{Fe(o)}\right] + \left[V'''_{Fe(o)}\right] + \left[V'''_{Fe(t)}\right] - \left[Fe^{\bullet\bullet}_{i}\right] - \left[Fe^{3\bullet}_{i}\right]$

If the concentration of chromium ions $\left[Cr^{\bullet}_{Fe}\right]$ is small, their presence will affect the values of the deviation from the stoichiometry mainly near the stoichiometric composition. Thus, when adjusting the values of ΔG^{o}_{i} of defects' formation, we should consider mainly the largest values of the deviation from the stoichiometry, at low and high oxygen pressures (near the limit oxygen pressures). Concentrations of defects are calculated as a function of the concentration of vacancies $\left[V'''_{Fe(o)}\right]$, and the equilibrium pressure is determined according to the relation 1.10b, using the values of $\Delta G^{o}_{V'''_{Fe(o)}}$ and the concentration of vacancies $V'''_{Fe(o)}$ and electron holes. Methods of calculations are analogous as for titanium-doped magnetite (see Section 3.2).

4.3 RESULTS OF CALCULATION OF THE DIAGRAMS OF THE CONCENTRATIONS OF DEFECTS, AND DISCUSSION

Figure 4.1a–f show diagrams of the concentrations of ionic defects for the spinel $(Fe_{1-x}Cr_x)_{3\pm\delta}O_4$ with chromium content of $x_{Cr} = 0.1$, 0.2, 0.333, 0.4 and 0.5 mol at 1473 K, (solid lines), calculated using the results of the deviation from the stoichiometry taken from (Töpfer et al. 1995) ((o) points). The dotted line shows the dependence of δ on p_{O_2} obtained using "equilibrium constants" of reactions of formation of cation vacancies K_V and interstitial ions K_I, determined in the work (Töpfer et al. 1995).

As can be seen in Figure 4.1a, b, for the spinel $(Fe_{1-x}Cr_x)_{3\pm\delta}O_4$ with chromium content of $x_{Cr} = 0.1$ and 0.2 mol, a very good match was obtained between the dependence of deviation from the stoichiometry on p_{O_2} and the experimental values

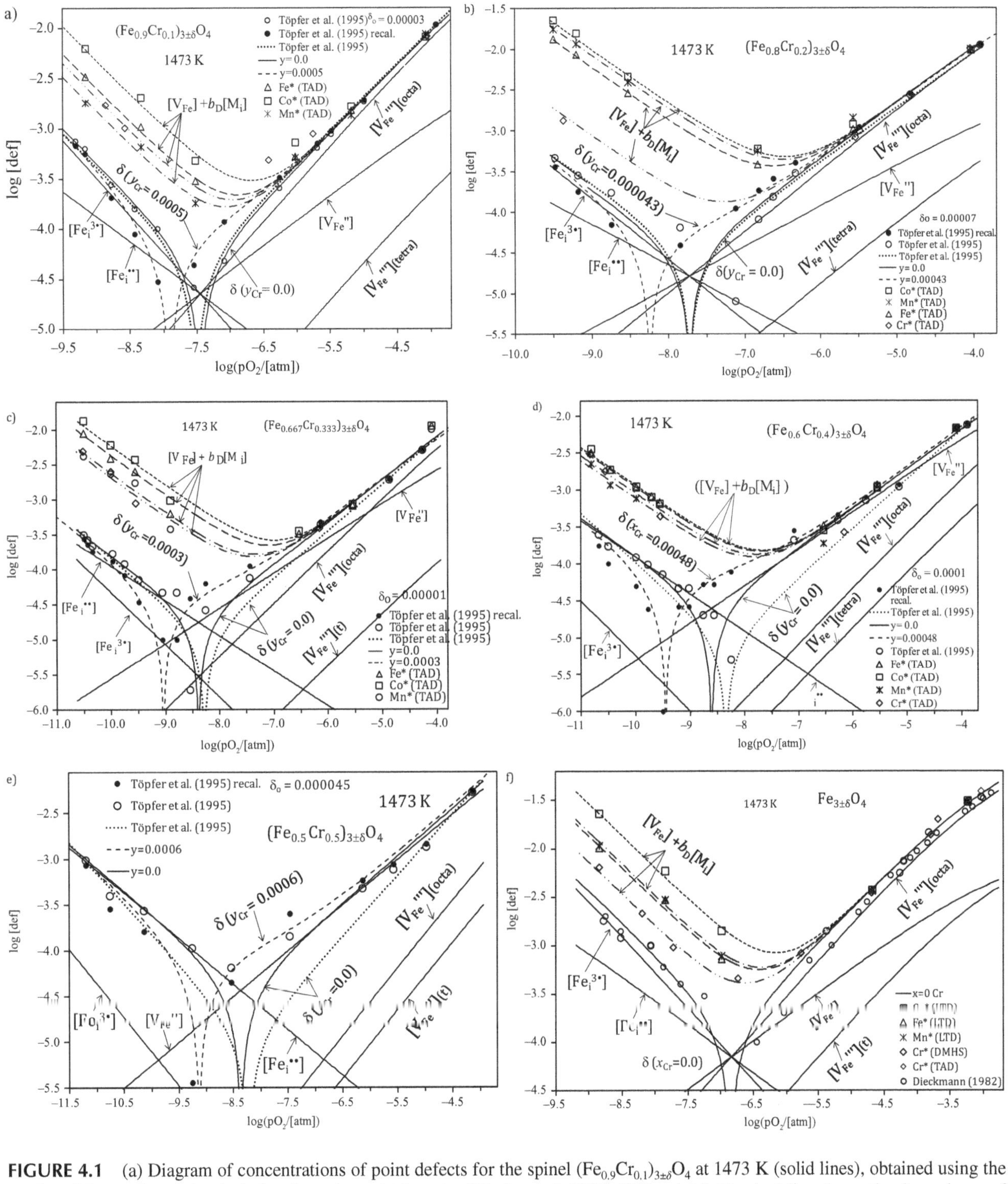

FIGURE 4.1 (a) Diagram of concentrations of point defects for the spinel $(Fe_{0.9}Cr_{0.1})_{3\pm\delta}O_4$ at 1473 K (solid lines), obtained using the results of the studies of the deviation from the stoichiometry (Töpfer et al. 1995) ((○) points). The (····) line shows the dependence of the deviation from the stoichiometry on the oxygen pressure, determined using the values of "equilibrium constants" of reactions of formation of cation vacancies K_V and interstitial cations K_I (Töpfer et al. 1995). Points (●) are the recalculated values of the deviation δ relative to the value $\delta_o = 3{\cdot}10^{-5}$. Line (---) represents the dependence of δ on p_{O_2} for the chromium concentration $y_{Cr}^{\bullet} = 5{\cdot}10^{-4}$ mol/mol. The remaining points represent the values of the sum of concentrations of ionic defects ($[V_{Fe}] + b_D[M_i]$), calculated according to Equation 1.47a and 1.47b for fitted values of $D^{o}_{V(M)}$ for the highest oxygen pressures and when using coefficients of self-diffusion of tracers: Fe* – (△) points, Co* – (□), Mn* – (✱), Ti* – (◇), taken from (Töpfer et al. 1995). Dashed lines denote the dependence of the sum of concentrations of defects ($[V_{Fe}] + b_D[M_i]$) on p_{O_2} obtained for the fitted value of b_D. (b) for the spinel $(Fe_{0.8}Cr_{0.2})_{3\pm\delta}O_4$, at 1473 K (solid lines). Points (●) are the recalculated values of the deviation from the stoichiometry relative to the value $\delta_o = 7{\cdot}10^{-5}$, and (---) line represents the dependence of δ on p_{O_2} for concentration of chromium $y_{Cr}^{\bullet} = 4.3{\cdot}10^{-5}$ mol/mol.

of δ. Dominating defects are cation vacancies $V'''_{Fe(o)}$ and interstitial cations $Fe_i^{3\bullet}$. Concentrations of vacancies $V''_{Fe(o)}$ and interstitial cations $Fe_i^{\bullet\bullet}$ are much lower and they can be considered the maximum possible concentrations of these defects.

For the content of $x_{Cr} = 0.333$, 0.4 and 0.5, in order to obtain a good match between the dependence of the deviation from the stoichiometry and the experimental values of δ, it was necessary to choose much lower values of $\Delta G^o_{F_2}$ of formation of Frenkel defects of ions with the charge of 2 than $\Delta G^o_{F_1}$ for the defects with the charge of 3 (see Figure 4.7). This causes that, as can be seen in Figure 4.1c–e, the concentration of cation vacancies $\left[V''_{Fe(o)}\right]$ and interstitial cations $\left[Fe_i^{\bullet\bullet}\right]$ is higher than the concentration $\left[V'''_{Fe(o)}\right]$ and $\left[Fe_i^{3\bullet}\right]$. (c) Diagram of concentrations of point defects for the spinel (Fe0.667Cr0.333)3±δO4, at 1473 K (solid lines). Points (●) are the recalculated values of the deviation from the stoichiometry relative to the value $\delta_o = 1{\cdot}10^{-5}$, and (---) line represents the dependence of δ on p_{O_2} for concentration of chromium $y_{Cr}^{\bullet} = 3{\cdot}10^{-4}$ mol/mol (other details as in caption to Figure 4.1a). (d) Diagram of concentrations of point defects for the spinel (Fe0.6Cr0.4)3±δO4, at 1473 K (solid lines). Points (●) are the recalculated values of the deviation from the stoichiometry relative to the value $\delta_o = 1{\cdot}10^{-4}$, and (---) line represents the dependence of δ on p_{O_2} for concentration of chromium $y_{Cr}^{\bullet} = 4.8{\cdot}10^{-4}$ mol/mol (other details as in caption to Figure 4.1a). (e) Diagram of concentrations of point defects for the spinel (Fe0.5Cr0.5)3±δO4, at 1473 K (solid lines). Points (●) are the recalculated values of the deviation from the stoichiometry relative to the value $\delta_o = 4.5{\cdot}10^{-5}$, and (---) line represents the dependence of δ on p_{O_2} for concentration of chromium $y_{Cr}^{\bullet} = 6{\cdot}10^{-4}$ mol/mol (other details as in caption to Figure 4.1a). (f) Diagram of concentrations of point defects for Fe3±δO4 at 1473 K (solid lines) (other details as in caption to Figure 4.1a).

Figure 4.2 compares the oxygen pressure dependences of deviation from the stoichiometry in the spinel $(Fe_{1-x}Cr_x)_{3\pm\delta}O_4$ with chromium content of $x_{Cr} = 0.0$, 0.1, 0.2, 0.333, 0.4 and 0.5 mol presented in Figure 4.1.

As can be seen in Figure 4.2, chromium doping in magnetite influences the concentrations of ionic defects in a different way than titanium ions (see Figure 3.3). The chromium doping only slightly changes the concentration of cation vacancies and the limit oxygen pressure at which spinel is transformed into hematite. Furthermore, the values of the deviation from the stoichiometry (concentration of cation vacancies) at which spinel is transformed into hematite are slightly lower (they are slightly lower than for pure magnetite). When chromium content increases, the range where cation vacancies dominate widens. This indicates that chromium doping does not practically affect the concentration of cation vacancies in the spinel. When the chromium content in the spinel $(Fe_{1-x}Cr_x)_{3\pm\delta}O_4$ increases, the concentration of interstitial cations significantly decreases (at the same oxygen pressure). The range of existence of spinel $(Fe_{1-x}Cr_x)_{3\pm\delta}O_4$ is similar to the range of existence of pure magnetite, and at the chromium content of $x_{Cr} = 0.333$–0.5 mol it is much wider, by about one order of magnitude.

Chromium doping, similarly as for titanium, causes a significant shift in the oxygen pressure at which the spinel $(Fe_{1-x}Cr_x)_3O_4$ reaches the stoichiometric composition towards lower pressures. The extent of the shift in this oxygen pressure is significantly lower than that for titanium. For example, in the case of chromium content $x_{Cr} = 0.2$ this is a shift by one order of magnitude, and in the case of titanium doping this was a shift by two orders of magnitude. In the case of chromium content $x_{Cr} = 0.1$–0.333 mol, the shift is higher than for $x_{Cr} = 0.4$ and 0.5 mol, where it is very small.

In the case of chromium content of $x_{Cr} = 0.1$, 0.2 mol, the range where interstitial cations dominate is shortened a little. As a result, despite the increase in chromium content, the transformation of spinel into doped oxide occurs at oxygen pressures that are only slightly lower than those for the transformation of pure magnetite, which is consistent with the

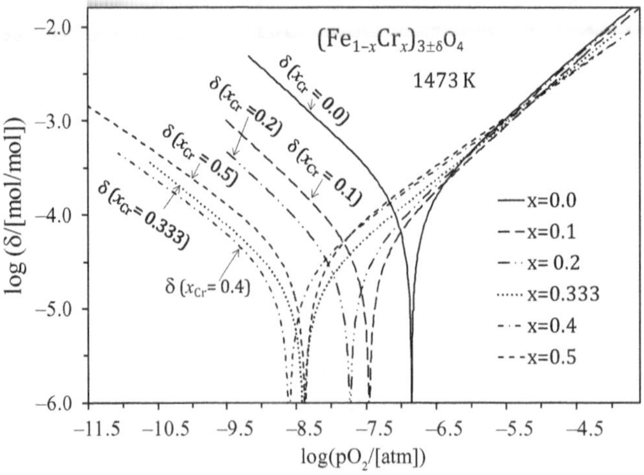

FIGURE 4.2 Dependences of the deviation from the stoichiometry on the oxygen pressure in the spinel $(Fe_{1-x}Cr_x)_{3\pm\delta}O_4$ with the chromium content $x_{Cr} = 0.0$–0.5 mol, at 1473 K, presented in Figure 4.1.

published results (Snethlage and Klemm 1975). Therefore, Cr^{3+} chromium ions have smaller effect on the increase in bonding energy than Ti^{4+} ions, although, when compared to Fe^{3+} iron ions, the influence is significant.

The chromium content equal $x_{Cr} = 0.333$ mol means that the spinel reached the composition $Fe(CrFe)O_4$, where 1 mol of Fe^{3+} ions was substituted. If we assume that chromium ions at lower concentrations incorporate into the sublattice with octahedral coordination, then beyond $x_{Cr} = 0.333$ mol (above 1 mol) chromium ions should be incorporated into the sublattice of cations with tetrahedral coordination. Chromium content above $x_{Cr} \geq 0.333$ mol has the influence inverse to that at lower concentrations to interactions between ions and concentration of ionic defects. As can be seen in Figure 4.2, the dependences of δ on p_{O_2} for the chromium content $x_{Cr} = 0.333$, 0.4 and 0.5 are very similar, which indicates that the influence of chromium ions on the range of existence of the spinel $(Fe_{1-y}Cr_yFeCr)_{1-\Delta}O_4$ and on the concentration of ionic defects is much smaller, or even inverse, than their influence at a lower content.

At the chromium content $x_{Cr} = 0.333$ mol, the spinel reaches the composition $CrFe_2O_4$ and it forms already at significantly lower pressure (in relation to the concentration of 0.2). In turn, further incorporation of the dopant, above the concentration of $x_{Cr} = 0.333$ mol, especially 0.5 mol, causes a significant decrease in the pressure of formation of the spinel. As can be seen in Figure 4.2, at the chromium content $x_{Cr} = 0.4$ mol, there occurs a slight decrease in the deviation from the stoichiometry, and at the content $x_{Cr} = 0.5$ mol – its increase (concentration of interstitial cations increases).

As mentioned earlier, in the spinel $(Fe_{1-x}Cr_x)_{3\pm\delta}O_4$ with the chromium content above $x_{Cr} = 0.333$ mol, the character of the dependence of δ on p_{O_2} is different than at lower concentrations, and in order to obtain an agreement of the dependence of the deviation from the stoichiometry and the experimental values of δ it was necessary to adjust lower values of $\Delta G^{o}_{F_2}$ of formation of Frenkel defects (V''_{Fe} and $Fe_i^{\bullet\bullet}$) and higher values of $\Delta G^{o}_{F_1}$. Due to this, concentrations of vacancies $\left[V''_{Fe}\right]$ and interstitial ions $\left[Fe_i^{\bullet\bullet}\right]$ are higher, and the concentrations of $\left[V'''_{Fe}\right]$ and $\left[Fe_i^{3\bullet}\right]$ are much lower. There arises a question whether the above discrepancies can be related to the incorporation of chromium ions in the form of Cr^{3+} into the sublattice of Fe^{2+} ions $\left(Cr^{\bullet}_{Fe}\right)$. As mentioned in the case of titanium doping, the absolute values of the deviation from the stoichiometry depend on the assumed (fitted) value of δ_o. The performed calculations indicate that a small correction of the value of δ_o allows obtaining the dependence of δ on p_{O_2} specific for doping with $Cr^{\bullet}_{Fe}$ ions.

Figure 4.1 presents the dependences of the deviation from the stoichiometry on the oxygen pressure for such a concentration of chromium ions $Cr^{\bullet}_{Fe}$ ($y_{Cr^{\bullet}}$) incorporated into the sublattice of Fe^{2+} ions (according to the reaction 4.2) ((---) line), which is consistent with the recalculated values of the deviation from the stoichiometry (Töpfer et al. 1995) ((●) points). Calculations were performed for the same values of $\Delta G^{o}_{F_1} \cong \Delta G^{o}_{F_2}$ as for the spinel not doped with $\left[Cr^{\bullet}_{Fe}\right]$ ions. However, in the case of a spinel with content above $x_{Cr} = 0.2$ mol, for the values $\Delta G^{o}_{F_1} \cong \Delta G^{o}_{F_2}$ ((*) points in Figure 4.7), a good match between the dependence of δ on p_{O_2} and the experimental values of δ was not obtained. Figure 4.3 presents the dependence of the concentration of ions $\left[Cr^{\bullet}_{Fe}\right] = 3/8\ y_{Cr}$ on the chromium content x_{Cr} in the spinel $(Fe_{1-x}Cr_x)_{3\pm\delta}O_4$.

As can be seen in Figure 4.3, the concentration of ions $\left[Cr^{\bullet}_{Fe}\right]$ decreases when the chromium content in the spinel increases and it reaches the minimum value for the spinel $CrFe_2O_4$. Further increase in chromium content in the spinel causes an increase in the concentration of ions $\left[Cr^{\bullet}_{Fe}\right]$. As can be seen in Figure 4.1, incorporation of chromium $Cr^{\bullet}_{Fe}$ causes a shift in the oxygen pressure at which spinel reaches the stoichiometric composition towards lower pressures. The

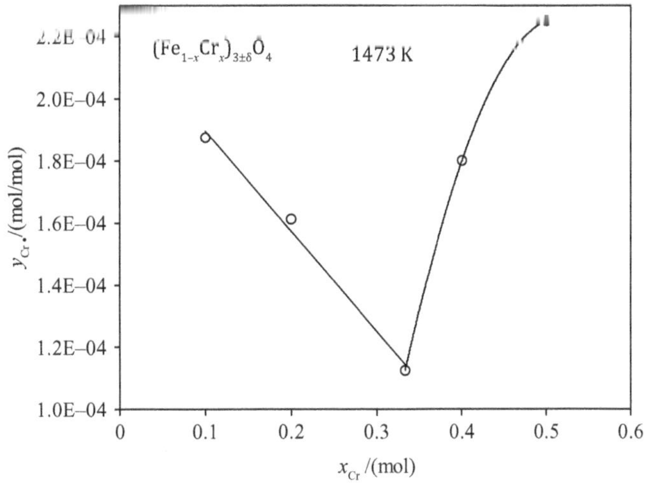

FIGURE 4.3 Dependence of the concentration of chromium ions $\left[Cr^{\bullet}_{Fe}\right] = 3/8 y_{Cr^{\bullet}}$ incorporated into the sublattice of ions Fe^{2+}, on the chromium content x_{Cr} in the spinel $(Fe_{1-x}Cr_x)_{3\pm\delta}O_4$, at 1473 K.

oxygen pressure at which concentrations of cation vacancies and interstitial cations are equal (Frenkel defects are present) changes only slightly. At such low concentrations of $Cr^{\bullet}_{Fe}$ ($y_{Cr}{}^{\bullet}$), changes in the concentration of ionic defects are practically insignificant (solid line and dashed line overlap). However, the character of the dependence of the deviation from the stoichiometry on p_{O_2} changes significantly, which is an effect of a logarithmic change in small values of δ (small difference between the concentrations of defects).

From the presented calculations it results that in chromium-doped magnetite, chromium ions form a solid solution (occupying nodes of Fe^{3+} ions), which causes a change in interionic interactions (change in lattice energy), a change in the range of spinel existence and a change in other properties related to the change in distribution of ions with different charges in spinel sublattices. However, it is likely that in the formed spinel structure, a certain small amount of chromium ions is incorporated into the sublattice of Fe^{2+} ions as ions $Cr^{\bullet}_{Fe}$ (positive in relation to the lattice). The above process is a result of setting of the equilibrium state between Cr_2O_3 oxide and the spinel.

Figure 4.4 presents the dependence of the derivative $(d\log p_{O_2}/d\log\delta) = n$ on the oxygen pressure, for the dependences of the deviation from the stoichiometry on p_{O_2} shown in Figure 4.1 (without considering $Cr^{\bullet}_{Fe}$ ions).

As can be seen in Figure 4.4, for the chromium content $x_{Cr} = 0.1$ and 0.2 mol at the lowest oxygen pressures the parameter n reaches the value of −1.5 ($1/n = -2/3$), and at the highest oxygen pressures it reaches the value of n = 1.5. For the chromium content $x_{Cr} = 0.333$ mol the values are higher, and for the concentration $x_{Cr} = 0.4$ and 0.5 mol the values of the parameter n are close, respectively, to −2 and 2, which is related to the domination of defects $\left[V''_{Fe(o)}\right]$ and $\left[Fe_i^{\bullet\bullet}\right]$.

Changes in the limit oxygen pressure depending on the chromium content in the spinel, discussed above, are consistent with the dependences presented in Figure 4.5.

As can be seen in Figure 4.5, the oxygen pressures at which spinels $(Fe_{1-x}Cr_x)_3O_4$ reach the stoichiometric composition decrease when the chromium content increases, and the dependence only slightly deviates from linear relation ((o) points). The pressure at which the concentrations of cation vacancies and interstitial cations are equal $[V_{Fe}] = [M_i]$ is slightly higher and it similarly slightly decreases, because the values of $y_{Cr}{}^{\bullet}$ are low. The decrease in the above oxygen pressures is a result of a slight increase in bonding energy due to a higher chromium content in the spinel $(Fe_{1-x}Cr_x)_3O_4$.

In turn, Figure 4.6 presents the dependence of the concentration of Frenkel defects on the chromium content x_{Cr} in the spinel $(Fe_{1-x}Cr_x)_3O_4$.

As can be seen in Figure 4.6, at the chromium content $x_{Cr} = 0.1$ mol the concentration of Frenkel defects significantly decreases and further increase in chromium content causes only a slight change in this concentration, and above $x_{Cr} > 0.333$ mol – a slight increase. This is a consequence of a rise in the value of $\Delta G^o_{F_1}$ when the chromium content in the spinel increases.

Figures 4.7–4.9 present the dependence of adjusted standard Gibbs energies: $\Delta G^o_{F_1}$, $\Delta G^o_{F_2}$, $\Delta G^o_{F_3}$, $\Delta G^o_{V''_{Fe(o)}}$, ΔG^o_I on the chromium content x_{Cr} in the spinel $(Fe_{1-x}Cr_x)_3O_4$ at 1473 K. Figures 4.8–4.10 show the dependence of the standard Gibbs energy $\Delta G^o_{V''_{Fe}}$ of formation of vacancies $V''_{Fe(o)}$ calculated according to Equation (1.35d), $\Delta G^o_{Fe_i^{3\bullet}}$ of formation of interstitial iron ions $Fe_i^{3\bullet}$ (calculated according to Equation (1.34c)) and ($\Delta G^o_{Fe_i^{\bullet\bullet}}$) of formation of ions $Fe_i^{\bullet\bullet}$ (calculated according to Equation (1.36c)) depending on the chromium content x_{Cr} in the spinel $(Fe_{1-x}Cr_x)_3O_4$.

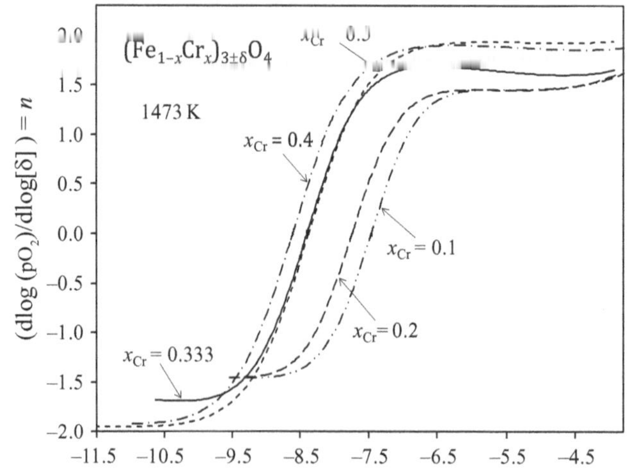

FIGURE 4.4 Dependence of the derivative: $n_\delta = d\log(p_{O_2})/d\log\delta$ on the oxygen pressure, obtained for the dependence of the deviation from the stoichiometry in the spinel $(Fe_{1-x}Cr_x)_{3\pm\delta}O_4$ with the chromium content $x_{Cr} = 0.1$–0.5 mol, at 1473 K (presented in Figure 4.1).

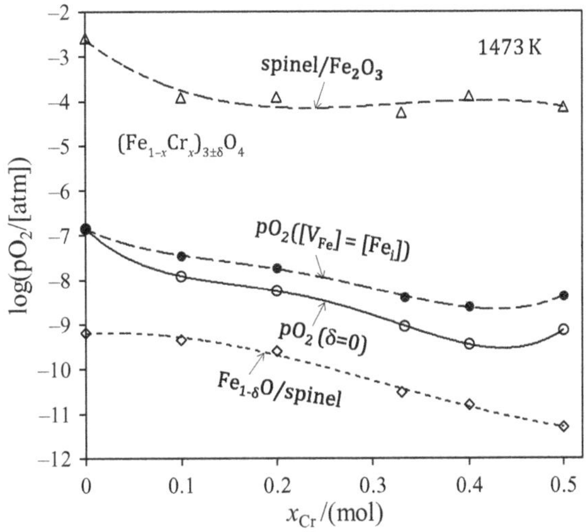

FIGURE 4.5 Dependence of the oxygen pressure at which the spinel $(Fe_{1-x}Cr_x)_3O_4$ reaches the stoichiometric composition ((○) points) on the chromium content x_{Cr}, at 1473 K, and the oxygen pressures at which the concentrations of iron vacancies and interstitial iron ions are equal ((●) points). Lines (---) and points (◇) denote oxygen pressures at the phase boundary doped wüstite/spinel and spinel/doped hematite taken from (Töpfer et al. 1995).

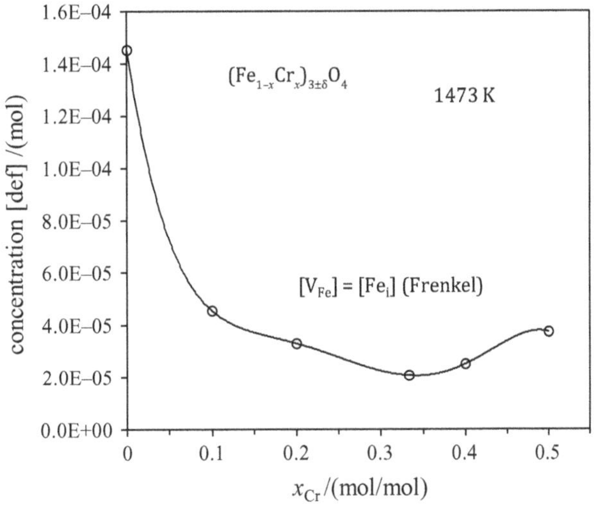

FIGURE 4.6 Dependence of the concentration of Frenkel defects on the chromium content x_{Cr} in the spinel $(Fe_{1-x}Cr_x)_{3\pm\delta}O_4$, at 1473 K.

As can be seen in Figure 4.7, the standard Gibbs energy of the formation of Frenkel defects $\Delta G^o_{F_1}$ increases with the chromium content in magnetite, to a smaller extent for the chromium content up to $x_{Cr} = 0.2$ mol, to a larger extent for the content above $x_{Cr} > 0.333$ mol. This results in a decrease in the concentration of Frenkel defects. Similarly, as $\Delta G^o_{F_1}$, to the chromium content of $x_{Cr} = 0.2$ mol, the value of $\Delta G^o_{F_2}$ changes; it decreases for the chromium content above $x_{Cr} > 0.333$ mol. A result of these changes is a small increase in the concentration of Frenkel defects with the increase in chromium content in the spinel. As it was mentioned, assuming the same values of $\Delta G^o_{F_1} = \Delta G^o_{F_2}$ for the chromium content of $x_{Cr} = 0.333–0.5$ mol ((✳) points) caused a worse match between the dependence of δ on p_{O_2} and the experimental results.

In turn, as can be seen in Figures 4.8 and 4.10, changes in the values of $\Delta G^o_{V''_{Fe}}$ and $\Delta G^o_{V'_{Fe}}$ are small up to the chromium content of $x_{Cr} = 0.2$ mol, and when the chromium content increases further, they increase and decrease respectively. The character of the change in values of $\Delta G^o_{Fe^{3\cdot}_i}$ and $\Delta G^o_{Fe^{\cdot\cdot}_i}$ is also complex, which is strictly related to the role of chromium doping at the chromium content of $x_{Cr} = 0.333$ mol and above.

Figures 4.8 and 4.10 show also the values of $\Delta G^o_{V_{Fe}}$ and $\Delta G^o_{Fe_i}$ calculated using "equilibrium constants" of reaction of formation of cation vacancies K_V and interstitial cations K_I extracted from (Töpfer et al. 1995). The values differ from these determined in the work, because they describe another model (equation) of defects.

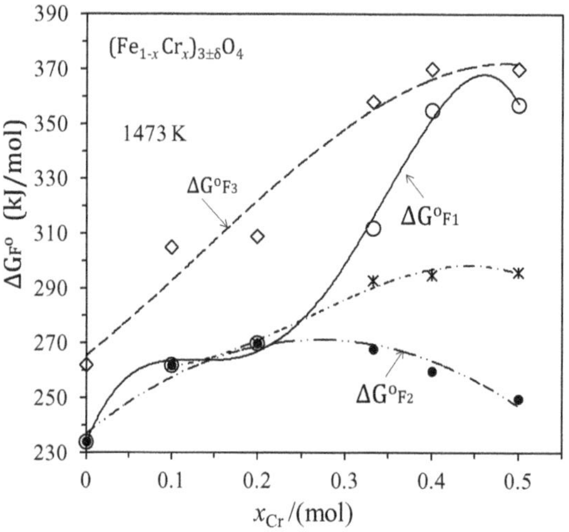

FIGURE 4.7 Dependence of the standard Gibbs energy $\Delta G^o_{F_1}$ ((○) points) $\Delta G^o_{F_2}$ (●) and $\Delta G^o_{F_3}$ (◇) of formation of Frenkel defects on the chromium content x_{Cr} in the spinel $(Fe_{1-x}Cr_x)_{3\pm\delta}O_4$, at 1473 K, obtained using the results of the studies of the deviation from the stoichiometry taken from (Töpfer et al. 1995). Points (✱) represent the values $\Delta G^o_{F_1} = \Delta G^o_{F_2}$.

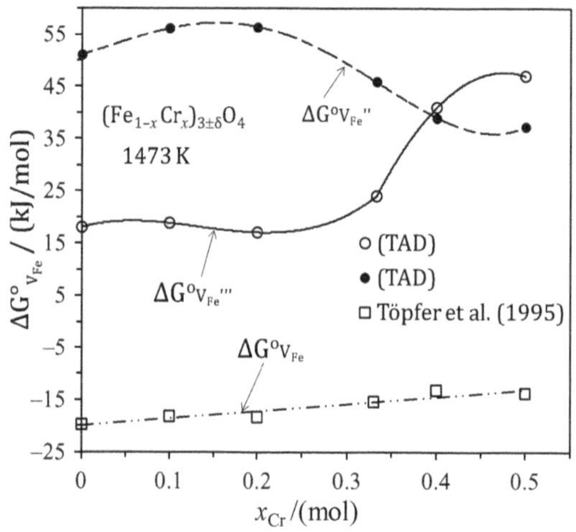

FIGURE 4.8 Dependence of the standard Gibbs energy $\Delta G^o_{V'''_{Fe}}$ and $\Delta G^o_{V''_{Fe}}$ of formation of vacancies $\left(V'''_{Fe(o)}\right)\left(V''_{Fe(o)}\right)$ on the chromium content x_{Cr} in the spinel $(Fe_{1-x}Cr_x)_{3\pm\delta}O_4$, at 1473 K and the values $\Delta G^o_{V_{Fe}}$ calculated using the values of "equilibrium constants" of reaction of formation of cation vacancies K_{V}, taken from (Töpfer et al. 1995).

4.4 DIFFUSION IN CHROMIUM-DOPED MAGNETITE

Similarly as for pure magnetite, using the concentration of ionic defects resulting from the diagram of defects' concentrations and the values of coefficients of diffusion of radioisotope elements in the spinel $(Fe_{1-x}Cr_x)_{3\pm\delta}O_4$, coefficients of diffusion (mobility) of metal ions via cation vacancies and via interstitial cations were determined (see: Section 1.6.2).

Figure 4.1 presents the dependence of the "sum" of concentrations of ionic defects ($[V_{Fe}] + b_D[Fe_i]$) on the oxygen pressure in the spinel $(Fe_{1-x}Cr_x)_{3\pm\delta}O_4$ for the fitted value of b_D (dashed lines) and the values of this sum, calculated according to Equation (1.47a and 1.47b), using the values of diffusion coefficients of iron ^{59}Fe ((△) points) and cobalt ^{60}Co (□), manganese ^{53}Mn (✱) and chromium ^{51}Cr (◇) taken from (Töpfer et al. 1995). For the highest oxygen pressures, the values of the coefficient of diffusion of cations $D^o_{V(M)}$ via cation vacancies and the parameter b_D, being the ratio of coefficients of diffusion via defects ($D^o_{I(M)}/D^o_{V(M)}$) were adjusted in such a way as to obtain a match between the dependence of the sum of the concentrations of defects ($[V_{Fe}] + b_D[Fe_i]$) with the values calculated using the coefficients of self-diffusion of tracers. As can be seen in Figure 4.1 a very good agreement was obtained between the dependence of the sum of the concentrations of defects ($[V_{Fe}] + b_D[Fe_i]$) on p_{O_2} and the values of this sum calculated using the coefficients of self-diffusion of tracers.

Figure 4.11 presents the dependence of the parameter b_D on the chromium content in the spinel $(Fe_{1-x}Cr_x)_{3\pm\delta}O_4$ for tracer ions: Fe* ((△) points) and Co* (□), Mn* (○) and Cr* (◇).

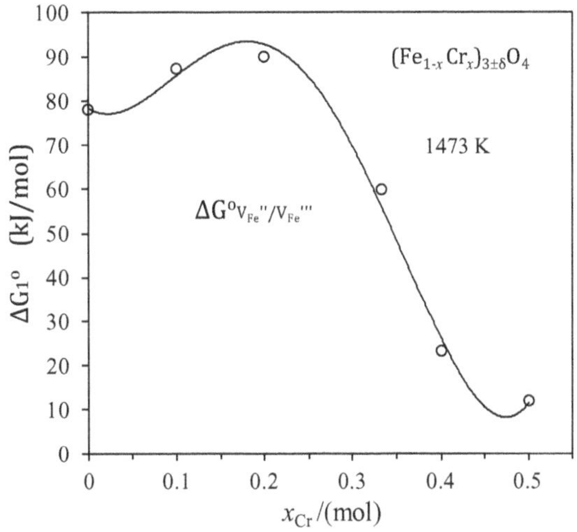

FIGURE 4.9 Dependence of the standard Gibbs energy ΔG_1^o on the chromium content x_{Cr} in the spinel $(Fe_{1-x}Cr_x)_{3\pm\delta}O_4$ at 1473 K, permitting determination of the ratio of concentrations of iron vacancies $\left[V''_{Fe(o)}\right]/\left[V'''_{Fe(o)}\right]$.

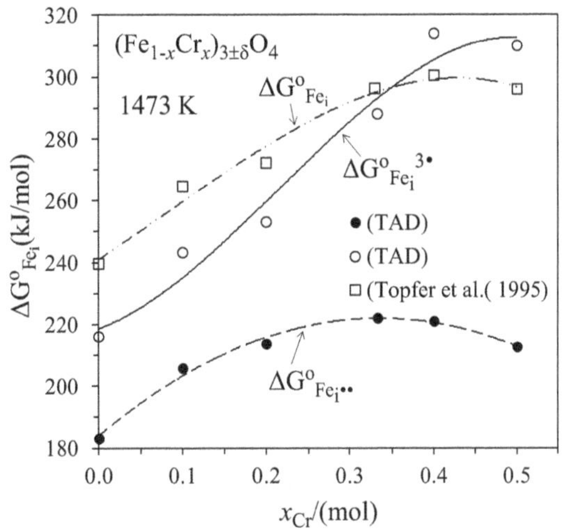

FIGURE 4.10 Dependence the of standard Gibbs energy $\Delta G^o_{Fe_i^{3\bullet}}$ and $\Delta G^o_{Fe_i^{\bullet\bullet}}$ of formation of interstitial ions $Fe_i^{3\bullet}$ and $Fe_i^{\bullet\bullet}$ on the chromium content x_{Cr} in the spinel $(Fe_{1-x}Cr_x)_{3\pm\delta}O_4$ at 1473 K (calculated from Equations 1.34c and 1.36c) and the values $\Delta G^o_{Fe_i}$ calculated using "equilibrium constants" of reaction of formation of interstitial cations, taken from (Töpfer et al. 1995).

As can be seen in Figure 4.11, the ratio of the coefficients of diffusion via interstitial ions and the coefficients of diffusion via cation vacancies changes with the chromium content. A small increase in the value of the parameter b_D occurs at the chromium content $x_{Cr} = 0.1$ mol. The highest values of the parameter b_D are present for the chromium content of $x_{Cr} = 0.2$ mol, for cobalt $b_D = 45$, for manganese 40, for iron 25. However, the value of the parameter b_D changes only slightly in the case of diffusion of chromium and the highest value $b_D = 13$ occurs at the content of $x_{Cr} = 0.333$ mol. At a higher content of chromium, the values of the parameter b_D decrease.

In turn, Figure 4.12 presents the dependences of the coefficients of diffusion of ions via cation vacancies $D^o_{V(M)}$ (solid points) and via interstitial ions $\left(D^o_{I(M)}\right)$ (empty points) on the chromium content in the spinel $(Fe_{1-x}Cr_x)_{3\pm\delta}O_4$.

As can be seen in Figure 4.12, in the case of diffusion of ions Fe*, Co* and Mn*, their mobilities via cation vacancies and via interstitial ions differ only slightly. At the chromium content of $x_{Cr} = 0.1$ mol the values $D^o_{V(M)}$ are higher than for pure magnetite, however, when chromium content increases further, they decrease, and above the concentration of $x_{Cr} = 0.333$ mol they increase.

In turn, according to the values of the parameter b_D, the coefficients of diffusion via interstitial cations are higher than the coefficients of diffusion via vacancies. Similarly, the mobilities of Fe, Co and Mn ions $\left(D^o_{I(M)}\right)$ are practically identical.

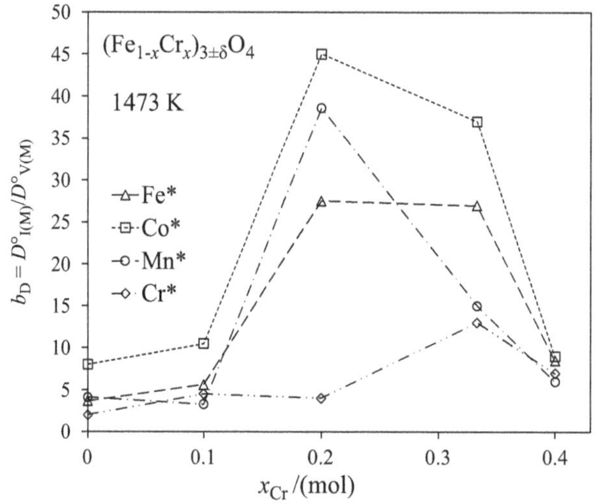

FIGURE 4.11 Dependence of the ratio of the coefficient of diffusion of ions via interstitial cations to the coefficient of diffusion of cations via cation vacancies $D^o_{I(M)}/D^o_{V(M)} = b_D$ on the chromium content x_{Cr} in the spinel $(Fe_{1-x}Cr_x)_{3\pm\delta}O_4$ at 1473 K, obtained using the values of coefficients of self-diffusion of tracers: Fe* ((△) points), Co* (□), Mn* (○), Cr* (◇), taken from (Töpfer et al. 1995) and concentrations of ionic defects.

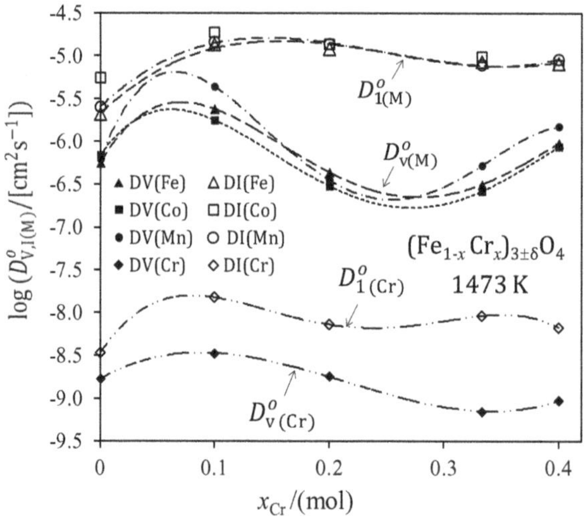

FIGURE 4.12 Dependence of the coefficient of diffusion of ions via cation vacancies $D^o_{V(M)}$ (solid points) and via interstitial cations $D^o_{I(M)}$ (empty points) on the chromium content x_{Cr} in the spinel $(Fe_{1-x}Cr_x)_{3\pm\delta}O_4$ at 1473 K, obtained using the values of coefficients of self-diffusion of tracers: Fe* ((▲,△) points), Co* (■,□), Mn* (●,○), Cr* (◆,◇), taken from (Töpfer et al. 1995) and concentrations of ionic defects.

At the chromium content of $x_{Cr} = 0.1$ mol the values $D^o_{I(M)}$ are higher than for pure magnetite, however, when chromium content increases further, they slightly decrease, practically within the error limits.

However, the coefficients of diffusion of chromium $D^o_{V(Cr)}$ and $D^o_{I(Cr)}$ are clearly lower, via cation vacancies as well as via interstitial cations; they only slightly depend on the chromium content, although the character of the change is similar to that for Fe*, Co* and Mn* ions. This might be related to the radius of chromium ions diffusing via defects and their effective charge (with electrostatic field of a chromium ion interacting with adjacent ions during their movement).

4.5 CONCLUSIONS

Using the results of the studies of deviation from the stoichiometry for the spinel $(Fe_{1-x}Cr_x)_{3\pm\delta}O_4$ with the chromium content of $x_{Cr} = 0.1–0.5$ mol, obtained at 1473 K (Töpfer et al. 1995), diagrams of the concentrations of defects were determined. It was found that chromium doping in magnetite only slightly affects the character of the dependence of the deviation from the stoichiometry on the oxygen pressure compared to that for the pure magnetite. In the spinel $(Fe_{1-x}Cr_x)_{3\pm\delta}O_4$ with chromium content up to $x_{Cr} > 0.333$ mol, in the range of its existence, the stoichiometric composition is reached, while there occurs a shift in the oxygen pressure at which the spinel reaches the stoichiometric composition towards lower values, by about one order of magnitude at the content of 0.333 mol.

For the chromium content up to $x_{Cr} > 0.2$ mol, the structure of point defects practically does not change. In the range of higher oxygen pressures, cation vacancies $\left[V_{Fe}'''\right]$ dominate; their concentration changes only slightly with the chromium content in the spinel. In the range of low oxygen pressures, interstitial cations $\left[Fe_i^{3\bullet}\right]$ dominate; their concentration decreases with the chromium content. For spinels with chromium content $x_{Cr} = 0.333$ mol and higher, the dominating defects are vacancies $\left[V_{Fe}''\right]$ and $\left[Fe_i^{\bullet\bullet}\right]$; their concentration is only slightly affected by chromium doping.

It was shown that it was possible to adjust a reference value δ_0 which causes a change in the character of the dependence of the deviation from the stoichiometry on the oxygen pressure indicating the incorporation of Cr^{3+} ions into the sublattice of Fe^{2+}, which are positive in relation to the lattice ($Cr_{Fe}^{\bullet}$). The maximum concentration of these ions at the chromium content of $x_{Cr} = 0.1$ mol can be $y_{Cr^\bullet} = 2{\cdot}10^{-4}$ mol/mol and it decreases at the content of $x_{Cr} = 0.1$–0.333 mol, and then it increases when the chromium content increases.

It was found that when the chromium content increases up to $x_{Cr} = 0.2$ mol in the spinel $(Fe_{1-x}Cr_x)_{3-\delta}O_4$, the standard Gibbs energies of formation of Frenkel defects $\Delta G_{F_1}^{o} = \Delta G_{F_2}^{o}$ increase. At higher concentrations of chromium, $\Delta G_{F_2}^{o}$ of formation of Frenkel defects ($\left[V_{Fe}''\right] = \left[Fe_i^{\bullet\bullet}\right]$) decreases. In consequence, it causes a decrease in the concentration of these defects at the stoichiometric composition to the value of $2{\cdot}10^{-4}$ mol and its slight change when the chromium content in the spinel increases.

The standard Gibbs energies of formation of cation vacancies $\Delta G_{V_{Fe}'''}^{o}$ and $\Delta G_{V_{Fe}''}^{o}$ and interstitial cations $\Delta G_{Fe_i^{3\bullet}}^{o}$ and $\Delta G_{Fe_i^{\bullet\bullet}}^{o}$ depend on the values of ΔG_{eh}^{o} of formation of electronic defects and they were determined for the value of $\Delta G_{eh}^{o} = 30$ kJ/mol, the same as the value assumed for pure magnetite. Their dependence, depending on the chromium content in the spinel, is complex, due to a change in the type of dominating defects

Using the determined concentration of ionic defects and the values of coefficients of self-diffusion of tracers M* (^{59}Fe, ^{60}Co, ^{53}Mn and ^{51}Cr) published in (Töpfer et al. 1995), the ratio of the coefficient of diffusion (mobility) of tracer ions via interstitial cations to the coefficient of diffusion via cation vacancies ($D_{I(M)}^{o}/D_{V(M)}^{o} = b_D$) was determined. It was found that when the chromium content increases up to $x_{Cr} = 0.2$ mol in the spinel $(Fe_{1-x}Cr_x)_{3\pm\delta}O_4$, the mobility via interstitial cations is significantly higher than the mobility via cation vacancies. At 1473 K, the value of the parameter b_D for Co ions reaches $b_D \approx 45$, for manganese $b_D \approx 40$, and for iron ions $b_D \approx 25$. The values of b_D for chromium ions are smaller, about 5, and for the chromium content of $x_{Cr} = 0.333$ mol they increase up to about $b_D \approx 15$.

At the chromium content $x_{Cr} = 0.1$ mol in the spinel $(Fe_{1-x}Cr_x)_{3\pm\delta}O_4$ there occurs an increase in the mobility of tracer ions compared to pure magnetite. In the case of diffusion via interstitial cations, further increase in the chromium content in the spinel causes a small decrease, within the error limits, in the value of $D_{I(M)}^{o}$. In the case of diffusion via cation vacancies, as to the values of $D_{V(M)}^{o}$, an increase in the chromium content above $x_{Cr} = 0.1$ mol causes their decrease, and above the content of $x_{Cr} = 0.333$ mol they increase.

REFERENCES

Birchenall, C. E. 1971. *Oxidation of Metals and Alloys*. Metals Park, OH: ASM Monograph.

Dieckmann, R. 1982. Defects and Cation Diffusion in Magnetite (IV): Nonstoichiometry and Point Defect Structure of Magnetite ($Fe_{3-\delta}O_4$). *Ber. Bunsenges. Phys. Chem.* 86: 112–118.

Katsura, T., and A. Muan. 1964. Experimental Study of Equilibria in the System FeO-Fe_2O_3-Cr_2O_3 at 1300 °C. *Trans. Metal. Soc. AIME.* 230: 77–84.

Katsura, T., M. Wakihara, S. I. Hara, and T. Sugihara. 1975. Some Thermodynamic Properties in Spinel Solid Solutions with the Fe_3O_4 Component. *J. Solid State Chem.* 13: 107–113.

Pelton, A. D., H. Schmaltzried, and J. Sticher. 1979. Computer-assisted Analysis and Calculation of Phase Diagrams of the Fe-Cr-O, Fe-Ni-O and Cr-Ni-O Systems. *J. Phys. Chem. Solids.* 40: 1103–1122.

Snethlage, R., and D. D. Klemm. 1975. Das System Fe-Cr-O bei 1 000, 1 095 und 1 200 °C. *N. Jb. Miner. Abh.* 125: 227–242.

Töpfer, J., S. Aggarwal, and R. Dieckmann. 1995. Point Defects and Cation Tracer Diffusion in $(Cr_xFe_{1-x})_{3-\delta}O_4$ Spinels. *Solid State Ionics.* 81: 251–266.

5 Aluminium-Doped Magnetite – $(Fe_{1-x}Al_x)_{3-\delta}O_4$

5.1 INTRODUCTION

Aluminium-doped magnetite was studied mainly due to its functional properties at room temperature. However, there is only a small number of studies at high temperatures, at the thermodynamic equilibrium, at a specific oxygen pressure. Aluminium dopant, due to a stable oxidation state of +3, incorporates into the sublattice of Fe^{3+} ions with octahedral or tetrahedral coordination. The range of existence of magnetite doped with aluminium was published in (Meyers et al. 1980) and the distribution of ions in the spinel $(Fe_{1-x}Al_x)_{3-\delta}O_4$ in (Mason and Bowen 1981a, 1981b). The studies of the deviation from the stoichiometry in the spinel $(Fe_{1-x}Al_x)_{3-\delta}O_4$ with the aluminium content of x_{Al} = 0.0–0.2 mol, in the temperature range of 1573–1673 K and in the oxygen pressure range from 0.1 × 10^{-4} atm can be found in (Yamauchi et al. 1983). The studies were limited to the range where cation vacancies dominate. In turn, the studies on chemical diffusion with the relaxation method, as a function of the oxygen pressure were published in (Petuskey 1977, Yamauchi et al. 1983). The studies on the self-diffusion of iron in iron-aluminium spinel at high temperatures were described in (Halloran and Bowen 1980). From the studies on diffusion it results that the defects' structure in the spinel $(Fe_{1-x}Al_x)_{3\pm\delta}O_4$ is similar to that in magnetite; at low oxygen pressures diffusion of ions occurs via interstitial cations, and at higher oxygen pressures it occurs via iron vacancies.

Electrical conductivity of aluminium-doped magnetite is described in (Gillot et al. 1976). The oxygen pressure dependence of the thermoelectric power and electrical conductivity in the temperature range of 1200–1800 K for the spinel $(Fe_{1-x}Al_x)_{3-\delta}O_4$ with the aluminium content of x_{Al} = 0.0, 0.163, 0.323 and 0.487 mol was reported in (Mason and Bowen 1981a, 1981b). It was shown that the electrical conductivity decreases with the aluminium content from about 200–250 $\Omega^{-1}cm^{-1}$ for pure magnetite to the value of about 10–50 Ω^{-1} cm^{-1} for the content of x_{Al} = 0.487 mol. The Seebeck coefficient values are negative and they decrease when the oxygen pressure increases (the values are more negative) (Mason and Bowen 1981a, 1981b). The character of the oxygen pressure dependence of electrical conductivity is similar to that for the pure magnetite. In a wide range of oxygen pressures, the electrical conductivity is constant and it decreases at high oxygen pressures. This indicates that the dominating defects are electrons and their transport occurs according to the mechanism of small polarons and that the mobility of electrons is much higher than the mobility of electron holes. Only at the aluminium content of x_{Al} = 0.487 mol in the spinel $(Fe_{1-x}Al_x)_{3-\delta}O_4$ at higher oxygen pressures electrical conductivity increases when oxygen pressure increases, which indicates different properties of the spinel with this composition.

5.2 METHODS AND RESULTS OF CALCULATION OF THE DIAGRAMS OF THE CONCENTRATIONS OF DEFECTS, AND DISCUSSION

The calculations of diagrams of point defects concentrations of in aluminium-doped magnetite were conducted with a method analogous to the one used for pure magnetite (see Section 1.5), using the results of the deviation from the stoichiometry (Yamauchi et al. 1983). It was assumed that Al^{3+} aluminium ions incorporate into sites with octahedral coordination or tetrahedral coordination, thus replacing Fe^{3+} ions, analogously to the case of chromium. Therefore, the incorporation of aluminium ions does not perturb electroneutrality of the spinel and it does not change the structure of ionic defects. Due to the lack of the value of ΔG^o_{eh} of formation of electronic defects, it was assumed that it only slightly depends on the temperature and on aluminium concentration. Due to that, in the calculations, a constant value of ΔG^o_{eh} = 30.3 kJ/mol was assumed, the same as for pure magnetite. As shown by initial calculations, a constant value of ΔG^o_{eh} in the range from 30 to 50 kJ/mol does not affect the degree of match between the dependence of δ on p_{O_2} and the experimental results δ. The assumed value of ΔG^o_{eh} affects only the changes of the values of $\Delta G^o_{V'''_{Fe(o)}}$ of formation of vacancies $V'''_{Fe(o)}$ and other defects (analogous to the case of pure magnetite). Due to the lack of results of the studies of the deviation from the stoichiometry in the range where interstitial cations dominate, as a first approximation, the same values of formation of Frenkel defects $\Delta G^o_{F_1} = \Delta G^o_{F_2}$, $\Delta G^o_{F_3}$ and ΔG^o_I were assumed as for pure magnetite (see Figures 1.11 and 1.13). The calculations performed in the range of their change up to 20 kJ/mol did not alter the character of the dependence of deviation from the stoichiometry on p_{O_2} in the analysed range of oxygen pressures.

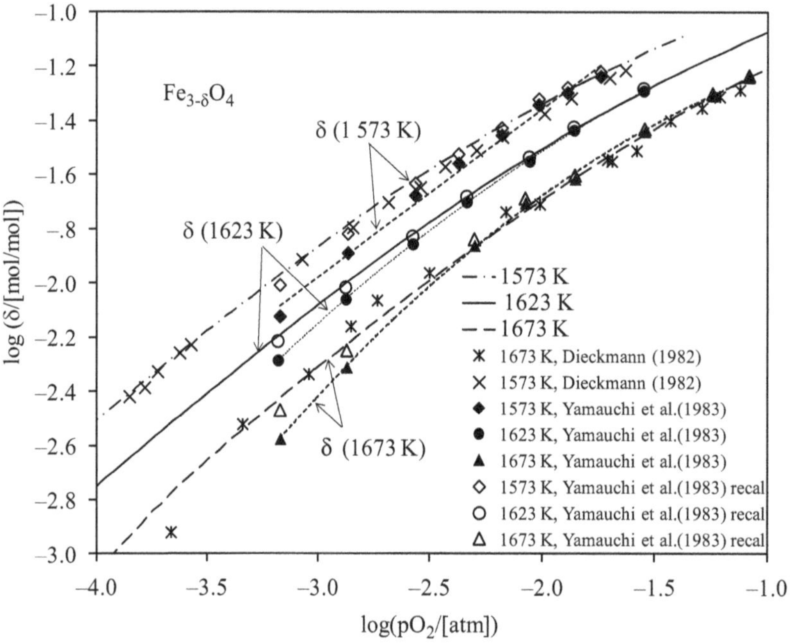

FIGURE 5.1 Dependences of the deviation from the stoichiometry on the oxygen pressure in magnetite at 1573–1673 K, obtained using the results of the studies on the deviation from the stoichiometry (Yamauchi et al. 1983) (solid points) and (Dieckmann 1982) ((✳,x) points). Empty points – the recalculated values of the deviation from the stoichiometry from (Yamauchi et al. 1983).

Figure 5.1 presents the dependence of the deviation from the stoichiometry on p_{O_2} for pure magnetite (Dieckmann 1982) ((✳,x) points) and (Yamauchi et al. 1983) ((□,●,▲) points) at 1573–1673 K. The lines show the dependence of deviation from the stoichiometry on p_{O_2} calculated in the present work based on (Dieckmann 1982).

As can be seen in Figure 5.1, the results of the deviation from the stoichiometry (Yamauchi et al. 1983) (solid points) and the character of the dependence of δ on p_{O_2} ((---) lines) differ from the dependence obtained using other results (Dieckmann 1982) ((✳,x) points). A similar character of the dependence of δ on p_{O_2} is present for aluminium-doped magnetite (Yamauchi et al. 1983) shown in Figure 5.2.

The dependence of δ on p_{O_2} is stronger and when the oxygen pressure decreases, the differences increase significantly. Such a strong change in the character of the dependence of δ on p_{O_2} cannot be explained by the influence of the concentration of ionic defects with lower ionisation degrees on the values of the deviation from the stoichiometry (the influence is inverse). The discrepancies can be related to the value of the assumed reference mass of magnetite sample, in relation to which the absolute values of the deviation from the stoichiometry were calculated. The lack of measurement results at low oxygen pressures (in the range where interstitial cations dominate) makes it difficult to appropriately determine the reference value δ_o by fitting. Due to that, it was decided to correct the results for pure magnetite and aluminium-doped magnetite, by fitting such a value of δ_o that the dependence of deviation from the stoichiometry on the oxygen pressure has a similar character as that of the results obtained by Dieckmann for pure magnetite. Thus, for the model where the defects with the highest ionisation degree dominate and the exponent in the dependence of δ on p_{O_2} is close to $1/n \cong 2/3$.

Figure 5.3 shows the values of δ_o correcting the results reported in (Yamauchi et al. 1983) at the given temperatures for pure magnetite and aluminium-doped magnetite, assuming that vacancies $V'''_{Fe(o)}$ dominate and that the deviation from the stoichiometry depends on the oxygen pressure with an exponent of 2/3 ($\delta \approx p_{O_2}^{2/3}$).

As can be seen in Figure 5.2, the obtained dependences of the deviation from the stoichiometry on p_{O_2} for spinels with the aluminium content of $x_{Al} = 0.0–0.2$ mol at temperatures 1573–1673 K matches the recalculated experimental δ values (empty points) quite well.

In turn, Figure 5.4 presents the fitted values of standard Gibbs energies: $\Delta G^o_{V'''_{Fe(o)}}$ of formation of vacancies $V'''_{Fe(o)}$ and $\Delta G^o_{V''_{Fe(o)}}$ calculated according to the Equation 1.35c as a function of the aluminium content x_{Al} in the spinel $(Fe_{1-x}Al_x)_{3-\delta}O_4$ at 1573–1673 K.

Figure 5.5 shows the dependence of standard Gibbs energy $\Delta G^o_{Fe_i^{3\bullet}}$ of formation of interstitial iron ions $Fe_i^{3\bullet}$ (calculated according to Equation 1.34c) and $\Delta G^o_{Fe_i^{\bullet\bullet}}$ of formation of ions $Fe_i^{\bullet\bullet}$ (calculated according to Equation 1.36c) depending on aluminium content in the spinel $(Fe_{1-x}Al_x)_{3-\delta}O_4$.

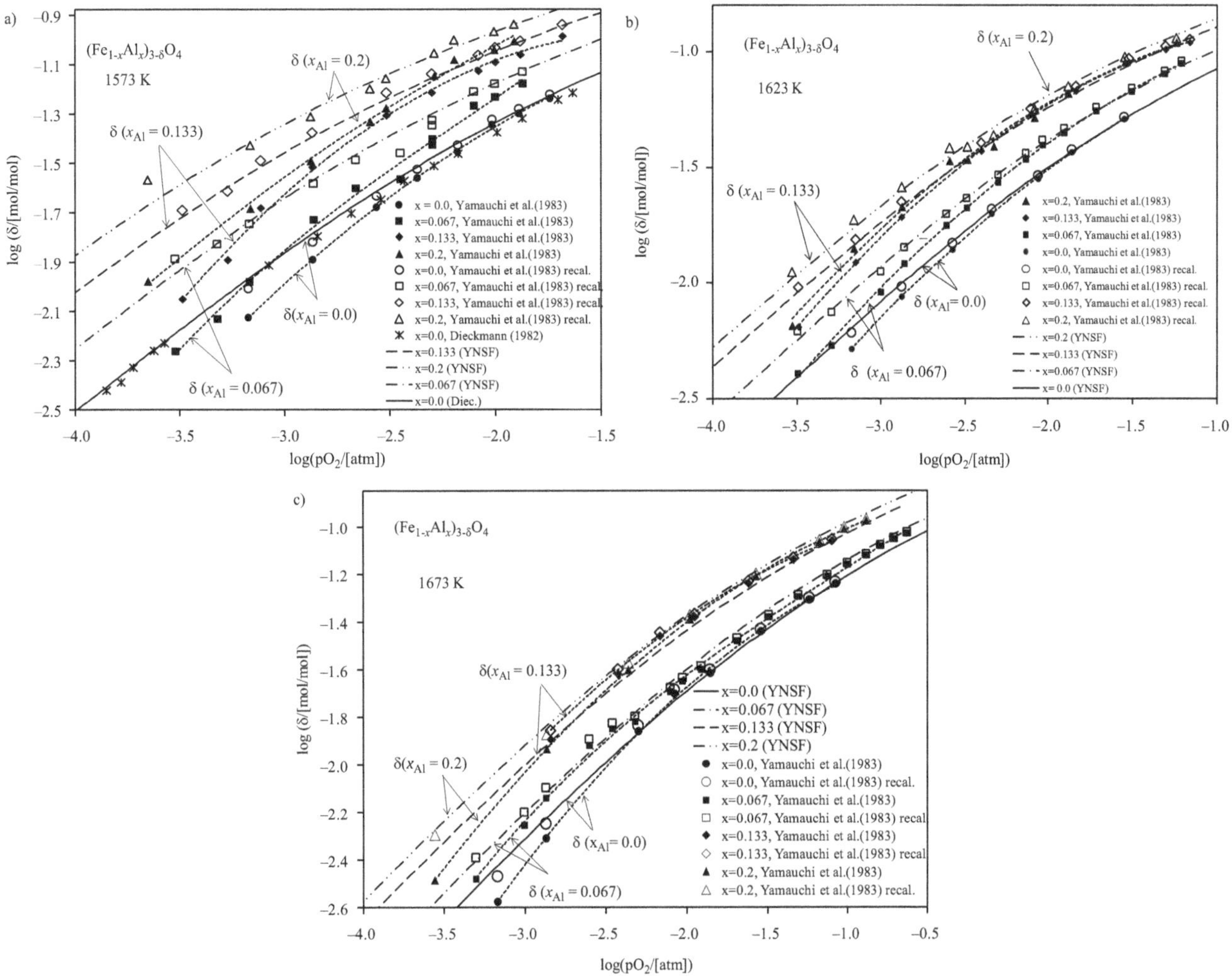

FIGURE 5.2 (a) Oxygen pressure dependences of the deviation from the stoichiometry in the spinel $(Fe_{1-x}Al_x)_{3-\delta}O_4$ with the aluminium content of $x_{Al} = 0.0, 0.067, 0.133$ and 0.2 mol, at 1573 K, obtained using the recalculated values (empty points) of the deviation from the stoichiometry (Yamauchi et al. 1983) (solid points) and for non-doped magnetite, obtained using the results reported in (Dieckmann 1982) ((∗) points). (b) Oxygen pressure dependences of the deviation from the stoichiometry in the spinel $(Fe_{1-x}Al_x)_{3-\delta}O_4$ with the aluminium content of $x_{Al} = 0.0, 0.067, 0.133$ and 0.2 mol, at 1623 K, obtained using the corrected values (empty points) of the deviation from the stoichiometry (Yamauchi et al. 1983) (solid points). (c) Oxygen pressure dependences of the deviation from the stoichiometry in the spinel $(Fe_{1-x}Al_x)_{3-\delta}O_4$ with the aluminium content of $x_{Al} = 0.0, 0.067, 0.133$ and 0.2 mol, at 1673 K, obtained using the corrected values (empty points) of the deviation from the stoichiometry (Yamauchi et al. 1983) (solid points).

Figures 5.6 and 5.7 show the linear temperature dependences of $\Delta G^{o}_{V'''_{Fe(o)}}$ and $\Delta G^{o}_{V^{\bullet}_{Fe(o)}}$ and of $\Delta G^{o}_{Fe_i^{\bullet\bullet}}$ and $\Delta G^{o}_{Fe_i^{3\bullet}}$ for the spinel $(Fe_{1-x}Al_x)_{3-\delta}O_4$ with the aluminium content of $x_{Al} = 0.0–0.2$ mol.

Figure 5.8 shows diagrams of the concentrations of ionic defects for the spinel $(Fe_{1-x}Al_x)_{3-\delta}O_4$ with aluminium content of $x_{Al} = 0.0–0.2$ mol at the temperatures of 1573–1673 K, in the range where cation vacancies dominate (solid lines), calculated using the recalculated values of the deviation from the stoichiometry (Yamauchi et al. 1983). As can be seen in Figure 5.8, when the temperature increases, the concentration of aluminium vacancies in the spinel decreases.

Figure 5.9 compares the oxygen pressure dependences of deviation from the stoichiometry in the spinel $(Fe_{1-x}Al_x)_{3-\delta}O_4$ with the aluminium content of $x_{Al} = 0.0–0.2$ mol at 1573 K, which, as can be seen, when the aluminium content increases, the curves are shifted towards lower oxygen pressures. When the aluminium content increases, the concentration of cation vacancies (deviation from the stoichiometry) increases, at the same oxygen pressure.

Figure 5.10 presents the dependence of oxygen pressure, at which the spinel $(Fe_{1-x}Al_x)_3O_4$ reaches the stoichiometric composition.

As can be seen in Figure 5.10, at higher temperatures these equilibrium oxygen pressures are higher. At a constant temperature, when the aluminium content in the spinel increases, these pressures become lower. Aluminium doping of the magnetite, similarly as titanium and chromium, causes an increase in bonding energy comparing to Fe^{3+} ions.

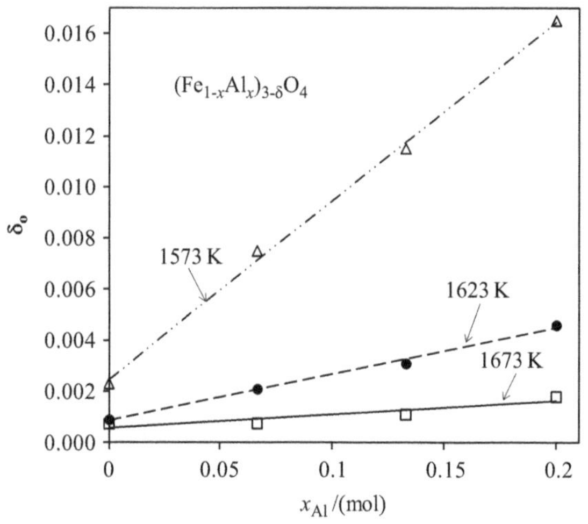

FIGURE 5.3 Values of the parameter δ_o, which was used for recalculation of the values of the deviation from the stoichiometry taken from (Yamauchi et al. 1983) in the spinel $(Fe_{1-x}Al_x)_{3-\delta}O_4$ with the aluminium content of $x_{Al} = 0.0$, 0.067, 0.133 and 0.2 mol at 1573, 1623 and 1673 K.

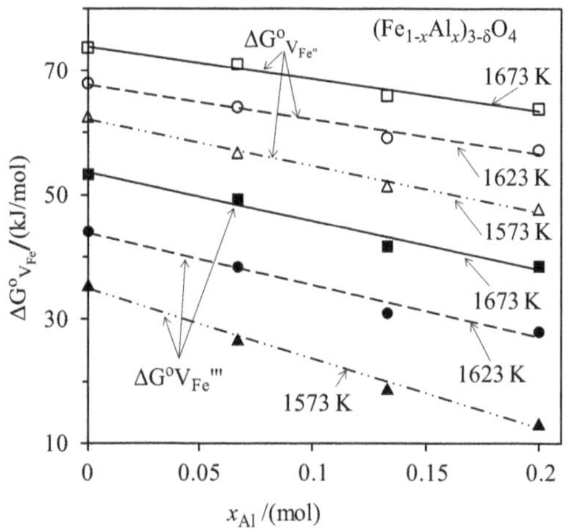

FIGURE 5.4 Dependence of standard Gibbs energy $\Delta G^o_{V_{Fe}'''}$ and $\Delta G^o_{V_{Fe}''}$ of formation of vacancies ($V'''_{Fe(o)}$ – solid points) and ($V''_{Fe(o)}$ – empty points) on the aluminium content x_{Al} in the spinel $(Fe_{1-x}Al_x)_{3-\delta}O_4$, at 1573, 1623 and 1673 K, obtained using the recalculated values of the deviation from the stoichiometry taken from (Yamauchi et al. 1983).

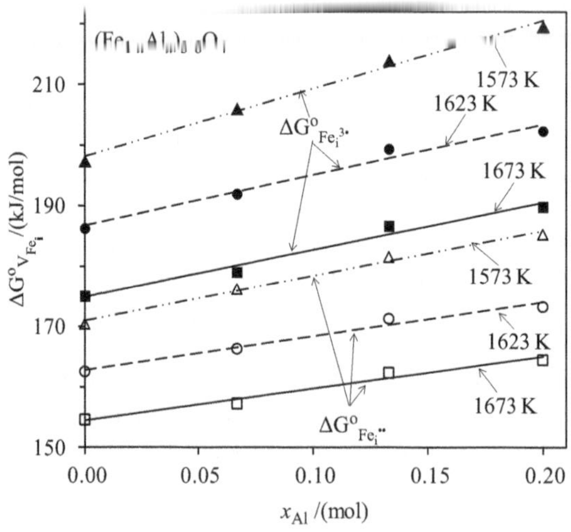

FIGURE 5.5 Dependence of standard Gibbs energy $\Delta G^o_{Fe_i^{3\bullet}}$ and $\Delta G^o_{Fe_i^{\bullet\bullet}}$ of formation of interstitial ions $Fe_i^{3\bullet}$ and $Fe_i^{\bullet\bullet}$ on the aluminium content x_{Al} in the spinel $(Fe_{1-x}Al_x)_{3-\delta}O_4$ at 1573–1673 K (calculated according to Equation. 1.34c and 1.36c).

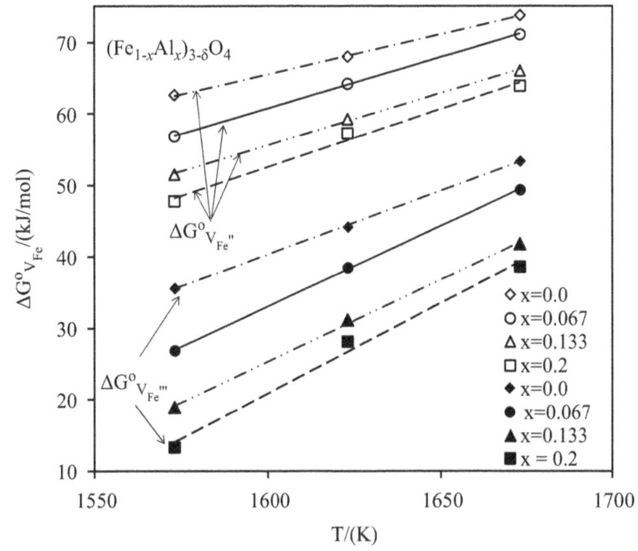

FIGURE 5.6 Temperature dependence of standard Gibbs energy $\Delta G^{o}_{V_{Fe}}$ of formation of vacancies $\left(V'''_{Fe(o)}\right)$ and $\left(V''_{Fe(o)}\right)$ in the spinel $(Fe_{1-x}Al_x)_{3-\delta}O_4$ with the aluminium content x_{Al} = 0.0, 0.067, 0.133 and 0.2 mol.

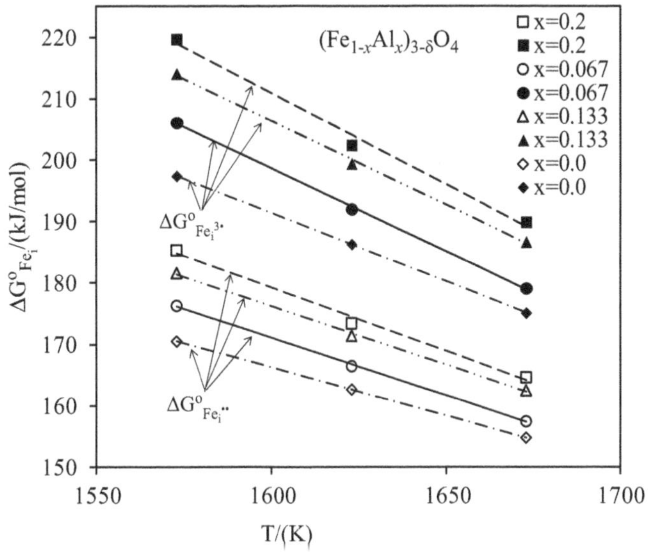

FIGURE 5.7 Temperature dependence of standard Gibbs energy $\Delta G^{o}_{Fe_i^{3\bullet}}$ and $\Delta G^{o}_{Fe_i^{\bullet\bullet}}$ of formation of interstitial ions $Fe_i^{3\bullet}$ (solid points) and $Fe_i^{\bullet\bullet}$ (empty points) in the spinel $(Fe_{1-x}Al_x)_{3-\delta}O_4$ with the aluminium content of x_{Al} = 0.0, 0.067, 0.133 and 0.2 mol.

5.3 CONCLUSIONS

Using the published results of the studies of deviation from the stoichiometry for the spinel $(Fe_{1-x}Al_x)_{3-\delta}O_4$ for the aluminium content of x_{Al} = 0.0, 0.067, 0.133 and 0.2 mol (Yamauchi et al. 1983), diagrams of defects' concentrations were determined at the temperatures of 1573–1673 K. Analysing the results of the deviation from the stoichiometry and the character of the oxygen pressure dependence of δ for pure magnetite (Yamauchi et al. 1983), it was found that the reference values should be corrected. Due to a narrow range of oxygen pressures in which the studies were performed (from 0.1–10^{-4} atm), the effect of aluminium doping in magnetite on the character of the dependence of δ on p_{O_2} was not found; owing to that, it was not possible to determine the effect on the value of standard Gibbs energy ΔG^{o}_{F} of formation of Frenkel defects. Due to that, for the determination of diagrams of the concentrations of defects, the same values of ΔG^{o}_{F} as for pure magnetite were assumed. It was found that in the studied range of higher oxygen pressures, the dominating defects are cation vacancies $\left[V'''_{Fe}\right]$; their concentration increases when the aluminium dopant content increases (at a constant oxygen

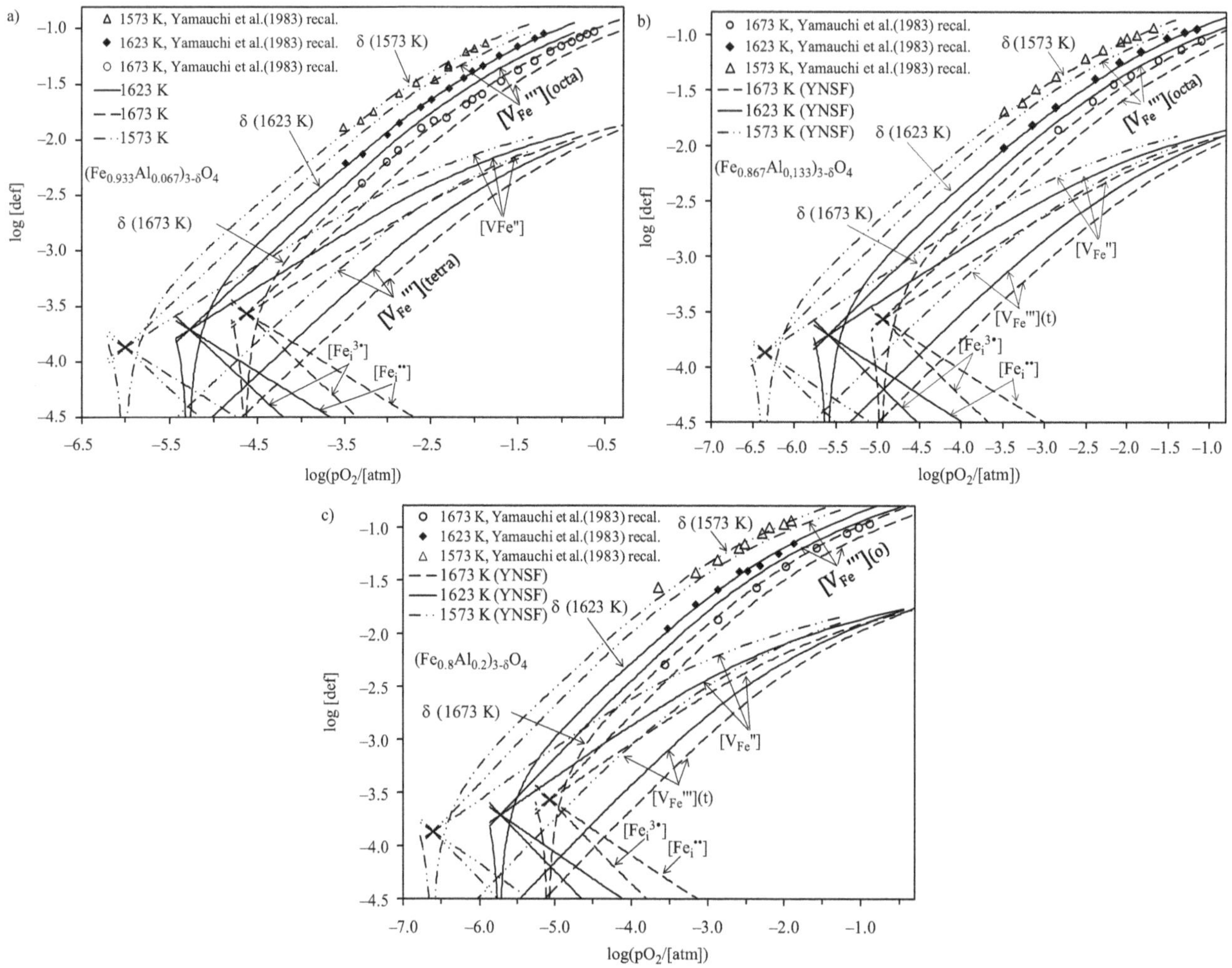

FIGURE 5.8 (a) Diagrams of concentrations of point defects for the spinel $(Fe_{0.93}Al_{0.067})_{3-\delta}O_4$ at 1573–1673 K, obtained using the recalculated values of the deviation from the stoichiometry ((△,◆,○) points) taken from (Yamauchi et al. 1983). (b) Diagrams of the concentrations of point defects for the spinel $(Fe_{0.867}Al_{0.133})_{3-\delta}O_4$ at 1573–1673 K. (c) Diagrams of the concentrations of point defects for the spinel $(Fe_{0.8}Al_{0.2})_{3-\delta}O_4$ at 1573–1673 K.

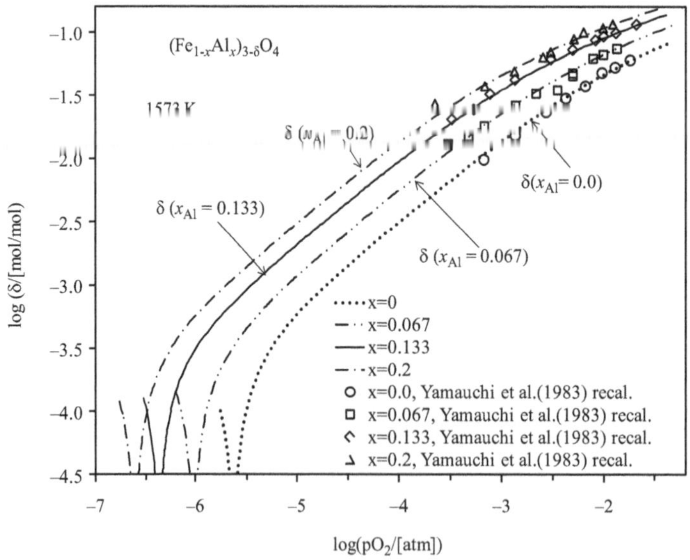

FIGURE 5.9 Oxygen pressure dependences of the deviation from the stoichiometry in the spinel $(Fe_{1-x}Al_x)_{3-\delta}O_4$ with the aluminium content of x_{Al} = 0.0–0.2 mol, at 1573 K, obtained using the corrected values of the deviation from the stoichiometry (Yamauchi et al. 1983).

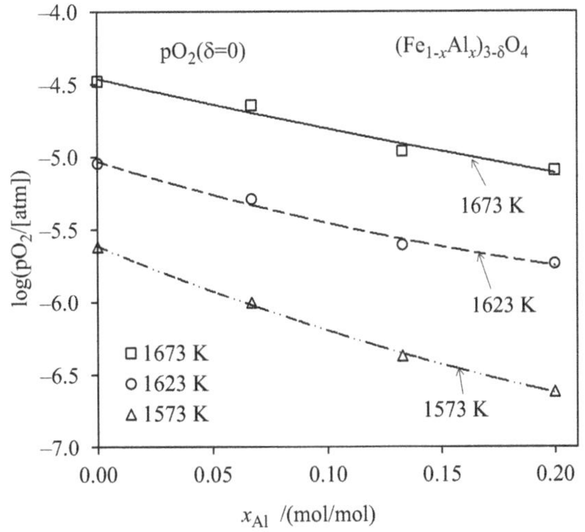

FIGURE 5.10 Dependence of the oxygen pressure at which the spinel $(Fe_{1-x}Al_x)_{3-\delta}O_4$ reaches the stoichiometric composition on the aluminium content x_{Al}, at 1573–1673 K.

pressure) and it decreases when the temperature increases. When the aluminium content in the spinel $(Fe_{1-x}Al_x)_{3-\delta}O_4$ increases, there occurs a shift towards lower values in the oxygen pressure at which the spinel reaches the stoichiometric composition, and when the temperature increases, towards higher oxygen pressures. For the assumed value of ΔG^{o}_{eh} of formation of electronic defects, equal $\Delta G^{o}_{eh} = 30$ kJ/mol, standard enthalpies of formation of cation vacancies $\Delta G^{o}_{V''_{Fe}}$ and $\Delta G^{o}_{V'''_{Fe}}$ and interstitial cations $\Delta G^{o}_{Fe_i^{3\bullet}}$ and $\Delta G^{o}_{Fe_i^{\bullet\bullet}}$ were determined, depending on the temperature and on the aluminium content in the spinel $(Fe_{1-x}Al_x)_{3-\delta}O_4$.

REFERENCES

Dieckmann, R. 1982. Defects and Cation Diffusion in Magnetite (IV): Nonstoichiometry and Point Defect Structure of Magnetite $(Fe_{3-\delta}O_4)$. *Ber. Bunsenges. Phys. Chem.* 86: 112–118.

Gillot, B., J. F. Ferriot, and A. Roussel. 1976. Conductivité electrique des magnetites finement divisées substituées a l'aluminium et au chrome. *J. Phys. Chem. Solids.* 37: 857–862.

Halloran, J. W., and H. K. Bowen. 1980. Iron Diffusion in Iron-Aluminate Spinels. *J. Am. Ceram. Soc.* 63: 58–65.

Mason, T. O., and H. K. Bowen. 1981a. Cation Distribution and Defect Chemistry of Iron-Aluminate Spinels. *J. Am. Ceram. Soc.* 64: 86–90.

Mason, T. O., and H. K. Bowen. 1981b. Electronic Conduction and Thermopower of Magnetite and Iron-Aluminate Spinels. *J. Am. Ceram. Soc.* 64: 237–242.

Meyers, C. E., T. O. Mason, W. T. Petuskey, J. W. Halloran, and H. K. Bowen. 1980. Phase Equilibria in the System Fe-Al-O. *J. Am. Ceram. Soc.* 63: 659–663.

Petuskey, W. T. 1977. Interdiffusion and Thermal Migration in Iron Aluminate Spinels. PhD Thesis. MIT Cambridge, Massachussetts.

Yamauchi, S., A. Nakamura, T. Shimizu, and K. Fuekl. 1983. Vacancy Diffusion in Magnetite hercynite Solid Solution. *J. Solid State Chem.* 50: 20–32.

6 Cobalt-Doped Magnetite – $(Fe_{1-x}Co_x)_{3\pm\delta}O_4$

6.1 INTRODUCTION

An example of magnetite doped with metal ions with the second ionisation degree is magnetite doped with cobalt, $(Fe_{1-x}Co_x)_{3\pm\delta}O_4$, where cobalt ions incorporate into nodes of Fe^{2+} ions, forming a solid solution. In the limit case, the composition of $CoFe_2O_4$ is reached. The range of existence of spinel $(Fe_{1-x}Co_x)_{3\pm\delta}O_4$ depending on the cobalt concentration was reported in the literature (Smiltens 1957, Carter 1960, Aukrust and Muan 1964, Schmalzried and Tretjakow 1966, Zhang and Chen 2013, Sahu, et al. 2015). The studies on the deviation from the stoichiometry, using the coulometric titration of the spinel $(Fe_{1-x}Co_x)_{3\pm\delta}O_4$, at 1473 K were performed by Schmalzried group (Sockel and Schmalzried 1968). Extensive studies of the deviation from the stoichiometry in dependence on the oxygen pressure, at 1473 K, in the spinel $(Fe_{1-x}Co_x)_{3\pm\delta}O_4$ with the cobalt content of $x_{Co} = 0.1$–0.333 mol were done by Dieckmann group (Lu et al. 1993). They conducted also studies on the oxygen pressure dependence of diffusion of radioisotope metals: iron ^{59}Fe, cobalt ^{60}Co, manganese ^{53}Mn (Lu et al. 1993). The studies of diffusion in cobalt-doped magnetite at a constant oxygen pressure were performed in a couple of laboratories (Borisenko and Morozov 1955, Müller and Schmalzried 1964). It was shown that in the range of existence of spinels $(Fe_{1-x}Co_x)_{3\pm\delta}O_4$ there are defects of the type similar to those in pure magnetite (Lu et al. 1993). At higher oxygen pressures, cation vacancies dominate, while at low pressures interstitial cations dominate. Studies on the diffusion showed, similarly as for magnetite, that the coefficients of self-diffusion depend on the type of metal and they are different in the range in which cation vacancies and interstitial cations dominate. Changes in the character of the dependence of coefficients of self-diffusion of tracer ions on the oxygen pressure are fully consistent with the model of the structure of ionic defects.

Electrical conductivity of the spinel $(Fe_{1-x}Co_x)_{3\pm\delta}O_4$ was studied mainly at room temperature and at higher temperatures (Constantin and Rosenberg 1971, Erickson and Mason 1985, Gillot 1982, Gillot and Jemmali 1983, Iida et al. 1958, Jonker 1959, Kimmet and Poplawsky 1974, Kumashiro 1967). It was found that the electrical conductivity increases when the temperature increases. Activation energy of electrical conductivity is about 0.1–0.2 eV depending on the cobalt content. When cobalt content in magnetite increases, the electrical conductivity decreases. The studies of thermoelectric power showed, similarly as for magnetite, that the Seebeck coefficient value is negative (Erickson and Mason 1985, Yamada 1973). The studies on electrical conductivity and thermoelectric power of the spinel $Fe_{3-x}Co_xO_4$ with cobalt content of $x_{Co} = 0.0$–0.8 mol, in the temperature range of 873–1673 K, were also reported (Erickson and Mason 1985). They confirmed that the charge transfer occurs according to the hopping mechanism of small polarons. Based on the values of the Seebeck coefficient, the distribution of metal ions in doped magnetite was determined (Erickson and Mason 1985); it implies that cobalt incorporates into nodes of Fe^{2+} ions in the sublattice with octahedral and tetrahedral coordination, which can be described with the formula.

$$Fe_a^{3+}Fe_{b-z}^{2+}Co_z^{2+}\left(Fe_{c-y}^{2+}Co_y^{2+}Fe_d^{3+}\right)O_4$$

Depending on the temperature, an equilibrium state and a determined distribution of ions is attained (specific values of coefficients *a*, *b*, *c*, *d*, *z* and *y* are set). The suggested distribution is qualitatively consistent with the studies of magnetisation (Murray and Linnett 1976, Pelton et al. 1979, Sawatzky et al. 1969) and the studies on the Mössbauer effect (Franke and Rosenberg 1977, Sorescu 2011), as well as with theoretical calculations (Subramanian et al. 1994).

6.2 METHODS OF THE CALCULATIONS OF THE DIAGRAMS OF THE CONCENTRATIONS OF DEFECTS

Cobalt-doped magnetite can be obtained from a solid-phase reaction, by annealing appropriate quantities of oxides: FeO, CoO and Fe_2O_3 according to the resultant reaction:

$$\left(1-x\right)Fe0 + xCoO + Fe_2O_3 \rightarrow Fe_{(t)}^{3+}\left(Fe_{1-x}^{2+}Co_x^{2+}Fe_{(o)}^{3+}\right)O_4 \tag{6.1}$$

It was assumed that cobalt ions Co^{2+} incorporate mainly into the nodes of Fe^{2+} ions with octahedral or tetrahedral coordination (at high temperatures, an equilibrium distribution of ions in spinel sublattices is set). Therefore, the incorporation of cobalt ions does not perturb electroneutrality of the spinel and the structure of ionic defects does not change. A change in the oxygen pressure causes an increase or decrease in the concentration of point defects according to Equations 1.10–1.14.

Analysis of the results of the deviation from the stoichiometry (Lu et al. 1993), plotted in a log–log scale, showed that in the range of low oxygen pressures, where interstitial cations should dominate, measured values significantly deviate from the dependence of δ on p_{O_2} with an exponent of −2/3, as suggested in (Lu et al. 1993) (see dotted line in Figure 6.1a). Large discrepancies are present especially in the spinel $(Fe_{1-x}Co_x)_{3\pm\delta}O_4$ with cobalt content of $x_{Co} = 0.2$ mol. Such character of the dependence of δ on p_{O_2} indicates that cobalt ions not only incorporate into nodes of Fe^{2+} ions, but also a small quantity of these ions occupies nodes of the sublattice of ions Fe^{3+} $\left(Co'_{Fe}\right)$, becoming ions that are negative in relation to the sublattice. The doping reaction, resulting in building-in of a molecule into the spinel crystal, assumes the form of:

$$4CoO + V_i \rightarrow Co^x_{Fe(2+)(o)} + Co'_{Fe(3+)(o)} + Co'_{Fe(3+)(t)} + 4O^x_O + Co_i^{\bullet\bullet} \tag{6.2}$$

Therefore, as a result of incorporation of four CoO “molecules”, four oxygen sites and three cobalt sites are created, as well as one interstitial cobalt ion. Thus, cobalt ions Co^{2+}, incorporated into nodes of the sublattice of Fe^{3+} ions with octahedral and tetrahedral coordination will have the effective negative charge in relation to the lattice $\left(Co'_{Fe}\right)$ and the fourth cobalt ion will incorporate into an interstitial void $\left(Co_i^{\bullet\bullet}\right)$, causing an increase in the concentration of the interstitial cations. As a result of diffusion of cobalt ions, an equilibrium distribution of cobalt ions and an equilibrium concentration of ionic defects will be set between the individual sublattices of the spinel. In consequence, in the spinel $(Fe_{1-x}Co_x)_{3\pm\delta}O_4$, apart from electroneutral cobalt ions Co^{2+}, there will be a small concentration of ions $\left[Co'_{Fe}\right]$ negative in relation to the lattice, which causes an increase in the concentration of interstitial cations. Therefore, according to the reaction (6.2), the incorporation, into the structure of spinel $(Fe_{1-x}Co_x)_3O_4$ with the stoichiometric composition, of cobalt ions $\left(Co'_{Fe}\right)$ in the amount of y_{Co}' mol will cause an increase in the concentration of interstitial cations by $1/4y_{Co}'$:

$$\left[Fe_i^{\bullet\bullet}\right] = y^o_{Fe_i^{\bullet\bullet}} + 1/4y_{Co'}$$

where $y^o_{Fe_i^{\bullet\bullet}}$ denotes the concentration of interstitial cations at the stoichiometric composition $\left[Fe_i^{\bullet\bullet}\right]$, as well as $\left[Co_i^{\bullet\bullet}\right]$. Incorporation of ions $\left(Co'_{Fe}\right)$ causes perturbation in the equilibrium between the concentration of point defects and the equilibrium pressure over such a doped spinel will be shifted towards lower oxygen pressures in relation to non-doped spinel. An increase in the oxygen pressure will cause a change in the concentration of point defects and, at a determined oxygen pressure, a new equilibrium state will be attained (a new equilibrium concentration of ionic defects and electronic defects will be set). As a result, the stoichiometric composition of the spinel doped with ions Co'_{Fe} will be reached at a higher oxygen pressure compared to non-doped spinel.

In calculations of concentrations of ionic defects, it is necessary to take into account, under the electroneutrality condition, the concentration of ions $\left[Co'_{Fe}\right] = 1/2y_{Co}'$:

$$\left[e'\right] + 2\left[V''_{Fe(o)}\right] + 3\left[V'''_{Fe(o)}\right] + 3\left[V'''_{Fe(t)}\right] + 1/2y_{Co} = \left[h^\bullet\right] + 2\left[Fe_i^{\bullet\bullet}\right] + 3\left[Fe_i^{\bullet\bullet\bullet}\right] \tag{6.3}$$

Also, when calculating the deviation from the stoichiometry, it is necessary to consider an increase in the number of sites due to incorporation of oxygen ions and cobalt ions:

$$\frac{\left[Fe\right] + \left[Co\right]}{\left[O\right]} = \frac{3 + y_{Co} - \Sigma\left[def\right]}{4 + y_{Co}} \tag{6.4}$$

where $\Sigma\left[def\right] = \left[V''_{Fe(o)}\right] + \left[V'''_{Fe(o)}\right] + \left[V'''_{Fe(t)}\right] - \left[Fe_i^{\bullet\bullet}\right] - \left[Fe_i^{3\bullet}\right]$

The method for calculating the diagrams of the concentrations of point defects presented in Chapter 1.5 for the spinel $(Fe_{1-x}Co_x)_{3\pm\delta}O_4$ should be modified, taking into account concentration of ions $\left[Co'_{Fe}\right]$ (similarly as in the case of titanium and chromium doping; see Chapters 3 and 4). Due to the lack of the value of ΔG^o_{eh} of formation of electronic defects, it was assumed that it depends only slightly on the content of cobalt in spinel $(Fe_{1-x}Co_x)_{3\pm\delta}O_4$. Due to this, in the calculations, a constant value of ΔG^o_{eh} = 30.3 kJ/mol was assumed, the same as for pure magnetite. As shown by initial calculations, a change in the value ΔG^o_{eh} in the range of 25–50 kJ/mol does not affect the degree of match between the dependence of δ on p_{O_2} to the experimental results δ, but only the values of ΔG^o_i of formation of individual defects change. As it was decided to take the concentration of ions $\left[Co'_{Fe}\right]$ into account in the calculations, the relative concentrations of defects were calculated as a function of the concentration of vacancies $\left[V''_{Fe}\right]$, considering an increase in the concentration of interstitial ions

$\left[Co_i^{\bullet\bullet}\right]$. Therefore, according to the method of calculations, the following were adjusted: the values of ΔG^o_{F2} of formation of Frenkel defects (in the first step of calculations it was assumed that $\Delta G^o_{F2} = \Delta G^o_{F1}$) and the values of $\Delta G^o_{V''_{Fe(o)}}$ of formation of cation vacancies $\left(V''_{Fe(o)}\right)$. Simultaneously, the values of ΔG^o_I were adjusted in order to obtain similar values of cation vacancies at the stoichiometric composition, $\left[V''_{Fe}\right] \cong \left[V'''_{Fe(o)}\right]$ (see Equation 1.35a) and the concentration of cobalt ions y_{Co}' was adjusted in such a way that the dependence of δ on p_{O_2} fitted the experimental results δ as much as possible. Then such values of ΔG^o_{F3} were chosen, that the concentration of vacancies $\left[V'''_{Fe(t)}\right]$ did not deteriorate the quality of match between the dependence of δ on p_{O_2} and the experimental values of δ. The equilibrium oxygen pressure was calculated according to Equation 1.10a, using the values of $\Delta G^o_{V''_{Fe}}$, concentration of vacancies V''_{Fe} and concentration of electron holes.

6.3 RESULTS OF CALCULATION OF THE DIAGRAMS OF THE CONCENTRATIONS OF DEFECTS, AND DISCUSSION

Figure 6.1 shows diagrams of the concentrations of ionic defects for the spinel $(Fe_{1-x}Co_x)_{3\pm\delta}O_4$, with cobalt content of x_{Co} = 0.1, 0.2, 0.3, and 0.333 mol at 1473 K (solid lines), considering the concentration of ions $\left[Co'_{Fe}\right]$ incorporated into the nodes of Fe^{3+}ions, obtained using the results of the deviation from the stoichiometry ((O) points). The (---) line denotes the

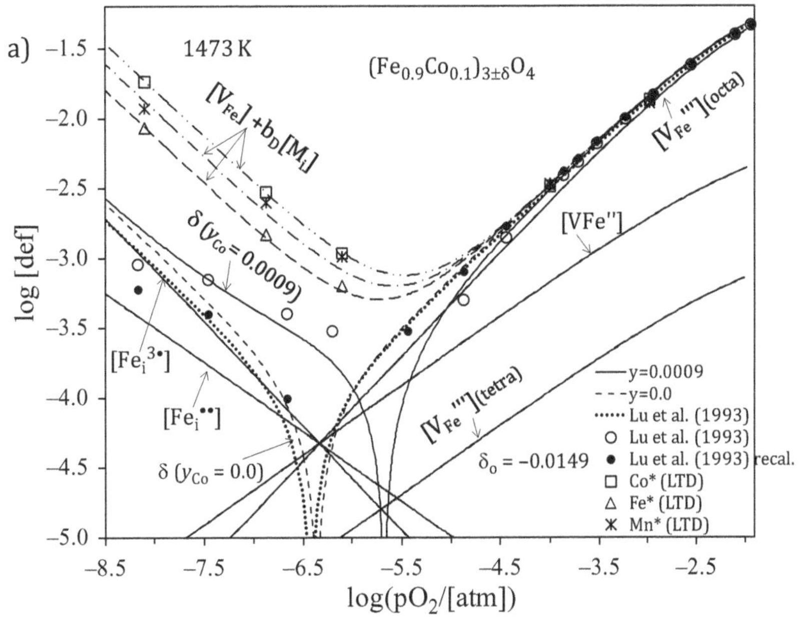

FIGURE 6.1 (a) Diagram of concentrations of point defects for the spinel $(Fe_{0.9}Co_{0.1})_{3\pm\delta}O_4$ at 1473 K (solid lines) obtained using the results of the studies of the deviation from the stoichiometry (Lu et al. 1993) ((O) points) and taking into account cobalt ions Co'_{Fe} incorporated into the sublattice of Fe^{3+} ions in the amount of $y_{Co}' = 9\cdot10^{-4}$ mol/mol. The (····) line shows the dependence of δ on p_{O_2} determined using "equilibrium constants" of reactions of formation of cation vacancies K_V and interstitial cations K_I (Lu et al. 1993). Points (●) are the recalculated values of the deviation from the stoichiometry relative to the value $\delta_o = {}^-1.49\cdot10^{-2}$, and (---) line represents the dependence of δ on p_{O_2} for $y_{Co}' = 0.0$. The remaining points represent the values of the sum of concentrations of ionic defects ($[V_{Fe}] + b_D[M_i]$), calculated according to Equation (1.47) for fitted values of $D^o_{V(M)}$ for the highest oxygen pressures and when using the values of coefficients of self-diffusion of tracers: Fe* – (△) points, Co* – (□), Mn* – (✳) (Lu et al. 1993). Lines (- -), (-·-·), (-··-·) denote, respectively, dependence of the sum of concentrations of ionic defects ($[V_{Fe}] + b_D[M_i]$) on p_{O_2} for the fitted value of b_D. (b) for the spinel $(Fe_{0.867}Co_{0.133})_{3\pm\delta}O_4$ at 1473 K when taking into account cobalt ions Co'_{Fe} in the amount of $y_{Co}' = 1.5\cdot10^{-3}$ mol/mol (solid lines), obtained using the recalculated values of the deviation from the stoichiometry ((●) points) relative to the value $\delta_o = -1.45\cdot10^{-3}$ (Sockel and Schmalzried 1968) ((□) points). Points (O) are the values recalculated relative to the value $\delta_o = -1.1\cdot10^{-3}$, and (---) line represents the dependence of δ on p_{O_2} for $y_{Co}' = 0.0$. Line (····) shows the dependence of δ on p_{O_2} determined using the extrapolated values of "equilibrium constants" K_V and K_I (Lu et al. 1993). (c) for the spinel $(Fe_{0.8}Co_{0.2})_{3\pm\delta}O_4$ at 1473 K when taking into account cobalt ions Co'_{Fe} in the amount of $y_{Co}' = 2.8\cdot10^{-3}$ mol/mol (solid lines), obtained using the results of the studies of the deviation from the stoichiometry (Lu et al. 1993) ((O) points). Points (●) are the recalculated values of δ relative to the value $\delta_o = -0.0019$. (d) for the spinel $(Fe_{0.7}Co_{0.3})_{3\pm\delta}O_4$, at 1473 K, when taking into account cobalt ions Co'_{Fe} in the amount of $y_{Co}' = 3\cdot10^{-4}$ mol/mol (solid lines). Points (●) are the recalculated values of δ relative to the value $\delta_o = -4.9\cdot10^{-4}$. (e) for the spinel $(Fe_{0.667}Co_{0.333})_{3\pm\delta}O_4$, at 1473 K (solid lines). Points (●) denote the recalculated values of the deviation from the stoichiometry δ relative to the value $\delta_o = -1.23\cdot10^{-4}$.

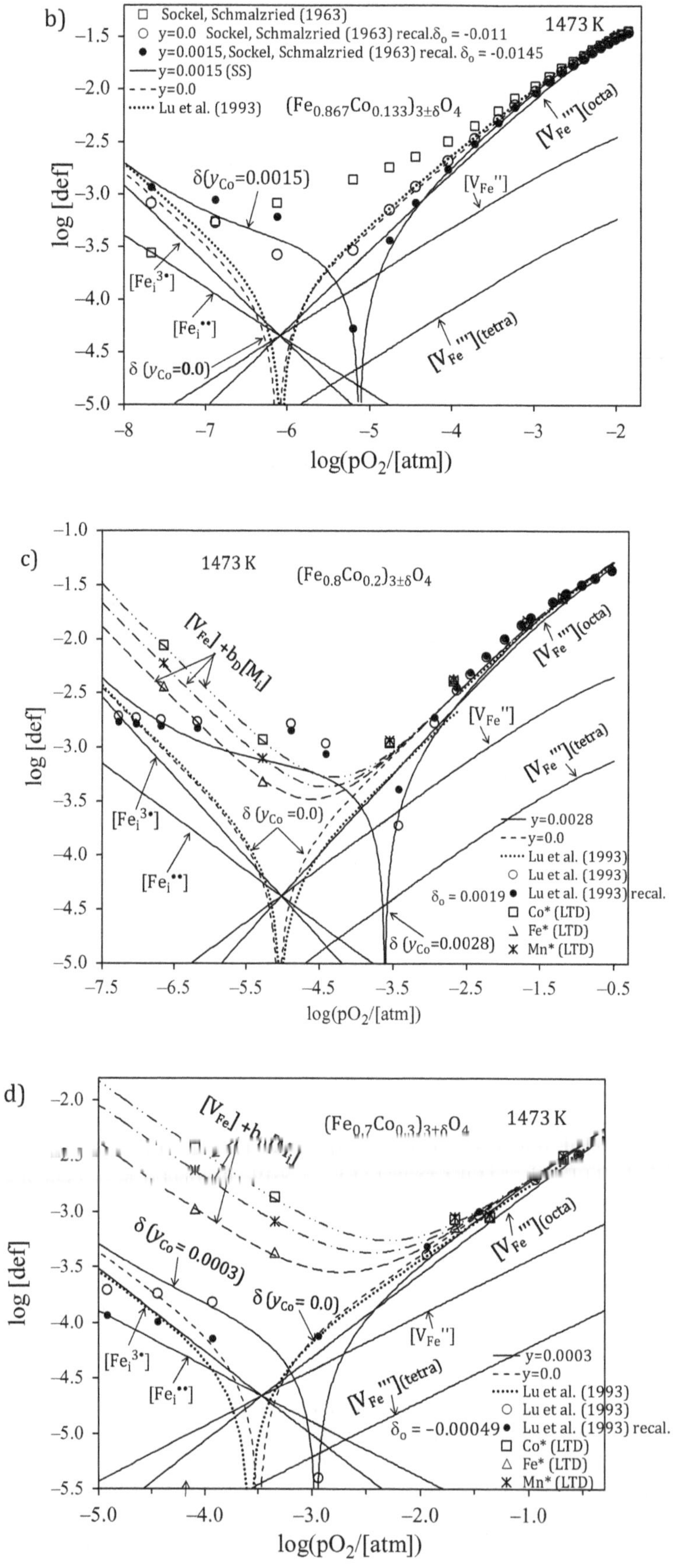

FIGURE 6.1 (Continued)

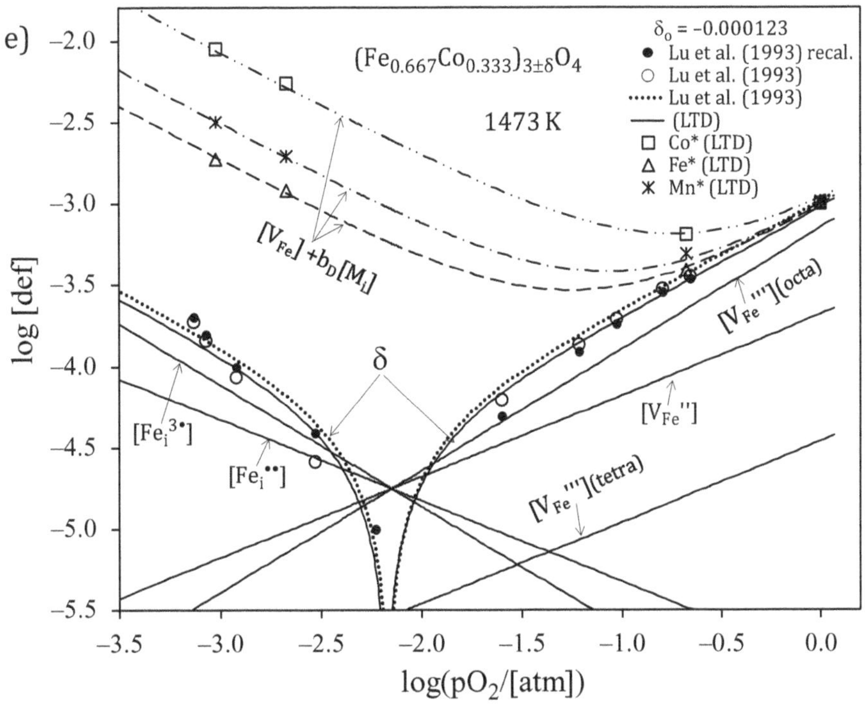

FIGURE 6.1 (Continued)

dependence of δ on p_{O_2} without considering the concentration of ions $[Co'_{Fe}]$. The dotted line shows the dependence of δ on p_{O_2} obtained using "equilibrium constants" of reactions of formation of cation vacancies K_V and interstitial ions K_I (Lu et al. 1993).

As can be seen in Figure 6.1, in the range of low oxygen pressures there are large differences between the experimental values ((○) points) and the calculated dependence of δ on p_{O_2} without considering the concentration of ions $[Co'_{Fe}]$ ((---) line), also in the case of the dependence suggested in (Lu et al. 1993) ((···) line). Only in the case of cobalt content of x_{Co} = 0.333 mol is the calculated dependence of δ on p_{O_2} consistent with the experimental values δ (Figure 6.1e). On the other hand, as can be seen in Figure 6.1a, for the cobalt content of = x_{Co} = 0.1 mol, the correction of the reference value (δ_o = −0.0149) causes that recalculated values of deviation δ ((●) points) are consistent with the dependence of δ on p_{O_2}, not considering the concentration of ions $[Co'_{Fe}]$.

Figure 6.1b presents the diagram of the concentrations of defects for the spinel $(Fe_{1-x}Co_x)_{3\pm\delta}O_4$, for the cobalt content of x_{Co} = 0.133 mol, using in the calculations the results (Sockel and Schmalzried 1968) ((□) points).

As can be seen in Figure 6.1b, the differences are larger, because in (Sockel and Schmalzried 1968) the stoichiometric composition was assumed at the decay pressure of the spinel. Correction of values by introducing the reference value δ_o = −0.011 ((○) points) only slightly changed the character of the dependence of δ on p_{O_2} at low oxygen pressures. At higher oxygen pressures, a match was obtained between the dependence of δ on p_{O_2}, determined using the extrapolated values of "equilibrium constants" K_V and K_I given in (Lu et al. 1993). A correction of the value of the deviation from the stoichiometry reported in (Sockel and Schmalzried 1968) at the value of δ_o = −0.0145 ((●) points) causes that the dependence of δ on p_{O_2}, taking into account the concentration of ions $[Co'_{Fe}]$ in the amount of y_{Co} = 0.0015 mol/mol is consistent with the recalculated values of the deviation δ.

Therefore, the experimental character of the deviation from the stoichiometry on p_{O_2} in spinels $(Fe_{1-x}Co_x)_{3\pm\delta}O_4$, for the cobalt content of x_{Co} = 0.1–0.2 (Lu et al. 1993, Sockel and Schmalzried 1968), presented in Figure 6.1, indicates that the sublattice of ions Fe^{3+} incorporates ions Co^{2+} (Co'_{Fe}). Due to large values of the deviation from the stoichiometry at low oxygen pressures, the correction of values δ only slightly affects the character of the dependence ((●) points). The result of incorporation of ions Co'_{Fe} into the spinel structure is an increase in the deviation at low oxygen pressures and a shift in the oxygen pressure at which the spinel reaches the stoichiometric composition towards higher pressures (compared with spinel non-doped with ions Co'_{Fe}).

Figure 6.2 presents the dependence of the concentration of cobalt ions $[Co'_{Fe}] = 1/2y_{Co'}$ on the cobalt content (x_{Co}) in the spinel $(Fe_{1-x}Co_x)_{3\pm\delta}O_4$.

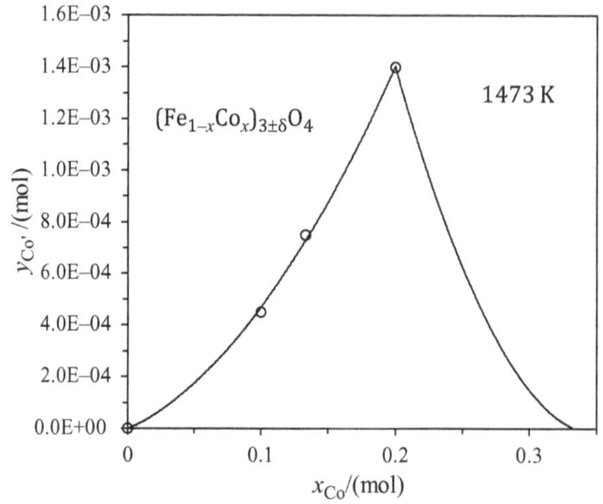

FIGURE 6.2 Dependence of the concentration of cobalt ions $\left[Co'_{Fe}\right] = 1/2y_{Co'}$ incorporated into the sublattice of ions Fe^{3+} on the cobalt content x_{Co} in the spinel $(Fe_{1-x}Co_x)_{3\pm\delta}O_4$, at 1473 K.

As can be seen in Figure 6.2, the concentration of ions $\left[Co'_{Fe}\right]$ increases when the cobalt content in the spinel increases and it reaches the value of $1.4 \cdot 10^{-3}$ mol/mol at the cobalt content of $x_{Co} = 0.2$ mol, and then it decreases down to zero at the cobalt content of $x_{Co} = 0.333$ mol, thus, in the spinel, in which all the Fe^{2+} sites were substituted with cobalt ions. The spinel $CoFe_2O_4$ should be therefore treated as a heterocation spinel, different from magnetite doped with cobalt, despite the fact that cobalt ions in the spinel $(Fe_{1-x}Co_x)_{3\pm\delta}O_4$ with a lower content ($x_{Co} < 0.333$ mol) form a solid solution.

As can be seen in Figure 6.1, the incorporation of cobalt in the form of Co'_{Fe} ions causes a shift in the oxygen pressure at which the spinel reaches the stoichiometric composition towards higher pressures by one order of magnitude, compared to spinel non-doped with Co'_{Fe} ions, and compared to the oxygen pressure at which the concentrations of defects are equal $[V_{Fe}] = [Fe_i]$ (Frenkel defects). Due to small concentrations of ion $\left[Co'_{Fe}\right]$ $(1/2y_{Co'})$, the concentrations of ionic defects practically do not change (the lines overlap), differently from the change in the character of the dependence of the deviation from the stoichiometry on p_{O_2}. This effect is not visible in a semi-logarhitmic scale (δ on log p_{O_2}) presented in (Sockel and Schmalzried 1968), but the differences are visible in a log–log plot.

Therefore, from the presented calculations it results that in cobalt-doped magnetite formed in a reaction between appropriate amounts of oxides: FeO, CoO and Fe_2O_3, Co^{2+} ions form a solid solution (occupying the nodes of Fe^{2+} ions), which causes a change in interionic interactions (a change in lattice energy of the spinel). However, a small concentration of cobalt ions, up to about $\left[Co'_{Fe}\right] = 0.001$ mol/mol incorporates into the formed spinel structure into the sublattice of Fe^{3+} ions that are negative in relation to the lattice. The effect is a specific change in the dependence of δ on p_{O_2} (perturbed symmetry of branches of the dependence of δ on p_{O_2}).

Figure 6.3 compares the dependences of deviation from the stoichiometry on the oxygen pressure in spinels $(Fe_{1-x}Co_x)_{3\pm\delta}O_4$ with the cobalt content of $x_{Co} = 0.0–0.333$ mol at 1473 K.

As can be seen in Figure 6.3, cobalt ions in the doped magnetite have an opposite effect on the oxygen pressures where the spinel exists than doping metal ions M^{3+}. When the cobalt content in magnetite increases, the range of existence of spinel is shifted towards higher oxygen pressures. Therefore, there occurs a shift in the transition of cobalt-doped wüstite into doped magnetite and the transition of spinel into doped hematite. It should be noted that the transition of spinel $(Fe_{1-x}Co_x)_{3\pm\delta}O_4$ into doped wüstite occurs at practically the same concentration of defects. The oxygen pressure at which spinels $(Fe_{1-x}Co_x)_3O_4$ reach the stoichiometric composition is also shifted towards higher oxygen pressures. The value of this pressure depends on the concentration of ions Co'_{Fe} and it is different from the oxygen pressure at which the concentration of cation vacancies and interstitial ions are equal $[V_{Fe}] = [Fe_i]$ (Frenkel defects).

Figure 6.4 presents the dependence of oxygen pressure, at which the spinel $(Fe_{1-x}Co_x)_3O_4$ reaches the stoichiometric composition, on the cobalt content x_{Co}.

As can be seen in Figure 6.4, these pressures increase when the cobalt content increases and the dependence only slightly deviates from linearity. The oxygen pressures at which Frenkel type defects are present, are significantly lower and they also increase when the cobalt content in the spinel increases. The differences present in the above oxygen pressures are a result of presence of a determined concentration of ions $\left[Co'_{Fe}\right]$.

In turn, Figure 6.5 presents the dependence of the concentration of Frenkel defects in the spinel $(Fe_{1-x}Co_x)_{3\pm\delta}O_4$ on the cobalt content x_{Co}.

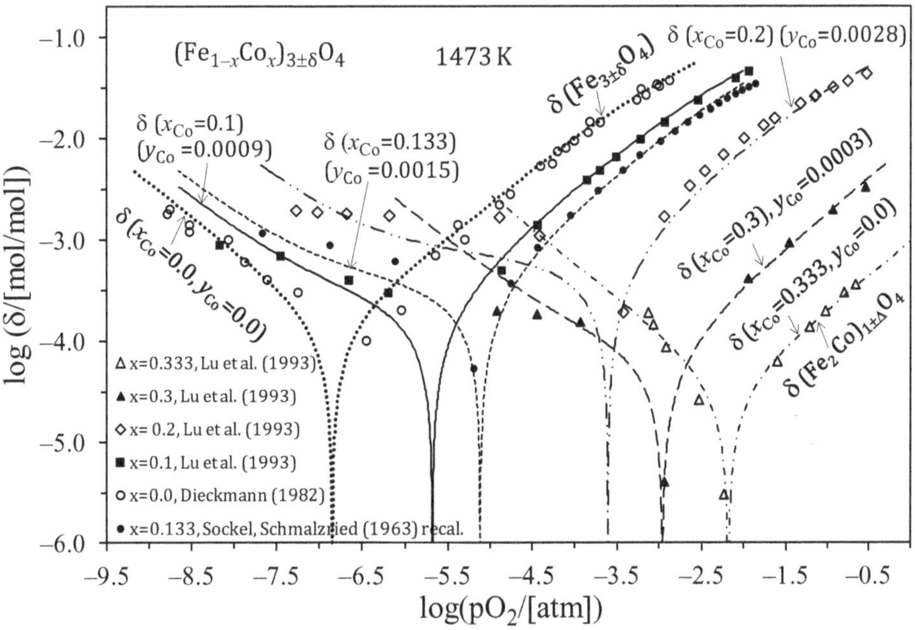

FIGURE 6.3 Oxygen pressure dependences of the deviation from the stoichiometry in the spinel $(Fe_{1-x}Co_x)_{3\pm\delta}O_4$ with the cobalt content of x_{Co} = 0.0, 0.1, 0.133, 0.2 and 0.333 mol, at 1473 K (presented in Figure 6.1).

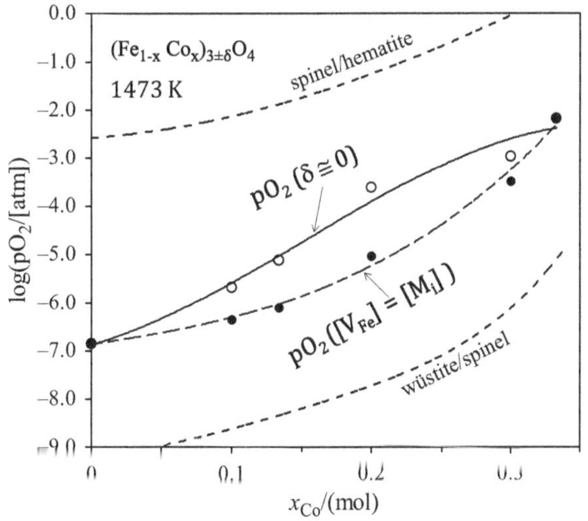

FIGURE 6.4 Dependence of the oxygen pressure at which the spinel $(Fe_{1-x}Co_x)_3O_4$ reaches the stoichiometric composition ((○) points) on the cobalt content x_{Co}, at 1473 K, and the oxygen pressures at which the concentrations of cation vacancies and interstitial cations are equal ((●) points). Dashed lines (——) denote pressures at the phase boundary doped wüstite/spinel and spinel/doped hematite (Zhang and Chen 2013).

As can be seen in Figure 6.5, at the cobalt content of x_{Co} = 0.1 mol there occurs a significant decrease in the concentration of Frenkel defects, and at the content of x_{Co} = 0.1–0.2 mol their concentration slightly decreases. The concentration of Frenkel defects decreases significantly at the cobalt content of x_{Co} = 0.3 mol. This is a result of a significant increase in the value of ΔG^{o}_{Fi} when the cobalt content in the spinel increases (see Figure 6.7). The obtained changes of the oxygen pressures at which Frenkel defects are present, and the decrease in their concentration with the increases of cobalt ion content in the spinel $(Fe_{1-x}Co_x)_{3\pm\delta}O_4$ indicate that cobalt ions Co^{2+} cause a decrease in the bonding energy and their effect is inverse than that in the case of doping ions Cr^{3+}, Al^{3+}, Ti^{3+}.

Figure 6.6 shows the dependence of the derivative $(d\log p_{O_2}/d\log\delta) = n$ on the oxygen pressure, for the dependences of the deviation from the stoichiometry on p_{O_2} shown in Figure 6.1.

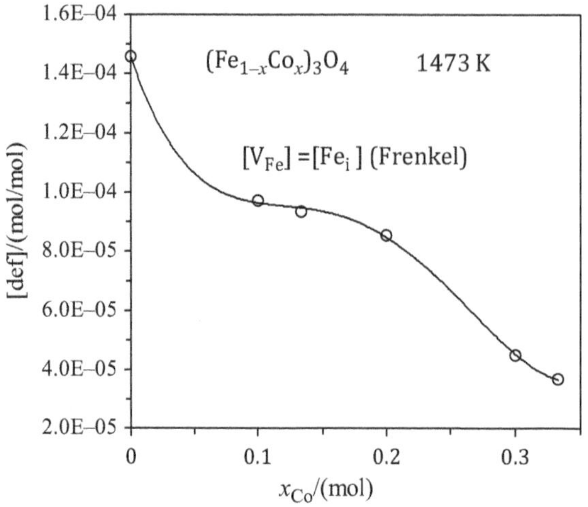

FIGURE 6.5 Dependence of the concentration of Frenkel defects on the cobalt content x_{Co} in the spinel $(Fe_{1-x}Co_x)_{3\pm\delta}O_4$ at 1473 K.

As can be seen in Figure 6.6, for the cobalt content of x_{Co} = 0.1 and 0.2 mol/mol, at the highest oxygen pressures, in a certain narrow range, the derivative reaches the value n = 1.5 and then it increases. At the lowest oxygen pressures the parameter n tends to the value of –1.5 ($1/n = -2/3$). The fact that the parameter n reaches high negative values is related to the character of the dependence of δ on p_{O_2}, resulting from the presence of Co'_{Fe}. ions.

Figures 6.7–6.9 present the dependence of adjusted standard Gibbs energies: $\Delta G^o_{F_1}$, $\Delta G^o_{F_2}$, $\Delta G^o_{F_3}$, $\Delta G^o_{V''_{Fe(o)}}$, ΔG^o_I on the cobalt content in the spinel $(Fe_{1-x}Co_x)_{3\pm\delta}O_4$ at 1473 K.

Figures 6.8–6.10 show the dependence of $\Delta G^o_{V'''_{Fe}}$ of formation of vacancies $V'''_{Fe(o)}$ (calculated according to Equation 1.35d), $\Delta G^o_{Fe_i^{3\bullet}}$ of formation of interstitial iron ions $Fe_i^{3\bullet}$ (calculated according to Equation 1.34c) and $\Delta G^o_{Fe_i^{\bullet\bullet}}$ of formation of ions $Fe_i^{\bullet\bullet}$ (calculated according to Equation 1.36c).

As can be seen in Figure 6.7, standard Gibbs energy $\Delta G^o_{F_1}$ of formation of Frenkel defects increases when the cobalt content in the spinel $(Fe_{1-x}Co_x)_{3\pm\delta}O_4$ increases. A significant increase in $\Delta G^o_{F_1}$ occurs at the cobalt content of x_{Co} = 0.1 mol, and a small increase at the content of x_{Co} = 0.1–0.2 mol. A significant increase in $\Delta G^o_{F_1}$ occurs at the cobalt content of x_{Co} = 0.3 mol. The consequence of these changes is a decrease in the concentration of Frenkel defects.

As can be seen in Figure 6.8 and 6.10, the values of ($\Delta G^o_{V'''_{Fe}}$) and ($\Delta G^o_{V''_{Fe}}$) monotonically decrease when the concentration of cobalt in magnetite increases, and the values of $\Delta G^o_{Fe_i^{3\bullet}}$ and $\Delta G^o_{Fe_i^{\bullet\bullet}}$, inversely, decrease.

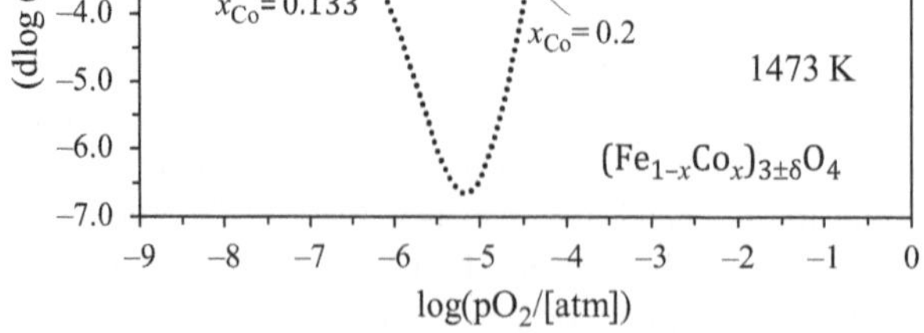

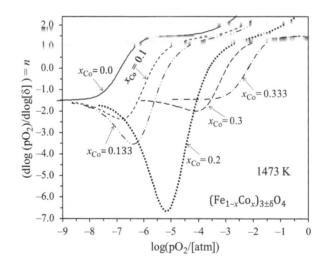

FIGURE 6.6 Dependences of the derivative $n_\delta = dlog(p_{O_2})/dlog\delta$ on p_{O_2}, obtained for the dependence of the deviation from the stoichiometry on p_{O_2} in the spinel $(Fe_{1-x}Co_x)_{3\pm\delta}O_4$ with the cobalt content x_{Co} = 0.0–0.333 mol, at 1473 K (presented in Figure 6.1).

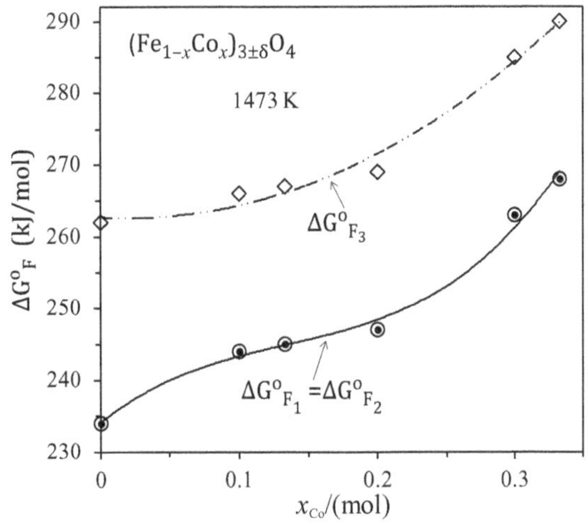

FIGURE 6.7 Dependence of standard Gibbs energy $\Delta G^o_{F_1}$ ((○) points), $\Delta G^o_{F_2}$ (●) and $\Delta G^o_{F_3}$ (◇) of formation of Frenkel defects on the cobalt content x_{Co} in the spinel $(Fe_{1-x}Co_x)_{3\pm\delta}O_4$, at 1473 K, obtained using the results of the studies of the deviation from the stoichiometry (Lu et al. 1993).

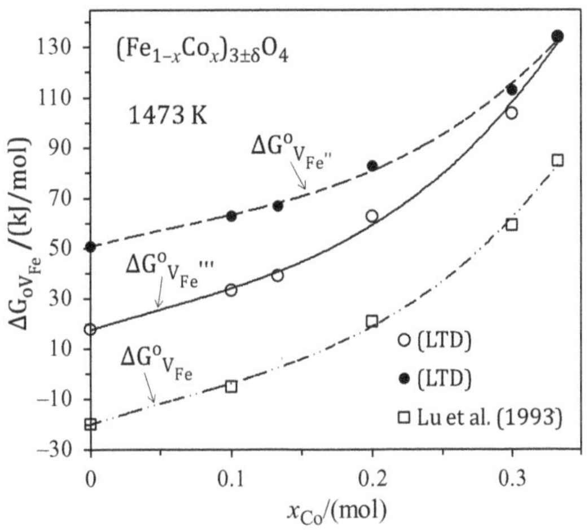

FIGURE 6.8 Dependence of standard Gibbs energy $\Delta G^o_{V'''_{Fe}}$ ((○) points) and $\Delta G^o_{V''_{Fe}}$ ((●) points) of formation of vacancies $\left(V'''_{Fe(o)}\right)$ and $\left(V''_{Fe(o)}\right)$ on the cobalt content x_{Co} in the spinel $(Fe_{1-x}Co_x)_{3\pm\delta}O_4$, at 1473 K and the values $\Delta G^o_{V_{Fe}}$ calculated using the values of "equilibrium constants" of reaction of formation of vacancies K_V (Lu et al. 1993) ((□) points).

Figures 6.8 and 6.10 show also the values of $\Delta G^o_{V_{Fe}}$ and $\Delta G^o_{Fe_i}$ calculated using "equilibrium constants" of reaction of formation of cation vacancies K_V and interstitial ions K_I (Lu et al. 1993). The values are different from those determined in the present work, as they describe another model of defects, but the character of the dependence on the cobalt content in the spinel is similar.

6.4 DIFFUSION IN COBALT-DOPED MAGNETITE

Similarly as for pure magnetite, using the concentration of ionic defects resulting from the diagram of defects' concentrations and the values of coefficients of self-diffusion of radioisotope elements in the spinel $(Fe_{1-x}Co_x)_{3\pm\delta}O_4$, the coefficients of diffusion of defects, and specifically the mobility of metal ions via cation vacancies $\left(D^o_{V(M)}\right)$ and via interstitial cations $\left(D^o_{I(M)}\right)$, were determined (see: Chapter 1.6.2).

Figure 6.1 shows the oxygen pressure dependence of the "sum" of concentrations of ionic defects ($[V_{Fe}] + b_D[Fe_i]$) and the values of this sum, calculated according to Equation 1.47a, when using the values of coefficients of self-diffusion of

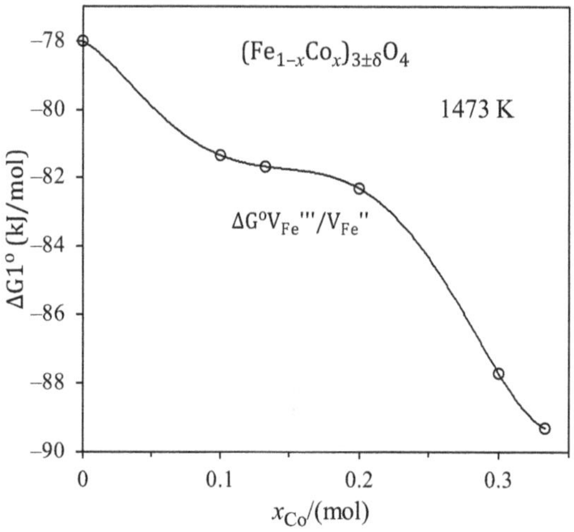

FIGURE 6.9 Dependence of standard Gibbs energy ΔG_I^o on the cobalt content x_{Co} in the spinel $(Fe_{1-x}Co_x)_{3\pm\delta}O_4$ at 1473 K, permitting to determine the ratio of concentrations of iron vacancies $\left[V''_{Fe(o)}\right]/\left[V'''_{Fe(o)}\right]$ (from Equation 1.35b).

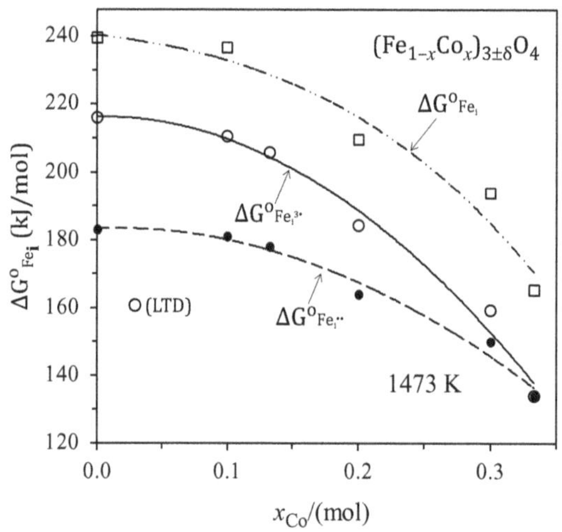

FIGURE 6.10 Dependence of standard Gibbs energy $\Delta G^o_{Fe_i^{3\bullet}}$ ((O) points) of formation of interstitial ions $Fe_i^{3\bullet}$ and $\Delta G^o_{Fe_i^{\bullet\bullet}}$ ((●) points) of formation of ions $Fe_i^{\bullet\bullet}$ on the cobalt content x_{Co} in the spinel $(Fe_{1-x}Co_x)_{3\pm\delta}O_4$, at 1473 K (calculated from Equation 1.34c and 1.36c and the values $\Delta G^o_{Fe_i}$ calculated using "equilibrium constants" of reaction of formation of interstitial cations K_I (Lu et al. 1993) ((□) points).

iron ^{59}Fe ((△) points), cobalt ^{60}Co - (□) and manganese ^{53}Mn - (✻) in the spinel $(Fe_{1-x}Co_x)_{3\pm\delta}O_4$ (Lu et al. 1993). The values of the coefficient of diffusion of cations $D^o_{V(M)}$ via iron vacancies at the highest oxygen pressures and of the parameter b_D, being the ratio of the coefficients of diffusion via defects $(D^o_{I(M)}/D^o_{V(M)})$, were adjusted in such a way as to obtain a match between the dependence of the sum of the concentrations of defects ($[V_{Fe}] + b_D[Fe_i]$) on the oxygen pressure (dashed lines) and the values calculated using the coefficients of diffusion of metal ions (tracers). As can be seen in Figure 6.1, a very good agreement was obtained between the dependence of the sum of the concentrations of defects ($[V_{Fe}] + b_D[Fe_i]$) on p_{O_2} (dashed lines (---)) and the values calculated using the coefficients of self-diffusion of tracers ((△,□,✻) points).

Figure 6.11 presents the dependence of the parameter b_D on the cobalt content x_{Co} in the spinel $(Fe_{1-x}Co_x)_{3-\delta}O_4$ for individual tracer ions.

As can be seen in Figure 6.11, the mobility via interstitial cations is higher than the mobility via cation vacancies. Their ratio $(D^o_{I(M)}/D^o_{V(M)} = b_D$ is different depending on the diffusing ion and it only slightly increases with the cobalt content up to $x_{Co} = 0.1$ mol (it is slightly higher than in pure magnetite). The parameter b_D significantly increases in the range of cobalt content of $x_{Co} = 0.2–0.333$ mol, for cobalt up to the value $b_D = 60$, for manganese up to 25, and for iron up to 15.

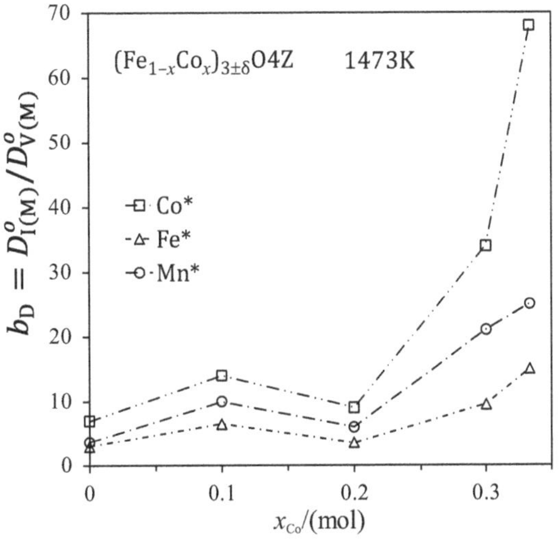

FIGURE 6.11 Dependence of the ratio of the coefficient of diffusion of ions via interstitial cations to the coefficient of diffusion via cation vacancies $D^o_{I(M)}/D^o_{V(M)} = b_D$ on the cobalt content x_{Co} in the spinel $(Fe_{1-x}Co_x)_{3\pm\delta}O_4$ at 1473 K, obtained using the of coefficients of self-diffusion of tracers: Fe* ((△) points), Co* (□), Mn* (○) (Lu et al. 1993) and concentrations of ionic defects calculated from diagrams.

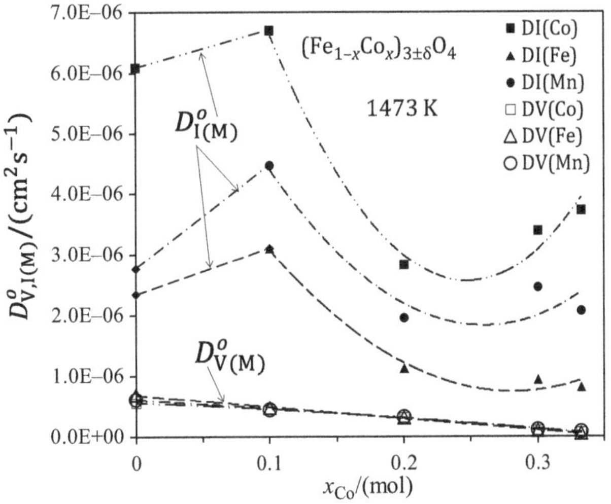

FIGURE 6.12 Dependence of the coefficient of diffusion of tracer ions via cation vacancies $D^o_{V(M)}$ (empty points) and via interstitial cations $D^o_{I(M)}$ (solid points) on the cobalt content x_{Co} in the spinel $(Fe_{1-x}Co_x)_{3\pm\delta}O_4$ at 1473 K, obtained using the of coefficients of self-diffusion of tracers: Fe* ((△,▲) points), Co* (□,■), Mn* (○,●) (Lu et al. 1993) and concentrations of ionic defects calculated from diagrams.

In turn, Figure 6.12 presents the dependence of the values of coefficients of diffusion of ions (of their mobility) via cation vacancies $D^o_{V(M)}$ (empty points) and via interstitial ions $D^o_{I(M)}$ (solid points) on the cobalt content x_{Co} in the spinel $(Fe_{1-x}Co_x)_{3\pm\delta}O_4$.

As can be seen in Figure 6.12, for the diffusion of tracers Fe*, Co* and Mn*, the coefficients of diffusion via cation vacancies $D^o_{V(M)}$ only slightly differ among them and they monotonically decrease with the increase in cobalt content in the spinel $Fe_{1-x}Co_x)_{3\pm\delta}O_4$.

According to the values of the parameter b_D, the coefficients of diffusion via interstitial cations $D^o_{I(M)}$ are significantly higher than those via vacancies and they differ significantly depending on the diffusing ion. The fastest diffusion occurs for cobalt, and the slowest for iron. The nature of the change is complex and dependent on the diffusing ion. For the cobalt content of $x_{Co} = 0.1$ mol in the spinel $(Fe_{1-x}Co_x)_{3\pm\delta}O_4$, the coefficient of diffusion via interstitial cations $D^o_{I(M)}$ is higher than that in magnetite. In the spinel with the cobalt content of $x_{Co} = 0.2$ mol they are significantly lower, and at the cobalt content above $x_{Co} = 0.3$ mol they slightly increase (within the measurement error limit).

6.5 CONCLUSIONS

Using the results of the studies of deviation from the stoichiometry for the spinel $(Fe_{1-x}Co_x)_{3\pm\delta}O_4$ with the cobalt content of x_{Co} = 0.1–0.333 mol, for the temperature of 1473 K (Lu et al. 1993, Sockel and Schmalzried 1968), diagrams of the concentrations of defects were determined. It was found that cobalt doping in magnetite affects the character of the oxygen pressure dependence of the deviation from the stoichiometry. The stoichiometric composition is reached in the range of existence of the spinel $(Fe_{1-x}Co_x)_{3\pm\delta}O_4$, while as the cobalt content increases, the oxygen pressure at which the spinel reaches the stoichiometric composition is shifted towards higher values, by about four orders of magnitude, at the cobalt content of x_{Co} = 0.333 mol.

In the range of higher oxygen pressures, cation vacancies $\left[V_{Fe}'''\right]$ dominate, and the character of the dependence of their concentration on the oxygen pressure is analogous to the case of pure magnetite (exponent about n = 2/3). When the concentration of cobalt in magnetite increases, the concentration of cation vacancies significantly decreases. In the range of low oxygen pressures, where interstitial cations $\left[Fe_i^{3\bullet}\right]$ dominate, the character of the dependence of deviation from the stoichiometry significantly changes when the cobalt content increases. It was shown that the above character of the dependence of δ on p_{O_2} is an effect of the incorporation of Co^{2+} ions into the spinel structure, into the sublattice of Fe^{3+} ions that are negative in relation to the lattice (Co_{Fe}'). The concentration of these ions increases with the cobalt content in the spinel, and at the content of x_{Co} = 0.2 mol reaches the value of y_{Co} = 0.0014 mol/mol, and then at the content of x_{Co} = 0.333 mol it decreases to zero.

It was found that when the cobalt content in the spinel $(Fe_{1-x}Co_x)_{3\pm\delta}O_4$ increases, standard Gibbs energies of formation of Frenkel defects increase, which results in a decrease in the concentration of Frenkel defects ($\left[V_{Fe}'''\right]=\left[Fe_i^{3\bullet}\right]$) from about $1.4{\cdot}10^{-4}$ to $4{\cdot}10^{-5}$ mol/mol at the cobalt content of x_{Co} = 0.333 mol. The standard Gibbs energies of formation of cation vacancies $\Delta G^o_{V_{Fe}'''}$ and $\Delta G^o_{V_{Fe}''}$ increase when the content of cobalt in the spinel increases, while $\Delta G^o_{Fe_i^{3\bullet}}$ and $\Delta G^o_{Fe_i^{\bullet\bullet}}$ of formation of interstitial cations decrease. The above values depend on the value of ΔG^o_{eh} of formation of electronic defects and they were determined for the value ΔG^o_{eh} = 30 kJ/mol, the same as the value assumed for pure magnetite.

Using the determined concentration of ionic defects and the values of coefficients of self-diffusion of tracers M* (^{59}Fe, ^{60}Co and ^{53}Mn) (Lu et al. 1993), the ratio of the coefficient of diffusion (mobility) of tracer ions via interstitial cations to the coefficient of diffusion via cation vacancies $\left(D^o_{I(M)}/D^o_{V(M)}=b_D\right)$ was determined. It was found that when the cobalt content in the spinel $(Fe_{1-x}Co_x)_{3\pm\delta}O_4$ increases, the mobility of ions via interstitial cations is higher than the mobility via cation vacancies and at the cobalt content of x_{Co} = 0.333 mol the parameter b_D reaches the highest values. At 1473 K, for the diffusion of cobalt, the parameter b_D reaches the value$b_D \approx 70$, for manganese $b_D \approx 25$, and for iron ions $b_D \approx 15$.

Cobalt doping of spinel $(Fe_{1-x}Co_x)_{3\pm\delta}O_4$ causes a decrease in the mobility of tracer ions via cation vacancies $\left(D^o_{V(M)}\right)$ and they are practically independent on the ion type. The mobilities of ions via interstitial cations $\left(D^o_{I(M)}\right)$ depend on the type of diffusing ions. The highest mobility is that of cobalt ions, the lowest is that of iron ions, and their values depend on the content of cobalt in the spinel. For the cobalt content of x_{Co} = 0.1 mol in the spinel $(Fe_{1-x}Co_x)_{3\pm\delta}O_4$, the coefficient of diffusion via interstitial cations $D^o_{I(M)}$ is higher than in magnetite. At the cobalt content of x_{Co} = 0.2 mol the values of $D^o_{I(M)}$ are significantly lower, and at the cobalt content above x_{Co} = 0.3 mol they slightly increase (within the measurement error limit).

REFERENCES

Aukrust, E., and A. Muan. 1964. Thermodynamic Properties of Solid Solutions With Spinel-Type Structure 2 System Co_3O_4-Fe_3O_4 At 1200 °C. *Trans. Met. Soc. AIME.* 230:1395–1399.

Borisenko, A. I., and E. I. Morozov. 1955. The Cobalt-60 Diffusion in Cobalt Ferrites. *Dokl. Akad. Nauk S. S. S. R.* 105: 1274–1277.

Carter, R. E. 1960. Dissociation Pressures of Solid Solutions from Fe_3O_4 To $0.4Fe_3O_4{\cdot}0.6CoFe_2O_4$. *J. Am. Ceram. Soc.* 43: 448–451.

Constantin, C., and M. Rosenberg. 1971. Thermoelectric Power of Pure and Substituted Magnetite Above and Below the Verwey Transmission. *Solid State Commun.* 9: 675–677.

Dieckmann, R. 1982. Defects and Cation Diffusion in Magnetite (Iv): Nonstoichiometry and Point Defect Structure Of Magnetite ($Fe_{3-\Delta}O_4$). *Ber. Bunsenges. Phys. Chem.* 86: 112–118.

Erickson, D. S., and T. O. Mason. 1985. Nonstoichiometry, Cation Distribution, and Electrical Properties in Fe_3O_4-$CoFe_2O_4$ at High Temperature. *J. Solid State Chem.* 59: 42–53.

Franke, H., and M. Rosenberg. 1977. Influence of Fe^{2+} Concentration on the Mössbauer Spectra of Iron-Cobalt and Iron-Nickel Ferrites. *Physica B+C.* 86–88:965–967.

Gillot, B. 1982. Thermopower-Composition Dependence in Iron-Chromium, Iron-Zinc, and Iron–Cobalt Ferrites. *Phys. Status Solidi (a).* 69: 719–725.

Gillot, B., and F. Jemmali. 1983. Dependence of Electrical Properties in Iron-Cobalt, Iron-Zinc Ferrites Near Stoichiometry on Firing Temperature and Atmosphere. *Phys. Status Solidi (a).* 76: 601–608.

Iida, S., H. Sekizawa, and Y. Aiyama. 1958. Uniaxial Anisotropy in Iron-Cobalt Ferrites. *J. Phys. Soc. Jpn.* 13: 58–71.

Jonker, G. H. 1959. Analysis of the Semiconducting Properties of Cobalt Ferrite. *J. Phys. Chem. Solids.* 9: 165–175.

Kimmet, S. G., and R. P. Poplawsky. 1974. Image Furnace Growth of Cobalt Ferrous Ferrite Single Crystals and their Electrical Resistivity at Low Temperatures. *Mater. Sci. Eng.* 16: 265–267.

Kumashiro, Y. 1967. Relationship between Magnetic Semiconducting Properties of $Co_x Fe_{3-x}O_4$ Polycrystals. *J. Electrochem. Soc. Jpn.* 35: 210–215.

Lu, F. H., S. Tinkler, and R. Dieckmann. 1993. Point Defects and Cation Tracer Diffusion in $(Co_x Fe_{1-x})_{3-\Delta}O_4$ Spinels. *Solid State Ionics.* 62: 39–52.

Müller, W., and H. Schmalzried. 1964. Fehlordnung in Kobaltferrit. *Ber. Bunsenges. Phys. Chem.* 68: 270–276.

Murray, P. J., and J. W. Linnett. 1976. Cation Distribution in the Spinels $Co_xFe_{3-x}O_4$. *J. Phys. Chem. Solids.* 37: 1041–1042.

Pelton, A. D., H. Schmalzried, and J. Sticher. 1979. Thermodynamics of Mn_3O_4-Co_3O_4, Fe_3O_4-Mn_3O_4 and Fe_3O_4-Co_3O_4 Spinels by Phase Diagram Analysis. *Ber. Bunsenges. Phys. Chem.* 83: 241–252.

Sahu, S. K., B. Huang, K. Lilova, B. F. Woodfield, and A. Navrotsky. 2015. Thermodynamics of Fe_3O_4-Co_3O_4 and Fe_3O_4-Mn_3O_4 Spinel Solid Solutions at the Bulk and Nanoscale. *Phys. Chem. Chem. Phys.* 17: 22286–22295.

Sawatzky, G. A., F. Van Der Woude, and A. H. Morrish. 1969. Mössbauer Study of Several Ferrimagnetic Spinels. *Phys. Rev.* 187: 747–757.

Schmalzried, H., and J. D. Tretjakow. 1966. Fehlordnung in Ferriten. *Ber. Bunsenges. Phys. Chem.* 70:180–189.

Smiltens, J. 1957. Investigation of the Ferrite Region of the Phase Diagram Fe-Co-O. *J. Am. Chem. Soc.* 79: 4881–4884.

Sockel, H. G., and H. Schmalzried. 1968. Coulometrische Titration an Übergangsmetalloxiden. *Ber. Bunsenges. Phys. Chem.* 72: 745–754.

Sorescu, M. 2011. Recoilless Fraction of Cobalt-Doped Magnetite. *Nuclear Instrum. Methods Phys. Res. B.* 269: 590–596.

Subramanian, R., R. Dieckmann, G. Eriksson, and A. D. Pelton. 1994. Model Calculations of Phase Stabilities of Oxide Solid Solutions in the Co-Fe-Mn-O System at 1200°C. *J. Phys. Chem. Solids.* 55: 391–404.

Yamada, T. 1973. Electrical Conduction in $CoFe_{2-x}Ti_xO_4$ and $Co_{1-x}Fe_{2+x}O_4$ Single Crystals. *J. Phys. Soc. Jpn.* 35: 130–133.

Zhang, W. W., and M. Chen. 2013. Thermodynamic Modeling of the Co-Fe-O System. *CALPHAD Computer Coupling Phase Diagrams Thermochem.* 41: 76–88.

7 Manganese-Doped Magnetite – $(Fe_{1-x}Mn_x)_{3\pm\delta}O_4$

7.1 INTRODUCTION

An important magnetite from the practical point of view, but also chronologically, is the magnetite doped with manganese, in which Mn^{2+} ions incorporate into nodes of Fe^{2+} ions and in the limit case the spinel reaches the composition $MnFe_2O_4$ (jacobsite). At a higher manganese content, the spinels should be treated as hausmannite doped with iron; this will be discussed in Chapter 12. Many works concerned the range of existence of spinels $(Fe_{1-x}Mn_x)_{3\pm\delta}O_4$, mainly at 1273 K and 1473 K, (Bergstein 1963, Bergstein and Kleinert 1964, Bulgakova and Rozanov 1970, Duquesnoy et al. 1975, Franke 1987, Franke and Dieckmann 1989, Franke and Dieckmann 1990, Komarov et al. 1965, Muan and Somiya 1962, Ono et al. 1971, Roethe et al. 1970, Schwerdtfeger and Muan 1967, Subramanian and Dieckmann 1993, Terayama et al. 1983, Wickham 1969, Yoo and Tuller 1988). The discussion and analysis of results on the range of existence of manganese ferrites and their enthalpies of formation can be found in (Kjellqvist and Selleby 2010). For the study of cation distribution in the sublattice of the spinel $MnFe_2O_4$ and $(Fe_{1-x}Mn_x)_{3\pm\delta}O_4$, various methods were applied, mainly neutron diffraction, X-ray diffraction (Hastings and Corliss 1956, Jirak and Vratislav 1974, Murthy et al. 1971, Rieck and Driessens 1966, Sawatzky et al. 1967) and Mössbauer spectroscopy (Becker et al. 1994, Bonsdorf et al. 1997). It was found that $MnFe_2O_4$ spinel at room temperature has the structure of normal spinel. Thus, in cation sites of the sublattice with tetrahedral coordination there are Mn^{2+} ions. At high temperatures, there occurs an exchange of ions and change in the oxidation state, and, depending on the temperature, an equilibrium distribution of cations is set. Partial studies of the deviation from the stoichiometry in a narrow range of oxygen pressures were presented in (Bergstein 1963, Bulgakova and Rozanov 1970, Duquesnoy et al. 1975, Komarov et al. 1965, Ono et al. 1971, Roethe et al. 1970, Terayama et al. 1983, Wickham 1969, Yoo and Tuller 1988).

Wide studies on the deviation from the stoichiometry and diffusion of radioisotope elements, ^{59}Fe, in the spinel $(Fe_{1-x}Mn_x)_{3\pm\delta}O_4$, at 1473 K, as a function of the oxygen pressure, were performed (Franke 1987, Franke and Dieckmann 1989, Franke and Dieckmann 1990). The equilibrium constants for the reaction of the formation of cation vacancies and interstitial cations were reported in (Lu and Dieckmann 1995). In their range of existence, the spinels $(Fe_{1-x}Mn_x)_{3\pm\delta}O_4$ reach the stoichiometric composition. Therefore, at higher oxygen pressures, cation vacancies dominate, while at low pressures interstitial cations dominate. The studies on the diffusion of iron ions showed, similarly as for magnetite, that the coefficients of self-diffusion are different in the range where cation vacancies and interstitial cations dominate and their dependence on the oxygen pressure is fully consistent with the model of ionic defects' structure. The studies on diffusion are consistent with the results obtained in other laboratories (Hohmann et al. 1967).

7.2 METHODS AND RESULTS OF CALCULATION OF THE DIAGRAMS OF THE CONCENTRATIONS OF DEFECTS AND DISCUSSION

As already mentioned, the defects' structure and the character of the dependence of the deviation from the stoichiometry on the oxygen pressure reported in (Franke and Dieckmann 1989, Lu and Dieckmann 1995) are similar as for pure magnetite. Due to that, the calculations of diagrams of concentrations of point defects in manganese-doped magnetite were conducted with a method analogous to the one used for pure magnetite (see Section 1.5), using the values of "equilibrium constants" of reactions of formation of cation vacancies K_V and interstitial ions K_I, given in (Lu and Dieckmann 1995).

Due to the lack of the value of ΔG^o_{eh} of formation of electronic defects, it was assumed that it only slightly depends on the concentration of doping ions. Furthermore, in the calculations, a constant value of $\Delta G^o_{eh} = 30.3$ kJ/mol was assumed, the same as for pure magnetite.

Figure 7.1a–e present the diagrams of the concentrations of ionic defects for spinels $(Fe_{1-x}Mn_x)_{3\pm\delta}O_4$ with the manganese content of $x_{Mn} = 0.0, 0.1, 0.2, 0.333$ and 0.4 mol, at 1473 K (solid lines), calculated using the values of "equilibrium constants" of reactions of formation of cation vacancies K_V and interstitial ions K_I, taken from (Lu and Dieckmann 1995). The dotted line (····) shows the dependence of δ on p_{O_2} obtained using the values of "equilibrium constants" K_V and K_V (Lu and Dieckmann 1995). When calculating the dependence of δ on p_{O_2}, it was assumed that at the stoichiometric composition, the concentrations of cation vacancies are equal, $\left[V'''_{Fe(o)}\right] = \left[V''_{Fe}\right]$.

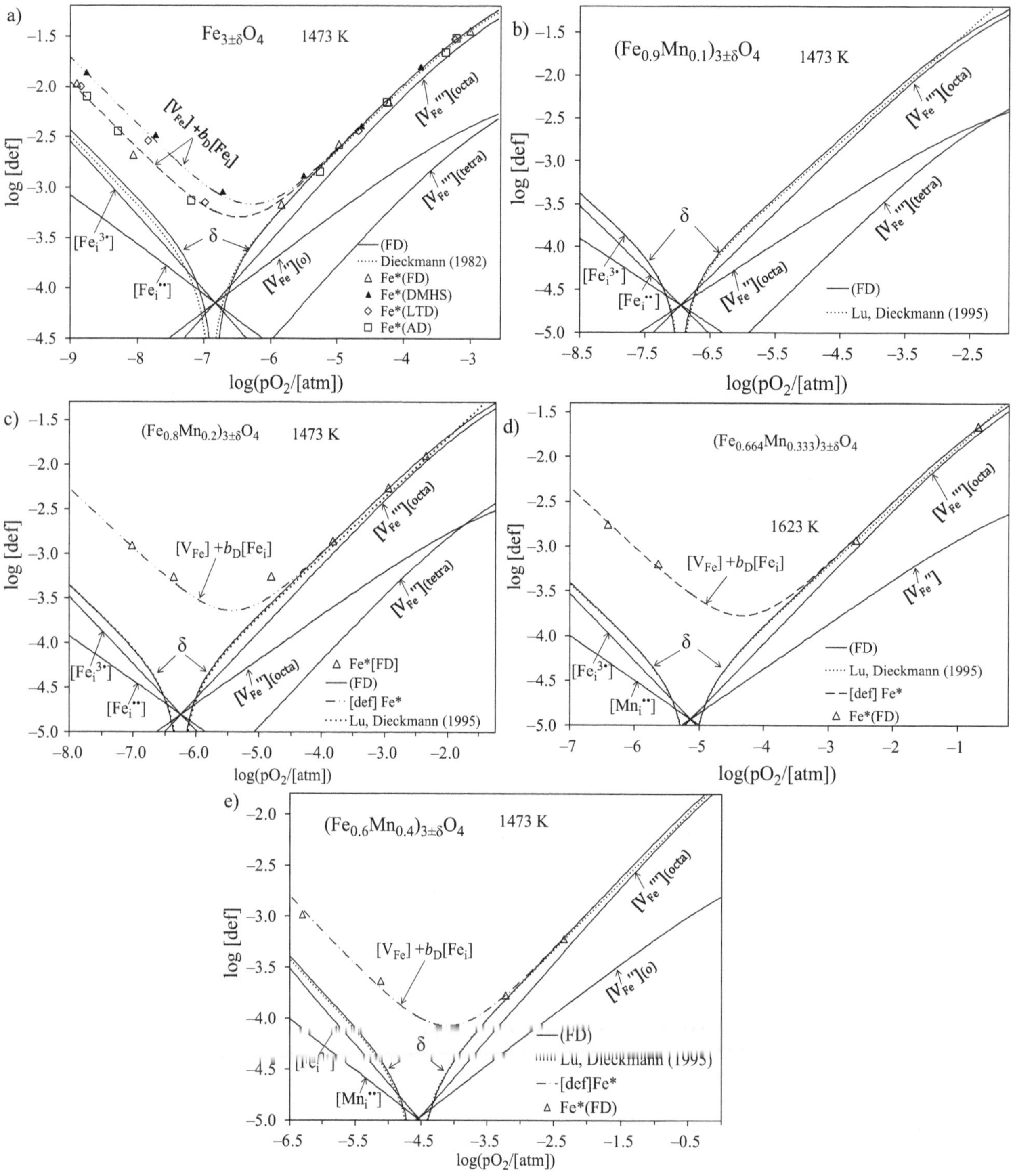

FIGURE 7.1 (a) Diagram of the concentrations of point defects for magnetite $Fe_{3\pm\delta}O_4$ at 1473 K (solid lines). The (····) line shows the oxygen pressure dependence of the deviation from the stoichiometry, determined using the "equilibrium constants" of reactions of formation of cation vacancies K_V and interstitial cations K_I (Dieckmann 1982). The points (△,▲,□,◇) represent the values of the sum of concentrations of ionic defects ($[V_{Fe}] + b_D[Fe_i]$), calculated according to Equation 1.47a and 1.47b for the fitted values of $D^o_{V(Fe)}$ for the highest oxygen pressures, and using the coefficients of self-diffusion of iron (Fe*) (Dieckmann et al. 1978, Franke and Dieckmann 1989, Lu et al. 1993). Lines (-···-) and (---) denote, respectively, dependence of the sum of concentrations of defects ($[V_{Fe}] + b_D[M_i]$) on p_{O_2} for the fitted values of b_D. (b) for the spinel $(Fe_{0.9}Mn_{0.1})_{3\pm\delta}O_4$, at 1473 K (solid lines). The (···) line shows the dependence of the deviation from the stoichiometry on p_{O_2}, calculated using the "equilibrium constants" K_V and K_I (Lu and Dieckmann 1995). (c) for the spinel $(Fe_{0.8}Mn_{0.2})_{3\pm\delta}O_4$ at 1473 K (solid lines). (d) for the spinel $(Fe_{0.664}Mn_{0.333})_{3\pm\delta}O_4$ at 1473 K (solid lines). (e) for the spinel $(Fe_{0.6}Mn_{0.4})_{3\pm\delta}O_4$ at 1473 K (solid lines).

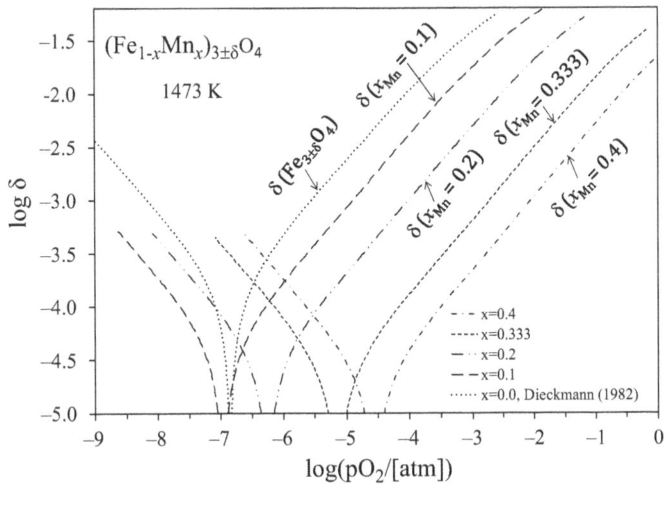

FIGURE 7.2 Dependences of the deviation from the stoichiometry on the oxygen pressure in the spinel $(Fe_{1-x}Mn_x)_{3\pm\delta}O_4$ with the manganese content of $x_{Mn} = 0.0, 0.1, 0.2, 0.333$ and 0.4 mol, at 1473 K (presented in Figure 7.1a–e).

As can be seen in Figure 7.1a–e, good agreement was obtained between the dependence of the deviation from the stoichiometry and the dependence resulting from the values of "equilibrium constants" K_V and K_I. Dominating defects are cation vacancies $V'''_{Fe(o)}$ and interstitial cations $Fe_i^{3\bullet}$. Concentrations of vacancies $\left[V''_{Fe(o)}\right]$ and interstitial ions $\left[Fe_i^{\bullet\bullet}\right]$ are much lower and do not affect the character of the dependence of δ on p_{O_2} .

Figure 7.2 compares the oxygen pressure dependences of the deviation from the stoichiometry obtained at 1473 K for pure magnetite and for $(Fe_{1-x}Mn_x)_{3\pm\delta}O_4$ spinel, presented in Figure 7.1.

As can be seen in Figure 7.2, manganese doping of magnetite, above $x_{Mn} > 0.1$ mol, causes a shift in the range of existence of spinel towards higher oxygen pressures. When the manganese content increases, the oxygen pressure at which the spinel $(Fe_{1-x}Mn_x)_3O_4$ reaches the stoichiometric composition is also shifted towards higher oxygen pressures, by about two orders of magnitude, at the content of $x_{Mn} = 0.333$ mol, therefore, to a much smaller extent than in the case of cobalt doping.

Figure 7.3 presents the dependence of oxygen pressure at which the spinel $(Fe_{1-x}Mn_x)_3O_4$ reaches the stoichiometric composition, on the manganese content x_{Mn}.

As can be seen in Figure 7.3, the oxygen pressure at which the stoichiometric composition is present increases monotonically when the manganese content increases. This is a result of a small change in bonding energy in the spinel, due to the incorporation of manganese into the nodes of Fe^{2+} ions.

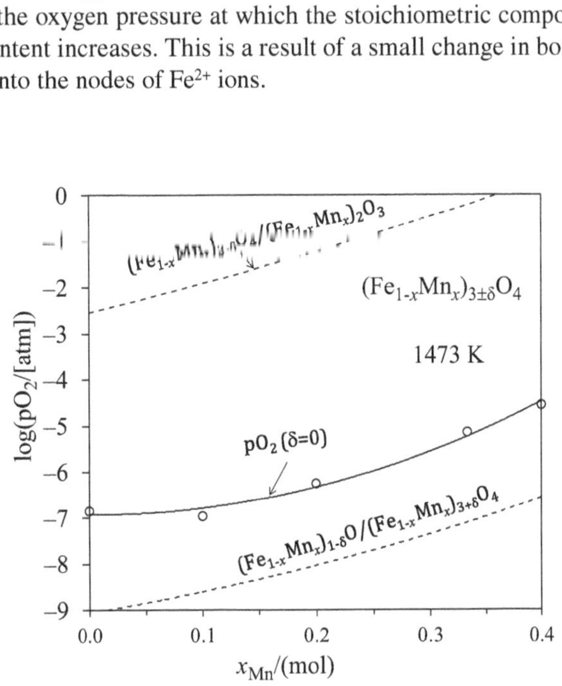

FIGURE 7.3 Dependence of the oxygen pressure at which the spinel $(Fe_{1-x}Mn_x)_{3\pm\delta}O_4$ reaches the stoichiometric composition ((O) points) on the manganese content x_{Mn}, at 1473 K. Dashed lines denote oxygen pressures at the phase boundary doped wüstite/spinel and spinel/doped hematite taken from (Franke 1987, Schwerdtfeger and Muan 1967, Terayama et al. 1983).

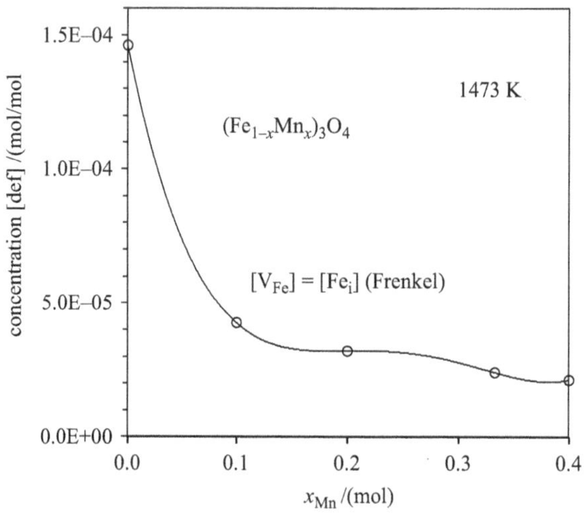

FIGURE 7.4 Dependence of the concentration of Frenkel defects on the manganese content x_{Mn} in the spinel $(Fe_{1-x}Mn_x)_{3\pm\delta}O_4$, at 1473 K.

In turn, Figure 7.4 presents the dependence of the concentration of Frenkel defects in the spinel $(Fe_{1-x}Mn_x)_3O_4$, which decreases significantly at the manganese content $x_{Mn} = 0.1$ mol. A further increase in the manganese content decreases their concentration only slightly. This is a consequence of a rise in the value of ΔG^o_{F1} when the manganese content increases.

Figure 7.5 shows the oxygen pressure dependence of the derivative (dlog p_{O_2} /logδ) = n, for the dependence of the deviation from the stoichiometry on p_{O_2} shown in Figure 7.1.

As can be seen in Figure 7.5, at the highest oxygen pressures, in a certain narrow range of pressure, the derivative reaches the value $n = 1.5$, which, however, increases, when the oxygen pressure increases. At the lowest oxygen pressures the parameter n tends to the value of –1.5 ($1/n = -2/3$).

Figures 7.6–7.8 present the dependence of the adjusted standard Gibbs energies: ΔG^o_{F1}, ΔG^o_{F2}, ΔG^o_{F3}, $\Delta G^o_{V'''_{Fe(o)}}$, ΔG^o_I on the manganese content in $(Fe_{1-x}Mn_x)_{3\pm\delta}O_4$ at 1473 K.

Figures 7.7 and 7.9 show the dependence of standard Gibbs energy $\Delta G^o_{V''_{Fe}}$ of formation of vacancies $V''_{Fe(o)}$ (calculated according to Equation 1.35d) and $\Delta G^o_{Fe_i^{3\bullet}}$ of the formation of interstitial iron ions $Fe_i^{3\bullet}$ (calculated according to Equation 1.34c) and ($\Delta G^o_{Fe_i^{\infty}}$) of formation of ions $Fe_i^{\bullet\bullet}$ (calculated according to Equation 1.36c).

As can be seen in Figure 7.6, the standard Gibbs energy of the formation of Frenkel defects ΔG^o_{F1} increases when the manganese content in magnetite increases.

As can be seen in Figures 7.7 and 7.9, the values ($\Delta G^o_{V'''_{Fe}}$) and ($\Delta G^o_{V''_{Fe}}$) increase when the manganese content x_{Mn} in the spinel $(Fe_{1-x}Mn_x)_{3\pm\delta}O_4$ increases, while the values $\Delta G^o_{Fe_i^{3\bullet}}$ and $\Delta G^o_{Fe_i^{\bullet\bullet}}$ after an increase for the manganese content $x_{Mn} = 0.1$ mol, decrease.

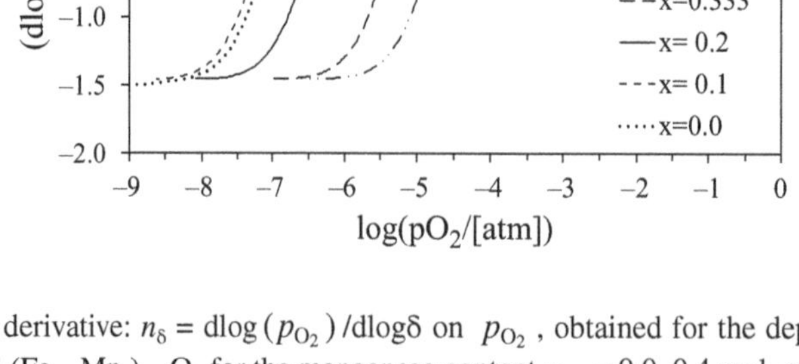

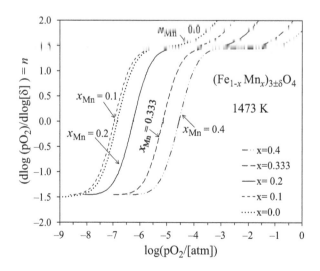

FIGURE 7.5 Dependence of the derivative: n_δ = dlog (p_{O_2}) /dlogδ on p_{O_2}, obtained for the dependence of the deviation from the stoichiometry on p_{O_2} in the spinel $(Fe_{1-x}Mn_x)_{3\pm\delta}O_4$ for the manganese content x_{Mn} = 0.0–0.4 mol, at 1473 K (presented in Figure 7.1).

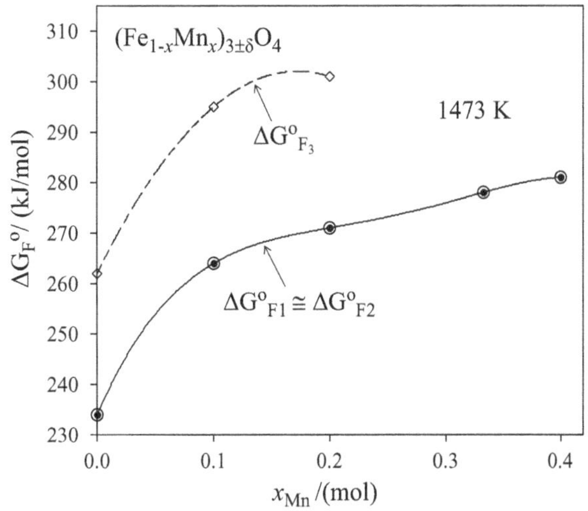

FIGURE 7.6 Dependence standard Gibbs energy $\Delta G^o_{F_1}$ ((O) points) and $\Delta G^o_{F_2}$ (●) and $\Delta G^o_{F_3}$ (◇) on the manganese content x_{Mn} in the spinel $(Fe_{1-x}Mn_x)_{3\pm\delta}O_4$, at 1473 K.

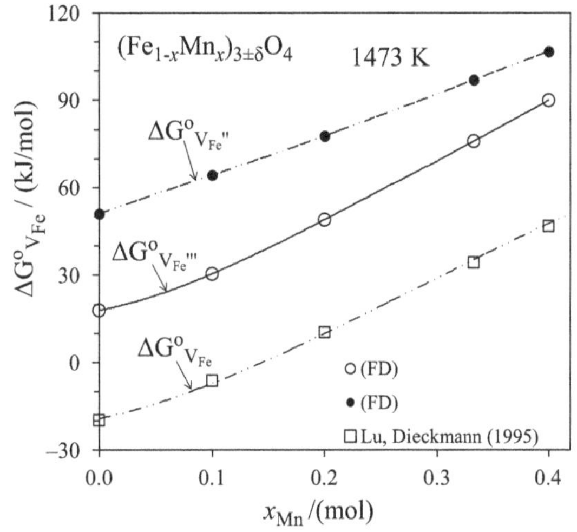

FIGURE 7.7 Dependence of standard Gibbs energy $\Delta G^o_{V_{Fe}'''}$ of formation of vacancies $V'''_{Fe(o)}$ ((O) points) and $\Delta G^o_{V_{Fe}''}$ of formation of $V''_{Fe(o)}$ ((●) points) on the manganese content x_{Mn} in the spinel $(Fe_{1-x}Mn_x)_{3\pm\delta}O_4$, at 1473 K and the values $\Delta G^o_{V_{Fe}}$ calculated using the values of "equilibrium constants" of reaction of formation of vacancies K_V, given in (Lu and Dieckmann 1995) ((□) points).

Figures 7.7 and 7.9 show also the values of $\Delta G^o_{V_{Fe}}$ and $\Delta G^o_{Fe_i}$ calculated using "equilibrium constants" of reactions of formation of cation vacancies K_V and interstitial ions K_I taken from (Lu and Dieckmann 1995). The values differ from those determined in the work, because they describe another model of defects, but they have a similar character of the dependence when the manganese content in the spinel increases.

7.3 DIFFUSION IN MANGANESE-DOPED MAGNETITE

Similarly as for pure magnetite, using the concentration of ionic defects resulting from the diagram of defects' concentrations and the values of coefficients of self-diffusion of iron in the spinel $(Fe_{1-x}Mn_x)_{3\pm\delta}O_4$, the ratio of coefficients of diffusion (mobility of metal ions) via cation vacancies and via interstitial cations was determined (see: Section 1.6.2).

Figure 7.1 shows the dependence of the "sum" of concentrations of ionic defects ($[V_{Fe}] + b_D[Fe_i]$) on the oxygen pressure for the fitted values of the parameter b_D and the values of this sum, calculated according to Equation 1.47a and 1.47b using the coefficients of self-diffusion of iron ^{59}Fe (Franke and Dieckmann 1989) and the fitted values of the coefficient of diffusion of iron ions $D^o_{V(Fe)}$ via iron vacancies at the highest oxygen pressures.

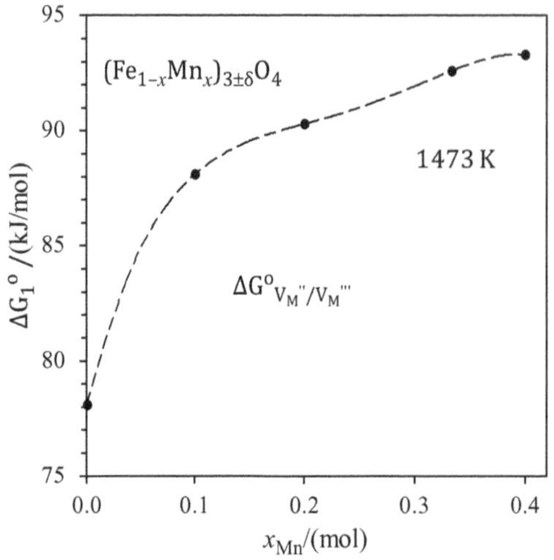

FIGURE 7.8 Dependence of standard Gibbs energy ΔG^o_I on the manganese content x_{Mn} in the spinel $(Fe_{1-x}Mn_x)_{3\pm\delta}O_4$ at 1473 K, permitting to determine the ratio of concentrations of iron vacancies $\left[V''_{Fe(o)}\right]/\left[V'''_{Fe(o)}\right]$, according to Equation 1.35b.

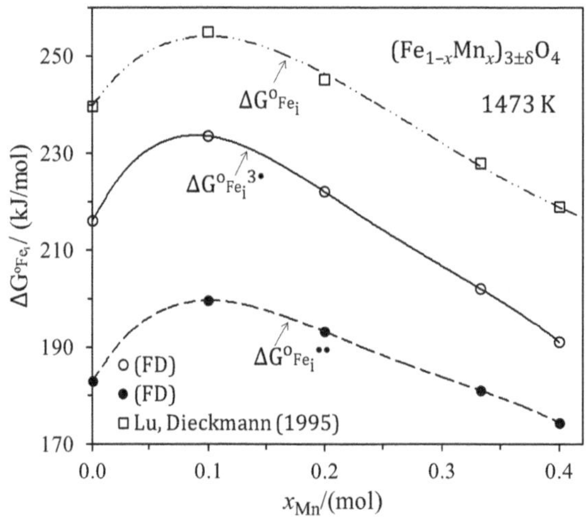

FIGURE 7.9 Dependence of the standard Gibbs energy $\Delta G^o_{Fe_i^{3\bullet}}$ of formation of interstitial ions $Fe_i^{3\bullet}$ ((○) points) calculated according to Equation 1.34c and $\Delta G^o_{Fe_i^{\bullet\bullet}}$ of formation of ions $Fe_i^{\bullet\bullet}$ ((●) points), calculated according to Equation 1.36c on the manganese content x_{Mn} in the spinel $(Fe_{1-x}Mn_x)_{3\pm\delta}O_4$, at 1473 K and the values $\Delta G^o_{Fe_i}$ calculated using the values of "equilibrium constants" of reaction of formation of interstitial cations K_I reported in (Lu and Dieckmann 1995) ((□) points).

As can be seen in Figure 7.1, good agreement was obtained between the dependence of the sum of the concentrations of ionic defects ($[V_{Fe}] + b_D[Fe_i]$) on p_{O_2}, determined from the defects' diagram, and the values of this sum determined using the coefficients of diffusion of iron.

Figure 7.10 presents the dependence of the ratio of coefficients of diffusion via defects ($D^o_{I(Fe)}/D^o_{V(Fe)} = b_D$) on the manganese content x_{Mn} in the spinel $(Fe_{1-x}Mn_x)_{3\pm\delta}O_4$.

As can be seen in Figure 7.10, the mobility of iron ions via interstitial ions is higher than the mobility via cation vacancies and their ratio increases when the manganese content increases from $x_{Mn} = 0.0$–0.333 mol, and then, for the content of 0.4 mol, it decreases.

Figure 7.11 presents the dependence of the values of coefficients of diffusion of iron ions via cation vacancies $D^o_{V(Fe)}$ and via interstitial ions $D^o_{I(Fe)}$ on the manganese content x_{Mn} in the spinel $(Fe_{1-x}Mn_x)_{3\pm\delta}O_4$.

As can be seen in Figure 7.11, the coefficients of diffusion of iron ions via cation vacancies decrease when the manganese content increases (for the manganese content of $x_{Mn} = 0.4$ they slightly increase).

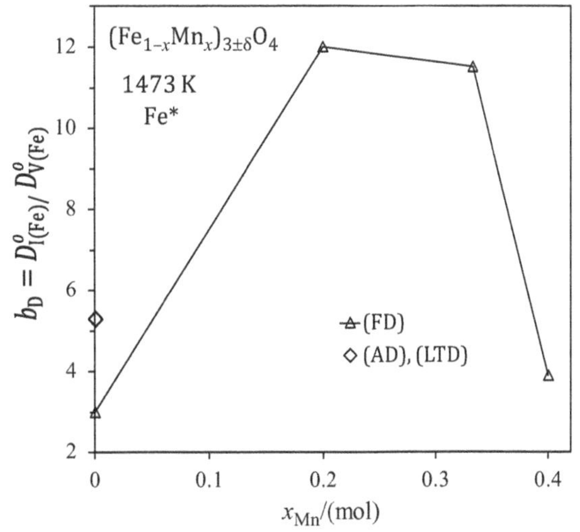

FIGURE 7.10 Dependence of the ratio of the coefficient of diffusion of iron ions via interstitial cations to the coefficient of diffusion via cation vacancies $D^o_{I(Fe)}/D^o_{V(Fe)} = b_D$ on the manganese content x_{Mn} in the spinel $(Fe_{1-x}Mn_x)_{3\pm\delta}O_4$ at 1473 K, obtained using the of coefficients of self-diffusion of iron ^{59}Fe (Aggarwal and Dieckmann 2002a, Aggarwal and Dieckmann 2002b, Franke and Dieckmann 1989, Lu et al. 1993) and concentrations of ionic defects calculated from diagrams.

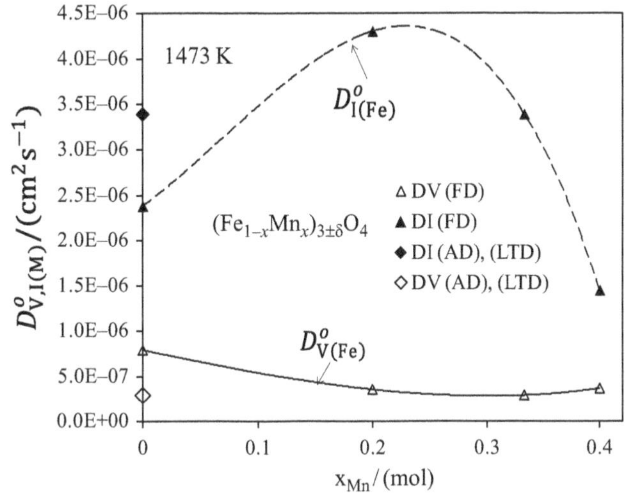

FIGURE 7.11 Dependence of the coefficient of diffusion of iron ions via cation vacancies $D^o_{V(Fe)}$ ((△) points) and the coefficient of diffusion via interstitial cations $D^o_{I(Fe)}$ ((▲) points) on the manganese content x_{Mn} in the spinel $(Fe_{1-x}Mn_x)_{3\pm\delta}O_4$ at 1473 K, obtained using the of coefficients of self-diffusion of iron ^{59}Fe (Aggarwal and Dieckmann 2002a, Aggarwal and Dieckmann 2002b, Franke and Dieckmann 1989, Lu et al. 1993) and concentrations of ionic defects calculated from diagrams.

According to the values of the parameter b_D, the mobility of iron ions via interstitial cations is significantly higher than the mobility via cation vacancies. At the manganese content $x_{Mn} = 0.2$ mol, the values of $D^o_{I(Fe)}$ are higher than in pure magnetite, but when the manganese content in the spinel $(Fe_{1-x}Mn_x)_{3\pm\delta}O_4$ further increases, they decrease.

7.4 CONCLUSIONS

Using the results of the studies of the deviation from the stoichiometry in the spinel $(Fe_{1-x}Mn_x)_{3\pm\delta}O_4$ with the manganese content $x_{Mn} = 0.1$–0.4 mol (Franke 1987, Lu and Dieckmann 1995), diagrams of the concentrations of defects at 1473 K were determined. It was found that the range of existence of spinel $(Fe_{1-x}Mn_x)_{3\pm\delta}O_4$ shifts towards higher oxygen pressures when the manganese content increases. Manganese doping of magnetite does not affect the character of the oxygen pressure dependence of the deviation from the stoichiometry. In their range of existence, spinels $(Fe_{1-x}Mn_x)_{3-\delta}O_4$ reach the stoichiometric composition, while when the manganese content in the spinel increases, the oxygen pressure at which the spinel reaches the stoichiometric composition is shifted towards higher values, by about two orders of magnitude, at the manganese content of $x_{Mn} = 0.333$ mol.

Manganese doping of spinel $(Fe_{1-x}Mn_x)_{3\pm\delta}O_4$ practically does not affect the point defects structure. In the range of higher oxygen pressures, cation vacancies $\left[V_{Fe}'''\right]$ dominate; on the other hand, in a range of low oxygen pressures, interstitial cations $\left[Fe_i^{3\bullet}\right]$ dominate.

It was found that when the manganese content in the spinel $(Fe_{1-x}Mn_x)_{3\pm\delta}O_4$ increases, standard Gibbs energy of formation of Frenkel defects increases, which results in a decrease in the concentration of Frenkel defects from the value of $1.5 \cdot 10^{-4}$ to $3 \cdot 10^{-5}$ mol/mol.

The standard Gibbs energies of formation of cation vacancies $\Delta G^{o}_{V_{Fe}'''}$ and $\Delta G^{o}_{V_{Fe}''}$ increase when the manganese content in the spinel $(Fe_{1-x}Mn_x)_{3\pm\delta}O_4$ increases, while $\Delta G^{o}_{Fe_i^{3\bullet}}$ and $\Delta G^{o}_{Fe_i^{\bullet\bullet}}$ of formation of interstitial cations, after a slight increase at the manganese content $x_{Mn} = 0.1$, decrease. The above standard enthalpies of formation of defects depend on the value of ΔG^{o}_{eh} of formation of electronic defects and they were determined for the value $\Delta G^{o}_{eh} = 30$ kJ/mol, the same as the value assumed for pure magnetite.

Using the determined concentration of ionic defects and the values of coefficients of self-diffusion of iron ions ^{59}Fe, the ratio of the coefficients of diffusion (mobility) of iron ions via interstitial cations to the coefficient of diffusion via vacancies $(D^{o}_{I(Fe)} / D^{o}_{V(Fe)} = b_D)$ was determined (Franke and Dieckmann 1989). It was found that at the manganese content $x_{Mn} = 0.2$–0.3 mol in the spinel $(Fe_{1-x}Mn_x)_{3\pm\delta}O_4$, the mobility via interstitial cations is higher than the mobility via cation vacancies and their ratio reaches the value of $b_D \approx 12$ and then it decreases.

At the manganese content $x_{Mn} = 0.2$ mol in the spinel $(Fe_{1-x}Mn_x)_{3\pm\delta}O_4$, the mobility of iron ions via interstitial cations $\left(D^{o}_{I(Fe)}\right)$ is higher than in magnetite, and when the manganese content increases further, this mobility decreases. The mobility of iron ions via cation vacancies $\left(D^{o}_{V(Fe)}\right)$ decreases when the manganese content in the spinel increases.

REFERENCES

Aggarwal, S., and R. Dieckmann. 2002a. Point Defects and Cation Tracer Diffusion in $(Ti_xFe_{1-x})_{3-\delta}O_4$. I. Non-stoichiometry and Point Defects. *J. Phys. Chem. Miner.* 29: 695–706.

Aggarwal, S., and R. Dieckmann. 2002b. Point Defects and Cation Tracer Diffusion in $(Ti_xFe_{1-x})_{3-\delta}O_4$. II. Cation Tracer Diffusion. *J. Phys. Chem. Miner.* 29: 707–718.

Becker, K. D., J. Pattanayak, and S. Wißmann. 1994. A High-Temperature Mössbauer Study of the Cation Distribution in $(Fe,Mn)_3O_4$ Spinels. *Solid State Ionics.* 70–71: 497–502.

Bergstein, A. 1963. Manganese Magnesium Ferrites. Iv. Oxygen Nonstoichiometry of Spinels $Mn_xFe_{3-x}O_{4+\gamma}$. *Collect. Czech. Chem. Commun.* 28: 2381–2386.

Bergstein, A., and P. Kleinert. 1964. Partial Phase Diagram of the System $Mn_xFe_{3-x}O_y$. *Collect. Czech. Chem. Commun.* 29:2549–2551.

Bonsdorf, G., M. A. Denecke, K. Schäfer, S. Christen, H. Langbein, and W. Gunßer. 1997. X-Ray Absorption Spectroscopic and Mössbauer Studies of Redox and Cation-Ordering Processes in Manganese Ferrite. *Solid State Ionics.* 101–103: 351–357.

Bulgakova, T. I., and A. G. Rozanov. 1970. Phase Equilibria in the Ferrite Region of the Mn-Fe-O System. *Russ. J. Phys. Chem.* 44: 385–388.

Dieckmann, R. 1982. Defects and Cation Diffusion in Magnetite (IV): Nonstoichiometry and Point Defect Structure of Magnetite $(Fe_{3-\delta}O_4)$. *Ber. Bunsenges. Phys. Chem.* 86: 112–118.

Dieckmann, R., T. O. Mason, J. D. Hodge, and H. Schmalzried. 1978. Defects and Cation Diffusion in Magnetite (III.) Tracer Diffusion of Foreign Tracer Cations as a Function of Temperature and Oxygen Potential. *Ber. Bunsenges. Phys. Chem.* 82: 778–783.

Duquesnoy, A., J. Couzin, and P. Gode. 1975. Isothermal Representation of Ternary Phase Diagrams ABO: *Study of the System Mn-Fe-O. C. R. Acad. Sci. Paris C.* 281: 107–109.

Franke, P. 1987. Disorder and Transport in Mixed Oxides of the Types (Fe, Mn) O and $(Fe, Mn)_3O_4$. PhD Thesis, University of Hannover, Germany.

Franke, P., and R. Dieckmann. 1989. Defect Structure and Transport Properties of Mixed Iron-manganese Oxides. *Solid State Ionics.* 32–33: 817–823.

Franke, P., and R. Dieckmann. 1990. Thermodynamics of Iron Manganese Mixed Oxides at High Temperatures. *J. Phys. Chem. Solids.* 51: 49–57.

Hastings, J. M., and L. M. Corliss. 1956. Neutron Diffraction Study of Manganese Ferrite. *Phys. Rev.* 104: 328–331.

Hohmann, H. H., W. Müller, H. Schmalzried, and Yu. D. Tretyakov. 1967. *Proc. Brit. Ceram Soc.* 8: 91.

Jirak, Z., and S. Vratislav. 1974. Temperature Dependence of Distribution of Cations in $Mnfe_2o_4$. *Czech. J. Phys. B.* 24: 642–647.

Kjellqvist, L., and M. Selleby. 2010. Thermodynamic Assessment of the Fe-Mn-O System. *J. Phase Equilib. Diff.* 31: 113–134.

Komarov, V. F., N. N. Oleinikov, Yu. G. Saksonov, and Yu. D. Tretyakov. 1965. Methods for the Calculation of Thermodynamic Properties of Ferrites and Solid Solutions. *Inorg. Mater.* 1: 365–382.

Lu, F. H., and R. Dieckmann. 1995. Point Defects in Oxide Spinel Solid Solutions of the Type. *J. Phys. Chem. Solids.* 56: 725–733.

Lu, F. H., S. Tinkler, and R. Dieckmann. 1993. Point Defects and Cation Tracer Diffusion in $(Co_xFe_{1-x})_{3-\Delta}O_4$ Spinels. *Solid State Ionics.* 62: 39–52.

Muan, A., and S. Somiya. 1962. The System Iron Oxide-Manganese Oxide in Air. *Am. J. Sci.* 260: 230–240.

Murthy, N. S. S., L. M. Rao, R. J. Begum, M. G. Natera, and S. I. Youssef. 1971. Neutron Scattering Studies of Some Spinels. *J. Phys. Colloq. (Suppl C1).* 32: 318–319.
Ono, K., T. Ueda, T. Ozaki, Y. Ueda, A. Yamaguchi, and J. Moriyama. 1971. Thermodynamic Study of the Iron-Manganese-Oxygen System. *J. Japan Inst. Metals.* 35: 757–763.
Rieck, G. D., and F. C. M. Driessens. 1966. The Structure of Manganese–Iron–Oxygen Spinels. *Acta Cryst.* 20: 521–525.
Roethe, A., K. P. Roethe, and G. H. Jerschkewitz. 1970. Gleichgewichtsmessungen an Oxidischen Mehrstoffsystemen. Ii Gleichgewichtsmessungen an Einer Me_3O_4-Phase im Quaternären System Fe-Mn-Zn-O. *Z. Anorg. Allg. Chem.* 378: 14–21.
Sawatzky, G. A., F. Van der Woude, and A. H. Morrish. 1967. Note on Cation Distribution of $Mnfe_2o_4$. *Phys. Lett. A.* 25: 147–148.
Schwerdtfeger, K., and A. Muan. 1967. Equilibria in System Fe-Mn-O Involving (Fe, Mn)O and $(Fe, Mn)_3O_4$ Solid Solutions. *Trans. Met. Soc. AIME.* 239: 1114–1119.
Subramanian, R., and R. Dieckmann. 1993. Nonstoichiometry and Thermodynamics of the Solid Solution $(Fe, Mn)_{1-\Delta}O$ at 1200°C. *J. Phys. Chem. Solids.* 54: 991–1000.
Terayama, K., M. Ikeda, and M. Taniguchi. 1983. Phase Equilibria in the Mn-Fe-O System in CO_2–H_2 Mixtures. *Trans. Jap. Inst. Met.* 24: 514–517.
Wickham, D. G. 1969. The Chemical Composition of Spinels in the System Fe_3O_4-Mn_3o_4. *J. Inorg. Nucl. Chem.* 31: 313–320.
Yoo, H. I., and H. L. Tuller. 1988. In Situ Phase Equilibria Determination of a Manganese Ferrite by Electrical Means. *J. Mater. Res.* 3: 552–556.

8 Magnetite Doped with Manganese and Cobalt – $(Co_xMn_zFe_{2z})_{3\pm\delta}O_4$

8.1 INTRODUCTION

Interesting useful properties are obtained as a result of doping the spinel $MnFe_2O_4$ with cobalt ions. In line with the expectations, at high temperatures ions Mn^{2+} and Fe^{2+} are substituted with cobalt ions Co^{2+} in the octahedral and tetrahedral sublattice. This type of distribution of ions in magnetite doped with Co and Mn was confirmed by the studies of thermoelectric power for the spinel $(Co_{1-x}Mn_x)_{0.4}Fe_{2.6}O_4$, performed at 1473 K (Carter and Mason 1988). It was demonstrated, based on the model of small polarons and the values of the Seebeck coefficient, that in the spinel $(Co_{1-x}Mn_x)_{0.4}Fe_{2.6}O_4$, manganese ions Mn^{2+} occupy the nodes of the tetrahedral sublattice and the octahedral sublattice to the same extent, independent of the manganese content. In turn, simultaneously, the concentration of Co^{2+} ions decreases; their concentration is higher in the octahedral sublattice than in the tetrahedral sublattice. The ratio of the concentration of ions Fe^{2+} and Fe^{3+} in the octahedral and the tetrahedral sublattices practically does not change. The above distribution is consistent with the thermodynamic calculations (Subramanian et al. 1994) and with the results of the studies of Mössbauer effect (Singh et al. 1981).

Extensive studies of the oxygen pressure dependence of the deviation from the stoichiometry and the diffusion of radioisotope elements: iron ^{59}Fe, cobalt ^{60}Co, manganese ^{54}Mn in the spinel $(Co_xMn_zFe_{2z})_{3\pm\delta}O_4$, (with composition fulfilling the condition ($x + 3z = 1$) with the cobalt content of $x_{Co} = 0.1–0.333$ mol, at 1473 K were done (Lu and Dieckmann 1992). It was demonstrated that the spinel $(Co_xMn_zFe_{2z})_{3\pm\delta}O_4$ reaches the stoichiometric composition in the range of its existence. Therefore, at higher oxygen pressures, cation vacancies dominate, while at low oxygen pressures interstitial cations dominate. The studies on the diffusion showed, similarly as for magnetite, that the coefficients of self-diffusion of tracers depend on the type of metal and they are different in the range where cation vacancies and interstitial cations dominate; the character of their change in dependence on the oxygen pressure is fully consistent with the model of ionic defects' structure. As the results of the studies of the deviation from the stoichiometry performed by Lu and Dieckmann include all the range of existence of spinels, it was decided to use them in the determination of diagram of the concentrations of defects in dependence on the content of cobalt and manganese.

8.2 METHODS AND RESULTS OF CALCULATION OF THE DIAGRAMS OF THE CONCENTRATIONS OF DEFECTS, AND DISCUSSION

As it was shown before, for the spinel $MnFe_2O_4$, Mn^{2+} ions are present mainly in the cation sites with tetrahedral coordination, although some amount of them will be present in the sites with octahedral coordination. Due to doping the spinel $(Co_xMn_zFe_{2z})_{3\pm\delta}O_4$, (where ($x + 3z = 1$)) with cobalt, the content of manganese and iron decreases, respectively, but the ratio of their concentrations is preserved, [Mn]/[Fe] = 0.5. As a result, the sum of the concentrations of cobalt ions and manganese ions is higher than one mole and it is ([Co] + [Mn]) = 1.2, 1.4 and 1.666 mol/mol, respectively. Simultaneously, the content of iron is lower than two, thus, ions Co^{2+} and Mn^{2+} incorporate into some nodes of Fe^{3+} ions. The character of the oxygen pressure dependence of deviation from the stoichiometry is similar as in the case of magnetite doped with cobalt or manganese (Lu and Dieckmann 1992). It can be expected that excess manganese ions will be oxidised during the synthesis of the spinel. However, this fact does not affect the structure of point defects. Due to that, the calculations of diagrams of concentrations of point defects in magnetite doped with cobalt and manganese were conducted with a method analogous to the one used for magnetite doped with cobalt ions (see: Section 6.2), using the results of the deviation from the stoichiometry (Lu and Dieckmann 1992).

Due to the lack of the value of ΔG^o_{eh} of formation of electronic defects, it was assumed that it only slightly depends on the concentration of doping ions. Due to that, in the calculations, a constant value of ΔG^o_{eh} = 30.3 kJ/mol was assumed, the same as for pure magnetite.

Figure 8.1a–d present diagrams of concentrations of ionic defects for spinels $(Co_xMn_zFe_{2z})_{3-\delta}O_4$, for the cobalt content x_{Co} = 0.0, 0.1, 0.2, 0.333 mol, at 1473 K (solid lines), calculated using the results of the deviation from the stoichiometry (Lu and Dieckmann 1992). In the calculations of the dependence of δ on p_{O_2} it was assumed that at the stoichiometric composition, the concentrations of vacancies are equal $\left[V'''_{Fe(o)}\right] = \left[V''_M\right]$. The dotted line shows the dependence of δ on p_{O_2}

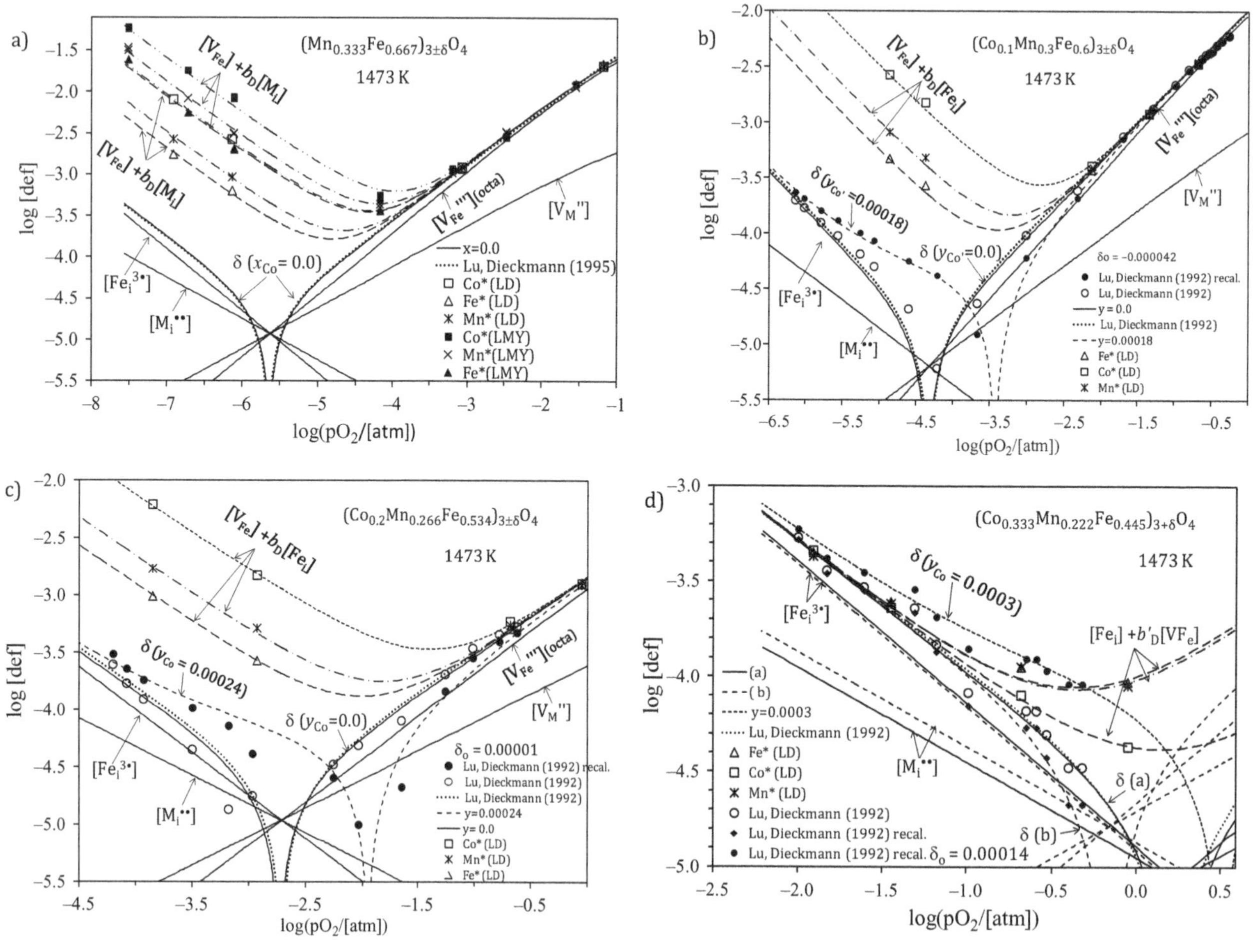

FIGURE 8.1 (a) Diagram of concentrations of point defects for the spinel $(Mn_{0.333}Fe_{0.667})_{3\pm\delta}O_4$, at 1473 K (solid lines). The (····) line shows the dependence of the deviation from the stoichiometry on the oxygen pressure, determined using the values of "equilibrium constants" of reactions of formation of cation vacancies K_V and interstitial cations K_I (Lu and Dieckmann 1995). The points represent the values of the sum of concentrations of ionic defects ($[V_{Fe}] + b_D[M_i]$), calculated according to Equation 1.47a, 1.47b for fitted values of $D^o_{V(M)}$ for the highest oxygen pressures and the values of coefficients of self-diffusion of tracers: Fe* ((△) points), Co* - (□), Mn* - (✳) (Lu and Dieckmann 1992) and (Lee et al. 2000) ((▲,■,x) points). Lines (- -), (---), (-··-) denote, respectively, the dependence of the sum of concentrations of defects ($[V_{Fe}] + b_D[M_i]$) on p_{O_2} obtained for the fitted value of b_D. (b) for the spinel $(Co_{0.1}Mn_{0.3}Fe_{0.6})_{3\pm\delta}O_4$ at 1473 K (solid lines), obtained using the results of the studies of the deviation from the stoichiometry (Lu and Dieckmann 1992) ((○) points). The (···) line shows the dependence of the deviation from the stoichiometry on p_{O_2}, determined using the values of "equilibrium constants" of reactions of formation of cation vacancies K_V and interstitial cations K_I extracted from (Lu and Dieckmann 1992). Points (●) show the recalculated values of the deviation from the stoichiometry δ published in (Lu and Dieckmann 1992) relative to the value $\delta_o = -4.2\cdot10^{-5}$, and line (---) represents the dependence of δ on p_{O_2} which taking into account cobalt ions Co'_{Fe} incorporated into the sublattice of ions Fe^{3+} in the amount of $y_{Co}' = 1.8\cdot10^{-4}$ mol/mol. (c) for the spinel $(Co_{0.2}Mn_{0.266}Fe_{0.534})_{3\pm\delta}O_4$, at 1473 K (solid lines). Points (●) show the recalculated values of the deviation from the stoichiometry δ for the value $\delta_o = 1\cdot10^{-5}$, and line (---) represents the dependence of δ on p_{O_2} when taking into account cobalt ions Co'_{Fe} incorporated into the sublattice of ions Fe^{3+} in the amount of $y_{Co}' = 2.4\cdot10^{-4}$ mol/mol. (d) for the spinel $(Co_{0.333}Mn_{0.222}Fe_{0.445})_{3\pm\delta}O_4$ at 1473 K (solid lines), obtained using the results of the studies of the deviation from the stoichiometry (Lu and Dieckmann 1992) ((○) points). The (- -) line represents the dependence of δ on p_{O_2} for the value $\Delta G^o_{F_1} = \Delta G^o_{F_2} = 270$ kJ/mol. Points (◆) denote the recalculated values of the deviation from the stoichiometry δ for the value $\delta_o = 7\cdot10^{-5}$. The line (---) represents the dependence of δ on p_{O_2} when taking into account cobalt ions Co'_{Fe} incorporated into the sublattice of ions Fe^{3+} in the amount of $y_{Co}' = 3\cdot10^{-4}$ mol/mol and points (●) denote the recalculated values of the deviation from the stoichiometry δ for the value $\delta_o = 1.4\cdot10^{-4}$.

obtained using "equilibrium constants" of reactions of formation of cation vacancies K_V and interstitial ions K_I (Lu and Dieckmann 1992).

As can be seen in Figure 8.1, a good match was obtained between the dependence of the deviation from the stoichiometry on p_{O_2} and the experimental values of δ ((○) points). Dominating defects are cation vacancies $V'''_{Fe(o)}$ and interstitial cations $Fe_i^{3\bullet}$. Concentrations of cation vacancies $[V''_M]$ and interstitial ions $[M_i^{\bullet\bullet}]$ are much lower and do not affect the character of the dependence of δ on p_{O_2}. The increase in the concentration of cobalt and the decrease in the concentration of manganese and iron causes a shift in the range of existence of spinel towards low oxygen pressures, while the character of the dependence of the deviation from the stoichiometry on the oxygen pressure practically does not change. It is analogous as in the case of spinel $(MnFe_2)_{1-\delta/3}O_4$.

However, it was decided to check if a change in the value δ_o assumed in (Lu and Dieckmann 1992) causes a change in the character of the dependence of δ on p_{O_2}, typical of the incorporation of cobalt ions, negative in relation to the lattice (Co'_{Fe}), according to the Section 6.2, analogously to the case of magnetite doped with cobalt (see Section 6.2). In the case of spinel $(Co_xMn_zFe_{2z})_{3\pm\delta}O_4$, the reaction of incorporation of cobalt ions into the spinel structure assumes the form of:

$$4CoO + V_i \rightarrow Co^{x}_{Fe(2+)} + 2Co'_{Fe(3+)(o)} + 4O^{x}_{O} + Co^{\bullet\bullet}_{i} \tag{8.1}$$

Figure 8.1a and b presents recalculated values of the deviation from the stoichiometry (Lu and Dieckmann 1992) for the value $\delta_o = 4.2 \cdot 10^{-6}$ at the cobalt content of $x_{Co} = 0.1$ mol and for $\delta_o = 1 \cdot 10^{-6}$ at the content of $x_{Co} = 0.2$ mol ((●) points) and the oxygen pressure dependence of the deviation from the stoichiometry for the concentration of ions $[Co'_{Fe}]$ equal $y_{Co}{}' = 1.8 \cdot 10^{-4}$ and $2.4 \cdot 10^{-4}$ mol/mol. As can be seen in Figure 8.1a and b, a good match was obtained between the dependence of δ on p_{O_2}, taking into account the concentration of ions $[Co'_{Fe}]$, and the recalculated values of the deviation δ ((●) points). The incorporation of cobalt ions Co'_{Fe} causes a shift in the oxygen pressure at which spinel reaches the stoichiometric composition towards higher pressures. At such low concentration $[Co'_{Fe}] = ½\, y_{Co}{}'$, changes in the concentrations of defects are practically insignificant, differently from the change in the character of the dependence of the deviation from the stoichiometry. The obtained results indicate that, similarly as in the case of magnetite doped with cobalt, cobalt ions Co^{2+} incorporate into the formed spinel structure $(Co_xMn_zFe_{2z})_{3\pm\delta}O_4$ into the sublattice of Fe^{3+} ions that are negative in relation to the lattice $[Co'_{Fe}]$. Figure 8.2 presents the dependence of the concentration of ions $[Co'_{Fe}]$ on the cobalt content in the spinel $(Co_xMn_zFe_{2z})_{3\pm\delta}O_4$.

As can be seen in Figure 8.1c, for the cobalt content of $x_{Co} = 0.333$ mol in the spinel $(Co_xMn_zFe_{2z})_{3\pm\delta}O_4$, the diagram of the concentration of ionic defects (solid line) is different and the spinel reaches the stoichiometric composition near the oxygen pressure of 1 atm. In the studied range of oxygen pressures, interstitial cations dominate. The range of oxygen pressures where cation vacancies dominate is located above the pressure of 1 atm and it was not studied, which makes it difficult to unambiguously fit the values of $\Delta G^{o}_{F_1}$ and $\Delta G^{o}_{F_2}$ of formation of Frenkel defects and to determine the concentration of ions $[Co'_{Fe}]$. Due to that, the results of the calculations were presented for the value $\Delta G^{o}_{F_1} = \Delta G^{o}_{F_2} = 286$ kJ/mol, for which a relatively good match was obtained between the dependence of δ on p_{O_2} and the experimental values of δ reported in (Lu and Dieckmann 1992) ((○) points). Moreover, the calculations were performed for the value $\Delta G^{o}_{F_1} = \Delta G^{o}_{F_2} = 270$ kJ/mol; this value could be more likely, as it allows obtaining a monotonic character of the dependence of standard Gibbs energy of formation of defects on the cobalt content (see Figures 8.7–8.9). The obtained character of the dependence of δ on p_{O_2} is consistent with the recalculated values of the deviation from the stoichiometry ((◇) points) for the value $\delta_o = 7 \cdot 10^{-5}$. The calculations were also performed when taking into account the concentration of ions $[Co'_{Fe}]$. As can be seen in Figure 8.1d

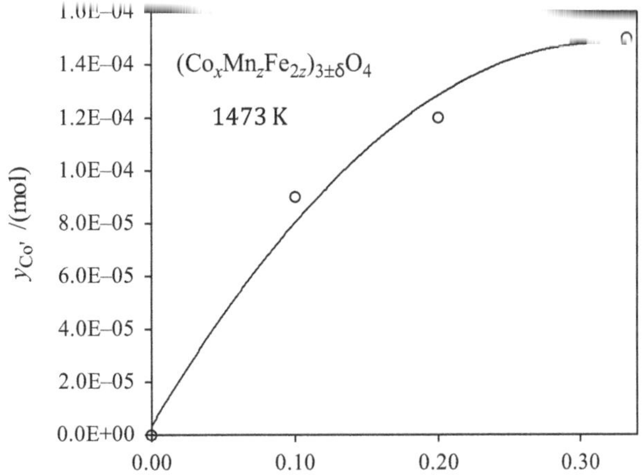

FIGURE 8.2 Dependence of the concentration of cobalt ions $[Co'_{Fe}] = 1/2 y_{Co}$ incorporated into the sublattice of ions Fe^{3+} on the cobalt content x_{Co} in the spinel $(Co_xMn_zFe_{2z})_{3\pm\delta}O_4$, at 1473 K.

the correction of the value $\delta_o = 1.4{\cdot}10^{-4}$ changes the values of the deviation from the stoichiometry ((●) points), which match well the dependence of δ on p_{O_2} taking into account the concentration of cobalt ions $y_{Co}' = 3{\cdot}10^{-4}$ mol/mol.

From the presented calculations it results that in the spinel $(Co_xMn_zFe_{2z})_{3\pm\delta}O_4$ (where $(x + 3z = 1)$), cobalt and manganese ions form a solid solution. They occupy the nodes of Fe^{2+} ions and partially the sites of Fe^{3+} ions, which causes a change in interionic interactions (a change in lattice energy). As a result, the range of existence of the spinel changes and other properties related to the change in distribution of ions with different charges are also changed. It cannot be excluded that some small number of cobalt ions (Co_{Fe}') negative in relation to the lattice, incorporate into the spinel structure in the sublattice of Fe^{3+} ions (despite the fact that the total concentration of manganese and cobalt is higher than 1 mol/mol).

Figure 8.3 compares the oxygen pressure dependence of deviation from the stoichiometry for the spinel $(Co_xMn_zFe_{2z})_{3\pm\delta}O_4$ with the cobalt content of $x_{Co} = 0.0$–333 mol (shown in Figure 8.1 (without taking into account Co_{Fe}' ions).

As can be seen in Figure 8.3, the increase in the cobalt content and the decrease in the content of manganese and iron in the spinel $(Co_xMn_zFe_{2z})_{3\pm\delta}O_4$ causes a practically parallel shift in the range of existence of the spinel towards higher oxygen pressures. There also occurs a shift towards higher oxygen pressures at which the spinel $(Co_xMn_zFe_{2z})_{3\pm\delta}O_4$ reaches the stoichiometric composition. Therefore, as can be seen in Figure 8.3, the substitution of iron ions Fe^{2+} with manganese ions Mn^{2+} in magnetite causes a two orders of magnitude shift towards higher values in the oxygen pressure at which the spinel $MnFe_2O_4$ reaches the stoichiometric composition. Cobalt doping of the spinel $(Co_xMn_zFe_{2z})_{3\pm\delta}O_4$ shifts this pressure by five more orders of magnitude (at the cobalt content of $x_{Co} = 0.333$ mol).

Figure 8.4 presents the dependence of the oxygen pressure at which the spinel $(Co_xMn_zFe_{2z})_{3\pm\delta}O_4$ reaches the stoichiometric composition on the cobalt content (taking into account the concentration of ions $[Co_{Fe}'] = 1/2y_{Co}'$, ((○) points), which, as can be seen, increases practically linearly when the cobalt content rises. The pressures at which concentrations of cation vacancies and interstitial cations are equal (Frenkel defects are present) are lower ((●) points). An increase in the above oxygen pressure $p_{O_2}^{(\delta=0)}$, as mentioned earlier, is a result of a decrease in bonding energy due to the increase in cobalt content when the content of manganese and iron in the spinel $(Co_xMn_zFe_{2z})_{3\pm\delta}O_4$ decreases.

Figure 8.5 presents the dependence of the concentration of Frenkel defects on the cobalt content x_{Co} in the spinel $(Co_xMn_zFe_{2z})_{3\pm\delta}O_4$, which decreases at the cobalt content of $x_{Co} = 0.1$ mol, and it increases when the cobalt content further increases. The character of the change in the concentration of Frenkel defects is consistent with the change in the value of $\Delta G_{F_1}^o$ of formation of Frenkel defects when the cobalt content in the spinel $(Co_xMn_zFe_{2z})_{3\pm\delta}O_4$ increases.

Figure 8.6 shows the oxygen pressure dependence of the derivative $(d\log p_{O_2}/\log\delta) = n$, for the dependence of the deviation from the stoichiometry on p_{O_2} shown in Figure 8.1 (without taking into account Co_{Fe}' ions). As can be seen in Figure 8.6, at the highest pressures the value of the derivative reaches the value $n = 1.5$. At the lowest oxygen pressures the parameter n reaches the value of -1.5 $(1/n = -2/3)$. When ions Co_{Fe}' are taken into account, the dependence of the derivative n on p_{O_2} is more complex, analogously to the case of magnetite doped with cobalt (see Figure 6.6), however, at limit pressures, the same values are reached.

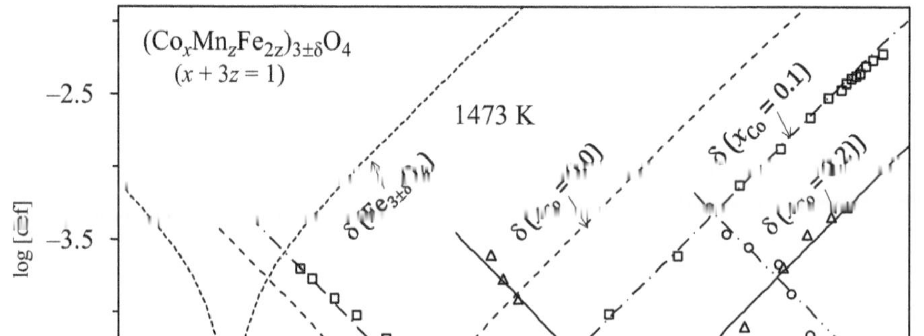

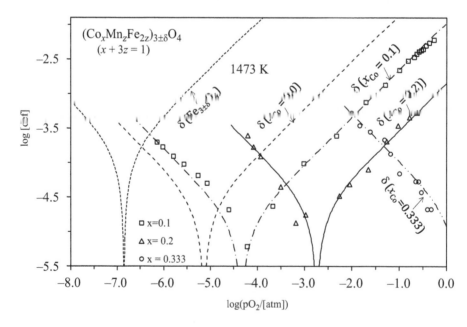

FIGURE 8.3 Oxygen pressure dependence of the deviation from the stoichiometry in the spinel $(Co_xMn_zFe_{2z})_{3\pm\delta}O_4$, (where $x + 3z = 1$) with the cobalt content of $x_{Co} = 0.0$–0.333 mol, at 1473 K, presented in Figure 8.1 (without taking into account the concentration of ions $[Co_{Fe}']$).

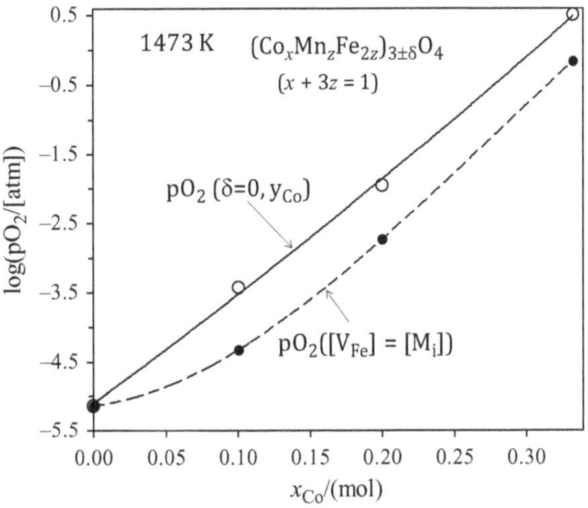

FIGURE 8.4 Dependence of the oxygen pressure at which the spinel $(Co_xMn_zFe_{2z})_3O_4$ reaches the stoichiometric composition on the cobalt content x_{Co} at 1473 K ((○) points), and the oxygen pressures at which the concentrations of cation vacancies and interstitial cations are equal ((●) points).

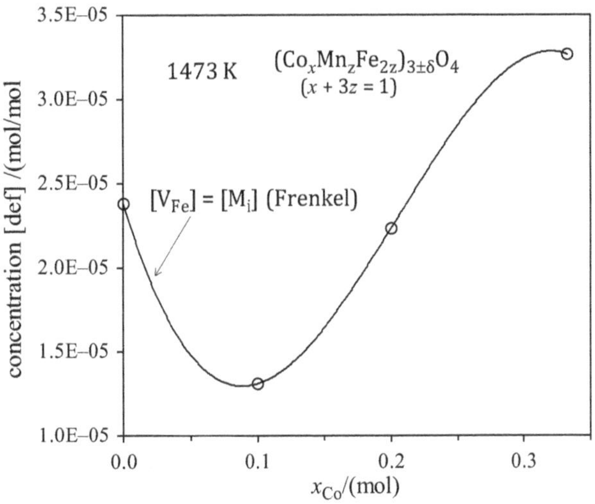

FIGURE 8.5 Dependence of the concentration Frenkel defects on the cobalt content x_{Co} in the spinel $(Co_xMn_zFe_{2z})_3O_4$ at 1473 K.

Figures 8.7–8.9 present the dependence of adjusted standard Gibbs energies: ΔG^o_{F1}, ΔG^o_{F2}, ΔG^o_{VM}, ΔG^o_I on the cobalt content x_{Co} in the spinel $(Co_xMn_zFe_{2z})_{3\pm\delta}O_4$ at 1473 K.

Figures 8.8–8.10 show the dependence of $\Delta G^o_{V''_{Fe}}$ of formation of vacancies $V'''_{Fe(o)}$ (calculated according to Equation 1.35A) and $\Delta G^o_{Fe_i^{3\bullet}}$ of the formation of interstitial iron ions $Fe_i^{3\bullet}$ (calculated according to Equation 1.34a) and $\Delta G^o_{M_i^{\bullet\bullet}}$ of formation of ions $M_i^{\bullet\bullet}$ (calculated according to Equation 1.36c).

As can be seen in Figure 8.7, the standard Gibbs energy of formation of Frenkel defects ΔG^o_{F1} increases at the cobalt content of x_{Co} = 0.1 mol in the spinel $(Co_xMn_zFe_{2z})_{3\pm\delta}O_4$, and when the cobalt content further increases, it decreases.

As can be seen in Figures 8.8 and 8.10, the values $\Delta G^o_{V''_{Fe}}$ and $\Delta G^o_{V'_M}$ increase when the cobalt content x_{Co} in the spinel $(Co_xMn_zFe_{2z})_{3\pm\delta}O_4$ increases, while the values $\Delta G^o_{Fe_i^{3\bullet}}$ and $\Delta G^o_{M_i^{\bullet\bullet}}$, inversely, decrease. As marked in Figure 8.1d, the values of ΔG^o_i of formation of defects in the spinel $(Co_xMn_zFe_{2z})_{3\pm\delta}O_4$ at the cobalt content of x_{Co} = 0.333 mol were determined for the values δ taken from Lu and Dieckmann (1992) ((a) curve, and for the recalculated values, (b) curve), which are consistent with the monotonic change in the character of the dependence of ΔG^o_i on the spinel composition.

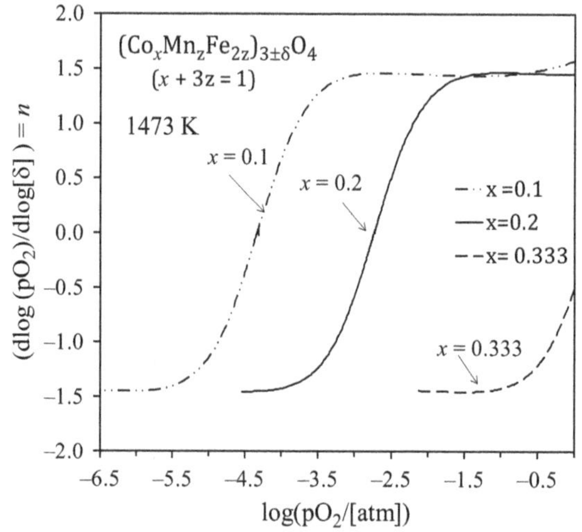

FIGURE 8.6 Oxygen pressure dependence of the derivative: $n_\delta = d\log(p_{O_2})/d\log\delta$, obtained for dependences of the deviation from the stoichiometry on p_{O_2} in the spinel $(Co_xMn_zFe_{2z})_3O_4$ with the cobalt content of $x_{Co} = 0.1–0.333$ mol, at 1473 K (presented in Figure 8.1 without taking into account the concentration of ions $\left[Co'_{Fe}\right]$).

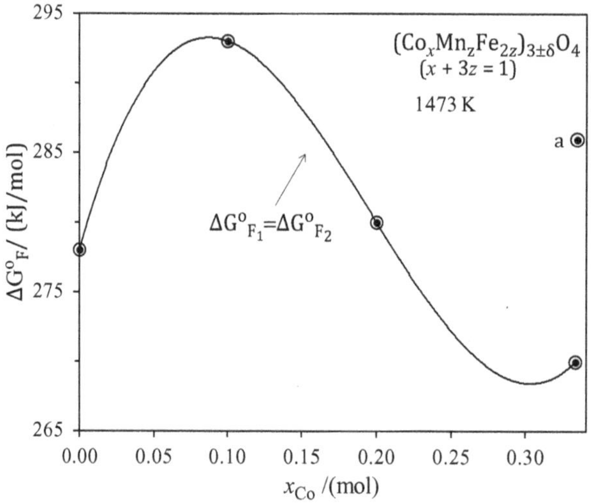

FIGURE 8.7 Dependence of standard Gibbs energy $\Delta G^o_{F_1}$ ((O) points) and $\Delta G^o_{F_2}$ (●) of formation of Frenkel defects on the cobalt content x_{Co} in the spinel $(Co_xMn_zFe_{2z})_{3\pm\delta}O_4$, at 1473 K, obtained using the results of the studies of the deviation from the stoichiometry (Lu and Dieckmann 1992).

Figures 8.7 and 8.10 show also the values of $\Delta G^o_{V_{Fe}}$ and $\Delta G^o_{Fe_i}$ calculated using the values of "equilibrium constants" of reaction of formation of cation vacancies K_V and interstitial cations K_I (Lu and Dieckmann 1992). The values differ from those determined in the work, because they describe another model of defects, but they have a similar character in dependence on the cobalt content.

8.3 DIFFUSION IN MAGNETITE DOPED WITH COBALT AND MANGANESE

Similarly, as for pure magnetite, using the concentration of ionic defects resulting from the diagram of defects' concentrations and the values of coefficients of self-diffusion of tracers in the spinel $(Co_xMn_zFe_{2z})_{3\pm\delta}O_4$, the coefficients of diffusion of defects, and specifically the mobility of tracer ions via cation vacancies and via interstitial cations, were determined (see Section 1.6.2).

Figure 8.1 shows the dependence of the "sum" of concentrations of ionic defects ($[V_{Fe}] + b_D[Fe_i]$) on p_{O_2} for the fitted values of the parameter b_D and the values of this sum, calculated according to Equation 1.47a, 1.47b using the values of diffusion coefficients for: iron ^{59}Fe, cobalt ^{60}Co, manganese ^{54}Mn, in the spinel $(Co_xMn_zFe_{2z})_{3\pm\delta}O_4$ (Lu and Dieckmann

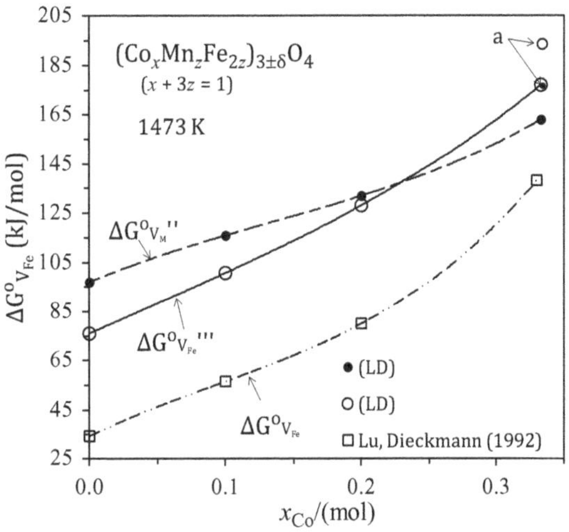

FIGURE 8.8 Dependence of the standard Gibbs energy $\Delta G^o_{V'''_{Fe}}$ and $\Delta G^o_{V''_{Fe}}$ of formation of vacancies $V'''_{Fe(o)}$ and $V''_{Fe(o)}$ on the cobalt content x_{Co} in the spinel $(Co_xMn_zFe_{2z})_{3\pm\delta}O_4$, at 1473 K and the values $\Delta G^o_{V_{Fe}}$ calculated using the values of "equilibrium constants" of reaction of formation of cation vacancies K_V (Lu and Dieckmann 1992).

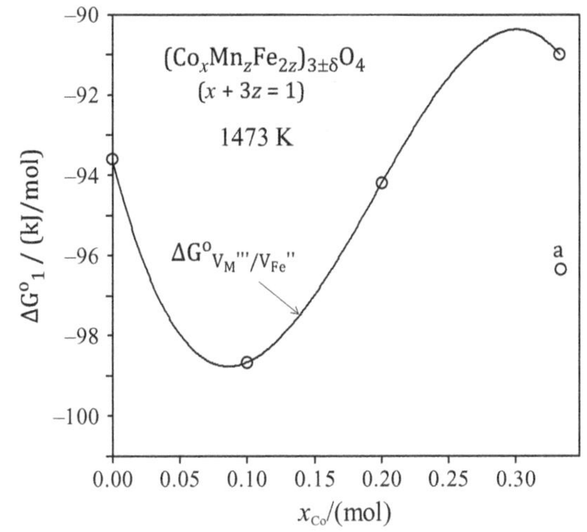

FIGURE 8.9 Dependence of standard Gibbs energy ΔG^o_1 on the cobalt content x_{Co} in the spinel $(Co_xMn_zFe_{2z})_{3\pm\delta}O_4$ at 1473 K, permitting the determination of the ratio of concentrations of iron vacancies $[V'''_{Fe(o)}]/[V''_{Fe(o)}]$ (according to Equation 1.35b).

1992) and in the spinel $(Mn_{0.333}Fe_{0.667})_{3\pm\delta}O_4$ taken from (Lu and Dieckmann 1992, Lee et al. 2000). The values of the coefficient of diffusion of cations $D^o_{V(M)}$ via cation vacancies at the highest oxygen pressures and of the parameter b_D, being the ratio of the coefficients of diffusion via defects $(D^o_{I(M)}/D^o_{V(M)})$, were adjusted in such a way as to obtain a match between the dependence of the sum of ionic defects $([V_{Fe}] + b_D[M_i])$ on p_{O_2} with the values of this sum calculated using the coefficients of self-diffusion of tracers. As in the case of spinel $(Co_xMn_zFe_{2z})_{3\pm\delta}O_4$ with the cobalt content of x_{Co} = 0.333 mol only interstitial cations dominate, for the lowest oxygen pressures, only the values of $D^o_{I(M)}$ and $b'_D = (D^o_{V(M)}/D^o_{I(M)}$ were adjusted, using a transformation of Equation 1.46a:

$$\left[M_i\right] + b'_D\left[V_{Mn}\right] = \frac{D^*_M}{D^o_{I(M)}} \tag{8.2}$$

As can be seen in Figure 8.1, a very good agreement was obtained between the dependence of the sum of the concentrations of ionic defects $([V_{Fe}] + b_D[Fe_i])$ on p_{O_2} for the fitted value of b'_D and the values of this sum calculated using the coefficients of self-diffusion of tracers (Lu and Dieckmann 1992). As can be seen in Figure 8.1a, there are quite significant

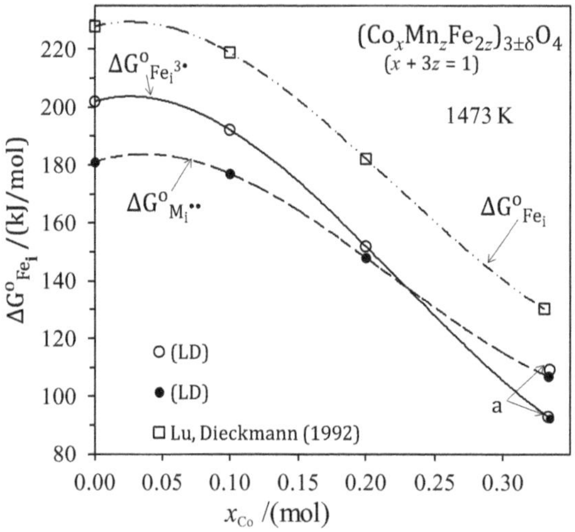

FIGURE 8.10 Dependence of the standard Gibbs energy $\Delta G^o_{Fe_i^{3\bullet}}$ and $\Delta G^o_{Fe_i^{\bullet\bullet}}$ of formation of interstitial ions $Fe_i^{3\bullet}$ and $Fe_i^{\bullet\bullet}$, calculated according to Equations 1.34c and 1.36c on the cobalt content x_{Co} in the spinel $(Co_xMn_zFe_{2z})_{3\pm\delta}O_4$, at 1473 K and the values $\Delta G^o_{Fe_i}$ calculated using the values of "equilibrium constants" of reaction of formation of interstitial cations K_I (Lu and Dieckmann 1992).

differences when using the results of the coefficients of diffusion of ions in the spinel $(Fe_{0.667}Mn_{0.333})_{3\pm\delta}O_4$ reported in Lu and Dieckmann (1992) and Lee et al. (2000).

Figure 8.11 presents the dependence of the parameter b_D on the cobalt content in the spinel $(Co_xMn_zFe_{2z})_{3\pm\delta}O_4$ for tracer ions: Fe* ((△,▲) points) and Co* (□,■), Mn* (○,●), using the values of coefficients of self-diffusion taken from Lu and Dieckmann (1992) (empty points) and Lee et al. (2000) (solid points).

As can be seen in Figure 8.11, the ratio of the coefficients of diffusion of tracer ions via interstitial ions to the coefficients of diffusion via cation vacancies for the content of $x_{Co} = 0.1$ mol in the spinel $(Co_xMn_zFe_{2z})_{3\pm\delta}O_4$ is higher than for the spinel $MnFe_2O_4$, and then, when the cobalt content increases, it decreases. The values of b_D for the spinel $MnFe_2O_4$ obtained when using the values of the coefficients of self-diffusion of tracers (Lee et al. 2000) are higher and they cause an essential change in the character of the dependence of b_D on the cobalt content in the spinel $(Co_xMn_zFe_{2z})_{3\pm\delta}O_4$.

Figure 8.12 presents the dependence of the values of coefficients of diffusion of tracer ions via cation vacancies $D^o_{V(M)}$ and via interstitial ions $D^o_{I(M)}$ on the cobalt content x_{Co} in the spinel $(Co_xMn_zFe_{2z})_{3\pm\delta}O_4$.

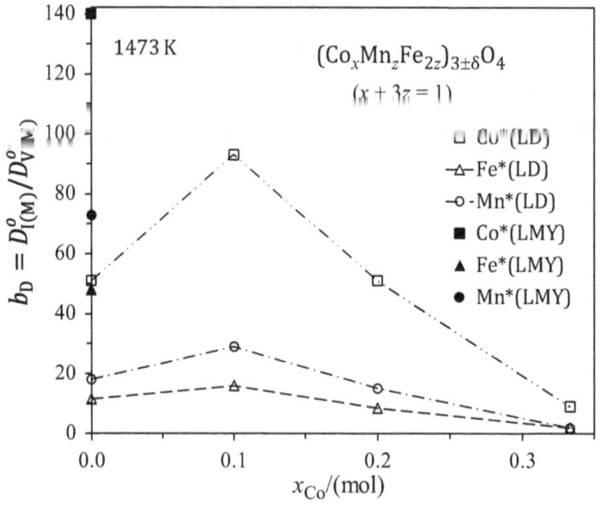

FIGURE 8.11 Dependence of the ratio of the coefficient of diffusion of tracer ions via interstitial cations to the coefficient of diffusion of cations via vacancies $D^o_{I(M)}/D^o_{V(M)}$, (parameter b_D) on the cobalt content x_{Co} in the spinel $(Co_xMn_zFe_{2z})_{3\pm\delta}O_4$ at 1473 K, obtained using the coefficients of self-diffusion of tracers: Fe* (△) points), Co* – (□), Mn* – (○), taken from (Lu and Dieckmann 1992) and (Lee et al. 2000) ((▲,■,●) points) and concentrations of ionic defects.

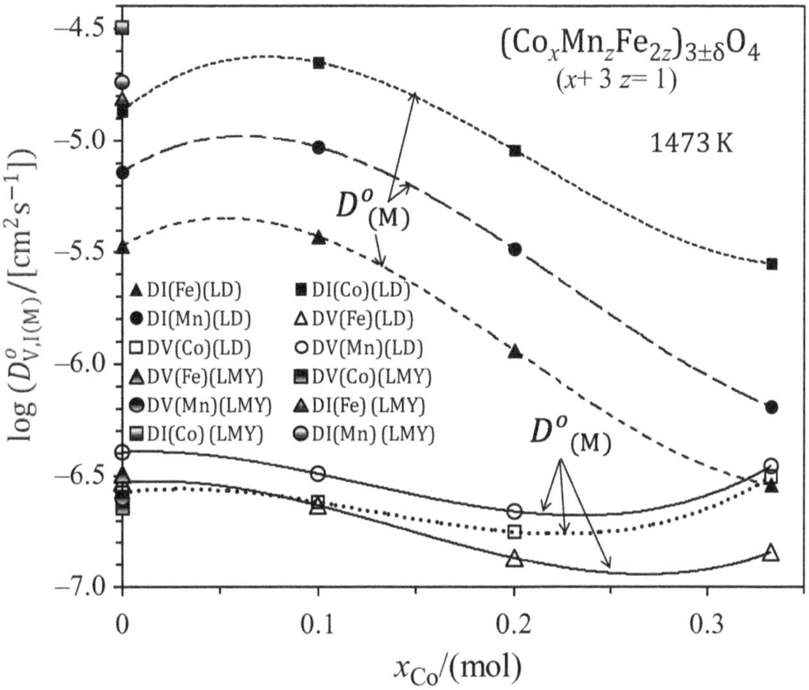

FIGURE 8.12 Dependence of the coefficient of diffusion via cation vacancies $D^o_{V(M)}$ (empty points) and via interstitial cations $D^o_{I(M)}$ (solid points) on the cobalt content x_{Co} in the spinel $(Co_xMn_zFe_{2z})_{3\pm\delta}O_4$ at 1473 K, obtained using the coefficients of self-diffusion of tracers: Fe* ((◆,▲) points), Co* – (□,■), Mn* – (○,●), taken from (Lu and Dieckmann 1992) and (Lee et al. 2000) ((◬,⬒,◒,◬,⬓,◓) points) and concentrations of ionic defects.

As can be seen in Figure 8.12, in the case of diffusion of tracer ions Fe*, Co* and Mn* in the spinel $(Co_xMn_zFe_{2z})_{3\pm\delta}O_4$, the coefficients of diffusion (mobility) of ions via cation vacancies only slightly differ among them and they decrease to the cobalt content up to $x_{Co} = 0.2$ mol, and at the cobalt content up to $x_{Co} = 0.333$ mol they increase significantly.

According to the values of the parameter b_D, the mobilities of tracer ions via interstitial cations are significantly higher than the mobilities via cation vacancies and the differences between them are significant. The highest mobility is that of cobalt ions, the lowest is that of iron ions. At the content of $x_{Co} = 0.1$ mol, the values of $D^o_{I(M)}$ are only slightly higher than for the spinel $MnFe_2O_4$ (using the results published in (Lee et al. 2000)). And then, as the content of cobalt in spinel $(Co_xMn_zFe_{2z})_{3\pm\delta}O_4$ increases, these values decrease. The values of $D^o_{I(M)}$ determined when using the results of the studies on the diffusion in the spinel $MnFe_2O_4$ (Lee et al. 2000) are significantly higher and, as can be seen in Figure 8.12, the values of $D^o_{I(M)}$ decrease when the cobalt content x_{Co} in the spinel $(Co_xMn_zFe_{2z})_{3\pm\delta}O_4$ increases.

8.4 CONCLUSIONS

Using the results of the studies of deviation from the stoichiometry for the spinel $(Co_xMn_zFe_{2z})_{3\pm\delta}O_4$ (where the composition is determined by the condition $x + 3y = 1$) with the cobalt content of $x_{Co} = 0.1$–0.333 mol (Lu and Dieckmann 1992), diagrams of defects' concentrations were determined, at 1473 K. It was found that cobalt doping of the spinel $(Co_xMn_zFe_{2z})_{3\pm\delta}O_4$ only slightly affects the character of the dependence of the deviation from the stoichiometry on the oxygen compared to the spinel $MnFe_2O_4$. The range of existence of the spinel is shifted practically in parallel towards higher oxygen pressures. In its range of existence, the spinel $(Co_xMn_zFe_{2z})_{3\pm\delta}O_4$ reaches the stoichiometric composition at a determined oxygen pressure, while when the cobalt content increases, this pressure is shifted towards higher pressures, by about five orders of magnitude (at the cobalt content of $x_{Co} = 0.333$ mol (compared to the spinel $MnFe_2O_4$).

In the range of higher oxygen pressures cation vacancies $\left[V_{Fe}'''\right]$ dominate, and the character of the dependence of their concentration on the oxygen pressure is analogous to the case of pure magnetite and the spinel $MnFe_2O_4$ (exponent about 2/3). When the cobalt content in $(Co_xMn_zFe_{2z})_{3\pm\delta}O_4$ increases, the concentration of cation vacancies significantly decreases (at the same oxygen pressure). In the range of low oxygen pressures, interstitial cations $\left[Fe_i^{3\bullet}\right]$ dominate.

It was shown that it is possible to adjust a reference value δ_o which causes a change in the character of the oxygen pressure dependence of the deviation from the stoichiometry indicating the occurrence of the incorporation of Co^{2+} ions in the sublattice of Fe^{3+} (Co_{Fe}') ions, which are negative in relation to the lattice. The maximum concentration of these ions at the cobalt content of $x_{Co} = 0.2$ mol can be $y_{Co'} = 1.2 \cdot 10^{-4}$ mol/mol.

It was found that for the cobalt content of x_{Co} = 0.1 mol in the spinel $(Co_xMn_zFe_{2z})_{3\pm\delta}O_4$, the standard Gibbs energy of the formation of Frenkel defects ΔG_F^o increases, and at a higher cobalt content it decreases. The result of the change in ΔG_F^o is a decrease in the concentration of Frenkel defects ($\left[V_{Fe}'''\right]=\left[Fe_i^{3\bullet}\right]$) to about $1.5\cdot10^{-5}$ and then, when the cobalt content increases, this concentration increases to $3\cdot10^{-5}$ mol/mol.

Standard Gibbs energies of formation of cation vacancies $\Delta G_{V_{Fe}''}^o$ and $\Delta G_{V_M''}^o$ increase when the content of cobalt in the spinel $(Co_xMn_zFe_{2z})_{3\pm\delta}O_4$ rises, while $\Delta G_{Fe_i^{3\bullet}}^o$ and $\Delta G_{M_i^{\bullet\bullet}}^o$ of formation of interstitial cations decrease. The above values depend on the value of ΔG_{eh}^o of formation of electronic defects adopted in the calculations and they were determined for the value of ΔG_{eh}^o = 30 kJ/mol, the same as the value assumed for pure magnetite.

Using the calculated concentration of ionic defects and the values of coefficients of self-diffusion of tracers M* (^{59}Fe, ^{60}Co and ^{53}Mn) in the spinel $(Co_xMn_zFe_{2z})_{3\pm\delta}O_4$ (Lu and Dieckmann 1992), the ratio of the coefficient of diffusion (mobility) of tracer ions via interstitial cations to the coefficient of diffusion via vacancies $\left(D_{I(M)}^o/D_{V(M)}^o=b_D\right)$ was determined. It was found that when the cobalt content in the spinel $(Co_xMn_zFe_{2z})_{3\pm\delta}O_4$ increases, the mobility of tracer ions via interstitial cations is higher than the mobility via cation vacancies and at the cobalt content of x_{Co} = 0.1 mol the parameter b_D reaches the highest values. At 1473 K, the values of the parameter b_D for cobalt ions is $b_D \approx 95$, for manganese $b_D \approx 31$, and for iron ions $b_D \approx 16$.

Cobalt doping of the spinel $(Co_xMn_zFe_{2z})_{3\pm\delta}O_4$, causes a decrease in the mobility of tracer ions, while the coefficients of their diffusion (mobility) via cation vacancies only slightly differ between them and they decrease to the cobalt content up to x_{Co} = 0.2 mol, and at the cobalt content up to x_{Co} = 0.333 mol they increase significantly.

For the diffusion via interstitial cations, the mobilities of tracer ions via interstitial cations are significantly higher than the mobilities via cation vacancies and the differences between them are significant. The highest mobility is that of cobalt ions, the lowest is that of iron ions. At the content of x_{Co} = 0.1 mol, the values of $D_{I(M)}^o$ are slightly higher than for the spinel $MnFe_2O_4$. And then, as the content of cobalt in spinel $(Co_xMn_zFe_{2z})_{3\pm\delta}O_4$ increases, these values decrease. The values of $D_{I(M)}^o$ determined when using the results of the studies on the diffusion in the spinel $MnFe_2O_4$ (Lee et al. 2000) are significantly higher and, when they are taken into account, the values of $D_{I(M)}^o$ decrease when the cobalt content x_{Co} in the spinel $(Co_xMn_zFe_{2z})_{3\pm\delta}O_4$ increases.

REFERENCES

Carter, D. C., and T. O. Mason. 1988. Electrical Properties and Site Distribution of Cations in $(Mn_yCo_{1-y})_{0.4}Fe_{2.6}O_4$. *J. Am. Ceram. Soc.* 71: 213–218.

Lee, J. H., M. Martin, and H.-I. Yoo. 2000. Self- And Impurity Cation Diffusion in Manganese–Zinc-Ferrite, $Mn_{1-x-y}Zn_xFe_{2+y}O_4$. *J. Phys. Chem. Solids.* 61: 1597–1605.

Lu, F. H., and R. Dieckmann. 1992. Point Defects and Cation Tracer Diffusion in $(Co, Fe,Mn)_{3-\delta}O_4$ Spinels: I. Mixed Spinels $(Co_xFe_{2y}Mn_y)_{3-\delta}O_4$. *Solid State Ionics.* 53-56: 290–302.

Lu, F. H., and R. Dieckmann. 1995. Point Defects in Oxide Spinel Solid Solutions of the Type $(Co,Fe,Mn)_{3-\Delta}O_4$ at 1200°C. *J. Phys. Chem. Solids.* 56: 725–733.

Singh, V. K., N. K. Khatri, and S. Lokanathan. 1981. Mössbauer Study of Ferrite Systems $Co_xMn_{1-x}Fe_2O_4$ and $Ni_xMn_{1-x}Fe_2O_4$. *Pramana.* 16: 273–280.

Subramanian, R., R. Dieckmann, G. Eriksson, and A. D. Pelton. 1994. Model Calculations of Phase Stabilities of Oxide Solid Solutions in the Co-Fe-Mn-O System at 1200°C. *J. Phys. Chem. Solids.* 55: 391–404.

9 Magnetite Doped with Zinc and Manganese – (Zn, Mn, Fe)$_{3\pm\delta}$O$_4$

9.1 INTRODUCTION

Zinc–manganese ferrites, due to their magnetic (soft magnetic materials) and electrical properties, have been an object of many works and analyses (McCurrie 1994, Valenzuela 1994). They are spinels that are formally similar to cobalt–manganese ferrites, discussed earlier, where manganese ions in the spinel $MnFe_2O_4$ are substituted with zinc ions. Studies of manganese–zinc ferrites were performed mainly at room temperature on frozen samples, due to their application. The range of existence of spinels as a function of the zinc content was studied with thermogravimetric method (Blank 1961, Morineau 1976, Morineau and Paulus 1973, Morineau and Paulus 1975, Tanaka 1981) and by electrical conductivity measurements (Jang and Yoo 1996, Tuller et al. 1985, Yoo and Tuller 1987, Yoo and Tuller 1988, Kim et al. 1990). However, there is no complete phase diagram, especially covering the range of existence of the spinel at higher oxygen pressures, where the spinel transition is possible. Beyond 1373 K, zinc–manganese ferrites should be stable up to the oxygen pressure of 1 atm. The studies of electrical conductivity of the spinel $Zn_{0.35}Mn_{0.54}Fe_{2.11}O_4$ as a function of the oxygen pressure and temperature were reported by Jang and Yoo (1996), and it was found that it is high, but lower than in magnetite, similar to that of other doped spinels. In the temperature range of 1373–1573 K, the electrical conductivity increases from 25 to 40 $\Omega^{-1}cm^{-1}$. When the oxygen pressure decreases, electrical conductivity decreases slightly, while it strongly decreases at high oxygen pressures as a result of decay and the transition of the spinel into doped hematite. The character of the oxygen pressure dependence of electrical conductivity was found to be analogous to that of pure magnetite, which would indicate that the mobility of electrons is much higher than the mobility of electron holes. Based on electrotransport measurements depending on the oxygen pressure the transport numbers of ions were determined (Jang and Yoo 1996) and were used for the calculations of cation diffusion coefficients. The character of the oxygen pressure dependence of the transport numbers indicates that the rates of diffusion of ions in the range of low and high oxygen pressures are different. Similar character of the dependence of coefficients of self-diffusion of tracers: ^{57}Co, ^{59}Fe, ^{54}Mn and ^{65}Zn in the spinel $MnFe_2O_4$ and $Zn_{0.35}Mn_{0.54}Fe_{2.11}O_4$ was reported in (Lee et al. 2000). The obtained results are analogous to the case of diffusion of pure and doped magnetite obtained by Dieckmann et al. (see Chapters 1–8).

Systematic studies of the oxygen pressure dependence of the deviation from the stoichiometry in the spinel $(Zn_xMn_{1-x}Fe_2)_{1\pm\delta/3}O_4$ with zinc content of $x_{Zn} = 0.1–0.5$ at 1473 K were published in (Töpfer et al. 2003). A second group of spinels studied were $Zn_{0.25-w/4}Mn_{0.75-3w/4}Fe_{2+w}O_4$ spinels, in which the iron content increased above two ($2 + w$), where w increased from $w = 0.04–0.1$ mol (Töpfer and Dieckmann 2004). For these spinels, the oxygen pressure dependence of the deviation from the stoichiometry was determined at 1473 K. For the spinel $Zn_{0.23}Mn_{0.69}Fe_{2.08}O_4$ the oxygen pressure dependence of the deviation from the stoichiometry was determined in the range of temperatures 1173–1473 K. The obtained results of the studies demonstrated that the range of existence of the spinel slightly changes depending on the composition. In the range of their existence the spinels attain the stoichiometric composition, and, similarly to other doped spinels, at low oxygen pressures interstitial cations dominate, and at high oxygen pressures cation vacancies dominate. The above defects' structure is consistent with the studies of self-diffusion (Jang and Yoo 1996, Lee et al. 2000). Using the results of the studies of the deviation from the stoichiometry (Töpfer et al. 2003, Töpfer and Dieckmann 2004), it was decided to determine the diagrams of defects' concentrations.

9.2 METHODS AND RESULTS OF CALCULATION OF THE DIAGRAMS OF THE CONCENTRATIONS OF DEFECTS, AND DISCUSSION

Spinels $(Zn_xMn_{1-x}Fe_2)_{1\pm\delta/3}O_4$ are obtained through a reaction of defined amounts of oxides: MnO, ZnO and Fe_2O_3. Zinc ions, Zn^{2+}, (with a stable oxidation state), as well as Mn^{2+} ions, as a result of the synthesis of the spinel will occupy cation sites, mainly with tetrahedral coordination. Thus, in the spinel $MnFe_2O_4$, manganese ions are substituted with zinc ions. The presence of zinc ions should not, therefore, affect the structure of point defects in the spinel. This is fully consistent with the character of the oxygen pressure dependence of the deviation from the stoichiometry reported in (Töpfer et al. 2003, Töpfer and Dieckmann 2004). Due to that, the calculations of diagrams of concentrations of point defects in magnetite doped with zinc and manganese were conducted with a method analogous to the one used for pure magnetite and cobalt-doped magnetite (see Sections 1.5 and 6.2), using the results of the deviation from the stoichiometry from (Töpfer et al. 2003, Töpfer and Dieckmann 2004).

Due to the lack of the value of ΔG°_{eh} of formation of electronic defects, it was assumed that it only slightly depends on the concentration of doping ions. Owing to that, in the calculations, a constant value of $\Delta G^{\circ}_{eh} = 30.3$ kJ/mol was assumed, the same as for pure magnetite and other doped spinels.

The analysis of the oxygen pressure dependence of the deviation from the stoichiometry from (Töpfer et al. 2003), plotted in a log–log scale, demonstrated that there are quite significant discrepancies (see Figure 9.1). A good match was obtained between the dependence of δ on p_{O_2} and the experimental values of δ in the range of higher oxygen pressures where cation vacancies dominate. There is no agreement between the dependence of δ on p_{O_2} and the experimental values of δ in the range of lower oxygen pressures, where interstitial cations dominate, similarly as in the case of spinels $(Fe_{1-x}Co_x)_{3\pm\delta}O_4$ and $(Co_xMn_zFe_{2z})_{3\pm\delta}O_4$. The character of the dependence of δ on p_{O_2} indicates that some zinc ions Zn^{2+} incorporate into the spinel structure into the sublattice of Fe^{3+} ions (Zn'_{Fe}), according to the reaction (6.2), which, for ZnO doping, assumes the form of:

$$4ZnO + V_i \rightarrow Zn^{x}_{Fe(2+)(o)} + 2Zn'_{Fe(3+)(o)} + 4O^{x}_{O} + Zn^{\bullet\bullet}_{i} \tag{9.1}$$

Due to that, according to the methods of calculations presented in Section 6.2, the values of $\Delta G^{\circ}_{F_1}$ and $\Delta G^{\circ}_{F_2}$ of formation of Frenkel defects were adjusted, assuming at a first approximation that these values are ual ($\Delta G^{\circ}_{F_1} = \Delta G^{\circ}_{F_2}$). Then, the

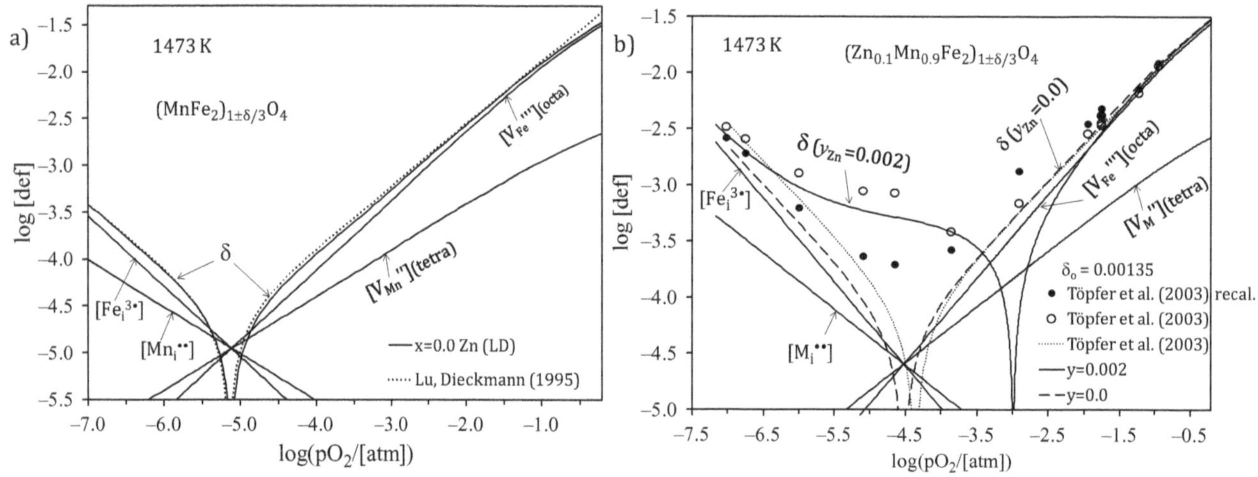

FIGURE 9.1 (a) Diagram of concentrations of point defects for the spinel $(Mn_{0.333}Fe_{0.667})_{3\pm\delta}O_4$, at 1473 K (solid lines). The (····) line shows the oxygen pressure dependence of the deviation from the stoichiometry, determined using the values of "uilibrium constants" of reactions of formation of cation vacancies K_V and interstitial cations K_I, given in Lu and Dieckmann (1995). (b) for the spinel $(Zn_{0.1}Mn_{0.9}Fe_2)_{1\pm\delta/3}O_4$ at 1473 K, obtained using the results of the studies of the deviation from the stoichiometry (Töpfer et al. 2003) ((○) points) and taking into account zinc ions Zn'_{Fe} incorporated into the sublattice of Fe^{3+} ions in the amount of $y'_{Zn} = 2{\cdot}10^{-3}$ mol/mol (solid lines). The (···) line shows the dependence of the deviation from the stoichiometry on p_{O_2}, determined using the "uilibrium constants" of reactions of formation of cation vacancies K_V and interstitial cations K_I, extracted from Töpfer et al. (2003). Points (●) denote the recalculated values of the deviation from the stoichiometry relative to the value $\delta_o = -1.35{\cdot}10^{-3}$, and (---) line represents the dependence of δ on p_{O_2} for $y'_{Zn} = 0.0$. (c) for the spinel $(Zn_{0.25}Mn_{0.75}Fe_2)_{1\pm\delta/3}O_4$, at 1473 K, when taking into account zinc ions Zn'_{Fe} incorporated into the sublattice of Fe^{3+} ions in the amount of $y'_{Zn} = 9{\cdot}10^{-4}$ mol/mol (solid lines). Points (●) denote the recalculated values of the deviation from the stoichiometry δ relative to $\delta_o = 6{\cdot}10^{-4}$. (d) for the spinel $(Zn_{0.2}Mn_{0.8}Fe_2)_{1\pm\delta/3}O_4$, at 1473 K, when taking into account zinc ions Zn'_{Fe} incorporated into the sublattice of Fe^{3+} ions in the amount of $y'_{Zn} = 1.4{\cdot}10^{-4}$ mol/mol (solid lines). Points (●) denote the recalculated values of the deviation from the stoichiometry δ relative to $\delta_o = -5{\cdot}10^{-4}$. (e) for the spinel $(Zn_{0.3}Mn_{0.7}Fe_2)_{1\pm\delta}/_3O_4$, at 1473 K, when taking into account zinc ions Zn'_{Fe} incorporated into the sublattice of Fe^{3+} ions in the amount of $y'_{Zn} = 7{\cdot}10^{-4}$ mol/mol (solid line). Points (●) denote the recalculated values of the deviation from the stoichiometry δ relative to $\delta_o = 4.8{\cdot}10^{-4}$. (f) for the spinel $(Zn_{0.4}Mn_{0.6}Fe_2)_{1\pm\delta}/_3O_4$, at 1473 K, when taking into account zinc ions Zn'_{Fe} incorporated into the sublattice of Fe^{3+} ions in the amount of $y'_{Zn} = 1.5{\cdot}10^{-3}$ mol/mol (solid line). Points (●) denote the recalculated values of the deviation from the stoichiometry δ relative to $\delta_o = 6{\cdot}10^{-4}$. (g) for the spinel $(Zn_{0.5}Mn_{0.5}Fe_2)_{1\pm\delta}/_3O_4$, at 1473 K, when taking into account zinc ions Zn'_{Fe} incorporated into the sublattice of Fe^{3+} ions in the amount of $y'_{Zn} = 2.5{\cdot}10^{-3}$ mol/mol (solid line). Points (●) denote the recalculated values of the deviation from the stoichiometry δ relative to the value $\delta_o = 1{,}0{\cdot}10^{-5}$.

values of $\Delta G^{o}_{V_M^{o}}$ of formation of cation vacancies V_M'' were fitted, along with such values of ΔG_I^{o} to obtain similar concentrations of iron vacancies at the stoichiometric composition $\left[V_M''\right] \cong \left[V_{Fe}'''\right]$ (according to Equation 1.35b). Therefore, such values of ΔG_i^{o} of the formation of defects and the value of the concentration of ions $\left[Zn_{Fe}'\right]$ (y_{Zn}') were adjusted at which for the highest values of the deviation from the stoichiometry, at high and low oxygen pressures, the dependence of δ on p_{O_2} matched the experimental values of the deviation δ as much as possible. Due to high concentrations of ions Mn^{2+} and

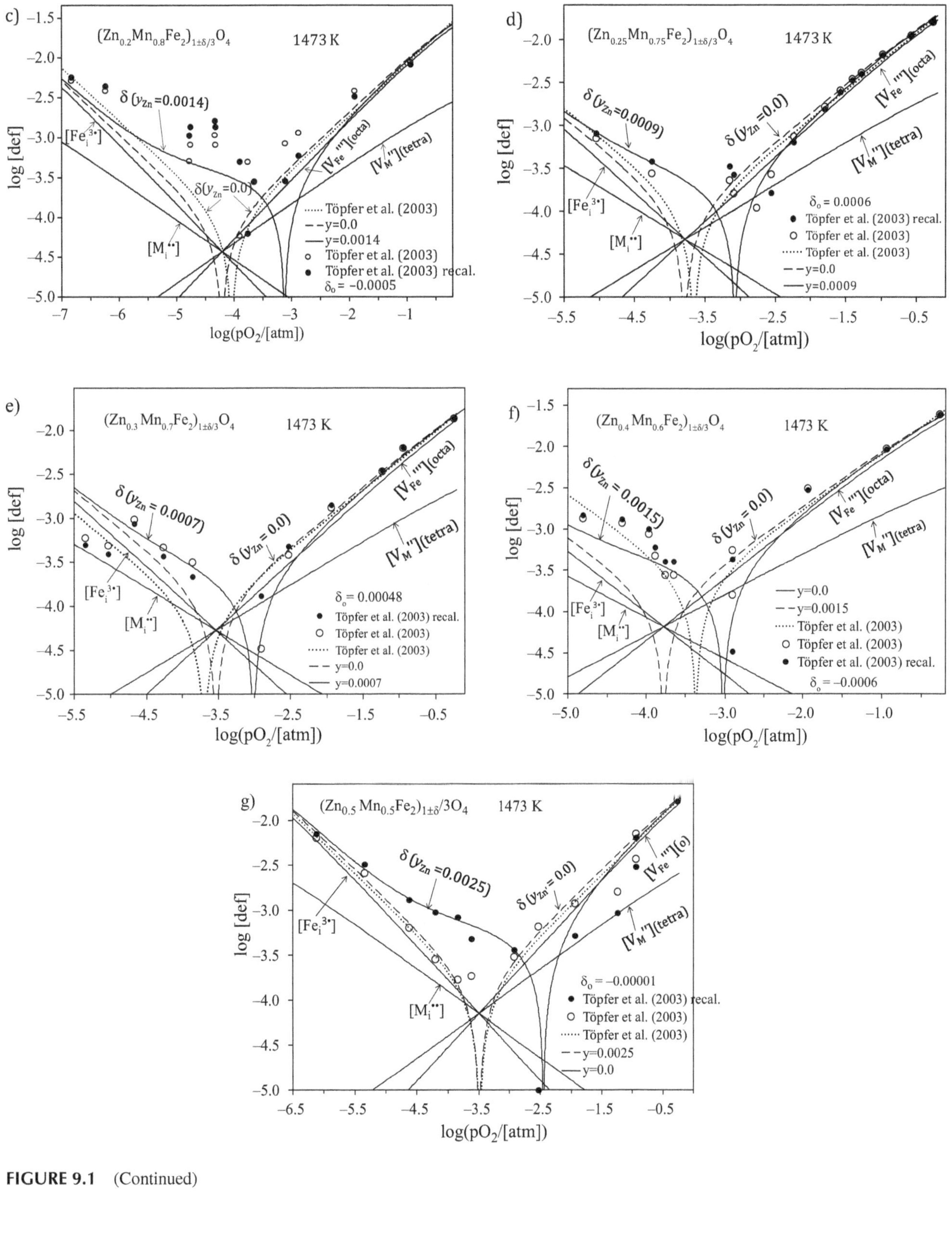

FIGURE 9.1 (Continued)

Zn^{2+}, which occupy cation sites with tetrahedral coordination, cation vacancies $V'''_{Fe(t)}$ in this sublattice were not considered.

Figure 9.1a–g show diagrams of the concentrations of ionic defects, for spinels $(Zn_xMn_{1-x}Fe_2)_{1\pm\delta/3}O_4$, with the zinc content of $x_{Zn} = 0.0–0.5$ mol at 1473 K (solid lines), calculated using the results of the deviation from the stoichiometry from Töpfer et al. (2003) ((○) points). The dotted line (···) shows the dependence of δ on p_{O_2} obtained using "uilibrium constants" of reactions of formation of cation vacancies K_V and interstitial ions K_I taken from Töpfer et al. (2003), and, for the spinel $(Mn_{0.333}Fe_{0.667})_{3\pm\delta}O_4$, given in Lu and Dieckmann (1995).

As can be seen in Figure 9.1a–g, the obtained dependence of δ on p_{O_2}, without considering the concentration of ions $[Zn'_{Fe}]$ ((---) line) is close to the values of the deviation from the stoichiometry reported in Töpfer et al. (2003) (dotted line (···)). The experimental values δ taken from the same work. ((○) points) near the stoichiometric composition deviate from the proposed dependence of δ on p_{O_2}. Attempts to fit other value of δ_o than that adopted by Töpfer (Töpfer et al. 2003), due to relatively high values of the deviation from the stoichiometry at low oxygen pressures, did not change significantly the values of the deviation from the stoichiometry ((●) points) or the character of the dependence of δ on p_{O_2}, which would be consistent with the character suggested in (Kim et al. 1990). As can be seen in Figure 9.1g, only in the case of spinel $Zn_{0.5}Mn_{0.5}Fe_2)_{1\pm\delta/3}O_4$ are the experimental values of δ ((○) points) consistent with the dependence of δ on p_{O_2}. However, it is possible to adjust a value of the parameter $\delta_o = 1.0·10^{-5}$ in such a way that the recalculated values of the deviation δ ((●) points) are consistent with the dependence of δ on p_{O_2}, taking into account the concentration of ions $\left(Zn'_{Fe}\right)$ in the amount of $y'_{Zn} = 0.0025$ mol/mol.

Therefore, the obtained character of the dependence of the deviation from the stoichiometry on p_{O_2} for spinels $(Zn_xMn_{1-x}Fe_2)_{1\pm\delta/3}O_4$ indicates that zinc ions, apart from the fact that they occupy Mn^{2+} nodes, in a small amount incorporate also into the nodes of Fe^{3+} ions $\left(Zn'_{Fe}\right)$ and they are negative in relation to the lattice. Only for the spinel $(Zn_{0.5}Mn_{0.5}Fe_2)_{1\pm\delta/3}O_4$ the result is ambiguous. It should be noted that in the case of spinels $(Zn_xMn_{1-x}Fe_2)_{1\pm\delta/3}O_4$ compared to other doped ferrites there are quite large discrepancies between the calculated dependence of δ on p_{O_2} and the values of the deviation from the stoichiometry, especially in the middle of the range of existence of the spinel (there is a large scatter of measured values). This might result from measurement difficulties for small changes of the deviation from the stoichiometry in this range of oxygen pressures. Slight differences in the values of the deviation δ in a log–log scale show a large scatter of measurement points.

Figure 9.2 presents the dependence of the concentration of ions $\left[Zn'_{Fe}\right]$ (y'_{Zn}) on the zinc content x_{Zn} in the spinel $(Zn_{0.5}Mn_{0.5}Fe_2)_{1\pm\delta/3}O_4$.

As can be seen in Figure 9.2, in the spinel with the zinc content of $x_{Zn} = 0.1$ mol the concentration of ions $\left[Zn'_{Fe}\right]$ reaches $7·10^{-4}$ mol/mol. A further increase in the zinc content, up to $x_{Zn} = 0.3$ mol, causes a decrease in the concentrations of these ions, and at higher zinc contents in the spinel the concentration of ions $\left[Zn'_{Fe}\right]$ increases again, up to $1.2·10^{-3}$ mol/mol.

As can be seen in Figure 9.1, similarly as in the case of other doped ferrites, in the range of high oxygen pressures cation vacancies $\left[V'''_{Fe}\right]$ dominate, and in the range of low oxygen pressures, interstitial cations $\left[Fe_i^{3\bullet}\right]$ dominate. The concentration of cation vacancies $\left[V''_{Fe}\right]$ and interstitial cations $\left[M_i^{\bullet\bullet}\right]$ is significantly lower.

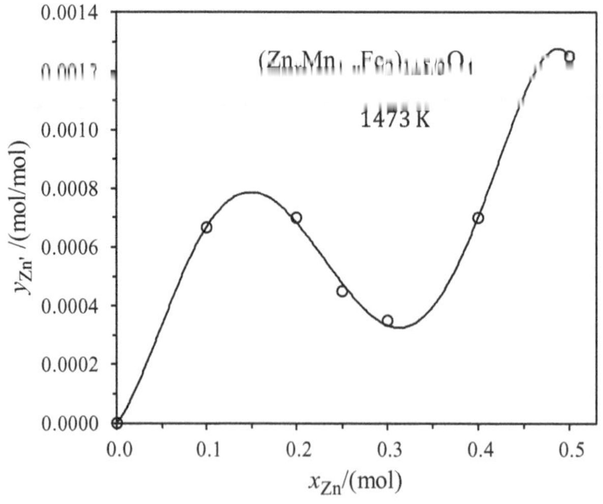

FIGURE 9.2 Dependence of the concentration of zinc ions $\left[Zn'_{Fe}\right]$ (y'_{Zn}) incorporated into the sublattice of Fe^{3+} on the zinc content x_{Zn} in the spinel $(Zn_xMn_{1-x}Fe_2)_{1\pm\delta/3}O_4$, at 1473 K, obtained using the results of the studies of the deviation from the stoichiometry from (Töpfer et al. 2003).

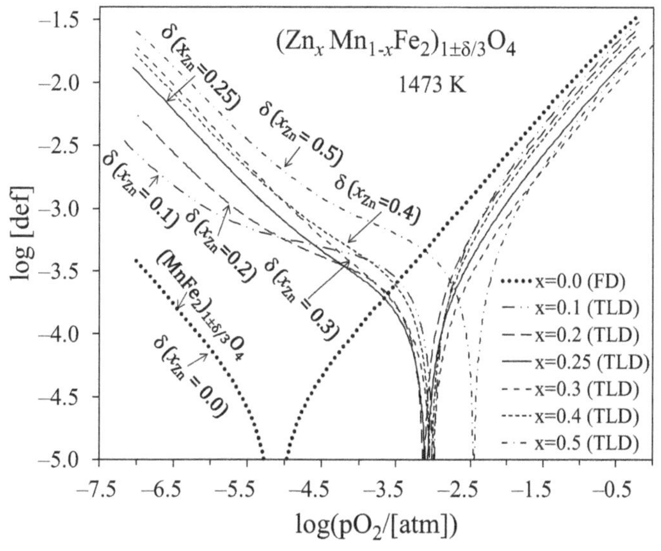

FIGURE 9.3 Dependences of the deviation from the stoichiometry on the oxygen pressure in the spinel $(Zn_xMn_{1-x}Fe_2)_{1\pm\delta/3}O_4$ with the zinc content x_{Zn} = 0.0–0.5 mol, at 1473 K, (presented in Figure 9.1 and 7.1d).

Figure 9.3 compares the oxygen pressure dependences of the deviation from the stoichiometry for the zinc content of x_{Zn} = 0.0–0.5 mol in the spinel $(Zn_xMn_{1-x}Fe_2)_{1\pm\delta/3}O_4$, obtained at 1473 K (presented in Figure 9.1). As can be seen in Figure 9.3, the range of existence of spinels $(Zn_xMn_{1-x}Fe_2)_{1\pm\delta/3}O_4$ slightly changes when the zinc content increases compared to the spinel $(MnFe_2)_{1\pm\delta/3}O_4$). When the zinc content increases in the spinel, in the range of higher oxygen pressures, the concentration of cation vacancies slightly decreases. The deviation from the stoichiometry increases in the range of low oxygen pressures, therefore, the concentration of interstitial cations increases.

In Figure 9.3, we can see that at the zinc content of x_{Zn} = 0.1 mol, in the spinel $(Zn_xMn_{1-x}Fe_2)_{1\pm\delta/3}O_4$ the oxygen pressure at which the spinel reaches the stoichiometric composition significantly shifts in relation to the spinel $MnFe_2O_4$. This is caused not only by the presence of zinc ions Zn^{2+}, but also the concentration of incorporated ions $\left[Zn'_{Fe}\right]$. A further increase in the zinc content up to x_{Zn} = 0.4 mol practically does not change this pressure $p_{O_2}^{(\delta=0)}$. The dependence that deviates from others is the dependence of δ on p_{O_2} for the spinel $(Zn_{0.5}Mn_{0.5}Fe_2)_{1\pm\delta/3}O_4$, which is consistent with the experimental values (Töpfer et al. 2003), but it seems correct also for the recalculated values ((●) points) (see Figure 9.1f).

Figure 9.4 shows the oxygen pressure dependence of the derivative $(d\log p_{O_2}/\log\delta) = n$, for the dependences of the deviation from the stoichiometry on p_{O_2} in the spinel $(Zn_xMn_{1-x}Fe_2)_{1\pm\delta/3}O_4$, shown in Figure 9.1.

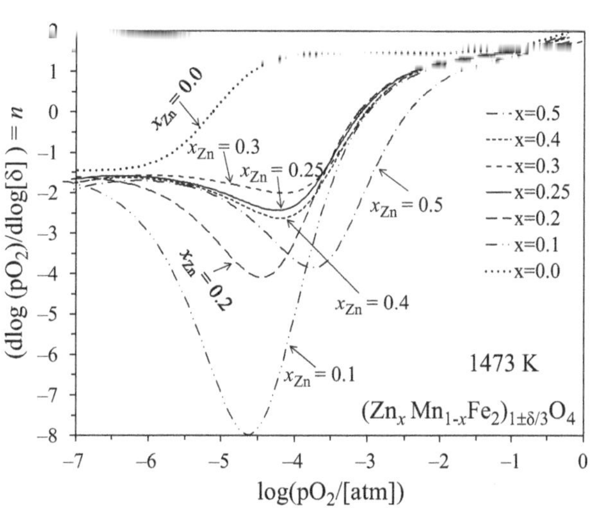

FIGURE 9.4 Dependence of the derivative: $n_\delta = d\log(p_{O_2})/d\log\delta$ on p_{O_2}, obtained for the dependence of the deviation from the stoichiometry on p_{O_2} in the spinel $Zn_xMn_{1-x}Fe_2)_{1\pm\delta/3}O_4$ with the zinc content x_{Zn} = 0.0–0.5 mol, at 1473 K (presented in Figure 9.1).

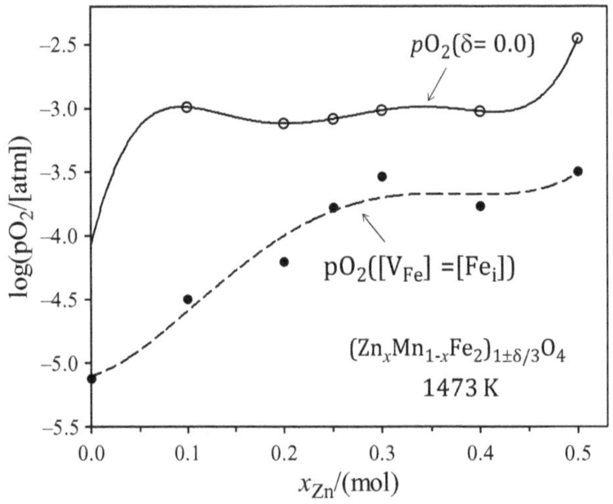

FIGURE 9.5 Dependence of the oxygen pressure at which the spinel $(Zn_xMn_{1-x}Fe_2)_{1\pm\delta/3}O_4$ reaches the stoichiometric composition on the zinc content x_{Zn} at 1473 K, when taking into account the concentration of zinc ions $\left[Zn'_{Fe}\right] = 1/2y_{Zn'})$ incorporated into the sublattice of Fe^{3+} ions ((○) points), and the oxygen pressure at which the concentrations of cation vacancies and interstitial cations are ual ((●) points).

As can be seen in Figure 9.4, at the highest pressures, in a certain narrow range, the value of the derivative reaches $n = 1.5$. At the lowest oxygen pressures the parameter n approaches the value of −1.5 ($1/n = -2/3$). The fact that the parameter n reaches high negative values is related to the character of the dependence of δ on p_{O_2}.

Figure 9.5 shows the dependence of the oxygen pressure at which spinels $(Zn_xMn_{1-x}Fe_2)_{1\pm\delta/3}O_4$ reach the stoichiometric composition on the zinc content x_{Zn} ((○) points, taking into account the concentration of ions $\left[Zn'_{Fe}\right]$).

As shown in Figure 9.5, the oxygen pressure at which the stoichiometric composition is present increases for the zinc content of $x_{Zn} = 0.1$ mol. When the zinc content in the spinel increases further, the changes are small. The pressure at which the concentrations of cation vacancies and interstitial ions are ual ($[V_{Fe}] = [Fe_i]$) monotonically increases when the zinc content increases ((●) points); due to the presence of $\left[Zn'_{Fe}\right]$ it occurs at lower oxygen pressures.

Figure 9.6 presents the dependence of the concentration of Frenkel defects on the zinc content x_{Zn} in the spinel $Zn_xMn_{1-x}Fe_2O_4$, which monotonically increases. This is a result of a decrease in the value of $\Delta G^o_{F_1}$ when the zinc content in the spinel increases (see Figure 9.7).

Figures 9.7–9.10 present the dependence of adjusted standard Gibbs energies: $\Delta G^o_{F_1}$, $\Delta G^o_{F_2}$, $\Delta G^o_{V''_M}$, ΔG^o_I on the zinc content x_{Zn} in the spinel $(Zn_xMn_{1-x}Fe_2)_{1\pm\delta/3}O_4$ at 1473 K. Figures 9.8 and 9.10 show the dependence of standard Gibbs energy of formation of vacancies $V'''_{Fe(o)}$ ($\Delta G^o_{V''_{Fe}}$) calculated according to Equation 1.35d, $\Delta G^o_{Fe_i^{3\bullet}}$ of formation of interstitial iron ions $Fe_i^{3\bullet}$ calculated according to Equation 1.34c and ($\Delta G^o_{M_i^{\infty}}$) of formation of ions $M_i^{\bullet\bullet}$ calculated according to Equation 1.36c.

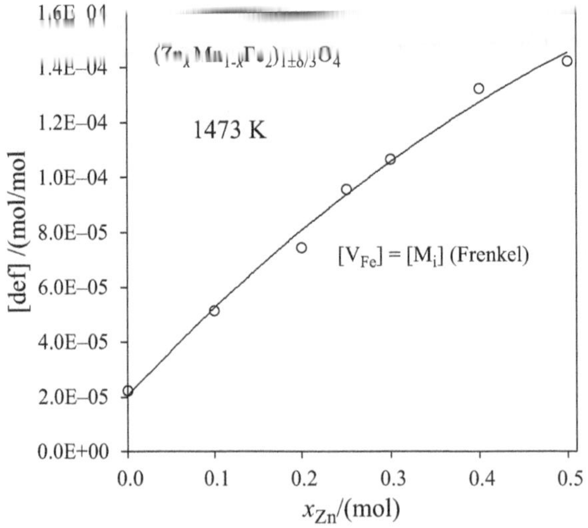

FIGURE 9.6 Dependence of the concentration of Frenkel defects on the zinc content x_{Zn} in the spinel $(Zn_xMn_{1-x}Fe_2)_{1\pm\delta/3}O_4$, at 1473 K.

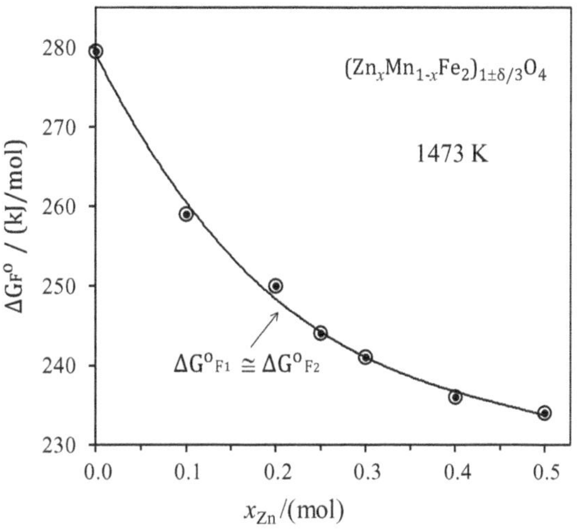

FIGURE 9.7 Dependence of standard Gibbs energy $\Delta G^o_{F_1}$ and $\Delta G^o_{F_2}$ of formation of Frenkel defects with the charge of two and three in the spinel $(Zn_xMn_{1-x}Fe_2)_{1\pm\delta/3}O_4$ on the zinc content x_{Zn}, at 1473 K, obtained using the results of the studies of the deviation from the stoichiometry (Töpfer et al. 2003).

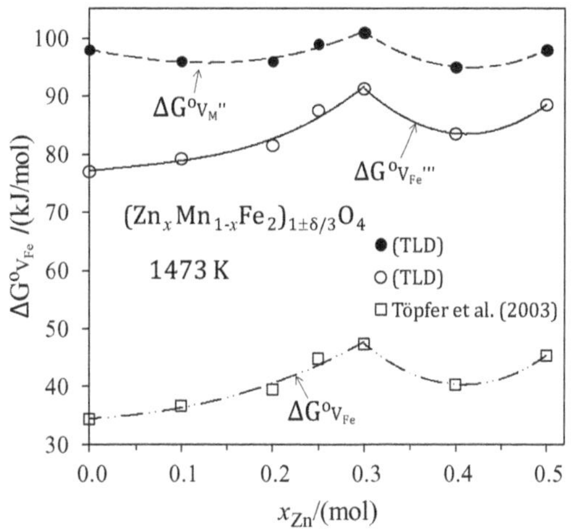

FIGURE 9.8 Dependence of standard Gibbs energy $\Delta G^o_{V'''_{Fe}}$ of formation of cation vacancies $\left(V'''_{Fe(o)}\right)$ and $\Delta G^o_{V''_{Fe}}$ of formation of vacancies $\left(V''_{Fe(o)}\right)$ in the spinel $(Zn_xMn_{1-x}Fe_2)_{1\pm\delta/3}O_4$ on the zinc content x_{Zn}, at 1473 K, and the values $\Delta G^o_{V_{Fe}}$ calculated using the "uilibrium constants" of reaction of formation of vacancies K_V from Töpfer et al. (2003).

As seen in Figure 9.7, the standard Gibbs energy of the formation of Frenkel defects $\Delta G^o_{F_1} = \Delta G^o_{F_2}$ decreases when the zinc content in the spinel increases and the manganese concentration decreases.

Figures 9.8 and 9.10 show that the values of $\Delta G^o_{V'''_{Fe}}$ and $\Delta G^o_{V''_M}$ of formation of vacancies slightly increase when the zinc content in the spinel increases up to $x_{Zn} = 0.3$ mol, and then, for $x_{Zn} = 0.4$ they decrease and at $x_{Zn} = 0.5$ they increase again. The values of $\Delta G^o_{Fe_i^{3\bullet}}$ and $\Delta G^o_{M_i^{\bullet\bullet}}$ decrease when the zinc content in the spinel increases.

Figures 9.7 and 9.10 also show the values of $\Delta G^o_{V_{Fe}}$ and $\Delta G^o_{Fe_i}$ calculated using "uilibrium constants" of reaction of formation of cation vacancies K_V and interstitial ions K_I taken from (Töpfer et al. 2003). The values differ from those determined in the present work, because they describe another model of defects, but they have a similar nature in dependence on the zinc content in the spinel.

Ferrites $Zn_{0.25-w/4}Mn_{0.75-3z/4}Fe_{2+w})_{1\pm\delta/3}O_4$

A second series of ferrites studied (Töpfer and Dieckmann 2004) were spinels $(Zn_{0.25-w/4}Mn_{0.75-3w/4}Fe_{2+w})_{1\pm\delta/3}O_4$, in which the iron content $(2 + w)$ increases, where $w = 0.0$–0.1 mol. Figure 9.11 presents the diagrams of the concentrations of ionic defects for spinels $(Zn_{0.25-w/4}Mn_{0.75-3w/4}Fe_{2+w})_{1\pm\delta/3}O_4$ at the temperature of 1473 K (solid lines), calculated using the results of the deviation from the stoichiometry taken from (Töpfer and Dieckmann 2004) ((○) points); the diagram for the spinel

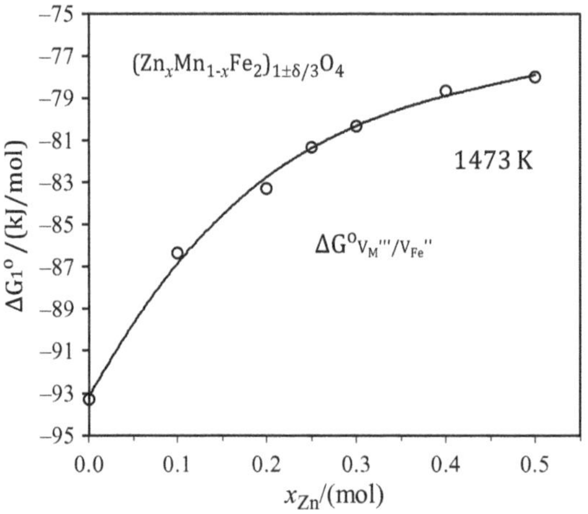

FIGURE 9.9 Dependence of standard Gibbs energy ΔG_I^o on the zinc content x_{Zn} in the spinel $(Zn_xMn_{1-x}Fe_2)_{1\pm\delta/3}O_4$ at 1473 K, permitting determination of the ratio of concentrations of iron vacancies $\left[V'''_{Fe(o)}\right]/\left[V''_M\right]$, according to Equation 1.35b.

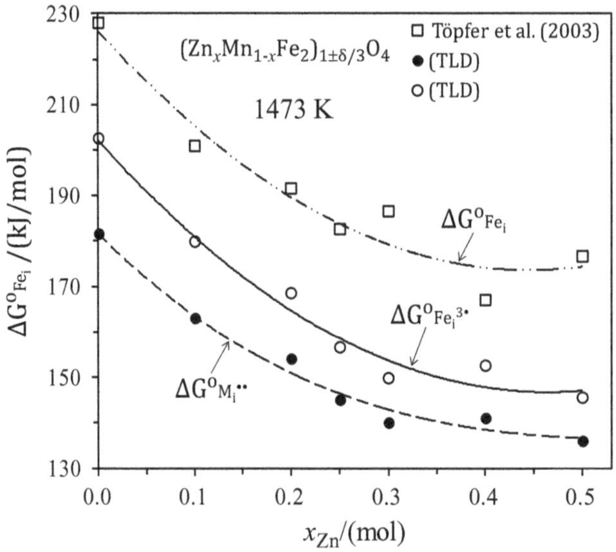

FIGURE 9.10 Dependence of the standard Gibbs energy $\Delta G^o_{Fe_i^{3\bullet}}$ and $\Delta G^o_{Fe_i^{\bullet\bullet}}$ of formation of interstitial ions $Fe_i^{3\bullet}$ and $Fe_i^{\bullet\bullet}$, calculated according to Equation (1.34c) and (1.30c) in the spinel $(Zn_xMn_{1-x}Fe_2)_{1\pm\delta/3}O_4$ on the zinc content x_{Zn}, at 1473 K and the values $\Delta G^o_{Fe_i}$ calculated using the values of "uilibrium constants" of reaction of formation of interstitial iron ions K_I from (Töpfer et al. 2003).

$(Zn_{0.25}Mn_{0.75}Fe_2)_{1\pm\delta/3}O_4$ ($w = 0.0$) is shown in Figure 9.1c. The dotted line (···) shows the dependence of δ on p_{O_2} obtained using the "uilibrium constants" K_V and K_I (Töpfer and Dieckmann 2004).

As can be seen in Figure 9.11a–d, in the spinel $(Zn_{0.25-w/4}Mn_{0.75-3w/4}Fe_{2+w})_{1\pm\delta/3}O_4$, in which the iron content only slightly increases above 2, diagrams of concentrations of defects also change only slightly in relation to the spinel $(Zn_{0.25}Mn_{0.75}Fe_2)_{1\pm\delta/3}O_4$. Similarly, in order to obtain a match between the dependence of δ on p_{O_2} with the experimental values, it was necessary to take into account in the calculations the concentrations of ions $\left[Zn'_{Fe}\right]$ according to methods outlined earlier. Attempts to fit other value of δ_o than that adopted by Töpfer and Dieckmann (2004), due to high values of the deviation from the stoichiometry at low oxygen pressures, did not change significantly their values ((●) points); also the character of the dependence of δ on p_{O_2} did not change – it deviates from the dependence suggested in Töpfer and Dieckmann (2004) (dotted line (···)).

Figure 9.12 presents the dependence of the fitted values of concentrations of ions $\left[Zn'_{Fe}\right] = y_{Zn'}$ on the values w_{Fe}.

As shown in Figure 9.12, an increase in the iron content above 2 mol causes a significant increase in the concentration of ions (Zn'_{Fe}) in the spinel.

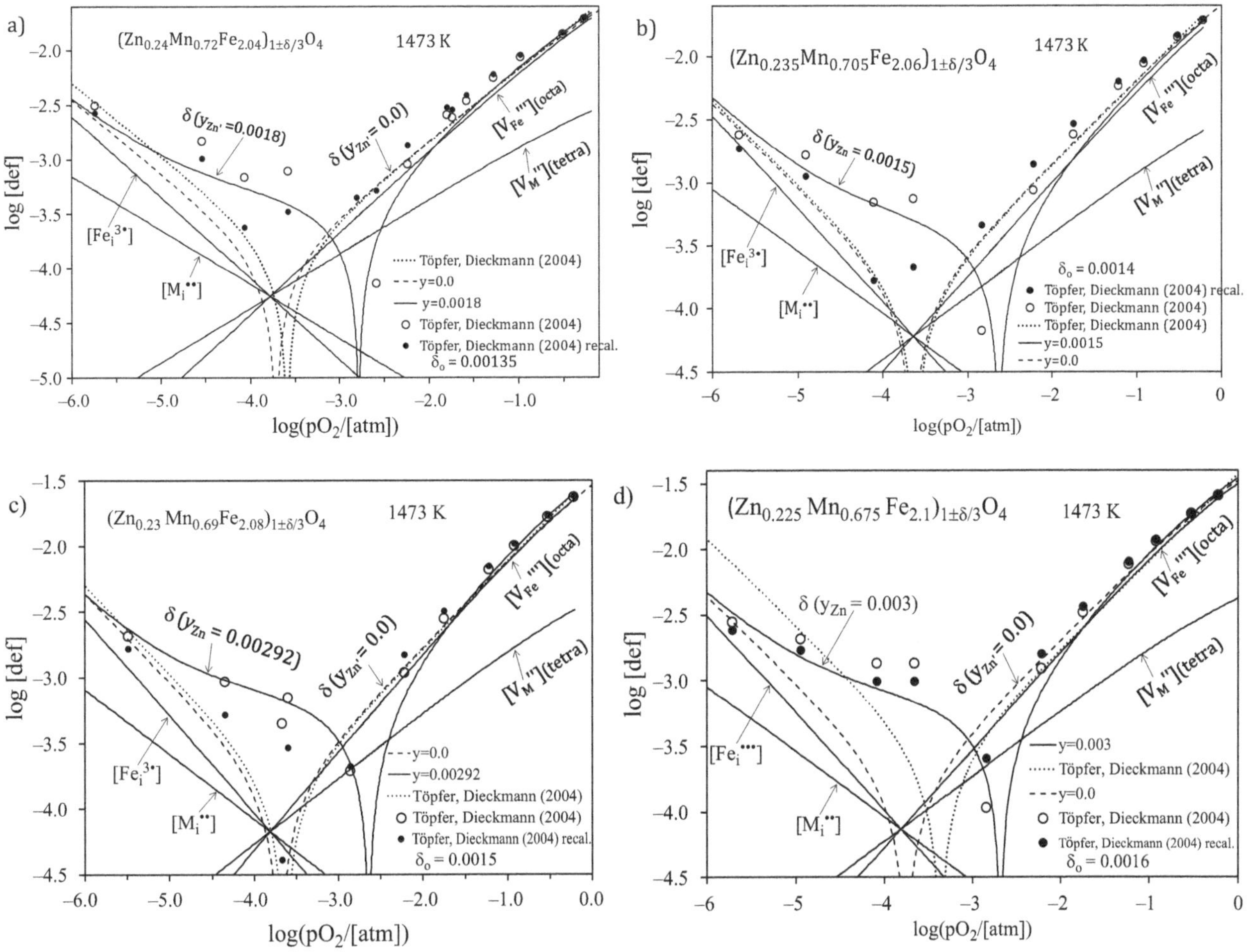

FIGURE 9.11 (a) Diagram of concentrations of point defects for the spinel $(Zn_{0.24}Mn_{0.72}Fe_{2.04})_{1\pm\delta/3}O_4$, at 1473 K, taking into account zinc ions Zn'_{Fe} incorporated into the sublattice of Fe^{3+} ions in the amount of $y'_{Zn} = 1.8\cdot10^{-3}$ mol/mol (solid lines). obtained using the results of the studies of the deviation from the stoichiometry taken from Töpfer and Dieckmann (2004) ((○) points). The (···) line shows the dependence of the deviation from the stoichiometry on p_{O_2}, determined using the "uilibrium constants" K_V and K_I given in Töpfer and Dieckmann (2004). Points (●) denote the recalculated values of the deviation from the stoichiometry relative to the value $\delta_o = 1.35\cdot10^{-3}$, and (---) line represents the dependence of δ on p_{O_2} for $y'_{Zn} = 0.0$. (b) for the spinel $(Zn_{0.235}Mn_{0.705}Fe_{2.06})_{1\pm\delta/3}O_4$, at 1473 K, when taking into account zinc ions Zn'_{Fe} incorporated into the sublattice of Fe^{3+} ions in the amount of $y'_{Zn} = 1.5\cdot10^{-3}$ mol/mol (solid lines). Points (●) show the recalculated values of the deviation from the stoichiometry (Töpfer and Dieckmann 2004) relative to the value $\delta_o = 1.4\cdot10^{-3}$. (c) for the spinel $(Zn_{0.23}Mn_{0.69}Fe_{2.08})_{1\pm\delta/3}O_4$, at 1473 K, when taking into account zinc ions Zn'_{Fe} incorporated into the sublattice of Fe^{3+} ions in the amount of $y'_{Zn} = 2.92\cdot10^{-3}$ mol/mol (solid lines). Points (●) show the recalculated values of the deviation from the stoichiometry (Töpfer and Dieckmann 2004) relative to the value $\delta_o = 1.5\cdot10^{-3}$. (d) for $(Zn_{0.225}Mn_{0.675}Fe_{2.1})_{1\pm\delta/3}O_4$, at 1473 K, when taking into account zinc ions Zn'_{Fe} incorporated into the sublattice of Fe^{3+} ions in the amount of $y'_{Zn} = 3\cdot10^{-3}$ mol/mol (solid lines). Points (●) represent the recalculated values of the deviation from the stoichiometry (Töpfer and Dieckmann 2004) relative to the value of $\delta_o = 1.6\cdot10^{-3}$.

In turn, Figure 9.13 compares the dependences of the deviation from the stoichiometry on the oxygen pressure, obtained for the spinel $(Zn_{0.25-w/4}Mn_{0.75-3w/4}Fe_{2+w})_{1\pm\delta/3}O_4$ for the values of $w_{Fe} = 0.0$–0.1 mol at the temperature of 1473 K.

As can be seen in Figure 9.13, an increase in the iron content above 2 mol causes a small increase in the concentration of cation vacancies and interstitial cations. Also, as can be seen in Figure 9.14, when the iron content (w) increases, the oxygen pressure at which spinels reach the stoichiometric composition increases only slightly. The change in the oxygen pressure at which the concentrations of cation vacancies and interstitial cations are ual, ($[V_{Fe}] = [Fe_i]$)), is even smaller. As can be seen in Figure 9.15, when the iron content (w) in the spinel increases, the concentration of Frenkel defects increases from $1\cdot10^{-4}$ to $1.6\cdot10^{-4}$ mol/mol).

Figures 9.16–9.18 present the dependence of adjusted standard Gibbs energies: $\Delta G^o_{F_1}$, $\Delta G^o_{F_2}$, $\Delta G^o_{V_M''}$, ΔG^o_I on the excess amount of iron w in the spinel $(Zn_{0.25-w/4}Mn_{0.75-3w/4}Fe_{2+w})_{1\pm\delta/3}O_4$ at 1473 K. Figures 9.17 and 9.19 present the dependence of

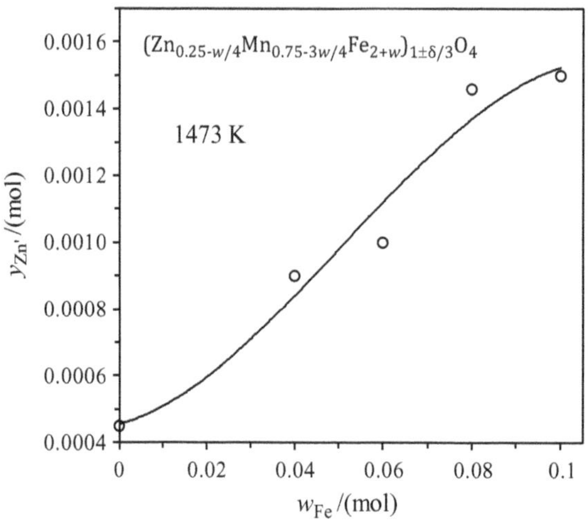

FIGURE 9.12 Dependence of the concentration of zinc ions $\left[Zn'_{Fe}\right]$ (y'_{Zn}) incorporated into the sublattice of Fe^{3+} ions, in the spinel $(Zn_{0.25-w/4}Mn_{0.75-3w/4}Fe_{2+w})_{1\pm\delta/3}O_4$, on the concentration of excess iron w_{Fe}, at 1473 K.

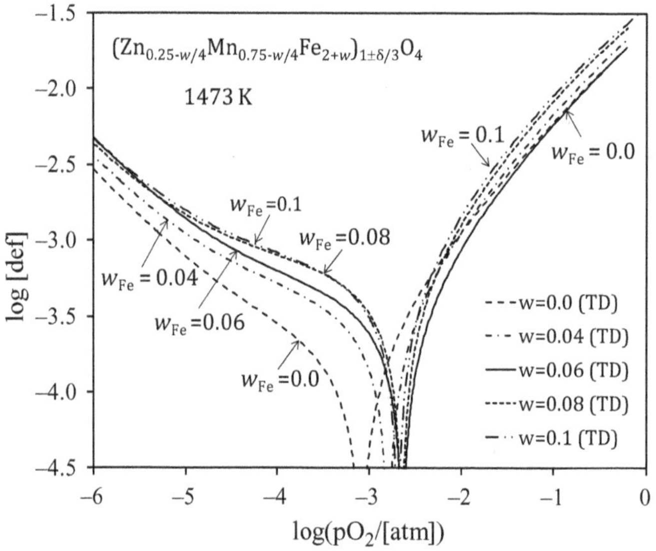

FIGURE 9.13 Oxygen pressure dependences of the deviation from the stoichiometry in the spinel $(Zn_{0.25-w/4}Mn_{0.75-3w/4}Fe_{2+w})_{1\pm\delta/3}O_4$ on the concentration of excess iron w_{Fe} = 0.0–0.1 mol, at 1473 K obtained using the results of the studies of the deviation from the stoichiometry (Töpfer and Dieckmann 2004).

the standard Gibbs energy $\Delta G^o_{V'''_{Fe}}$ of formation of vacancies $V'''_{Fe(o)}$ and $\Delta G^o_{Fe_i^{3\bullet}}$ of formation of interstitial iron ions $Fe_i^{3\bullet}$ and ($\Delta G^o_{M_i^{\infty}}$) of formation of interstitial cations $M_i^{\bullet\bullet}$.

Shown in Figure 9.16, the standard Gibbs energy of the formation of Frenkel defects $\Delta G^o_{F_1} = \Delta G^o_{F_2}$ decreases linearly when the iron content in the spinel increases.

As can be seen in Figures 9.17 and 9.19, the values of ($\Delta G^o_{V'''_{Fe}}$) and ($\Delta G^o_{V''_M}$), but also $\Delta G^o_{Fe_i^{3\bullet}}$ and $\Delta G^o_{M_i^{\bullet\bullet}}$, only slightly decrease when the iron content increases. A similar decrease is present for the values of $\Delta G^o_{V_{Fe}}$ and $\Delta G^o_{Fe_i}$ calculated using "uilibrium constants" of reaction of formation of cation vacancies K_V and interstitial cations K_I (Töpfer and Dieckmann 2004).

Ferrite $(Zn_{0.23}Mn_{0.68}Fe_{2.08})_{1-\delta/3}O_4$

In the third series of studies (Töpfer and Dieckmann 2004), the oxygen pressure dependence of the deviation from the stoichiometry for the spinel $(Zn_{0.23}Mn_{0.68}Fe_{2.08})_{1\pm\delta/3}O_4$ in the temperature range of 1173–1473 K was determined.

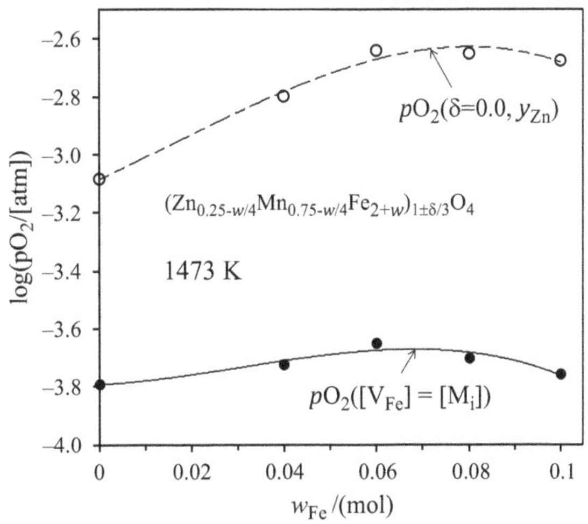

FIGURE 9.14 Dependence of the oxygen pressure at which the spinel $(Zn_{0.25-w/4}\,Mn_{0.75-3w/4}Fe_{2+w})O_4$ reaches the stoichiometric composition on the concentration of excess iron w_{Fe}, at 1473 K ((O) points) and the oxygen pressure at which the concentrations of cation vacancies and interstitial cations are ual ((●) points).

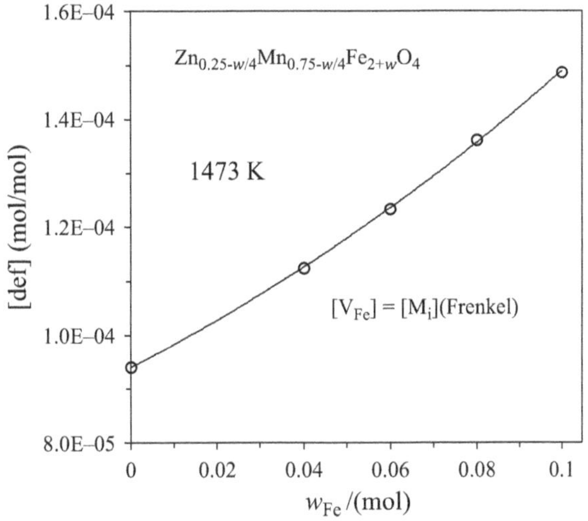

FIGURE 9.15 Dependence of the concentration of Frenkel defects on the concentration of excess iron w_{Fe} in the spinel $(Zn_{0.25-w/4}\,Mn_{0.75-3w/4}Fe_{2+w})_{1\pm\delta/3}O_4$ at 1473 K.

Figure 9.20a–c present diagrams of concentrations of ionic defects for the spinel $(Zn_{0.23}Mn_{0.68}Fe_{2.08})_{1\pm\delta/3}O_4$ determined at the temperatures of 1173–1373 K (solid lines), calculated using the results of the deviation from the stoichiometry taken from (Töpfer and Dieckmann 2004) ((O) points). The diagram for the temperature of 1473 K is presented in Figure 9.11c.

Figure 9.20a–c show that at every temperature in the range of higher oxygen pressures, where cation vacancies dominate, a good agreement was obtained between the oxygen pressure dependence of the deviation from the stoichiometry and the experimental values of δ. There is no match between the dependence of δ on p_{O_2} and the experimental values of δ in the range of lower oxygen pressures where interstitial cations dominate. Quite a good match between the dependence of δ on p_{O_2} and the experimental values δ was obtained when assuming, similarly as before, that some zinc ions Zn^{2+} incorporate into the spinel structure into the sublattice of Fe^{3+} ions $\left(Zn'_{Fe}\right)$. Attempts to fit other value of δ_o than that adopted by Töpfer and Dieckmann (2004) did not significantly change the values of the deviation from the stoichiometry ((●) points).

Figure 9.21 presents the dependence of the concentration of ions $\left[Zn'_{Fe}\right] = 1/2y_{Zn}$ on the temperature in the spinel $(Zn_{0.23}Mn_{0.68}Fe_{2.08})_{1\pm\delta/3}O_4$, which, as can be seen, increase when the temperature increases, from $1{\cdot}10^{-3}$ $1.5{\cdot}10^{-3}$ mol/mol.

Figure 9.22 compares the oxygen pressure dependences of the deviation from the stoichiometry at temperatures 1173–1473 K (presented in Figure 9.20). As can be seen in Figure 9.22, when the temperature increases, the range of existence of the spinel $(Zn_{0.23}Mn_{0.68}Fe_{2.08})_{1\pm\delta/3}O_4$ is shifted practically in parallel towards higher oxygen pressures.

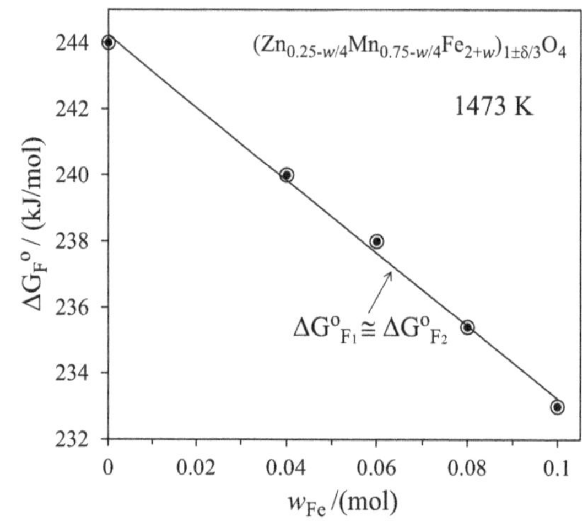

FIGURE 9.16 Dependence of standard Gibbs energy $\Delta G^o_{F_1} = \Delta G^o_{F_2}$ ((O,●) points) of formation of Frenkel defects with charge 2 and 3 on the concentration of excess iron w_{Fe} in the spinel $(Zn_{0.25-w/4}Mn_{0.75-3w/4}Fe_{2+w})_{1\pm\delta/3}O_4$ at 1473 K, obtained using the results of the studies of the deviation from the stoichiometry (Töpfer and Dieckmann 2004).

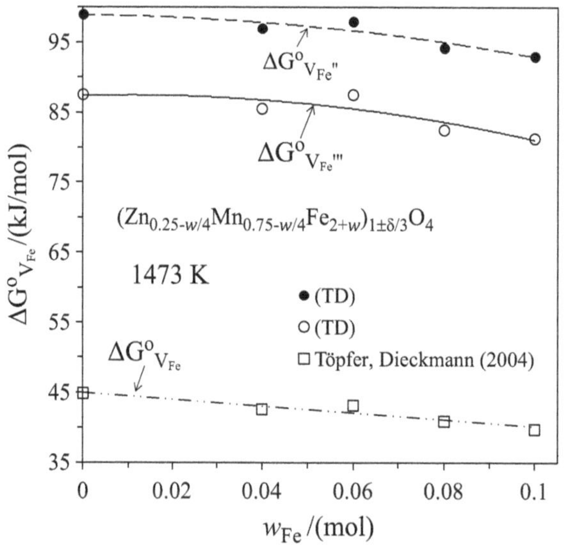

FIGURE 9.17 Dependence of standard Gibbs energy $\Delta G^o_{V_{Fe}'''}$ and $\Delta G^o_{V_{Fe}''}$ of formation of vacancies $(V_{Fe(o)})$ ((O) points), (V_M'') - (●) on the concentration of excess iron w_{Fe} in the spinel $(Zn_{0.25-w/4}Mn_{0.75-3w/4}Fe_{2+w})_{1\pm\delta/3}O_4$, at 1473 K and the values $\Delta G^o_{V_{Fe}}$ calculated using the values of "uilibrium constants" of reactions of formation of cation vacancies K_V (Töpfer and Dieckmann 2004) ((□) points).

Figure 9.23 presents the oxygen pressure at which the doped ferrite reaches the stoichiometric composition ((O) points) and the pressure at which the concentrations of cation vacancies and interstitial ions ($[V_{Fe}] = [Fe_i]$) are ual ((●) points), depending on the temperature. As can be seen in Figure 9.23, the oxygen pressures at which the spinel reaches the stoichiometric composition and at which Frenkel defects are present increase when the temperature increases.

In turn, Figure 9.24 presents the temperature dependence of the concentration of Frenkel defects in the spinel $(Zn_{0.23}Mn_{0.68}Fe_{2.08})O_4$; their concentration increases when the temperature increases, from $2 \cdot 10^{-5}$ to $1.4 \cdot 10^{-4}$ in the temperature range of 1173–1473 K.

Figures 9.25–9.28 present temperature dependences of standard Gibbs energies ΔG_i^o of formation of individual defects, and Table 9.1 compares enthalpies and entropies of the reaction of formation of the individual defects, resulting from the parameters of the obtained linear reactions. It should be noted that the above determined standard Gibbs energies of formation of defects depend on the assumed value of ΔG^o_{eh} and its temperature dependence, which was not taken into account in the present calculations. Thus, they can be burdened with a systematic error.

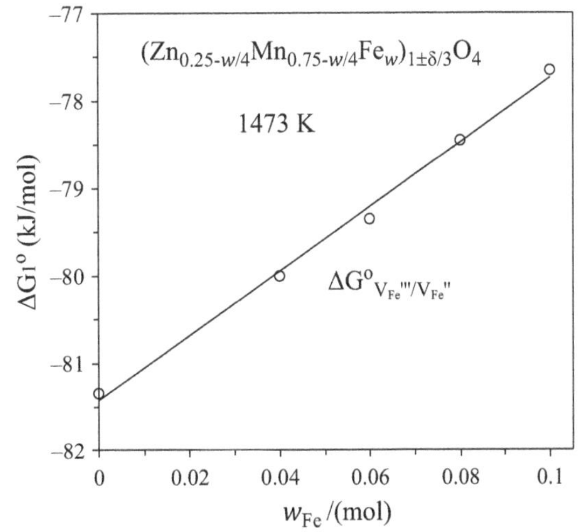

FIGURE 9.18 Dependence of standard Gibbs energy ΔG_1^o on the concentration of excess iron w_{Fe} in the spinel $(Zn_{0.25-w/4}Mn_{0.75-3w/4}Fe_{2+w})_{1\pm\delta/3}O_4$ at 1473 K, permitting determination of the ratio of concentrations of iron vacancies $\left[V_{Fe(o)}'''\right]/\left[V_{Fe(o)}''\right]$, according to Equation 1.35a.

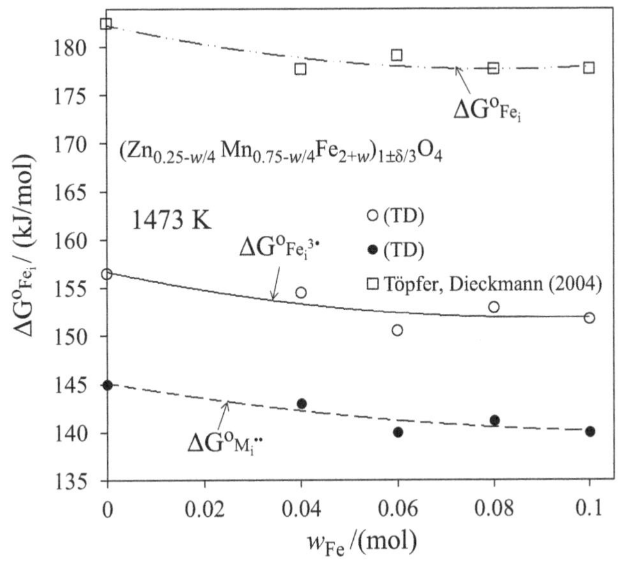

FIGURE 9.19 Dependence of the standard Gibbs energy $\Delta G^o_{Fe_i^{3\bullet}}$ and $\Delta G^o_{M_i^{\bullet\bullet}}$ of formation of interstitial ions $M_i^{3\bullet}$ ((○) points) and $M_i^{\bullet\bullet}$ - (●), calculated according to Equations 1.34c and 1.36c) on the concentration of excess iron w_{Fe} in the spinel $(Zn_{0.25-w/4}Mn_{0.75-3w/4}Fe_{2+w})_{1\pm\delta/3}O_4$, at 1473 K and the values $\Delta G^o_{Fe_i}$ calculated using the "uilibrium constants" of reactions of formation of interstitial cations K_I (Töpfer and Dieckmann 2004) ((□) points).

9.3 DIFFUSION IN MAGNETITE DOPED WITH ZINC AND MANGANESE

Similarly for pure magnetite, using the concentration of ionic defects resulting from the diagram of defects' concentrations and the values of coefficients of self-diffusion of tracers: ^{59}Fe, ^{60}Co, ^{53}Mn and ^{65}Zn in the spinel $(Zn_{0.35}Mn_{0.54}Fe_{2,11})_{1\pm\delta/3}O_4$ (Lee et al. 2000), it was decided to determine the coefficients of diffusion of defects, and specifically the mobility of tracer metal ions via cation vacancies and via interstitial cations (see Section 1.6.2). As can be seen in Figures 9.3 and 9.13, the dependences of the deviation from the stoichiometry on p_{O_2} in the spinel $(Zn_xMn_{1-x}Fe_2)_{1\pm\delta/3}O_4$ with the zinc content of x_{Zn} = 0.3 and 0.4 mol, and for the spinel $(Zn_{0.225}Mn_{0.675}Fe_{2.1})_{1\pm\delta/3}O_4$ are very similar. Due to that, for the spinel $(Zn_{0.35}Mn_{0.54}Fe_{2.11})_{1\pm\delta/3}O_4$, the values of ΔG_i^o of formation of defects were extrapolated and it was assumed that $\Delta G^o_{F_1} = \Delta G^o_{F_2}$ = 230 kJ/mol and $\Delta G^o_{V_{Fe}''}$ = 79.5 kJ/mol, at which the spinel reaches the stoichiometric composition at the oxygen pressure p_{O_2} = 1.53·10^{-4} atm and the concentration of Frenkel defects is the same as the one present in the spinels mentioned above, with a similar composition.

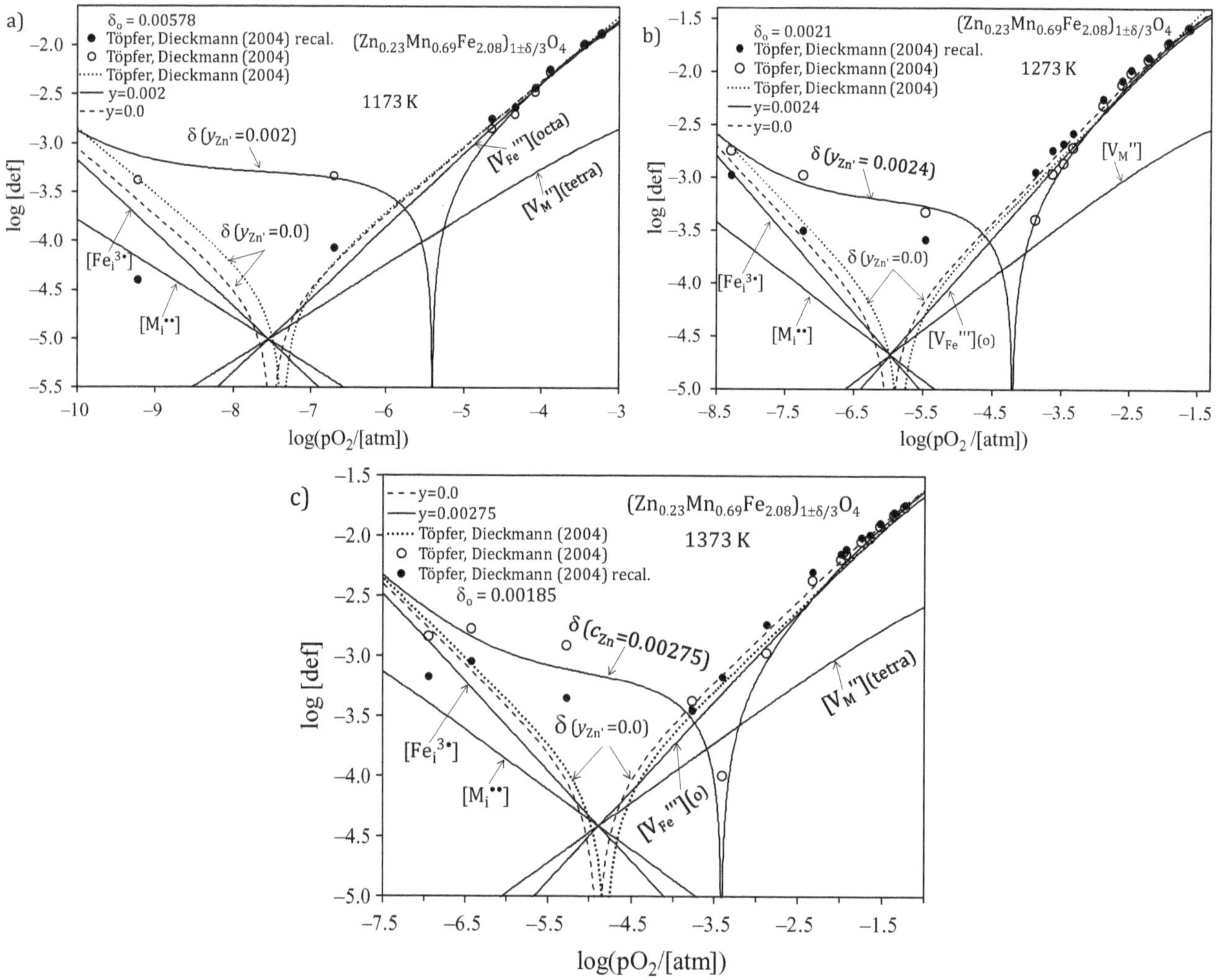

FIGURE 9.20 (a) Diagram of concentrations of point defects for the spinel $(Zn_{0.23}Mn_{0.69}Fe_{2.08})_{1\pm\delta/3}O_4$, at 1173 K, taking into account zinc ions Zn'_{Fe} incorporated into the sublattice of Fe^{3+} ions in the amount of $y'_{Zn} = 2{\cdot}10^{-3}$ mol/mol (solid line). obtained using the results of the studies of the deviation from the stoichiometry (Töpfer and Dieckmann 2004) ((○) points). The (···) line shows the dependence of the deviation from the stoichiometry on p_{O_2}, determined using the "uilibrium constants" K_V and K_I (Töpfer and Dieckmann 2004). Points (●) represent the recalculated values of the deviation from the stoichiometry δ taken from (Töpfer and Dieckmann 2004) relative to the value $\delta_o = 5.78{\cdot}10^{-3}$, and (---) line represents the dependence of δ on p_{O_2} for $y'_{Zn} = 0.0$. (b) at 1473 K, when taking into account zinc ions Zn'_{Fe} incorporated into the sublattice of Fe^{3+} ions in the amount of $y'_{Zn} = 2.4{\cdot}10^{-3}$ mol/mol (solid lines). Points (●) show the recalculated values of the deviation from the stoichiometry δ (Töpfer and Dieckmann 2004) relative to the value $\delta_o = 2.1{\cdot}10^{-3}$. (c) at 1473 K when taking into account zinc ions Zn'_{Fe} incorporated into the sublattice of Fe^{3+} ions in the amount of $y'_{Zn} = 2.75{\cdot}10^{-3}$ mol/mol (solid line). Points (●) show the recalculated values of the deviation from the stoichiometry δ from (Töpfer and Dieckmann 2004) relative to the value $\delta_o = 1.85{\cdot}10^{-3}$.

Figure 9.1a presents the dependence of the "sum" of concentrations of ionic defect ($[V_{Fe}] + b_D[Fe_i]$) on p_{O_2} for the spinel $(MnFe_2)_{1\pm\delta/3}O_4$, and Figure 9.29 – for the spinel $(Zn_{0.35}Mn_{0.54}Fe_{2.1})_{1\pm\delta/3}O_4$ for the fitted value of b_D and the values of this sum, calculated according to Equation 1.47a using the values of coefficients of self-diffusion of: Fe*, Co* Mn* and Zn* (Lee et al. 2000) and for the spinel $(MnFe_2)_{1\pm\delta/3}O_4$ (Lu and Dieckmann 1992). The values of the coefficient of diffusion of cations $D^o_{V(M)}$ via iron vacancies at the highest oxygen pressures and of the parameter b_D, being the ratio of the coefficients of diffusion via defects ($D^o_{I(M)}/D^o_{V(M)}$), were adjusted in such a way as to obtain a match between the dependence of the sum of concentrations of ionic defects ($[V_{Fe}] + b_D[M_i]$) on p_{O_2} with the values of this sum calculated using the coefficients of self-diffusion of tracers.

As can be seen in Figures 9.1a and 9.29, a good agreement was obtained between the dependence of the sum ($[V_{Fe}] + b_D[Fe_i]$) on p_{O_2} determined from the diagram of concentrations of defects and the values of this sum calculated using the coefficients of self-diffusion of tracers. However, as can be seen in Figure 9.1a, there is a large difference in the

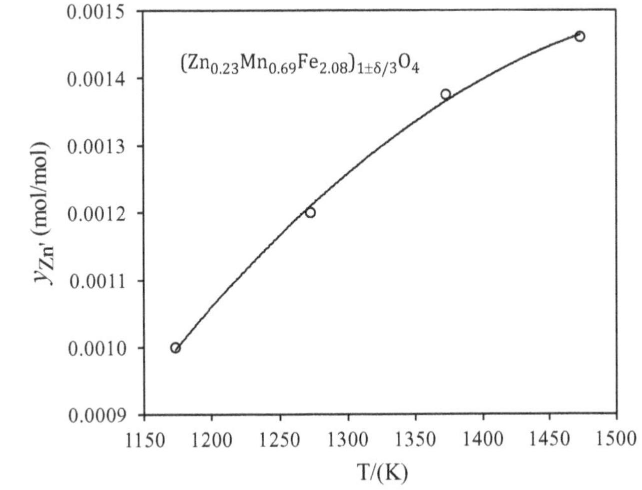

FIGURE 9.21 Temperature dependence of the concentration of zinc ions $\left[Zn'_{Fe}\right]$ (y'_{Zn}) incorporated into the sublattice of Fe^{3+} ions in the spinel $(Zn_{0.23}Mn_{0.69}Fe_{2.08})_{1\pm\delta/3}O_4$, obtained using the results of the studies of the deviation from the stoichiometry (Morineau and Paulus 1973).

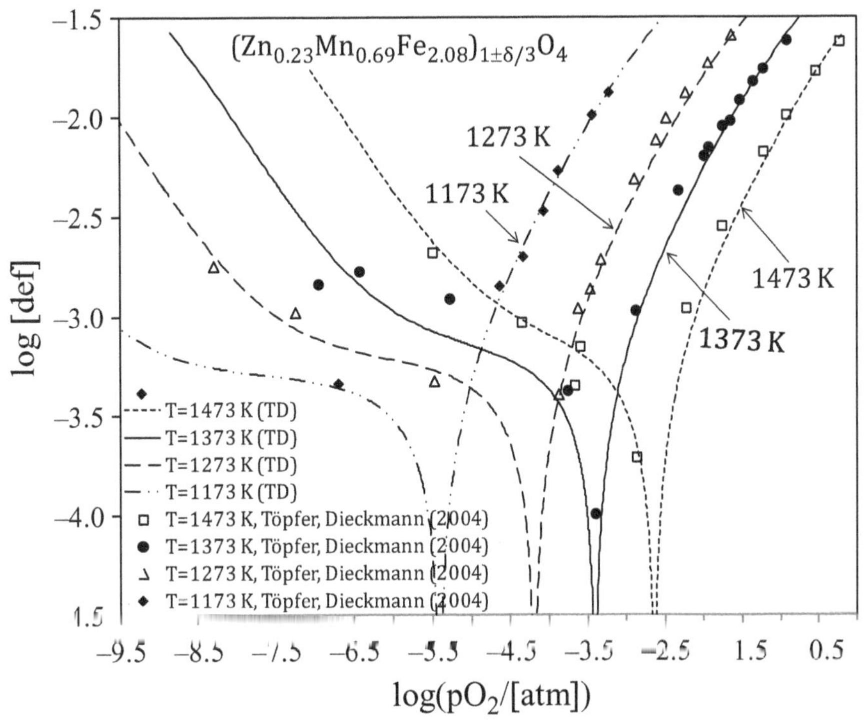

FIGURE 9.22 Dependences of the deviation from the stoichiometry on the oxygen pressure in the spinel $(Zn_{0.23}Mn_{0.69}Fe_{2.08})_{1\pm\delta/3}O_4$ at temperatures 1173–1473 K (presented in Figure 9.20).

dependence of the sum ($[V_{Fe}] + b_D[Fe_i]$) on p_{O_2} obtained when using the results from Lu and Dieckmann (1992) and Lee et al. (2000) . Figure 9.29 presents also the dependence of the sum of concentrations of defects ($[V_{Fe}] + b_D[Fe_i]$) on p_{O_2}, determined for the fitted parameter $b_D = t_{I(M)}/t_{V(M)}$, being the ration of transport numbers of ions via interstitial cations and via cation vacancies and the values of this sum calculated using the values of transport numbers of ions (Jang and Yoo 1996), determined by electrotransport method in the spinel $(Zn_{0.35}Mn_{0.54}Fe_{2.11})_{1\pm\delta/3}O_4$. The obtained value $b_D = 0.06$ indicates, contrary to the conclusions from the studies on diffusion on tracers, that the mobility of ions via cation vacancies is 17 times higher than the mobility via interstitial cations.

Figure 9.30 presents the dependence of the parameter b_D for the zinc content of $x_{Zn} = 0.0$ and 0.35 mol for tracers Fe*, Co*, Mn* and Zn*. As can be seen in the diagram, there is a large difference between the values of the ratio $\left(D^o_{I(M)}/D^o_{V(M)}\right)$

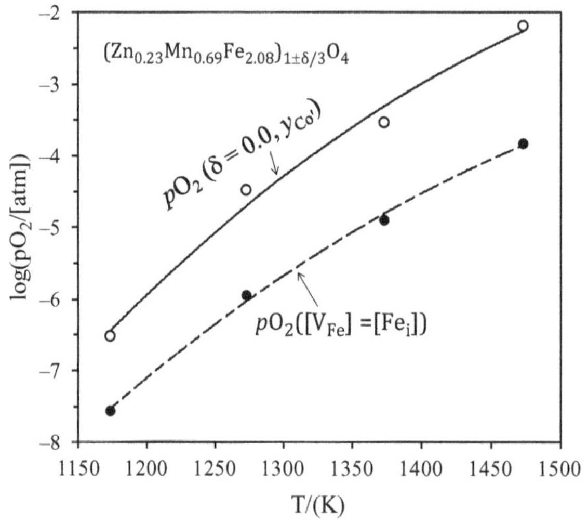

FIGURE 9.23 Temperature dependence of the oxygen pressure at which the spinel $(Zn_{0.23}Mn_{0.69}Fe_{2.08})$ O_4 reaches the stoichiometric composition ((○) points) and the oxygen pressure at which the concentrations of cation vacancies and interstitial ions are ual ((●) points).

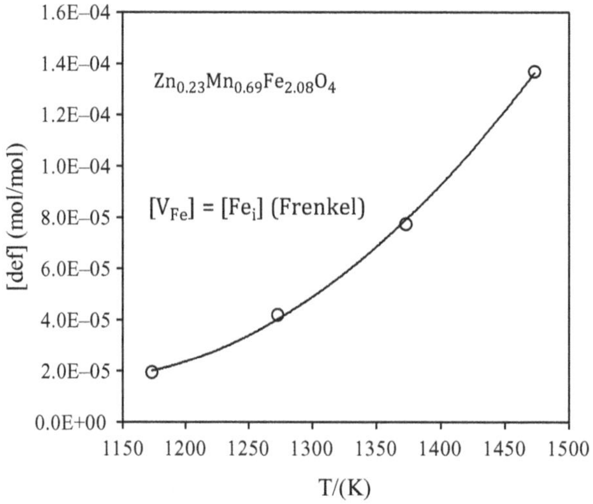

FIGURE 9.24 Temperature dependence of the concentration of Frenkel defects in the spinel $(Zn_{0.23}Mn_{0.69}Fe_{2.08})_{1\pm\delta/3}O_4$.

in the case of the spinel $(MnFe_2)_{1\pm\delta/3}O_4$ obtained when using the values of the coefficients of self-diffusion of tracer ions from Lu and Dieckmann (1992) and Lee et al. (2000).

The values of b_D are positive, thus, the mobilities of tracer ions Fe*, Co*, Mn* via interstitial ions in the spinel $MnFe_2O_4$ are many times higher than the mobilities via cation vacancies. On the other hand, in the spinel $(Zn_{0.35}Mn_{0.54}Fe_{2,1})_{1\pm\delta/3}O_4$ the mobilities of ions are similar (b_D is slightly higher than one).

Figure 9.31 shows the values of the coefficients of diffusion of ions Fe*, Co*, Mn* and Zn* via cation vacancies $D^o_{V(M)}$ and interstitial ions $D^o_{I(M)}$ in spinels $(MnFe_2)_{1\pm\delta/3}O_4$ and $(Zn_{0.35}Mn_{0.54}Fe_{2.1})_{1\pm\delta/3}O_4$. As can be seen in Figure 9.31, for the diffusion of tracer ions Fe*, Co* and Mn* in the spinel $(MnFe_2)_{1\pm\delta/3}O_4$ their coefficients of diffusion via cation vacancies $D^o_{V(M)}$ obtained using the results taken from Lu and Dieckmann (1992) and Lee et al. (2000) are similar. The values of coefficients of tracer diffusion via interstitial cations $D^o_{I(M)}$ are much higher and also the differences between them are higher, when using the results of the studies published in Lu and Dieckmann (1992) and Lee et al. (2000).

In turn, as can be seen in Figure 9.31, the mobility of ions Fe*, Co*, Mn* and Zn* via cation vacancies in the spinel $(Zn_{0.35}Mn_{0.64}Fe_{2.1})_{1\pm\delta/3}O_4$ are similar to the values in the spinel $(MnFe_2)_{1\pm\delta/3}O_4$. According to the values of the parameter b_D, the mobilities of tracer ions via interstitial cations in the spinel $(MnFe_2)_{1\pm\delta/3}O_4$ are much higher, while in the spinel $(Zn_{0.35}Mn_{0.64}Fe_{2.1})_{1\pm\delta/3}O_4$ they decreased significantly. Therefore, if the diagram of concentrations of defects for the spinel $(Zn_{0.35}Mn_{0.54}Fe_{2.1})_{1\pm\delta/3}O_4$, when using the extrapolated values of ΔG^o_i of formation of defects, is determined correctly, then the obtained results indicate that zinc ion doping of the spinel $(Zn_{0.35}Mn_{0.64}Fe_{2.1})_{1\pm\delta/3}O_4$ only slightly changes the concentration of cation vacancies; also the mobility of ions via cation vacancies does not change.

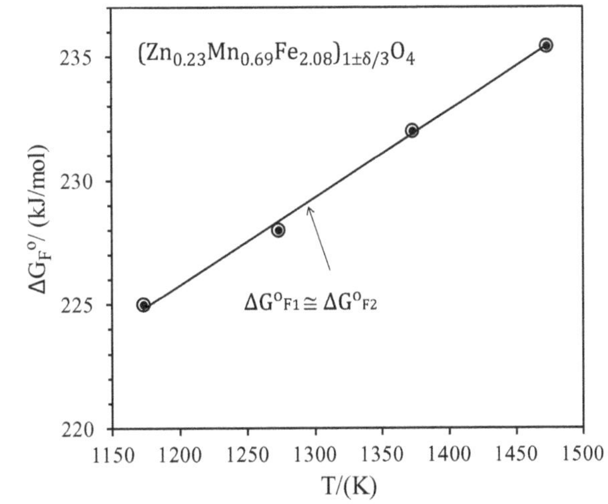

FIGURE 9.25 Temperature dependence of standard Gibbs energy $\Delta G^o_{F_1} \cong \Delta G^o_{F_2}$ ((O,●) points) of formation of Frenkel defects with charge two and three in the spinel $(Zn_{0.23}Mn_{0.69}Fe_{2.08})_{1\pm\delta/3}O_4$, obtained using the results of the studies of the deviation from the stoichiometry (Töpfer and Dieckmann 2004).

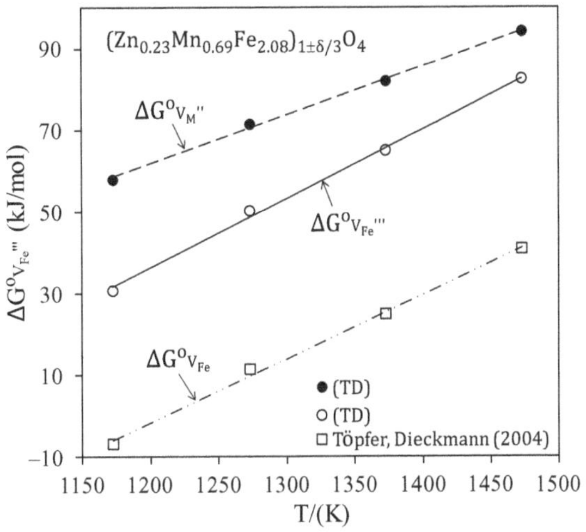

FIGURE 9.26 Temperature dependence of standard Gibbs energy $\Delta G^o_{V_{Fe}'''}$ ((O) points) and $\Delta G^o_{V_{Fe}''}$ ((●) points) of formation of cation vacancies $\left(V_{Fe}'''\right)$ and $\left(V_M''\right)$ in the spinel $(Zn_{0.23}Mn_{0.69}Fe_{2.08})_{1\pm\delta/3}O_4$ and the values $\Delta G^o_{V_{Fe}}$ calculated using the values of "uilibrium constants" of reactions of formation of cation vacancies K_V (Töpfer and Dieckmann 2004).

Zinc ion doping causes an increase in the concentration of interstitial cations and a significant decrease in the mobility of tracer ions (in relation to the spinel $(MnFe_2)_{1\pm\delta/3}O_4$). A decrease in the mobility via interstitial cations can be therefore related to an increase in the concentration of these defects with the increase in zinc content in the spinel $(Zn_xMn_{0.99-x}Fe_{2.11})_{1\pm\delta/3}O_4$. The above conclusion, as well as differences in relation to the results of the studies on electrotransport, ruire verification and more studies on diffusion in spinels with different zinc content.

9.4 CONCLUSIONS

Using the results of the studies on the deviation from the stoichiometry for the spinel $(Zn_xMn_{1-x}Fe_2)_{1\pm\delta/3}O_4$ with the zinc content of x_{Zn} = 0.1–0.5 mol and for spinels with the formula of $Zn_{0.25-w/4}Mn_{0.75-3w/4}Fe_{2+w})_{1\pm\delta/3}O_4$, in which the composition depended on the iron content (2 + w), where w = 0.0–0.1 mol (Töpfer et al. 2003), the diagrams of concentrations of defects at 1473 K were drawn. It was found that zinc doping in the spinel $(Zn_xMn_{1-x}Fe_2)_{1\pm\delta/3}O_4$ affects its range of existence and the character of the dependence of the deviation from the stoichiometry on the oxygen pressure to a much smaller extent than cobalt doping in the spinel $(Co_xMn_zFe_{2z})_{3\pm\delta}O_4$. The oxygen pressure at which the spinel $Zn_xMn_{1-x}Fe_2O_4$ reaches the stoichiometric composition at the zinc content of x_{Zn} = 0.1 mol is shifted by two orders of magnitude towards higher

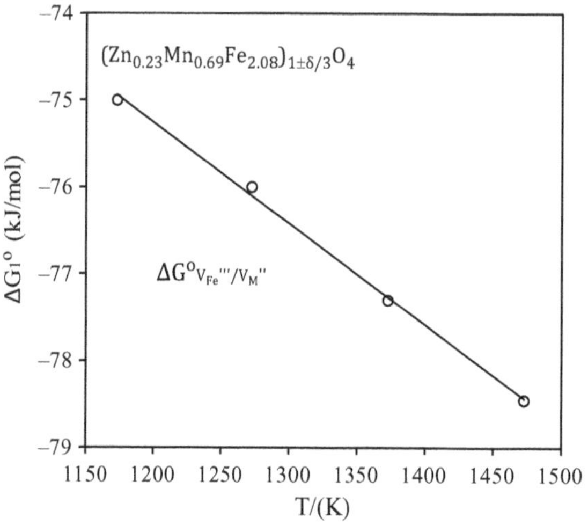

FIGURE 9.27 Temperature dependence of standard Gibbs energy ΔG_i^o in the spinel $(Zn_{0.23}Mn_{0.69}Fe_{2,08})_{1\pm\delta/3}O_4$, permitting determination of the ratio of concentrations of iron vacancies $\left[V'''_{Fe(o)}\right]/\left[V''_{Fe(o)}\right]$, according to Equation 1.35b.

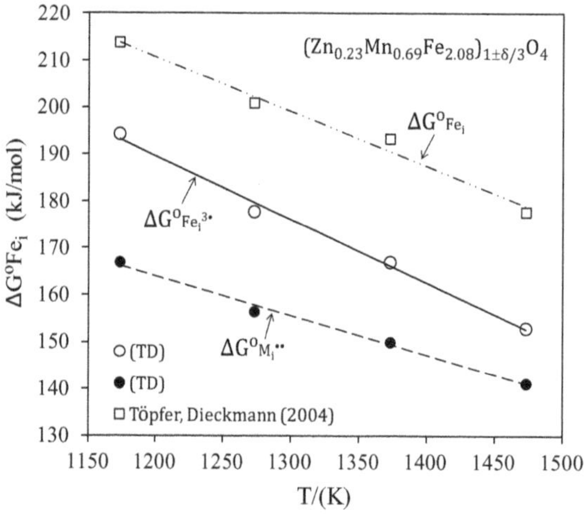

FIGURE 9.28 Temperature dependence of the standard Gibbs energy $\Delta G^o_{Fe_i^{3\bullet}}$ and $\Delta G^o_{M_i^{\bullet\bullet}}$ of formation of interstitial ions $Fe_i^{3\bullet}$ and $M_i^{\bullet\bullet}$, calculated according to Equations 1.34c and 1.36c) in the spinel $(Zn_{0.23}Mn_{0.69}Fe_{2,08})_{1\pm\delta/3}O_4$ and the values $\Delta G^o_{Fe_i}$ calculated using the values of "uilibrium constants" K_I of reaction of formation of interstitial cations (Töpfer and Dieckmann 2004).

TABLE 9.1

Values of enthalpies ΔH_i^o and entropies ΔS_i^o of formation of intrinsic Frenkel defects, cation vacancies V''_{Fe} and V'''_{Fe}, interstitial cations $M_i^{\bullet\bullet}$ and $Fe_i^{3\bullet}$ in the spinel $(Zn_{0.23}Mn_{0.68}Fe_{2.08})_{1\pm\delta/3}O_4$ in the temperature range of 1173–1473 K, determined based on the dependences shown in Figures 9.25–9.28.

Equation	ΔH (kJ/mol)	ΔS (J/(mol·K))
$\left[V'''_{Fe}\right]\left[Mn_i^{3\bullet}\right] = K_{F_1} = K_{F_2}$ (Equation 1.34b)	184 ± 2	−35 ± 1
$V'''_{Fe(o)}$ (Equation 1.9b)	−168 ± 8	−170 ± 8
V''_{Fe} (Equation 1.9a)	−81 ± 5	−119 ± 4
$Fe_i^{3\bullet}$ (Equation 1.13b)	351 ± 10	135 ± 7
$M_i^{\bullet}$ (Equation 1.13a)	265 ± 7	84 ± 5
$K_{V'''_{Fe}/V''_{Fe}} = K_1$ (Equation 1.35b)	61 ± 1	−12 ± 1
V_{Mn} (Morineau and Paulus 1973)	−189 ± 8	−156 ± 6
Mn_i (Morineau and Paulus 1973)	350 ± 14	116 ± 10

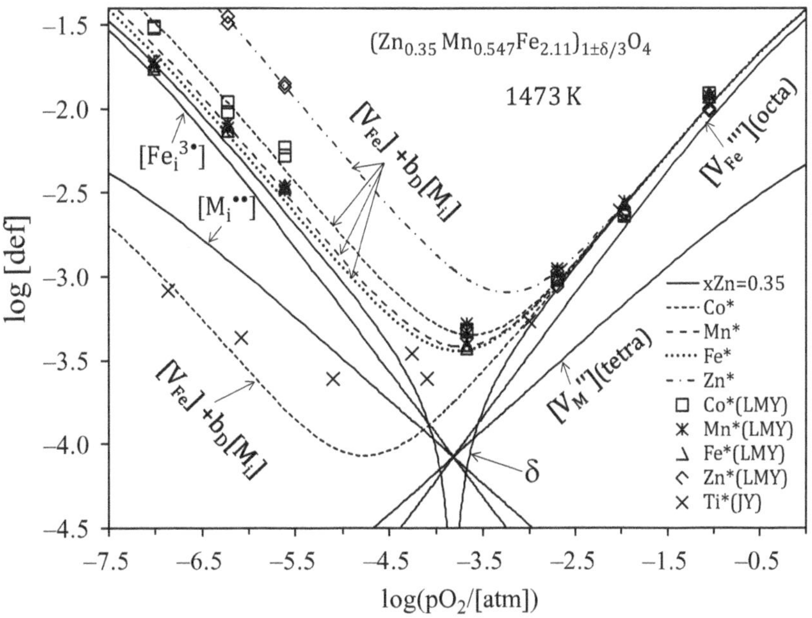

FIGURE 9.29 Diagram of concentrations of point defects for the spinel $(Zn_{0.35}Mn_{0.54}Fe_{2.11})_{1\pm\delta/3}O_4$ at 1373 K (solid lines) determined using the extrapolated values $\Delta G^o_{F_1} = \Delta G^o_{F_2} = 230$ kJ/mol of formation of Frenkel defects and $\Delta G^o_{V_{Fe}} = 79.5$ kJ/mol of formation of cation vacancies. The points represent the values of the sum of concentrations of ionic defects ($[V_{Fe}] + b_D[M_i]$) calculated according to Equation 1.47a, 1.47b for fitted values of $D^o_{V(M)}$ for the highest oxygen pressures and when using coefficients of self-diffusion of tracers: Fe* – (△) points, Co* – (□), Mn* – (✳), Zn* – (◇), (Lee et al. 2000) and points (×) – values obtained when using the transport numbers of ions, (Jang and Yoo 1996). Lines (···), (- - -), (---), (-·-·-), (-···-) denote, respectively, the dependence of the sum of concentrations of defects ($[V_{Fe}] + b_D[M_i]$), obtained for the fitted value of b_D.

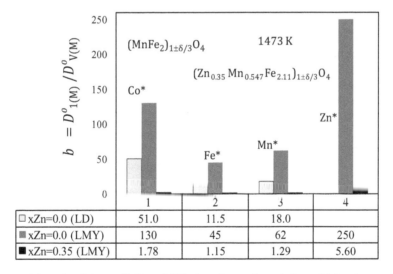

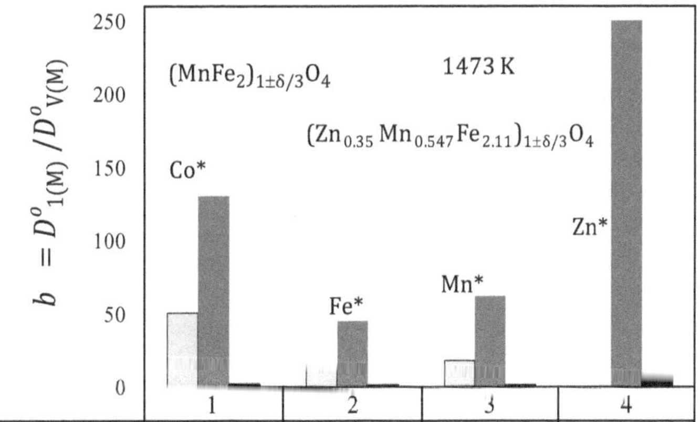

	1	2	3	4
□ xZn=0.0 (LD)	51.0	11.5	18.0	
■ xZn=0.0 (LMY)	130	45	62	250
■ xZn=0.35 (LMY)	1.78	1.15	1.29	5.60

FIGURE 9.30 Dependence of the ratio of the coefficient of diffusion of tracer ions via interstitial cations to the coefficient of diffusion via cation vacancies $D^o_{I(M)}/D^o_{V(M)} = b_D$ in the spinel $(MnFe_2)_{3\pm\delta}O_4$ and $(Zn_{0.35}Mn_{0.54}Fe_{2.11})_{1\pm\delta/3}O_4$ at 1473 K, obtained using the coefficients of self-diffusion of tracers Fe*, Co* Mn* and Zn*(Lu and Dieckmann 1992, Lee et al. 2000) and concentration of ionic defects calculated from the diagrams.

pressures compared with the spinel $MnFe_2O_4$ and a further increase in the zinc content practically does not change it. Only at the content of $x_{Zn} = 0.5$ mol there occurs a shift, by half an order of the magnitude.

In the range of higher oxygen pressures cation vacancies $\left[V'''_{Fe}\right]$ dominate, and their concentration only slightly decreases when the zinc content in the spinel $(Zn_xMn_{1-x}Fe_2)_{1\pm\delta/3}O_4$ increases. It is slightly lower than that in the spinel $(MnFe_2)_{1\pm\delta/3}O_4$. In the range of low oxygen pressures, interstitial cations $\left[Fe_i^{3\bullet}\right]$ dominate and the character of the dependence of deviation from the stoichiometry is significantly different from that in the spinel $(MnFe_2)_{1\pm\delta/3}O_4$. Similarly as in the case of spinels

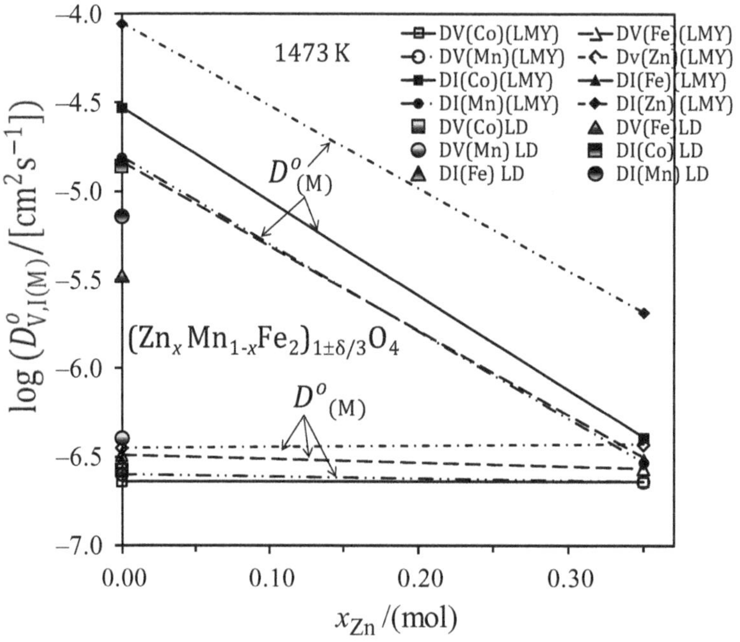

FIGURE 9.31 Dependence of the coefficient of diffusion of tracer ions via cation vacancies $D^o_{V(M)}$ (empty points) and via interstitial cations $D^o_{I(M)}$ (solid points) on the zinc content x_{Zn} in the spinel $(MnFe_2)_{3\pm\delta}O_4$ and $(Zn_{0.35}Mn_{0.54}Fe_{2.11})_{1\pm\delta/3}O_4$, at 1473 K, obtained using the coefficients of self-diffusion of tracers Fe* ((△,▲,◈,◈) points), Co* (□,■,◬,◬), Mn*(○,●,⊟,⊟), Zn* (◇,◆,⊖,⊖) (Lu and Dieckmann 1992, Lee et al. 2000) and concentrations of ionic defects and the calculated parameter b_D.

$(Fe_{1-x}Co_x)_{3\pm\delta}O_4$ and $(Co_xMn_zFe_{2z})_{3\pm\delta/3}O_4$, it was demonstrated that the above character of the dependence of δ on p_{O_2} is an effect of the incorporation of Zn^{2+} ions into the spinel structure, into the sublattice of Fe^{3+} ions, which are negative in relation to the lattice (Zn'_{Fe}). At the zinc content of x_{Zn} = 0.1 mol, in the spinel $(Zn_xMn_{1-x}Fe_2)_{1\pm\delta/3}O_4$, the concentration of $\left[Zn'_{Fe}\right]$ ions is 0.001 mol/mol, and then, when the zinc content increases to x_{Zn} = 0.333 mol, it decreases to $3\cdot10^{-4}$, and when the zinc content increases further, it increases again up to $1.3\cdot10^{-3}$ mol/mol.

It was found that when the zinc content in the spinel $(Zn_xMn_{1-x}Fe_2)_{1\pm\delta/3}O_4$ increases, the standard Gibbs energy of formation of Frenkel defects decreases, which results in a decrease in the concentration of Frenkel defects ($\left[V'''_{Fe}\right]=\left[Fe^{3\bullet}_i\right]$) from about $2\cdot10^{-5}$ to $1.4\cdot10^{-4}$ mol/mol at the zinc content of x_{Zn} = 0.5 mol.

The standard Gibbs energies of formation of cation vacancies $\Delta G^o_{V'''_{Fe}}$ and $\Delta G^o_{V''_M}$ increase when the zinc content in the spinel increases up to x_{Zn} = 0.3 mol, and then, for x_{Zn} = 0.4 they decrease, and at x_{Zn} = 0.5 they increase again. The values of $\Delta G^o_{Fe^{3\bullet}_i}$ and $\Delta G^o_{M^{\bullet\bullet}_i}$ decrease when the zinc content increases. The above values depend on the value of ΔG^o_{eh} of formation of electronic defects adopted in the calculations and they were determined for the value of ΔG^o_{eh} = 30 kJ/mol, the same as the value assumed for pure magnetite.

The studies on spinels $(Zn_{0.25-w/4}Mn_{0.75-3w/4}Fe_{2+w})_{1\pm\delta/3}O_4$, in which the iron content increases 2–2.1 ((2 + w), where w = 0.0–0.1 mol), indicate that an increase in the iron content causes a slight increase in the concentration of vacancies and Frenkel defects ($1\cdot10^{-4}$–$1.5\cdot10^{-4}$ mol/mol) and that the concentration of zinc ions incorporated into the sublattice of iron ions Fe^{3+}, $\left[Zn'_{Fe}\right]$, increases from about $4\cdot10^{-4}$ to $1.5\cdot10^{-3}$ mol/mol. When the iron content increases, the standard Gibbs energy of formation of Frenkel defects decreases. The values of $\Delta G^o_{V'''_{Fe}}$ and $\Delta G^o_{V''_M}$ of formation of vacancies and $\Delta G^o_{Fe^{3\bullet}_i}$ and $\Delta G^o_{M^{\bullet\bullet}_i}$ of formation of interstitial cations slightly decrease when the iron content increases (*w*).

Diagrams of concentrations of ionic defects were also prepared for the second group of spinels, $(Zn_{0.23}Mn_{0.68}Fe_{2.08})_{1\pm\delta/3}O_4$, in the range of temperatures of 1173–1473 K, based on the published results of the deviation from the stoichiometry (Töpfer and Dieckmann 2004). It was found that when the temperature increases, the range of existence of the spinel $(Zn_{0.23}Mn_{0.68}Fe_{2.08})_{1\pm\delta/3}O_4$ is shifted practically in parallel towards higher oxygen pressures. When the temperature increases, the oxygen pressures at which the spinel reaches the stoichiometric composition and at which Frenkel defects are present also increase.

Enthalpies of formation of defects were determined (Table 9.1) based on temperature dependences of standard Gibbs energies of formation of defects ΔG^o_i. The above values depend on the value of ΔG^o_{eh} of formation of electronic defects, assumed in the calculations, and on its dependence on the temperature, which was not taken into account; therefore, they are burdened with a systematic error.

Using the determined concentrations of ionic defects in spinels $(MnFe_2)_{1\pm\delta/3}O_4$ and $(Zn_{0.35}Mn_{0.64}Fe_{2.1})_{1\pm\delta/3}O_4$ and the values of coefficients of self-diffusion of tracers M* (^{59}Fe, ^{57}Co, ^{54}Mn, ^{65}Zn) taken from Lu and Dieckmann (1992) and Lee et al. (2000), the ratio of the coefficient of diffusion (mobility) of tracer ions via interstitial cations to the coefficient of diffusion via vacancies $\left(D^o_{I(M)}/D^o_{V(M)} = b_D\right)$ was determined. It was fund that the mobilities of tracer ions Fe*, Co*, Mn* via interstitial ions in the spinel $MnFe_2O_4$ are many times higher than the mobilities via cation vacancies. In the spinel $(Zn_{0.35}Mn_{0.64}Fe_{2.1})_{1\pm\delta/3}O_4$ the mobilities of ions are similar (b_D is slightly higher than one).

It was found that the coefficients of diffusion of tracer ions Fe*, Co* and Mn* via cation vacancies $D^o_{V(M)}$ in the spinel $(MnFe_2)_{1\pm\delta/3}O_4$ are similar. The values of coefficients of tracer diffusion via interstitial cations $D^o_{I(M)}$ are much higher.

The mobility of tracer ions Fe*, Co*, Mn* and Zn* via cation vacancies in the spinel $(Zn_{0.35}Mn_{0.64}Fe_{2.1})_{1\pm\delta/3}O_4$ are similar to the values in the spinel $(MnFe_2)_{1\pm\delta/3}O_4$. The mobility of tracer ions via interstitial cations significantly decreases compared to the spinel $(MnFe_2)_{1\pm\delta/3}O_4$ and it is similar to the mobility via cation vacancies; for Zn*, the value is higher. The obtained results indicate that zinc doping of the spinel $(Zn_{0.35}Mn_{0.64}Fe_{2.1})_{1\pm\delta/3}O_4$ only slightly changes the concentration of cation vacancies and the mobility of ions via cation vacancies. Doping with zinc ions causes a significant increase in the concentration of interstitial cations and a significant decrease in the mobility of tracer ions via these defects. A decrease in the mobility via interstitial cations can be therefore related to an increase in the concentration of these defects with the increase in zinc content in the spinel $(Zn_xMn_{0.99-x}Fe_{2.11})_{1\pm\delta/3}O_4$.

REFERENCES

Blank, J. 1961. Equilibrium Atmosphere Schedules for the Cooling of Ferrites. *J. Appl. Phys.* 32: S378.

Jang, Y.-J., and H. I. Yoo. 1996. Phase Stability and Ionic Transference Number of a Ferrite Spinel, $Mn_{0.54}Zn_{0.35}Fe_{2.11}O_4$. *Solid State Ionics.* 84: 77–88.

Kim, J. H., H. I. Yoo, and H. L. Tuller. 1990. Electrical Properties and Phase Stability of a Zinc Ferrite. *J. Am. Ceram. Soc.* 73: 258–262.

Lee, J. H., M. Martin, and H.-I. Yoo. 2000. Self- and Impurity Cation Diffusion in Manganese–Zinc-Ferrite, $Mn_{1-x-y}Zn_xFe_{2+y}O_4$. *J. Phys. Chem. Solids.* 61: 1597–1605.

Lu, F. H., and R. Dieckmann. 1992. Point Defects and Cation Tracer Diffusion in $(Co, Fe, Mn)_{3-\delta}O_4$ Spinels: I. Mixed Spinels $(Co_xFe_{2y}Mn_y)_{3-\delta}O_4$. *Solid State Ionics.* 53–56: 290–302.

Lu, F. H., and R. Dieckmann. 1995. Point Defects in Oxide Spinel Solid Solutions of the Type $(Co, Fe, Mn)_{3-\delta}O_4$ at 1200°C. *J. Phys. Chem. Solids.* 56: 725–733.

McCurrie, R. A. 1994. *Ferromagnetic Materials: Structure and Properties.* San Diego: Academic Press.

Morineau, R. 1976. Structural Defects and Oxidation-Reduction uilibrium in Mn, Zn Ferrites. *Phys. Status Solidi (a).* 38: 559–568.

Morineau, R., and M. Paulus. 1973. Oxygen Partial Pressures of Mn-Zn Ferrites. *Phys. Status Solidi (a).* 20: 373–380.

Morineau, R., and M. Paulus. 1975. Chart Of p_{O_2} Versus Temperature and Oxidation Degree for Mn-Zn Ferrites in the Composition Range: 50 < Fe_2O_3< 54; 20 < MnO < 35; 11 < ZnO < 30 (Mole %). *IEEE Trans. Magn. Mag.* 11: 1312–1314.

Tanaka, T. 1981. Equilibrium Oxygen Pressures of Mn-Zn Ferrites. *J. Am. Ceram. Soc.* 64: 419–421.

Töpfer, J., and R. Dieckmann. 2004. Point Defects and Deviation From Stoichiometry in $(Zn_{x-y/4}Mn_{1-x-3y/4}Fe_{2+y})_{1-\Delta/3}O_4$. *J. Europ. Ceram. Soc.* 24: 603–612.

Töpfer, J., L. Liu, and R. Dieckmann. 2003. Deviation from Stoichiometry and Point Defects in $(Zn_xMn_{1-x}Fe_2)_{1-\Delta/3}O_4$. *Solid State Ionics.* 159: 397–404.

Tuller, H. L., H. I. Yoo, W. Kehr, and R. W. Scheidecker. 1985. Electrical Property-Phase uilibria Correlations in Manganese–Zinc Ferrites. *Adv. Ceram.* 15: 315–324.

Valenzuela, R. 1994. *Magnetic Ceramics.* Cambridge: Cambridge University Press.

Yoo, H. I., and H. L. Tuller. 1987. Iron-Excess Manganese Ferrite: Electrical Conductivity and Cation Distributions. *J. Am. Ceram. Soc.* 70: 388–392.

Yoo, H. I., and H. L. Tuller. 1988. In Situ Phase uilibria Determination of a Manganese Ferrite by Electrical Means. *J. Mat. Res.* 3: 552–556.

10 Effect of Dopants on the Concentrations of Point Defects and Their Mobilities in Magnetite

A Summary

10.1 EFFECT OF DOPANTS ON THE CONCENTRATIONS OF POINT DEFECTS IN MAGNETITE

As already discussed, the features of spinel structure in magnetite make it possible to substitute iron ions with other cations, which form, practically in the whole range of concentrations, a solid solution. These possibilities result from the fact that in a regular structure with densely packed oxygen sublattice, cations occupy half of the voids with octahedral coordination, and they occupy part of the void with tetrahedral coordination. Therefore, compared to MO oxides with cubic structure, the spinel structure is much "looser" and, due to that, heterocation spinels are formed, with practically continuous variation in composition.

Magnetite doped, for example, with M^{3+} and/or M^{2+} ions is obtained by annealing, at a high temperature, appropriate quantities of oxides: FeO, Fe_2O_3, M_2O_3 or MO. According to the mechanism of reactions between solids, a spinel layer is formed between the reacting oxides. The growth of layer thickness depends on the mechanism of reaction and on the rate of diffusion of ions through the product layer. As a result, at phase boundaries, the oxides are rebuilt into a spinel structure. Contact of different oxide grains causes the formation of spinels with different composition. However, as a result of mutual diffusion of cations, a spinel structure with a determined composition is formed, where doped cations in octahedral and tetrahedral sublattices form a solid solution. Reactions of powdered reactants significantly shorten diffusion paths, during the formation of spinel as well as during mutual diffusion between crystallites with different composition. As a result, at a high temperature, an equilibrium state is set, as well as a determined distribution of doping ions and iron ions in the sublattices of spinel structure. It results from mutual interactions between ions and from the presence of a determined concentration of electronic defects, localised at metal cations.

Dopant ions, M^{3+} or M^{2+}, substituting iron ions with the same oxidation state, do not perturb the electroneutrality of the crystal (they are electroneutral relative to the lattice), which does not cause significant changes in the character of the dependence of the deviation from the stoichiometry on the oxygen pressure (structure of point defects). Dopant atoms change the energy of interactions between ions (lattice energy), which affects the range of existence of spinel (the oxygen pressure at which phase transitions occur) and the values of standard Gibbs energies of formation of point defects. Thus, the concentration of point defects will change, but not so much as in the case of dopants where the oxidation state is different from that of parental atoms of the sublattice (as in the case of A^+ or D^{3+} dopants in oxides of MO type).

The type of cation sites occupied by doping ions will depend on their radius, stable oxidation state and the energy of electrostatic interactions, which is dependent on their effective charge. In the spinel, a change in the oxidation state of the doping ion can occur, similarly as for iron ions in magnetite. At high temperatures during the annealing, the distribution of iron ions and doping ions can change. As a result, after an appropriately long time, an equilibrium state sets in between all ions with different charges (atoms with different oxidation states), present in different cation sublattices of the spinel. Therefore, depending on the temperature, the distribution of ions in the spinel structure changes. Also, defined concentration of intrinsic electronic defects is attained; they cause a change in the oxidation state (ion charge). Charge and spin of ions and their distribution in the spinel structure will affect magnetic and electric properties. High electrical conductivity of magnetite at high temperatures, only slightly dependent on the type and concentration of the dopant, indicates that the concentration of electronic defects is high and the dopant should not affect the mechanism of transport of electronic defects, but it will influence their mobility.

Using the results of the studies of the deviation from the stoichiometry and metal ions diffusion in magnetite doped with ions: Ti^{4+} and Cr^{3+} and with Co^{2+}, Mn^{2+}, Zn^{2+}, obtained by Dieckmann et al. (Dieckmann 1982, Lu and Dieckmann 1992, Lu et al. 1993, Lu and Dieckmann 1995, Töpfer et al. 1995, Aggarwal and Dieckmann 2002, Töpfer et al. 2003, Töpfer and Dieckmann 2004), diagrams of defects' concentrations were determined; they allow determining the influence of dopant ions on the range of concentration of spinels with a determined composition and on the structure and concentrations of point defects. According to the conclusions of Dieckmann et al, and of other authors, it was found that as a result of the annealing of a mixture of oxides, heterocation spinels are formed, in which the dopant ions occupy cation sites in the sublattices of spinel structure, becoming node ions (electroneutral in relation to the lattice). Their distribution in the sublattices of spinel structure at high temperatures is random, but it results from attaining an equilibrium state. Dopant ions cause a change in interactions between ions, which affects the range of oxygen pressures where spinels exist. Dopant ions only slightly affect the structure of point defects, thus, the character of the dependence of the deviation from the stoichiometry on the oxygen pressure.

Similarly, as pure magnetite, heterocation spinels, independent of the composition, reach the stoichiometric composition in the range of their existence and there is a range of oxygen pressures where at low oxygen pressures interstitial iron cations, $Fe_i^{3\bullet}$, and dopant cations dominate $M_i^{3\bullet}$. In the range of higher oxygen pressures, cation vacancies V_{Fe}''' in the sublattice of iron ions Fe^{3+} with octahedral coordination dominate. The maximum concentration of cation vacancies at the highest oxygen pressures can reach 0.15 mol/mol, and the concentration of interstitial cations at the lowest oxygen pressures can be below 0.1 mol/mol; they depend on the type and concentration of the dopant. The values of $\Delta G_{F_1}^{o}$ of formation of Frenkel defects adopted in the calculations unambiguously determine the concentration of ionic defects.

It was also shown that at the values of $\Delta G_{F_2}^{o}$ of formation of Frenkel defects (V_{Fe}'' and $Fe_i^{\bullet\bullet}$), equal $\Delta G_{F_1}^{o}$, the concentration of vacancies V_{Fe}'' and interstitial cations $Fe_i^{\bullet\bullet}$ does not affect the degree of match between the dependence of the deviation from the stoichiometry on p_{O_2} and the experimental values δ. Thus, the concentration of these defects can be assumed as the maximum concentration in the spinel; in reality it could be lower. Adopting in the calculations the same values of $\Delta G_{F_1}^{o} = \Delta G_{F_2}^{o}$ assumes that, despite the difference in charges of cations the energies of formation of these defects are approximatively similar, which can be justified by similar surroundings of ions Fe^{3+} and Fe^{2+} with octahedral coordination in the spinel structure (similar coordination polyhedrons).

In the case of magnetite doped with cobalt ions Co^{2+} (spinels $(Fe_{1-x}Co_x)_{3\pm\delta}O_4$ and $(Co_xMn_zFe_{2z})_{3\pm\delta}O_4$) and with zinc ions Zn^{2+} (spinel $(Zn_xMn_{0.333-x}Fe_{0.667})_{3\pm\delta}O_4$) it was shown that a small quantity of ions Co^{2+} or Zn^{2+} incorporates into the formed spinel structure, into the sublattice of ions Fe^{3+}, as ions M_{Fe}' electroneutral in relation to the lattice. Their presence causes an increase in the concentration of interstitial cations, and, as a result, in the range of low pressures there occurs a change in the character of the oxygen pressure dependence of the deviation from the stoichiometry. There also occurs a shift in the oxygen pressure at which spinel reaches the stoichiometric composition. It is possible to observe this effect experimentally in the case of a high concentration of interstitial cations at low oxygen pressures and at relatively high concentration of incorporated ions M_{Fe}'. An analogous change in the character of the dependence of δ on p_{O_2} occurs in the case of doping A^+ ions in oxides $M_{1-\delta}O$ (see: Stokłosa 2015). It was shown that also for the remaining spinels it is possible that ions M^{3+} and M^{4+} incorporate into the magnetite structure, into the sublattice of iron ions Fe^{2+} and ions $M_{Fe}^{\bullet}$ are created.

Therefore, at high temperatures, as a result of contact of oxides of type MO, M_2O_3 or MO_2 with crystals of the formed spinel, ions are incorporated into the sublattice of ions with lower or higher oxidation state than that of the parent ions in the individual sublattices, which is fully consistent with the fact that a state of thermodynamic equilibrium is set. The determined concentrations of ions $[M_{Fe}']$ or $[M_{Fe}^{\bullet}]$ are the maximum concentrations that equal from $2 \cdot 10^{-4}$ to $1 \cdot 10^{-3}$ mol/mol. Figure 10.1 presents the maximum concentrations of ions $M_{Fe}^{\bullet}$ for magnetite doped with ions Ti^{4+} and Cr^{3+} and concentrations of ions M_{Fe}' for doping with ions Co^{2+} and Zn^{2+}, which incorporate into the structure of the analysed spinels, depending on the dopant content x_M.

As can be seen in Figure 10.1, a high concentration of ions $[M_{Fe}']$ is incorporated in the case of the spinel $(Fe_{1-x}Co_x)_{3\pm\delta}O_4$ and $(Zn_xMn_{0.333-x}Fe_{0.667})_{3\pm\delta}O_4)$ and ions $[Ti_{Fe}^{\bullet}]$ in the case of the spinel $(Fe_{1-x}Ti_x)_{3\pm\delta}O_4$. Significantly lower concentrations of ions $[Co_{Fe}']$ and $[Cr_{Fe}^{\bullet}]$ are present in spinels $(Co_xMn_zFe_{2z})_{3\pm\delta}O_4)$ and $(Fe_{1-x}Cr_x)_{3\pm\delta}O_4$. For the spinel $(Fe_{1-x}Co_x)_{3\pm\delta}O_4$ and $(Fe_{1-x}Ti_x)_{3\pm\delta}O_4$, the maximum concentrations of ions $[Co_{Fe}']$ and $[Ti_{Fe}^{\bullet}]$ is reached at the dopant content of $x_{Fe} = 0.2$ mol and at a higher content of doping ions it decreases. This indicates that the spinels with the content of $x_M = 0.333$ mol (MFe_2O_4) have a more ordered structure and different properties compared to the spinels with a lower dopant content. A complex relation is present in the case of the dependence of the concentration of ions $[Zn_{Fe}']$ on the zinc content x_{Zn} in the spinel $(Zn_xMn_{0.333-x}Fe_{0.667})_{3\pm\delta}O_4)$; the character of this change indicates a different effect of zinc on the interactions between ions in the spinel structure depending on its content. This effect is not present in a similar spinel $(Co_xMn_zFe_{2z})_{3\pm\delta}O_4)$, which might be related to the fact of a significant decrease in the content of iron and manganese ions.

Thus, the equilibrium distribution of cations in the individual sublattices of doped magnetite results from the quantity of doping ions in a given cation sublattice. These ions preserve their oxidation state, or partly change it as an effect of exchange of electrons (due to internal oxidation or reduction) and these ions are electroneutral relative to the lattice. Their content can reach 1 mol in a sublattice (up to the total exchange of iron ions in the sublattice). A portion of the iron ions have an oxidation state higher or lower than that of parent ions in the sublattice, due to the presence of a significant concentrations of electron holes and electrons, which are localised on these cations ($Fe_{Fe}^{\bullet}$, Fe_{Fe}'), or the concentration of these ions is an equivalent of the concentration of electronic defects. Finally, there exist doping ions ($M_{Fe}^{\bullet}$, M_{Fe}') charged in relation to the lattice, which were incorporated into the formed spinel structure into the appropriate sublattice and their

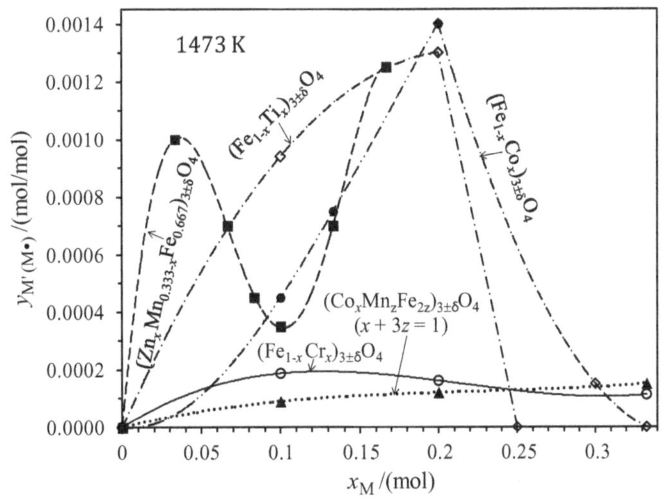

FIGURE 10.1 Dependence of the concentration of M^{3+} ions ($M^{\bullet}_{Fe}$) incorporated into the sublattice of Fe^{2+} ions or M^{2+} ions (Me'_{Fe}) incorporated into the sublattice of Fe^{3+} ions on the content of x_M ions in magnetite doped with metal ions: Ti ((◇) points), Cr, – (○), Co – (●), Co and Mn – (▲), Zn and Mn – (■) at 1473 K, obtained using the results of the studies on the deviation from the stoichiometry obtained by Dieckmann et al. (Lu and Dieckmann 1992, Lu and Dieckmann 1995, Lu et al. 1993, Töpfer et al. 1995, Aggarwal and Dieckmann 2002, Töpfer et al. 2003).

concentration can reach 0.001 mol/mol. Therefore, despite the fact that in the individual sublattices of doped magnetite, iron ions and doping ions have the oxidation state of 2+ or 3+, they play different roles.

As already mentioned, the range of existence and the character of the dependence of the deviation from the stoichiometry on the oxygen pressure are affected by the type and concentration of the dopant in magnetite. Figure 10.2 compares the oxygen pressure dependences of the deviation from the stoichiometry at 1473 K, for spinels $(Fe_{1-x}Ti_x)_{3\pm\delta}O_4$ (solid line) and $(Fe_{1-x}Cr_x)_{3\pm\delta}O_4$ (dashed lines) with the dopant content from x_M = 0.0–0.333 mol, which, as can be seen, are essentially different.

As can be seen in Figure 10.2, titanium doping of magnetite causes that the range of existence of the spinel changes – the curves of the deviation from the stoichiometry shift "in parallel" towards lower oxygen pressures. This indicates that when the titanium content increases, the bonding energy increases. When the titanium content of the spinel $(Fe_{1-x}Ti_x)_{3\pm\delta}O_4$ increases, the range of oxygen pressures where the spinel exists gets shorter. Above all, the range of oxygen pressures where interstitial cations dominate is shortened. As can be seen in Figure 10.2, titanium doping of magnetite causes that the range of existence of the spinel changes – the curves of the deviation from the stoichiometry shift "in parallel" towards lower oxygen pressures. This indicates that when the titanium content increases, the bonding energy increases. For the

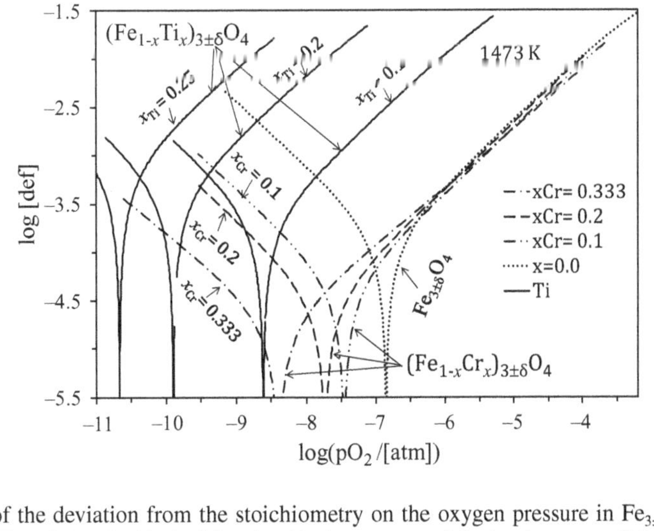

FIGURE 10.2 Dependences of the deviation from the stoichiometry on the oxygen pressure in $Fe_{3\pm\delta}O_4$ (dotted line) and in spinels $(Fe_{1-x}Ti_x)_{3\pm\delta}O_4$ (solid line) and $(Fe_{1-x}Cr_x)_{3\pm\delta}O_4$ (dashed lines) with the content of doping ions x_M = 0.0–0.333 mol, at 1473 K, obtained using the results of the studies of the deviation from the stoichiometry (Dieckmann 1982, Töpfer et al. 1995, Aggarwal and Dieckmann 2002).

spinel $(Fe_{1-x}Cr_x)_{3\pm\delta}O_4$, the effect of chromium is significantly smaller. The range of existence is shifted, to a smaller extent, towards lower oxygen pressures. Above all, the concentration of cation vacancies virtually does not change. Also, the range where interstitial cations dominate gets only slightly shorter. At the same oxygen pressure, when the chromium content of the spinel $(Fe_{1-x}Cr_x)_{3\pm\delta}O_4$ increases, the concentration of the interstitial cations decreases significantly.

In turn, Figure 10.3 compares the dependences of the deviation from the stoichiometry on the oxygen pressure at 1473 K, for spinels $(Fe_{1-x}Mn_x)_{3\pm\delta}O_4$ (solid line) and spinels $(Fe_{1-x}Co_x)_{3\pm\delta}O_4$ (dashed lines) with the dopant content of $x_M = 0.0$–0.333 mol.

As can be seen in Figure 10.3, doping cations M^{2+} in magnetite cause an almost parallel shift towards higher oxygen pressures in the range of existence of spinel, while the effect of manganese is significantly smaller than that of cobalt. In the case of the spinel $(Fe_{1-x}Mn_x)_{3\pm\delta}O_4$ with manganese content of $x_{Mn} = 0.0$–0.333 mol, there occurs a practically parallel shift towards higher oxygen pressures in the dependence of the deviation from the stoichiometry on p_{O_2}. For the spinel $(Fe_{1-x}Co_x)_{3\pm\delta}O_4$ with cobalt content $x_{Co} = 0.1$ and 0.2 mol, a change in the character of the dependence of the deviation from the stoichiometry is visible, in the range of low oxygen pressures, caused by an incorporation of cobalt ions Co^{2+} into the spinel structure, into the sublattice of iron ions Fe^{3+} (Co'_{Fe}).

Figure 10.4 presents the oxygen pressure dependence of the deviation from the stoichiometry at 1473 K, for the spinels $(Co_xMn_zFe_{2z})_{3\pm\delta}O_4$ (where $x + 3z = 1$) (solid line) with cobalt content of $x_{Co} = 0.0$–0.333 mol and for the spinel $(Zn_xMn_{0.333-x}Fe_{0.667})_{3\pm\delta}O_4$ with zinc content from $x_{Zn} = 0.0$–0.167 mol (dashed lines). For the above spinels, cobalt and zinc doping modifies the properties of the spinel $(Mn_{0.333}Fe_{0.667})_{3-\delta}O_4$.

As can be seen in Figure 10.4, manganese doping of magnetite at the content of $x_{Mn} = 0.333$ mol (1 mol/mol) shifts the range of existence of the spinel towards higher oxygen pressures by two orders of magnitude ((---) line). The increase in the cobalt content in the spinel $(Co_xMn_zFe_{2z})_{3-\delta}O_4$ (solid lines) causes a significant shift in the range of existence of the spinel compared to the spinel $(Mn_{0.333}Fe_{0.667})_{3\pm\delta}O_4$ $(MnFe_2)_{1\pm\Delta}O_4)$. When cobalt content of the spinel $(Co_xMn_zFe_{2z})_{3\pm\delta/3}O_4$ increases, the content of manganese and iron decreases, with $2z$ dropping from 0.667to 0.445 mol, which is the reason for such a big shift in the range of existence of the spinel. The influence of zinc in the spinel $(Zn_xMn_{0.333-x}Fe_{0.667})_{3\pm\delta}O_4$ is significantly smaller and already at zinc content of $x_{Zn} = 0.033$ mol a significant shift occurs in the oxygen pressure at which the spinel reaches the stoichiometric composition. Further increase in zinc content up to $x_{Zn} = 0.133$ mol in the spinel only slightly affects its range of existence and only slightly changes the oxygen pressure dependence of deviation from the stoichiometry. A significant shift in the dependence of δ on p_{O_2} occurs only at zinc content of $x_{Zn} = 0.167$ mol.

Figure 10.5 presents the influence of doping ions Ti^{4+} and Cr^{3+} and M^{2+} on the oxygen pressure at which Frenkel defects ($[V_{Fe}] = [M_i]$) are present; it indicates, more precisely, the extent of the shift in the range of existence of individual spinels. At the above pressure, the spinel reaches the stoichiometric composition if ions M'_{Fe} or $M^{\bullet}_{Fe}$ are not incorporated. It is different from the pressure at which the spinel is doped with the above ions.

As can be seen in Figure 10.5, manganese and cobalt ions shift the oxygen pressures at which Frenkel defects are present, ($[V_{Fe}] = [M_i]$), towards higher pressures, while ions M^{4+} and M^{3+} cause a decrease in these pressures. This effect indirectly indicates the influence of doping ions on the energy of interaction between ions in the doped magnetite. Ions M^{2+} cause its decrease; Mn^{2+} ions to a small degree, Co^{2+} ions to a larger degree. A significant increase in the bonding energy is caused by Ti^{4+} (or Ti^{3+}) ions and a smaller one – by Cr^{3+} ions.

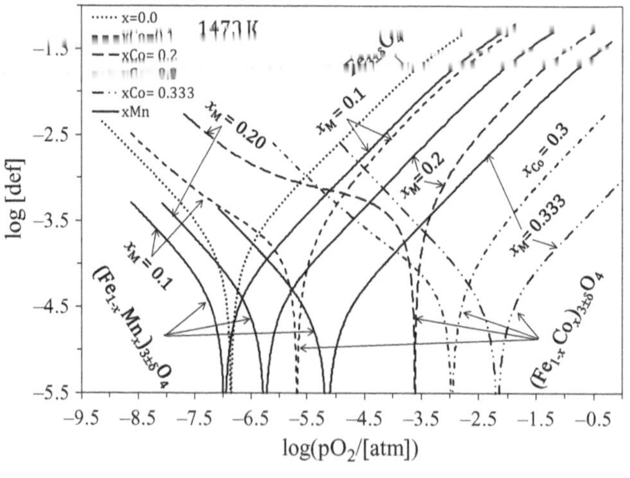

FIGURE 10.3 Oxygen pressure dependences of the deviation from the stoichiometry in $Fe_{3\pm\delta}O_4$ (dotted line), in the spinel $(Fe_{1-x}Co_x)_{3\pm\delta}O_4$ (dashed lines) and $(Fe_{1-x}Mn_x)_{3\pm\delta}O_4$ (solid lines) with the dopant content of $x_M = 0.0$–0.333 mol, at 1473 K, obtained using the results of the studies of the deviation from the stoichiometry (Dieckmann 1982, Lu and Dieckmann 1995, Lu et al. 1993).

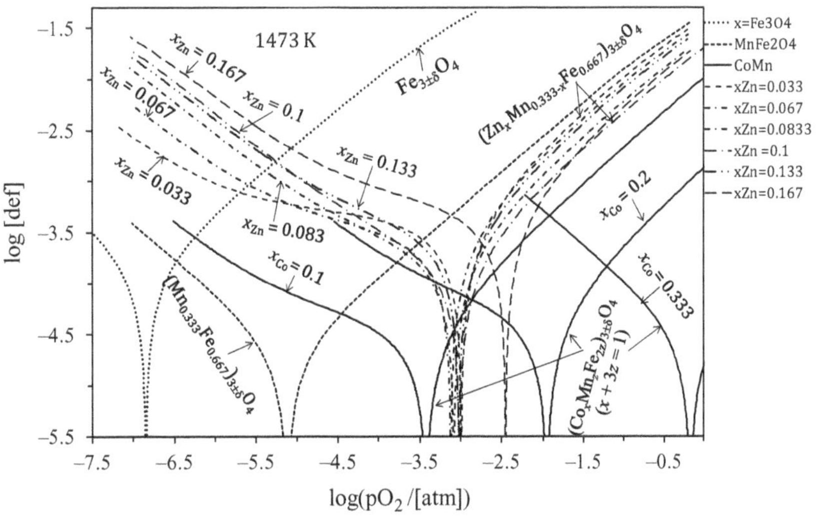

FIGURE 10.4 Dependence of the deviation from the stoichiometry on the oxygen pressure in $Fe_{3\pm\delta}O_4$, (dotted line), in the spinel $(Mn_{0.333}Fe_{0.667})_{3\pm\delta}O_4$ ((---) line), in $(Co_xMn_zFe_{2z})_{3\pm\delta}O_4$ (solid lines) and in $(Zn_xMn_{0.33-x}Fe_{0.667})_{3\pm\delta}O_4$ (dashed lines) with the dopant content of $x_M = 0.0$–0.333 mol, at 1473 K, obtained using the results of the studies of the deviation from the stoichiometry (Dieckmann 1982, Lu and Dieckmann 1992, Lu and Dieckmann 1995, Töpfer et al. 2003).

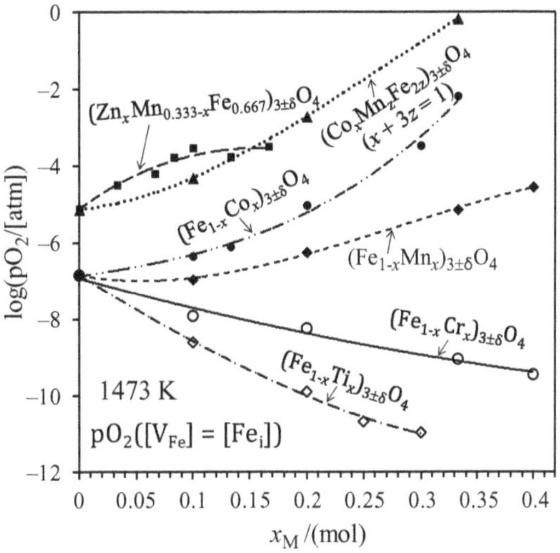

FIGURE 10.5 Dependence of the oxygen pressure at which the concentrations of cation vacancies and interstitial cations in the analysed spinels are equal on the dopant content x_M at 1473 K.

In turn, Figure 10.6 presents the dependence of the standard Gibbs energy $\Delta G^o_{F_1} = \Delta G^o_{F_2}$ of formation of Frenkel defects (V'''_{Fe} and $Fe^{3\bullet}_i$) as well as (V''_{Fe} and $Fe^{\bullet\bullet}_i$). Quite large values of the deviation from the stoichiometry at the limit oxygen pressures, where individual spinels exist, allow unambiguous determination of the values of $\Delta G^o_{F_1}$.

As can be seen in Figure 10.6, when the content of ions Co, Mn and Cr in magnetite increases, values of $\Delta G^o_{F_1}$ increase, while titanium doping causes a significant decrease in these values. On the other hand, ions Co^{2+} in the spinel $(Co_xMn_zFe_{2z})_{3\pm\delta}O_4$ (where $(x + 3z = 1)$) and ions Zn^{2+} in the spinel $(Zn_xMn_{1-x}Fe_2)_{1\pm\delta/3}O_4$ have a different effect on the value of $\Delta G^o_{F_1}$; these ions modify the properties of the spinel $MnFe_2O_4$. Zinc doping of the spinel $MnFe_2O_4$ causes a decrease in the value of $\Delta G^o_{F_1}$, while cobalt doping, after a slight increase at the content of $x_{Co} = 0.1$ mol, when the cobalt content increases further, causes a decrease in this value. As demonstrated in Stokłosa (2015), the value of $\Delta G^o_{F_1}$ of formation of Frenkel defects is related to the energy of incorporation of oxygen into the spinel structure and it depends on the difference between the energy of formation of cation vacancies and interstitial cations. Thus, it depends on several values that are a consequence of energy of interactions between ions and it is difficult to interpret the change in ΔG^o_F.

A result of the change in the value of $\Delta G^o_{F_1}$ is a change in the concentration of Frenkel defects. Figure 10.7 shows the dependence of the concentration of Frenkel defects on the dopant content in analysed spinels.

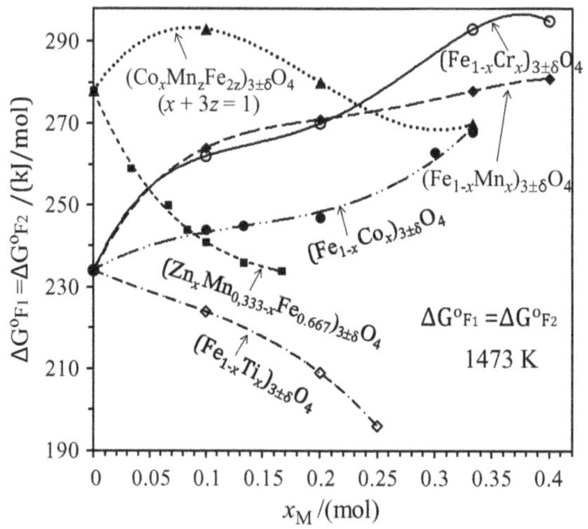

FIGURE 10.6 Dependence of standard Gibbs energy $\Delta G^o_{F_1} \cong \Delta G^o_{F_2}$ of formation of Frenkel defects with charge 2 and 3 on the dopant content x_M in magnetite doped with metal ions: Ti ((◇) points), Cr – (○), Co – (●), Mn – (◆), Co and Mn – (▲), Zn and Mn – (■) at 1473 K, obtained using the results of the studies on the deviation from the stoichiometry obtained Dieckmann et al. (Dieckmann 1982, Lu and Dieckmann 1992, Lu et al. 1993, Lu and Dieckmann 1995, Töpfer et al. 1995, Aggarwal and Dieckmann 2002, Töpfer et al. 2003).

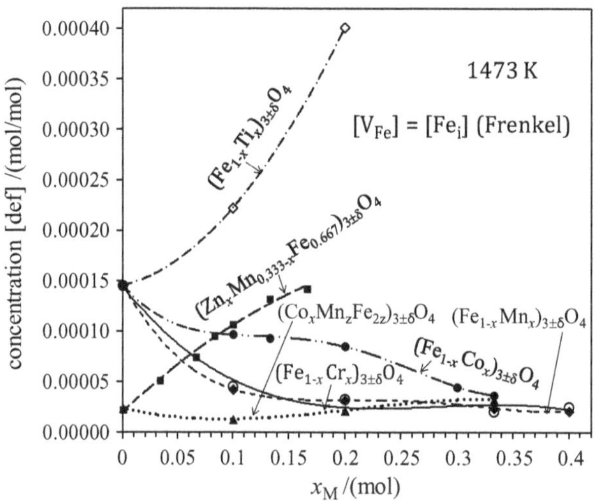

FIGURE 10.7 Dependence of the concentration of Frenkel defects on the dopant content x_M in analysed spinels, at 1473 K.

As can be seen in Figure 10.7, the concentration of Frenkel defects are small and the nature of their change is inverse in relation to the nature of the change in $\Delta G^o_{F_1}$ of their formation.

In turn, Figure 10.8 presents the dependence of $\Delta G^o_{V_{Fe}'''}$ of formation of cation vacancies V_{Fe}''' and $\Delta G^o_{Fe_i^{3\bullet}}$ of formation of interstitial cations $Fe_i^{3\bullet}$, and Figure 10.9 – the dependence of $\Delta G^o_{V_{Fe}''}$ and $\Delta G^o_{Fe_i^{\bullet\bullet}}$ of formation of V_{Fe}'' and $Fe_i^{\bullet\bullet}$ on the dopant content in analysed spinels.

As shown earlier, the fitted values of $\Delta G^o_{V_{Fe}'''}$ or $\Delta G^o_{V_{Fe}''}$, as well as the calculated values of $\Delta G^o_{Fe_i^{3\bullet}}$ and $\Delta G^o_{Fe_i^{\bullet\bullet}}$, depend on the value ΔG^o_{eh} of formation of electronic defects adopted in the calculations. In the calculations it was approximatively assumed that ΔG^o_{eh} does not depend on the concentration of dopant in magnetite and it is equal $\Delta G^o_{eh} = 30.3$ kJ/mol. The real values of ΔG^o_i of formation of defects can differ slightly, but the character of their dependence on the dopant concentration should be similar to the one presented in the above figures. Furthermore, the values $\Delta G^o_{V_{Fe}''}$ and $\Delta G^o_{Fe_i^{\bullet\bullet}}$ depend on the value of $\Delta G^o_{F_2}$, which also can be different in reality.

As can be seen in Figure 10.8, when the content of Co, Mn and Cr in magnetite increases, the values of $\Delta G^o_{V_{Fe}'''}$ increase monotonically, while in the case of titanium they decrease. The character of the dependence of $\Delta G^o_{Fe_i^{3\bullet}}$ on the dopant content is inverse than $\Delta G^o_{V_{Fe}'''}$, but it is not symmetrical, which indicates that the dopant affects the standard Gibbs energy $\Delta G^o_{V_{Fe}'''}$ and $\Delta G^o_{Fe_i^{3\bullet}}$ of formation of these defects differently. The effect of the sum of their values is ΔG^o_F of formation of

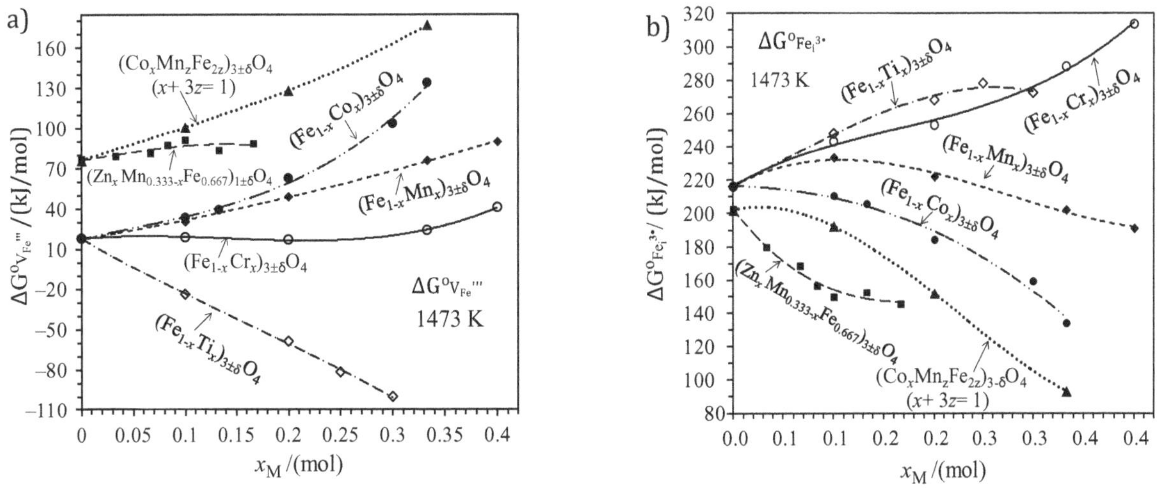

FIGURE 10.8 Dependence of standard Gibbs energy a) $\Delta G^o_{V_{Fe}'''}$ of formation of cation vacancies $V''_{Fe(o)}$, b) $\Delta G^o_{Fe_i^{3\bullet}}$, of formation of interstitial ions $Fe_i^{3\bullet}$ on the dopant content x_M in magnetite doped with metal ions: Ti ((◇) points), Cr – (○), Co – (●), Mn – (◆), Co and Mn – (▲), Zn and Mn – (■) at 1473 K, obtained using the results of the studies on the deviation from the stoichiometry (Dieckmann 1982, Lu and Dieckmann 1992, Lu et al. 1993, Lu and Dieckmann 1995, Töpfer et al. 1995, Aggarwal and Dieckmann 2002, Töpfer et al. 2003, Töpfer and Dieckmann 2004).

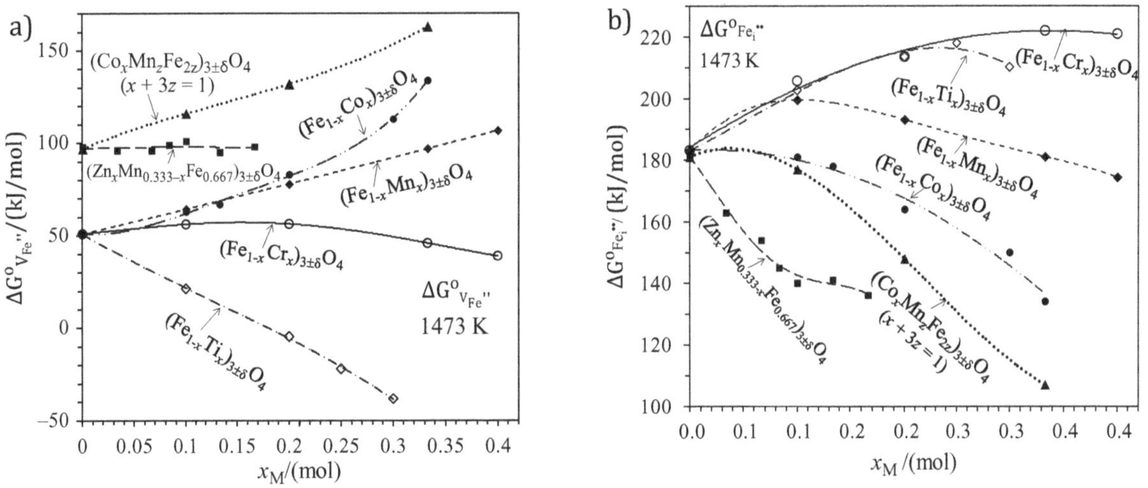

FIGURE 10.9 Dependence of standard Gibbs energy a) $\Delta G^o_{V_{Fe}''}$ of formation of cation vacancies $V''_{Fe(o)}$, b) $\Delta G^o_{Fe_i^{\bullet\bullet}}$ of formation of interstitial ions $Fe_i^{\bullet\bullet}$ in magnetite doped with metal ions: Ti ((◇) points), Cr – (○), Co – (●), Mn – (◆), Co and Mn – (▲), Zn and Mn – (■) on the dopant content x_M, at 1473 K, obtained using the results of the studies on the deviation from the stoichiometry (Dieckmann 1982, Lu and Dieckmann 1992, Lu et al. 1993, Lu and Dieckmann 1995, Töpfer et al. 1995, Aggarwal and Dieckmann 2002, Töpfer et al. 2003, Töpfer and Dieckmann 2004).

Frenkel defects and the nature of its change depending on the dopant content (see Equation 1.34c). As can be seen in Figure 10.9, the values of $\Delta G^o_{V_{Fe}''}$ and $\Delta G^o_{Fe_i^{\bullet\bullet}}$ change in a similar way depending on the dopant content in spinels.

10.2 EFFECT OF DOPANTS ON THE MOBILITY OF IONS Fe, Co, Mn, Ti AND Cr IN MAGNETITE

The studies on the diffusion of tracers in doped magnetite (Franke and Dieckmann 1989, Lu and Dieckmann 1992, Lu et al. 1993, Lu and Dieckmann 1995, Töpfer et al. 1995, Aggarwal and Dieckmann 2002) demonstrated, similarly as for pure magnetite (Dieckmann et al. 1978), that the diffusion of metal ions (tracers) strictly depends on the concentration of cation vacancies and interstitial cations. Analogously to the case of pure magnetite, using the concentration of ionic defects and the values of coefficients of self-diffusion of tracers (^{59}Fe, ^{60}Co and ^{53}Mn, ^{51}Cr, ^{44}Ti) the coefficients of diffusion (mobility) of ions via cation vacancies $D^o_{V(M)}$ and via interstitial cations $D^o_{I(M)}$ were determined.

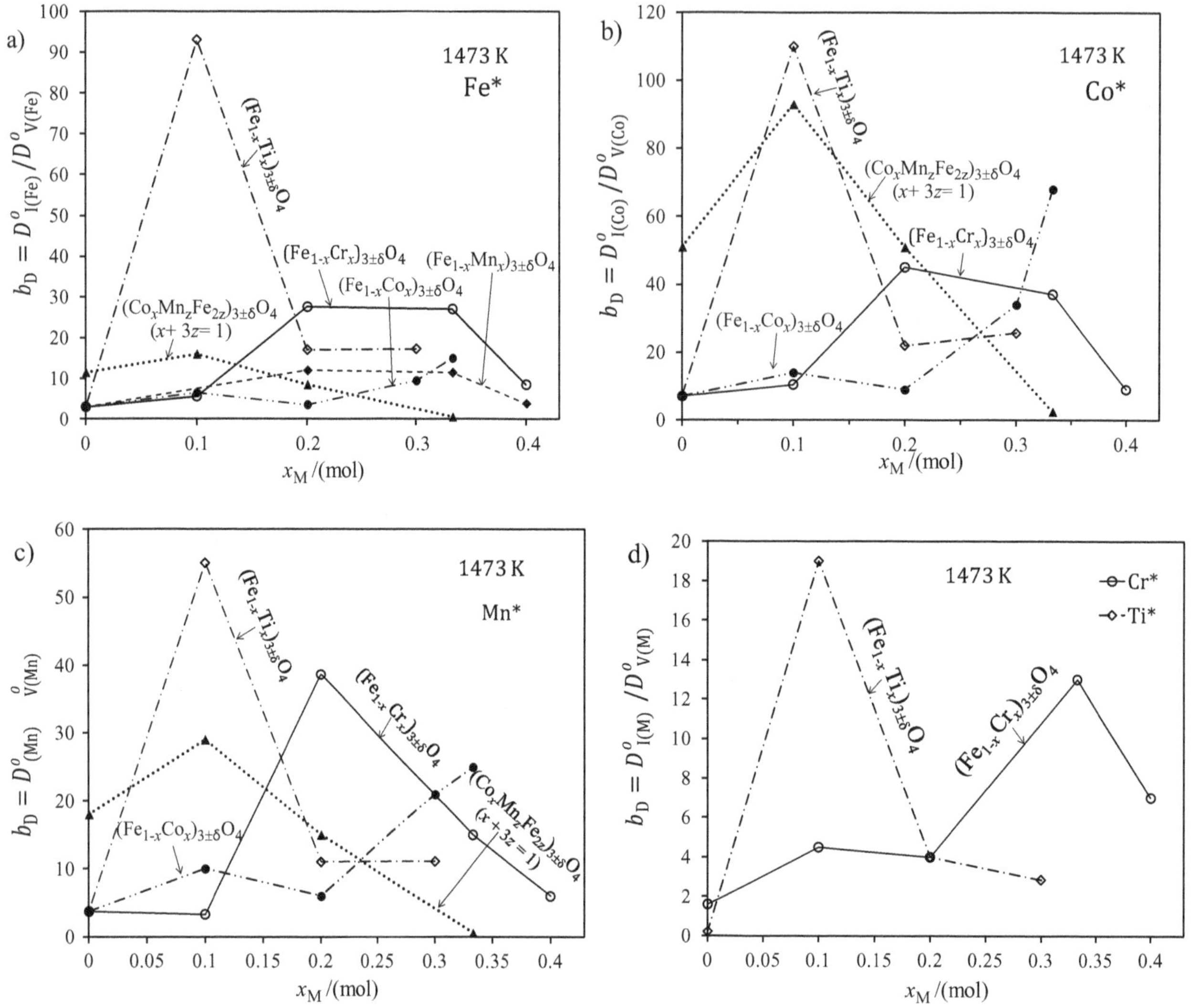

FIGURE 10.10 Dependence of the ratio of the coefficient of diffusion of tracer ions via interstitial cations to the coefficient of diffusion of cations via vacancies $D^o_{I(M)}/D^o_{V(M)} = b_D$ on the dopant content x_M in spinels $(Fe_{1-x}Ti_x)_{3\pm\delta}O_4$, $(Fe_{1-x}Cr_x)_{3\pm\delta}O_4$, $(Fe_{1-x}Co_x)_{3\pm\delta}O_4$, $(Fe_{1-x}Mn_x)_{3\pm\delta}O_4$ and $(Co_xMn_zFe_{2z})_{3\pm\delta}O_4$, at 1473 K, obtained based on the values of coefficients of self-diffusion of: a) Fe*, b) Co*, c) Mn*, d) Cr* and Ti* (Dieckmann et al. 1978, Franke and Dieckmann 1989, Lu and Dieckmann 1992, Lu et al. 1993, Lu and Dieckmann 1995, Töpfer et al. 1995, Aggarwal and Dieckmann 2002) and concentrations of ionic defects.

The performed calculations demonstrated that in doped magnetite, the mobility of ions via interstitial cations is higher than the mobility via vacancies, while their ratio ($D^o_{I(M)}/D^o_{V(M)} = b_D$) depend on the type of spinel and on the concentration of the dopant. Figures 1.19, 3.20, 4.11, 6.11, 7.10, 8.11 present the dependences of the parameter b_D on the dopant content for individual tracers in analysed spinels. From these dependences it results that the ratio $\left(D^o_{I(M)}/D^o_{V(M)}\right)$ for the individual tracers in spinels is different and the largest differences are present in the case of cobalt diffusion, smaller ones for manganese, and the smallest ones for iron. Figure 10.10, compares the dependence of the ratio $\left(D^o_{I(M)}/D^o_{V(M)}\right)$ on the dopant content x_M in analysed spinels, resulting from the diffusion of tracers Fe*, Co*, Mn*, Ti* and Cr*, respectively.

As can be seen in Figure 10.10, the character of the dependence of b_D on the dopant content x_M in individual spinels is complex and it depends on the diffusing ion, on the spinel type and on the concentration of doping ions. There is no general trend in the changes. Usually, at the dopant content of $x_M = 0.1$ or 0.2 mol the parameter b_D increases, and when the dopant content in the spinel further increases, this parameter decreases. This indicates that the dopant affects the ratio of mobilities of ions $\left(D^o_{I(M)}/D^o_{V(M)}\right)$ in spinels in a significant way. The fact that the mobility of ions decreases at the composition MFe_2O_4 indicates that in these spinels there is the highest degree of order compared to spinels with a lower dopant content and they can be treated as new heterocation spinels, which is widely adopted. The only deviating dependence is the

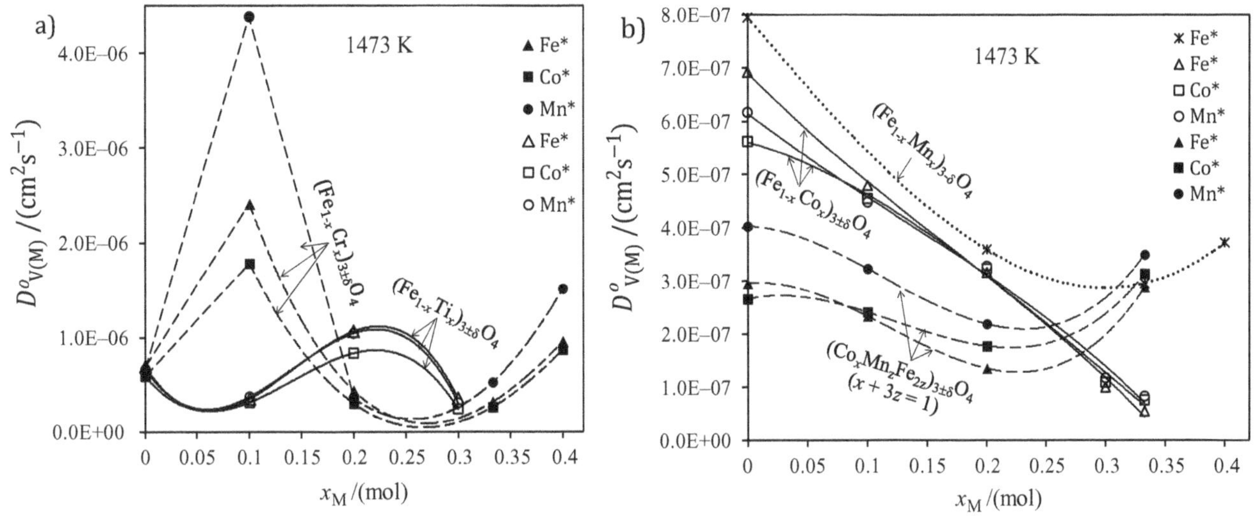

FIGURE 10.11 Dependence of the coefficient of diffusion of tracer ions via cation vacancies $D^o_{V(Fe)}$ on the dopant content x_M in spinels a) $(Fe_{1-x}Ti_x)_{3\pm\delta}O_4$, $(Fe_{1-x}Cr_x)_{3\pm\delta}O_4$, b) $(Fe_{1-x}Co_x)_{3\pm\delta}O_4$, $(Fe_{1-x}Mn_x)_{3\pm\delta}O_4$ and $(Co_xMn_zFe_{2z})_{3\pm\delta}O_4$, at 1473 K, obtained using the values of coefficients of self-diffusion of Fe*, Co*, Mn* (Dieckmann et al. 1978, Franke and Dieckmann 1989, Lu and Dieckmann 1992, Lu et al. 1993, Lu and Dieckmann 1995, Töpfer et al. 1995, Aggarwal and Dieckmann 2002) and concentrations of ionic defects resulting from the diagram of defects.

dependence of b_D on x_{Co} in the case of diffusion of tracer ions in the spinel $(Fe_{1-x}Co_x)_{3\pm\delta}O_4$, which is practically monotonically increasing up to $b_D = 70$.

Figure 10.11a compares the dependence of $D^o_{V(M)}$ on the dopant content in spinels $(Fe_{1-x}Ti_x)_{3-\delta}O_4$, $(Fe_{1-x}Cr_x)_{3\pm\delta}O_4$, and in Figure 10.11b – in spinels $(Fe_{1-x}Co_x)_{3-\delta}O_4$, $(Fe_{1-x}Mn_x)_{3\pm\delta}O_4$ and $(Co_xMn_zFe_{2z})_{3\pm\delta}O_4$, resulting from the diffusion of tracers Fe*, Co* and Mn*.

As can be seen in Figure 10.11, the character of the dependence of $D^o_{V(M)}$ on the dopant content is complex and different depending on the dopant content in the spinel. The mobility of tracer ions Fe*, Co*, Mn* via cation vacancies $\left(D^o_{V(M)}\right)$ differ only slightly depending on the type of diffusing ion. Larger differences between the coefficients of diffusion $D^o_{V(M)}$ are present in the case of the spinel $(Co_xMn_zFe_{2z})_{3\pm\delta}O_4$. The occurring differences in the values of $D^o_{V(M)}$ are only slightly higher than the limit of the measurement error. As can be seen in Figure 10.11a, the character of the dependence of $D^o_{V(M)}$ on the dopant content is independent of the type of diffusing ions but it is specific for the individual spinels.

As can be seen in Figure 10.11a, in the case of spinels $(Fe_{1-x}Ti_x)_{3\pm\delta}O_4$ and $(Fe_{1-x}Cr_x)_{3\pm\delta}O_4$ the characters of the dependences of $D^o_{V(M)}$ on x_M are opposite. In turn, as can be seen in Figure 10.11b, in the case of spinels $(Fe_{1-x}Co_x)_{3\pm\delta}O_4$, $(Fe_{1-x}Mn_x)_{3\pm\delta}O_4$, when the dopant content increases, the mobility of ions via cation vacancies decreases. The character of the dependence of $D^o_{V(Co)}$ on x_{Co} is slightly different for the spinel $(Co_xMn_zFe_{2z})_{3\pm\delta}O_4$ up to the content of $x_{Co} = 0.2$ mol, a similar decrease in the value $D^o_{V(Co)}$ occurs, and at the content of $x_M = 0.333$ mol, there occurs a strong increase.

On the other hand, Figure 10.12 shows the dependence of the coefficients of diffusion of tracer ions: Fe*, Co*, Mn* via interstitial cations $D^o_{I(M)}$ on the dopant content in analysed spinels.

As can be seen in Figure 10.12, the character of the dependence of $D^o_{I(M)}$ on the dopant content in the individual spinels is different, while the mobility of tracer ions Fe*, Co*, Mn* via interstitial cations is significantly different depending on the type of diffusing ion. As can be seen in Figure 10.12a, in the case of spinels $(Fe_{1-x}Ti_x)_{3\pm\delta}O_4$, $(Fe_{1-x}Cr_x)_{3\pm\delta}O_4$, the coefficients $D^o_{I(M)}$ only slightly differ between them, although in the spinel $(Fe_{1-x}Ti_x)_{3\pm\delta}O_4$ the mobility of manganese ions is much lower than that of Fe* and Co*. As can be seen in Figure 10.12b, clear differences are present in the case of mobilities of ions in spinels $(Fe_{1-x}Co_x)_{3\pm\delta}O_4$, $(Fe_{1-x}Mn_x)_{3\pm\delta}O_4$ and $(Co_xMn_zFe_{2z})_{3\pm\delta}O_4$. In all the analysed spinels, at the dopant content of $x_M = 0.1$ mol, the coefficients $D^o_{I(M)}$ are higher that in pure magnetite, but when the dopant content in the individual spinels increases further, they decrease.

Figure 10.13 presents the dependence of coefficients of diffusion of ions Ti* and Cr* via interstitial ions $D^o_{I(M)}$ and via interstitial cations $D^o_{V(M)}$ on the dopant content in spinels $(Fe_{1-x}Ti_x)_{3\pm\delta}O_4$ and $(Fe_{1-x}Cr_x)_{3\pm\delta}O_4$ and, for comparison, the dependences of coefficients of diffusion of cations via cation vacancies $D^o_{V(M)}$ in the spinel $(Fe_{1-x}Al_x)_{3-\delta}O_4$ are based on the measurements of chemical diffusion at 1573 K (Yamauchi et al. 1983).

As can be seen in Figure 10.13, there are large differences in the mobility of ions: Ti* in the spinel $(Fe_{1-x}Ti_x)_{3\pm\delta}O_4$ and Cr* in the spinel $(Fe_{1-x}Cr_x)_{3\pm\delta}O_4$; the character of the dependence of $D^o_{V(M)}$ on the dopant content is also different. Similarly,

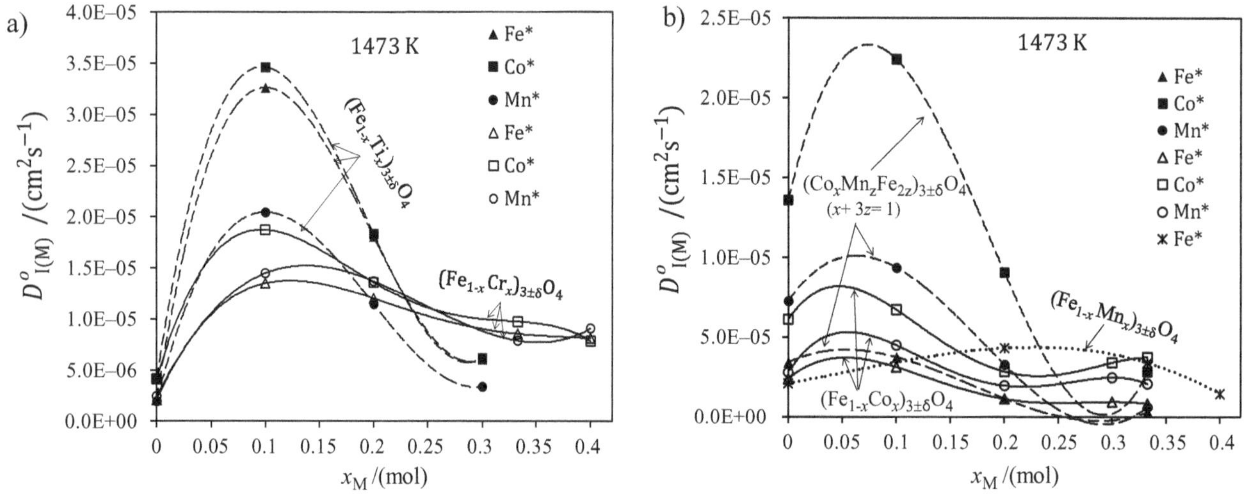

FIGURE 10.12 Dependence of the coefficient of diffusion of ions via interstitial cations $D^o_{I(M)}$ on the dopant content x_M in spinels a) $(Fe_{1-x}Ti_x)_{3\pm\delta}O_4$, $(Fe_{1-x}Cr_x)_{3\pm\delta}O_4$, b) $(Fe_{1-x}Co_x)_{3\pm\delta}O_4$, $(Fe_{1-x}Mn_x)_{3\pm\delta}O_4$ and $(Co_xMn_zFe_{2z})_{3\pm\delta}O_4$, at 1473 K, obtained using the values of coefficients of self-diffusion of Fe*, Co*, Mn* (Dieckmann et al. 1978, Franke and Dieckmann 1989, Lu and Dieckmann 1992, Lu et al. 1993, Lu and Dieckmann 1995, Töpfer et al. 1995, Aggarwal and Dieckmann 2002) and concentrations of ionic defects.

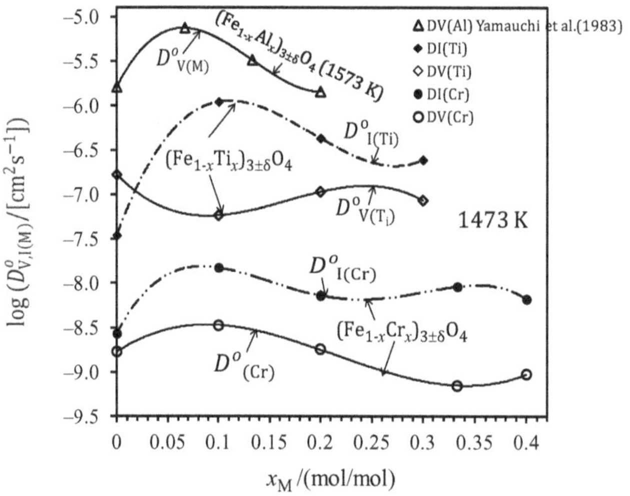

FIGURE 10.13 Dependence of the coefficient of diffusion of ions Ti, Cr via cation vacancies $D^o_{V(M)}$ and via interstitial cations $D^o_{I(M)}$ on the dopant content x_M in spinels $(Fe_{1-x}Ti_x)_{3\pm\delta}O_4$ and $(Fe_{1-x}Cr_x)_{3\pm\delta}O_4$ at 1473 K, obtained using the values of coefficients of self-diffusion of Ti*, Cr* (Dieckmann et al. 1978, Franke and Dieckmann 1989, Töpfer et al. 1995, Aggarwal and Dieckmann 2002) and of diffusion of ions via cation vacancies in the spinel $(Fe_{1-x}Al_x)_{3-\delta}O_4$ at 1573 K, calculated based on the coefficients of chemical diffusion (Yamauchi et al. 1983).

as for the ions discussed earlier, in the case of the mobilities of Ti* and Cr* in spinels, it is also difficult to determine a global trend of the changes in coefficients $D^o_{V(M)}$ and $D^o_{I(M)}$. In the case of diffusion of chromium in the spinel $(Fe_{1-x}Cr_x)_{3\pm\delta}O_4$ and cations in the spinel $(Fe_{1-x}Al_x)_{3-\delta}O_4$, the mobility of Cr ions and M ions via cation vacancies increases at the content of $x_M = 0.1$ mol, and when the dopant content increases further, it decreases. The mobility of titanium ions in the spinel $(Fe_{1-x}Ti_x)_{3\pm\delta}O_4$, inversely, slightly decreases at the content of $x_{Ti} = 0.1$ mol, and at higher concentration of titanium it increases. In turn, the mobilities of ions Ti and Cr via interstitial cations, in the spinel $(Fe_{1-x}Ti_x)_{3\pm\delta}O_4$ as well as in the spinel $(Fe_{1-x}Cr_x)_{3\pm\delta}O_4$, at the content of $x_M = 0.1$ mol, increase, and when the dopant content increases further, they slightly decrease, similarly as the mobilities of ions Fe*, Co* and Mn*. The differences in the values of coefficients of diffusion of tracer ions $D^o_{V(M)}$ and $D^o_{I(M)}$ in the individual spinels are so large that the character of their dependence on the dopant content rather cannot be related to a measurement error. It results, above all, from the error in the measurement of the coefficients of self-diffusion of tracers and it depends on the calculated concentration of ionic defects, which clearly impedes the interpretation of their dependence on the dopant content. However, the obtained results indicate that the changes in mobilities of ions via cation vacancies and interstitial cations depend on the concentration of dopant in magnetite, causing changes in the interactions between ions. The reason for quite complex dependences of mobilities of ions via cation vacancies can be the distribution of dopant ions in the individual sublattices of spinel structure, affecting the mutual interactions of ions. It should be noted that the mobilities of ions via cation vacancies $\left[V_{Fe}'''\right]$ and $\left[V_{Fe}''\right]$ will be different; their contribution to ion transport cannot be determined. As can be seen in Figure 10.12, more regular changes of mobilities of ions are present for the diffusion of ions via interstitial cations. At the dopant content in spinels of about $x_M = 0.1$ mol, the mobility of ions increases, and when the dopant content increases further, it decreases. This might indicate that dopant ions cause a decrease in the mobility of ions. A decrease in the mobility of ions via interstitial cations can be also related with an increase in the concentration of interstitial cations in the range of low concentrations and thus with a decrease in the concentrations of available voids.

It should be noted that the extent of changes of mobilities of tracer ions in doped magnetite, the character of the dependence of the change in the parameter b_D on the dopant concentration, as well as the values of concentration of ions M_{Fe}' or $M_{Fe}^{\bullet}$ in spinels indicate that spinels in which the concentration of dopant reaches the value of one (MFe_2O_4) can be treated as new compounds with different properties (iron ions are practically substituted with doping ions in a given sublattice). In these spinels, the highest degree of order is reached. The above conclusion is fully consistent with the fact of treating the spinels with such a composition as new heterocation spinels MFe_2O_4. A change in the content of the dopant M (composition) between the magnetite and the spinel MFe_2O_4 is continuous, but usually at the dopant content within the range x_M = 0.1–0.2 there occur maximums or minimums of the change in parameters (quantities) related to specific properties.

REFERENCES

Aggarwal, S., and R. Dieckmann. 2002. Point Defects and Cation Tracer Diffusion in $(Ti_xFe_{1-x})_{3-\delta}O_4$ I. Non-Stoichiometry and Point Defects. *J. Phys. Chem. Miner.* 29: 695–706.

Dieckmann, R. 1982. Defects and Cation Diffusion in Magnetite (IV): Nonstoichiometry and Point Defect Structure of Magnetite ($Fe_{3-\delta}O_4$). *Ber. Bunsenges. Phys. Chem.* 86: 112–118.

Dieckmann, R., T. O. Mason, J. D. Hodge, and H. Schmalzried. 1978. Defects and Cation Diffusion in Magnetite (III.) Tracer Diffusion of Foreign Tracer Cations as a Function of Temperature and Oxygen Potential. *Ber. Bunsenges. Phys. Chem.* 82: 778–783.

Franke, P., and R. Dieckmann. 1989. Defect Structure and Transport Properties of Mixed Iron-Manganese Oxides. *Solid State Ionics.* 32–33: 817–823.

Lu, F. H., and R. Dieckmann. 1992. Point Defects and Cation Tracer Diffusion in $(Co, Fe, Mn)_{3-\delta}O_4$ Spinels: I. Mixed Spinels $(Co_xFe_{2y}Mn_y)_{3-\delta}O_4$. *Solid State Ionics.* 53–56: 290–302.

Lu, F. H., and R. Dieckmann. 1995. Point Defects in Oxide Spinel Solid Solutions of the type $(Co, Fe, Mn)_{3-}O_4$ at 1200 °C. *J. Phys. Chem. Solids.* 56: 725–733.

Lu, F. H., S. Tinkler, and R. Dieckmann. 1993. Point Defects and Cation Tracer Diffusion in $(Co_xFe_{1-x})_{3-\delta}O_4$ Spinels. *Solid State Ionics.* 62: 39–52.

Stokłosa, A. 2015. *Non-Stoichiometric Oxides of 3d Metals.* Pfäffikon: Trans Tech Publications Ltd.

Töpfer, J., S. Aggarwal, and R. Dieckmann. 1995. Point Defects and Cation Tracer Diffusion in $(Cr_xFe_{1-x})_{3-\delta}O_4$ Spinels. *Solid State Ionics.* 81: 251–266.

Töpfer, J., and R. Dieckmann. 2004. Point Defects and Deviation from Stoichiometry in $(Zn_{x-y/4}Mn_{1-x-3y/4}Fe_{2+y})_{1-\delta/3}O_4$. *J. Europ. Ceram. Soc.* 24: 603–612.

Töpfer, J., L. Liu, and R. Dieckmann. 2003. Deviation from Stoichiometry and Point Defects in $(Zn_xMn_{1-x}Fe_2)_{1-\Delta/3}O_4$. *Solid State Ionics.* 159: 397–404.

Yamauchi, S., A. Nakamura, T. Shimizu, and K. Fueki. 1983. Vacancy Diffusion in Magnetite Hercynite Solid Solution. *J. Solid State Chem.* 50: 20–32.

Part III

Hausmannite Doped with Cobalt, Iron and Lithium

11 Cobalt-Doped Hausmannite – $(Mn_{1-x}Co_x)_{3\pm\delta}O_4$

11.1 INTRODUCTION

Hausmannite with a spinel structure, similarly as magnetite, can be doped with ions of other metals. Cobalt ions, Co^{2+}, in the spinel $(Mn_{1-x}Co_x)_{3\pm\delta}O_4$, form a solid solution and they occupy mainly cation sites in the tetrahedral sublattice, substituting Mn^{2+} ions and, in the limit case, the composition of $CoMn_2O_4$ is reached. At higher temperatures, due to attaining the equilibrium state, part of these ions occupies also cation sites in the sublattice with octahedral coordination. The range of existence of spinels $(Mn_{1-x}Co_x)_{3\pm\delta}O_4$ depending on cobalt concentration was reported (Aukrust and Muan 1964, Keller and Dieckmann 1985, Lu and Dieckmann 1993, Subramanian et al. 1994). Studies of the deviation from the stoichiometry and diffusion with tracer ions, using radioisotope elements: ^{60}Co, ^{59}Fe, ^{54}Mn, at 1473 K, were also performed (Lu and Dieckmann 1993). The studies on the deviation from the stoichiometry of the spinel $(Mn_{1-x}Co_x)_{3\pm\delta}O_4$ with the cobalt content of x_{Co} = 0.1–0.333 mol showed that interstitial cations dominate practically the whole range of existence. The studies of diffusion indicate that apart from diffusion via interstitial cation, diffusion via cation vacancies is also significant. Using the results of the studies on the deviation from the stoichiometry (Lu and Dieckmann 1993), it was decided to determine the diagrams of point defects and the coefficients of diffusion of defects in spinels $(Mn_{1-x}Co_x)_{3\pm\delta}O_4$.

11.2 METHODS AND RESULTS OF CALCULATION OF THE DIAGRAMS OF THE CONCENTRATIONS OF DEFECTS, AND DISCUSSION

The spinel $(Mn_{1-x}Co_x)_{3\pm\delta}O_4$ is generally obtained through a reaction of specific quantities of oxides: MnO, CoO and Mn_2O_3. In the spinel structure, analogous to the case of hausmannite, cobalt ions Co^{2+} and manganese ions Mn^{2+} will occupy cation sites with tetrahedral coordination. Therefore, the presence of cobalt ions does not affect the structure of point defects, which is consistent with the character of the oxygen pressure dependence of the deviation from the stoichiometry in the spinel $(Mn_{1-x}Co_x)_{3\pm\delta}O_4$ (Lu and Dieckmann 1993). Due to that, the calculations of diagrams of concentrations of point defects in cobalt-doped hausmannite were conducted with a method analogous to the one used for cobalt-doped magnetite (see Section 1.5, 6.2), using the results of the of deviation from the stoichiometry for the spinel $(Mn_{1-x}Co_x)_{3\pm\delta}O_4$ with the cobalt content of x_{Co} = 0.0–0.333 mol (Lu and Dieckmann 1993).

Due to the lack of the value of ΔG^o_{eh} of formation of electronic defects, it was assumed that it only slightly depends on the concentration of doping ions in hausmannite. Owing to that, in the calculations, a constant value of ΔG^o_{eh} = 60 kJ/mol was assumed, the same as for pure hausmannite at 1473 K. In the calculations of diagrams of concentrations of defects, it was assumed that standard Gibbs energies of formation of Frenkel defects have the same values: $\Delta G^o_{F_2} = \Delta G^o_{F_1}$. Thus, at the stoichiometric composition, the concentrations of vacancies are equal $[V''_{Mn(A)}]$ $[V'''_{Mn(O)}]$

Figure 11.1a–d present the diagrams of the concentrations of ionic defects for spinels $(Mn_{1-x}Co_x)_{3\pm\delta}O_4$, with the cobalt content of x_{Co} = 0.0–0.333 mol, at the temperature of 1473 K (solid lines), determined using the values of the deviation from the stoichiometry (Lu and Dieckmann 1993) ((○) points) and for pure hausmannite (Keller and Dieckmann 1985). The dotted line shows the dependence of δ on p_{O_2} obtained using "equilibrium constants" of reactions of formation of cation vacancies K_V and interstitial ions K_I (Keller and Dieckmann 1985, Lu and Dieckmann 1993).

As can be seen in Figure 11.1a–d, the experimental values and the calculated dependences of δ on p_{O_2} are different compared to pure hausmannite. Cobalt doping shifted the range of existence of spinels to such extent that in the studied range of oxygen pressures, interstitial cations $[Mn_i^{3\bullet}]$ dominate, and the concentration of ions $[Mn_i^{\bullet\bullet}]$ is much lower; this concentration must be assumed the maximum concentration of these defects. Spinels $(Mn_{1-x}Co_x)_{3\pm\delta}O_4$ reach the stoichiometric composition near the oxygen pressure of 1 atm. The range of pressures where cation vacancies dominate is above the oxygen pressure of 1 atm. The obtained dependences of δ on p_{O_2} for spinels with the cobalt content of x_{Co} = 0.1, 0.2 mol ((---) line), apart from the lowest values of the deviation δ, are consistent with the values of the deviation δ (Lu and Dieckmann 1993), which were determined with the assumption that the stoichiometric composition is reached near the oxygen pressure of 1 atm. The match quality is worse for the spinel $(Mn_{1-x}Co_x)_{3\pm\delta}O_4$ with the cobalt content of x_{Co} = 0.333 mol ($(CoMn_2)_{1\pm\delta/3}O_4$). The lack of results in the range of oxygen pressures where cation vacancies dominate makes it

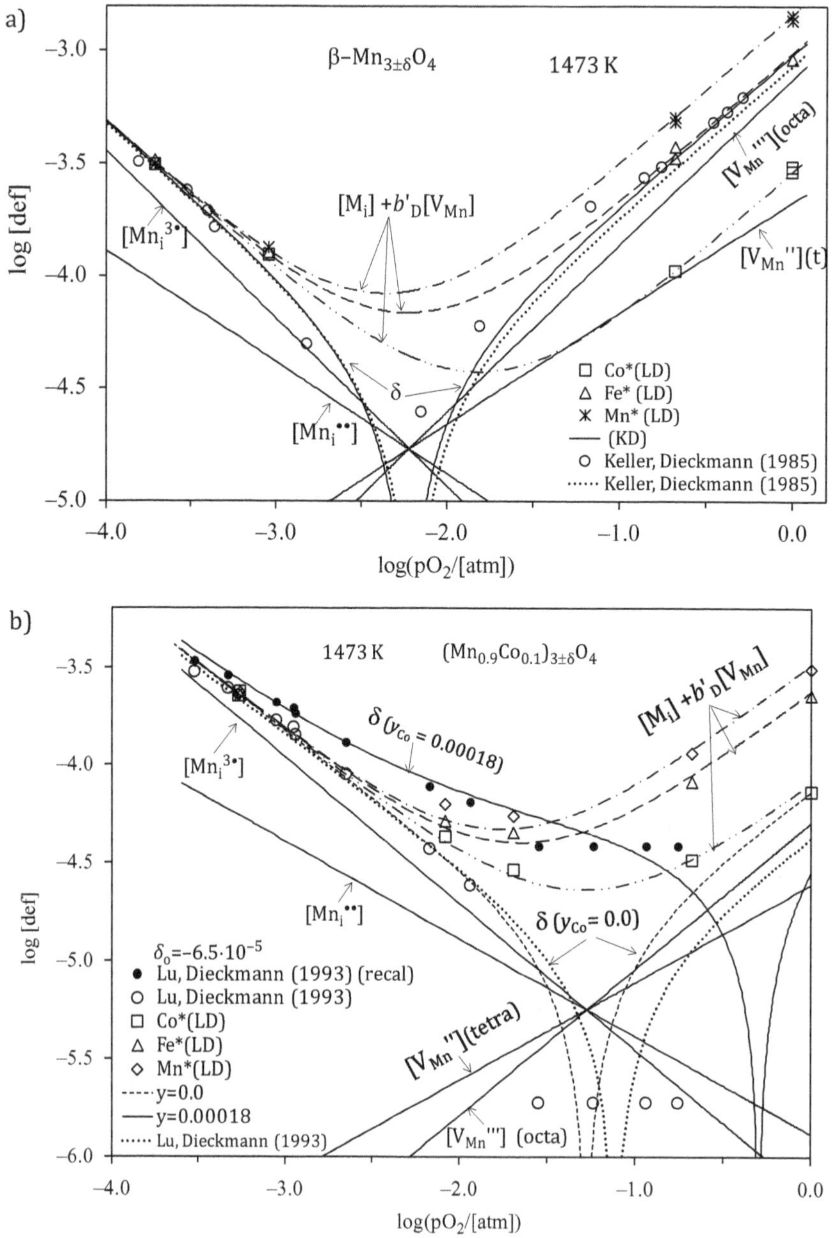

FIGURE 11.1 (a) Diagram of concentrations of point defects for $\beta\text{-}Mn_{3\pm\delta}O_4$ at 1473 K (solid lines), obtained using the values of the deviation from the stoichiometry (Keller and Dieckmann 1985) ((○) points). The (····) line shows the dependence of the deviation δ on p_{O_2} determined using "equilibrium constants" of reactions of formation of cation vacancies K_V and interstitial cations K_I (Keller and Dieckmann 1985). The points (△,□,✳) represent the values of the sum of concentrations of ionic defects $\left(\left[M_i\right]+b'_D\left[V_M\right]\right)$ calculated according to Equation 11.1 for fitted values of $D^o_{I(M)}$ for the lowest oxygen pressures and the values of coefficients of self-diffusion of tracers: Fe* ((△) points), Co* – (□), Mn* – (✳) (Lu and Dieckmann 1993). Lines (---), (-··-), (-···-) denote, respectively, dependence of the sum of concentrations of ionic defects $\left(\left[M_i\right]+b'_D\left[V_{Fe}\right]\right)$ on p_{O_2} obtained for the fitted value of b_D. (b) for the spinel $(Mn_{0.9}Co_{0.1})_{3\pm\delta}O_4$ at 1473 K (solid lines) obtained using the results of the studies of the deviation from the stoichiometry (Lu and Dieckmann 1993) ((○) points) and taking into account cobalt ions Co'_{Mn} incorporated into the sublattice of Mn^{3+} ions in the amount of $y_{Co}' = 1.8\cdot10^{-4}$ mol/mol. Points (●) represent the recalculated values of the deviation from the stoichiometry for $\delta_o = -6.5\cdot10^{-5}$. The (····) line shows the dependence of the deviation from the stoichiometry p_{O_2}, determined using the "equilibrium constants" of reactions of formation of cation vacancies K_V and interstitial ions K_I (Lu and Dieckmann 1993). Line (---) represents the dependence of δ on p_{O_2} for $y_{Co} = 0.0$. (c) for the spinel $(Mn_{0.8}Co_{0.2})_{3\pm\delta}O4$ at 1473 K (solid lines) obtained using the results of the studies of the deviation from the stoichiometry (Lu and Dieckmann 1993) ((○) points) and taking into account cobalt ions Co'_{Mn} incorporated into the sublattice of Mn^{3+} ions in the amount of $y_{Co}' = 1\cdot10^{-4}$ mol/mol. Points (●) represent the recalculated values of the deviation from the stoichiometry relative to the value of $\delta_o = -4.5\cdot10^{-5}$. (d) for $(Mn_{0.667}Co_{0.333})_{3\pm\delta}O_4$ at 1473 K (solid lines) obtained using the values of the deviation from the stoichiometry (Lu and Dieckmann 1993) ((○) points) and taking into account the concentration of cobalt ions Co'_{Mn} incorporated into the sublattice of Mn^{3+} ions in the amount of $y_{Co}' = 8\cdot10^{-5}$ mol/mol. Points (●) represent the recalculated values of the deviation from the stoichiometry relative to the value of $\delta_o = -2.75\cdot10^{-5}$.

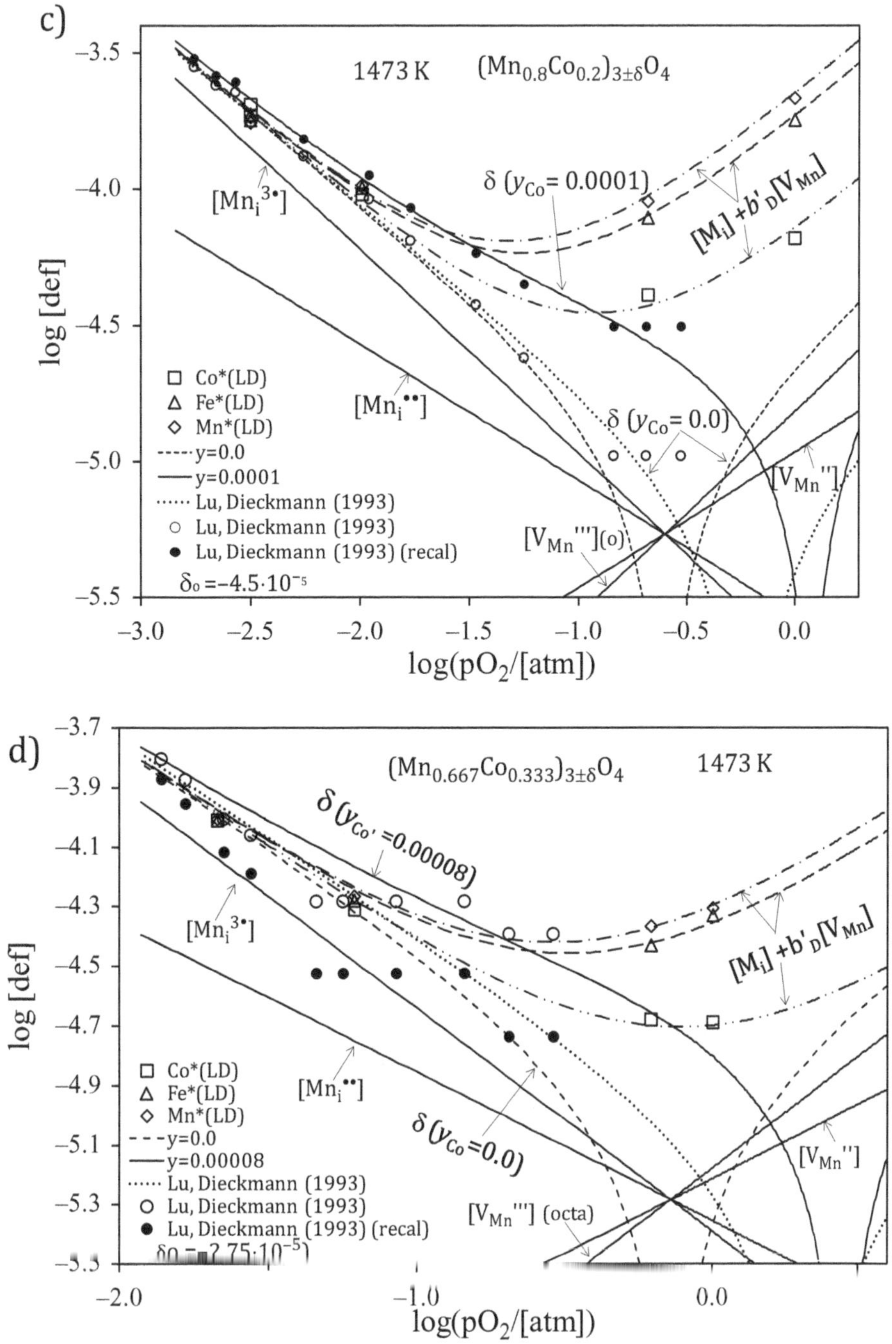

FIGURE 11.1 (Continued)

difficult to unambiguously determine the values of δ_o and the absolute values of the deviation from the stoichiometry, especially near the oxygen pressure > 1 atm, where no changes in the sample mass were observed.

Despite a good match between the calculated dependence of the deviation from the stoichiometry on p_{O_2} and the experimental values δ, it was decided to check if the occurring discrepancies of the values of the deviation δ are related to the incorporation of ions Co^{2+} into the sublattice of Mn^{3+} ions $\left(Co'_{Mn}\right)$, and if the change in the value δ_o (Lu and Dieckmann 1993) causes a change in the dependence of δ on p_{O_2}, specific for the incorporation of these ions. Similarly, as for magnetite doped with M^{2+} ions (see Section 6.2, reaction 6.2). The equation for doping of cobalt ions into the structure of the spinel $(Mn_{1-x}Co_x)_{3\pm\delta}O_4$ will assume the form of:

$$4CoO + V_i \rightarrow Co^{x}_{Mn(2+)(o)} + 2Co'_{Mn(3+)(o)} + 4O^{x}_{O} + Co_i^{\bullet\bullet} \quad (11.1)$$

Figure 11.1b and c presents the dependence of the deviation from the stoichiometry on the oxygen pressure, taking into account the concentration of ions $\left[Co'_{Mn}\right] = 1/2y_{Co'}$ incorporated into the spinel structure according to the reaction 10.1 (solid line). As can be seen in Figure 11.1b and c, there is a good match between the recalculated values of the deviation from the stoichiometry reported in (Lu and Dieckmann 1993) ((●) points) and the dependence of δ on p_{O_2}. As Figure 11.1d shows, for the spinel $(Mn_{1-x}Co_x)_{3\pm\delta}O_4$ with the cobalt content of $x_{Co} = 0.333$ mol, the values of the deviation from the stoichiometry (Lu and Dieckmann 1993) ((○) points) and the character of their dependence on p_{O_2} indicate that ions (Co'_{Mn}) are incorporated into the spinel structure. The solid line in Figure 11.1d presented the results of the calculations for the value of $y_{Co}' = 8 \cdot 10^{-5}$ mol/mol, which, as can be seen, describes the experimental results better than the dependence suggested in (Lu and Dieckmann 1993) (dotted line). The recalculated values of the deviation from the stoichiometry for the value $\delta_o = -2.75 \cdot 10^{-5}$ ((●) points) are close to the dependence obtained when using "equilibrium constants" K_V and K_I (Lu and Dieckmann 1993) ((···) line), as well as to the one determined in the present work for a spinel not doped with Co'_{Mn} ions ($y_{Co} = 0.0$, (---) line).

Figure 11.2 presents the dependence of the concentration of ions $\left[Co'_{Mn}\right] = 1/2y_{Co}'$ on the content of cobalt ions x_{Co} in the spinel $(Mn_{1-x}Co_x)_{3\pm\delta}O_4$.

As can be seen in Figure 11.2, introduction of cobalt in the amount of $x_{Co} = 0.1$ mol causes also a significant incorporation of ions Co'_{Mn} into the spinel structure. A further increase in the cobalt content causes a decrease in the concentration of these ions.

In turn, Figure 11.3 compares the dependences of the deviation from the stoichiometry on the oxygen pressure at 1473 K, presented in Figure 11.1.

As can be seen in Figure 11.3, the oxygen pressure at which there occurs a transition of $Mn_{1-\delta}O$ oxide into a spinel, shifts towards higher oxygen pressures in relation to pure hausmannite. The oxygen pressure at which the spinel reaches the stoichiometric composition is also shifted; the extent of this shift depends on the concentration of ions $\left[Co'_{Mn}\right]$. When the cobalt content in hausmannite increases, the concentration of interstitial cations increases (at a constant oxygen pressure).

Figure 11.4 presents the dependence of the oxygen pressure at which the spinel $(Mn_{1-x}Co_x)_3O_4$ reaches the stoichiometric composition on the cobalt content x_{Co} ((○) points), which, as can be seen, increases when the cobalt content in the spinel increases. The oxygen pressures at which the concentrations of cation vacancies and interstitial ions are equal, $([V_{Mn}] = [Mn_i])$, is lower ((●) points).

In turn, Figure 11.5 presents the dependence of the concentration of Frenkel defects $([V_{Mn}] = [Mn_i])$ at the cobalt content of x_{Co} in the spinel $(Mn_{1-x}Co_x)_{3\pm\delta}O_4$. As can be seen in Figure 11.5, at the cobalt content of $x_{Co} = 0.1$ mol, the concentration of Frenkel defects decreases, and when the content increases further, it is practically unchanged.

Figure 11.6 shows the oxygen pressure dependence of the derivative $(d \log p_{O_2}/\log \delta) = n$ for the calculated dependence of the deviation from the stoichiometry on p_{O_2} shown in Figure 11.1a–d (taking into account the concentration of ions $\left[Co'_{Mn}\right]$). As can be seen in Figure 11.6, at the lowest oxygen pressures the parameter n tends to the value of -1.5 ($1/n = -2/3$). The presence of ions Co'_{Fe} changes the character of the dependence of the derivative n on p_{O_2} at higher oxygen pressures.

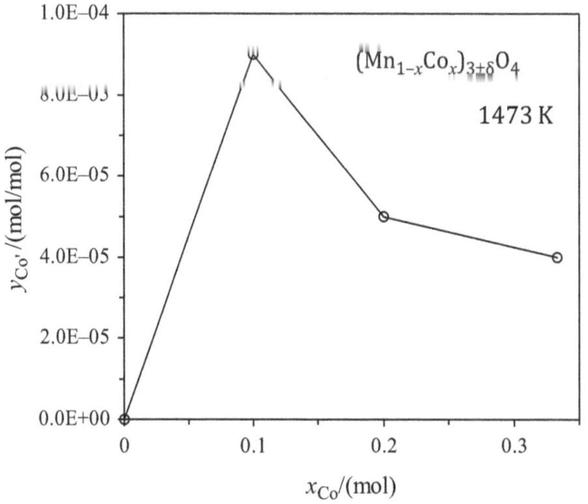

FIGURE 11.2 Dependence of the concentration of cobalt ions $\left[Co'_{Mn}\right] = 1/2y_{Co}'$ incorporated into the sublattice of ions Mn^{3+}, on the cobalt content x_{Co} in the spinel $(Mn_{1-x}Co_x)_{3\pm\delta}O_4$ at 1473 K.

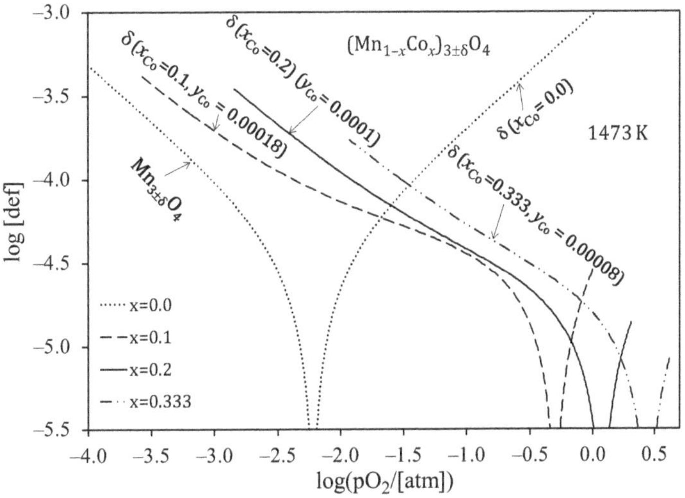

FIGURE 11.3 Oxygen pressure dependences of the deviation from the stoichiometry in the spinel $(Mn_{1-x}Co_x)_{3\pm\delta}O_4$, at 1473 K, at the cobalt content of $x_{Co} = 0.0–0.333$ mol, when taking into account the incorporation of cobalt ions Co'_{Mn} into the sublattice of Mn^{3+} ions (presented in Figure 11.1).

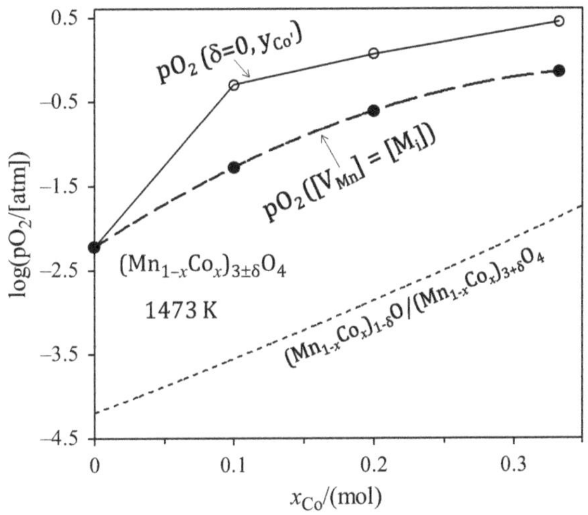

FIGURE 11.4 Dependence of the oxygen pressure at which the spinel $(Mn_{1-x}Co_x)_3O_4$ reaches the stoichiometric composition on the cobalt content x_{Co}, at 1473 K ((○) points), and the oxygen pressure at which the concentrations of cation vacancies and interstitial cations are equal ((●) points). Line (---) denotes oxygen pressures at the phase boundary of doped manganese oxide/spinel (Lu and Dieckmann 1993).

Figures 11.7–11.9 present the dependence of adjusted standard Gibbs energies: ΔG^o_{F1}, ΔG^o_{F2}, $\Delta G^o_{V''_{Mn}}$, ΔG^o_I on the cobalt content x_{Co} of the spinel $(Mn_{1-x}Co_x)_{3\pm\delta}O_4$ at 1473 K. Figures 11.8 and 11.10 show the dependence of the calculated standard Gibbs energies $\Delta G^o_{V'''_{Mn}}$ of formation of vacancies $V'''_{Mn(o)}$ and $\Delta G^o_{Mn_i^{3\bullet}}$ of formation of interstitial iron ions $Mn_i^{3\bullet}$ and $(\Delta G^o_{Mn_i^{\bullet\bullet}})$ of formation of ions $Mn_i^{\bullet\bullet}$.

As can be seen in Figure 11.7, the standard Gibbs energy of formation of Frenkel defects $\left(\Delta G^o_{F1} = \Delta G^o_{F2}\right)$ significantly increases at the cobalt content of $x_{Co} = 0.1$ mol in the spinel $(Mn_{1-x}Co_x)_{3\pm\delta}O_4$ and when the cobalt content increases further, the increase is slight. This results in a decrease in the concentration of Frenkel defects.

As can be seen in Figures 11.8 and 11.10, the values of $(\Delta G^o_{V'''_{Mn}})$ and $(\Delta G^o_{V''_{Mn}})$ monotonically increase when the cobalt content in the spinel increases, the values of $\Delta G^o_{Mn_i^{3\bullet}}$ and $\Delta G^o_{Mn_i^{\bullet\bullet}}$, inversely, decrease linearly.

Figures 11.8 and 11.10 show also the values of $\Delta G^o_{V_{Mn}}$ and $\Delta G^o_{Mn_i}$ calculated using the values of "equilibrium constants" of reaction of formation of cation vacancies K_V and interstitial cations K_I (Keller and Dieckmann 1985, Lu and Dieckmann

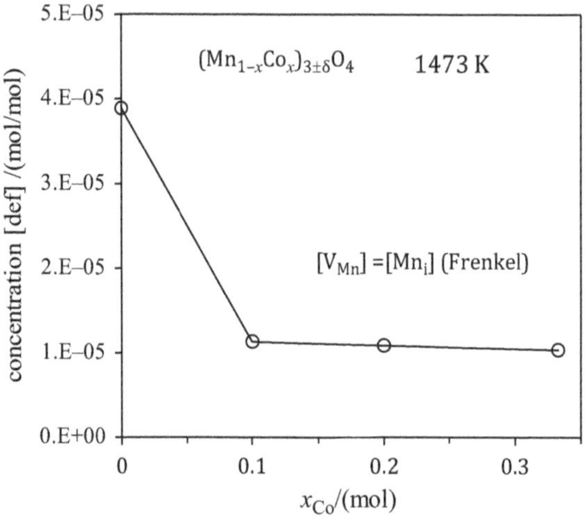

FIGURE 11.5 Dependence of the concentration of Frenkel defects in the spinel $(Mn_{1-x}Co_x)_3O_4$ on the cobalt content x_{Co}, at 1473 K.

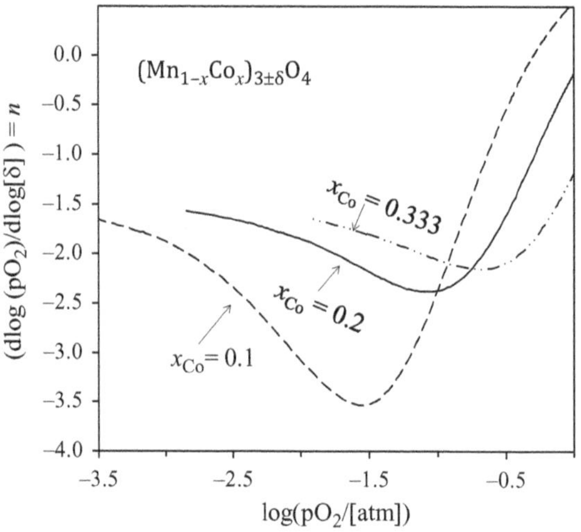

FIGURE 11.6 Dependences of the derivative $n_\delta = d\log(p_{O2})/d\log\delta$ on p_{O_2}, obtained for the dependence of the deviation from the stoichiometry on p_{O_2} in the spinel $(Mn_{1-x}Co_x)_{3\pm\delta}O_4$ with the cobalt content of $x_{Co} = 0.0–0.333$ mol, at 1473 K presented in Figure 11.1.

1993). The values differ from those determined in the present work, because they describe another model of defects, but they have a similar nature in dependence on the cobalt content in hausmannite.

11.3 MOBILITY OF IONS IN COBALT-DOPED HAUSMANNITE

Similarly as for pure magnetite, using the concentration of ionic defects resulting from the diagram of defects' concentrations and the values of coefficients of self-diffusion of tracers Fe*, Co* and Mn* in the spinel $(Mn_{1-x}Co_x)_{3\pm\delta}O_4$, coefficients of diffusion (mobility) of metal ions via cation vacancies ($D^o_{V(M)}$ and via interstitial cations $D^o_{I(M)}$ were determined (see: Section 1.6.2, Equation 1.46a, 1.46b). Due to the fact that in the spinel $(Mn_{1-x}Co_x)_{3\pm\delta}O_4$, interstitial cations dominate, by transforming Equation 1.46a we obtain:

$$\left[M_i\right] + b'_D\left[V_{Mn}\right] = \frac{D^*_M}{D^o_{I(M)}} \tag{11.2}$$

where the parameter b'_D is the ratio of the coefficient of diffusion of ions via cation vacancies $D^o_{V(M)}$ to the coefficient of diffusion via interstitial cations $(D^o_{V(M)}/D^o_{I(M)})$.

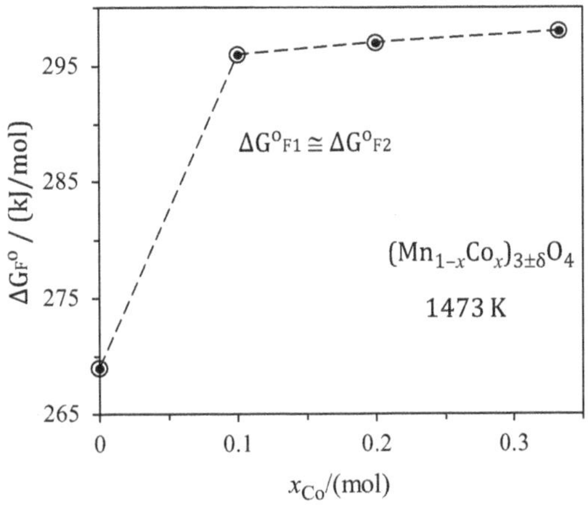

FIGURE 11.7 Dependence of standard Gibbs energy $\Delta G^o_{F_1}$ ((○) points) and $\Delta G^o_{F_2}$ – (●) of formation of Frenkel defects with charge two and three on the cobalt content x_{Co} in the spinel $(Mn_{1-x}Co_x)_{3\pm\delta}O_4$, at 1473 K, obtained using the results of the studies of the deviation from the stoichiometry (Keller and Dieckmann 1985, Lu and Dieckmann 1993).

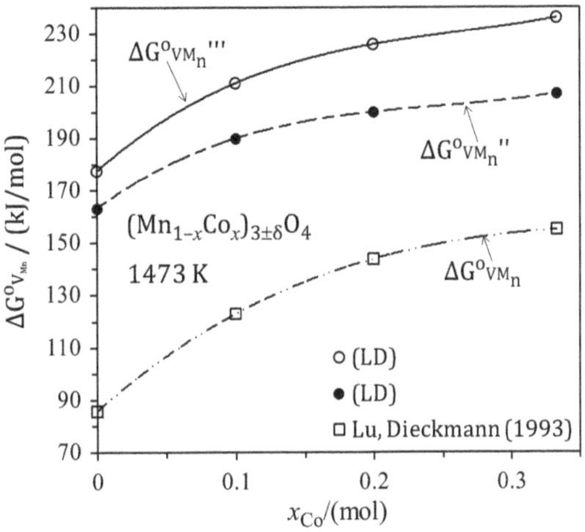

FIGURE 11.8 Dependence of standard Gibbs energy $\Delta G^o_{V_{Fe}'''}$ ((○) points) and $\Delta G^o_{V_{Fe}''}$ – (●) of formation of vacancies $V'''_{Fe(o)}$ and $V''_{Fe(o)}$ on the cobalt content x_{Co} in the spinel $(Mn_{1-x}Co_x)_{3\pm\delta}O_4$, at 1473 K and the values $\Delta G^o_{V_{Fe}}$ calculated using the values of "equilibrium constants" of reaction of formation of vacancies K_V (Lu and Dieckmann 1993) ((□) points).

Figure 11.11a–d shows the dependence of the "sum" of concentrations of ionic defects $\left([Mn_i] + b'_D[V_{Mn}]\right)$ on p_{O_2} for the fitted parameters b'_D and the values of this sum, calculated according to Equation (11.1), using the values of coefficients of self-diffusion of cobalt ^{60}Co ((□) points), manganese ^{54}Mn (◇) and iron ^{59}Fe (△) (Lu and Dieckmann 1993). For the highest values of the deviation from the stoichiometry, at the lowest oxygen pressures, the values of the coefficients of diffusion of ions via interstitial cations $\left(D^o_{I(M)}\right)$ and of the parameter b'_D were fitted is such a way as to obtain a match of the dependence of the sum of concentrations of ionic defects $\left([Mn_i] + b'_D[V_{Mn}]\right)$ on p_{O_2} with the values calculated using the coefficients of self-diffusion of tracers. As can be seen in Figure 11.1a–d, a good agreement was obtained between the dependence of the sum of the concentrations of defects $\left([Mn_i] + b'_D[V_{Mn}]\right)$ on p_{O_2} for the fitted value of b'_D and the values calculated using the coefficients of self-diffusion of tracers in the spinel $(Mn_{1-x}Co_x)_{3\pm\delta}O_4$.

Figure 11.11 presents the dependence of the parameter b'_D on the cobalt content x_{Co} in the spinel $(Mn_{1-x}Co_x)_{3\pm\delta}O_4$ for tracers: Fe* ((△) points) and Co* (□), Mn* (○). As can be seen in Figure 11.11, the mobility of tracer ions via cation vacancies is larger than the mobility via interstitial cations (apart from the mobility of cobalt in pure hausmannite). The ratio of the coefficients of diffusion $(D^o_{V(M)}/D^o_{I(M)})$ in the range of x_{Co} = 0.0–0.2 mol increases when the cobalt content in

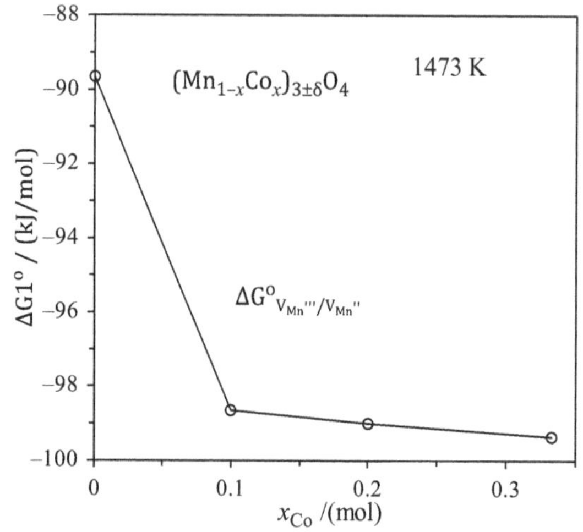

FIGURE 11.9 Dependence of standard Gibbs energy ΔG_I^o on the cobalt content x_{Co} in the spinel $(Mn_{1-x}Co_x)_{3\pm\delta}O_4$ at 1473 K, permitting determination of the ratio of concentrations of manganese vacancies $\left[V_{Mn}'''\right]/\left[V_{Mn}''\right]$ (according to Equation 1.35a).

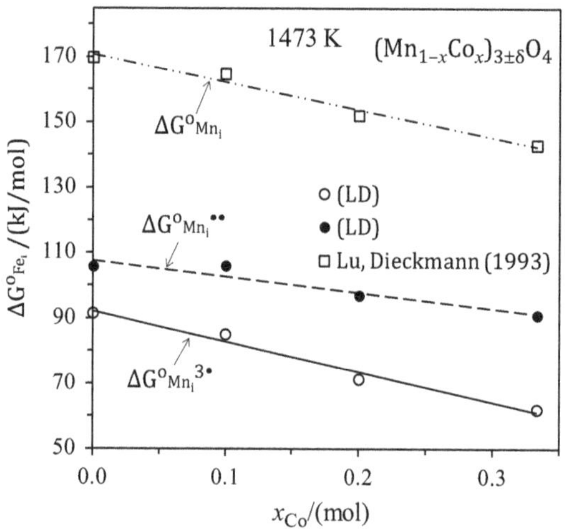

FIGURE 11.10 Dependence of standard Gibbs energy $\Delta G^o_{Mn_i^{3\bullet}}$ ((○) points) and $\Delta G^o_{Mn_i^{\bullet\bullet}}$ - (●) of formation of interstitial cations $Me_i^{3\bullet}$ and $Me_i^{\bullet\bullet}$ on the cobalt content x_{Co} in the spinel $(Mn_{1-x}Co_x)_{3\pm\delta}O_4$, at 1473 K and the values $\Delta G^o_{Fe_i}$ calculated using the values of "equilibrium constants" of reaction of formation of interstitial cations K_I (Lu and Dieckmann 1993) ((□) points).

the spinel increases, and when the cobalt content increases further, it decreases. The highest values of the parameter b_D' are present in the case of diffusion of manganese and iron, and the lowest ones – for cobalt.

Figure 11.12 presents the dependence of the values of coefficients of diffusion of ions via interstitial ions $D^o_{I(M)}$ and via cation vacancies $D^o_{V(M)}$ on the cobalt content x_{Co} in the spinel $(Mn_{1-x}Co_x)_{3\pm\delta}O_4$. According to the values of the parameter b_D', the coefficients of diffusion via interstitial cations are lower than the coefficients of diffusion via vacancies. Therefore, despite the fact that in the range of existence of the spinel, interstitial cations dominate, the mobility via cation vacancies is higher.

As can be seen in Figure 11.2, the differences between the coefficients of diffusion $D^o_{V(M)}$ for the diffusion of Co and Mn are small, but for iron they are much smaller. Larger differences in coefficients $D^o_{I(M)}$ are present for the diffusion via interstitial cations. In the case of diffusion via interstitial cations, at the cobalt content of $x_{Co} = 0.1$ mol, a decrease in the value of $D^o_{I(M)}$ occurs, and then, when the cobalt content increases, the mobilities of tracer ions increase in a linear way. In the case of diffusion via cation vacancies, up to the cobalt content of $x_{Co} = 0.2$ mol, $D^o_{V(M)}$ increase, and at the cobalt content of $x_{Co} = 0.333$ mol they decrease. The highest mobility via defects in the spinel $(Mn_{1-x}Co_x)_{3\pm\delta}O_4$ is the mobility of cobalt, for manganese it is lower, and the lowest mobility is that of iron.

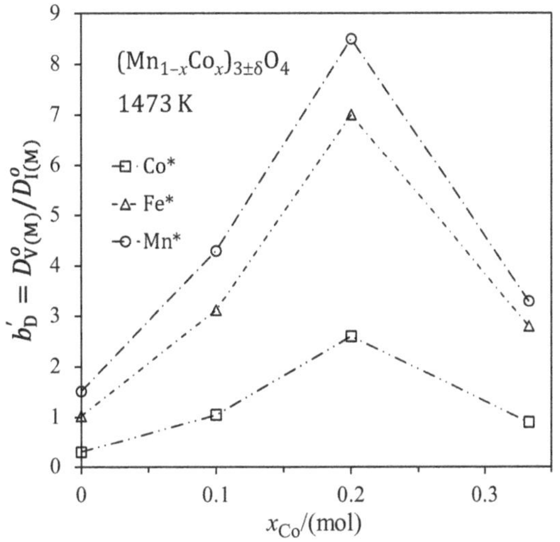

FIGURE 11.11 Dependence of the ratio of the coefficient of diffusion of ions via cation vacancies to the coefficient of diffusion via interstitial cations $D^o_{V(M)}/D^o_{I(M)} = b'_D$ on the cobalt content x_{Co} in the spinel $(Mn_{1-x}Co_x)_{3\pm\delta}O_4$ at 1473 K, obtained using the of coefficients of self-diffusion of tracers: Fe* ((△) points), Co* – (□), Mn* – (○) (Lu and Dieckmann 1993).

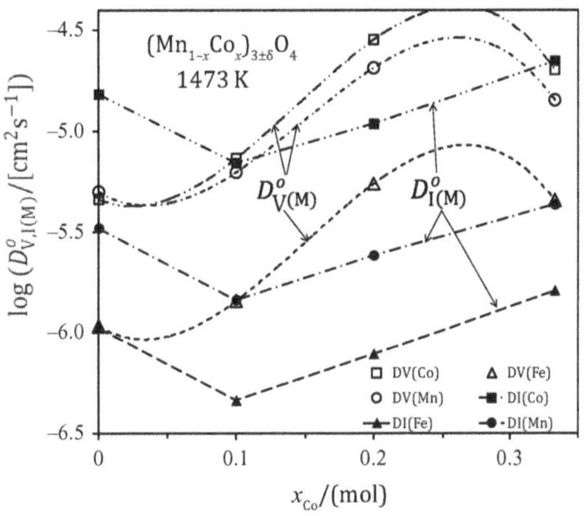

FIGURE 11.12 Dependence of the coefficient of diffusion of tracer ions via cation vacancies $D^o_{V(M)}$ and via interstitial cations $D^o_{I(M)}$ on the cobalt content x_{Co} in the spinel $(Mn_{1-x}Co_x)_{3\pm\delta}O_4$ at 1473 K, obtained using the of coefficients of self-diffusion of tracers: Fe* ((△) points), Co* – (□), Mn* – (○) (Lu and Dieckmann 1993), concentrations of ionic defects and the fitted parameter b'_D.

11.4 CONCLUSIONS

Using the results of the studies of the deviation from the stoichiometry on the oxygen pressure in the spinel $(Mn_{1-x}Co_x)_{3\pm\delta}O_4$ with the cobalt content of $x_{Co} = 0.1–0.333$ mol (Lu and Dieckmann 1993), diagrams of concentrations of defects at 1473 K were determined. It was found that the cobalt doping in hausmannite causes a shift towards higher oxygen pressures in the range of existence of spinel. Spinels $(Mn_{1-x}Co_x)_3O_4$, depending on the cobalt content, reach the stoichiometric composition near, and likely above, the pressure of 1 atm.

In the range of low oxygen pressures, interstitial cations $\left[Mn_i^{3\bullet}\right]$ dominate, while cation vacancies $\left[V_{Mn}'''\right]$ dominate at oxygen pressures above 1 atm. A correction of the value of the deviation from the stoichiometry by fitting the reference value δ_o indicates that the character of the dependence of δ on p_{O_2} can be an effect of the incorporation into the structure of the spinel $(Mn_{1-x}Co_x)_{3\pm\delta}O_4$ of Co^{2+} ions, into the sublattice of Mn^{3+} ions (Co'_{Mn}), which are negatively charged in relation to the lattice. The concentration of these ions, for the content of $x_{Co} = 0.1$ mol reaches the value of $y_{Co}' = 9{\cdot}10^{-5}$ mol/mol and when the cobalt content increases, it decreases.

It was found that when the cobalt content in the spinel $(Mn_{1-x}Co_x)_{3\pm\delta}O_4$ increases, standard Gibbs energies of formation of Frenkel defects increase, which results in a decrease in the concentration of Frenkel defects $(\left[V_{Mn}'''\right] = \left[Mn_i^{3\bullet}\right])$ from about $4{\cdot}10^{-5}$ to $1{\cdot}10^{-5}$ mol/mol at the cobalt content of $x_{Co} = 0.1$ mol and when the cobalt content increases further, they are unchanged. Standard Gibbs energies of formation of cation vacancies $\Delta G^{o}_{V_{Mn}'''}$ and $\Delta G^{o}_{V_{Mn}''}$ increase when the content of cobalt in the spinel $(Mn_{1-x}Co_x)_{3\pm\delta}O_4$ increases, while $\Delta G^{o}_{Mn_i^{3\bullet}}$ and $\Delta G^{o}_{Mn_i^{\bullet}}$ of formation of interstitial cations decrease. The above values depend on the value of ΔG^{o}_{eh} of formation of electronic defects and they were determined for the value $\Delta G^{o}_{eh} = 60$ kJ/mol, the same as the value assumed for pure hausmannite.

Using the determined concentration of ionic defects and the values of coefficients of self-diffusion of tracers M* (^{59}Fe, ^{60}Co and ^{543}Mn) (Lu and Dieckmann 1993), the ratio of the coefficient of diffusion (mobility) of tracer ions via vacancies to the coefficient of diffusion via interstitial cations $\left(D^{o}_{V(M)}/D^{o}_{I(M)} = b'_D\right)$ was determined. It was found that when the cobalt content in the spinel $(Mn_{1-x}Co_x)_{3-\delta}O_4$ increases, the mobility via cation vacancies is higher than the mobility via interstitial cations and at the cobalt content of $x_{Co} = 0.2$ mol the parameter b'_D reaches the highest values. At the temperature of 1473 K, for the diffusion of cobalt ions, the parameter b'_D reaches the value $b'_D \approx 3$, for iron 7, and for manganese 9.

In the spinel $(Mn_{1-x}Co_x)_{3\pm\delta}O_4$, similarly as in pure hausmannite, the mobilities of tracer ions depend on ion type. When the cobalt content increases, the mobility of ions via cation vacancies $\left(D^{o}_{V(M)}\right)$ increases. For the diffusion of ions via interstitial cations, the coefficient $\left(D^{o}_{I(M)}\right)$ at the cobalt content of $x_{Co} = 0.1$ mol is lower than in hausmannite, and then it increases in a linear way. The highest mobility is that of cobalt ions, it is lower for manganese ions and the lowest mobility is that of iron ions.

REFERENCES

Aukrust, E., and A. Muan. 1964. Thermodynamic Properties of Solid Solutions with Spinel-Type Structure: I. The System Co_3O_4-Mn_3O_4. *Trans. Metall. Soc. AIME.* 230: 378–382.

Keller, M., and R. Dieckmann. 1985. Defect Structure and Transport Properties of Manganese Oxides: (II) The Nonstoichiometry of Hausmannite ($Mn_{3-\delta}O_4$). *Ber. Bunsenges. Phys. Chem.* 89: 1095–1104.

Lu, F. H., and R. Dieckmann. 1993. Point Defects and Cation Tracer Diffusion in $(Co_xMn_{1-x})_{3-\delta}O_4$ Spinels. *Solid State Ionics.* 67: 145–155.

Subramanian, R., R. Dieckmann, G. Eriksson, and A. D. Pelton. 1994. Model Calculations of Phase Stabilities Of Oxide Solid Solutions in the Co-Fe-Mn-O System at 1200°C. *J. Phys. Chem. Solids.* 55: 391–404.

12 Iron-Doped Hausmannite – $(Mn_{1-x}Fe_x)_{3\pm\delta}O_4$

12.1 INTRODUCTION

In Chapter 7, manganese-doped magnetite was discussed; in a limit case, it forms the spinel $MnFe_2O_4$. At higher manganese content the spinels should be treated as iron-doped hausmannite $(Mn_{1-x}Fe_x)_{3\pm\delta}O_4$, which at the limit concentration reaches the composition of $FeMn_2O_4$. Many works discussed the range of existence of the spinel $(Mn_{1-x}Fe_x)_{3\pm\delta}O_4$; mainly, the pressures at the phase boundary of doped oxide/$(Mn_{1-x}Fe_x)_{3\pm\delta}O_4$ were determined (Muan and Somiya 1962, Bergstein 1963, Bergstein and Kleinert 1964, Komarov et al. 1965, Tretyakov et al. 1965, Schwerdtfeger and Muan 1967, Wickham 1969, Bulgakova and Rozanov 1970, Roethe et al. 1970, Ono et al. 1971, Duquesnoy, et al. 1975, Terayama et al. 1983, Franke 1987, Franke and Dieckmann 1989, Franke and Dieckmann 1990, Subramanian and Dieckmann 1993). The most complete range of existence of iron-doped hausmannite was studied at 1273 K and 1473 K (Franke and Dieckmann 1989, Franke and Dieckmann 1990, Subramanian and Dieckmann 1993). The analysis of the studies of the Mn–Fe–O system, the enthalpy of spinel formation and the distribution of ions in sublattices of the spinel were also reported (Kjellqvist and Selleby 2010).

The deviation from the stoichiometry as a function of the oxygen pressure and diffusion of iron, ^{59}Fe, in the spinel $(Mn_{1-x}Fe_x)_{3\pm\delta}O_4$, at 1473 K was studied (Franke and Dieckmann 1989, Lu and Dieckmann 1995). It was demonstrated that spinels $(Mn_{1-x}Fe_x)_{3\pm\delta}O_4$ with the iron content of $x_{Fe} = 0.1–0.4$ mol, reach the stoichiometric composition in the range of their existence. Therefore, at higher oxygen pressures, cation vacancies dominate, while at low pressures interstitial cations dominate. The studies on the diffusion of iron showed that the coefficients of self-diffusion are different in the range where cation vacancies and interstitial cations dominate, and their change is fully consistent with the model of ionic defects' structure.

12.2 METHODS AND RESULTS OF CALCULATION OF THE DIAGRAMS OF THE CONCENTRATIONS OF DEFECTS, AND DISCUSSION

Iron-doped hausmannite can be obtained from solid phase reaction of appropriate amounts of oxides MnO, Mn_2O_3 and Fe_2O_3. Iron ions incorporate into the sites with mainly octahedral coordination, substituting ions Mn^{3+} and at the limit concentration, the composition $FeMn_2O_4$ is reached. At higher temperatures, due to the setting of the equilibrium state, there occurs a slight exchange of ions between sublattices of the spinel structure. As already mentioned, the defects' structure and the character of the dependence of the deviation from the stoichiometry on the oxygen pressure are similar to those for pure hausmannite (Franke and Dieckmann 1989, Lu and Dieckmann 1995). Due to that, the calculations of diagrams of the concentrations of point defects in the spinel $(Mn_{1-x}Fe_x)_{3\pm\delta}O_4$ were conducted with a method analogous to the one used for pure hausmannite (see Section 2.2), using the values of "equilibrium constants" of formation of cation vacancies K_V and interstitial ions K_I, given in (Lu and Dieckmann 1995).

Due to the lack of the value of ΔG^o_{eh} of formation of electronic defects, it was assumed that it only slightly depends on the concentration of doping ions in hausmannite. Owing to that, in the calculations, a constant value of $\Delta G^o_{eh} = 60$ kJ/mol was assumed, the same as for pure hausmannite. In the calculations of diagrams of concentrations of defects, it was assumed that standard enthalpies of formation of Frenkel defects with the charge of 2 and 3 are identical $\Delta G^o_{F_1} = \Delta G^o_{F_2}$. Thus, at the stoichiometric composition, the concentrations of cation vacancies are equal $\left[V'''_{Mn(o)}\right] = \left[V''_{Mn(t)}\right]$.

Figure 12.1a–d present the diagrams of the concentrations of ionic defects for spinels $(Mn_{1-x}Fe_x)_{3\pm\delta}O_4$ with the iron content of $x_{Fe} = 0.0–0.4$ mol, at 1473 K (solid lines), calculated using the "equilibrium constants" of formation of cation vacancies K_V and interstitial ions K_I (Lu and Dieckmann 1995). The dotted line presents the dependence of δ on p_{O_2} obtained using the values of "equilibrium constants" K_V and K_I (Lu and Dieckmann 1995).

As can be seen in Figure 12.1, a very good agreement was obtained between the dependence of the deviation from the stoichiometry and the dependence resulting from the values of "equilibrium constants" K_V and K_I. Dominating defects are cation vacancies $V'''_{Mn(o)}$ and interstitial cations $Mn_i^{3\bullet}$. Concentrations of vacancies $\left[V''_{Mn}\right]$ and interstitial ions $\left[Mn_i^{\bullet\bullet}\right]$ are much lower and do not affect the character of the dependence of δ on p_{O_2}.

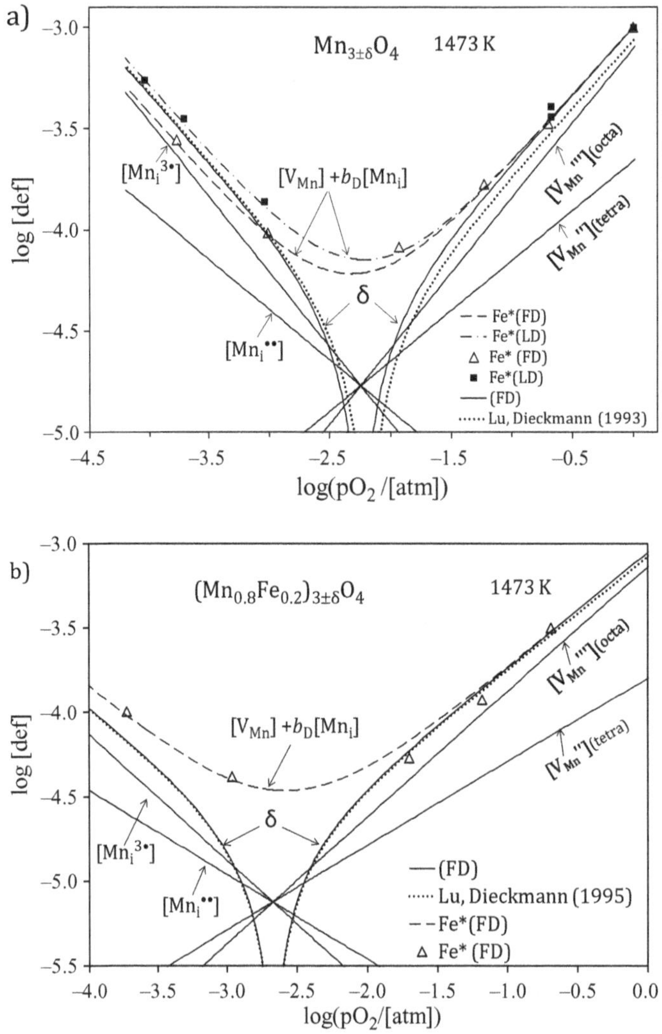

FIGURE 12.1 (a) Diagram of concentrations of point defects for β-$Mn_{3\pm\delta}O_4$ at 1473 K (solid lines). The (····) line shows the dependence of the deviation from the stoichiometry on p_{O_2}, determined using the "equilibrium constants" of reactions of formation of vacancies K_V and interstitial cations K_I (Keller and Dieckmann 1985). The points represent the sum of concentrations of ionic defects ($[V_{Mn}] + b_D[Mn_i]$) calculated according to Equation 1.47a, 1.47b for the fitted value of $D^o_{V(Fe)}$ for the highest oxygen pressures, and using the coefficients of self-diffusion of iron (^{59}Fe) (Franke and Dieckmann 1989) ((△)(FD) points) and (Lu and Dieckmann 1993) – (■) (LD). Lines (-··-) and (---) denote the dependence of the sum of concentrations of defects ($[V_{Fe}] + b_D[M_i]$) on p_{O_2} obtained for the fitted value of b_D. (b) for the spinel $(Mn_{0.8}Fe_{0.2})_{3\pm\delta}O_4$ at 1473 K (solid lines), determined using the "equilibrium constants" of reactions of formation of vacancies K_V and interstitial cations K_I (Lu and Dieckmann 1995). (c) for the spinel $(Mn_{0.664}Fe_{0.333})_{3\pm\delta}O_4$ at 1473 K (solid lines). (d) for the spinel $(Mn_{0.6}Fe_{0.4})_{3\pm\delta}O_4$ at 1473 K (solid lines).

Figure 12.2 compares the dependences of the deviation from the stoichiometry on the oxygen pressure obtained in the spinel $(Mn_{1-x}Fe_x)_{3\pm\delta}O_4$ at 1473 K, presented in Figure 12.1.

As can be seen in Figure 12.2, iron doping in hausmannite causes a shift in the range of existence of the spinel towards lower oxygen pressures. Simultaneously, in the range of higher oxygen pressures, the deviation from the stoichiometry (concentration of cation vacancies) is practically unchanged. In the spinel doped with iron of $x_{Fe} = 0.4$ mol, the concentration of vacancies increases. In turn, at lower oxygen pressures, when the iron content increases, the deviation from the stoichiometry decreases, thus, the concentration of interstitial cations decreases. When the iron content increases, a shift

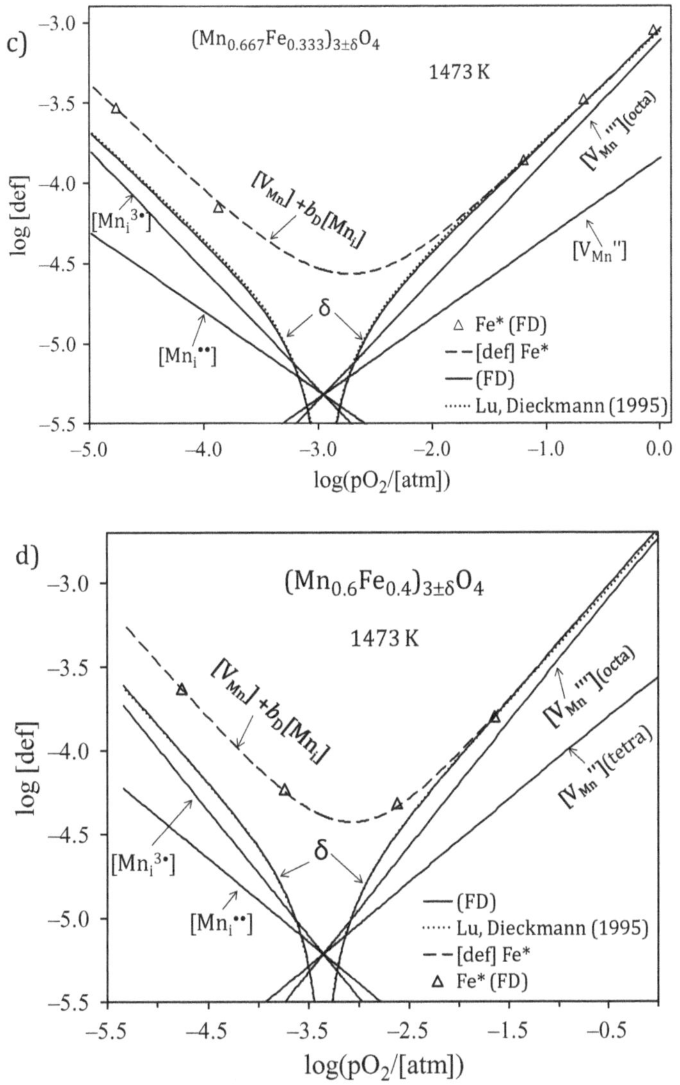

FIGURE 12.1 (Continued)

towards lower oxygen pressures occurs for the point at which the spinels $(Mn_{1-x}Fe_x)_3O_4$ reach the stoichiometric composition.

Figure 12.3 presents the dependence of the oxygen pressure at which the spinel $(Mn_{1-x}Fe_x)_3O_4$ reaches the stoichiometric composition on the iron content x_{Fe}.

As can be seen in Figure 12.3, the oxygen pressures at which the stoichiometric composition is present decrease when the iron content in the spinel increases, and the dependence only slightly deviates from linear relation. Also, the equilibrium oxygen pressure at the phase boundary of doped manganese oxide/spinel decreases. This is a result of an increase in bonding energy due to the incorporation of iron into nodes of manganese ions.

In turn, Figure 12.4 presents the dependence of the concentration of Frenkel defects, which decreases when the iron content x_{Fe} increases, which is a result of an increase in the value of $\Delta G^{o}_{F_1}$ when the dopant content increases.

Figure 12.5 compares the dependence of the derivative $(d\log p_{O_2}/d\log\delta) = n$ on the oxygen pressure, for the dependences of the deviation from the stoichiometry on p_{O_2} shown in Figure 12.1. As can be seen in Figure 12.5, at the highest oxygen pressures, the derivative reaches the value $n = 1.5$. At the lowest oxygen pressures the parameter n tends towards the value of -1.5 $(1/n = -2/3)$.

Figures 12.6–12.9 present the dependence of adjusted standard Gibbs energies: $\Delta G^{o}_{F_1}$, $\Delta G^{o}_{F_2}$, $\Delta G^{o}_{V_{Mn}''}$, ΔG^{o}_{I} on the iron content x_{Fe} in the spinel $(Mn_{1-x}Fe_x)_{3\pm\delta}O_4$ at 1473 K. Figures 12.7 and 12.9 show the dependence of standard Gibbs energy $\Delta G^{o}_{V_{Mn}''}$ of formation of vacancies V_{Mn}'' (calculated according to Equation 1.35a) and $\Delta G^{o}_{Mn_i^{3\bullet}}$ of the formation of interstitial

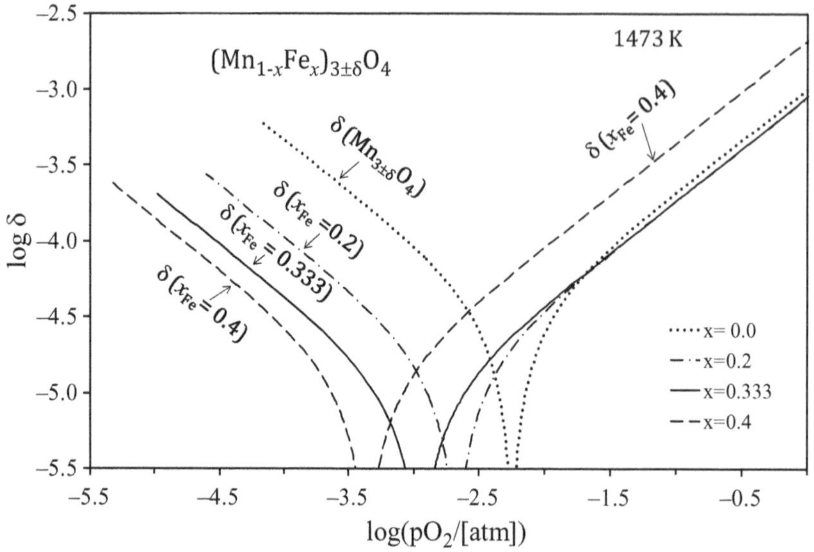

FIGURE 12.2 Dependences of the deviation from the stoichiometry on the oxygen pressure in the spinel $(Mn_{1-x}Fe_x)_{3\pm\delta}O_4$, with the iron content $x_{Fe} = 0.0$–0.4 mol at 1473 K, presented in Figure 12.1.

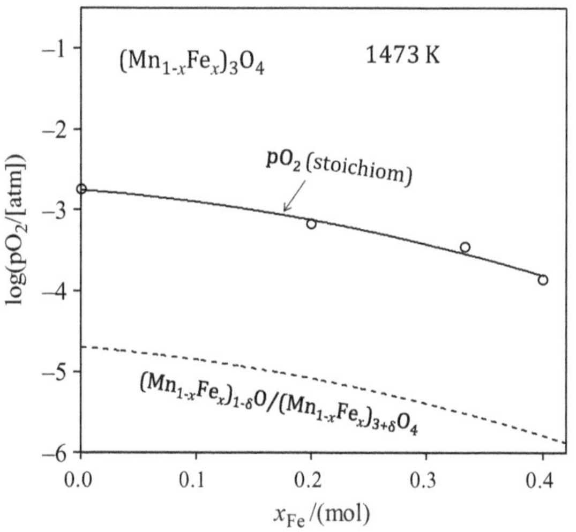

FIGURE 12.3 Dependence of the oxygen pressure at which the spinel $(Mn_{1-x}Fe_x)_3O_4$ reaches the stoichiometric composition on the iron content x_{Fe}, at 1473 K ((O) points). Dashed line denotes oxygen pressures at the phase boundary of iron-doped manganese oxide/spinel (Subramanian and Dieckmann 1993).

iron ions $Mn_i^{3\bullet}$(calculated according to Equation 1.34c) and $\Delta G^o_{Mn_i^{\bullet\bullet}}$ of formation of ions $Mn_i^{\bullet\bullet}$ (calculated according to Equation 1.36c).

As can be seen in Figure 12.6, standard enthalpy of formation of Frenkel defects $\Delta G^o_{F_I}$ increases when the iron content in the spinel $(Mn_{1-x}Fe_x)_{3\pm\delta}O_4$ increases. The result is a decrease in the concentration of Frenkel defects at the stoichiometric composition.

As can be seen in Figures 12.7 and 12.9, the values $\Delta G^o_{V_{Mn}'''}$ practically do not depend on the iron content in the spinel and the values of $\Delta G^o_{V_{Mn}''}$, increase only slightly. Also, the values of $\Delta G^o_{Mn_i^{3\bullet}}$ and $\Delta G^o_{Mn_i^{\bullet\bullet}}$ increase when the iron content in the spinel increases. As can be seen, the values of ΔG_i^o of defects' formation for the iron content of $x_{Fe} = 0.4$ mol, thus, after reaching the composition $FeMn_2O_4$, their behaviour deviates from that for lower iron contents.

Figures 12.7 and 12.9 show also the values of $\Delta G^o_{V_{Mn}}$ and $\Delta G^o_{Mn_i}$ calculated using "equilibrium constants" of reaction of formation of cation vacancies K_V and interstitial ions K_I, compared in the work (Lu and Dieckmann 1995). The values differ from those determined in the present work, because they describe another model of defects, but they have a similar character of the dependence on the iron content in the spinel $Mn_{1-x}Fe_x)_{3\pm\delta}O_4$.

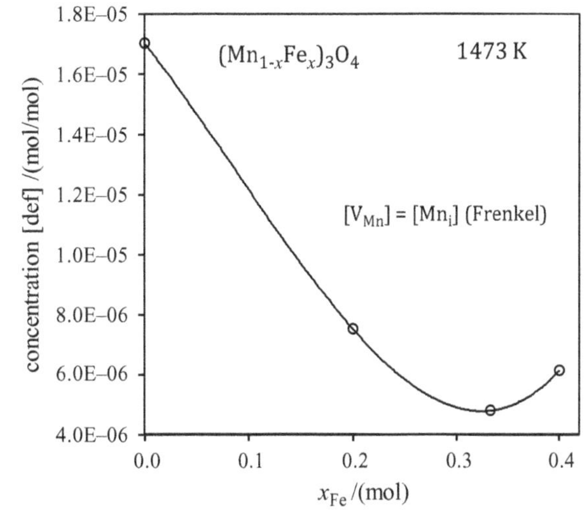

FIGURE 12.4 Dependence of the concentration of Frenkel defects on the iron content x_{Fe} in the spinel $(Mn_{1-x}Fe_x)_3O_4$ at 1473 K.

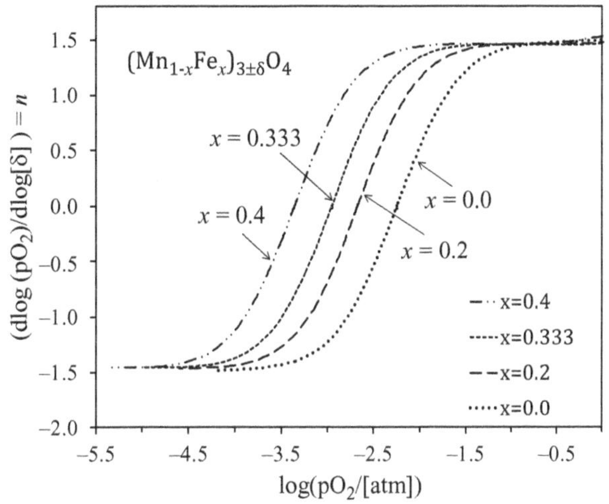

FIGURE 12.5 Dependence of the derivative: $n_\delta = \mathrm{d}\log(p_{O_2})/\mathrm{d}\log\delta$ on the oxygen pressure, obtained for the dependence of the deviation from the stoichiometry on p_{O_2} in the spinel $(Mn_{1-x}Fe_x)_{3\pm\delta}O_4$ with the iron content x_{Fe} = 0.0–0.4 mol, at 1473 K (presented in Figure 12.1).

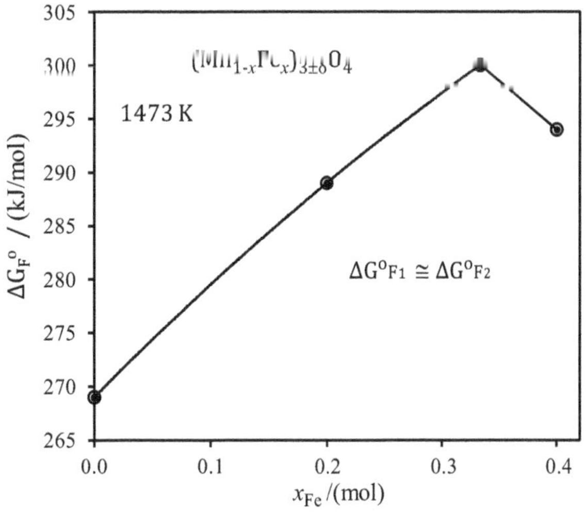

FIGURE 12.6 Dependence of standard Gibbs energy $\Delta G^{o}_{F_1}$ ((○) points) and $\Delta G^{o}_{F_2}$ (●) of formation of Frenkel defects in the spinel $(Mn_{1-x}Fe_x)_{3\pm\delta}O_4$ on the iron content x_{Fe}, at 1473 K.

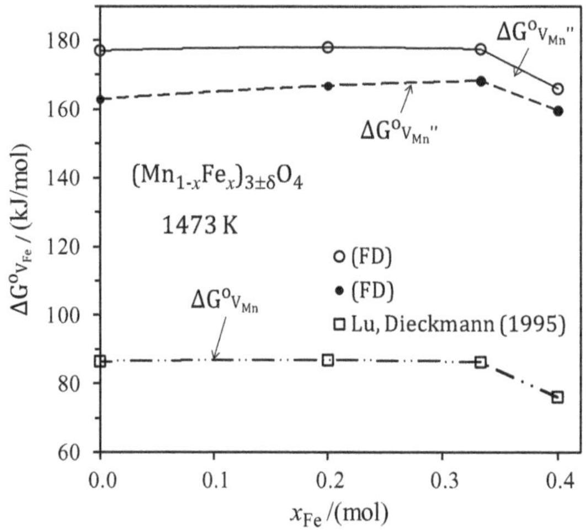

FIGURE 12.7 Dependence of standard Gibbs energy $\Delta G^o_{V'''_{Mn}}$ ((○) points) and $\Delta G^o_{V''_{Mn}}$ ((●) points) of formation of cation vacancies $V'''_{Mn(o)}$ and $V''_{Mn(o)}$ on the iron content x_{Fe} in the spinel $(Mn_{1-x}Fe_x)_{3\pm\delta}O_4$ at 1473 K and the values $\Delta G^o_{V_{Mn}}$ calculated using the "equilibrium constants" of reaction of formation of cation vacancies K_V (Lu and Dieckmann 1995) ((□) points).

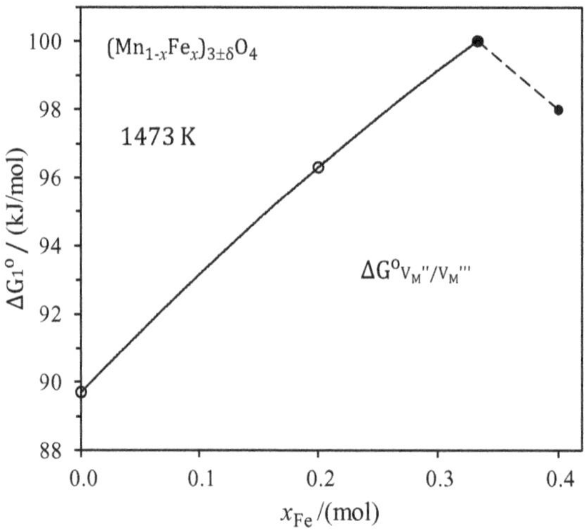

FIGURE 12.8 Dependence of standard Gibbs energy ΔG^o_I on the iron content x_{Fe} in the spinel $(Mn_{1-x}Fe_x)_{3\pm\delta}O_4$ at 1473 K, permitting determination of the ratio of concentrations of manganese vacancies $\left[V''_{Mn}\right]/\left[V'''\right]$, according to Equation 2.3.

12.3 DIFFUSION IN IRON-DOPED HAUSMANNITE

Similarly as for pure magnetite, using the concentration of ionic defects resulting from the diagram of defects' concentrations and the values of coefficients of diffusion of iron ^{59}Fe in the spinel $Mn_{1-x}Fe_x)_{3\pm\delta}O_4$ coefficients of diffusion of iron ions (mobility) via cation vacancies and via interstitial cations were determined (see Section 1.6.2).

Figure 12.1a–d show the dependence of the "sum" of concentrations of ionic defects ($[V_{Mn}] + b_D[Mn_i]$) on the oxygen pressure for the fitted values of the parameter b_D and the values of this sum, calculated according to Equation 1.47a, 1.47b using the values of the coefficients diffusion of iron ^{59}Fe (Franke and Dieckmann 1989, Lu and Dieckmann 1993) and the fitted value of the coefficient of diffusion of iron $D^o_{V(Fe)}$ via cations vacancies.

As can be seen in Figure 12.1a–d, a good agreement was obtained between the oxygen pressure dependence of the sum ($[V_{Mn}] + b_D[Mn_i]$), and the calculated values using the coefficients of self-diffusion of iron ((△) points). For pure hausmannite, the results taken from (Lu and Dieckmann 1993) were used ((■) points). Figure 12.10 presents the dependence of the ratio of mobility of iron ions via interstitial cations to the mobility via cation vacancies ($D^o_{I(Fe)}/D^o_{V(Fe)} = b_D$) on the iron content x_{Fe} in the spinel $(Mn_{1-x}Fe_x)_{3\pm\delta}O_4$.

As can be seen in Figure 12.10, the mobility of iron ions via interstitial cations is higher than the mobility via cation vacancies. The parameter b_D increases when the iron content increases, up to $b_D = 2.4$ at the iron content of $x_{Fe} = 0.4$ mol.

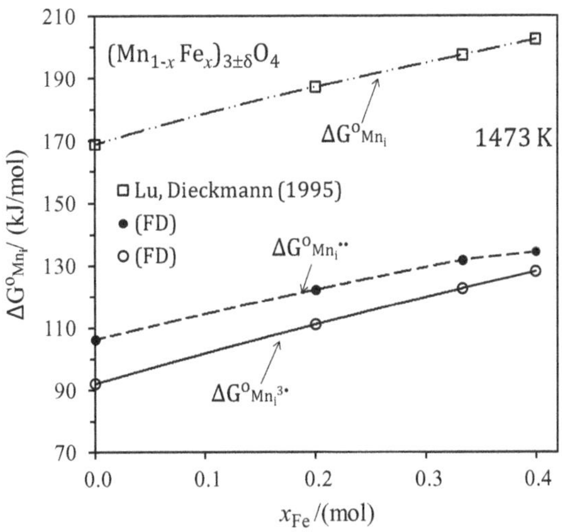

FIGURE 12.9 Dependence of standard Gibbs energy $\Delta G^o_{M_i^{3•}}$ ((○) points) and $\Delta G^o_{M_i^{••}}$ ((●) points) of formation of interstitial cations $M_i^{3•}$ and $M_i^{••}$, calculated according to Equation 2.3b and 2.3c on the iron content x_{Fe} in the spinel $(Mn_{1-x}Fe_x)_{3\pm\delta}O_4$, at 1473 K and the values $\Delta G^o_{M_i}$ calculated using the "equilibrium constants" of reaction of formation of interstitial cations K_I (Lu and Dieckmann 1995) ((□) points).

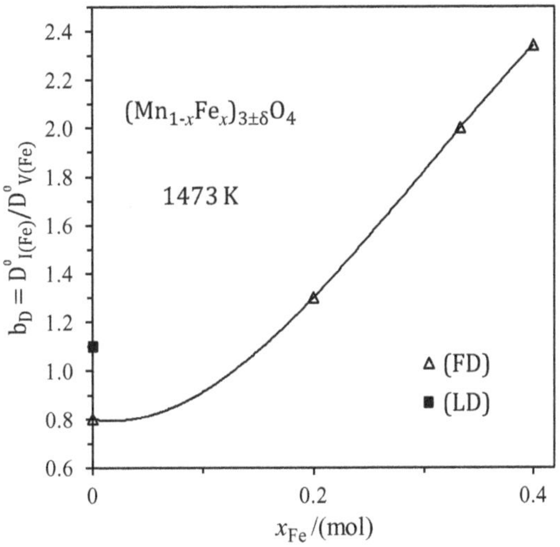

FIGURE 12.10 Dependence of the ratio of the coefficient of diffusion of iron ions via interstitial cations to the coefficient of diffusion via cation vacancies $D^o_{I(Fe)}/D^o_{V(Fe)} = b_D$ on the iron content x_{Fe} in the spinel $(Mn_{1-x}Fe_x)_{3\pm\delta}O_4$ at 1473 K, obtained using the of coefficients of self-diffusion of iron Fe* (Franke and Dieckmann 1989) – ((△)(FD) points) and (Lu and Dieckmann 1993) – (■)(LD) and concentrations of ionic defects.

In turn, Figure 12.11 presents the dependence of the mobilities of iron ions via cation vacancies $D^o_{V(Fe)}$ and via interstitial ions $D^o_{I(Fe)}$ on the iron content in the spinel $(Mn_{1-x}Fe_x)_{3\pm\delta}O_4$.

As can be seen in Figure 12.11, the mobilities of iron ions in the spinel $(Mn_{1-x}Fe_x)_{3\pm\delta}O_4$ via cation vacancies and via interstitial cations increase when the iron content increases. They decrease for the iron content $x_{Fe} = 0.4$ mol.

12.4 CONCLUSIONS

Using the results of the studies of deviation from the stoichiometry for the spinel $(Mn_{1-x}Fe_x)_{3\pm\delta}O_4$ with the iron content $x_{Fe} = 0.1–0.4$ mol (Lu and Dieckmann 1995), diagrams of concentrations of defects at 1473 K were determined. It was found that iron doping of hausmannite shifts the range of existence of the spinel $(Mn_{1-x}Fe_x)_{3\pm\delta}O_4$ towards low oxygen pressures. It does not affect the character of the dependence of the deviation from the stoichiometry on the oxygen pressure. In the range of its existence, the spinel $(Mn_{1-x}Fe_x)_{3\pm\delta}O_4$ reaches the stoichiometric composition, while the oxygen pressure at

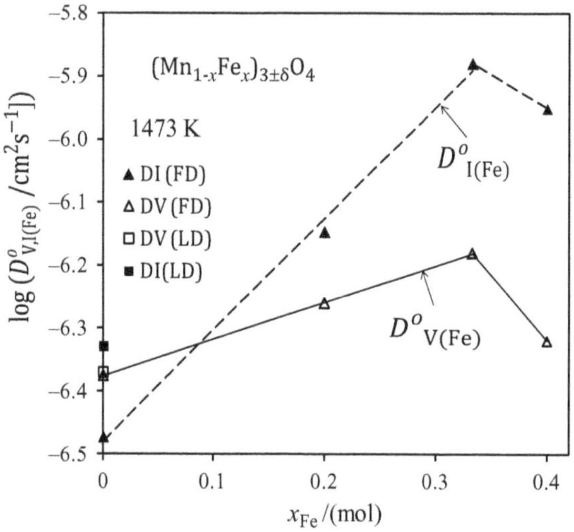

FIGURE 12.11 Dependence of the coefficient of diffusion of iron ions via cations vacancies $D^o_{V(Fe)}$ and via interstitial cations $D^o_{I(Fe)}$ on the iron content x_{Fe} in the spinel $(Mn_{1-x}Fe_x)_{3\pm\delta}O_4$ at 1473 K, obtained using the of coefficients of self-diffusion of iron Fe* (Franke and Dieckmann 1989) – ((△,▲)(FD)) points and (Lu and Dieckmann 1993) – (□,■)(LD), concentrations of ionic defects and the fitted value of b_D.

which it is reached, shifts, when the iron content increases, towards lower pressures, by about one order of magnitude at the iron content of $x_{Fe} = 0.4$ mol.

Iron ions in the spinel $(Mn_{1-x}Fe_x)_{3\pm\delta}O_4$ practically do not affect the point defect structure. In the range of higher oxygen pressures, cation vacancies $\left[V'''_{Mn}\right]$ dominate; their concentration, up to the iron content of $x_{Fe} = 0.333$ mol, does not change (it is the same as in hausmannite). An increase in the deviation from the stoichiometry (concentration of cation vacancies) was found only at the iron content of $x_{Fe} = 0.4$ mol. In the range of low oxygen pressures, interstitial cations $\left[Mn_i^{3\bullet}\right]$ dominate; their concentration at the same oxygen pressure decreases when the iron content increases. The limit concentration of interstitial cations, at which there occurs a transformation of spinel into doped manganese oxide are close, but lower than in pure hausmannite.

It was found that when the iron content in the spinel $(Mn_{1-x}Fe_x)_{3\pm\delta}O_4$ increases, the standard Gibbs energies of formation of Frenkel defects ΔG^o_F increase up to the iron content of $x_{Fe} = 0.333$ mol. At the content of $x_{Fe} = 0.4$ mol, the value of ΔG^o_F decreases slightly. This causes a decrease in the concentration of Frenkel defects at the stoichiometric composition from the value of $4{\cdot}10^{-5}$–$5{\cdot}10^{-6}$ mol/mol when the iron content increases.

The standard Gibbs energies of formation of cation vacancies $\Delta G^o_{V''_{Fe}}$ and $\Delta G^o_{V'''_{Fe}}$ slightly increase when the iron content in the spinel $(Mn_{1-x}Fe_x)_{3\pm\delta}O_4$ increases, and at the content of $x_{Fe} = 0.4$ mol they decrease. The increase in the values of $\Delta G^o_{Fe_i^{3\bullet}}$ and $\Delta G^o_{Fe_i^{\bullet\bullet}}$ of formation of interstitial oxygen cations, when the iron content in the spinel increases, is larger. The above values depend on the value of ΔG^o_{eh} of formation of electronic defects and they were determined for the value ΔG^o_{eh} = 60 kJ/mol, the same as the value assumed for pure hausmannite.

Using the determined concentration of ionic defects and the values of coefficients of self-diffusion of iron ions ^{59}Fe (Franke and Dieckmann 1989), the ratio of the coefficients of diffusion (mobility) of iron ions via interstitial cations to the coefficient of diffusion via vacancies ($D^o_{I(Fe)}/D^o_{V(Fe)} = b_D$) was determined. It was found that at the iron content $x_{Fe} = 0.2$–0.4 mol in the spinel $(Mn_{1-x}Fe_x)_{3\pm\delta}O_4$, the mobility via interstitial cations is higher than the mobility via cation vacancies and it increases to the value of $b_D \approx 2.5$ 12 at the iron content of $x_{Fe} = 0.4$ mol.

An increase in the iron content in the spinel $(Mn_{1-x}Fe_x)_{3\pm\delta}O_4$ causes an increase in the mobility of iron ions via interstitial cations $D^o_{I(Fe)}$, as well as, to a smaller extent, the mobility via cation vacancies $D^o_{V(Fe)}$.

REFERENCES

Bergstein, A. 1963. Manganese Magnesium Ferrites. IV. Oxygen Nonstoichiometry of Spinels $Mn_xFe_{3-x}O_{4+\gamma}$. *Collect. Czech. Chem. Commun.* 28: 2381–2386.

Bergstein, A., and P. Kleinert. 1964. Partial Phase Diagram of the System $Mn_xFe_{3-x}O_y$. *Collect. Czech. Chem. Commun.* 29: 2549–2551.

Bulgakova, T. I., and A. G. Rozanov. 1970. Phase Equilibria in the Ferrite Region of the Mn-Fe-O System. *Russ. J. Phys. Chem.* 44: 385–388.

Duquesnoy, A., J. Couzin, and P. Gode. 1975. Isothermal Representation of Ternary Phase Diagrams ABO. Study of the System Mn-Fe-O. *C. R. Acad. Sci. (Paris) Sér. C.* 281: 107–109.

Franke, P. 1987. Disorder and Transport in Mixed Oxides of the Types (Fe,Mn) O and $(Fe,Mn)_3O_4$. PhD Thesis, University of Hannover, Germnay.

Franke, P., and R. Dieckmann. 1989. Defect Structure and Transport Properties of Mixed Iron-Manganese Oxides. *Solid State Ionics.* 32–33: 817–823.

Franke, P., and R. Dieckmann. 1990. Thermodynamics of Iron Manganese Mixed Oxides at High Temperatures. *J. Phys. Chem. Solids.* 51: 49–57.

Keller, M., and R. Dieckmann. 1985. Defect Structure and Transport Properties of Manganese Oxides: (II) The Nonstoichiometry of Hausmannite $(Mn_{3-\delta}O_4)$. *Ber. Bunsenges. Phys. Chem.* 89: 1095–1104.

Kjellqvist, L., and M. Selleby. 2010. Thermodynamic Assessment of the Fe-Mn-O System. *J. Phase Equilib. Diff.* 31: 113–134.

Komarov, V. F., N. N. Oleinikov, Yu. G. Saksonov, and Yu. D. Tretyakov. 1965. *Inorg. Mater.* 1: 365–382.

Lu, F. H., and R. Dieckmann. 1993. Point Defects and Cation Tracer Diffusion in $(Co_xMn_{1-x})_{3-\delta}O_4$ Spinels. *Solid State Ionics.* 67: 145–155.

Lu, F. H., and R. Dieckmann. 1995. Point Defects in Oxide Spinel Solid Solutions of the Type $(Co,Fe,Mn)_{3-\delta}O_4$ at 1200 °C. *J. Phys. Chem. Solids.* 56: 725–733.

Muan, A., and S. Somiya. 1962. The System Iron Oxide-Manganese Oxide in Air. *Am. J. Sci.* 260: 230–240.

Ono, K., T. Ueda, T. Ozaki, Y. Ueda, A. Yamaguchi, and J. Moriyama. 1971. Thermodynamic Study of the Iron-Manganese-Oxygen System. *J. Japan Inst. Metals.* 35: 757–763.

Roethe, A., K. P. Roethe, and G. H. Jerschkewitz. 1970. Gleichgewichtsmessungen an oxidischen Mehrstoffsystemen. II Gleichgewichtsmessungen an einer Me_3O_4-Phase im quaternären System Fe-Mn-Zn-O. *Z. Anorg. Allg. Chem.* 378: 14–21.

Schwerdtfeger, K., and A. Muan. 1967. Equilibria in System Fe-Mn-O Involving (Fe, Mn)O and $(Fe, Mn)_3O_4$ Solid Solutions. *Trans. Metall. Soc. AIME.* 239: 1114–1119.

Subramanian, R., and R. Dieckmann. 1993. Nonstoichiometry and Thermodynamics of the Solid Solution $(Fe, Mn)_{1-\Delta}O$ at 1200°C. *J. Phys. Chem. Solids.* 54: 991–1000.

Terayama, K., M. Ikeda, and M. Taniguchi. 1983. Phase Equilibria in the Mn-Fe-O System in CO_2–H_2 Mixtures. *Trans. Jap. Inst. Metals.* 24: 514–517.

Tretyakov, Yu. D., Yu. G. Saksonov, and I. V. Gordeer. 1965. Phase Diagram of the System Fe_3O_4-Mn_3O_4-MnO-FeO at 1000 °C and Thermodynamic Properties of the Coexisting Phases. *Inorg. Mater.* 1: 382.

Wickham, D. G. 1969. The Chemical Composition of Spinels in the System Fe_3O_4-Mn_3O_4. *J. Inorg. Nucl. Chem.* 31: 313–320.

13 Hausmannite Doped with Iron and Cobalt – $(Co_xFe_zMn_{2z})_{3\pm\delta}O_4$

13.1 INTRODUCTION

In Chapter 8, we discussed the structure of point defects in the spinel $MnFe_2O_4$ doped with cobalt, $(Co_xMn_zFe_{2z})_{3\pm\delta}O_4$. An important issue is also an analogue of the above spinel $(Co_xFe_zMn_{2z})_{3\pm\delta}O_4$ (where $(x + 3z = 1)$) and the effect of cobalt ions on the structure of point defects in the spinel $FeMn_2O_4$ (or $(Fe_{0.333}Mn_{0.667})_{3\pm\delta}O_4$). The Co–Fe–Mn–O system was studied at 1473 K and in the range of existence of spinel phase (Aukrust and Muan 1964, Subramanian and Dieckmann 1992, Subramanian and Dieckmann 1993, Subramanian et al. 1994). Extensive studies of the deviation from the stoichiometry at 1473 K and the diffusion of radioisotope elements: cobalt ^{60}Co, iron ^{59}Fe, manganese ^{54}Mn in the spinel $(Co_xFe_zMn_{2z})_{3\pm\delta}O_4$ (where $(x + 3z = 1)$) with the cobalt content of $x_{Co} = 0.0–0.333$ mol as a function of the oxygen pressure were reported (Lu and Dieckmann 1993). They demonstrated that spinels $(Co_xFe_zMn_{2z})_{3\pm\delta}O_4$ reach the stoichiometric composition in the range of their existence. Therefore, at low oxygen pressures interstitial cations dominate, while in a narrow range of higher oxygen pressures, cation vacancies dominate. The studies on the diffusion of tracer ions showed, similarly as for hausmannite, that the coefficients of diffusion depend on the type of metal, and they are different in the range where cation vacancies and interstitial cations dominate; their change is fully consistent with the model of ionic defects' structure.

13.2 METHODS, RESULTS OF CALCULATION OF THE DIAGRAMS OF THE CONCENTRATIONS OF DEFECTS, AND DISCUSSION

For $(Co_xFe_zMn_{2z})_{3\pm\delta}O_4$, where $x + 3z = 1$, the original spinel is the spinel $(FeMn_2)_{1\pm\Delta}O_4$ $(Fe_{0.333}Mn_{0.667})_{3\pm\delta}O_4$) (Lu and Dieckmann 1995). It should be noted that incorporation of a defined quantity of cobalt ions into the spinel $(Co_xFe_zMn_{2z})_{3\pm\delta}O_4$ causes a decrease in the concentration of manganese ions below 2 mol and iron ions below 1 mol, while the ratio of iron ions to manganese ions is preserved, [Fe]/[Mn] = 0.5. Doping of the spinel $FeMn_2O_4$ with cobalt ions Co^{2+} will cause their incorporation into nodes of Mn^{2+} ions with tetrahedral coordination. At higher temperatures, an equilibrium state and a determined distribution of ions in the sublattices of the spinel is set, which, however, should not affect significantly the structure of point defects. This was confirmed by the studies of the deviation from the stoichiometry (Lu and Dieckmann 1993) that showed the character of the dependence of δ on p_{O_2} analogous to that for pure hausmannite or hausmannite doped with iron or cobalt. Due to that, the calculations of diagrams of the concentrations of point defects for spinels $(Co_xFe_zMn_{2z})_{3\pm\delta}O_4$ were conducted with a method analogous to the one used for cobalt-doped hausmannite (see Section 11.2), using the data on the deviation from the stoichiometry at 1473 K (Lu and Dieckmann 1993).

Owing to the lack of the value of ΔG^{o}_{eh} of formation of electronic defects, it was assumed that it only slightly depends on the concentration of doping ions. Due to that, in the calculations, a constant value of $\Delta G^{o}_{eh} = 60$ kJ/mol was assumed, the same as for pure hausmannite. It was assumed that $\Delta G^{o}_{F_2} = \Delta G^{o}_{F_1}$ of formation of Frenkel defects are identical, thus, at the stoichiometric composition, the concentrations of vacancies $\left[V''_{Mn}\right] \cong \left[V'''_{Mn(o)}\right]$ are similar.

Figure 13.1a–d show diagrams of the concentrations of ionic defects for the spinels $(Co_xFe_zMn_{2z})_{3\pm\delta}O_4$, with the cobalt content of $x_{Co} = 0.0, 0.1, 0.2, 0.333$ mol at 1473 K (solid lines), calculated using the results of the deviation from the stoichiometry (Lu and Dieckmann 1993, Lu and Dieckmann 1995). The dotted lines (···) show the dependence of δ on p_{O_2} obtained using "equilibrium constants" of reactions of formation of cation vacancies K_V and interstitial ions K_I (Lu and Dieckmann 1993, Lu and Dieckmann 1995).

As can be seen in Figure 13.1b, in the spinel $(Co_xFe_zMn_{2z})_{3\pm\delta}O_4$, with the cobalt content of $x_{Co} = 0.1$ mol, a good agreement was obtained between the dependence of the deviation from the stoichiometry and the experimental results δ. Dominating defects are cation vacancies V'''_{Mn} and interstitial cations $Mn_i^{3\bullet}$, and the concentration of V''_{Mn} and $Mn_i^{\bullet\bullet}$ is lower. However, it was decided to check if a change in the reference value δ_o taken from Lu and Dieckmann (1993) causes a change in the dependence of δ on p_{O_2} to the one characteristic of the incorporation of cobalt ions in the sublattice of Mn^{3+} (Co'_{Mn}) ions, negative in relation to the lattice (analogous to the doping of the hausmannite with Co^{2+} ions, see reaction 10.1).

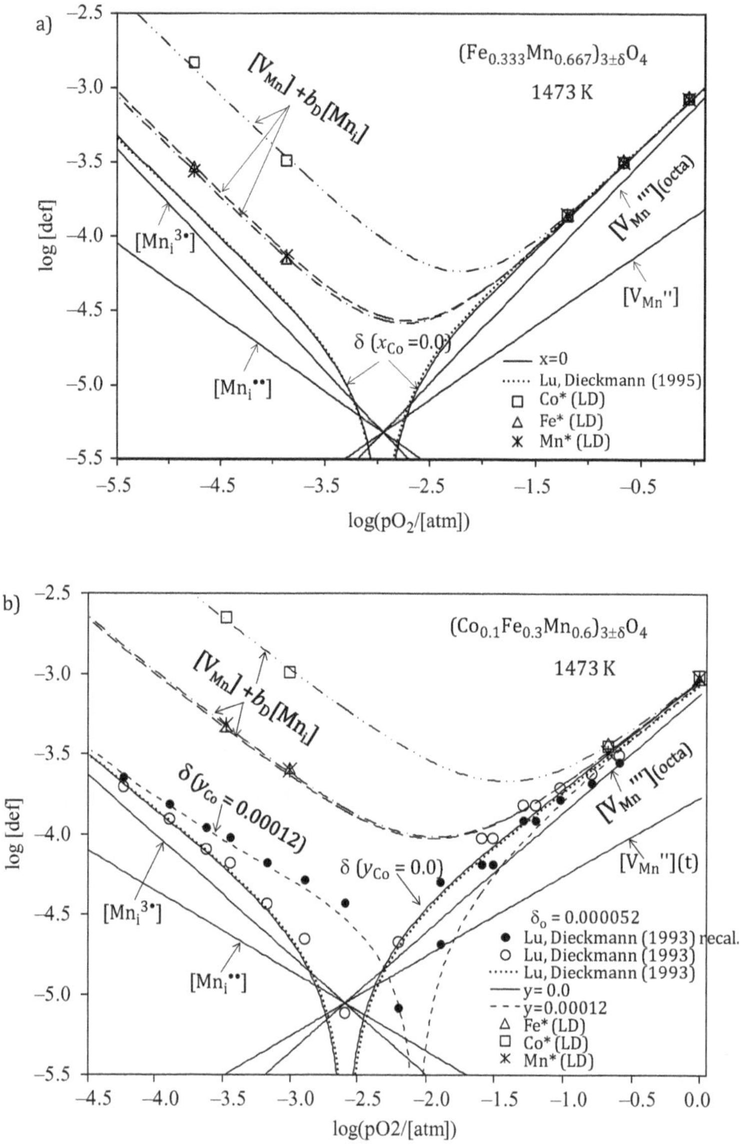

FIGURE 13.1 (a) Diagram of concentrations of point defects for the spinel $(Fe_{0.333}Mn_{0.667})_{3\pm\delta}O_4$ at 1473 K (solid line). Line (····) represents the dependence of the deviation from the stoichiometry on the oxygen pressure, determined using the values of "equilibrium constants" of reactions of formation of cation vacancies K_V and interstitial cations K_I (Lu and Dieckmann 1995). The points represent the values of the sum of concentrations of ionic defects ($[V_{M}] + b_D[M_i]$) calculated according to Equation 1.17a, 1.47b for fitted values of $D_{V(M)}$ for the highest oxygen pressures and the values of coefficients of self-diffusion of tracers: Fe* ((△) points), Co* (□), Mn* (✻) (Lu and Dieckmann 1993). Lines (---), (-·-), (-······-) denote, respectively, dependence of the sum of concentrations of ionic defects ($[V_{MnFe}] + \boldsymbol{b}_D[M_i]$) on p_{O_2} for the fitted value of $\boldsymbol{b}_D$. (b) for the spinel $(Co_{0.1}Fe_{0.3}Mn_{0.6})_{3\pm\delta}O_4$ at 1473 K (solid lines), obtained using the deviation from the stoichiometry data (Lu and Dieckmann 1993) ((○) points). The (···) line shows the oxygen pressure dependence of the deviation from the stoichiometry, calculated using the "equilibrium constants" K_V and K_I of formation of defects (Lu and Dieckmann 1995). Points (●) show the recalculated values of the deviation from the stoichiometry δ (Lu and Dieckmann 1993) relative to the value $\delta_o = 5.2 \cdot 10^{-5}$, and line (---) represents the dependence of δ on p_{O_2} when taking into account cobalt ions Co'_{Mn} incorporated into the sublattice of ions Mn^{3+} in the amount of $y_{Co}' = 1.2 \cdot 10^{-4}$ mol/mol. (c) for the spinel $(Co_{0.2}Fe_{0.266}Mn_{0.534})_{3\pm\delta}O_4$, at 1473 K, when taking into account cobalt ions Co'_{Mn} incorporated into the sublattice of Mn^{3+} ions in the amount of $y_{Co}' = 3.2 \cdot 10^{-4}$ mol/mol (solid lines). Points (●) show the recalculated values of the deviation from the stoichiometry δ (Lu and Dieckmann 1993) relative to the value $\delta_o = 9.9 \cdot 10^{-5}$. Line (---) (a) represents the dependence of the deviation from the stoichiometry on p_{O_2} obtained using the values of δ at the lowest oxygen pressures (Lu and Dieckmann 1993) ((○) points) (without taking into account Co'_{Mn} ions) and line (---) (b) represents the dependence of δ on p_{O_2} consistent with the values of the deviation δ at limit oxygen pressures. (d) for the spinel $(Co_{0.333}Fe_{0.222}Mn_{0.445})_{3\pm\delta}O_4$, at 1473 K (solid lines), when taking into account cobalt ions Co'_{Mn} incorporated into the sublattice of Mn^{3+} ions in the amount of $y_{Co}' = 6 \cdot 10^{-5}$ mol/mol. Points (●) show the recalculated values of the deviation from the stoichiometry δ (Lu and Dieckmann 1993) relative to the value $\delta_o = 5.6 \cdot 10^{-5}$. Line (---) represents the dependence of δ on p_{O_2} obtained using the values δ at the lowest oxygen pressures (Lu and Dieckmann 1993) ((○)) (without taking into account ions Co'_{Mn}.

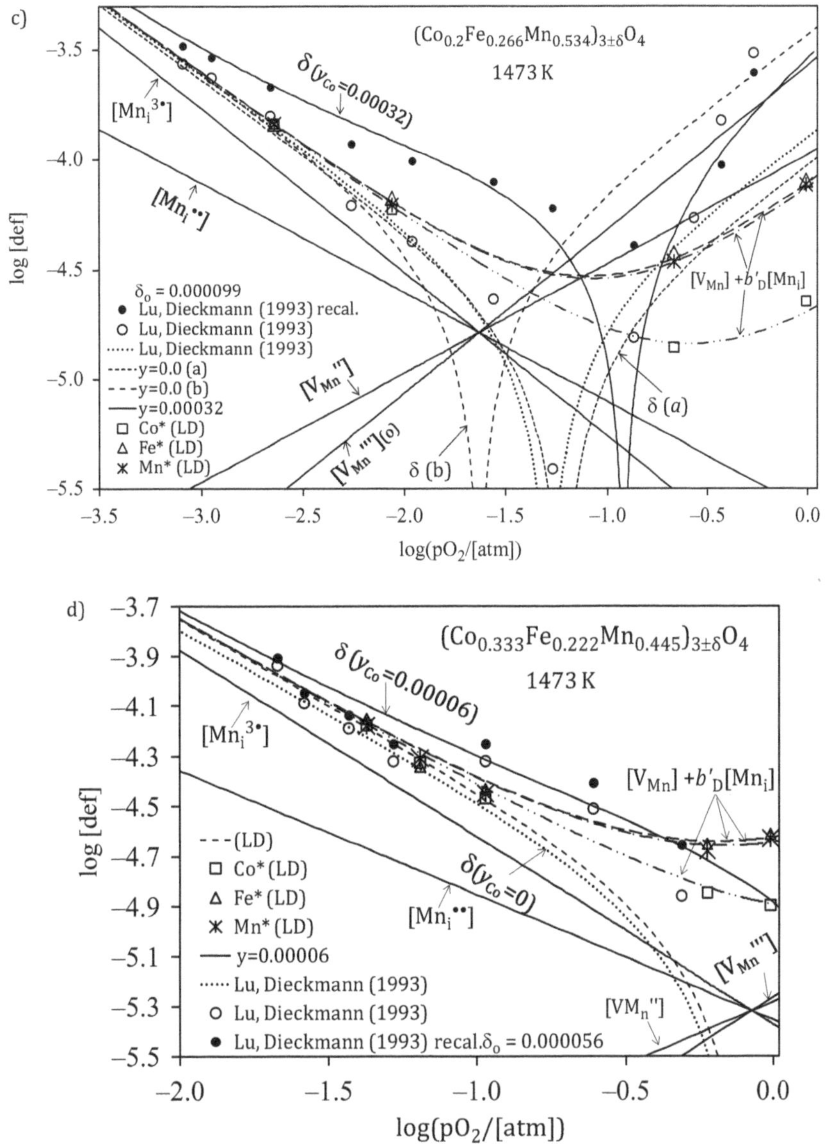

FIGURE 13.1 (Continued)

Figure 13.1b presents the oxygen pressure dependence of the deviation from the stoichiometry, taking into account the concentration of cobalt ions $\left(Co'_{Mn}\right)$ equal $y_{Co}{}' = 1.2\cdot10^{-4}$ mol/mol (dashed line). As can be seen in Figure 13.1b, a good match was obtained between the dependence of δ on p_{O_2} and the recalculated values of the deviation from the stoichiometry for the value of $\delta_o = 5.2\cdot10^{-5}$ ((●) points).

In turn, as can be seen in Figure 13.1c, in the spinel $(Co_xFe_zMn_{2z})_{3\pm\delta}O_4$, with the cobalt content of $x_{Co} = 0.2$ mol there are significant discrepancies between the dependence of δ on p_{O_2} ((---) line (δ (a)) and the experimental values of the deviation δ, especially in the range of higher oxygen pressures where cation vacancies dominate. Similar discrepancies are present between the dependence of δ on p_{O_2} obtained when using "equilibrium constants" K_V and K_I (Lu and Dieckmann 1993, Lu and Dieckmann 1995) (dotted line). This is quite strange, as at such large values of the deviation from the stoichiometry the measurement error should have been the lowest. The character of the dependence of δ on p_{O_2} indicates that discrepancies might be an effect of the incorporation of ions $\left(Co'_{Mn}\right)$. Figure 13.1c presents the dependence of δ on p_{O_2} which was calculated considering only the values of the deviation at limit, extreme oxygen pressures ((---) line δ(b)), which also deviates from the experimental values δ. As seen in Figure 13.1c, taking into account in the calculations the concentration of ions $\left[Co'_{Mn}\right]$ equal $y_{Co}{}' = 3.2\ 10^{-4}$ mol/mol causes that a good match is obtained between the dependence of δ on p_{O_2} and the recalculated values of the deviation from the stoichiometry for $\delta_o = 9.9\cdot10^{-6}$ ((●) points).

In turn, as can be seen in Figure 13.1d, for the cobalt content of $x_{Co} = 0.333$ mol, the range of oxygen pressures where cation vacancies dominate is above the oxygen pressure of 1 atm, which makes it difficult to adjust the value of $\Delta G^o_{F_I}$.

Figure 13.1d presents the results of the calculations of the diagram of the concentrations of defects, for the value of $\Delta G^{o}_{F_1} = \Delta G^{o}_{F_2} = 300$ kJ/mol, at which a relatively good match was obtained between the dependence of δ on p_{O_2} ((---) line) and the values of the deviation from the stoichiometry (Lu and Dieckmann 1993) ((○) points), at the lowest oxygen pressures. When taking into account the concentration of ions $[Co'_{Mn}]$ equal $y_{Co}' = 6 \cdot 10^{-5}$ mol/mol (solid line), a good match was obtained with the recalculated experimental values δ for $\delta_o = 5.6 \cdot 10^{-5}$ ((●) points).

Figure 13.2 presents the dependence of the concentration of ions $[Co'_{Mn}] = 1/2 y_{Co}'$ on the cobalt content in the spinel $(Co_xFe_zMn_{2z})_{3\pm\delta}O_4$, which, as can be seen, increases when the cobalt content increases, up to $[Co'_{Mn}] = 0.8\ 10^{-4}$ mol/mol for $x_{Co} = 0.2$ mol, and at the content of $x_{Co} = 0.333$ mol they possibly decrease.

From the presented calculations it results that in the spinel $(Co_xFe_zMn_{2z})_{3\pm\delta}O_4$, cobalt ions form a solid solution (they occupy nodes of Mn^{2+} ions), which causes a change in interionic interactions (change in lattice energy), a change in the range of spinel existence and a change in other properties related to the change in distribution of ions with different charges. In the case of a spinel with the content of $x_{Co} = 0.1$ and 0.2 mol, a certain small quantity of cobalt ions is possibly incorporated into the sublattice of Mn^{3+} or Fe^{3+} ions, as ions Co'_{Mn} negative in relation to the lattice.

Figure 13.3 compares the oxygen pressure dependence of deviation from the stoichiometry at 1473 K for the spinel $(Co_xFe_zMn_{2z})_{3\pm\delta}O_4$, with the cobalt content of $x_{Co} = 0.0$–0.333 mol, presented in Figure 13.1 (taking into account the concentration of ions Co'_{Mn}).

As can be seen in Figure 13.3, the range of existence of the spinel $(Fe_{0.333}Mn_{0.667})_{3\pm\delta}O_4$ is shifted towards lower oxygen pressures compared to pure hausmannite. Cobalt doping of the spinel $(Co_xFe_zMn_{2z})_{3\pm\delta}O_4$, where $(x + 3z = 1)$) causes a shift towards higher oxygen pressures in the range of existence of spinel. Also, the oxygen pressure at which the spinel $Co_xFe_zMn_{2z})_{3\pm\delta}O_4$ reaches the stoichiometric composition is shifted towards higher oxygen pressures compared to the spinel $(Fe_{0.333}Mn_{0.667})_{3\pm\delta}O_4$. At the cobalt content up to $x_{Co} = 0.1$ mol in the spinel, the deviation from the stoichiometry only slightly decreases in the range of higher oxygen pressures (the concentration of cation vacancies is unchanged) in relation to hausmannite and to the spinel $(Fe_{0.333}Mn_{0.667})_{3\pm\delta}O_4$.

Figure 13.4 presents the dependence of the oxygen pressure on the cobalt content x_{Co}, at which the spinel $(Co_xFe_zMn_{2z})_3O_4$ reaches the stoichiometric composition ((○) points), which, as can be seen, increases when the cobalt content increases. The oxygen pressure at which the concentrations of cation vacancies and interstitial ions are equal changes in a similar way ((●) points).

Figure 13.5. presents the dependence of the concentration of Frenkel defects at the stoichiometric composition, which increases when the cobalt content increases up to $x_{Co} = 0.2$ mol in the spinel $(Co_xFe_zMn_{2z})_{3-\delta}O_4$, and then it probably decreases.

Figures 13.6–13.8 present the dependence of adjusted standard Gibbs energies: $\Delta G^{o}_{F_1}$, $\Delta G^{o}_{F_2}$, $\Delta G^{o}_{V''_{Fe(o)}}$, ΔG^{o}_{I} on the cobalt content in the spinel $(Co_xFe_zMn_{2z})_{3\pm}O_4$ at 1473 K. Figures 13.7–13.9 show the dependence standard Gibbs energy $(\Delta G^{o}_{V_{Mn}})$ of formation of vacancies V'''_{Mn} (calculated according to Equation 1.35c) and $\Delta G^{o}_{Mn_i^{3\bullet}}$ of formation of interstitial manganese

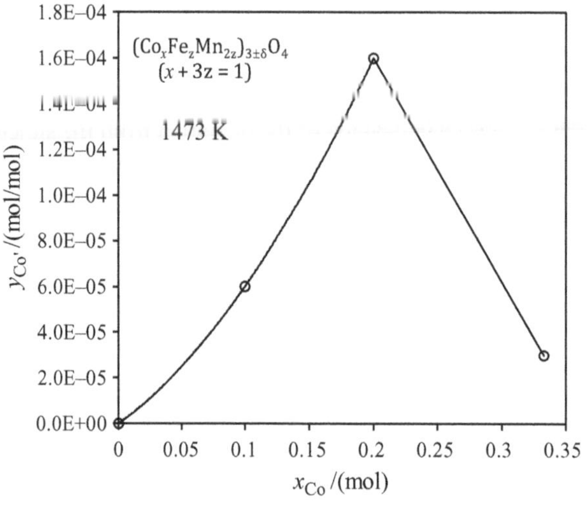

FIGURE 13.2 Dependence of the concentration of cobalt ions $[Co'_{Mn}] = 1/2 y_{Co'}$ incorporated into the sublattice of ions Mn^{3+} on the cobalt content x_{Co} in the spinel $(Co_xFe_zMn_{2z})_{3\pm\delta}O_4$ (where $x + 3z = 1$) at 1473 K.

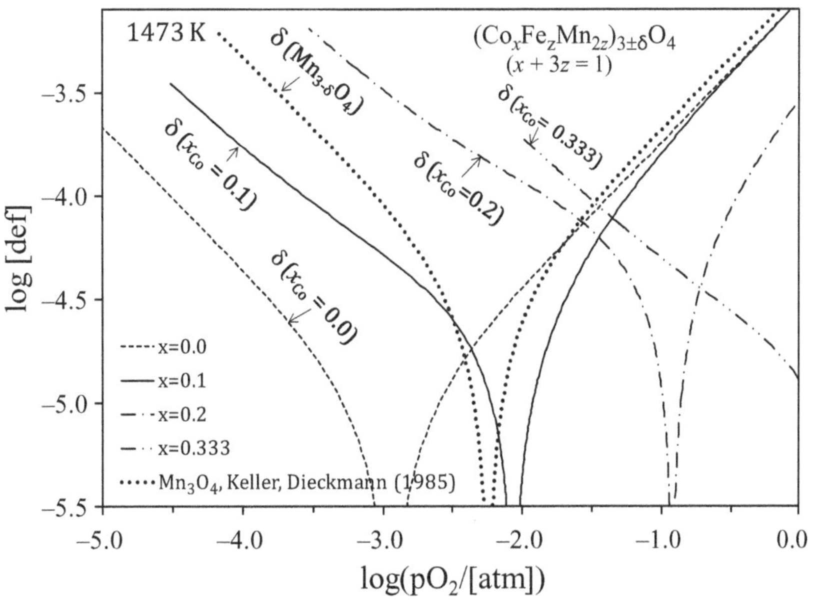

FIGURE 13.3 Dependences of the deviation from the stoichiometry on the oxygen pressure in the spinel $(Co_xFe_zMn_{2z})_{3\pm\delta}O_4$ with the cobalt content of $x_{Co} = 0.0–0.333$ mol at 1473 K presented in Figure 13.1 (without taking into account ions Co'_{Mn}).

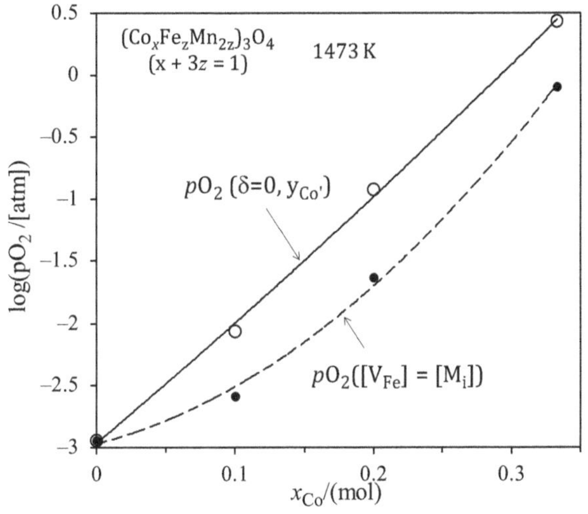

FIGURE 13.4 Dependence of the oxygen pressure at which the spinel $(Co_xFe_zMn_{2z})_3O_4$ reaches the stoichiometric composition ((○) points)), in the case of incorporation of cobalt ions Co'_{Mn} into the sublattice of Mn^{3+}, on the cobalt content x_{Co}, at 1473 K, and the oxygen pressure at which the concentrations of cation vacancies and interstitial cations are equal ((●) points).

ions $Mn_i^{3\bullet}$ (calculated according to Equation 1.34c) and ($\Delta G^o_{Mn_i^{\bullet\bullet}}$) of formation of ions $Mn_i^{\bullet\bullet}$ (calculated according to Equation 1.36c). The points (a) refer to values at which the oxygen pressure dependence of δ(a), shown in Figure 13.1c, was determined.

As can be seen in Figure 13.6, the standard Gibbs energy $\Delta G^o_{F_I}$ of formation of Frenkel defects decreases when the cobalt content increases up to $x_{Co} = 0.2$ mol, and then, at the content of $x_{Co} = 0.333$ mol, it increases.

As can be seen in Figures 13.7 and 13.10, the values ($\Delta G^o_{V''_{Mn}}$) and ($\Delta G^o_{V'''_{Mn}}$) increase when the cobalt content in the spinel $(Co_xFe_zMn_{2z})_{3\pm}O_4$ increases, while the values $\Delta G^o_{Mn_i^{3\bullet}}$ and $\Delta G^o_{Mn_i^{\bullet\bullet}}$, inversely, decrease.

Figures 13.7 and 13.9 show also the values of $\Delta G^o_{V_{Mn}}$ and $\Delta G^o_{Mn_i}$ calculated using “equilibrium constants” of reaction of formation of cation vacancies K_V and interstitial ions K_I (Lu and Dieckmann 1993). The values differ from those

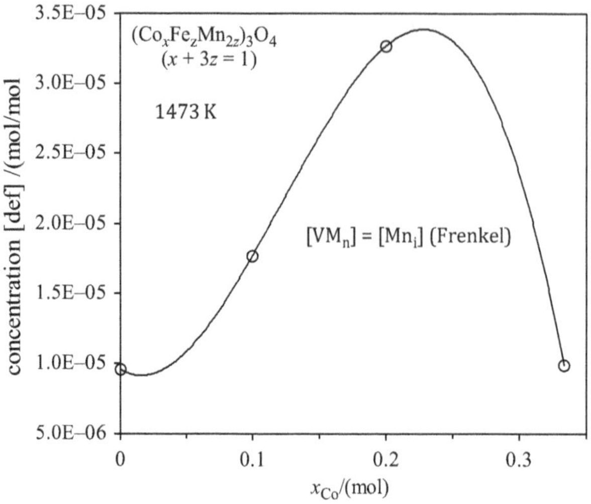

FIGURE 13.5 Dependence of the concentration of Frenkel defects on the cobalt content x_{Co} in the spinel $(Co_xFe_zMn_{2z})_{3\pm\delta}O_4$ at 1473 K.

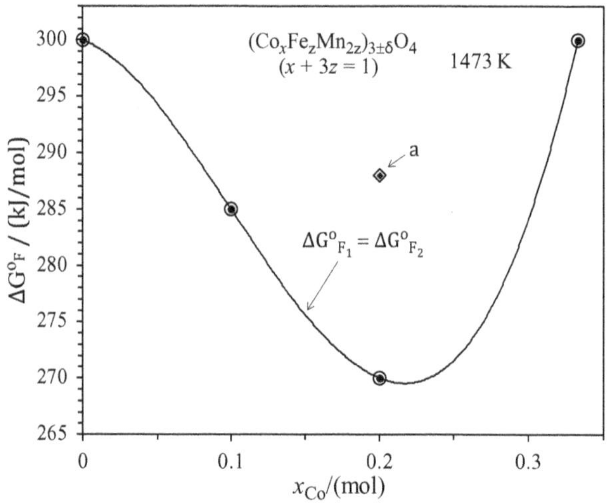

FIGURE 13.6 Dependence of the standard Gibbs energy $\Delta G^o_{F_1}$ ((○) points) and $\Delta G^o_{F_2}$ – (●) of formation of Frenkel defects in the spinel $(Co_xFe_zMn_{2z})_{3\pm\delta}O_4$ on the cobalt content x_{Co}, at 1473 K, obtained using the results of the studies of the deviation from the stoichiometry (Lu and Dieckmann 1993). Points *a* – values for the curve δ(a) in Figure 13.1c.

determined in the work, because they describe another model of defects, but they have a similar character of the dependence on the manganese content in the spinel.

13.3 DIFFUSION IN HAUSMANNITE DOPED WITH COBALT AND IRON

Similarly as for pure magnetite, using the concentration of ionic defects resulting from the diagram of defects' concentrations and the values of coefficients of self-diffusion of tracers in the spinel $(Co_xFe_zMn_{2z})_{3\pm\delta}O_4$, the coefficients of diffusion (mobility) of metal ions via cation vacancies $D^o_{V(M)}$ and via interstitial cations $D^o_{I(M)}$ were determined (see Section 1.6.2).

Figure 13.1 shows the dependence of the "sum" of concentrations of ionic defects ($[V_{Mn}] + b_D[Mn_i]$) on p_{O_2} for the fitted values of b_D and the values of this sum, calculated according to Equation 1.47a using the values of coefficients of self-diffusion of: iron ^{59}Fe, cobalt ^{60}Co and manganese ^{54}Mn (Lu and Dieckmann 1993). For the spinel $(Co_xFe_zMn_{2z})_{3\pm\delta}O_4$ with the cobalt content of x_{Co} = 0.0, 0.1 mol, the values of the coefficient of diffusion of cations $D^o_{V(M)}$ via cation vacancies and the parameter b_D, being the ratio of coefficients of diffusion via defects ($D^o_{I(M)}/D^o_{V(M)}$), were fitted is such a way as to obtain a match of the dependence resulting from the diagram of defects and from the values of this sum calculated using the

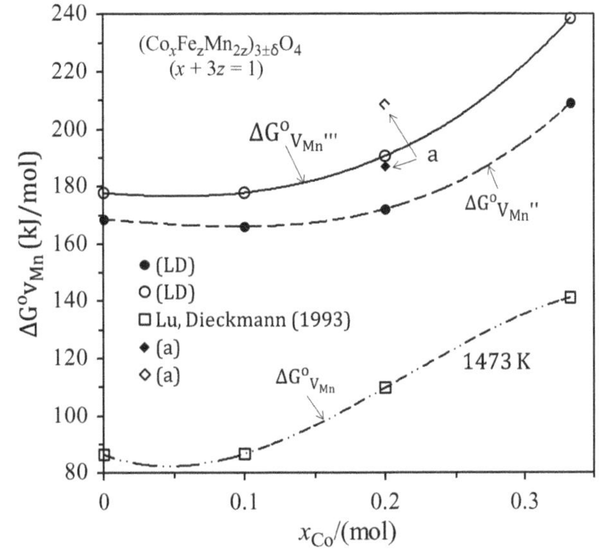

FIGURE 13.7 Dependence of standard Gibbs energy $\Delta G^o_{V'''_{Mn}}$ ((O) points) and $\Delta G^o_{V''_{Mn}}$ ((●) points) of formation of vacancies $\left(V'''_{Mn}\right)$ and $\left(V''_{Mn}\right)$ in the spinel $(Co_xFe_zMn_{2z})_{3\pm\delta}O_4$, on the cobalt content x_{Co}, at 1473 K and the values $\Delta G^o_{V_{Fe}}$ calculated using the "equilibrium constants" of reaction of formation of vacancies K_V (Lu and Dieckmann 1993). Points *a* – values for the curve δ(a) in Figure 13.1c.

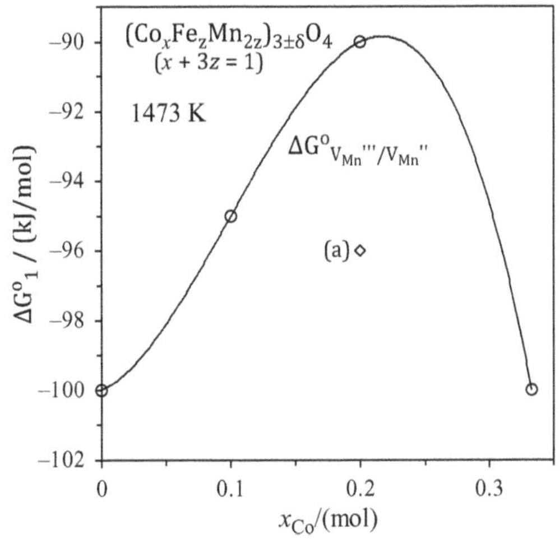

FIGURE 13.8 Dependence of standard Gibbs energy ΔG^o_1 on the cobalt content x_{Co} in the spinel $(Co_xFe_zMn_{2z})_{3\pm\delta}O_4$ at 1473 K, permitting determination of the ratio of concentrations of iron vacancies $\left[V'''_{Mn}\right]/\left[V''_{Mn}\right]$, according to Equation (1.35a). Point *a* – values for the curve δ(a) in Figure 13.1c.

coefficients of tracer self-diffusion. For the cobalt content of $x_{Co} = 0.2$ and 0.333 mol the sums $\left(\left[Mn_i\right]+b'_D\left[V_{Mn}\right]\right)$ were calculated, fitting the value of $b'_D = \left(D^o_{V(M)}/D^o_{I(M)}\right)$, and, for the lowest oxygen pressures, the values of the coefficient of diffusion $D^o_{I(M)}$ via interstitial cations was fitted (see Equation 11.1). As can be seen in Figure 13.1, a very good agreement was obtained between the dependence of the sum of the concentrations of ionic defects ($[V_{Mn}] + b_D[Mn_i]$) and the values calculated using the coefficients of self-diffusion of tracers.

Figure 13.10 presents the dependence of the ratio of mobilities $\left(D^o_{V(M)}/D^o_{I(M)} = b_D\right)$ on the cobalt content in the spinel $(Co_xFe_zMn_{2z})_{3\pm\delta}O_4$ for ions: Fe*, Co* and Mn*.

As can be seen in Figure 13.10, the mobility of tracer ions via interstitial ions is higher than the mobility via cation vacancies, especially for cobalt diffusion. The ratio of the coefficients of diffusion $b_D = \left(D^o_{V(M)}/D^o_{I(M)}\right)$ at the cobalt content

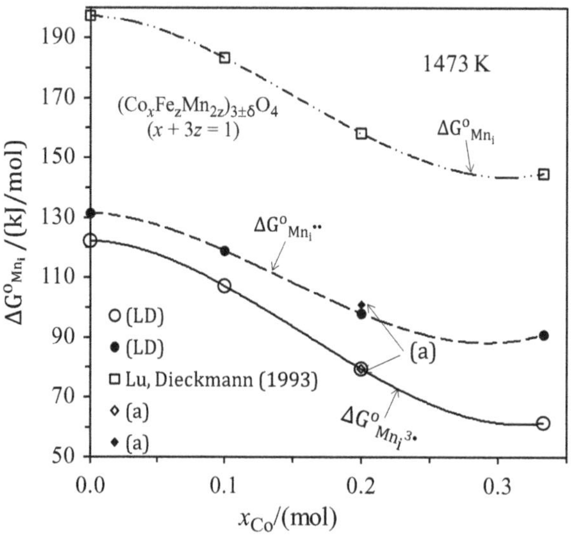

FIGURE 13.9 Dependence of the standard Gibbs energy $\Delta G^o_{Mn_i^{3\bullet}}$ ((○) points) and $\Delta G^o_{Mn_i^{\bullet\bullet}}$ ((●) points) of formation of interstitial ions $Mn_i^{3\bullet}$ and $Mn_i^{\bullet\bullet}$ on the cobalt content x_{Co} in the spinel $(Co_xFe_zMn_{2z})_{3\pm\delta}O_4$, at 1473 K and the values $\Delta G^o_{Mn_i}$ calculated using the "equilibrium constants" of reaction of formation of interstitial cations K_I (Lu and Dieckmann 1993). Points a – values for the curve δ(a) in Figure 13.1c.

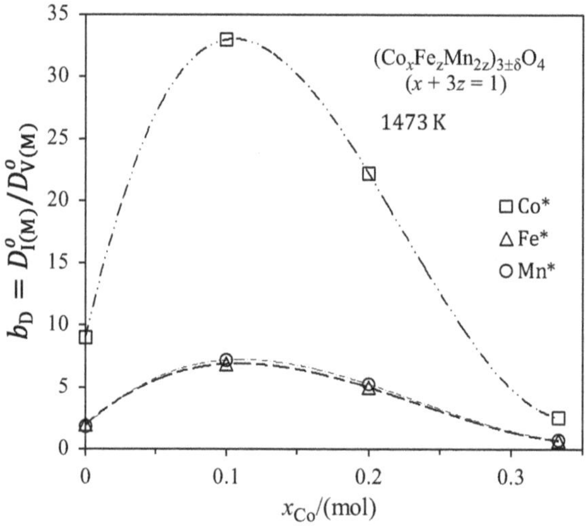

FIGURE 13.10 Dependence of the ratio of the coefficient of diffusion of tracer ions via interstitial cations to the coefficient of diffusion of cations via vacancies $D^o_{I(M)}/D^o_{V(M)}$, (parameter b_D) on the cobalt content x_{Co} in the spinel $(Co_xFe_zMn_{2z})_{3\pm\delta}O_4$ at 1473 K, obtained using the coefficients of self-diffusion of tracers: Fe* ((△) points), Co* – (□), Mn* – (○) (Lu and Dieckmann 1993) and concentrations of ionic defects.

of $x_{Co} = 0.1$ increases significantly especially for cobalt diffusion, and then, when the cobalt content increases further, it decreases.

In turn, Figure 13.11 presents the dependence of the mobilities of ions via cation vacancies $D^o_{V(M)}$ and via interstitial ions $D^o_{I(M)}$ on the cobalt content in the spinel $(Co_xFe_zMn_{2z})_{3\pm\delta}O_4$.

As can be seen in Figure 13.11, for the diffusion of ions Fe*, Co* and Mn*, the coefficients of diffusion $D^o_{I(M)}$ via interstitial cations change only slightly when the cobalt content in the spinel increases, but they significantly differ between them. The fastest diffusion occurs for cobalt, and the slowest for iron.

The mobilities of ions via cation vacancies $D^o_{V(M)}$ decrease up to the cobalt content of $x_{Co} = 0.2$ mol, and then, for the content of $x_{Co} = 0.333$ mol they increase significantly. The mobilities of cobalt ions and manganese ions via cation vacancies $\left(D^o_{V(M)}\right)$ are similar, while the mobility of iron ions is much lower.

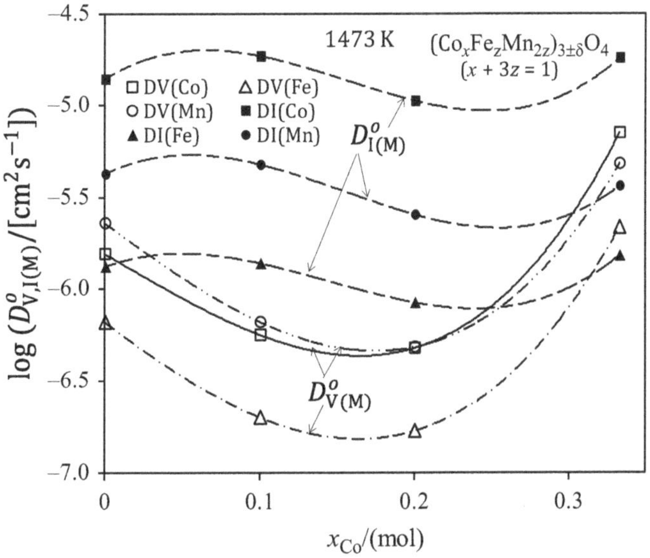

FIGURE 13.11 Dependence of the coefficient of diffusion of ions via cation vacancies $D^o_{V(M)}$ and via interstitial cations $D^o_{I(M)}$ on the cobalt content x_{Co} in the spinel $(Co_xFe_zMn_{2z})_{3\pm\delta}O_4$ at 1473 K, obtained using the coefficients of self-diffusion of tracers: Fe* ((△,▲), points), Co* – (□,■), Mn* – (○,●) (Lu and Dieckmann 1993), concentrations of ionic defects and the fitted parameter b_D or b'_D.

13.4 CONCLUSIONS

Using the results of the studies of the deviation from the stoichiometry in the spinel $(Co_xFe_zMn_{2z})_{3\pm\delta}O_4$ (where the composition is determined by the condition $x+3z = 1$) with the cobalt content of $x_{Co} = 0.1$–0.333 mol at 1473 K (Lu and Dieckmann 1993), the diagrams of concentrations of defects were determined. It was found that cobalt doping of the spinel $(Co_xMn_zFe_{2z})_{3\pm\delta}O_4$ causes a shift towards higher oxygen pressures in the range of existence of spinel. Also, the oxygen pressure at which the spinel $(Co_xFe_zMn_{2z})_{3\pm\delta}O_4$ reaches the stoichiometric composition is shifted towards higher oxygen pressures (compared to the spinel $(Fe_{0.333}Mn_{0.667})_{3\pm\delta}O_4$). At the cobalt content of $x_{Co} = 0.1$ mol in the spinel, in the range of higher oxygen pressures, the deviation from the stoichiometry (the concentration of cation vacancies) is practically unchanged in relation to hausmannite and to the spinel $(Fe_{0.333}Mn_{0.667})_{3\pm\delta}O_4$. In the range of low oxygen pressures, interstitial cations $\left[Mn_i^{3\bullet}\right]$ dominate; their concentration at the same oxygen pressure increases when the cobalt content in the spinel increases.

It was demonstrated that it is possible to fit a reference value δ_o which causes a change in the character of the dependence of the deviation from the stoichiometry on the oxygen pressure indicating the incorporation of ions Co^{2+} into the sublattice of ions Mn^{3+}, (Co'_{Mn}) negative in relation to the lattice. The maximum concentration of these ions at the cobalt content of $x_{Co} = 0.2$ mol can be $y_{Co'} = 1.6 \cdot 10^{-4}$ mol/mol.

It was found that when the cobalt content in the spinel $(Co_xMn_zFe_{2z})_{3\pm\delta}O_4$ increases, standard Gibbs energies of formation of Frenkel defects, up to the cobalt content of $x_{Co} = 0.2$ mol, decrease, which results in an increase in the concentration of Frenkel defects $\left(\left[V'''_{Mn}\right] = \left[Mn_i^{3\bullet}\right]\right)$ from about $1 \cdot 10^{-5}$–$1.6 \cdot 10^{-5}$ mol/mol. In turn, standard Gibbs energies of formation of cation vacancies $\Delta G^o_{V''_{Mn}}$ and $\Delta G^o_{V'''_{Mn}}$ increase when the content of cobalt in the spinel $(Co_xMn_zFe_{2z})_{3\pm\delta}O_4$ increases, while $\Delta G^o_{Mn_i^{3\bullet}}$ and $\Delta G^o_{Mn_i^{\bullet\bullet}}$ of formation of interstitial cations decrease. The above values depend on the value of ΔG^o_{eh} of formation of electronic defects and they were determined for the value $\Delta G^o_{eh} = 60$ kJ/mol, the same as the value assumed for pure magnetite.

Using the determined concentration of ionic defects and the values of coefficients of self-diffusion of tracers M* (^{59}Fe, ^{60}Co and ^{54}Mn) in the spinel $(Co_xMn_zFe_{2z})_{3\pm\delta}O_4$ (Lu and Dieckmann 1993), the ratio of the coefficient of diffusion (mobility) of tracer ions via cation vacancies to the coefficient of diffusion via interstitial cations $\left(D^o_{I(M)}/D^o_{V(M)} = b_D\right)$ was determined. It was found that when the cobalt content in the spinel $(Co_xMn_zFe_{2z})_{3\pm\delta}O_4$ increases, the mobility of via interstitial cations is higher than the mobility via cation vacancies and at the cobalt content of $x_{Co} = 0.1$ mol the parameter b_D reaches the highest values. At the temperature of 1473 K, the parameter b_D reaches for the of cobalt ions the value $b_D \approx 33$, and for iron and manganese $b_D \approx 7$.

It was found that the mobilities of tracer ions Fe*, Co* and Mn* via interstitial cations $\left(D^o_{I(M)}\right)$ slightly change when the cobalt content in the spinel increases, but they significantly differ between them. The fastest diffusion occurs for cobalt, and the slowest for iron.

The mobilities of ions via cation vacancies $D^o_{V(M)}$ decrease to the cobalt content of $x_{Co} = 0.2$ mol, and then, for the content of $x_{Co} = 0.333$ they increase significantly. The mobilities of cobalt ions and manganese ions via cation vacancies are similar, while the mobility of iron ions is much lower.

REFERENCES

Aukrust, E., and A. Muan. 1964. Thermodynamic Properties of Solid Solutions with Spinel Type Structure. The System Co_3O_4-Mn_3O_4. *Trans. Metall. Soc. AIME.* 230: 378–382.

Lu, F. H., and R. Dieckmann. 1993. Point Defects and Cation Tracer Diffusion in $(Co,Fe,Mn)_{3-\delta}O_4$ Spinels: II. Mixed Spinels $(Co_x Fe_zMn_{2z})_{3-\delta}O_4$. *Solid State Ionics.* 59: 71–82.

Lu, F. H., and R. Dieckmann. 1995. Point Defects in Oxide Spinel Solid Solutions of the Type $(Co,Fe,Mn)_{3-\delta}O_4$ at 1200 °C. *J. Phys. Chem. Solids.* 56: 725–733.

Subramanian, R., and R. Dieckmann. 1992. Limits of the Thermodynamic Stability of Cobalt-Iron-Manganese Mixed Oxides at 1200°C. *J. Am. Ceram. Soc.* 75: 382–391.

Subramanian, R., and R. Dieckmann. 1993. Nonstoichiometry and Thermodynamics of the Solid Solution $(Fe,Mn)_{1-\Delta}O$ At 1200°C. *J. Phys. Chem. Solids.* 54: 991–1000.

Subramanian, R., R. Dieckmann, G. Eriksson, and A. D. Pelton. 1994. Model Calculations of Phase Stabilities of Oxide Solid Solutions in the Co–Fe–Mn–O System at 1200°C. *J. Phys. Chem. Solids.* 55: 391–444.

14 Lithium-Manganese Spinel – $LiMn_2O_4$

14.1 INTRODUCTION

The spinel $LiMn_2O_4$ was the subject of many studies, as it revealed to be a very good cathode material in reversible lithium batteries, where it is possible to electrochemically intercalate and deintercalate lithium ions (Thackeray et al. 1983, Ohzuku et al. 1990, Tarascon and Guyomard 1991, Scrosati 1992, Thackeray 1995, Thackeray 1997). It can be obtained through a solid-phase reaction between lithium carbonate and MnO_2 (Palacín et al. 2000, Thackeray 1995) or obtained with sol--gel method (Cao et al. 2009Dziembaj et al. 2003, Fei et al. 2008). $LiMn_2O_4$ has two polymorphic variants, a variant with tetragonal structure (point group $I4_1$), where the ratio of lattice parameters is c/a = 1.011 (Yamada and Tanaka 1995, Yamada et al. 1999); it is stable up to 280 K. This structure is a deformed spinel structure (Jahn–Teller distortion is present in oxygen octahedrons ($O–Mn^{3+}$)). To this variation, also an orthorhombic structure is assigned (point group Fddd) (Rodriguez-Carvajal et al. 1998, Rousse et al. 1999). The temperature of phase transition depends on the composition, lithium content and method of production. At the transition temperature, a small leap in electrical conductivity or a change in the slope of the dependence of σ on T is observed and there is a hysteresis of conductivity upon rise and decrease of temperature (Azzoni et al. 1999, Iguchi et al. 1998Marzec et al. 2002, Massarotti et al. 1997, Molenda et al. 2003, 2005Shimakawa et al. 1997). The size and shape of hysteresis depends on the lithium content (Scrosati 1992, Thackeray 1995, 1997). At the transition temperature, there is also an anomaly of thermoelectric power (Thackeray 1995, 1997), indicating that it depends on the ratio of ions Mn^{3+}/Mn^{4+} (Palacín et al. 2000). This was confirmed also by EPR studies, from which it was concluded that the ratio of ions Mn^{3+}/Mn^{4+} depends on the composition and on the preparation method and it changes from 0.5–1.1 (Dziembaj et al. 2003).

Above the transition temperature, $LiMn_2O_4$ has a cubic spinel structure ($Fd\bar{3}m$) (Miura et al. 1996, Sugiyama et al. 1995). In the spinel structure lithium ions occupy cation sites with tetrahedral coordination, and manganese ions Mn^{3+} and Mn^{4+} occupy sites with octahedral coordination. Manganese–lithium spinel can be also considered as hausmannite doped with a maximum amount of lithium ions. Lithium doping of hausmannite caused a shift towards higher oxygen pressures in the range of existence of the spinel $LiMn_2O_4$ and at 1023 K the spinel is stable above the oxygen pressure of 0.01 atm. Thermogravimetric studies TG demonstrated that the spinel $LiMn_2O_{4-x}$ shows a mass deficit in the oxygen sublattice (Paulsen and Dahn 1999, Richard et al. 1994,Sugiyama et al. 1996, Tarascon et al. 1994, Thackeray et al. 1996, Yamada et al. 1995). This was confirmed by equilibrium studies on the deviation from the stoichiometry as a function of the oxygen pressure (Hosoya et al. 1997). The obtained dependence of δ on p_{O_2} is not typical compared to other spinels, discussed earlier. In particular, it was found that in the range of the oxygen pressure from 1 to 0.03 atm there is no mass change of the spinel sample, and below this pressure, to 0.01 atm, it decreases. It was assumed (Hosoya et al. 1997) that the observed deficit of sample mass is related to the existence of oxygen vacancies in $LiMn_2O_{4-x}$. However, it might be related to the presence of defects in the cation sublattice. Further studies (Goodenough et al. 1993, Masquelier et al. 1996), as well as quantum mechanical calculations (Hoang 2014), indicate the domination of cation defects.

The studies on electrical conductivity and thermoelectric power demonstrated that the spinel $LiMn_2O_{4-x}$ in the temperature range 850–1090 K shows high electrical conductivity, increasing with temperature, from 0.1 to 2 $\Omega^{-1}cm^{-1}$ (Molenda et al. 1999, Molenda and Kucza 1999). It is practically independent of the oxygen pressure. The spinel $LiMn_2O_{4-x}$ shows negative values of the Seebeck coefficient in the range of –30 to –15 μV/K at 850–1090 K; these values slightly decrease when the oxygen content decreases. This indicates that the charge carriers in the spinel $LiMn_2O_{4-x}$ are electrons and their transport occurs according to the mechanism of small polarons.

Using the results of the studies of deviation from the stoichiometry (Hosoya et al. 1997) and the results of electrical conductivity (Molenda et al. 1999), it was decided to attempt to determine the diagram of concentrations of point defects in $LiMn_2O_{4-x}$.

14.2 REACTIONS OF FORMATION OF POINT DEFECTS

As previously mentioned, the oxygen pressure dependence of the deviation from the stoichiometry (Hosoya et al. 1997) is not typical of spinels and of oxides showing defected oxygen sublattice. The presence of so wide range of oxygen pressures where the spinel would have the stoichiometric composition is typical of oxides with significantly higher concentration of ionic defects than electronic defects in the so-called stoichiometric compounds (e.g. in solid electrolytes), where the dominating defects are Frenkel type or Schottky type and the concentration of electronic defects is by a few orders of magnitude lower and depends exponentially on the oxygen pressure ($p_{O_2}^{1/n}$ on $p_{O_2}^{-1/n}$). High electrical conductivity $LiMn_2O_{4-x}$ excludes this type of defects. For the spinels discussed earlier, that have high electrical conductivity, the oxygen pressure dependence of deviation from the stoichiometry (concentration of ionic defects) quite strongly depends on the oxygen pressure ($p_{O_2}^{2/3}$ or $p_{O_2}^{-2/3}$).

If we assume that the dominating ionic defects in the spinel $LiMn_2O_{4-x}$ are cation defects, then the spinel formula will be $(LiMn_2)_{1+\Delta}O_4$ or $(Li_{0.333}Mn_{0.667})_{3+\delta}O_4$. Analogously to the case of hausmannite, in the spinel $LiMn_2O_4$ with the stoichiometric composition we can distinguish three different cation sublattices: the sublattice of lithium ions with tetrahedral coordination and two sublattices of manganese ions Mn^{3+} and Mn^{4+} with octahedral coordination. Analogously to the case of magnetite (see Section 1.4.1), in order to write defect reactions, it is necessary to adopt the following formula for the spinel:

$$\left(Li_{(t)}^{(+)}O_{1/2}\right)\left(Mn_{(o)}^{(3+)}O_{3/2}\right)\left(Mn_{(o)}^{(4+)}O_2\right)$$

Therefore, the formation of lithium vacancies occurs according to the reaction:

$$1/4O_2 = 1/2O_O^x + V_{Li}' + h^\bullet \qquad \Delta G_{V_{Li}'}^o \qquad (14.1a)$$

The formation of manganese vacancies occurs according to the following reactions:

$$3/4O_2 = 3/2O_O^x + V_{Mn(o)}''' + 3h^\bullet \qquad \Delta G_{V_{Mn}'''}^o \qquad (14.1b)$$

$$O_2 = 2O_O^x + V_{Mn(o)}^{4'} + 4h^\bullet \qquad \Delta G_{V_{Mn}^{4'}}^o \qquad (14.1c)$$

In turn, the formation of interstitial cations will occur according to the reactions:

$$Li_{Li(t)}^x + 1/2O_O^x + h^\bullet + V_i = Li_i^\bullet + 1/4O_2 \qquad \Delta G_{Li_i^\bullet}^o \qquad (14.2a)$$

$$Mn_{Mn(o)}^{x(3+)} + 3/2O_O^x + 3h^\bullet + V_i = Mn_i^{3\bullet} + 3/4O_2 \qquad \Delta G_{Mn_i^{3\bullet}}^o \qquad (14.2b)$$

$$Mn_{Mn(o)}^{x(3+)} + 2O_O^x + 4h^\bullet + V_i = Mn_i^{4\bullet} + O_2 \qquad \Delta G_{Mn_i^{4\bullet}}^o \qquad (14.2c)$$

There will be also an equilibrium between electronic defects. If we assume, analogously to the case of hausmannite, that electronic defects are present in the sublattice of ions with octahedral coordination, then in the sublattice of ions Mn^{3+} the presence of electron holes will be related to the presence of manganese ions Mn^{4+}, which will be positive in relation to ions of the sublattice $\left(Mn_{Mn}^\bullet\right)$. In the sublattice of ions Mn^{4+}, Mn^{3+} ions will be present, negative in relation to ions of the sublattice $\left(Mn_{Mn}'\right)$. The concentration of these ions will be equivalent to the concentration of electron holes and electrons (see Section 1.4.4). Therefore, in the sublattice of manganese ions with octahedral coordination, apart from ions Mn^{3+} and Mn^{4+}, which will be electroneutral relative to the lattice, ions with the same oxidation state will be present, but charged in relation to the lattice. A change in the concentration of electronic defects will cause a change in the concentration of ions Mn^{3+} and Mn^{4+}. An equilibrium will be set between the electronic defects, according to the reaction:

$$Mn_{Mn(4+)}^x + Mn_{Mn(3+)}^x = Mn_{Mn}^\bullet + Mn_{Mn}' \equiv h^\bullet + e' \qquad \Delta G_{eh}^o \qquad (14.3)$$

Expressions for the equilibrium constants of reactions 14.1a–14.3 assume the form of:

$$K_{V_{Li}'} = \frac{\left[V_{Li}'\right]\left[h^\bullet\right]}{p_{O_2}^{1/4}} \qquad (14.4a)$$

$$K_{V_{Mn}'''} = \frac{\left[V_{Mn(o)}'''\right]\left[h^\bullet\right]^3}{p_{O_2}^{3/4}} \qquad (14.4b)$$

$$K_{V_{Mn}^{4'}} = \frac{\left[V_{Mn}^{4'}\right]\left[h^\bullet\right]^4}{p_{O_2}} \qquad (14.4c)$$

$$K_{Li_i^{\bullet}} = \frac{\left[Li_i^{\bullet}\right] p_{O_2}^{1/4}}{\left[h^{\bullet}\right]} \quad (14.4d)$$

$$K_{Mn_i^{3\bullet}} = \frac{\left[Mn_i^{3\bullet}\right] p_{O_2}^{3/4}}{\left[h^{\bullet}\right]^3} \quad (14.4e)$$

$$K'_{Mn_i^{4\bullet}} = \frac{\left[Mn_i^{4\bullet}\right] p_{O_2}}{\left[h^{\bullet}\right]^4} \quad (14.4f)$$

$$K_{eh} = \left[h^{\bullet}\right][e'] \quad (14.5)$$

Reactions 14.1a–14.3 are mutually related and at a constant temperature, an equilibrium state will be set, where, as it results from Equation 14.4a–14.4f, mutual concentrations of ionic defects wizzll depend on the oxygen pressure and on the concentration of electronic defects.

At high temperatures, in the structure of the spinel $LiMn_2O_4$, similarly as for pure and doped hausmannite, migration of ions between sublattices might occur and, as a result, a determined equilibrium distribution of lithium ions and manganese ions in the octahedral and tetrahedral sublattices is set. The distribution of ions with different charges (oxidation states) results from optimum interactions between ions and from the presence of electronic defects, and, indirectly, ionic defects, with the concentration depending on the oxygen pressure.

As it was mentioned, the character of the dependence of the deviation from the stoichiometry on the oxygen pressure (Hosoya et al. 1997) is not typical compared to the spinels discussed earlier and to other oxides. Figure 14.1 presents the dependence of the deviation from the stoichiometry (Hosoya et al. 1997), assuming defected cation sublattice $(Li_{0.333}Mn_{0.667})_{3+\delta}O_4$ ((○) points). In the range of oxygen pressures where the sample mass does not change it was assumed that the deviation from the stoichiometry is zero (Hosoya et al. 1997). For the logarithm to have a finite value, it was assumed that the deviation from the stoichiometry is $1{\cdot}10^{-4}$ (below the measurement error value). If we assume that the published results (Hosoya et al. 1997) determine the change in the deviation from the stoichiometry $\Delta\delta$, then, in order to get the absolute value of the deviation from the stoichiometry, it is necessary to determine the reference composition, which was not done experimentally. According to the method proposed by Dieckmann (1982) and used for magnetite and other spinels, in order to determine the absolute values of the deviation from the stoichiometry it is necessary to adjust such a reference value that the character of the dependence δ of p_{O_2} consisted of two branches (ranges of oxygen pressure) where cation vacancies and interstitial cations dominate, respectively. In such a case the absolute value of the deviation from the stoichiometry is:

$$\delta = \delta_o + \Delta\delta$$

For magnetite and other spinels, the value of δ_o can be unambiguously specified (fitted), because the character of the dependence of δ on p_{O_2} consists of two symmetrical branches, divided with the oxygen pressure at which the spinel reaches the stoichiometric composition (see the dependences of δ on p_{O_2} for other spinels). The performed attempts of fitting the value of δ_o demonstrated that the character of the dependence of deviation from the stoichiometry is different than that for magnetite or hausmannite. The points (●) in Figure 14.1 present the recalculated values δ when assuming the value of δ_o = 0.005. The obtained character of the dependence indicates that the spinel, at low oxygen pressures, reaches the stoichiometric composition and there is a narrow range where the deviation from the stoichiometry increases when the oxygen pressure decreases. The obtained character of the dependence, with a wide range of a constant value of the deviation from the stoichiometry, is typical of spinels (also for other MO oxides (Stokłosa 2015), in which there occurred an incorporation of doping ions with a charge higher than the charge of ions of the parent sublattice. The presence of doping of this type was shown for magnetite doped with cobalt, with Co and Mn and with Zn and Mn. It probably also occurs in the case of magnetite doped with chromium and titanium (see Chapters 3–9). Therefore, the presence of a small excess of MnO_2 oxide will cause the incorporation of ions Mn^{4+} into the formed structure of the spinel $(LiMn_2)_{1+\Delta}O_4$, into cation sites of the sublattice of lithium and manganese Mn^{3+}. These ions should be treated as doping ions with a charge relative to the sublattice of lithium ions $\left(Mn_{Li}^{3\bullet}\right)$ and manganese ions Mn^{3+} $\left(Mn_{Mn(3+)}^{\bullet}\right)$. The doping will occur in all the sublattices, and, as a result, a molecule of spinel is built-in (as a result of the doping, the ratio of oxygen sites to cation sites must be preserved). Analogously to the case of magnetite, hausmannite and other spinels, the reaction of incorporation of ions Mn^{4+} (MnO_2) into the structure of $LiMn_2O_4$ can be written with the following reactions:

$$2MnO_2 \rightarrow 4O_O^x + Mn_{Li}^{3\bullet} + Mn_{Mn(4+)}^x + V_{Mn}''' \quad (14.6a)$$

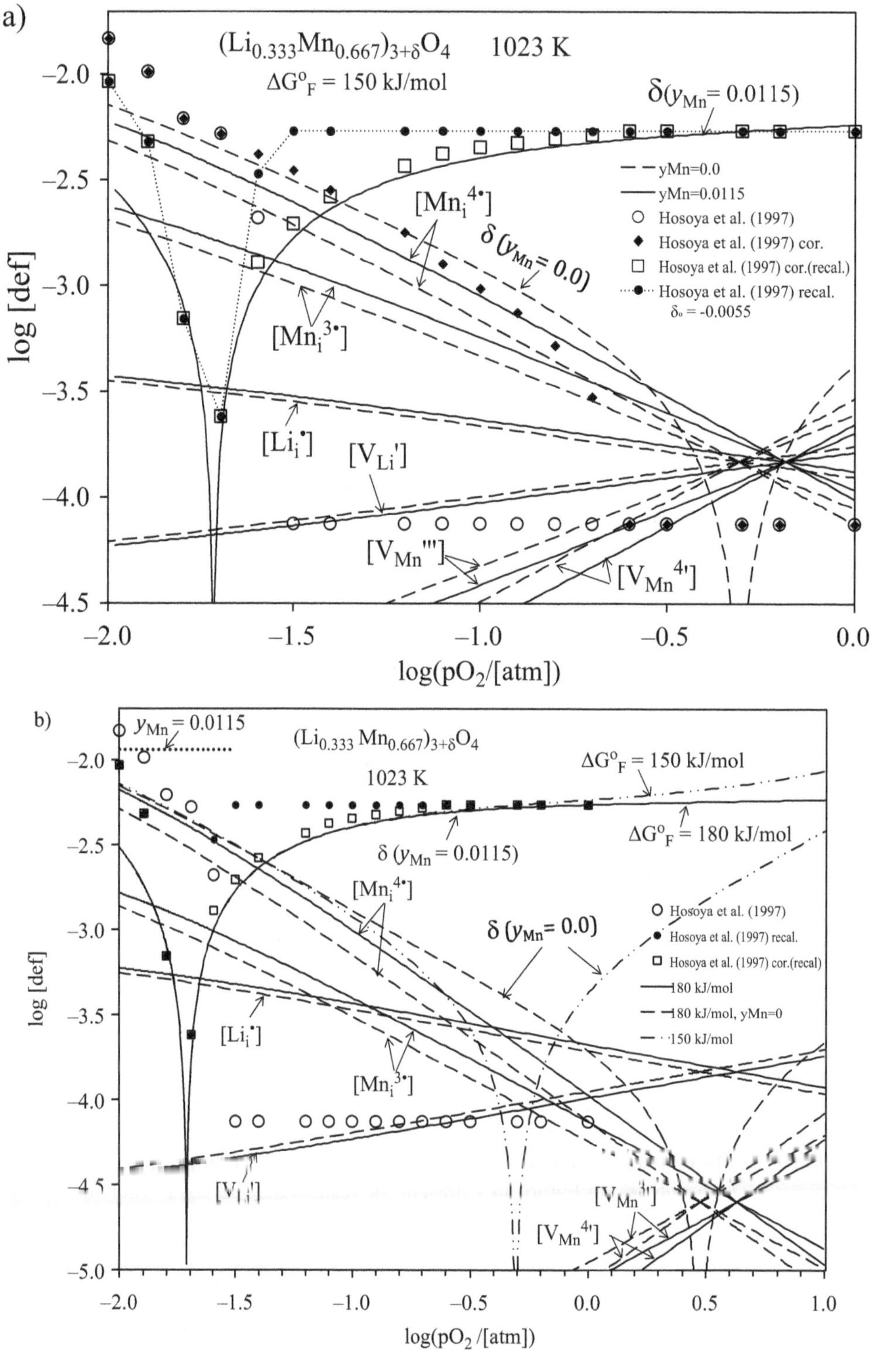

FIGURE 14.1 (a) Diagram of concentrations of point defects for the spinel $(Li_{0.333}Mn_{0.667})_{3+\delta}O_4$ at 1023 K (solid lines) obtained using the recalculated values of the deviation from the stoichiometry (Hosoya et al. 1997), when assuming doping of the spinel structure with Mn^{4+} ions in the amount of $y_{Mn} = 0.0115$ mol/mol, for the value $\Delta G^o_{F_1} = \Delta G^o_{F_2} = \Delta G^o_{F_{Li}} = 150$ kJ/mol. Lines (---) represent the deviation from the stoichiometry and the concentration of ionic defects at the value of $y_{Mn} = 0.0$ mol/mol. Points (○) represent the results of the studies on the deviation δ (Hosoya et al. 1997), and points (◆) – the corrected values. Points (●) are the values recalculated relative to the value $\delta_o = 0.0055$, points (□) – the values recalculated relative to the corrected values of the deviation from the stoichiometry (◆). (b) at 1023 K (solid line) when assuming doping of the spinel structure with Mn^{4+} ions in the amount of $y_{Mn} = 0.0115$ mol/mol, for the values $\Delta G^o_{F_{Li}} = 150$ kJ/mol, $\Delta G^o_{F_1} = \Delta G^o_{F_2} = 180$ kJ/mol. Line (---) represents the dependence of the deviation from the stoichiometry on p_{O_2} when $y_{Mn} = 0.0$ mol/mol. The (-···) line represents the dependence of δ on p_{O_2} when $\Delta G^o_{F_{Li}} = \Delta G^o_{F_1} = \Delta G^o_{F_2} = 150$ kJ/mol, presented in Figure 14.1a.

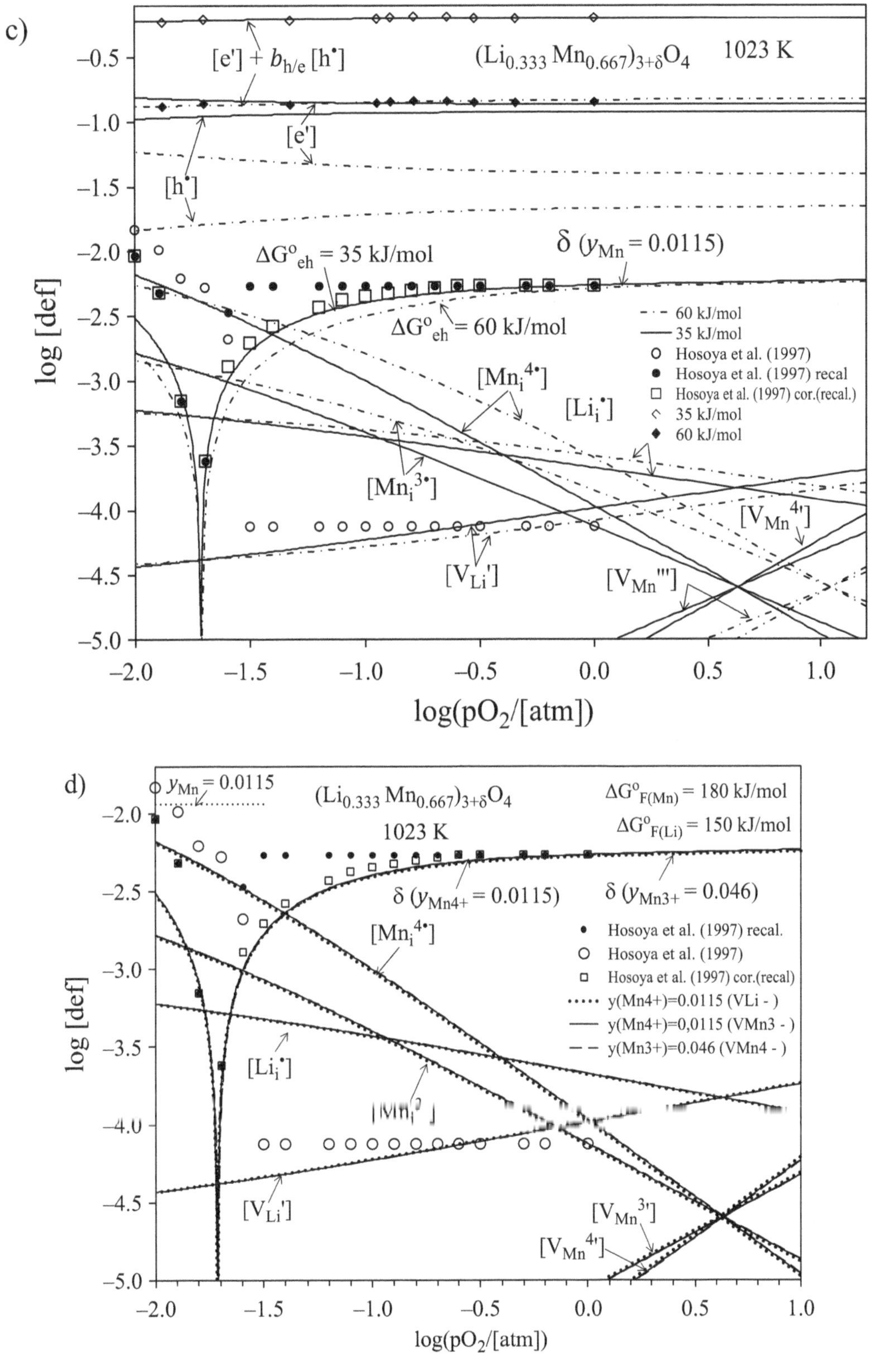

FIGURE 14.1 (Continued) (c) Diagram of concentrations of point defects for the spinel $(Li_{0.333}Mn_{0.667})_{3+\delta}O_4$ at 1023 K (solid lines) when assuming doping of the spinel structure with Mn^{4+} ions in the amount of $y_{Mn} = 0.0115$ mol/mol, for the values $\Delta G^o_{F_{Li}} = 150$ kJ/mol, $\Delta G^o_{F_1} = \Delta G^o_{F_2} = 180$ kJ/mol and for the value $\Delta G^o_{eh} = 35$ kJ/mol) (solid lines) and $\Delta G^o_{eh} = 60$ kJ/mol ((-·-) lines) (other details as in caption to Figure 14.1a. (d) at 1023 K, when assuming doping of the spinel structure with Mn^{4+} ions in the amount of $y_{Mn} = 0.0115$ mol/mol, according to Equation 14.6a ((- - -) lines), according to Equation 14.6b (solid lines), and Mn^{3+} in the amount of $y_{Mn} = 0.046$ mol/mol according to Equation 14.7a (dotted lines) for the value $\Delta G^o_{F_{Li}} = 150$ kJ/mol, $\Delta G^o_{F_1} = \Delta G^o_{F_2} = 180$ kJ/mol.

If we assume that Mn^{4+} ions will incorporate only into the sublattice of cations with octahedral coordination, then the process will follow the reaction:

$$2MnO_2 \rightarrow 4O_O^x + Mn_{Mn(3+)}^{\bullet} + Mn_{Mn(4+)}^x + V_{Li}' \tag{14.6b}$$

As the reactions of formation of defects 14.1 and 14.2 are mutually related, the reactions 14.6a and 14.6b are equivalent and they formally determine the mechanism of incorporation of Mn^{4+} ions. Therefore, the incorporation of Mn^{4+} ions into the spinel $(LiMn_2)_{1+\Delta}O_4$ will cause an increase in the concentration of vacancies $\left[V_{Mn}'''\right]$ or/and $\left[V_{Li}'\right]$, which will result in a change in concentrations of the remaining point defects.

A similar effect of doping is possible for doping with Mn^{3+} ions (Mn_2O_3). In such a case the doping process would follow the reaction:

$$4/3Mn_2O_3 \rightarrow 4O_O^x + Mn_{Li}^{\bullet\bullet} + Mn_{Mn(3+)}^x + 2/3Mn_{Mn(4+)}' + 1/3V_{Mn}^{4'} \tag{14.7a}$$

or

$$4/3Mn_2O_3 \rightarrow 4O_O^x + Mn_{Mn(3+)}^x + Mn_{Mn(4+)}' + 2/3Mn_{Li}^{\bullet\bullet} + 1/3V_{Li}' \tag{14.7b}$$

As we can see from Equations 14.6a, 14.6b and 14.7a, 14.7b, the doping of the spinel sublattice with manganese ions causes an increase in the concentration of manganese and lithium vacancies, while the concentration of interstitial cations will decrease (concentrations of defects are mutually related). Thus, doping of the spinel ($LiMn_2O_4$) with the stoichiometric composition with manganese ions leads to an increase in the concentration of cation vacancies. The equilibrium oxygen pressure over the spinel with such a composition will be shifted towards higher oxygen pressures compared to non-doped spinel. A decrease in the pressure with the change in relative concentrations of ionic defects, and, as the concentration of cation vacancies increased, the stoichiometric composition will be reached at lower oxygen pressures than in the case of non-doped spinel.

The structure of the spinel $LiMn_2O_4$ can be also doped with lithium ions into the sublattice of manganese ions, which, due to their charge and size, seems less likely. The process of doping with lithium ions will follow the reaction:

$$4Li_2O + 5V_i \rightarrow 4O_O^x + Li_{Mn(3+)}'' + Li_{Mn(4+)}''' + Li_{Li}^x + 5Li_i^{\bullet} \tag{14.8}$$

From the reaction 14.8 it results that the incorporation, into the spinel with the stoichiometric composition, of excess lithium ions into the sublattice of manganese ions leads to an increase in the concentration of interstitial lithium ions $\left(Li_i^{\bullet}\right)$, which causes that the equilibrium oxygen pressure over such a composition will be shifted lower oxygen pressures than before the doping. An increase in the oxygen pressure will cause a change in mutual concentrations of ionic defects and due to an increase in the concentration of interstitial cations the stoichiometric composition will be reached at higher oxygen pressures than in non-doped spinel. Thus, doping with manganese and lithium ions, charged in relation to the lattice, have an opposite effect to the change in the character of the dependence of the deviation from the stoichiometry on p_{O_2}. Due to the compensation of the influence of the above doping ions, their effective concentration should be expected, being a difference of the concentrations of ions charged in relation to the lattice. Depending on their effective concentration, a change occurs in the character of dependence of the deviation from the stoichiometry and the concentration of point defects on the oxygen pressure.

14.3 METHODS OF THE CALCULATIONS OF THE DIAGRAMS OF THE CONCENTRATIONS OF DEFECTS

In order to determine the diagram of concentrations of point defects, calculation methods analogous to the case of doped magnetite were used (see Section 1.5 and 6.2), as well as the results of the deviation from the stoichiometry (Hosoya et al. 1997). The value of δ_o was fitted in order to obtain a range of oxygen pressure where there is a constant value of the deviation from the stoichiometry and a range of oxygen pressure where it increases. As can be seen in Figure 14.1, by fitting the value of $\delta_o = 0.055$, we caused a significant change in the value of the deviation from the stoichiometry ((●) points) and a change in the character of the dependence δ on p_{O_2} ((···) line). There is a range of higher oxygen pressures where the values of the deviation from the stoichiometry are constant, independent of the oxygen pressure. There is an oxygen pressure at which the spinel reaches the stoichiometric composition and a narrow range p_{O_2} where the deviation from the stoichiometry δ increases. At lower values of δ_o, the oxygen pressure at which the spinel reaches the stoichiometric composition shifts slightly and the values of the deviation from the stoichiometry decrease in the range of higher oxygen pressures and at the lowest pressures.

The above character of the dependence of δ on p_{O_2} indicates that a given concentration of manganese ions y_{Mn} is incorporated into the spinel $(Li_{0.333}Mn_{0.667})_{3+\delta}O_4$ (according to the reactions 14.6a, 14.6b or 14.7a, 14.7b); as a result, the concentration of cation vacancies increases.

A lack of a range of oxygen pressure, where a significant concentration of cation vacancies, would be present (there are no symmetrical branches of the dependence of δ in p_{O_2}), as in the case of other heterocation spinels, makes it difficult to unambiguously determine the diagram of the concentrations of defects. In such a case, apart from fitting the value of ΔG_i^o of formation of defects, it is necessary to simultaneously fit the value of the concentration of manganese ions incorporated into the spinel structure (y_{Mn}). Analogously to the case of magnetite, in order to calculate the concentration of ionic defects, the expression describing the equilibrium constants (Equation 14.4a–14.4f) is used, as well as the relations, resulting from these equations, between the concentrations of the individual defects at a constant oxygen pressure.

If the process of doping the spinel $(Li_{0.333}Mn_{0.667})_{3+\delta}O_4$ with Mn^{4+} ions follows Equation 14.6b, it will lead to an increase in the concentration of lithium vacancies. Due to that, we calculate the concentration of ionic defects depending on the concentration of lithium vacancies.

From Equation 14.4a and 14.4b, it results that at the same oxygen pressure there is a relation between the concentration of vacancies $\left[V_{Mn}'''\right]$ and $\left[V_{Li}'\right]$:

$$\frac{\left[V_{Li}'\right]\left[h^{\bullet}\right]^4}{K_{V_{Li}'}^4} = \frac{\left[V_{Mn}'''\right]^{4/3}\left[h^{\bullet}\right]^4}{K_{V_{Mn}'''}^{4/3}} \tag{14.9a}$$

The ratio of concentration of vacancies $\left[V_{Li}'\right]/\left[V_{Mn}'''\right]$ depends on the equilibrium constants $K_{V_{Li}'}$ and $K_{V_{Mn}'''}$, and, after transformation, we get:

$$\frac{\left[V_{Mn}'''\right]^{4/3}}{\left[V_{Li}'\right]^4} = \frac{K_{V_{Mn}'''}^2}{K_{V_{Li}'}^4} = K_{V_{Mn}'''/V_{Li}'} = K_1 \tag{14.9b}$$

The concentrations of manganese vacancies $\left[V_{Mn}'''\right]$ dependence on the concentration of lithium vacancies assumes the form of:

$$\left[V_{Mn}'''\right] = K_1^{3/4}\left[V_{Li}'\right]^3 \tag{14.9c}$$

According to Equation 14.9b, the standard Gibbs energy $\Delta G_{V_{Mn}'''}^o$ of the formation of vacancies V_{Mn}''' depends on ΔG_1^o and $\Delta G_{V_{Li}'}^o$ according to the equation:

$$\Delta G_{V_{Mn}'''}^o = 3/4\Delta G_1^o + 3\Delta G_{V_{Li}'}^o \tag{14.9d}$$

The concentration of vacancies $\left[V_{Mn}^{4'}\right]$, depending on the concentration $\left[V_{Li}'\right]$, according to Equation 14.4a and 14.4e assumes the form of:

$$\frac{\left[V_{Mn}^{4'}\right]^4}{\left[V_{Li}'\right]^4} = \frac{K_{V_{Mn}^{4'}}}{K_{V_{Li}'}^4} = K_{V_{Mn}^{4'}/V_{Li}'} = K_2 \tag{14.10a}$$

The dependence of the concentration of manganese vacancies $\left[V_{Mn}^{4'}\right]$ on the concentration of lithium vacancies assumes the form of:

$$\left[V_{Mn}^{4'}\right] = K_2\left[V_{Li}'\right]^4 \tag{14.10b}$$

while the standard Gibbs energy $\Delta G_{V_{Mn}^{4'}}^o$ of formation of vacancies $V_{Mn}^{4'}$ depends on ΔG_2^o and $\Delta G_{V_{Li}'}^o$, according to the equation:

$$\Delta G_{V_{Mn}^{4'}}^o = \Delta G_2^o + 4\Delta G_{V_{Li}'}^o \tag{14.10c}$$

In turn, the concentration of interstitial cations is calculated using the equilibrium constants of formation of Frenkel defects; these relations, at a constant oxygen pressure, result from Equation 14.4a and 14.4d etc.:

$$\left[V'_{Li}\right]\left[Li_i^\bullet\right] = K_{V'_{Li}} K_{Li_i^\bullet} = K_{F_{Li}} \tag{14.11a}$$

$$\left[V'''_{Mn}\right]\left[Mn_i^{3\bullet}\right] = K_{V^{3'}_{Mn}} K_{Mn_i^{3\bullet}} = K_{F_1} \tag{14.11b}$$

$$\left[V^{4'}_{Mn}\right]\left[Mn_i^{4\bullet}\right] = K_{V^{4'}_{Mn}} K_{Mn_i^{4\bullet}} = K_{F_2} \tag{14.11c}$$

According to Equation 14.6b, the incorporation of Mn^{4+} ions into the spinel with the stoichiometric composition causes an increase in the concentration of lithium vacancies by $1/2y_{Mn}$:

$$\left[V'_{Li}\right] = y^{o}_{V'_{Li}} + 1/2y_{Mn} \tag{14.12}$$

where $y^{o}_{V'_{Li}}$ denotes the concentration of lithium vacancies at the stoichiometric composition.

As it results from Equation (14.4b), the incorporation of ions Mn^{4+} in an amount of y_{Mn} causes the formation in the sublattice of Mn^{3+} ions, $Mn^\bullet_{Mn(3+)}$ ions in the amount of $\left[Mn^\bullet_{Mn}\right] = 1/2y_{Mn}$, which perturbs the electroneutrality of the lattice and the electroneutrality condition assumes the form of:

$$\left[e'\right]+\left[V'_{Li}\right]+3\left[V'''_{Mn}\right]+4\left[V^{4'}_{Mn}\right]=\left[h^\bullet\right]+\left[Li_i^\bullet\right]+3\left[Mn_i^{3\bullet}\right]+4\left[Mn_i^{4\bullet}\right]+1/2y_{Mn} \tag{14.13}$$

Thus, the concentration of electronic defects will depend on the concentration of ionic defects and on the concentration of ions $\left[Mn^\bullet_{Mn}\right]$. Calculating from Equation 14.13 e.g. electron concentration and inserting into Equation 14.5, relating the concentrations of electronic defects, we get a dependence of the concentration of electron holes on the concentration of defects (see Equation 1.41). Solving the quadratic equation, knowing the equilibrium constant K_{eh} which is derived from the value of ΔG^{o}_{eh}, we determine the dependence of the concentration of electron holes on the concentration of ionic defects and, next, the concentration of electrons.

As a result of doping the spinel $(Li_{0.333}Mn_{0.667})_{3+\delta}O_4$ with Mn^{4+} ions, there occurs a build-up of the crystal (the number of lattice nodes increases), which must be taken into account when calculating the deviation from the stoichiometry, which is calculated according to the relation:

$$\frac{\left[Mn\right]+\left[Li\right]}{\left[O\right]} = \frac{3+y_{Mn}+\Sigma\left[def\right]}{4+2y_{Mn}} = \frac{3+\delta}{4} \tag{14.14}$$

The equilibrium pressure that results from the concentration of ionic defects and electronic defects can be calculated from Equation 14.4a using the equilibrium constant $K_{V'_{Li}}$ $(\Delta G^{o}_{V'_{Li}})$, concentrations of lithium vacancies and electron holes:

$$p_{O_2} = K_{V'_{Li}}\left[V'_{Li}\right]^4\left[h^\bullet\right]^4 \tag{14.15}$$

Thus, in order to determine the diagram of the concentrations of defects it is necessary to assume the value ΔG^{o}_{eh} and fit the values of $\Delta G^{o}_{V'_{Li}}$, of formation of lithium vacancies, values $\Delta G^{o}_{F_{Li}}$, $\Delta G^{o}_{F_1}$ and $\Delta G^{o}_{F_2}$ of formation of Frenkel defects, and it is necessary to fit the value of ΔG^{o}_{1} which allows calculating the constant $K_1 = K_{V'''_{Mn}/V'_{Li}}$, relating the concentration of manganese vacancies $\left[V'''_{Mn(o)}\right]$ and $K_2 = K_{V^{4'}_{Mn}/V'_{Li}}$, relating the concentration of vacancies $\left[V^{4'}_{Mn}\right]$ with the concentration of lithium vacancies (see Equation 14.9c and 14.10b), and fit such a value of y_{Mn} that the dependence of deviation from the stoichiometry on p_{O_2} would be close to the experimental results δ. At every change in the value ΔG^{o}_{i} of formation of defects, we cause a perturbation of the electroneutrality condition (14.13). Consequently, the calculations should be performed until this condition has been fulfilled. In the first stage of the calculation, identical values of ΔG^{o}_{i} of formation of Frenkel defects were assumed ($\Delta G^{o}_{F_{Li}} = \Delta G^{o}_{F_1} = \Delta G^{o}_{F_2}$); as a result, at the stoichiometric composition, the concentrations of cation vacancies and interstitial cations, respectively, are the same. In next steps it is necessary to analyse the effect of the individual values of ΔG^{o}_{i} on the degree of match between the dependence of δ on p_{O_2} to the experimental values of δ. Analogous expressions for the dependence of the concentration of defects as Equations 14.9a–14.9d to 14.11a–14.11c can be obtained when the incorporation of Mn^{4+} ions occurs according to Equation 14.6a or Mn^{3+} ions, according to Equation 14.7a and 14.7b. In this case the relative concentrations are determined depending on the concentration of vacancies $\left[V_{Mn}^{'''}\right]$ or $\left[V_{Mn}^{4'}\right]$.

14.4 RESULTS OF CALCULATION OF THE DIAGRAMS OF THE CONCENTRATIONS OF POINTS DEFECTS

As mentioned, the experimental character of the oxygen pressure dependence of the deviation from the stoichiometry, and particularly the lack, at higher oxygen pressures, of a range where the concentration of cation vacancies would depend on p_{O_2}, does not permit unambiguous determination of the values of standard Gibbs energies of formation of Frenkel defects ΔG_F^o and the concentration of doped manganese ions (y_{Mn}). Figure 14.1a present the results of calculations of concentrations of ionic defects (solid lines) for the lowest values of $\Delta G_{F_{Li}}^o = \Delta G_{F_1}^o = \Delta G_{F_2}^o = 150$ kJ/mol and for the value $y_{Mn} = 0.0115$ mol/mol, which can be considered as optimum values, at which there is quite a good match between the dependence of δ on p_{O_2} and the recalculated values of the deviation from the stoichiometry ((●) points) in the range of higher oxygen pressures and quite a good consistency near the stoichiometric composition. When fitting the values of ΔG_F^o of formation of Frenkel defects, the highest values of the deviation from the stoichiometry, at the lowest oxygen pressures were considered. The lower the values of ΔG_F^o, the higher the concentration of Frenkel defects. As can be seen in Figure 14.1a, there are slight differences between the dependence of δ on p_{O_2} and the recalculated values of the deviation from the stoichiometry at the value of $\delta_o = 0.055$ ((●) points). Lower values of ΔG_F^o cause an increase in the value of the deviation from the stoichiometry at the lowest oxygen pressures, but the degree of match between the dependence of δ and p_{O_2} and the experimental values of δ gets worse at higher oxygen pressures. In turn, at lower number of doped ions $y_{Mn} < 0.0115$ mol/mol, the range where the deviation from the stoichiometry is independent of the oxygen pressure is narrower. The dashed lines denote the deviation and the concentrations of defects for $y_{Mn} = 0$. As can be seen in Figure 14.1a, near the oxygen pressure at which the doped spinel reaches the stoichiometric composition, there are differences between the calculated dependence of δ on p_{O_2} and the recalculated values of the deviation from the stoichiometry ((●) points).

A small correction of the experimental values (Hosoya et al. 1997) ((◆) points), with the extent shown in Figure 14.2, causes that the recalculated values of the deviation from the stoichiometry ((□) points) for the value of $\delta_o = 0.055$) are significantly closer to the dependence of δ on p_{O_2}. The size of the correction of the δ is practically within the measurement error limit. However, this error is difficult to assess, as it depends on the time of setting of the equilibrium state, which is not given in the work (Hosoya et al. 1997). Setting of a constant mass of spinel sample usually requires quite a long time, especially when its changes are small and the oxygen pressures, relatively high, differ slightly.

In turn, Figure 14.1b presents the results of the calculations at the values of $\Delta G_{F_{Li}}^o = 150$ kJ/mol of formation of Frenkel defects ($\left[V_{Li}'\right]i\left[Li_i^{\bullet}\right]$) in the lithium sublattice and $\Delta G_{F_1}^o = \Delta G_{F_2}^o = 180$ kJ/mol of formation of Frenkel defects in the manganese sublattice. As can be seen in Figure 14.1b, at the highest pressures, a better match of the dependence of δ on p_{O_2}

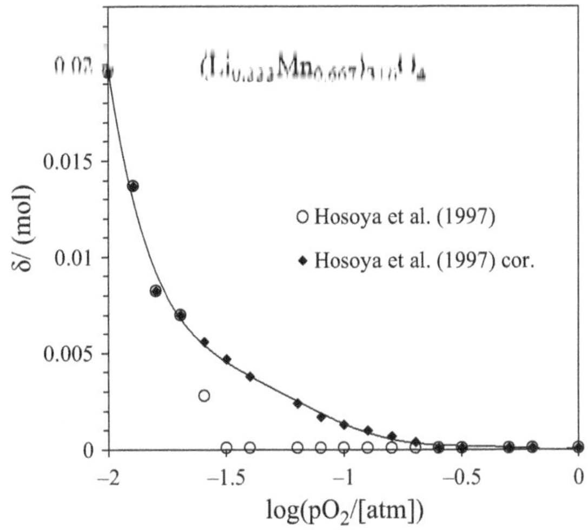

FIGURE 14.2 Dependence of the deviation from the stoichiometry on the oxygen pressure in the spinel $(Li_{0.333}Mn_{0.667})_{3+\delta}O_4$ (Hosoya et al. 1997) ((○) points) and the corrected values – (◆) points.

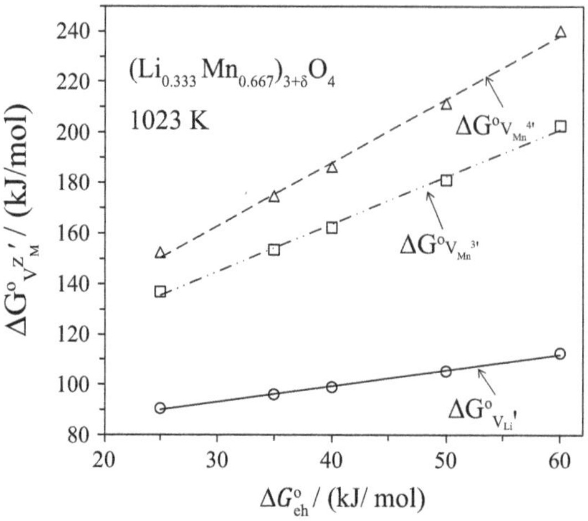

FIGURE 14.3 Dependence of standard Gibbs energy $\Delta G^o_{V_{Li}'}$ ((O) points) of formation of lithium vacancies V_{Li}', $\Delta G^o_{V_{Mn}'''}$ ((□) points) and $\Delta G^o_{V_{Mn}^{4'}}$ ((△) points) of formation of vacancies V_{Mn}''' and $V_{Mn}^{4'}$ on the value, assumed in the calculations, of ΔG^o_{eh} of formation of electronic defects in the spinel $(Li_{0.333}Mn_{0.667})_{3+\delta}O_4$ at 1023 K, at which there is a good degree of match between the dependence of the deviation from the stoichiometry on the oxygen pressure and the recalculated experimental values of δ.

and the value of the deviation δ was obtained; moreover, the range where the deviation from the stoichiometry is independent from the oxygen pressure gets longer.

As can be seen in Figure 14.1a,b, at the assumed values of $\Delta G^o_{F(Mn)}$ of formation of Frenkel defects in the manganese sublattice in the considered range of oxygen pressures (0.02–1 atm), the presence of an excess doping with Mn^{4+} ions, equal $y_{Mn} = 0.0115$ mol/mol, causes a very small increase in the concentration of interstitial cations and a decrease in the concentration of cation vacancies (solid lines) compared to the concentration of defects without taking into account $Mn^{\bullet}_{Mn}$ ions ((---) lines). However, in a log–log scale, there occurs a very important change in the character of the dependence of δ on p_{O_2}. When the value of ΔG^o_F of formation of Frenkel defects in the manganese sublattice increases, the oxygen pressure at which Frenkel defects are present ($[V_M] = [M_i]$) for the spinel doped with $Mn^{\bullet}_{Mn}$ ions and the pressure, at which the non-doped spinel ($y_{Mn} = 0$) reaches the stoichiometric composition ((---) line) shifts towards higher oxygen pressures ($p_{O_2} > 1$ atm). The (-···-) line denotes the dependence of δ on p_{O_2} for the value $\Delta G^o_{F_1} = \Delta G^o_{F_2} = 150$ kJ/mol presented in Figure 14.1a. The values of the deviation from the stoichiometry are practically unchanged at the lowest oxygen pressures. Larger values $\Delta G^o_{F_1} = \Delta G^o_{F_2} > 180$ do not change the degree of match between the dependence of δ on p_{O_2} and the value of the deviation δ, while the value of $\Delta G^o_{V_{Li}'}$ changes, as well as other values of ΔG^o_i of formation of defects. Also, the range of oxygen pressures where the deviation from the stoichiometry is constant increases, much above 1 atm.

Similarly, as in the case of other spinels, the values $\Delta G^o_{V_{Li}'}$ and ΔG^o_i of formation of the remaining defects depend on the assumed value of ΔG^o_{eh}. Figures 14.3 and 14.4 present the dependence of $\Delta G^o_{V_{Li}'}$, $\Delta G^o_{V_{Mn}^{4'}}$ and $\Delta G^o_{V_{Mn}'''}$ as well as $\Delta G^o_{Li_i^{\bullet}}$, $\Delta G^o_{Mn_i^{3\bullet}}$ and $\Delta G^o_{Mn_i^{4\bullet}}$ on the value ΔG^o_{eh} assumed in the calculations (in the range 25–60 kJ/mol), at which a relatively good match is obtained between the dependence of δ on p_{O_2} and the recalculated values of the deviation from the stoichiometry.

As can be seen in Figure 14.3, the dependence is linear up to the value of $\Delta G^o_{eh} = 60$ kJ/mol, with the correlation coefficient close to one ($R^2 = 0.9998$). However, at values above $\Delta G^o_{eh} = 40$ kJ/mol, the degree of match between the dependence of δ and p_{O_2} and the experimental results gets worse. Figure 14.1c compares the diagrams of concentrations of defects determined at the value of $\Delta G^o_{eh} = 35$ and 60 kJ/mol for the values $\Delta G^o_{F_{Li}} = 150$ kJ/mol and $\Delta G^o_{F_2} = \Delta G^o_{F_3} = 180$ kJ/mol.

As can be seen in Figure 14.1c, at the value of $\Delta G^o_{eh} = 60$ kJ/mol, the match between the dependence of δ on p_{O_2} ((-·-) line) and the recalculated values of the deviation ((□) points) is significantly worse. The oxygen pressure at which concentrations of cation vacancies and interstitial cations (Frenkel defects) are equal $[V_M] = [M_i]$ also shifts towards higher oxygen pressures. The concentration of electronic defects is significantly lower and the difference between the concentration of holes and electrons increases.

Analogous calculations of diagrams of concentrations of defects were performed for the spinel $(Li_{0.333}Mn_{0.667})_{3+\delta}O_4$ in which excess ions Mn^{4+} were incorporated, according to the reaction 14.6a, and identical concentrations of defects were obtained. Similarly, the calculations were performed for the case of incorporation of excess Mn^{3+} ions according to the

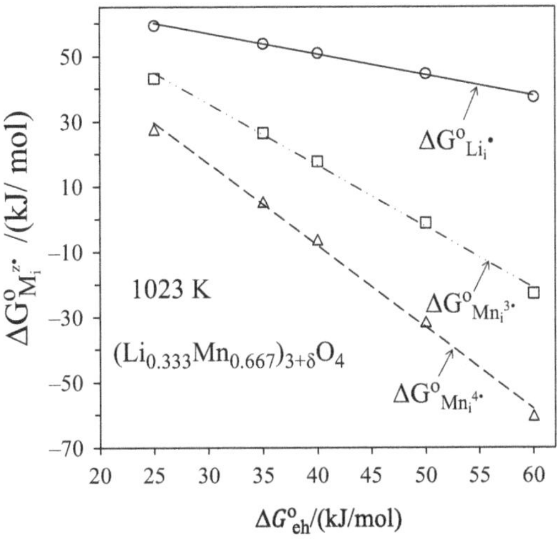

FIGURE 14.4 Dependence of the standard Gibbs energy $\Delta G^o_{Li_i^\cdot}$ ((○) points) of formation of interstitial lithium ions $Li_i^\cdot$ and $MnG^o_{Mn_i^{3\cdot}}$ ((□) points) and $\Delta G^o_{Mn_i^{4\cdot}}$ ((△) points) of formation of interstitial manganese ions $Mn_i^{3\cdot}$ and $Mn_i^{4\cdot}$ on the value, assumed in the calculations, of ΔG^o_{eh} of formation of electronic defects in the spinel $(Li_{0.333}Mn_{0.667})_{3+\delta}O_4$ at 1023 K, at which there is a good degree of match between the dependence of the deviation from the stoichiometry on the oxygen pressure and the recalculated experimental results δ.

reaction 14.7a and 14.7b. Figure 14.1d compares the diagrams of concentrations of defects calculated when taking into account the incorporation of excess Mn^{4+} ions according to the reaction 14.7a (solid line), according to Equation 14.9b ((···) lines), and Mn^{3+} ions, according to the reaction 14.8 ((---) line).

As can be seen in Figure 14.1d, in all the cases the same degree of match was obtained between the dependence of δ on p_{O_2} on and the recalculated values of the deviation ((□) points), as well as the same dependences of the concentration of defects on p_{O_2} (the lines overlap). However, in the case of the incorporation of Mn^{3+} ions, according to Equation 14.7a, the match of the dependence of δ on p_{O_2} was obtained for four times higher amount of the dopant (y_{Mn} = 0.046 mol/mol), which is due to different doping equations. The above results of calculations confirm that the patterns of the incorporation of Mn^{4+} and Mn^{3+}, according to Equation 14.6a, 14.6b, 14.8 are equivalent and they indicate an increase in one case of the concentrations of point defects, which causes a change in the concentrations of the remaining point defects.

However, the lack of results of studies at the oxygen pressure above 1 atm, where the deviation from the stoichiometry would have significant values, higher than the amount of the incorporated dopant (y_{Mn}), does not permit unambiguous determination of the values of ΔG_i^o of formation of defects. From the performed calculations it results that the minimum values $\Delta G^o_{F_1}$ and $\Delta G^o_{F_2}$ of formation of Frenkel defects are within the limits 160–180 kJ/mol, although in reality they could be much higher, as in the case of pure and doped hausmannite. Higher values $\Delta G^o_{F_1} = \Delta G^o_{F_2} > 170$ kJ/mol cause a shift in the oxygen pressures, at which Frenkel defects are present, towards higher oxygen pressures. Also, the concentration of Frenkel defects into the manganese sublattice decreases. The minimum value of $\Delta G^o_{F_1}$ of formation of Frenkel defects which does not deteriorate the match between the dependence of δ on p_{O_2} and the values of deviation δ is about 150 kJ/mol. At the above values, the concentration of defects in the lithium sublattice is about $5 \cdot 10^{-4}$ mol/mol at the oxygen pressure of 8 atm. The maximum concentration of the interstitial lithium ions at oxygen pressures 0.01 atm is about 10^{-3} mol/mol.

From the obtained calculations of diagrams of concentrations of defects in $(Li_{0.333}Mn_{0.667})_{3+\delta}O_4$ it results that the character of the dependence of δ on p_{O_2} is a result of incorporation of excess manganese ions Mn^{4+} or/and Mn^{3+} into the sublattices of lithium ions and manganese ions (y_{Mn}) Figure 14.1 Incorporation of a relatively high concentration of excess manganese ions into the spinel with the stoichiometric composition causes an increase in the concentration of lithium vacancies or/and manganese vacancies, which leads to a shift in the equilibrium oxygen pressure much above 1 atm. In the range of the studied oxygen pressures, 0.01–1 atm, there occurs a significant increase in the deviation from the stoichiometry and the doped spinel reaches the stoichiometric composition at low oxygen pressures. However, in the analysed range of oxygen pressures, the relative concentrations of point defects change only slightly. The concentration of the interstitial cations (dominating defects) increases; the oxygen pressure at which Frenkel defects are present, slightly decreases.

As can be seen in Figure 14.1, at the values of $\Delta G^o_{F_1} = \Delta G^o_{F_2} = 180$ kJ/mol, the dominating defects are interstitial ions $Mn_i^{4\cdot}$. The concentration of interstitial manganese ions $\left[Mn_i^{3\cdot}\right]$ is lower; it can be considered as the maximum concentration, as they do not deteriorate the match between the dependence of δ on p_{O_2} and the experimental results of the deviation δ (when $\Delta G^o_{F_1} = \Delta G^o_{F_2}$). Similarly, due to a weak dependence of the concentration of lithium defects on the oxygen pressure, their concentration can be significantly higher than that of manganese defects (exponent $n = -1/4$ when assuming a high concentration of electronic defects (see Equation 14.4d). As it was found, the concentration of interstitial lithium ions $\left[Li_i^\bullet\right]$, at the value of $\Delta G^o_{F_{Li}}$ = 150 kJ/mol, is the maximum concentration of these ions that does not deteriorate the match between the dependence of δ on p_{O_2} and the experimental values of δ. As it was already mentioned, the fitted concentration of doping manganese ions (y_{Mn}) is the effective concentration, resulting from the difference of the concentrations of excess manganese ions and possible excess lithium ions y_{Li}, as their influence on the concentration of ionic defects gets compensated. Thus, the real concentration of excess manganese ions $\left[Mn_{Li}^{3\bullet}\right]$, $\left[Mn_{Mn}^\bullet\right]$ or $\left[Mn_{Li}^{\bullet\bullet}\right]$, $\left[Mn'_{Mn}\right]$ see Equations 14.6a, 14.6b and 14.7a, 14.7b) could be higher.

14.5 CONCENTRATION OF ELECTRONIC DEFECTS AND THEIR MOBILITY

The value of the electrical conductivity and its dependence on the oxygen pressure can be used for the verification of the determined concentration of electronic defects. The results of the studies of electrical conductivity and thermoelectric power (Molenda et al. 1999) in the temperature range of 950–1090 K indicate that charge carriers are mainly electrons (negative value of the Seebeck coefficient) and their transport occurs according to the mechanism of small polarons. From the diagram of concentrations of point defects, it results that near the pressure of 1 atm (or above), the concentrations of cation vacancies and interstitial ions (Frenkel defects) should be equal. At the above oxygen pressure, also the concentrations of holes and electrons should be equal, and these defects should contribute to the total electrical conductivity. The character of the oxygen pressure dependence of the electrical conductivity should be affected by the concentration of ionic defects (interstitial cations).

The analysis of the character of the oxygen pressure dependence of the electrical conductivity indicates that at higher oxygen pressures it is practically independent of p_{O_2}. At the lowest oxygen pressures, the electrical conductivity systematically slightly decreases, despite the fact that when the oxygen pressure decreases, the concentration of electrons should increase, as the concentration of interstitial cations increase (see Figure 14.1, or Equation 14.4f and 14.5). In order to verify the above discrepancy, it was decided to use the calculated concentrations of electronic defects with the results of the studies of the electrical conductivity.

If we assume that in the spinel $(Li_{0.333}Mn_{0.667})_{3+\delta}O_4$, at 1023 K, mixed electrical conductivity occurs, and it is a sum of electron conductivity and hole conductivity (see Equation 1.48a, 1.48b), then, in such a case, analogous to the case of magnetite, the oxygen pressure dependence of the sum of the concentrations of electronic defects ($[e'] + b_{h/e}[h^\bullet]$) for the fitted value of the parameter $b_{h/e}$, which is the ratio of the mobility of holes to the mobility of electrons ($b_{h/e} = \mu_h/\mu_e$), should be consistent with the values of this sum calculated when using the values of the electrical conductivity (Molenda et al. 1999) for a fitted value of the mobility of electrons (see Equation 1.49):

$$\left[e'\right] + b_{h/e}\left[h^\bullet\right] = \frac{\sigma V_{LiMn_2O_4}}{F\mu_e} \tag{14.16}$$

where σ denotes the electrical conductivity, $b_{h/e}$ – the ratio of the mobility of holes to the mobility of electrons ($b_{h/e} = \mu_h/\mu_e$), F – Faraday constant, $V_{LiMn_2O_4}$ – molar volume of the spinel, which equals [illegible] cm³. Figure 14.4 presents the diagram of the concentrations of defects taking into account the concentration of electron holes and electrons and the sum of electronic defects ($[e'] + b_{h/e}[h^\bullet]$) obtained when assuming the value of ΔG^o_{eh} = 35 and 60 kJ/mol. In turn, Figure 14.5 presents only the concentration of electronic defects resulting from the diagram of concentrations of defects and the sum of concentrations of electronic defects ($[e'] + b_{h/e}[h^\bullet]$) for the value ΔG^o_{eh} = 35, 50 and 65 kJ/mol at a fitted value of $b_{h/e}$.

As can be seen in Figure 14.5, at all the values of ΔG^o_{eh} in the studied range of oxygen pressures, the concentration of electrons is higher than the concentration of electron holes and it increases when the oxygen pressure decreases, which is consistent with the negative values of the thermoelectric power. In turn, as can be seen, at the value of ΔG^o_{eh} = 35 kJ/mol assumed in the calculations, in the oxygen pressure up to 0.1 atm, the dependence of the concentration of electron holes and electrons only slightly depends on the oxygen pressure. When the value of ΔG^o_{eh} increases, this range is narrower and the difference between the concentrations of electrons and holes increases. The values of the sum of the concentrations of electronic defects ($[e'] + b_{h/e}[h^\bullet]$) at higher oxygen pressures practically do not depend on p_{O_2} and at low pressures they slightly decrease (analogously to the dependence of the electrical conductivity). Such a character of the above relationship indicates that in the range of low oxygen pressures, where the concentration of electrons is higher than the concentration of electron holes, the mobility of holes is higher than the mobility of electrons. For comparison, Figure 14.5 shows the dependence of ($[e'] + b_{h/e}[h^\bullet]$) on p_{O_2} for $b_{h/e} = 1$ (dotted line), determined when the value of ΔG^o_{eh} = 50 kJ/mol was assumed

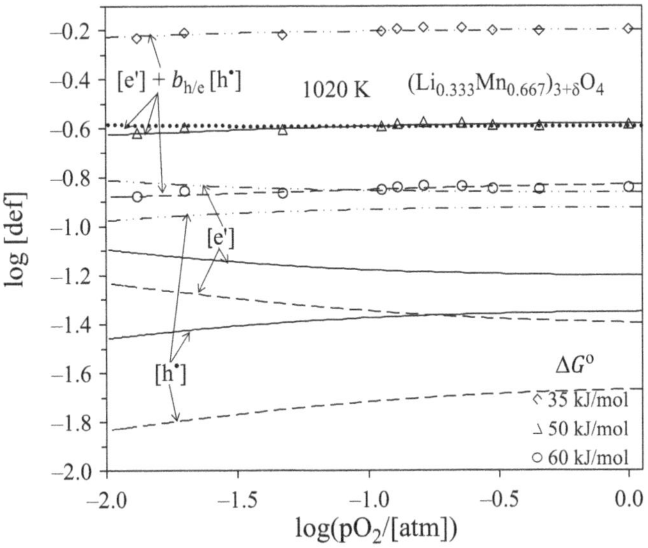

FIGURE 14.5 Oxygen pressure dependence of concentrations of electron holes and electrons and the sum of concentrations ($[e'] + b_{h/e}[h^•]$) in the spinel $(Li_{0.333}Mn_{0.667})_{3+\delta}O_4$ at 1023 K. Points denote the values of the sum ($[e'] + b_{h/e}[h^•]$) calculated according to Equation 14.16 using the values of the electrical conductivity (Molenda et al. 1999) and the fitted value of the mobility of electrons.

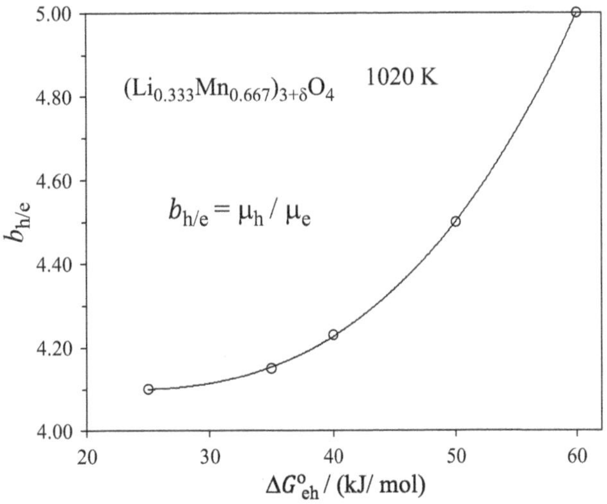

FIGURE 14.6 Dependence of the parameter $b_{h/e} = \mu_h/\mu_e$ (the ratio of mobility of holes to the mobility of electrons) in $(Li_{0.333}Mn_{0.667})_{3+\delta}O_4$ at 1023 K on the value of ΔG^o_{eh} of formation of electronic defects assumed in the calculations.

in the calculations. As can be seen, at low oxygen pressures the values deviate from a straight line. In turn, Figure 14.6 shows the dependence of the ratio of the mobility of electron holes to the mobility of electrons $b_{h/e} = \mu_h/\mu_e$ on the assumed values of ΔG^o_{eh} of formation of electronic defects that vary from 4.1 to 5.

Figure 14.7 shows the fitted values of the mobility of electrons; the value of the parameter $b_{h/e}$ was used to calculate the mobilities of electron holes, which, as can be seen, at the value of $\Delta G^o_{eh} = 35$ kJ/mol are equal about 0.1 and 0.5 cm^2/Vs.

14.6 CONCLUSIONS

The studies on the mass change of a $LiMn_2O_{4-\Delta}$ spinel sample (Hosoya et al. 1997) and the obtained character of the dependence indicated that the deviation from the stoichiometry cannot be related to oxygen vacancies. It results in effect from defects in the cation sublattice. Oxygen vacancies can be minority defects, but their concentration is so low that it does not

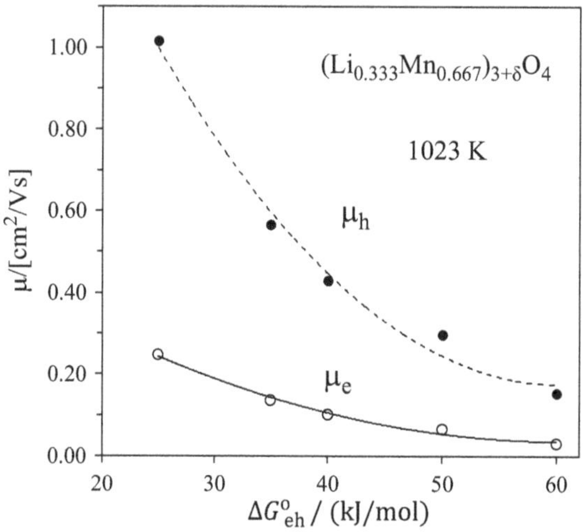

FIGURE 14.7 Dependence of the mobility of electrons μ_e ((○) points) and electron holes μ_h ((●) points) in the spinel $(Li_{0.333}Mn_{0.667})_{3+\delta}O_4$ on the value of ΔG^o_{eh} for formation of electronic defects, assumed in the calculations, at 1020 K.

affect the dependence of δ on p_{O_2}. It was found that when the reference value δ_o is fitted, the obtained character of dependence of the deviation from the stoichiometry is typical of a spinel with defects in the cation sublattice, where Mn^{4+} ions incorporate into the sublattice of Mn^{3+} ions (Mn_{Mn} with positive charge in relation to the sublattice of Mn^{3+}) and/or into the sublattice of lithium ions. Due to that, based on the proposed reactions of formation of point defects in sublattices of the spinel and using the recalculated values of the deviation from the stoichiometry (Hosoya et al. 1997), a diagram of the concentrations of point defects for the spinel $(Li_{0.333}Mn_{0.667})_{3+\delta}O_4$ was determined. From the obtained diagram of concentrations of point defects, it results that the spinel reaches the stoichiometric composition near the oxygen pressure of 0.01 atm. The pressure at which concentrations of cation vacancies and interstitial cations (Frenkel defects) are equal occurs near the oxygen pressure of 1 atm. In the analysed range of existence of the spinel (p_{O_2} = 1–0.01 atm), the dominating defects are interstitial manganese cations $\left(Mn_i^{4\bullet}\right)$ and ($Mn_i^{3\bullet}$), with a lower concentration. When the oxygen pressure decreases, the concentration of interstitial cations increases from $3{\cdot}10^{-5}$–0.01 mol/mol. The maximum concentration of interstitial lithium ions that does not affect the degree of match between the dependence of the deviation from the stoichiometry on p_{O_2} and the experimental values of the deviation δ can be from 10^{-4} to $3{\cdot}10^{-3}$ mol/mol (at $\Delta G^o_{F(Li)}$ of formation of Frenkel defects in the lithium sublattice about 150 kJ/mol). It was demonstrated that depending on the assumed value of ΔG^o_{eh} of formation of electronic defects (in the range from ΔG^o_{eh} = 25–60 kJ/mol), the values of ΔG^o_i of formation of ionic defects vary linearly. Simultaneously, above the value of ΔG^o_{eh} = 40 kJ/mol, the degree of match between the dependence of the deviation from the stoichiometry on p_{O_2} and the experimental values of the deviation δ deteriorates.

From the determined diagram of point defects, it results that the concentration of electrons is higher than the concentration of electron holes, which is consistent with the negative values of the thermoelectric power and with the lack of the dependence of electrical conductivity on the oxygen pressure. However, it was demonstrated that at low oxygen pressures, where there is a high concentration of interstitial cations, the mobility of electron holes can be higher than the mobility of electrons. The determined mobility of holes is μ_h = 0.5 cm²/Vs, and the mobility of electrons is about μ_e = 0.15 cm²/Vs (when ΔG^o_{eh} = 35 kJ/mol).

REFERENCES

Azzoni, C. B., M. C. Mozzati, A. Paleari, M. Bini, D. Capsoni, and V. Massarotti. 1999. Stoichiometry Effects on the Electrical Conductivity of Lithium-Manganese Spinels. *Z. Naturforsch. A.* 54: 579–584.

Cao, J., Y. Zhu, K. Bao, L. Shi, S. Liu, and Y. Qian. 2009. Microscale Mn_2O_3 Hollow Structures: Sphere, Cube, Ellipsoid, Dumbbell, and Their Phenol Adsorption Properties. *J. Phys. Chem. C.* 113: 17755–17760.

Dieckmann, R. 1982. Defects and Cation Diffusion in Magnetite (IV): Nonstoichiometry and Point Defect Structure of Magnetite ($Fe_{3-\delta}O_4$). *Ber. Bunsenges. Phys. Chem.* 86: 112–118.

Dziembaj, R., M. Molenda, D. Majda, and S. Walas. 2003. Synthesis, Thermal and Electrical Properties of $Li_{1+\Delta}mn_{2-\Delta}O_4$ Prepared by a Sol–Gel Method. *Solid State Ionics.* 157: 81–87.

Fei, J. B., Y. Cui, X. H. Yan, W. Qi, Y. Yang, K. W. Wang, Q. He, and J. B. Li. 2008. Controlled Preparation of MnO_2 Hierarchical Hollow Nanostructures and Their Application in Water Treatment. *Adv. Mater.* 20: 452–456.

Goodenough, J. B., A. Manthiran, and P. Wnętrzewski. 1993. Electrodes for Lithium Batteries. *J. Power Sources.* 43: 269–275.

Hoang, K. 2014. Understanding the Electronic and Ionic Conduction and Lithium Over-Stoichiometry in $Limn_2o_4$ Spinel. *J. Mater. Chem. A.* 2: 18271–18280.

Hosoya, M., H. Ikuta, T. Uchida, and and M. Wakihara. 1997. The Defect Structure Model in Nonstoichiometric $LiMn_2O_{4-\delta}$. *J. Electrochem. Soc.* 144: L52–L53.

Iguchi, E., N. Nakamura, and A. Aoki. 1998. Electrical Transport Properties in $Limn_2o_4$. *Phil. Mag. B.* 78: 65–77.

Marzec, J., K. Świerczek, J. Przewoźnik, J. Molenda, D. R. Simon, E. M. Kelder, and J. Schoonman. 2002. Conduction Mechanism in Operating a $Limn_2o_4$ Cathode. *Solid State Ionics.* 146: 225–237.

Masquelier, C., M. Tabuchi, K. Ado, R. Kanno, Y. Kobayashi, Y. Maki, O. Nakamura, and J. B. Goodenough. 1996. Chemical and Magnetic Characterization of Spinel Materials in the $LiMn_2O_4$-$Li_2Mn_4O_9$-$Li_4Mn_5O_{12}$ System. *J. Solid State Chem.* 123: 255–266.

Massarotti, V., D. Capsoni, M. Bini, C. B. Azzoni, and A. Paleari. 1997. Stoichiometry of Li_2MnO_3 and $LiMn_2O_4$ Coexisting Phases: XRD and EPR Characterization. *J. Solid State Chem.* 128: 80–86.

Massarotti, V., D. Capsoni, M. Bini, G. Chiodelli, C. B. Azzoni, M. C. Mozzati, and A. Paleari. 1997. Electric and Magnetic Properties of $LiMn_2O_4$ and Li_2MnO_3 Type Oxides. *J. Solid State Chem.* 131: 94–100.

Miura, K., A. Yamada, and M. Tanaka. 1996. Electric States of Spinel $Li_xMn_2O_4$ as a Cathode of the Rechargeable Battery. *Electrochim. Acta.* 41: 249–256.

Molenda, J., R. Dziembaj, E. Podstawka, and L. M. Proniewicz. 2005. Changes in Local Structure of Lithium Manganese Spinels (Li:Mn=1:2) Characterised by Xrd, Dsc, Tga, Ir, and Raman Spectroscopy. *J. Phys. Chem. Solids.* 66: 1761–1768.

Molenda, J., and W. Kucza. 1999. Transport Properties of $LiMn_2O_4$. *Solid State Ionics.* 117: 41–46.

Molenda, J., W. Ojczyk, M. Marzec, J. Marzec, J. Przewoźnik, R. Dziembaj, and M. Molenda. 2003. Electrochemical and Chemical Deintercalation of $LiMn_2O_4$. *Solid State Ionics.* 157: 73–79.

Molenda, J., K. Świerczek, W. Kucza, J. Marzec, and A. Stokłosa. 1999. Electrical Properties of $Limn_2o_{4-\Delta}$ at Temperatures 220–1100 K. *Solid State Ionics.* 123: 155–163.

Ohzuku, T., M. Kitagawa, and T. Hirai. 1990. Electrochemistry of Manganese Dioxide in Lithium Nonaqueous Cell: III. X-Ray Diffractional Study on the Reduction of Spinel-Related Manganese Dioxide. *J. Electrochem. Soc.* 137: 769–775.

Palacín, M. R., Y. Chabre, L. Dupont, M. Hervieu, P. Strobel, G. Rousse, C. Masquelier, and M. Anne. 2000. On the Origin of the 3.3 and 4.5 V Steps Observed in $LiMn_2O_4$- Based Spinels. *J. Electrochem. Soc.* 147: 845–853.

Paulsen, J. M., and J. R. Dahn. 1999. Phase Diagram of Li-Mn-O Spinel in Air. *Chem. Mater.* 11: 3065–3079.

Richard, M. N., E. W. Fuller, and J. R. Dahn. 1994. The Effect oAmmonia Reduction on the Spinel Electrode Materials, $LiMn_2O_4$ and $Li(Li_{13}Mn_{53})O_4$. *Solid State Ionics.* 73: 81–91.

Rodriguez-Carvajal, J., G. Rousse, C. Masquelier, and M. Hervieu. 1998. Electronic Crystallization in a Lithium Battery Material: Columnar Ordering of Electrons and Holes in the Spinel $LiMn_2O_4$. *Phys. Rev. Lett.* 81: 4660–4663.

Rousse, G., C. Masquelier, J. Rodriguez-Carvajal, E. Elkaim, J. P. Lauriat, and J. L. Martinez. 1999. X-ray Study of the Spinel $LiMn_2O_4$ at Low Temperatures. *Chem. Mater.* 11: 3629–3635.

Scrosati, B. 1992. Lithium Rocking Chair Batteries: An Old Concept? *J. Electrochem. Soc.* 139: 2776–2781.

Shimakawa, Y., T. Numata, and J. Tabuchi. 1997. Verwey-Type Transition and Magnetic Properties of the $LiMn_2O_4$ Spinels. *J. Solid State Chem.* 131: 138–143.

Stokłosa, A. 2015. *Non-Stoichiometric Oxides of 3d Metals.* Pfäffikon: Trans Tech Publications Ltd.

Sugiyama, J., T. Atsumi, T. Hioki, S. Noda, and N. Kamegashira. 1996. Oxygen Nonstoichiometry of Spinel $LiMn_2o_{4-\Delta}$. *J. Alloys Comp.* 235: 163–169.

Sugiyama, J., T. Tamura, and H. Yamauchi. 1995. Elastic/Anelastic Behaviour During The Phase Transition In Spinel $LiMn_2o_4$. *J. Phys. Condens. Mat.* 7: 9755–9764.

Tarascon, J. M., F. Coowari, T. N. Bowmei, and G. Amatucci. 1994. Synthesis Conditions and Oxygen Stoichiometry Effects on Li Insertion into the Spinel $LiMn_2O_4$. *J. Electrochem. Soc.* 141: 1421–1431.

Tarascon, J. M., and D. Guyomard. 1991. Li Metal-Free Rechargeable Batteries Based on $Li_{1+x}Mn_20_4$ Cathodes ($0 \leq x \leq 1$) and Carbon Anodes. *J. Electrochem. Soc.* 138: 2864–2868.

Thackeray, M. M. 1995. Structural Considerations of Layered and Spinel Lithiated Oxides for Lithium Ion Batteries. *J. Electrochem. Soc.* 142: 2558–2563.

Thackeray, M. M. 1997. Manganese Oxides For Lithium Batteries. *Prog. Solid State Chem.* 25: 1–71.

Thackeray, M. M., W. I. F. David, P. G. Bruce, and J. B. Goodenough. 1983. Lithium Insertion Into Manganese Spinels. *Mater. Res. Bull.* 18: 461–472.

Thackeray, M. M., M. F. Mansuetto, D. W. Dees, and D. Vissers. 1996. The Thermal Stability Of Lithium-Manganese-Oxide Spinel Phases. *Mater. Res. Bull.* 31: 133–140.

Yamada, A., K. Miura, K. Hinokuma, and M. Tanaka. 1995. Synthesis and Structural Aspects of $LiMn_2O_{4\pm\delta}$ as a Cathode for Rechargeable Lithium Batteries. *J. Electrochem. Soc.* 142: 2149–2154.

Yamada, A., and M. Tanaka. 1995. Jahn-Teller structural phase transition around 280 K in $LiMn_2O_4$. *Mater. Res. Bull.* 30: 715–721.

Yamada, A., M. Tanaka, K. Tanaka, and K. Sekai. 1999. Jahn–Teller Instability In Spinel Li-Mn-O. *J. Power Sources.* 81–82: 73–78.

15 Structure of Defects in Doped Hausmannite

Summary

15.1 EFFECT OF DOPANTS ON THE CONCENTRATION OF POINT DEFECTS IN HAUSMANNITE

Similarly, as in the case of magnetite, the spinel structure of hausmannite permits incorporation of quite a significant number of ions of other metals that form solid solutions in spinel sublattices. Using the results of the studies of the deviation from the stoichiometry (Keller and Dieckmann 1985, Lu and Dieckmann 1993a, Lu and Dieckmann 1993b, Lu and Dieckmann 1995), the diagrams of defect concentrations were determined for hausmannite doped with iron ions, $(Mn_{1-x}Fe_x)_{3\pm\delta}O_4$, and with cobalt ions, $(Mn_{1-x}Co_x)_{3\pm\delta}O_4$, and with Co and Fe $(Co_xFe_zMn_{2z})_{3\pm\delta}O_4$ (where $x + 3y = 1$) with the dopant content of $x_M = 0.0$–0.333 mol; which allowed determination of their effect on the concentration and structure of point defects and on the range of existence of the above spinels. Conclusions from this research (Keller and Dieckmann 1985, Lu and Dieckmann 1993a, Lu and Dieckmann 1993b, Lu and Dieckmann 1995) indicated that, analogously as in magnetite, heterocation spinels are formed, in which doping ions form a solid solution and occupy cation sites in the sublattices of spinel structure, becoming site ions that are electroneutral in relation to the lattice. The distribution of cations with different charge in the sublattices of spinel structure at high temperatures is random, but resulting from the setting of an equilibrium state. Doping ions mainly cause a change in interactions between ions, which affects the range of oxygen pressures where spinels exist. Doping ions only slightly affect the structure and the concentration of point defects, as well as the character of the oxygen pressure dependence of the deviation from the stoichiometry. Doped hausmannite reaches the stoichiometric composition in the range of its existence and there is a range of oxygen pressures where at low oxygen pressures interstitial manganese cations $Mn_i^{3\bullet}$ or/and doping ions $M_i^{3\bullet}$ dominate. In the range of higher oxygen pressures, cation vacancies V_{Mn}''' dominate.

Figure 15.1 compares the oxygen pressure dependence of deviation from the stoichiometry at 1473 K for the spinels $(Mn_{1-x}Fe_x)_{3\pm\delta}O_4$ (dashed lines), $(Mn_{1-x}Co_x)_{3\pm\delta}O_4$ ((-·-), (-··-) lines) and $(Co_xFe_zMn_{2z})_{3\pm\delta}O_4$ (their composition fulfils the relation ($x + 3z = 1$)) (solid lines) with the dopant content of $x_M = 0.0$–0.333 mol, which, as can be seen, are essentially different. As can be seen in Figure 15.1, doping iron of hausmannite, in the spinel $(Mn_{1-x}Fe_x)_{3\pm\delta}O_4$ causes that the range of existence of the spinel shifts towards lower oxygen pressures in relation to the pure hausmannite (dotted line (···)). Cobalt doping shifts the range of existence of the spinel $(Mn_{1-x}Co_x)_{3\pm\delta}O_4$ towards higher oxygen pressures ((-··-) lines). Similarly, doping Co^{2+} ions of the spinel $(Co_xFe_zMn_{2z})_{3\pm\delta}O_4$ (solid lines) causes a shift in the range of existence of the spinel towards higher oxygen pressures.

For the spinel $(Mn_{1-x}Fe_x)_{3\pm\delta}O_4$, iron doping practically does not change the character of the oxygen pressure dependence of the deviation from the stoichiometry (in relation to $Mn_{3\pm\delta}O_4$). The concentration of cation vacancies changes slightly (at the same oxygen pressure). When the iron content increases, the concentration of interstitial cations significantly decreases (at the same oxygen pressure). Also, the maximum concentration of interstitial cations at the lowest oxygen pressures is lower.

In turn, in spinels $(Mn_{1-x}Co_x)_{3\pm\delta}O_4$ and $(Co_xFe_zMn_{2z})_{3\pm\delta}O_4$, similarly as in magnetite, into the sublattice of Mn^{3+} ions in the spinel structure, incorporation of a small concentration of Co^{2+} ions occurs, which become negative in relation to the lattice Co_{Mn}'. They cause a shift towards higher oxygen pressures at which the spinel reaches the stoichiometric composition, while the character of the dependence of the deviation from the stoichiometry on the oxygen pressure changes. Owing to a smaller deviation at the lowest oxygen pressures, and, above all, due to a narrow range of oxygen pressures (or a lack of this range) in which cation vacancies dominate, the determination of absolute values of the deviation from the stoichiometry by fitting is ambiguous and requires experimental verification.

Figure 15.2 presents the dependence of the probable maximum concentration of ions Co_{Mn}' in spinels $(Mn_{1-x}Co_x)_{3\pm\delta}O_4$ and $(Co_xFe_zMn_{2z})_{3\pm\delta}O_4$, which can reach the concentration of $1.5 \cdot 10^{-4}$ mol/mol. In the spinel $LiMn_2O_4$, the concentration of excess manganese ions, charged in relation to the lattice, is about 0.01 mol/mol.

As can be seen in Figure 15.1, when the content of the cobalt increases in spinels $(Mn_{1-x}Co_x)_{3\pm\delta}O_4$ and $(Co_xFe_zMn_{2z})_{3\pm\delta}O_4$, the deviation from the stoichiometry (concentration of interstitial cations) increases at the same oxygen pressure, while the

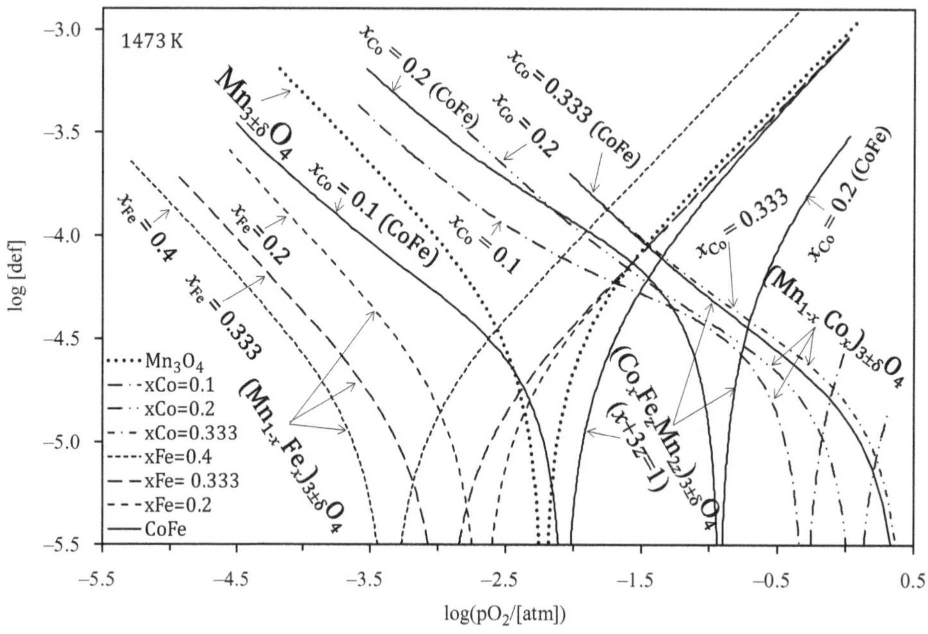

FIGURE 15.1 Dependence of the deviation from the stoichiometry on the oxygen pressure in β-$Mn_{3\pm\delta}O_4$ ((···) line), in spinels: $(Mn_{1-x}Fe_x)_{3\pm\delta}O_4$ (dashed lines), $(Mn_{1-x}Co_x)_{3\pm\delta}O_4$ (-·· lines) and $(Co_xFe_zMn_{2z})_{3\pm\delta}O_4$ (where $x + 3y = 1$) (solid line), with the content of ions x_M = 0.0–0.333 mol, at 1473 K, obtained using the results of the studies on the deviation from the stoichiometry (Keller and Dieckmann 1985, Lu and Dieckmann 1993a, Lu and Dieckmann 1993b, Lu and Dieckmann 1995).

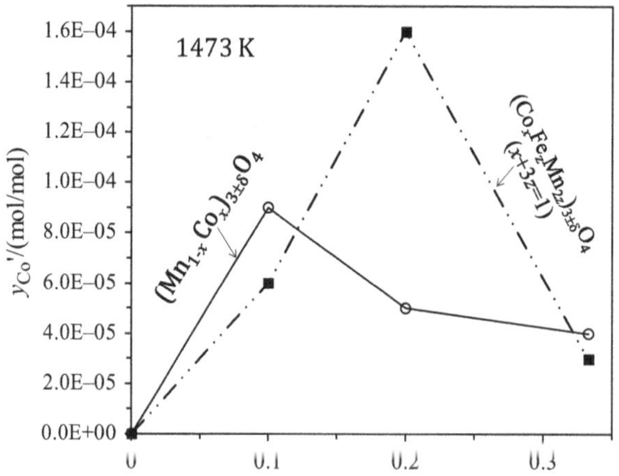

FIGURE 15.2 Dependence of the concentration of ions Co^{2+} (Co'_{Mn}) negative in relation to the lattice incorporated into the sublattice of Mn^{3+} ions on the content of doping ions x_M in spinels $(Mn_{1-x}Co_x)_{3\pm\delta}O_4$ ((○) points) and $(Co_xFe_zMn_{2z})_{3\pm\delta}O_4$ (where $x + 3y = 1$) ((■) points), at 1473 K, obtained using the results of the studies on the deviation from the stoichiometry (Lu and Dieckmann 1993b, Lu and Dieckmann 1995).

maximum concentration of interstitial cations, at which a transformation occurs of a spinel into a doped oxide, changes only slightly. This results from the shift towards higher oxygen pressures in the oxygen pressure dependence of the deviation from the stoichiometry.

Figure 15.3 presents the influence of doping iron ions and cobalt ions on the oxygen pressure at which, in the spinel, the concentrations of cation vacancies and interstitial cations – Frenkel defects – are equal, ($[V_M] = [M_i]$). These pressures indicate the real shift in the range of existence of the individual spinels. At the above pressure, the spinel reaches the stoichiometric composition, if ions M'_{Fe} are not incorporated.

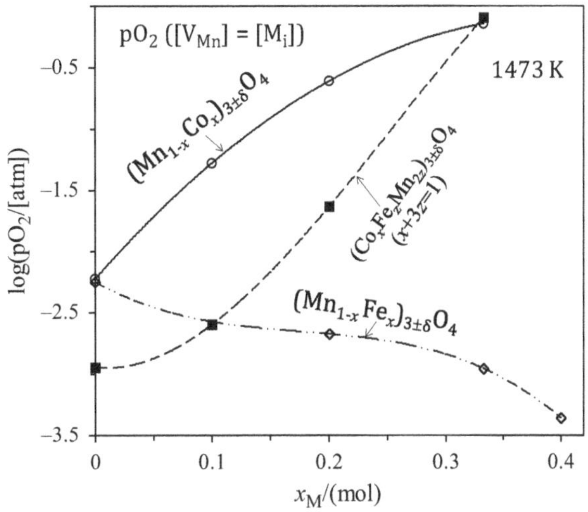

FIGURE 15.3 Dependence of the oxygen pressure at which the concentrations of cation vacancies and interstitial cations are equal on the doping ions content x_M in spinels: $(Mn_{1-x}Fe_x)_{3\pm\delta}O_4$ ((◇) points), $(Mn_{1-x}Co_x)_{3\pm\delta}O_4$ ((○) points) and $(Co_xFe_zMn_{2z})_{3\pm\delta}O_4$ (where $x + 3z = 1$) ((■) points) at 1473 K.

As can be seen in Figure 15.3, iron ions in hausmannite cause a slight decrease in the pressure at which the spinel reaches the stoichiometric composition. Co^{2+} ions in spinels $(Mn_{1-x}Co_x)_{3\pm\delta}O_4$ and $(Co_xFe_zMn_{2z})_{3\pm\delta}O_4$ cause an increase in the pressure at which the concentrations of defects are equal ($[V_{Fe}] = [M_i]$).

In turn, Figure 15.4 presents the dependence of the standard Gibbs energy of formation of Frenkel defects. Assuming in the calculations the same values of $\Delta G^{o}_{F_1} = \Delta G^{o}_{F_2}$ means that, despite the presence of doping ions and different charges of manganese cations, the energies of formation of Frenkel defects are similar. From the calculations it results that the values of $\Delta G^{o}_{F_2}$ of formation of Frenkel defects (V''_{Mn} and $Mn_i^{\bullet\bullet}$) are the lowest values that do not affect the degree of match between the dependence of the deviation from the stoichiometry on p_{O_2} and the experimental values of δ (the concentration of these defects can be considered as the maximum concentration in the spinel).

As can be seen in Figure 15.4, when the content of cobalt and iron in hausmannite increases, the values of $\Delta G^{o}_{F_1}$ increase. Cobalt doping, up to the content of $x_{Co} = 0.2$ mol in the spinel $(Co_xFe_zMn_{2z})_{3\pm\delta}O_4$ causes a decrease in the value of $\Delta G^{o}_{F_1}$, and then, at the content of $x_{Co} = 0.333$ mol, this value increases (the values were not determined in an unambiguous way). A decrease in the value of $\Delta G^{o}_{F_1}$ for the spinel $(Mn_{1-x}Fe_x)_{3\pm\delta}O_4$ with the iron content of $x_{Co} = 0.4$ mol indicates that its

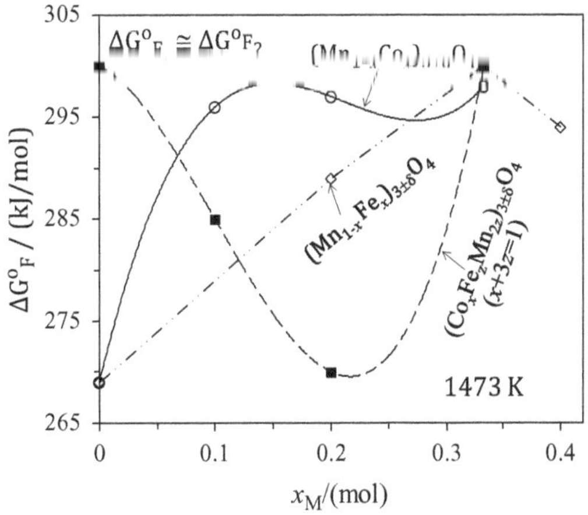

FIGURE 15.4 Dependence of standard Gibbs energy $\Delta G^{o}_{F_1} \cong \Delta G^{o}_{F_2}$ of formation of Frenkel defects on the dopant content x_M in spinels: $(Mn_{1-x}Fe_x)_{3\pm\delta}O_4$ ((◇) points), $(Mn_{1-x}Co_x)_{3\pm\delta}O_4$ ((○) points) and $(Co_xFe_zMn_{2z})_{3\pm\delta}O_4$ (where $x + 3z = 1$) ((■) points) at 1473 K, obtained using the results of the studies on the deviation from the stoichiometry (Keller and Dieckmann 1985, Lu and Dieckmann 1993a, Lu and Dieckmann 1993b, Lu and Dieckmann 1995).

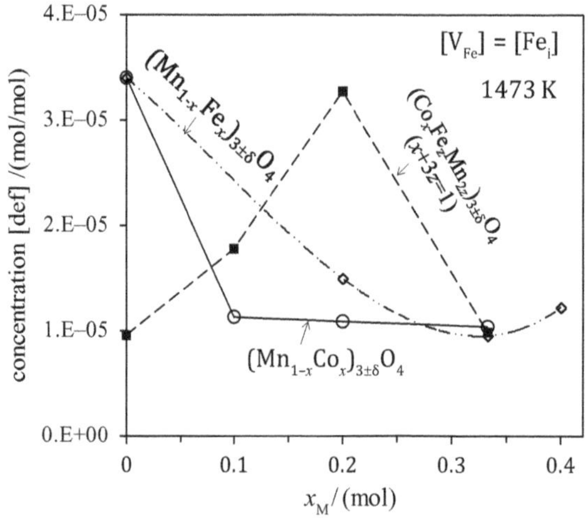

FIGURE 15.5 Dependence of the concentration of Frenkel defects on the dopant content x_M in spinels: $(Mn_{1-x}Fe_x)_{3\pm\delta}O_4$ ((◇) points), $(Mn_{1-x}Co_x)_{3\pm\delta}O_4$ ((○) points) and $(Co_xFe_zMn_{2z})_{3\pm\delta}O_4$ (where $x + 3z = 1$) ((■) points) at 1473 K, obtained using the results of the studies on the deviation from the stoichiometry (Keller and Dieckmann 1985, Lu and Dieckmann 1993a, Lu and Dieckmann 1993b, Lu and Dieckmann 1995).

properties are already different from those of spinels with a lower iron content (a transition range between the spinels $FeMn_2O_4$ and $MnFe_2O_4$).

The concentration of Frenkel defects depends on the value of $\Delta G^o_{F_1}$. Figure 15.5 presents the dependence of the concentration of Frenkel defects on the dopant content in spinels; as can be seen, they are opposite to the changes of the value of $\Delta G^o_{F_1}$ in individual spinels.

Figure 15.6 presents the dependence of $\Delta G^o_{V'''_{Mn}}$ and $\Delta G^o_{Mn_i^{3\bullet}}$ of formation of cation vacancies V'''_{Mn} and interstitial ions $Mn_i^{3\bullet}$, and Figure 15.7 – $\Delta G^o_{V''_{Mn}}$ and $\Delta G^o_{Mn_i^{\bullet\bullet}}$ of formation of V''_{Mn}, and $Mn_i^{\bullet\bullet}$ on the dopant content x_M in individual spinels. As demonstrated earlier, the values of the above quantities depend on the value of ΔG^o_{eh} of formation of electronic defects, assumed in the calculations. In the calculations it was approximatively assumed that the value of ΔG^o_{eh} does not depend on the dopant concentration in hausmannite and it is equal $\Delta G^o_{eh} = 60$ kJ/mol. It should be expected that the real values of ΔG^o_i of formation of defects can differ slightly, but the character of their dependence on the dopant concentration should be similar to the one presented in the above figures. Furthermore, the values $\Delta G^o_{V''_{Mn}}$ and $\Delta G^o_{Mn_i^{\bullet\bullet}}$ depend on the value of $\Delta G^o_{F_2}$, which also can be different in reality.

As can be seen in Figures 15.6 and 15.7, the character of the dependence of ΔG^o_i of formation of defects on the dopant content in spinels $(Mn_{1-x}Fe_x)_{3\pm\delta}O_4$, $(Mn_{1-x}Co_x)_{3\pm\delta}O_4$ and $(Co_xFe_zMn_{2z})_{3\pm\delta}O_4$ is different and it shows that the influence of the individual dopant ions on the interactions between ions is different.

15.2 INFLUENCE OF DOPANTS ON THE MOBILITY OF IONS VIA DEFECTS IN HAUSMANNITE

The studies on the diffusion of tracer ions Fe*, Co* and Mn* in doped hausmannite (Franke and Dieckmann 1989, Lu and Dieckmann 1993a, Lu and Dieckmann 1993b, Lu and Dieckmann 1995) demonstrated, similarly as for non-doped hausmannite, that the diffusion of metal ions (tracers) strictly depends on the concentration of cation vacancies and interstitial cations. Using the concentrations of ionic defects and the values of coefficients of self-diffusion of tracers (^{59}Fe, ^{60}Co and ^{54}Mn), the ratio of the coefficients of diffusion (mobilities) of ions via cation vacancies and via interstitial cations $\left(D^o_{V(M)}/D^o_{I(M)} = b'_D\right)$ was determined, as well as the mobilities of these ions in spinels $(Mn_{1-x}Fe_x)_{3\pm\delta}O_4$, $(Mn_{1-x}Co_x)_{3\pm\delta}O_4$ and $(Co_xFe_zMn_{2z})_{3\pm\delta}O_4$, depending on the dopant content x_M.

The performed calculations demonstrated that in pure hausmannite and in spinels $(Mn_{1-x}Fe_x)_{3\pm\delta}O_4$ and $(Co_xFe_zMn_{2z})_{3\pm\delta}O_4$ the mobility of ions via interstitial cations is higher than the mobility via cation vacancies, while their ratio $(D^o_{I(M)}/D^o_{V(M)})$ depends on the type of spinel and on the dopant concentration. In the spinel $(Mn_{1-x}Co_x)_{3\pm\delta}O_4$ the mobility of ions via cation vacancies is higher than the mobility via interstitial cations. Figure 15.8a compares the dependence of the ratio $\left(D^o_{I(M)}/D^o_{V(M)}\right)$,

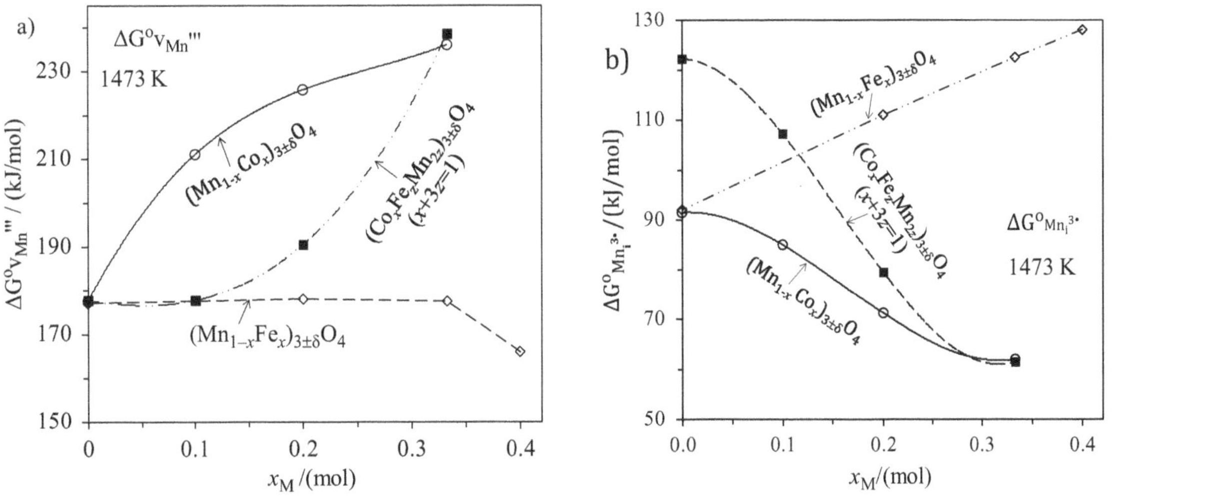

FIGURE 15.6 Dependence of standard Gibbs energy, a) $\Delta G^{o}_{V'''_{M}}$ of formation of cation vacancies $V'''_{M(o)}$, b) $\Delta G^{o}_{M_i^{3\bullet}}$, of formation of interstitial ions $M_i^{3\bullet}$ on the dopant content x_M in spinels: $(Mn_{1-x}Fe_x)_{3\pm\delta}O_4$ ((◇) points), $(Mn_{1-x}Co_x)_{3\pm\delta}O_4$ ((○) points) and $(Co_xFe_zMn_{2z})_{3\pm\delta}O_4$ (where $x + 3z = 1$) ((■) points) at 1473 K, obtained using the results of the studies on the deviation from the stoichiometry (Keller and Dieckmann 1985, Lu and Dieckmann 1993a, Lu and Dieckmann 1993b, Lu and Dieckmann 1995).

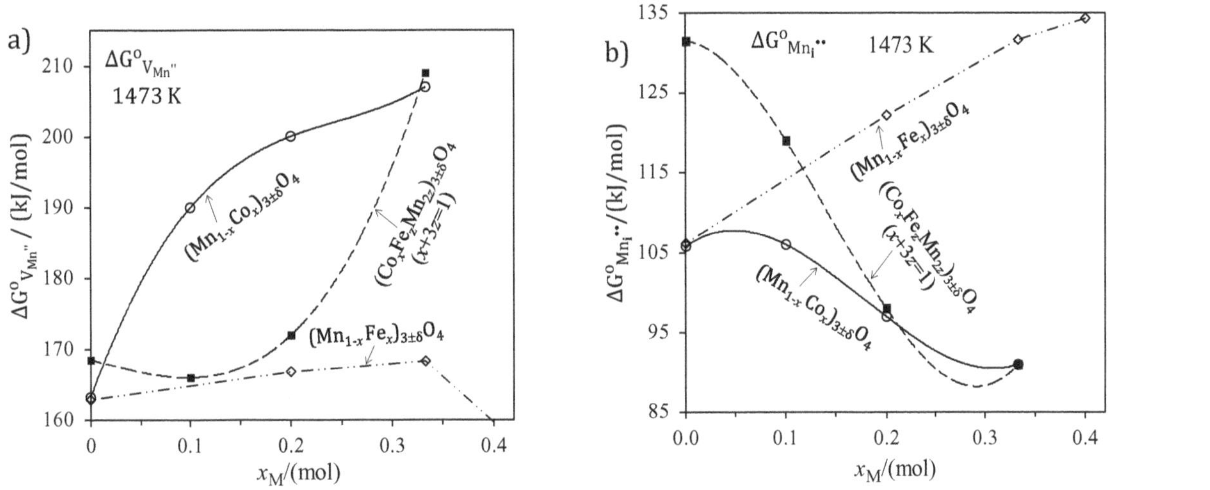

FIGURE 15.7 Dependence of standard Gibbs energy, a) $\Delta G^{o}_{V''_{M}}$ of formation of cation vacancies $V''_{M(o)}$, b) $\Delta G^{o}_{M_i^{\bullet\bullet}}$, of formation of interstitial ions $M_i^{\bullet\bullet}$ on the dopant content x_M in spinels: $(Mn_{1-x}Fe_x)_{3\pm\delta}O_4$ ((◇) points), $(Mn_{1-x}Co_x)_{3\pm\delta}O_4$ ((○) points) and $(Co_xFe_zMn_{2z})_{3\pm\delta}O_4$ (where $x + 3z = 1$) ((■) points) at 1473 K, obtained using the results of the studies on the deviation from the stoichiometry (Keller and Dieckmann 1985, Lu and Dieckmann 1993a, Lu and Dieckmann 1993b, Lu and Dieckmann 1995).

and Figure 15.8b – the dependence of $\left(D^{o}_{V(M)}/D^{o}_{I(M)} = b'_D\right)$ on the dopant content x_M for the diffusion of the individual tracers: Fe*, Co* and Mn* in analysed spinels.

As can be seen in Figure 15.8a and 15.8b, there is no general trend towards the change in the parameter b'_D or b_D depending on the type of diffusing ion and on the dopant content in the spinel. For the spinel $(Mn_{1-x}Fe_x)_{3\pm\delta}O_4$, the values b_D increase from 1 to 2, at the iron content of $x_{Fe} = 0.333$ mol. In turn, for the spinel $(Co_xFe_zMn_{2z})_{3\pm\delta}O_4$, at the cobalt content of $x_{Co} = 0.1$ mol, they reach the highest values, for the diffusion of cobalt, $b_D = 33$, and for the diffusion of Fe* and Mn* $b_D = 7$. For the spinel $(Mn_{1-x}Co_x)_{3\pm\delta}O_4$, at the dopant content of $x_{Co} = 0.2$ mol, the parameter b'_D reaches the highest values and it is, in the case of diffusion of iron ions, $b'_D = 7$, for manganese 9 and for cobalt 3.

In turn, Figure 15.9 presents the dependence of the coefficients of diffusion of ions Fe*, Co* and Mn* via cation vacancies $D^{o}_{V(M)}$ on the dopant content x_M in spinels $(Mn_{1-x}Fe_x)_{3\pm\delta}O_4$, $(Mn_{1-x}Co_x)_{3\pm\delta}O_4$ and $(Co_xFe_zMn_{2z})_{3\pm\delta}O_4$.

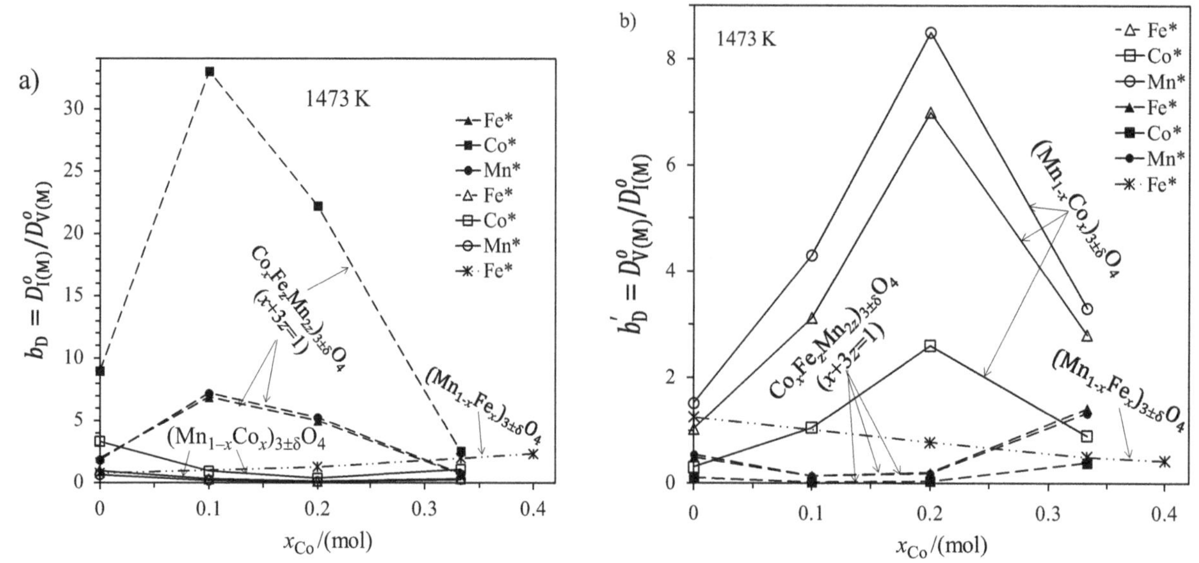

FIGURE 15.8 (a) Dependence of the ratio of the coefficient of diffusion of ions Fe*, Co* and Mn* via interstitial cations to the coefficient of diffusion via cation vacancies $\left(D^o_{I(M)}/D^o_{V(M)} = b_D\right)$ on the dopant content x_M in spinels: $(Mn_{1-x}Fe_x)_{3\pm\delta}O_4$ ((✱) points), $(Mn_{1-x}Co_x)_{3\pm\delta}O_4$ ((△,□,○) points) and $(Co_xFe_zMn_{2z})_{3\pm\delta}O_4$ (where $x + 3z = 1$) ((▲,■,●) points) at 1473 K, obtained when using the values of the coefficients of self-diffusion of tracers: Fe*, Co* and Mn* (Franke and Dieckmann 1989, Lu and Dieckmann 1993a, Lu and Dieckmann 1993b), and concentrations of ionic defects. (b) Dependence of the ratio of the coefficient of diffusion of ions Fe, Co and Mn via cation vacancies to the coefficient of diffusion via interstitial cations $\left(D^o_{V(M)}/D^o_{I(M)} = b'_D\right)$ on the dopant content x_M in spinels: $(Mn_{1-x}Fe_x)_{3\pm\delta}O_4$ ((✱) points), $(Mn_{1-x}Co_x)_{3\pm\delta}O_4$ ((△,□,○) points) and $(Co_xFe_zMn_{2z})_{3\pm\delta}O_4$ (where $x + 3z = 1$) ((▲,■,●) points) at 1473 K, obtained when using the values of the coefficients of self-diffusion of tracers: Fe*, Co* and Mn* (Franke and Dieckmann 1989, Lu and Dieckmann 1993a, Lu and Dieckmann 1993b), and concentrations of ionic defects.

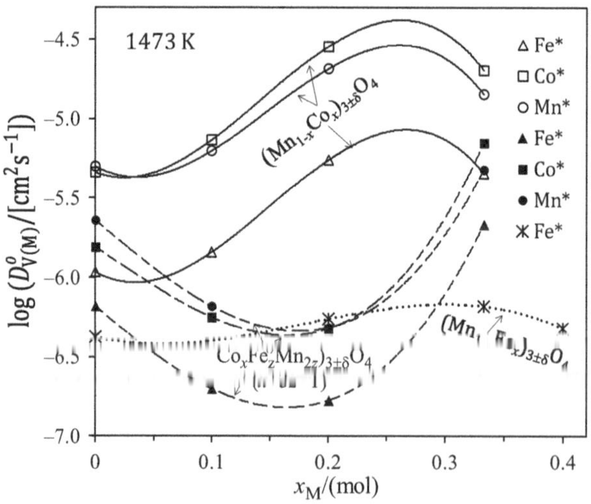

FIGURE 15.9 Dependence of coefficient of diffusion of tracer ions via cation vacancies $D^o_{V(M)}$ on the dopant content x_M in spinels: $(Mn_{1-x}Fe_x)_{3\pm\delta}O_4$ ((✱) points), $(Mn_{1-x}Co_x)_{3\pm\delta}O_4$ ((△,□,○) points) and $(Co_xFe_zMn_{2z})_{3\pm\delta}O_4$ (where $x + 3z = 1$) ((▲,■,●) points) at 1473 K, obtained based on the values of coefficients of self-diffusion of Fe*, Co* and Mn* (Franke and Dieckmann 1989, Lu and Dieckmann 1993a, Lu and Dieckmann 1993b) and the concentration of ionic defects.

As can be seen in Figure 15.9, in the case of spinel $(Fe_{1-x}Mn_x)_{3\pm\delta}O_4$, the coefficients of iron diffusion via cation vacancies $D^o_{V(Fe)}$ slightly increase when the iron content increases. On the other hand, for spinels $(Fe_{1-x}Co_x)_{3\pm\delta}O_4$, up to the cobalt content of $x_{Co} = 0.2$ mol an increase occurs in the value of the coefficient of diffusion $D^o_{V(M)}$, and at the content of $x_{Co} = 0.333$ mol – a slight decrease. In turn, cobalt doping of the spinel $(Co_xFe_zMn_{2z})_{3\pm\delta}O_4$ causes opposite changes, up to the content of $x_M = 0.2$ mol the values of $D^o_{V(M)}$ decrease, and then increase. Small differences are present for mobilities of ions Co* and Mn*; the mobility of iron ions is much lower.

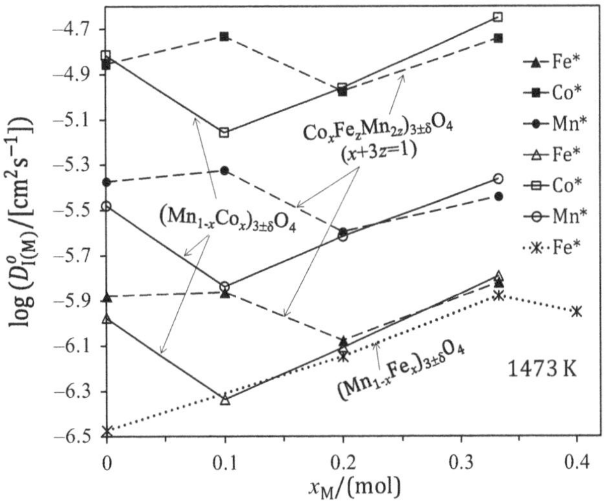

FIGURE 15.10 Dependence of coefficient of diffusion of tracer ions via interstitial cations $D^o_{I(M)}$ on the dopant content x_M in spinels: $(Mn_{1-x}Fe_x)_{3\pm\delta}O_4$ ((∗) points), $(Mn_{1-x}Co_x)_{3\pm\delta}O_4$ ((△,□,○) points) and $(Co_xFe_zMn_{2z})_{3\pm\delta}O_4$ (where $x + 3z = 1$) ((▲,■,●) points) at 1473 K, obtained based on the values of coefficients of self-diffusion of Fe*, Co* and Mn* (Franke and Dieckmann 1989, Lu and Dieckmann 1993a, Lu and Dieckmann 1993b) and the concentration of ionic defects.

Figure 15.10 compares the dependence of the coefficients of diffusion of ions Fe*, Co* and Mn* via interstitial ions $D^o_{I(M)}$ on the dopant content in the individual spinels. As can be seen in Figure 15.10, there are quite significant differences in the mobilities of the individual tracer ions via interstitial cations. The highest mobility is the mobility of cobalt ions, it is lower for manganese ions and the lowest mobility is that of iron ions. The character of the dependence of the mobility of ions on the dopant content in the individual spinels is complex. For spinels $(Mn_{1-x}Co_x)_{3\pm\delta}O_4$, with the cobalt content of $x_{Co} = 0.1$ mol, the mobility of ions via interstitial cations is lower than that in hausmannite, and then, when the cobalt content increases, it increases in a linear way. The mobility of ions in the spinel $(Co_xFe_zMn_{2z})_{3\pm\delta}O_4$ with the content of $x_M = 0.1$ mol is higher than that in the spinel $FeMn_2O_4$, and then, at the cobalt content of $x_M = 0.2$ mol it decreases to the value of $D^o_{I(M)}$ in the spinel $(Fe_{1-x}Co_x)_{3\pm\delta}O_4$, and then it further increases when the cobalt content increases. For the spinel $(Fe_{1-x}Mn_x)_{3\pm\delta}O_4$, when the iron content increases, the diffusion coefficient $D^o_{I(M)}$ quite significantly increases linearly. As can be seen in Figure 15.10, the mobilities of ions at the dopant content above $x_M = 0.2$ mol in the individual spinels have the same values, within the error limit.

REFERENCES

Franke, P., and R. Dieckmann. 1989. Defect Structure and Transport Properties of Mixed Iron-Manganese Oxides. *Solid State Ionics.* 32–33: 817–823.

Keller, M., and R. Dieckmann. 1985. Defect Structure and Transport Properties of Manganese Oxides: (II). The Nonstoichiometry of Hausmannite ($Mn_{3-\delta}O_4$). *Ber. Bunsenges. Phys. Chem.* 89: 1095–1104.

Lu, F. H., and R. Dieckmann. 1993a. Point Defects and Cation Tracer Diffusion in $(Co,Fe,Mn)_{3-\delta}O_4$ Spinels: II. Mixed Spinels $(Co_xFe_zMn_{2z})_{3-\delta}O_4$. *Solid State Ionics.* 59: 71–82.

Lu, F. H., and R. Dieckmann. 1993b. Point Defects and Cation Tracer Diffusion in $(Co_xMn_{1-x})_{3-\delta}O_4$ Spinels. *Solid State Ionics.* 67: 145–155.

Lu, F. H., and R. Dieckmann. 1995. Point Defects in Oxide Spinel Solid Solutions of the Type $(Co,Fe,Mn)_{3-\delta}O_4$ at 1200 °C. *J. Phys. Chem. Solids.* 56: 725–733.

Part IV

Diagrams of Concentrations of Point Defects in Oxide Solid Solutions $(Fe_{1-x}M_x)_{1-\delta}O$ (where M = Mn and Co)

16 Structure of Point Defects in Oxides $Fe_{1-\delta}O$, $Mn_{1-\delta}O$ and $Co_{1-\delta}O$

16.1 SHORT INTRODUCTION

As it was shown in the previous part of the work, a relatively "loose" crystallographic structure of magnetite and the presence of iron ions Fe^{2+} and Fe^{3+} causes that these ions can be substituted with metal ions M^{2+} and M^{3+}, which form solid solutions in the cation sublattice, in the whole range of concentrations. As it was shown, the presence of doping ions which have the same oxidation state as cations of the parent sublattice does not cause a change in the structure of point defects and their concentrations vary only slightly. Due to changes in interactions between ions, the range of oxygen pressures where spinel phases exists changes.

There arises a question about the change in the structure of point defects in the oxides $Fe_{1-\delta}O$, $Co_{1-\delta}O$ and $Mn_{1-\delta}O$ in the case of doping ions with the same oxidation state. These oxides have, similarly to spinels, a cubic crystallographic structure of NaCl-type. Therefore, in a densely-packed oxygen sublattice, all voids with octahedral coordination are occupied by cations, while the voids with tetrahedral coordination are vacant. Despite a small difference in the electronic structure of ions Fe^{2+}, Mn^{2+}, Co^{2+}, their interactions with oxygen cause that the range of oxygen pressure where the oxides of these metal exist changes significantly. These oxides also differ quite significantly as to the structure and concentrations of point defects.

The analysis presented in the work (Stokłosa 2015) implies that the iron oxide (wüstite) $Fe_{1-\delta}O$ exists only above the temperature of 830 K. The range of its existence, compared to other oxides of 3*d* metals, is very narrow. For example, at 1073 K, $Fe_{1-\delta}O$ exists in the range of the oxygen pressure from 10^{-20} to 10^{-18} atm, and at 1573 K, from 10^{-12} to 10^{-9} atm. In such a narrow range, a significant change in the deviation from the stoichiometry occurs. At the decomposition pressure at the $Fe/Fe_{1-\delta}O$ phase boundary, the deviation is $\delta = 0.045–0.05$ mol/mol (in the 1073–1573 K temperature range), and at the $Fe_{1-\delta}O/Fe_3O_4$ phase boundary, at 1073 K it reaches $\delta = 0.1$, and at 1573 K as much as 0.15 mol/mol. Dominating defects are iron vacancies with different ionisation degrees and several defect complexes are present. The range of existence of the manganese oxide $Mn_{1-\delta}O$ is much wider. For example, at 1173 K the manganese oxide exists in the range of the oxygen pressure from 10^{-27} to 10^{-8} atm, and at 1773 K, from 10^{-15} to 10^{-2} atm. Moreover, there is a significant range of the change in the concentration of defects, from 10^{-5} mol/mol at the decay pressure up to about 0.1 mol/mol at the $Mn_{1-\delta}O/Mn_3O_4$ phase boundary. The defects' structure is relatively simple; near the decay pressure the manganese oxide reaches the stoichiometric composition, and in a wide range of oxygen pressures, 2– manganese vacancies $[V''_{Mn}]$ dominate. At temperatures above 1473 K, in the range of higher oxygen pressures, vacancies with lower ionisation degrees start to dominate in $Mn_{1-\delta}O$. Cobalt oxide, $Co_{1-\delta}O$, exists in a much wider range than $Fe_{1-\delta}O$, from 10^{-12} to 1 atm at 1273 K, and the concentration of point defects (deviation from the stoichiometric composition) changes from 10^{-3} mol/mol at the decay pressure to 0.01 mol/mol at the oxygen pressure of 1 atm. The structure of point defects is similar, at low pressures, 2– cobalt vacancies $[V''_{Co}]$ dominate, but the concentration of interstitial cations is also significant. When the oxygen pressure increases, the concentration of vacancies with lower ionisation degrees increases significantly. A wider discussion of the results of the studies of deviation from the stoichiometry in the oxides $Fe_{1-\delta}O$, $Mn_{1-\delta}O$ and $Co_{1-\delta}O$, together with diagrams of concentrations of points defects in these oxides was presented in the work (Stokłosa 2015).

Wüstite, compared to oxides $Mn_{1-\delta}O$ and $Co_{1-\delta}O$, essentially differs as to the structure of point defects. It was found that in wüstite, defect complexes are present, formed due to interactions between iron vacancies and interstitial cations, The neurography studies have shown the presence of Fe^{3+} ions in octahedral and tetrahedral positions, analogously as in magnetite, Fe_3O_4 (Roth 1960). Further studies, X-ray diffraction and scattering (Manenc 1963, Manenc et al. 1963, Herai et al. 1964, Smuts 1966, Koch and Cohen 1969, Gartstein et al. 1986, Welberry and Christy 1997), neutron diffraction (Cheetham et al. 1971, Battle and Cheetham 1979, Gavarri et al. 1979, Schweika and Carlsson 1989, Gavarri and Carel 1990, Radler et al. 1990, Schweika et al. 1992, Schweika et al. 1995) and electron diffraction (Andersson and Sletnes 1977, Ijima 1974, Ishiguro and Nagakura 1985a, Ishiguro and Nagakura 1985b, Labidi and Monty 1991, Lebreton and Hobbs 1983, Yamamoto 1982), as well as Mössbauer spectroscopy (Checherskaya et al. 1973, Elias and Linnett 1969, Greenwood and Howe 1972a, Greenwood and Howe 1972b, Greenwood and Howe 1972c, Hope et al. 1982, Hrynkiewicz et al. 1972, Inglot et al. 1991, Johnson 1969, Pattek-Janczyk et al. 1986, Shirane et al. 1962) have confirmed that in $Fe_{1-\delta}O$ there are Fe^{3+} ions in tetrahedral positions, which, interacting with iron vacancies, form defect complexes. As a result, based on the above studies, Roth proposed 2:1-type complexes – $\left\{\left(V_{Fe}''\right)_2 Fe_i^{3\bullet}\right\}'$ (Roth 1960), Cheetham 4:1 complexes – $\left\{\left(V_{Fe}''\right)_4 Fe_i^{3\bullet}\right\}^{5\prime}$ (Cheetham et al. 1971). More complex clusters, of type 6:2 and 8:3, were suggested in (Catlow and Fender 1975), and 13:4 clusters in (Koch and Cohen 1969). Also, whole structures of complex defect clusters may also be formed (Gavarri et al. 1979, Presss and Ellis 1987, Radler et al. 1990). The performed computer simulations fully confirmed the possibility of existence of several complexes and defect clusters in $Fe_{1-\delta}O$ (Catlow and Fender 1975, Catlow and Stoneham 1981, Catlow et al. 1979, Grimes et al. 1986, Grimes et al. 1987, Presss and Ellis 1987). In the oxides $Mn_{1-\delta}O$ and $Co_{1-\delta}O$ defect complexes of this type were not found, which can indicate that their concentration is much smaller than in wüstite and difficult to confirm experimentally, but their concentration may be significant, as it results from calculations (Catlow and Fender 1975, Catlow and Stoneham 1981, Catlow et al. 1979, Grimes et al. 1986, Grimes et al. 1987, Presss and Ellis 1987, Schuster et al. 1989, Tomlison and Catlow 1989).

A second significant difference between wüstite and oxides $Mn_{1-\delta}O$ and $Co_{1-\delta}O$ is a complex character of dependence of the deviation from the stoichiometry in $Fe_{1-\delta}O$. A large change in the composition in wüstite depending on the oxygen pressure and a relatively short equilibration time caused that the studies of the deviation from the stoichiometry are burdened with a relatively small error and there is a good agreement between the results obtained by various authors (Ackermann and Sandford Jr. 1966, Barbi 1964, Bransky and Hed 1968, Fender and Riley 1969, Gerdanian 1988, Gerdanian and Dodé 1965, Giddings 1972, Giddings and Gordon 1973, Giddings and Gordon 1974, Marucco et al. 1970a, 1970b, Picard and Dodé 1970, Rizzo 1968, Rizzo and Smith 1968, Rizzo et al. 1968, Rizzo et al. 1969, Sockel and Schmalzried 1968, Swaroop and Wagner Jr. 1967, Takayama and Kimizuka 1980, Vallet and Raccah 1964, Vallet and Raccah 1965, Vallet et al. 1963). It was found that at temperatures above 1173 K, the dependence of the logarithm of the equilibrium oxygen pressure on the composition of the oxide FeO_x is well described by a linear relation:

$$\log p_{O_2} = Ax + B \tag{16.1}$$

However, the results of the studies of the deviation from the stoichiometry (Vallet and Raccah 1965) in the temperature range up to 1173 K in a semi-log scale demonstrated a more complex character of the dependence – they were described with three straight lines, and at higher temperatures – with two straight lines. Based on the analysis of constants *A* and *B* in Equation 16.1 depending on the temperature, Kleman suggested a hypothesis that in the range of existence of $Fe_{1-\delta}O$, different "phases" (pseudo-phases) of wüstite might be present (Kleman 1965). A wider analysis of *A* and *B* constants and of thermodynamic parameters of wüstite formation, presented in many works (Carel and Gavarri 1976, Carel and Vallet 1964, Carel et al. 1965, Gavarri and Carel 1981, Gavarri et al. 1976, Gavarri et al. 1981, Gavarri et al. 1988, Giddings 1972, Giddings and Gordon 1973, Giddings and Gordon 1974, Kleman 1965, Vallet 1965, Vallet 1975, Vallet 1977, Vallet and Carel 1979, Vallet and Carel 1986), showed a change in the character of the analysed values; as a result, a phase system was proposed, defining ranges of existence of three "pseudo-phases". The w_1 pseudo-phase should exist at the lowest deviations from the stoichiometry, up to about $\delta = 0.075$ mol/mol, while w_2 pseudo-phase – at higher values of δ, up to 1423 K. Next, the w_3 pseudo-phase should exist in the temperature range up to 1273 K, at the values above $\delta > 0.085$. The suggested pseudo-phases have the same crystallographic structure (cubic, of NaCl-type), but they can differ by the structure of point defects (type and concentration of defect complexes) and order of defects. The possibility of existence of several wüstite phases is indicated also by the electrochemical titration studies on the deviation from the stoichiometry (Fender and Riley 1969), where, for an oxide with constant deviation from the stoichiometry, a complex dependence of cell EMF on the temperature was obtained. The obtained dependence shows two or three straight lines, and the points where the character of the dependence changes were interpreted as a confirmation of the existence of several wüstite pseudo-phases in individual ranges of temperature and deviation from the stoichiometry (Fender and Riley 1969). The existence of pseudo-phases in wüstite is indicated by kinetics and oxidation of wüstite studies (changes of the deviation from the stoichiometry) in gas mixtures $^{13}CO/CO_2$ (Worrel and Coley 2013). The possibility of the existence of wüstite pseudo-phases at higher temperatures was also confirmed by the results of studies of deviation from the stoichiometry (Takayama, 1980), where a visible change in the character of the dependence of the oxide's composition on p_{O_2}, described with Equation 16.1 was seen. In the temperature range 1373–1573 K, the obtained results form two straight lines, which indicated that also above 1400 K, a w_2 pseudo-phase should exist. The results and the discussion concerning the existence of pseudo-phases in wüstite was summarised in (Gavarri and Carel 2019). Despite a good agreement of the results of the deviation from the stoichiometry described by Equation 16.1 it was not theoretically justified (based on thermodynamics of point defects), which indicates that this equation has no physical meaning.

The first attempts at the interpretation of the results of the deviation from the stoichiometry according to the defect theory have shown that the oxygen pressure dependence of the deviation from the stoichiometry can be described with an equation of the type $\delta = a(p_{O_2})^{1/n}$ with an exponent of about 1/6, which would indicate the domination of vacancies $\left[V_{Fe}''\right]$ (Darken and Gurry 1946, Hauffe and Pfeiffer 1953). However, the interpretation of the results of the studies of other authors has shown that the exponent $1/n$, when the deviation increases, changes from 1/4 to 1/6, and even to 1/10, which indicates a complex structure of defects (Swaroop and Wagner Jr. 1967, Kofstad 1972, Kofstad and Hed 1968, Smyth 1961, Sørensen 1981). The decrease in the exponent may indicate the presence of interactions between defects. Further attempts at relating the deviation from the stoichiometry only with the concentration of defect complexes did not give unambiguous results (Nowotny and Rekas 1989, Rekas and Mrowec 1987). In the work (Stokłosa 2015), interpretation of the results of the studies on the deviation from the stoichiometry was performed and an attempt was made to determine diagrams of concentrations of defects; the maximum concentration of defect complexes was also determined. As it results from the calculations, it can be about 0.001 mol/mol, it is much lower than the concentration of cations vacancies and does not affect the oxygen pressure dependence of the deviation from the stoichiometry. However, as it results from the calculations, at the highest oxygen pressures, the concentration of defect complexes is as high that it is possible that they are packed, with the maximum density, in the wüstite structure (Gavarri and Carel 1990, Gavarri et al. 1976, Gavarri et al. 1979, Gavarri and Carel 2019, Stokłosa 2015).

The problem of the existence of wüstite pseudo-phases was not solved by the X-ray crystallography of the $Fe_{1-\delta}O$ structure (Gavarri et al. 1976, Gavarri et al. 1988, Hayakawa et al. 1972, Hayakawa et al. 1973, Jette and Foote 1933, Levin and Wagner Jr. 1966, Touzelin 1974). However, an analysis concerning the change in lattice parameter a depending on the composition of $Fe_{1-\delta}O$ for frozen samples, has shown a possibility of the existence of discontinuities that are present at the composition corresponding to the "phase" transition in wüstite, w_1/w_2 (Carel et al. 1965, Carel and Vallet 1964). Similar results were obtained in dilatometry studies (Bars and Carel 1969, Carel and Vallet 1964, Gavarri and Carel 1981). In most X-ray crystallography studies of $Fe_{1-\delta}O$, there is no basis for assuming no other than the linear dependence of the change in the lattice parameter a on the composition of $Fe_{1-\delta}O$. This is mainly due to the measurement error, small number of studied oxides with different deviation from the stoichiometry and the fact that the studies were performed for samples with frozen structure. The problem of the existence of pseudo-phases in wüstite was neither solved by X-ray crystallographic studies performed at the equilibrium conditions, at a determined oxygen pressure, in the range of high temperatures; from these studies it transpires that a change in the lattice parameter with the change in composition of $Fe_{1-\delta}O$ is approximately linear (Hayakawa et al. 1972, Hayakawa et al. 1973, Jette and Foote 1933, Levin and Wagner Jr. 1966, Touzelin 1974).

The studies of the work function depending on the wüstite's composition, performed under equilibrium conditions, at 948–1223 K (Nowotny and Sikora 1978), did not provide a response to the above problem. The obtained monotonic character of the change in the work function depending on the pressure does not indicate the presence of phases in $Fe_{1-\delta}O$, however, at the temperatures of 1173 and 1223 K, near the $Fe_{1-\delta}O/Fe_3O_4$ phase boundary, a complex character of the dependence was obtained for a large deviation from the stoichiometry, indicating different surface properties of $Fe_{1-\delta}O$.

Therefore, interpretations of the results of the deviation from the stoichiometry and other properties of $Fe_{1-\delta}O$ did not provide convincing (hard) evidence of the presence of pseudo-phases in the range of existence of $Fe_{1-\delta}O$. Nevertheless, the change in the character of the dependence of the deviation from the stoichiometry on the oxygen pressure and, similarly, of the electrical conductivity of wüstite indicates a presence of two or three different "phases" (pseudo-phases), differing by structure of point defects (Ariya and Brach 1964, Aubry and Marion 1955, Bransky and Tannhauser 1967a, Bransky and Tannhauser 1967b, Geiger et al. 1966, Heikes et al. 1963, Hillegas 1968, Kozheurov and Mikhailov 1967, Molenda et al. 1987a, Molenda et al. 1987b, Neushutz and Towhidi 1970, Nowotny et al. 1982, Tannhauser 1962, Wagner and Koch 1936). The calculations performed in the work (Stokłosa 2015) indicate a possibility of formation of defect complexes, with concentrations that could reach 0.01–0.001 mol/mol. Simultaneously, the concentration of cation vacancies decreases. At higher oxygen pressures, when the concentration of defect complexes increases, they can start to be ordered, which could result in the aforementioned change in the character of the oxygen pressure dependence of the deviation from the stoichiometry. Also, larger defect clusters can be formed. The limit case of ordering of defect clusters is the previously discussed spinel structure (see Section 1.3, Figure 1.1).

Despite the differences in the energy of interactions of ions Fe^{2+}, Mn^{2+} and Co^{2+} and in their electronic structure, these ions in wüstite form a solid solution in the whole range of concentrations. Thus, a question arises – how do Mn^{2+} and Co^{2+} doping ions in wüstite affect the structure and concentrations of point defects, particularly the concentration of defect complexes and the existence of pseudo-phases, differing by the ordering of defects. Also, inversely, how do Fe^{2+} ions influence the concentration of defects in oxides $Mn_{1-\delta}O$ and $Co_{1-\delta}O$.

16.2 DEFECT EQUILIBRIA IN SIMPLE OXIDES $M_{1-\delta}O$ AND IN OXIDE SOLUTIONS $(Fe_{1-x}M_x)_{1-\delta}O$

As it was mentioned, in oxides of type $M_{1-\delta}O$, which show a real deficiency of metal, there are defects in the cation sublattice, where cation vacancies and interstitial ions are present. Formation of double-ionised cation vacancies V_M'' and interstitial ions $M_i^{\bullet\bullet}$ in the oxide $M_{1-\delta}O$, according to the Kröger–Vink notation, can be written with the following reactions:

$$1/2\,O_2 = O_O^x + V_M'' + 2h^\bullet \qquad \Delta G^o_{V_M''} \qquad (16.2)$$

$$M_M^x + O_O^x + V_i + 2h^\bullet = M_i^{\bullet\bullet} + 1/2\,O_2 \qquad \Delta G^o_{M_i^{\bullet\bullet}} \tag{16.3}$$

where V_i denotes an interstitial void occupied by a cation $M_i^{\bullet\bullet}$ (tetrahedral void in a NaCl-type structure).

Apart from the formation of ionic defects, electronic defects are present in the oxide, and their process of formation is described by the reaction:

$$(nil) = h^\bullet + e' \qquad \Delta G^o_{eh} \tag{16.4}$$

At higher concentrations of defects, it is possible that there are interactions between ionic and electronic defects and that defects with lower ionisation degrees are formed, which is described by the reactions:

$$V_M'' + h^\bullet = V_M' \qquad \Delta G^o_{V_M'} \tag{16.5}$$

$$V_M' + h^\bullet = V_M^x \qquad \Delta G^o_{V_M^x} \tag{16.6}$$

$$M_i^{\bullet\bullet} + e' = M_i^\bullet \qquad \Delta G^o_{M_i^\bullet} \tag{16.7}$$

$$M_i^\bullet + e' = M_i^x \qquad \Delta G^o_{M_i^x} \tag{16.8}$$

With the process of formation of defects, a specific change in the standard Gibbs energy of reaction ΔG_n^o is associated; the equilibrium constant, and in consequence the concentration of defects at a determined oxygen pressure, depend on this value.

For reactions 16.2–16.8, the expressions describing the equilibrium state assume the form of:

$$\frac{-\Delta G^o_{V_M''}}{RT} = \ln K_{V_M''} = \ln \frac{[V_M''][h^\bullet]^2}{p_{O_2}^{1/2}} \tag{16.9}$$

$$\frac{-\Delta G^o_{M_i^{\bullet\bullet}}}{RT} = \ln K_{M_i^{\bullet\bullet}} = \ln \frac{[M_i^{\bullet\bullet}]p_{O_2}^{1/2}}{[h^\bullet]^2} \tag{16.10}$$

$$\frac{-\Delta G^o_{eh}}{RT} = \ln K_{eh} = \ln [h^\bullet][e'] \tag{16.11}$$

$$\frac{-\Delta G^o_{V_M'}}{RT} = \ln K_{V_M'} = \ln \frac{[V_M']}{[V_M''][h^\bullet]} \tag{16.12}$$

$$\frac{-\Delta G^o_{V_M^x}}{RT} = \ln K_{V_M^x} = \ln \frac{[V_M^x]}{[V_M'][h^\bullet]} \tag{16.13}$$

$$\frac{-\Delta G^o_{M_i^\bullet}}{RT} = \ln K_{M_i^\bullet} = \ln \frac{[M_i^\bullet]}{[M_i^{\bullet\bullet}][e']} \tag{16.14}$$

$$\frac{-\Delta G^o_{M_i^x}}{RT} = \ln K_{M_i^x} = \ln \frac{[M_i^x]}{[M_i^\bullet][e']} \tag{16.15}$$

where K_n denote equilibrium constants of individual reactions of formation of defects, which depend on ΔG_i^o of the formation of defects (enthalpy and entropy of the reaction: $\Delta G_i^o = \Delta H_i^o - T\Delta S_i^o$).

From Equation 16.9 and 16.10 it results that at a determined oxygen pressure, the concentration of cation vacancies and interstitial ions are precisely defined, and at the stoichiometric composition, these concentrations are identical $\left[V_M''\right]=\left[M_i^{\bullet\bullet}\right]$. By adding Equation 16.2 and 16.3 we get the reaction of formation of intrinsic ionic Frenkel defects:

$$M_M^x + V_i = M_i^{\bullet\bullet} + V_M'' \qquad \Delta G_F^o \qquad (16.16)$$

The equilibrium constant for the reaction 16.16 is a product of equilibrium constants of reactions 16.2 and 16.3 and according to Equation 16.9 and 16.10 we get:

$$\frac{-\Delta G_F^o}{RT} = \ln\left(K_{V_M''}K_{M_i^{\circ\circ}}\right) = \ln K_F = \ln\left[V_M''\right]\left[M_i^{\bullet\bullet}\right] \qquad (16.17)$$

where K_F denotes the equilibrium constant of formation of Frenkel defects.

Vacancies with lower ionisation degrees formed, according to reactions 16.5 and 16.6, are defects with interactions between a cation vacancy and an electron hole $\left(V_M'',h^\bullet\right)'$, which is localized on a cation adjacent to the vacancy. A localisation of an electron hole is equivalent to an increase in oxidation state of metal ion; thus, it is an interaction with a metal ion M^{3+}, (M_M°). Therefore, a single-ionised vacancy (V_M') can be treated as the simplest defect complex $\left(V_M''M_M^\bullet\right)'$. An electroneutral cation vacancy $\left(V_M^x\right) \equiv \left(h^\bullet V_M'' h^\bullet\right)$, can be treated as a complex of a vacancy with two interstitial ions $\left(M_M^\bullet V_M'' M_M^\bullet\right)$; this complex, as can be seen, has a higher symmetry.

As discussed in the introduction, at significant concentrations of ionic defects and electron holes in the oxide $M_{1-\delta}O$, due to interactions between cation vacancies and interstitial cations, formation of complexes and larger defect clusters might occur. The simplest ionic complexes are complexes of type 2:1 $\equiv \left\{\left(V_M''\right)_2 M_i^{3\bullet}\right\}'$ (Roth 1960), (4:1) $\equiv \left\{\left(V_M''\right)_4 M_i^{3\bullet}\right\}^{5'}$ (Cheetham et al. 1971) and more complex clusters (6:2) $\equiv \left\{\left(V_M''\right)_6 \left(M_i^{3\bullet}\right)_2\right\}^{6'}$ (Catlow and Fender 1975). The performed computer simulations indicate that they can be present also in other oxides of type $M_{1-\delta}O$ (Catlow and Fender 1975, Catlow and Stoneham 1981, Catlow et al. 1979, Grimes et al. 1986, Grimes et al. 1987, Presss and Ellis 1987, Schuster et al. 1989, Tomlison and Catlow 1989).

The formation of complexes of this type can be presented with reactions analogous to reactions of formation of defects with lower ionisation degrees. Thus, the formation of the simplest complex of type 2:1 occurs as a result of interaction between two cation vacancies with an interstitial ion and an electron hole, which can be described by the reaction:

$$M_i^{\bullet\bullet} + 2V_M'' + h^\bullet = \left\{\left(V_M''\right)_2 M_i^{3\bullet}\right\}' \qquad \Delta G_{C_2'}^o \qquad (16.18)$$

Interaction of a 2:1 complex with two cation vacancies causes the formation of a complex with a higher symmetry, of 4:1 type:

$$\left\{\left(V_M''\right)_2 M_i^{3\bullet}\right\}' + 2V_M'' = \left\{\left(V_M''\right)_4 M_i^{3\bullet}\right\}^{5'} \qquad \Delta G_{C_4^{5'}}^o \qquad (16.19)$$

The above complexes could form more complex clusters, e.g.:

$$\left\{\left(V_M''\right)_2 M_i^{3\bullet}\right\}' + \left\{\left(V_M''\right)_4 M_i^{3\bullet}\right\}^{5'} = \left\{\left(V_M''\right)_6 \left(M_i^{3\bullet}\right)_2\right\}^{6'} \qquad \Delta G_{C_6^{5'}}^o \qquad (16.20)$$

where ΔG_i^o denote the standard Gibbs energies of reactions of formation of individual defect complexes. Indices C_2', $C_4^{5'}$ and $C_6^{6'}$ denote, respectively, clusters: $\left\{\left(V_M''\right)_2 M_i^{3\bullet}\right\}'$, $\left\{\left(V_M''\right)_4 M_i^{3\bullet}\right\}^{5'}$ and $\left\{\left(V_M''\right)_6 \left(M_i^{3\bullet}\right)_2\right\}^{6'}$. Therefore, several elementary reactions occur, which are mutually coupled.

For reactions 16.18–16.20, the expressions describing the equilibrium state assume the form of:

$$\frac{-\Delta G_{C_2'}^o}{RT} = \ln K_{C_2'} = \ln \frac{\left[\left\{\left(V_M''\right)_2 M_i^{3\bullet}\right\}'\right]}{\left[M_i^{\bullet\bullet}\right]\left[V_M''\right]^2\left[h^\bullet\right]} \qquad (16.21)$$

$$\frac{-\Delta G^{o}_{C_4^{5'}}}{RT} = \ln K_{C_4^{5'}} = \ln \frac{\left[\left\{\left(V''_M\right)_4 M_i^{3\bullet}\right\}^{5'}\right]}{\left[\left\{\left(V''_M\right)_2 M_i^{3\bullet}\right\}^{'}\right]\left[V''_M\right]^2} \tag{16.22}$$

$$\frac{-\Delta G^{o}_{C_6^{6'}}}{RT} = \ln K_{C_6^{6'}} = \ln \frac{\left[\left\{\left(V''_M\right)_6 \left(M_i^{3\bullet}\right)_2\right\}^{6'}\right]}{\left[\left\{\left(V''_M\right)_4 M_i^{3\bullet}\right\}^{5'}\right]\left[\left\{\left(V''_M\right)_2 M_i^{3\bullet}\right\}^{'}\right]} \tag{16.23}$$

Formation of more complex defect clusters that were found in wüstite can be presented with a series of reactions, where there occurs an interaction of more and more complex clusters, starting from $\left\{\left(V''_M\right)_6 \left(M_i^{3\bullet}\right)_2\right\}^{6'}$ with complexes $\left\{\left(V''_M\right)_2 M_i^{3\bullet}\right\}^{'}$ or $\left\{\left(V''_M\right)_4 M_i^{3\bullet}\right\}^{5'}$. The process of their formation (association) can be described, using the simplified notation, as:

(6:2) + (2:1) = (8:3)

(8:3) + (2:1) = (10:4)

(10:4) + (2:1) = (12:5)

(8:3) + (4:1) = (12:4)

(12:4) + (4:1) = (16:5) etc.

The formation of defect clusters of this type will cause a change in the concentration of simple complexes, but they will not affect the oxygen pressure dependence of the deviation from the stoichiometry. It should be noted that the above complexes or defect clusters will be in equilibrium with simple defects and their formation will cause a change in the concentration of cation vacancies $\left[V''_M\right]$, interstitial ions and electron holes. The presence of clusters of this type is possible in an oxide with a large deviation from the stoichiometry, as for example in $Fe_{1-\delta}O$. As it transpires from the calculations (Stokłosa 2015), the concentration of defect complexes is as high that it is possible that at higher oxygen pressures ordered cluster should form, as well as larger defect clusters which will form a "mixture" of structures with local order, or even "ordered phases" with the composition changing within a certain range. Therefore, at a specified oxygen pressure, an equilibrium state will be set and the oxide $M_{1-\delta}O$ will show a defined deviation from the stoichiometric composition. For the reasons mentioned above, clusters with a complex structure will not be considered in the further part of the work, as their concentration should be small and it does not affect the oxygen pressure dependence of the deviation from the stoichiometry.

If an interaction between cation vacancies and electron holes is analysed (Equation 16.5 and 16.6), in the model of defects one should also consider the possibility of interactions of electron holes, particularly of a node ion $M_M^{\bullet}$, (M^{3+}) with complexes and defect clusters. Reactions of the formation of the individual complexes with a lower ionisation degree are described as follows:

$$\left\{\left(V''_M\right)_2 M_i^{3\bullet}\right\}^{'} + h^{\bullet} = \left\{\left(V''\right)_2 M_i^{3\bullet}\right\}^{x} \qquad \Delta G^{o}_{C_2^{x}} \tag{16.24}$$

$$\left\{\left(V''_M\right)_4 M_i^{3\bullet}\right\}^{5'} + h^{\bullet} = \left\{\left(V''_M\right)_4 M_i^{3\bullet}\right\}^{4'} \qquad \Delta G^{o}_{C_4^{4'}} \tag{16.25}$$

$$\left(\left[V''_M\right]_6 \left[M_i^{3\bullet}\right]_2\right)^{6'} + h^{\bullet} = \left\{\left(V''_M\right)_6 \left(M_i^{3\bullet}\right)_2\right\}^{5'} \qquad \Delta G^{o}_{C_6^{5'}} \tag{16.26}$$

Expression describing the equilibrium constants for the above reactions will assume a form analogous to this for vacancies with lower ionisation degrees (Equation 16.12).

From the considerations presented above it results that the setting of an equilibrium state in the oxide will be related to quite a high number of reactions of formation of defects and their decay. The above reactions are mutually coupled, and the concentrations of defects depend on ΔG_i^o of formation of individual simple defects and defect complexes (equilibrium constants of reactions 16.2–16.8, 16.18–16.20 and 16.24–16.26). A change in the oxygen pressure causes a change in relative concentrations of simple defects and defect complexes and setting of a new equilibrium state. The equilibrium constants will depend on the deviation from the stoichiometry, as at high concentration of point defects, the energy of interaction between ions will change, thus the standard Gibbs energy of cohesion will change. This will cause a change in ΔG_i^o of formation of individual simple defects and defect complexes. Therefore, only in a specified (narrow) range of oxygen pressures, it is possible that the effective equilibrium constants have constant values, as they are a product of the equilibrium constant and coefficients of activity of components (effective activity component).

REFERENCES

Ackermann, R. J., and R. W. Sandford Jr. 1966. A Thermodynamic Study of the Wüstite Phase. USAEC, Argonne Nat. Lab., Rep. ANL-7250, p. 46.

Andersson, B., and J. O. Sletnes. 1977. Decomposition and Ordering in $Fe_{1-x}O$. *Acta Crystall. A.* 33:268–276.

Ariya, A. M., and B. Y. Brach. 1964. Electrical Conductivity of Ferrous Oxide. *Sov. Phys. Solid State.* 5:2565–2567.

Aubry, J., and F. Marion. 1955. Sur les variation de la conductivite électrique et de la température de fusion du protoxyde de fer en fonction de sa composition. *C. R. Acad. Sci. (Paris).* 241:1778–1781.

Barbi, G. B. 1964. The Stability of Wüstite by Electromotive Force Measurements on All-Solid Electrolytic Cells. *J. Phys. Chem.* 68:2912–2916.

Bars, J.-P., and C. Carel. 1969. Étude dilatométrique de la wüstite solide à l'intérieur de son domaine d'existence. *C. R. Acad. Sci. (Paris) Sér. C.* 269:1152–1154.

Battle, P. D., and A. K. Cheetham. 1979. The Magnetic Structure of Non-Stoichiometric Ferrous Oxide. *J. Phys. C, Solid State Phys.* 12:337–345.

Bransky, I., and A. Z. Hed. 1968. Thermogravimetric Determination of the Composition-Oxygen Partial Pressure Diagram of Wüstite ($Fe_{1-y}O$). *J. Am. Ceram. Soc.* 51:231–232.

Bransky, I., and D. S. Tannhauser. 1967a. High-Temperature Defect Structure of Ferrous Oxide. *Trans. Metall. Soc. AIME.* 239:75–80.

Bransky, I., and D. S. Tannhauser. 1967b. Mobility of Charge Carriers in Wüstite (Feo_{1+X}). *Physica.* 37:547–552.

Carel, C., and J. R. Gavarri. 1976. Introduction to Description of Phase Diagram of Solid Wüstite: I. Structural Evidence of Allotropic Varieties. *Mater. Res. Bull.* 11:745–756.

Carel, C., and P. Vallet. 1964. Dilatometric Study of Different Varieties of Solid Wüstite and the Existence of a Metastable Triple Point among the Three Varieties. *C. R. Acad. Sci. (Paris).* 258:3281–3284.

Carel, C., D. Weigel, and P. Vallet. 1965. Variations du paramètre cristallin des trois variétés de wüstite solide dans leurs domains d'existence respectifs. *C. R. Acad. Sci. (Paris).* 260:4325–4328.

Catlow, C. R. A., and B. E. F. Fender. 1975. Calculations of Defect Clustering in $Fe_{1-x}O$. *J. Phys. C: Solid State Phys.* 8:3267–3279.

Catlow, C. R. A., and A. M. Stoneham. 1981. Defect Equilibria in Transition Metal Oxides. *J. Am. Ceram. Soc.* 64:234–236.

Catlow, C. R. A., W. C. Mackrodt, M. J. Norgett, and A. M. Stoneham. 1979. The Basic Atomic Processes of Corrosion Ii. Defect Structures and Cation Transport in Transition-metal Oxides. *Phil. Mag. A.* 40:161–172.

Checherskaya, L. F., V. P. Romanov, and P. A. Tatsienko. 1973. Mössbauer Effect in Wüstite. *Phys. Status Solidi (a).* 19: K177–K182.

Cheetham, A. K., B. E. F. Fender, and R. I. Taylor. 1971. High Temperature Neutron Diffraction Study of $Fe_{1-x}O$. *J. Phys. C, Solid State Phys.* 4:2160–2165.

Elias, D. J., and J. W. Linnett. 1969. Oxidation of Metals and Alloys. Part 3. Mössbauer Spectrum and Structure of Wüstite. *Trans. Faraday Soc.* 65:2673–2676.

Fender, B. E. F., and F. D. Riley. 1969. Thermodynamic Properties of $Fe_{1-x}O$. Transitions in the Single Phase Region. *J. Phys. Chem. Solids.* 30:793–798.

Gartstein, E., O. Mason, and J. B. Cohen. 1986. Defect Agglomeration In Wüstite At High Temperatures - I: The Defect Arrangement. *J. Phys. Chem. Solids.* 47:759–773.

Gavarri, J. R., and C. Carel. 1981. Evolution structurale de monoxydes non-stoechiométriques du type wüstite: Simulations et relations entre paramètres structuraux. *J. Solids State Chem.* 38:368–380.

Gavarri, J. R., and C. Carel. 1990. Structural Evolution Study of Substituted Wustites $Fe_{1-Z-Y}(Ca, Mg)_YO$. *J. Phys. Chem. Solids.* 51:1131–1136.

Gavarri, J. R., and C. Carel. 2019. The Complex Nonstoichiometry of Wüstite $Fe_{1-Z}o$: Review and Comments. *Prog. Solid State Chem.* 53:27–49.

Gavarri, J. R., D. Weigel, and C. Carel. 1976. Introduction to Description of Phase Diagram of Solid Wüstite: II. Structural Review. *Mater. Res. Bull.* 11:917–926.

Gavarri, J. R., C. Carel, and D. Weigel. 1979. Contribution to the Structural Study of High Temperature Solid Wüstite. *J. Solid State Chem.* 29:81–95.

Gavarri, J. R., C. Carel, S. Jasienska, and J. Janowski. 1981. Morphology and Structures of Wüstite. Evolution of Complex Clusters and the Ordering. *Rev. Chim. Min.* 18:608–624.

Gavarri, J. R., C. Carel, and D. Weigel. 1988. Reexamination of the Cluster Structure of the P′ and P″ Quenched Wüstites. *C. R. Acad. Sci. (Paris) Sér. II.* 307:705–710.

Geiger, G. H., R. L. Levin, and J. B. Wagner Jr. 1966. Studies on the Defect Structure of Wustite Using Electrical Conductivity and Thermoelectric Measurements. *J. Phys. Chem. Solids.* 27:947–956.

Gerdanian, P. 1988. A New $\Delta h(O_2)$ Determination Method by High-Temperature Microcalorimetry Study of Oxide-Oxido-Reducing Gaseous Mixtures Reaction. Application to wüstite. *J. Phys. Chem. Solids.* 49:819–826.

Gerdanian, P., and M. Dodé. 1965. Étude Thermodynamique Des Oxydes Feo_{1+X} À 800°C. *J. Chim. Phys.* 62:1018–1022.

Giddings, R. G. 1972. PhD Thesis. University of Utah, Salt Lake City.

Giddings, R. G., and R. S. Gordon. 1973. Review of Oxygen Activities and Phase Boundaries in Wüstite as Determined by Electromotive Force and Gravimetric Methods. *J. Am. Ceram. Soc.* 56:111–116.

Giddings, R. G., and R. S. Gordon. 1974. Solid-State Coulometric Titration: Critical Analysis and Application to Wüstite. *J. Electrochem. Soc.* 121:793–800.

Greenwood, N. N., and A. T. Howe. 1972a. Mössbauer Studies of $Fe_{1-x}o$. Part I. The Defect Structure of Quenched Samples. *J. Chem. Soc, Dalton Trans.* 1:110–116.

Greenwood, N. N., and A. T. Howe. 1972b. Mössbauer Studies of $Fe_{1-x}O$. Part II. Disproportionation Between 300 and 700 K. *J. Chem. Soc, Dalton Trans.* 1:116–122.

Greenwood, N. N., and A. T. Howe. 1972c. Mössbauer Studies of $Fe_{1-x}O$. Part III. Diffusion Line Broadening at 1074 and 1173 K. *J. Chem. Soc, Dalton Trans.* 1:122–126.

Grimes, R. W., A. B. Anderson, and A. H. Heuer. 1986. Defect Clusters in Nonstoichiometric 3d Transition-Metal Monoxides. *J. Am. Ceram. Soc.* 69:619–623.

Grimes, R. W., A. B. Anderson, and A. H. Heuer. 1987. Interaction of Dopant Cations With 4:1 Defect Clusters in Non-Stoichiometric 3d Transition Metal Monoxides: A Theoretical Study. *J. Phys. Chem. Solids.* 48:45–50.

Darken, L. S., and R. W. Gurry. 1946. The System Iron-Oxygen II. Equilibrium and Thermodynamics of Liquid Oxide and Other Phases. *J. Am. Chem. Soc.* 68:798–816.

Hauffe, K., and P. Pfeiffer. 1953. The Kinetics of Wüstite Formation in the Oxidation of Iron. *Z. Metallk.* 44:27–36.

Hayakawa, M., J. B. Cohen, and T. B. Reed. 1972. Measurement of the Lattice Parameter of Wüstite at High Temperatures. *J. Am. Ceram. Soc.* 55:160–164.

Hayakawa, M., M. Morinaga, and J. B. Cohen. 1973. The Defect Structure of Transition-Metal Monoxides. In *Defects and Transport in Oxides*, ed. M.S. Seltzer and R.I. Feffee, 177–203. New York: Plenum Press.

Heikes, R. R., A. A. Maradudin, and R. C. Miller. 1963. Une étude des propriétés de transport des semiconducteurs de valence mixte. *Ann. Phys. Paris.* 13:733–746.

Herai, T., B. Thomas, J. Manenc, and J. Bénard. 1964. Étude au microscope électronique du stade de pré-précipitation dans FeO. *C. R. Acad. Sci. (Paris).* 258:4528–4530.

Hillegas, J. 1968. Seebeck Coefficient and Electrical Conductivity Measurements on Doped and Undoped Wüstite. PhD Thesis. Northwestern University, Evanston, Ill.

Hope, D. A., A. K. Cheetham, and G. J. Long. 1982. A Neutron Diffraction, Magnetic Susceptibility, and Moessbauer-Effect Study of the (Manganese Iron) Oxide ($(Mn_xfe_{1-X})_Yo$) Solid Solutions. *Inorg. Chem.* 21:2804–2809.

Hrynkiewicz, H. U., D. S. Kulawczuk, E. S. Mazanek, A. M. Pustowka, K. Tomala, and M. E. Wyderko. 1972. Mössbauer Effect Studies of Ferrous Oxides $Fe_{1-x}O$. *Phys. Status Solidi (a).* 9:611–616.

Ijima, S. 1974. High Resolution Em Study of Wustite. *32-th Annual Electron Microsc. (Canberra Australia) Proc.* 32:352–353.

Inglot, Z., D. Wiarda, K. P. Lieb, T. Wenzel, and M. Uhrmacher. 1991. Defects in $Fe_{1-x}O$ and the $Fe_{1-x}O$ to Fe_3o_4 Phase Transition Studied by the Perturbed Angular Correlation Method. *J. Phys. Condens. Matter.* 3:4569–4587.

Ishiguro, T., and S. Nagakura. 1985a. Structure of the Commensurate Phase P of Wüstite $Fe_{0.902}O$ Studied by High Resolution Electron Microscopy. *Jpn. J. Appl. Phys. Part 2.* 24: L723–L726.

Ishiguro, T., and S. Nagakura. 1985b. *Structure of the Commensurate Phase P″ of Wüstite Fe0.902o. H.R.E.M. Proc. XI-th International Congress on Electron Microscopy, (Kyoto Jpn. Proc)*, 963–964.

Jette, E. R., and F. Foote. 1933. An X-Ray Study of the Wüstite (Feo) Solid Solutions. *J. Chem. Phys.* 1:29–36.

Johnson, D. P. 1969. Mössbauer Study of the Local Environments of ^{57}fe in Feo. *Solid State Comm.* 7:1785–1788.

Kleman, M. 1965. Thermodynamic Properties of Iron Protoxide in Solid Form Application of Experimental Results to Outline the Equilibrium Diagram. *Mém. Sci. Rev. Métall.* 62:457–469.

Koch, F., and J. B. Cohen. 1969. The Defect Structure of $Fe_{1-x}O$. *Acta Crystall. B.* 25:275–287.

Kofstad, P. 1972. *Nonstoichiometry, Diffusion and Electrical Conductivity in Binary Metal Oxides*. New York: John Wiley.

Kofstad, P., and A. Z. Hed. 1968. Defect Structure Model for Wustite. *J. Electrochem. Soc.* 15:102–104.

Kozheurov, V. A., and G. G. Mikhailov. 1967. Electrical Conductivity of Wüstite. *Russ. J. Phys. Chem.* 11:1552–1555.

Labidi, M., and C. Monty. 1991. P′ And P″ Phase Structures in $Fe_{1-X}o$ and $Fe_{1-X-Y}ca_yo$. *Phase Trans.* 31:99–106.

Lebreton, C., and L. W. Hobbs. 1983. Defect Structure of $Fe_{1-x}O$. *Radiat. Eff.* 74:227–236.

Levin, R. L., and J. B. Wagner Jr. 1966. Lattice-Parameter Measurements of Undoped and Chromium-Doped Wüstite. *Trans. Metall. Soc. AIME.* 236:516–519.

Manenc, J. 1963. Existence of a Superstructure in the Iron Protoxide. *J. Phys. Radium.* 24:447–450.

Manenc, J., J. Bourgeot, and J. Bernard. 1963. Quelques observations concernant la structure du protoxyde de fer. *C. R. Acad. Sci. (Paris).* 256:931–933.

Marucco, J. F., P. Gerdanian, and M. Dodé. 1970a. Détermination des grandeurs molaires partielles de mélange de l'oxygène dans le protoxyde de fer à 1 075 °C. II.—Mesures des pressions partielles d'oxygène en équilibre avec le protoxyde. *J. Chim. Phys.* 67:906–913.

Marucco, J. F., C. Picard, P. Gerdanian, and M. Dodé. 1970b. Détermination des grandeurs molaires partielles de mélange de l'oxygène dans le protoxyde de fer à 1 075 °C. I. Mesures directes des enthalpies molaires partielles de mélange de l'oxygène à l'aide d'un microcalorimètre à haute température de type Tian-Calvet. *J. Chim. Phys.* 67:914–916.

Molenda, J., A. Stokłosa, and W. Znamirowski. 1987a. Transport Properties of Ferrous Oxide $Fe_{1-y}O$ at High Temperature. *Phys. Status Solidi (b).* 142:517–529.

Molenda, J., A. Stokłosa, and W. Znamirowski. 1987b. Electrical Properties of Manganese Doped Ferrous Oxide at High Temperatures. *Solid State Ionics.* 24:39–44.

Neushutz, D., and N. Towhidi. 1970. Die elektrische Leitfähigkeit von Wüstit. *Arch. Eisenhüttenwes.* 41:303–307.

Nowotny, J., and M. Rekas. 1989. Defect Structure and Thermodynamic Properties of the Wüstite Phase ($Fe_{1-y}O$). *J. Am. Ceram. Soc.* 72:1221–1228.

Nowotny, J., and I. Sikora. 1978. Surface Electrical Properties of the Wüstite Phase. *J. Electrochem. Soc.* 125:781–786.

Nowotny, J., M. Rekas, and M. Wierzbicka. 1982. Defect Structure and Electrical Properties of the Wüstite Phase. *Z. Phys. Chem. N. F.* 131:191–198.

Pattek-Janczyk, A., B. Sepioł, J. C. Grenier, and L. Fournes. 1986. Double Electron Exchange in $Fe_{1-x}O$: A Mössbauer Study. *Mater. Res. Bull.* 21:1083–1092.

Picard, C., and M. Dodé. 1970. Ferrous Oxide at 900° C. *Bull. Soc. Chim. Fr.* 7:2486.

Presss, M. R., and D. E. Ellis. 1987. Defect Clusters in Wüstite $Fe_{1-x}o$. *Phys. Rev. B.* 35:4438–4455.

Radler, M., J. B. Cohen, and J. Faber Jr. 1990. Point Defect Clusters in Wüstite. *J. Phys. Chem. Solids.* 51:217–228.

Rekas, M., and S. Mrowec. 1987. On Defect Clustering in the Wüstite Phase. *Solid State Ionics.* 22:185–197.

Rizzo, H. F. 1968. PhD Thesis. University of Utah, Salt Lake City.

Rizzo, H. F., and J. V. Smith. 1968. Coulometric Titration of Wüstite. *J. Phys. Chem.* 72:485–488.

Rizzo, H. F., R. S. Gordon, and I. B. Cutler. 1968. The Determination of Thermodynamic Properties in Single Phase Wüstite by Coulometric Titration in a High Temperature Galvanic Cell. In *Mass Transport in Oxides*, eds. J.B. Wachtman, Jr. and A.D. Franklin, 129–142. Washington: Nat. Bur. Stand. Spec. Publ, 296.

Rizzo, H. F., R. S. Gordon, and I. B. Cutler. 1969. The Determination of Phase Boundaries and Thermodynamic Functions in the Iron-Oxygen System by EMF Measurements. *J. Electrochem. Soc.* 116:266–274.

Roth, W. L. 1960. Defects in the Crystal and Magnetic Structures of Ferrous Oxide. *Acta Crystall.* 13:140–149.

Schuster, D., R. Dieckmann, and W. Schweika. 1989. The Question of Vacancy Clusters in Manganosite $Mn_{1-\Delta}O$. *Ber. Bunsenges. Phys. Chem.* 93:1347–1349.

Schweika, W., and A. E. Carlsson. 1989. Short-Range Order in Ising-Like Models With Many-Body Interactions: Description Via Effective Pair Interactions. *Phys. Rev. B.* 40:4990–4999.

Schweika, W., A. Hoser, and M. Martin. 1992. In-situ Study of the Defect Structure of Wüstite $Fe_{1-x}O$ by Diffuse Elastic Neutron Scattering. *Ber. Bunsenges. Phys. Chem.* 96:1541–1544.

Schweika, W., A. Hoser, M. Martin, and A. E. Carlsson. 1995. Defect Structure of Ferrous Oxide $Fe_{1-x}o$. *Phys. Rev. B.* 51:15771–15778.

Shirane, G., D. E. Cox, and L. Ruby. 1962. Mössbauer Study of Isomer Shift, Quadrupole Interaction, and Hyperfine Field in Several Oxides Containing Fe^{57}. *Phys. Rev.* 125:1158–1164.

Smuts, J. 1966. Structure of Wustite and the Variation of Its X-Ray Diffraction Intensities with Composition. *J. Iron Steel Inst. (London).* 204:237–239.

Smyth, D. M. 1961. Deviations from Stoichiometry in Mno and Feo. *J. Phys. Chem. Solids.* 19:167–169.

Sockel, H. G., and H. Schmalzried. 1968. Coulometrische Titration an Übergangsmetalloxiden. *Ber. Bunsenges. Phys. Chem.* 72:745–754.

Sørensen, O. T. 1981. Thermodynamics and Defect Structure of Nonstoichiometric Oxides. In *Nonstoichiometric Oxides*, ed. O. T. Sørensen, 28–56. New York: Academic Press.

Stokłosa, A. 2015. *Non-Stoichiometric Oxides of 3d Metals.* Pfäffikon: Trans Tech Publications Ltd.

Swaroop, B., and J. B. Wagner Jr. 1967. On the Vacancy Concentration of Wüstite (FeO_x) Near the p to n Transition. *Trans. Metall. Soc. AIME.* 239:1215–1218.

Takayama, E., and N. Kimizuka. 1980. Thermodynamic Properties and Subphases of Wüstite Field Determined by Means of Thermogravimetric Method in the Temperature Range of 1100–1300 °C. *J. Electrochem. Soc.* 127:970–976.

Tannhauser, D. S. 1962. Conductivity in Iron Oxides. *J. Phys. Chem. Solids.* 23:25–34.

Tomlison, S. M., and C. R. A. Catlow. 1989. Computer Symulation Studies of $Fe_{1-x}O$ and $Mn_{1-x}O$. In *Non-Stoichiometric Compounds: Surfaces, Grain Boundaries and Structural Defects*, ed. J. Nowotny, and J.W. Weppner, 53-76. Dordrecht: Kluver Academic Publishers.

Touzelin, B. 1974. High-Temperature X-Ray Determination of Iron Monoxide Lattice Parameters under Controlled Atmosphere. Decomposition of Iron Monoxide between 25 and 570 °c. *Rev. Hautes Temp. Refract.* 11:219–229.

Vallet, P. 1965. On Some New Boundaries of the Solid Wüstite Domain and the Three Triple Points Resulting at 910 °C. *C. R. Acad. Sci. (Paris).* 261:4396–4399.

Vallet, P. 1975. Sur les proprietes thermodynamiques de la wustite solide au-dessus de 911 °C. *C. R. Acad. Sci. (Paris) Sér. C.* 281:291–294.

Vallet, P. 1977. On the Thermodynamic Properties of Solid Wüstite. *C. R. Acad. Sci. (Paris) Sér. C.* 284:545–548.

Vallet, P., and C. Carel. 1979. Contribution to the Study of Solid Non-Stoichiometric Iron Monoxide.T-P-X Diagram. *Mater. Res. Bull.* 14:1181–1194.

Vallet, P., and C. Carel. 1986. Molar Thermodynamic Properties of Solid Wüstite from Their Equilibrium Thermogravimetric Study. *Rev. Chim. Min.* 23:362–377.

Vallet, P., and P. Raccah. 1964. On the Limits of the Domain of Solid Wüstite and the General Diagram Resulting. *C. R. Acad. Sci. (Paris).* 258:3679–3682.

Vallet, P., and P. Raccah. 1965. Contribution to the Study of the Thermodynamic Properties of Solid Iron Protoxide. *Mém. Sci. Rev. Métall.* 62:1–29.

Vallet, P., M. Kleman, and P. Raccah. 1963. On New Thermodynamic Properties and a New Diagram of Solid Wüstite. *C. R. Acad. Sci. (Paris).* 256:136–138.

Wagner, C., and E. Koch. 1936. Electrical Conductivity of Cobalt and Iron Oxides. *Z. Phys. Chem. B.* 32:439–446.

Welberry, T. R., and A. G. Christy. 1997. Defect Distribution and the Diffuse X-Ray Diffraction Pattern of Wüstite, $Fe_{1-x}o$. *Phys. Chem. Miner.* 24:24–38.

Worrel, E., and K. S. Coley. 2013. Defect Structure of Pseudo Phases of Wüstite. *Can. Metall. Quart.* 52:23–33.

Yamamoto, A. 1982. Modulated Structure of Wüstite ($Fe_{1-x}O$) (Three-Dimensional Modulation). *Acta Crystall. B.* 38:1451–1456.

17 Manganese–Wüstite – $(Fe_{1-x}Mn_x)_{1-\delta}O$

17.1 INTRODUCTION

Oxide solid solutions $(Fe_{1-x}Mn_x)_{1-\delta}O$ were the subject of many studies that allowed determining the range of existence of oxide phases and a phase system (Duquesnoy et al. 1975, Engell and Kohl 1962, Foster and Welch 1956,Franke 1987, Franke and Dieckmann 1989, Franke and Dieckmann 1990, Pelton and Thompson 1975, Schwerdtfeger and Muan 1967, Subramanian and Dieckmann 1993). Summary of these studies is presented by Kjellqvist and Selleby (2010). The studies on the deviation from the stoichiometry of the oxide $(Fe_{0.8}Mn_{0.2})_{1-\delta}O$ were reported in (Keller and Dieckmann 1985, Okinaka et al. 1968, Takeuchi et al. 1960). Systematic studies of the oxygen pressure dependence of the deviation from the stoichiometry at 1273 K and 1473 K in the oxides $(Fe_{1-x}Mn_x)_{1-\delta}O$ in the whole range of concentrations of manganese were performed by Dieckmann group (Franke and Dieckmann 1989, Franke and Dieckmann 1990, Keller and Dieckmann 1985, Subramanian and Dieckmann 1993). In turn, Soliman and Stokłosa performed studies of the deviation from the stoichiometry in the oxide $(Fe_{1-x}Mn_x)_{1-\delta}O$ with the manganese content of $x_{Mn} = 0.037$, 0.122 and 0.295 mol/mol (Soliman and Stokłosa 1990). From the above studies it results that at a low manganese content, when the oxygen pressure increases, there occurs a change in the character of dependence of the deviation, analogous to the case of pure wüstite, which indicates the existence of pseudo-phases w_1 and w_2, differing by the structure of point defects.

Iron diffusion in the oxides $(Fe_{1-x}Mn_x)_{1-\delta}O$ was studied as a function of manganese content and the temperature using a method of radioactive tracers (^{59}Fe) (Libanati et al. 1965), however, the work did not cover the effect of the diffusion rate on the deviation. The studies on the diffusion of iron in the oxides $(Fe_{1-x}Mn_x)_{1-\delta}O$ showed that the coefficient of self-diffusion depends on the oxygen pressure in an analogous way to the deviation from the stoichiometry (Franke and Dieckmann 1989). Similarly, the studies on the electrical conductivity also indicate its strict dependence on the concentration of defects (Franke and Dieckmann 1989, Molenda et al. 1991). Using the results of deviation from the stoichiometry studies (Franke and Dieckmann 1989, Subramanian and Dieckmann 1993) and Soliman and Stokłosa (Soliman and Stokłosa 1990), it was decided to determine the diagrams of defects' concentrations in the oxides $(Fe_{1-x}Mn_x)_{1-\delta}O$.

17.2 CALCULATIONS OF DEFECT CONCENTRATIONS DIAGRAMS

As already mentioned, the defects' structure and the character of the dependence of the deviation from the stoichiometry on the oxygen pressure at low manganese content, reported in (Franke and Dieckmann 1989, Soliman and Stokłosa 1990, Subramanian and Dieckmann 1993), have a similar nature as for pure wüstite, and at higher manganese content, similar as for pure manganese oxide. In the calculations it was assumed that there are mutual couplings between the concentrations of the defects, due to reactions of the formation of simple defects and defect complexes; as a result, determined concentrations of defects are set at a given oxygen pressure. Despite the fact that for wüstite with a low manganese content, a significant deviation from the stoichiometry is present even at the lowest oxygen pressures (near the decomposition pressure of the oxide), it was decided to attempt to determine the defects' concentrations. Due to that, the calculations of diagrams of concentrations of point defects in manganese-doped wüstite were conducted with a method analogous to the one used for pure oxides (Stokłosa 2015) and the one used for pure magnetite and doped magnetite (see Section 1.5), using the results of the studies of the deviation from the stoichiometry reported by Franke and Dieckmann (1989), Soliman and Stokłosa (1990), Subramanian and Dieckmann (1993). Similarly, as for other oxides showing smaller deviation from the stoichiometry, it was assumed that the point defects can be considered as quasi-particles, obeying the laws of chemical thermodynamics. As already mentioned, the method of adjusting the values of ΔG_i^o of the formation of defects takes into account a mean value of activity coefficients in a given range of oxygen pressures (see Section 1.5 and also 6.5 (Stokłosa 2015)). This fact, despite high concentrations of defects, allows for an attempt to determine of the defects' concentrations, according to the method presented in Section 1.5. For the calculations, it is necessary to know the value of ΔG_{eh}^o of the formation of electronic defects, which was not determined for pure wüstite and for manganese-doped wüstite. Based on the analysis of the results of electrical conductivity (Molenda et al. 1987) it was assumed that in wüstite it is $\Delta G_{eh}^o = 220$ kJ/mol and it is independent of the temperature. It is also similar to the value of ΔG_{eh}^o for manganese oxide, which is 227 kJ/mol (Stokłosa 2015). The initial calculations have shown that a change in the value of ΔG_{eh}^o within the range of 180–240 kJ/mol, does not cause a change in the character of the dependence of the deviation from the stoichiometry on p_{O_2}, in the range of oxygen pressures where $Fe_{1-\delta}O$ exists. The value of $\Delta G_{V_M''}^o$ of formation of cation vacancies and other defects changes only slightly; the oxygen pressure $p_{O_2}^{(s)}$ at which the oxide reaches the stoichiometric composition, changes. As this pressure is lower than the decomposition pressure by about 20 orders of magnitude, the error is insignificant. Considering the above facts, for the calculations of diagrams of concentrations of defects in manganese-doped wüstite, a constant value of ΔG^{o}_{eh} equal 220 kJ/mol was assumed.

As the concentrations of the individual defects are mutually related (they are dependent on the oxygen pressure, on the concentration of electron holes and on the concentration of defects out of which they are formed), during their calculation we can use the relations resulting from the expression for equilibrium constants (see Equation 16.9 and 16.10, Equation 16.12–16.15, Equation 16.21–16.23).

In the first stage we assume that the dominating defects are defects with the highest ionisation degree and we adjust the values of $\Delta G^{o}_{V_M''}$; we calculate the equilibrium constant according to Equation 16.9 and the concentration of vacancies $\left[V_M''\right]$ (dominating defects). With the change in the concentration of vacancies $\left[V_M''\right]$, the concentration of interstitial iron ions is associated. At the same oxygen pressure, from Equation 16.9 and 16.10, a relation between the concentration of vacancies $\left[V_M''\right]$ and the concentration of interstitial ions $\left[M_i^{\bullet\bullet}\right]$ is derived:

$$\frac{\left[V_M''\right]\left[h^{\bullet}\right]}{K_{V_M''}} = \frac{K_{M_i^{\bullet\bullet}}\left[h^{\bullet}\right]}{\left[M_i^{\bullet\bullet}\right]} \tag{17.1}$$

Transforming Equation 17.1 we get:

$$\left[V_M''\right]\left[M_i^{\bullet\bullet}\right] = K_{V_M''}K_{M_i^{\bullet\bullet}} = K_F \tag{17.2}$$

Adjusting the value of ΔG^{o}_{F} of formation of Frenkel-type defects, we calculate the value of K_F and determine the concentration of ions $\left[M_i^{\bullet\bullet}\right]$ depending on the concentration $\left[V_M''\right]$.

The standard Gibbs energy of the formation of interstitial $\left[M_i^{\bullet\bullet}\right]$ ions according to Equation 17.2 will be:

$$\Delta G^{o}_{M_i^{\bullet\bullet}} = \Delta G^{o}_{F} - \Delta G^{o}_{V_M''} \tag{17.3}$$

Therefore, we fit the values of $\Delta G^{o}_{V_M''}$ and ΔG^{o}_{F} in such a way that the calculated deviations from the stoichiometry resulting from the concentration of vacancies $\left[V_M''\right]$ and interstitial ions $\left[M_i^{\bullet\bullet}\right]$ are close to the experimental values of δ.

Due to the lack of small values of the deviation from the stoichiometry it was impossible to assume that there is a range where only iron vacancies $\left(V_{Fe}''\right)$ dominate and it was not possible to determine unambiguously the values of ΔG^{o}_{F} of the formation of intrinsic ionic defects, as for example for magnetite. In the calculations, the value of ΔG^{o}_{F} was assumed, presuming that it changes linearly when the manganese content in the oxide $(Fe_{1-x}Mn_x)_{1-\delta}O$ increases, using the values for pure oxides $Fe_{1-\delta}O$ and $Mn_{1-\delta}O$ determined in the work (Stokłosa 2015). The performed calculations showed that a change in the value ΔG^{o}_{F} within the range of 20 kJ/mol does not cause a change in the degree of match between the dependence of the deviation from the stoichiometry on the oxygen pressure and the experimental results δ; only the value of $\Delta G^{o}_{V_M''}$ slightly changes.

The performed calculations showed that at higher oxygen pressures there are large discrepancies between the dependence of the deviation from the stoichiometry and the experimental values of δ. Due to that, in the next stage, cation vacancies with lower ionisation degrees were taken into account; their concentration, according to Equation 16.12–16.15 depends on the values of equilibrium constant and on the concentrations of defects from which they are formed. The maximum values of $\Delta G^{o}_{V_M'}$ and $\Delta G^{o}_{V_M^{x}}$ were chosen in order to obtain a full consistency of the dependence of δ on p_{O_2} with the experimental values of δ. When considering the defects with lower ionisation degrees it was assumed that $\Delta G^{o}_{V_M'} = \Delta G^{o}_{M_i^{\bullet}}$ and $\Delta G^{o}_{V_M^{x}} = \Delta G^{o}_{M_i^{x}}$. The values of these constants were adjusted in such a way that the calculated dependence of deviation from the stoichiometry was consistent with the experimental values of δ.

At the next stage of calculations, the concentrations of individual defect complexes of type $\left\{\left(V_M''\right)_2 M_i^{3\bullet}\right\}' \equiv (C_2')$, $\left(\left\{\left(V_M''\right)_4 M_i^{3\bullet}\right\}^{5'}\right) \equiv (C_4^{5'})$ and clusters $\left(\left[V_M''\right]_6\left[M_i^{3\bullet}\right]_2\right)^{6'} \equiv (C_6^{6'})$ were taken into account, using the reactions of their formation (Equation 16.21–16.23). The calculations were performed with and without considering simple defects with lower ionisation degree. As an increase in the concentration of defect complexes caused a decrease in the concentration of cation vacancies $\left[V_M''\right]$, the maximum values of ΔG^{o}_{C} of the formation of defect complexes were chosen in such a way to allow obtaining the real values of defects' concentrations. At the limit value of ΔG^{o}_{C}, the concentration of cation vacancies $\left[V_M''\right]$ was independent of the oxygen pressure, which made further calculations impossible.

Therefore, the values of $\Delta G^{o}_{C_2'}$, $\Delta G^{o}_{C_4^{5'}}$, $\Delta G^{o}_{C_6^{6'}}$ of the formation of complexes were sequentially adjusted in order to obtain their real concentrations, and then $\Delta G^{o}_{V_M'} = \Delta G^{o}_{V_M^{x}}$ of formation of cation vacancies $\left[V_M'\right]$ and $\left[V_M^{x}\right]$ and complexes with a lower ionisation degrees were fitted in order to obtain a match between the dependence of the deviation from the stoichiometry on the oxygen pressure and the experimental values of δ.

This way, for pure wüstite and for wüstite doped with a small amount of manganese it was possible to obtain a good match only for the range of lower oxygen pressures, up to the value of the deviation below $\delta < 0.08$ mol/mol (in the range of existence of w_1 pseudo-phase).

For the range of higher oxygen pressures (in the range of existence of w_2 pseudo-phase) for pure wüstite and for an oxide with a low manganese content, the calculated values of the deviation from the stoichiometry resulting from the existence of vacancies V''_M were higher than the experimental ones. Such character of the dependence of δ on p_{O_2} may indicate the presence of a strong interaction between cation vacancies and defect complexes and the formation of highly ordered structures. As a result, this causes changes in interactions between ions, causing a change in the value of ΔG_i^o of formation of defects which is no more possible to be corrected with a mean activity coefficient. As shown in the work (Stokłosa 2015), for pure wüstite, for the range of higher oxygen pressures, the match between the calculated dependence of the deviation from the stoichiometry on p_{O_2} and the experimental results is obtained when assuming that $\Delta G^o_{V''_{Fe}}$ is a linear function of the deviation δ:

$$\Delta G^{o(\delta)}_{V''_{Fe}} = \Delta G^{o(\delta=0)}_{V''_{Fe}} + \alpha\delta \tag{17.4}$$

A change in the value of $\Delta G^{o(\delta)}_{V''_{Fe}}$ with the increase in the deviation from the stoichiometry is not only a computational (technical) operation, but it is also consistent with the thermodynamics of the system. The setting of equilibrium indicates that the structures of defects are thermodynamically reversible, as for each value of the deviation resulting from the oxygen pressure, a determined ratio of concentrations of defects (a determined deviation δ) is set. Therefore, due to a change in interactions, "equilibrium constants" are changed, also $\Delta G_i^{o(\delta)}$, which are not constant, because they depend on δ. The biggest change should be present for the value of $\Delta G^o_{V''_{Fe}}$ of the formation of vacancies $\left(V''_{Fe}\right)$ that are the dominating defects. At higher manganese contents in the oxide $(Fe_{1-x}Mn_x)_{1-\delta}O$ the above correction was not necessary, but in order to obtain higher concentrations of defect complexes, the dependence of δ on p_{O_2} ($\Delta G^o_{V''_{Fe}}$ values) was corrected with the parameter α.

Thus, in the case where the dependence of the deviation from the stoichiometry on the oxygen pressure consists of two curves with different slopes (ranges of existence of w_1 and w_2 pseudo-phases), fitting of the value of ΔG_i^o of formation of defects was performed independently for each range.

Therefore, the deviation from the stoichiometric composition in the oxide $(Fe_{1-x}Mn_x)_{1-\delta}O$ is the difference between the concentrations of cation vacancies and interstitial cations, also those present in defect complexes, which can be written as:

$$\delta = \left[V''_M\right] + \left[V'_M\right] + \left[V^x_M\right] + \sum_i C_i - \left[M_i^{\bullet\bullet}\right] \tag{17.5}$$

where

$$\sum_i \left[C_i\right] = \left[C'_2\right] + 3\left[C_4^{5'}\right] + 4\left[C_6^{6'}\right] + \left[C_2^x\right] + 3\left[C_4^{4'}\right] + 4\left[C_6^{5'}\right]$$

The concentrations of all ionic and electronic defects are bound by the electroneutrality condition, which assumes the form of:

$$\left[e'\right] + 2\left[V''_M\right] + \left[V'_M\right] + \left[C'_2\right] + 5\left[C_4^{5'}\right] + 6\left[C_6^{6'}\right] + 4\left[C_4^{4'}\right] + 5\left[C_6^{5'}\right] = \left[h^\bullet\right] + 2\left[M_i^{\bullet\bullet}\right] \tag{17.6}$$

Thus, the concentration of electronic defects is dependent on the concentration of ionic defects. Due to that, in order to determine the concentration of electronic defects it is necessary to know the value of the standard Gibbs energy of formation of electronic defects ΔG^o_{eh} (value of equilibrium constant). Calculating the concentration of electrons from Equation 17.6 and inserting into Equation 16.11, we get the dependence of the concentration of electron holes on the concentration of ionic defects:

$$K_{eh} = \left[h^\bullet\right]\left(\left[h^\bullet\right] + 2\left[M_i^{\bullet\bullet}\right] - \left(2\left[V''_M\right] + \left[V'_M\right] + \left[C'_2\right] + 5\left[C_4^{5'}\right] + 6\left[C_6^{6'}\right] + 4\left[C_4^{4'}\right] + 5\left[C_6^{5'}\right]\right)\right) \tag{17.7}$$

$$K_{eh} = \left[h^\bullet\right]\left(\left[h^\bullet\right] + \Sigma[\text{def}]\right) \tag{17.8}$$

Solving the quadratic Equation 17.7, knowing the equilibrium constant K_{ch} derived from the value of ΔG^o_{ch}, we determine the dependence of the concentration of electron holes on the concentration of ionic defects and, next, the concentration of electrons.

We calculate the relative concentrations of ionic defects according to Equation 17.2, Equations 16.12–16.15, 16.21–16.23, depending on the concentration of vacancies $\left[V''_M\right]$. The oxygen pressure at which there are equilibrium concentrations of point defects were calculated using the expression for the equilibrium constant of formation of vacancies V''_M (Equation 16.9), which after transformation has the form of:

$$p_{O_2} = \frac{\left[V''_{Fe}\right]^2\left[h^\bullet\right]^4}{K^2_{V''_M}} \tag{17.9}$$

A change in the value of ΔG^o_i of formation of one of the defects causes a change in the concentration of this defect and other defects. The calculations are performed by the method of successive approximation until the electroneutrality condition is fulfilled.

17.3 POINT DEFECTS CONCENTRATIONS DIAGRAMS – RESULTS

With the method presented above, Figures 17.1a–g and 17.2a–e show diagrams of concentrations of point defects (ionic and electronic) in pure oxides $Fe_{1-\delta}O$ and $Mn_{1-\delta}O$, using the results of the deviation from the stoichiometry reported by Bransky and Hed (1968Vallet and Raccah 1965) ((△,◆) points) and Keller and Dieckmann (1985) – (▲) and for oxide solutions $(Fe_{1-x}Mn_x)_{1-\delta}O$, using the results of the deviation from the stoichiometry taken from Franke and Dieckmann (1989) ((■,□) points) at 1273 K for the manganese content of x_{Mn} = 0.2, 0.4, 0.6 and 0.8 mol/mol at 1473 K and for the manganese content of x_{Mn} = 0.2, 0.5, 0.8 and 0.9 mol/mol given by Subramanian and Dieckmann (1993).

As can be seen in Figures 17.1 and 17.2, a good match was obtained between the dependence of the deviation from the stoichiometry and the experimental values of δ. Similarly for pure wüstite, also in the oxides $(Fe_{1-x}Mn_x)_{1-\delta}O$ with the manganese content up to x_{Mn} = 0.4 mol/mol at 1273 K, and at 1473 K up to x_{Mn} = 0.8 mol/mol, there occurs a change in the character of the dependence of δ on p_{O_2}, which indicates the existence of two wüstite pseudo-phases, w_1 and w_2 (see Figure 17.7).

From the performed calculations it results that in w_1 pseudo-phase, doubly charged negative cation vacancies $\left[V''_M\right]$ dominate, but there is also a significant concentration of vacancies with lower ionisation degrees, $\left[V'_M\right]$ and $\left[V^x_M\right]$. The

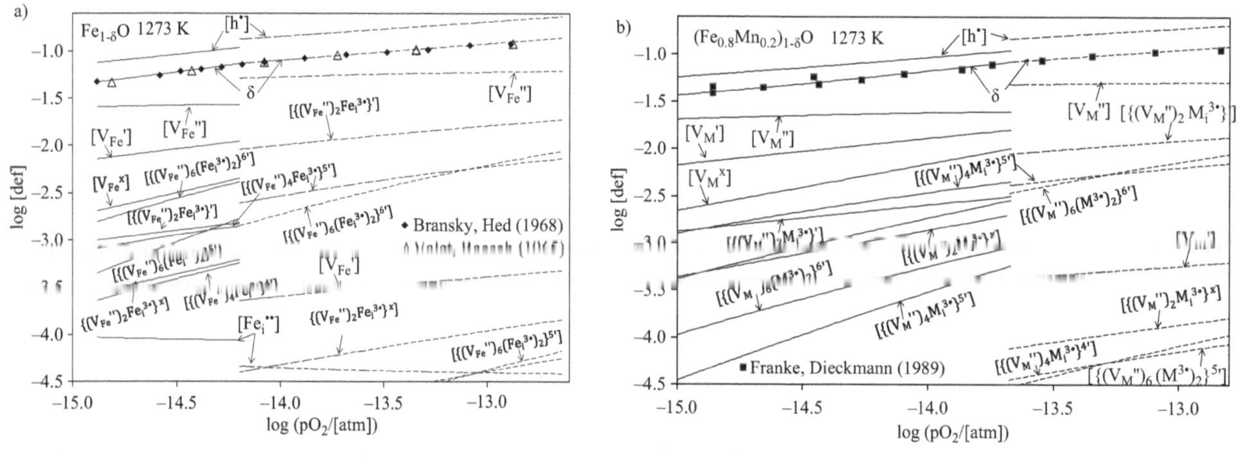

FIGURE 17.1 (a) Diagram of concentration of point defects for the oxide $Fe_{1-x}M_x)_{1-\delta}O$ Fe1–δO at 1273 K, taking into account simple defects and defect complexes $\left\{\left(V''_{Fe}\right)_2 Fe_i^{3\bullet}\right\}'$, $\left\{\left(V''_{Fe}\right)_4 Fe_i^{3\bullet}\right\}^{5'}$ and clusters $\left\{\left(V''_{Fe}\right)_6\left(Fe_i^{3\bullet}\right)_2\right\}^{6'}$, as well as defects with lower ionisation degrees, obtained using the results of the studies on the deviation from the stoichiometry taken from (Vallet and Raccah 1965) – (△) points, (Bransky and Hed 1968) – (◆). Solid lines denote the concentration of defects in w_1 "pseudo-phase", dashed lines (---) in w_2 "pseudo-phase". (b) for the oxide $(Fe_{0.8}Mn_{0.2})_{1-\delta}O$ obtained using the results of the studies on the deviation from the stoichiometry taken from (Franke and Dieckmann 1989) – (■). (c) for the oxide $(Fe_{0.6}Mn_{0.4})_{1-\delta}O$. (d) for the oxide $(Fe_{0.4}Mn_{0.6})_{1-\delta}O$. (e) for the oxide $(Fe_{0.2}Mn_{0.8})_{1-\delta}O$. (f) for the oxide $(Fe_{0.1}Mn_{0.9})_{1-\delta}O$. (g) for the oxide $Mn_{1-\delta}O$ obtained using the results of the studies on the deviation from the stoichiometry extracted from (Keller and Dieckmann 1985) – ((□) points).

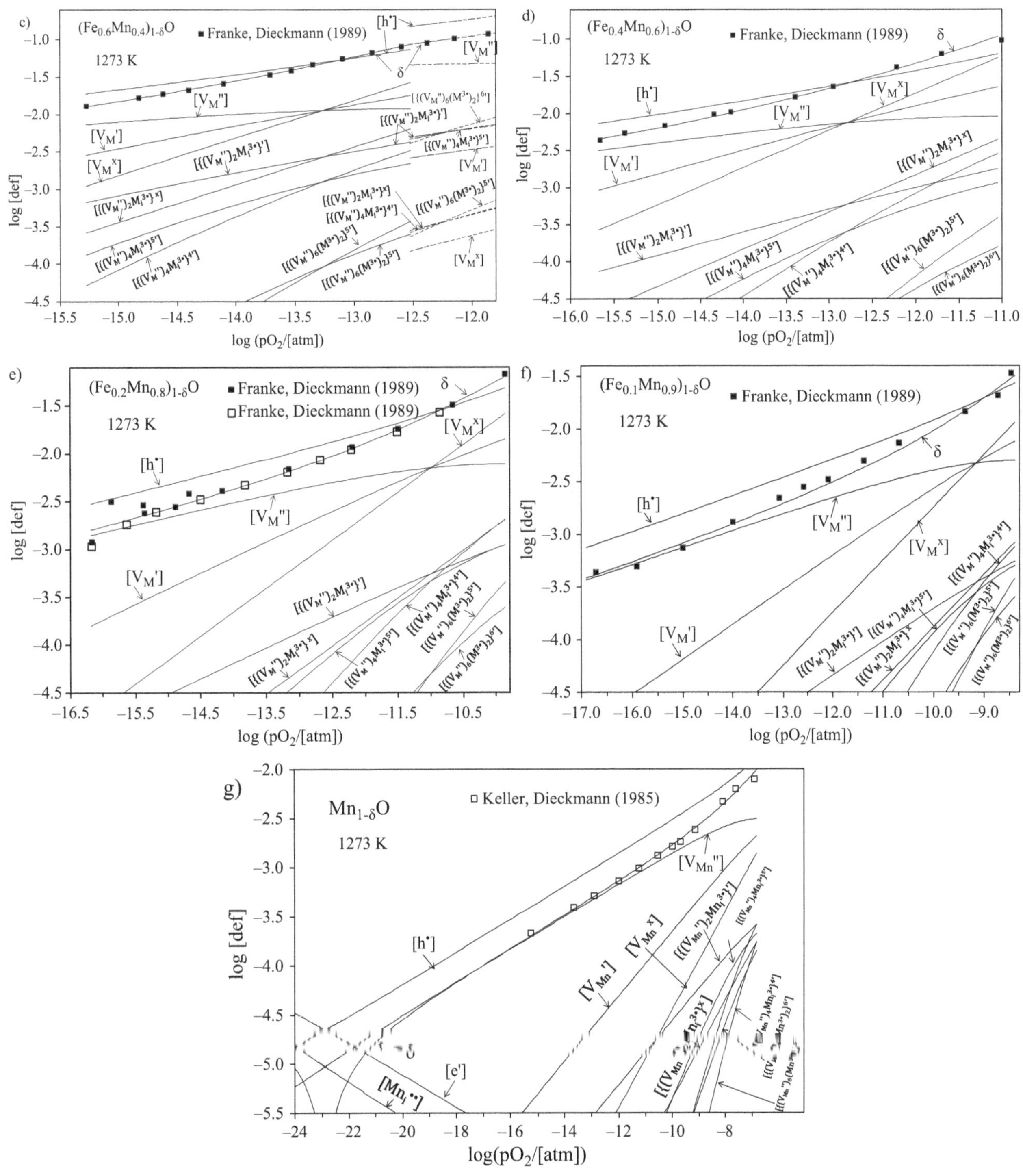

FIGURE 17.1 (Continued)

concentration of defect complexes in the oxides $(Fe_{1-x}Mn_x)_{1-\delta}O$ reaches 10^{-2} mol/mol and when the manganese content increases, it decreases to about 10^{-3} mol/mol. The calculated concentration of cation vacancies is the maximum one, as a further increase in their concentration causes a decrease in the concentration of cation vacancies $[V_M'']$, which in consequence becomes independent from the oxygen pressure, making the calculations impossible.

Due to a weaker dependence of the deviation from the stoichiometry on the oxygen pressure in in the range of existence of w_2 pseudo-phase, other values of ΔG_i^o of formation of defects were fitted. As can be seen in Figure 17.1, the dependence of the deviation from the stoichiometry on the oxygen pressure is dependent mainly on the concentration of vacancies $[V_M'']$. Due to that, such maximum concentrations of defect complexes and cation vacancies with lower ionisation degrees were fitted that do not deteriorate the degree of match between the dependence of δ on p_{O_2} and the experimental values of δ. An

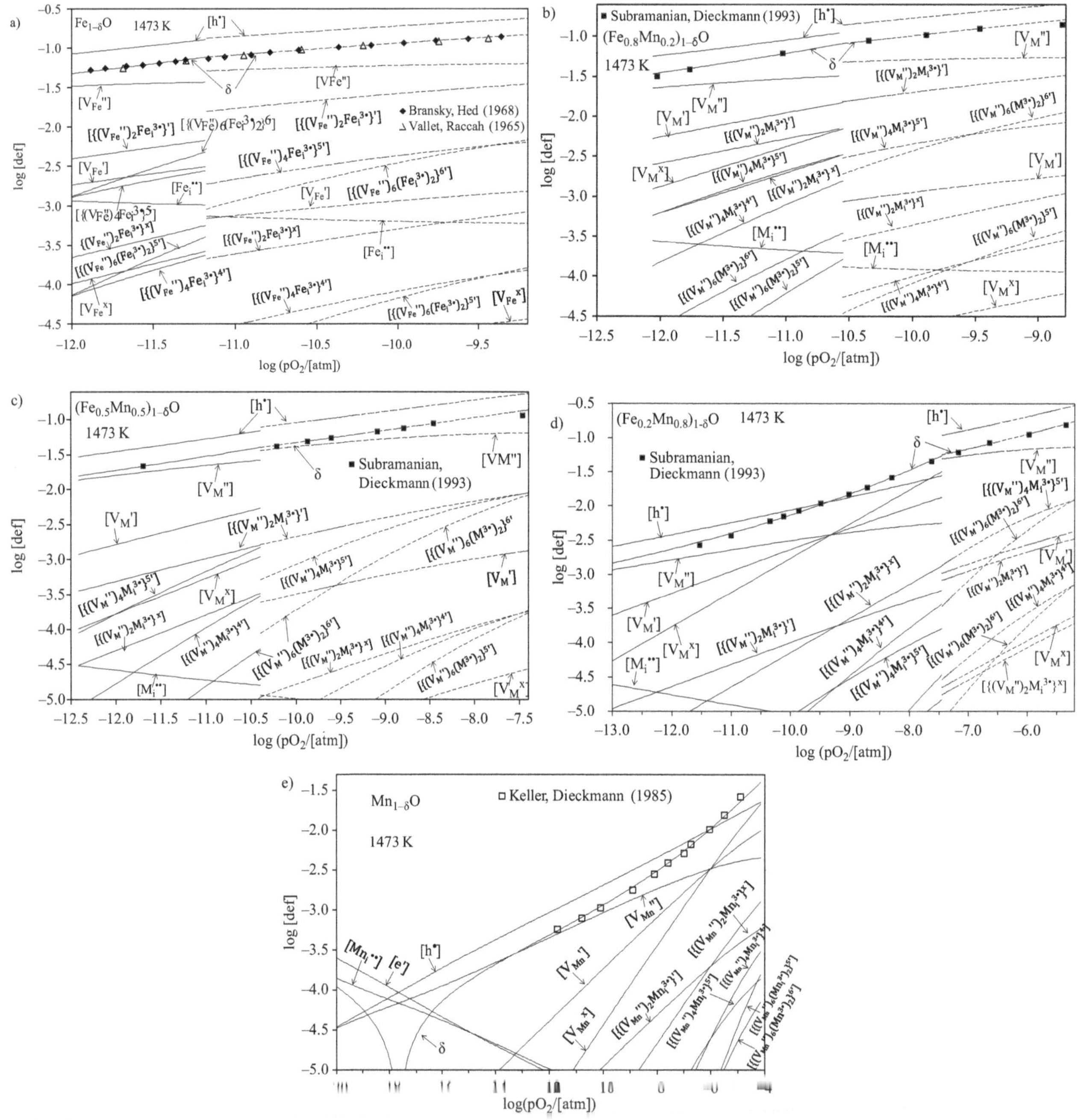

FIGURE 17.2 (a) Diagram of concentrations of point defects for the oxide $Fe_{1-\delta}O$ at 1473 K, taking into account simple defects, defect complexes: $\left\{\left(V_{Fe}''\right)_2 Fe_i^{3\bullet}\right\}'$, $\left\{\left(V_{Fe}''\right)_4 Fe_i^{3\bullet}\right\}^{5'}$ and clusters $\left\{\left(V_{Fe}''\right)_6 \left(Fe_i^{3\bullet}\right)_2\right\}^{6'}$, as well as defects with lower ionisation degrees, obtained using the results of the studies of the deviation from the stoichiometry reported by Vallet and Raccah (1965) – (△) points, (Bransky and Hed 1968) – (◆). Solid lines denote the concentration of defects in w_1 "pseudo-phase", dashed lines (---) in w_2 "pseudo-phase". (b) for the oxide $(Fe_{0.8}Mn_{0.2})_{1-\delta}O$ obtained using the results of the studies on the deviation from the stoichiometry taken from Subramanian and Dieckmann (1993) – (■) points. (c) for the oxide $(Fe_{0.5}Mn_{0.5})_{1-\delta}O$. (d) for the oxide $(Fe_{0.8}Mn_{0.8})_{1-\delta}O$. (e) K for the oxide $Mn_{1-\delta}O$ obtained using the results of the studies on the deviation from the stoichiometry given in (Keller and Dieckmann 1985) – (□) points.

increase in the concentration of defect complexes caused a decrease in the concentration of vacancies $[V_M'']$ and its independence of the oxygen pressure.

In turn, Figures 17.3–17.5 present diagrams of concentrations of defects for the oxides $(Fe_{1-x}Mn_x)_{1-\delta}O$ with the manganese content of x_{Mn} = 0.037, 0.122, 0.295 mol/mol, in the temperature range of 1173–1473 K, obtained when using the results reported in (Soliman and Stokłosa 1990).

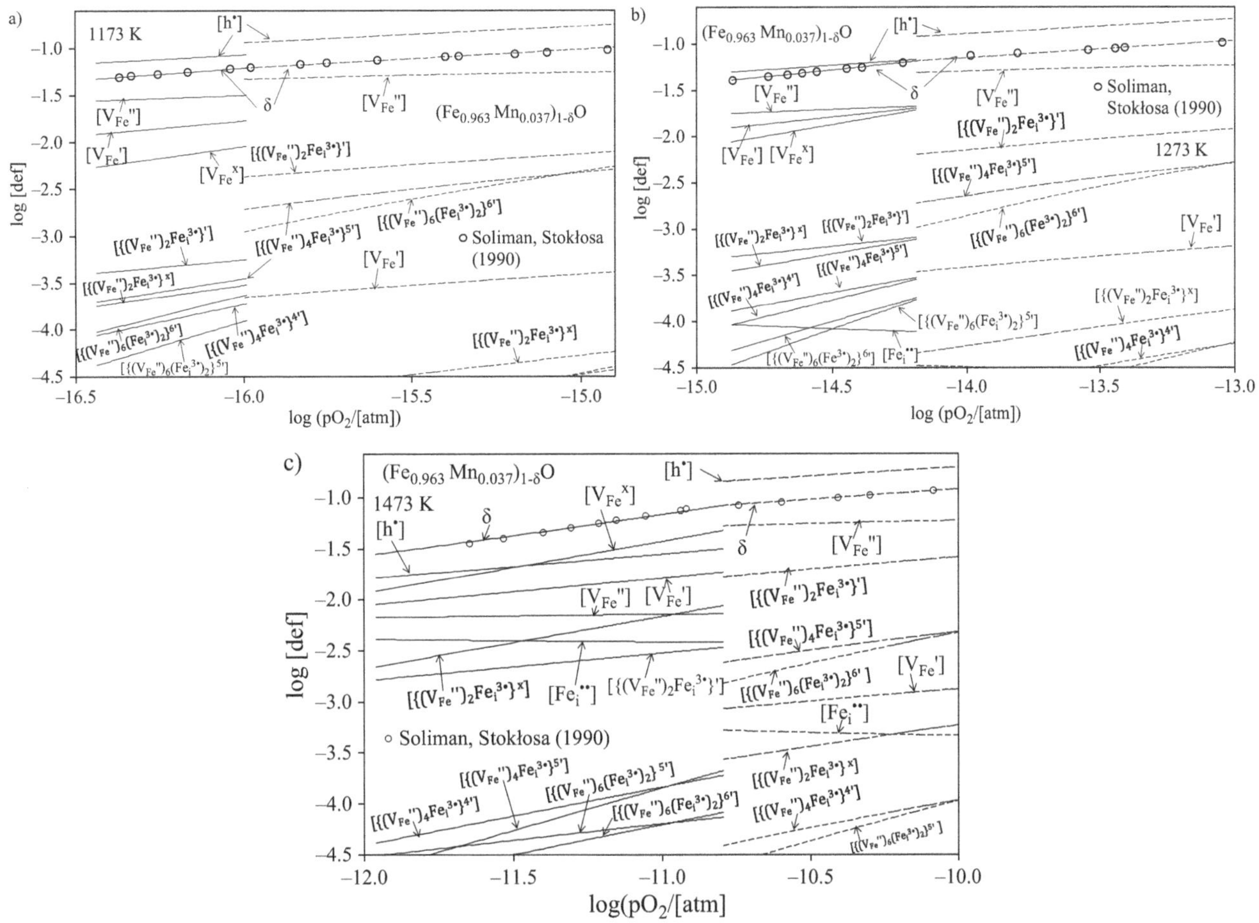

FIGURE 17.3 (a) Diagram of concentration of point defects for the oxide $(Fe_{0.963}Mn_{0.037})_{1-\delta}O$ at 1173 K, taking into account simple defects and defect complexes $\{(V''_{Fe})_2 Fe_i^{3\bullet}\}'$, $\{(V''_{Fe})_4 Fe_i^{3\bullet}\}^{5'}$ and clusters $\{(V''_{Fe})_6 (Fe_i^{3\bullet})_2\}^{6'}$, as well as defects with lower ionisation degrees, obtained using the results of the studies on the deviation from the stoichiometry (Soliman and Stokłosa 1990) ((O) points). Solid lines denote the concentration of defects in w_1 "pseudo-phase", dashed lines in w_2 "pseudo-phase". (b) for the oxide $(Fe_{0.963}Mn_{0.037})_{1-\delta}O$ at 1273 K. (c) for the oxide $(Fe_{0.963}Mn_{0.037})_{1-\delta}O$ at 1473 K.

As can be seen in Figures 17.3a–c, 17.4a–c and 17.5a–c, a good match was also obtained between the dependence of the deviation from the stoichiometry and the experimental results δ. At the manganese content of x_{Mn} = 0.037, 0.122 mol/mol there are two ranges of oxygen pressure, where the character of the dependence of δ on p_{O_2} is different, which indicates the existence of two wüstite pseudo-phases, w_1 and w_2. For the manganese content of x_{Mn} = 0.295 mol/mol, wüstite pseudo-phase w_2 probably exists at higher oxygen pressures.

In the case of wüstite with the manganese content of x_{Mn} = 0.037 mol/mol, at 1273 K and 1323 K, in the range of existence of w_2 pseudo-phase, the calculated values of the deviation from the stoichiometry, considering only vacancies $[V''_M]$, were higher than the experimental ones, similarly as in pure wüstite. The match was obtained when assuming that the standard Gibbs energy of formation of cation vacancies $\Delta G^o_{V''_M}$ is a linear function of the deviation from the stoichiometry according to Equation 17.4. In order to obtain a higher concentration of defect complexes in the oxide $(Fe_{1-x}Mn_x)_{1-\delta}O$ with a higher manganese content, a correction of the value of $\Delta G^o_{V''_M}$ was introduced, by fitting the parameter α (see Figure 17.13).

From the performed calculations of diagrams of concentrations of defects for the oxides $(Fe_{1-x}Mn_x)_{1-\delta}O$ it results that based on the dependence of the deviation from the stoichiometry it is impossible to conclude unambiguously on the concentration of defects complexes, because the determined concentrations of defect complexes do not affect the character of the dependence of the deviation δ on p_{O_2}. The calculated maximum concentration of complexes is of an order of 0.01–0.001 mol/mol, and, as it was shown by Stokłosa 2015), in the case of wüstite it is high enough for their dense-packing (ordering) in the structure.

Figure 17.6a–c compare the dependences of the deviation from the stoichiometry on the oxygen pressures (fragments of defects' diagrams) in the oxide $(Fe_{1-x}Mn_x)_{1-\delta}O$ at 1173–1473 K, obtained using the results taken from (Soliman and

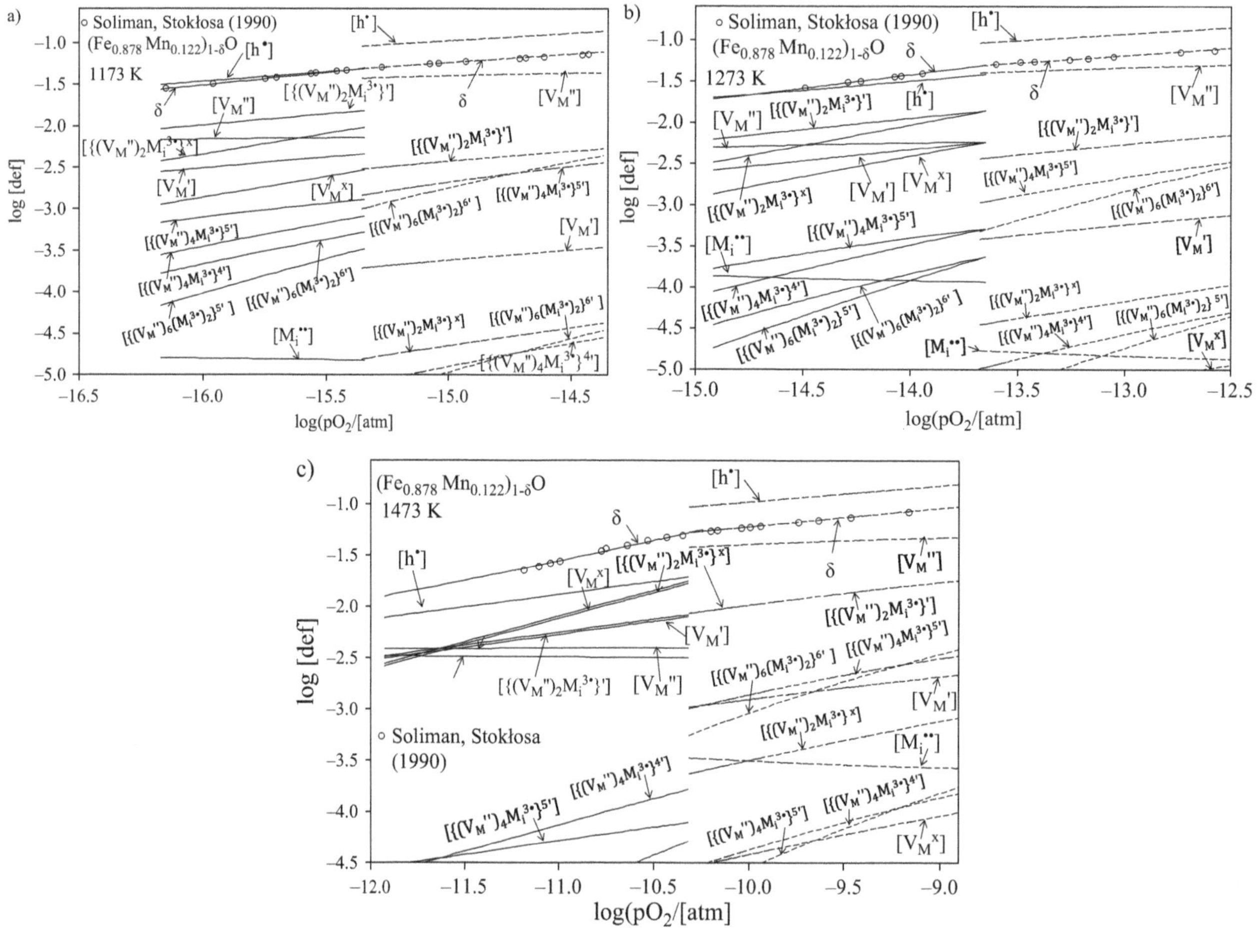

FIGURE 17.4 (a) Diagram of concentration of point defects for the oxide $(Fe_{0.878}Mn_{0.122})_{1-\delta}O$ at 1173 K, taking into account simple defects and defect complexes $\{(V''_{Fe})_2 Fe_i^{3\bullet}\}'$, $\{(V''_{Fe})_4 Fe_i^{3\bullet}\}^{5'}$ and clusters $\{(V''_{Fe})_6 (Fe_i^{3\bullet})_2\}^{6'}$, as well as defects with lower ionisation degrees, obtained using the results of the studies on the deviation from the stoichiometry (Soliman and Stokłosa 1990) ((O) points). Solid lines denote the concentration of defects in w_1 "pseudo-phase", dashed lines in w_2 "pseudo-phase". (b) for the oxide $(Fe_{0.878}Mn_{0.122})_{1-\delta}O$ at 1273 K. (c) for the oxide $(Fe_{0.878}Mn_{0.122})_{1-\delta}O$ at 1473 K.

Stokłosa 1990), and in Figure 17.7, extracted from (Franke and Dieckmann 1989, Subramanian and Dieckmann 1993). As can be seen in Figures 17.6a–c and 17.7a–d, at low manganese contents in wüstite there occurs a clear change in the character of the dependence of the deviation from the stoichiometry on p_{O_2}. Thus, similarly as for pure wüstite, also for doped wüstite w_1 and w_2 pseudo phases are present. When the manganese content increases, the range of oxygen pressures where the oxides $(Fe_{1-x}Mn_x)_{1-\delta}O$ exists increases, and, in the range of low oxygen pressures, the deviation from the stoichiometry decreases, causing an increase in the slope of the dependence of δ on p_{O_2}.

As can be seen in Figure 17.6a, lines of the dependence of δ on p_{O_2} at the manganese content of $x_{Mn} = 0.037$ and 0.2 mol/mol in the oxide $(Fe_{1-x}Mn_x)_{1-\delta}O$, obtained using the results reported by Franke and Dieckmann (1989) and Soliman and Stokłosa (1990), cross. Similarly, in Figures 17.7b and c, the values of the deviation from the stoichiometry for the manganese content of $x_{Mn} = 0.122$, 0.295 mol/mol (Soliman and Stokłosa 1990) differ significantly from the dependences for similar manganese contents reported elsewhere (Franke and Dieckmann 1989, Subramanian and Dieckmann 1993). The above discrepancies can be related to the evaporation of manganese from doped wüstite; this value was considered in the calculations presented in (Soliman and Stokłosa 1990). They can be also related to methods of measurement of low oxygen pressures. Due to difficulties in the determination of low oxygen activities with an oxygen cell, in the work of Soliman and Stokłosa (1990) the basis was the measurement of electrical conductivity of a calibrated $TiO_{2-\delta}$ sample.

In order to compare the character of the dependence from the stoichiometry on the oxygen pressure in the oxides $(Fe_{1-x}Mn_x)_{1-\delta}O$ (presented in Figures 17.6a–c and 17.7a–d), Figure 17.8a–c present the oxygen pressure dependence of the derivative $(d\log p_{O_2}/d\log\delta) = n_\delta$.

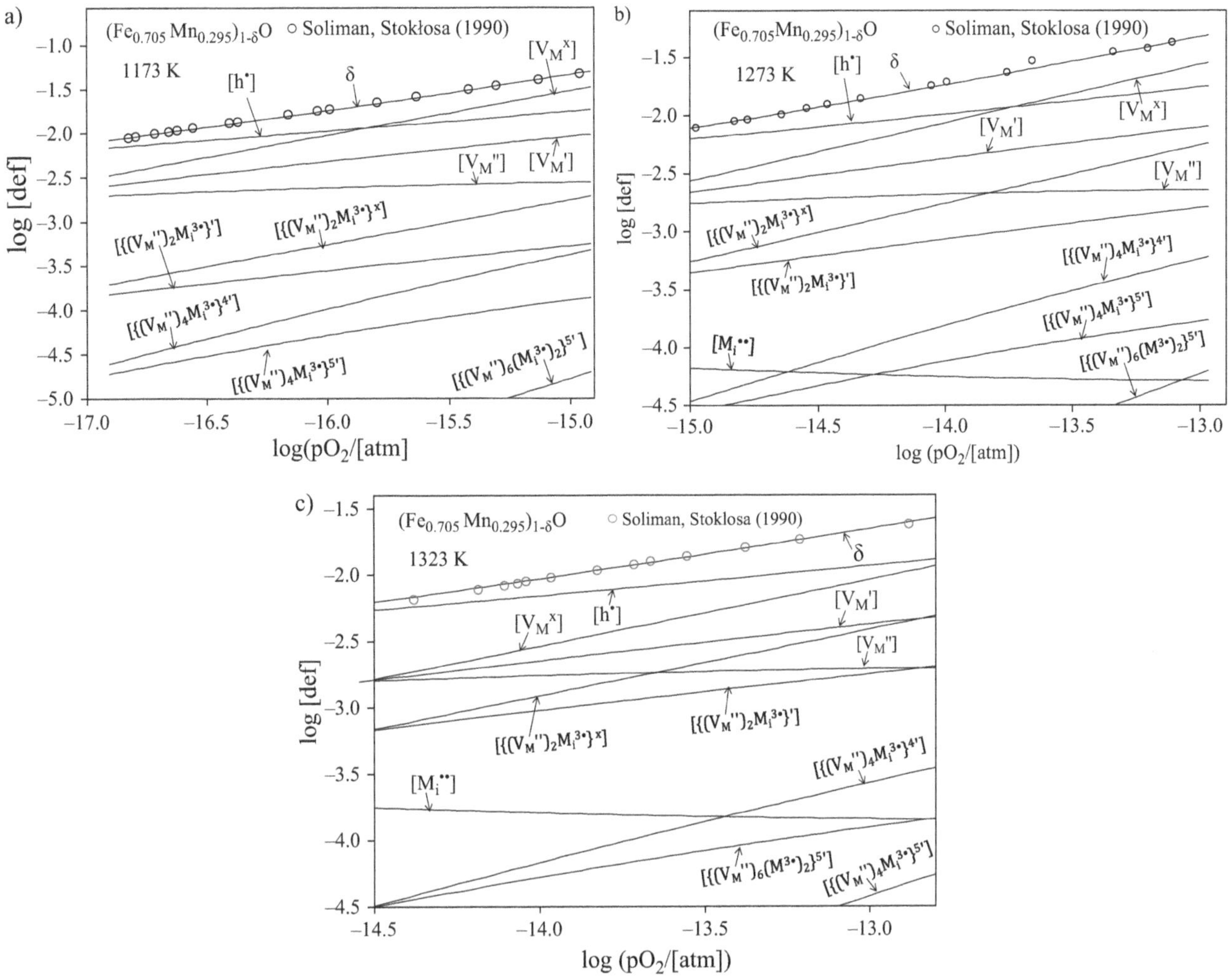

FIGURE 17.5 (a) Diagram of concentrations of point defects for the oxide $(Fe_{0.705}Mn_{0.295})_{1-\delta}O$ at 1173 K, taking into account simple defects and defect complexes $\left\{(V''_{Fe})_2 Fe_i^{3\bullet}\right\}'$, $\left\{(V''_{Fe})_4 Fe_i^{3\bullet}\right\}^{5'}$ and clusters $\left\{(V''_{Fe})_6 \left(Fe_i^{3\bullet}\right)_2\right\}^{6'}$, as well as defects with lower ionisation degrees, obtained using the results of the studies on the deviation from the stoichiometry (Soliman and Stokłosa 1990) ((○) points). Solid lines denote the concentration of defects in w_1 "pseudo-phase", dashed lines in w_2 "pseudo-phase". (b) at 1273 K. (c) at 1323 K.

As can be seen in Figure 17.8a–c, for all oxides $(Fe_{1-x}Mn_x)_{1-\delta}O$, there is no range of oxygen pressure where the exponent $1/n_\delta$ of the dependence of the deviation from the stoichiometry would be constant, thus, where one type of defects would dominate. As can be seen in Figure 17.8a, in the oxide $(Fe_{1-x}Mn_x)_{1-\delta}O$ with a low manganese content, in the range of existence of w_1 pseudo-phase, changes in the values of the derivative n_δ are small, but they significantly differ depending on the manganese content and on the temperature and they change within the range of $2 < n_\delta < 4$. In the case of w_2 pseudo-phase, the changes are smaller and the values are close to $n_\delta \cong 6$. In turn, as can be seen in Figure 17.8b and c, when the manganese concentration in the oxide $(Fe_{1-x}Mn_x)_{1-\delta}O$ increases, the range of the oxide's existence increases, and the values of the derivative (n_δ) change significantly. In the case of manganese oxide, when the oxygen pressure decreases, the values of n_δ increase from 3 and reach 6, and then they decrease towards zero as we approach the pressure at which the oxide reaches the stoichiometric composition. In the range of these pressures, the manganese vacancies V''_{Mn} dominate.

Figure 17.9 presents the range of oxygen pressures where the oxides $(Fe_{1-x}Mn_x)_{1-\delta}O$ exist, at 1273 K and 1473 K, as well as the oxygen pressure at which the phase transition w_1/w_2 occurs.

As can be seen in Figure 17.9, at 1273 K the oxygen pressure at which "phase" transition w_1/w_2 occurs, quite strongly increases when the manganese content increases, and above the content of $x_{Mn} = 0.4$ mol/mol, w_2 pseudo-phase does not exist anymore. At 1473 K, the "phase" transition w_1/w_2 occurs practically in the middle of the range of the oxide existence and it is still present at the manganese content of $x_{Mn} = 0.8$ mol/mol.

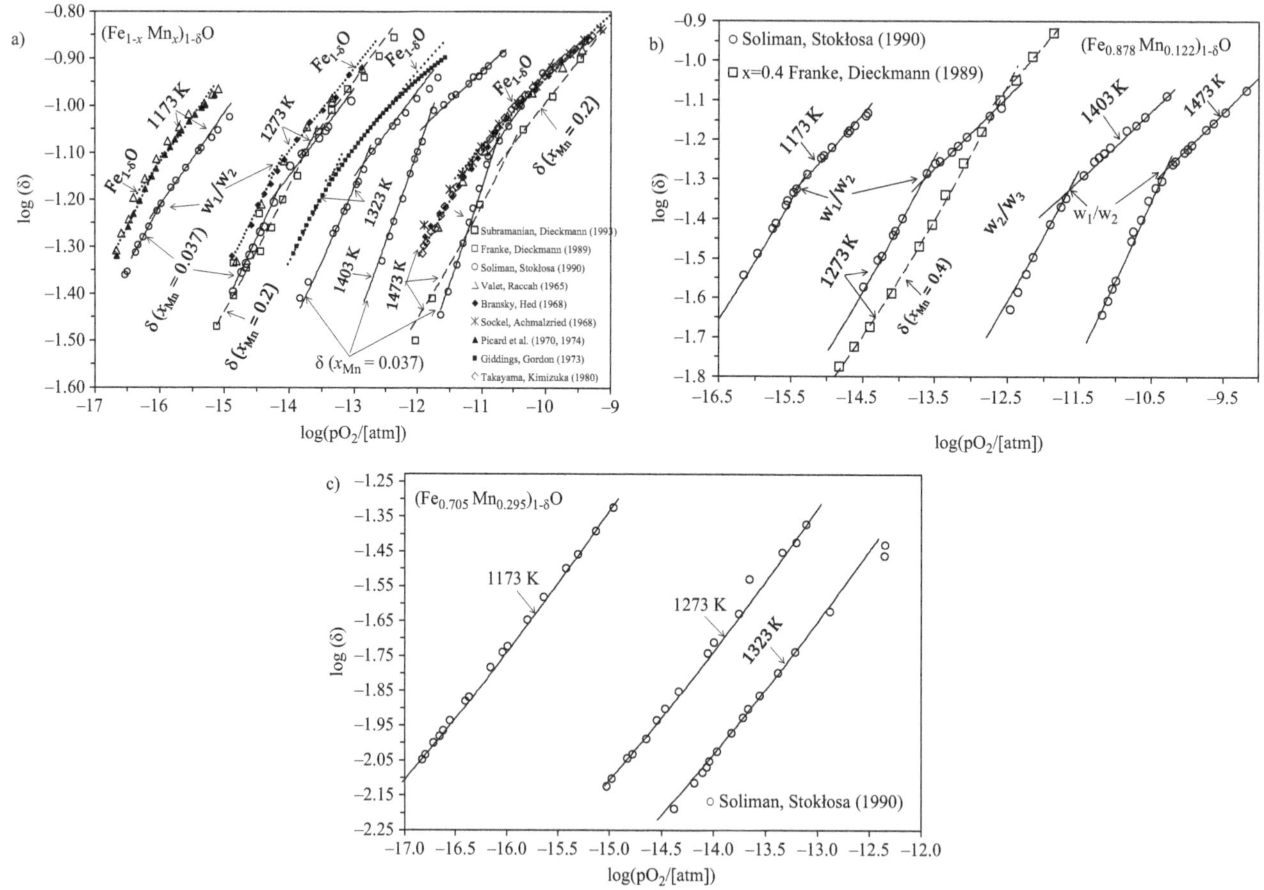

FIGURE 17.6 (a) Fragments of diagrams of concentrations of point defects, presenting the oxygen pressure dependence of the deviation from the stoichiometry at 1173–1473 K in the oxide $Fe_{1-\delta}O$ and $(Fe_{1-x}Mn_x)_{1-\delta}O$ with the manganese content of x_{Mn}= 0.037 and 0.2 mol/mol, obtained using the results of the studies on the deviation from the stoichiometry extracted from Franke and Dieckmann (1989), Subramanian and Dieckmann (1993) – (□), Soliman and Stokłosa (1990) – (○). Vallet and Raccah (1965) – (△) points, Bransky and Hed (1968) – (◆), Sockel and Schmalzried (1968) – (✳), Picard and Dodé (1970), Picard and Gerdanian (1974) – (▲), Giddings and Gordon (1973) – (■), Takayama and Kimizuka (1980) – (◆). (b). Dependences of the deviation from the stoichiometry on the oxygen pressure at temperatures 1173–1473 K in the oxide $(Fe_{1-x}Mn_x)_{1-\delta}O$ obtained using the results of the studies on the deviation from the stoichiometry reported by Soliman and Stokłosa (1990), for the manganese content of x_{Mn}= 0.122 mol/mol – (○) and Franke and Dieckmann (1989), for x_{Mn} = 0.4 mol/mol – (□) at 1273 K. (c) Dependences of the deviation from the stoichiometry on the oxygen pressure at temperatures 1173–1323 K in the oxide $(Fe_{0.705}Mn_{0.295})_{1-\delta}O$. Points (○) represent the results from (Soliman and Stokłosa 1990).

Figure 17.10 shows the dependence of the hypothetical oxygen activity ($p_{O_2}^{(s)}$), at which the oxides $(Fe_{1-x}Mn_x)_{1-\delta}O$ should reach the stoichiometric composition.

As can be seen, these pressures are by over 10 orders of magnitude lower than the decomposition pressure of oxides and when the manganese content in the oxide $(Fe_{1-x}Mn_x)_{1-\delta}O$ increases, they increase and approach the decomposition pressure of the oxide. Higher decomposition pressures than the pressure at which the oxide reaches the stoichiometric composition result from strong repulsive interactions between ions in a densely-packed cubic structure of NaCl type. When the manganese content in the oxide $(Fe_{1-x}Mn_x)_{1-\delta}O$ increases, these interactions decrease and in consequence the manganese oxide reaches the stoichiometric composition near the decay pressure. It should be noted that the calculated pressures ($p_{O_2}^{(s)}$) depend on the assumed value of ΔG^o_{eh} of formation of electronic defects and the fitted value of ΔG^o_F of formation of Frenkel defects. Moreover, the fitted values of ΔG^o_i of formation of defects and the calculated equilibrium constants are effective constants, being a product of the value and the mean activity coefficient of components in the range of higher oxygen pressures. Thus, the determined values of $p_{O_2}^{(s)}$ are approximative, however, as it results from the performed calculations, the differences should not be large, in the order of 1–2 orders of magnitude.

Figure 17.11a presents the dependence of the standard Gibbs energy of the formation of Frenkel defects ΔG^o_F) on the manganese content in the oxide $(Fe_{1-x}Mn_x)_{1-\delta}O$. It was assumed that these values change in a linear way with the

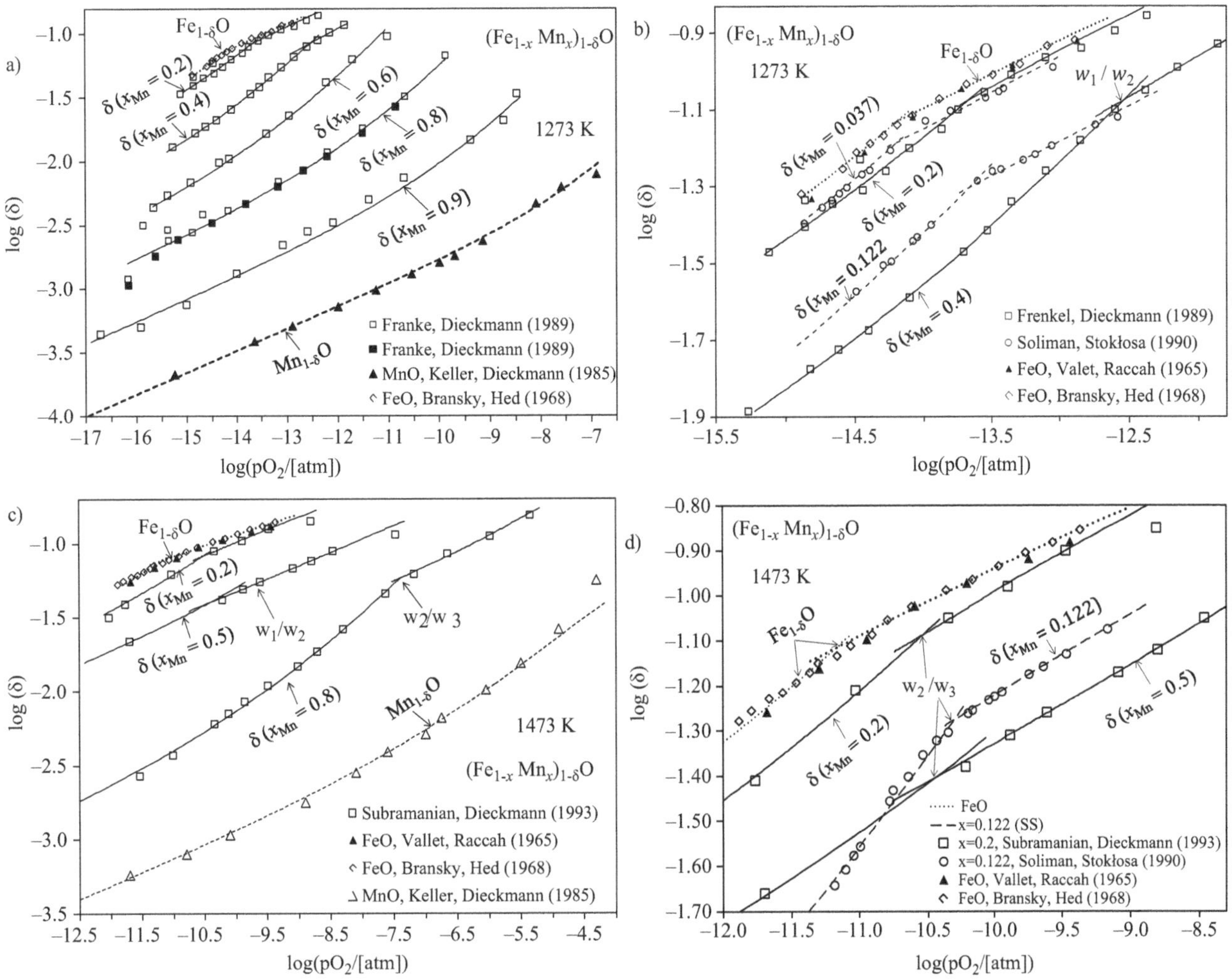

FIGURE 17.7 (a)(b) Fragments of diagrams of concentrations of point defects presenting the dependences of the deviation from the stoichiometry on the oxygen pressure at 1273 K, in the oxide $Fe_{1-\delta}O$, (···) line, and $Mn_{1-\delta}O$, (---) line, and in the oxide $(Fe_{1-x}Mn_x)_{1-\delta}O$ with the manganese content of: a) $x_{Mn} = 0.2–0.9$ mol/mol (solid lines), b) $x_{Mn} = 0.122$, 0.2 and 0.4 mol/mol, obtained using the results of the studies of the deviation from the stoichiometry reported in (Franke and Dieckmann 1989) – (□,■), (Keller and Dieckmann 1985) – (▲), (Soliman and Stokłosa 1990) – (○), (Vallet and Raccah 1965) – (▲), (Bransky and Hed 1968) – (◇). (c)(d) Dependences of the deviation from the stoichiometry on the oxygen pressure at 1473 K, in the oxide $Fe_{1-\delta}O$ ((···) line) and $(Fe_{1-x}Mn_x)_{1-\delta}O$ (solid lines) with the manganese content of c) $x_{Mn} = 0.0–0.8$ mol/mol, d) 0.122, 0.2 and 0.5 mol/mol. The points mark the results of the studies of the deviation from the stoichiometry taken from (Vallet and Raccah 1965) – (▲) points, (Bransky and Hed 1968) – (◇), (Subramanian and Dieckmann 1993) – (□), (Soliman and Stokłosa 1990) – (○), (Keller and Dieckmann 1985) – (△).

manganese concentration in wüstite. For the phase of pure oxides $Fe_{1-\delta}O$ and $Mn_{1-\delta}O$, the values determined in the work (Stokłosa 2015) were used. As it was already mentioned, a change in the value of ΔG^{o}_{F} within the range of 20 kJ/mol does not cause a change in the degree of match between the dependence of the deviation from the stoichiometry and the experimental values of δ; due to that, the same values of ΔG^{o}_{F} were assumed for w_1 and w_2 pseudo-phases.

Figure 17.11b. presents the temperature dependence of ΔG^{o}_{F} in the oxide $(Fe_{1-x}Mn_x)_{1-\delta}O$ with the manganese content of $x_{Mn} = 0.0$, 0.037, 0.122, 0.2, 0.295 and 0.8 mol/mol, which, as can be seen, are linear functions. Table 17.1 compares the values of enthalpy and entropy of formation of Frenkel defects, calculated based on the relations shown in Figure 17.11b. Enthalpies of formation of Frenkel defects increase when the manganese content in the oxide $(Fe_{1-x}Mn_x)_{1-\delta}O$ increases, within the range of $\Delta H^{o}_{F} = 216–309$ kJ/mol.

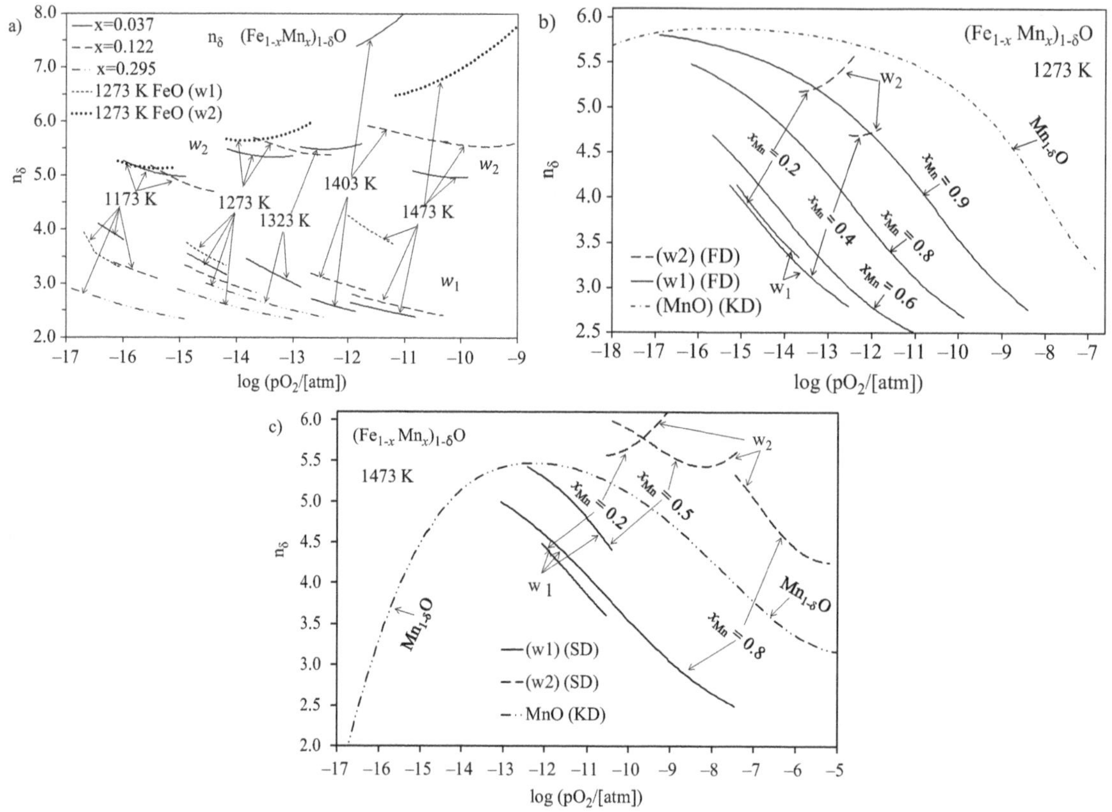

FIGURE 17.8 (a) Oxygen pressure dependence of the derivative: $(d\log p_{O_2}/d\log\delta) = n_\delta$ at 1173–1473 K, obtained for the oxygen pressure dependence of the deviation from the stoichiometry in the oxide $Fe_{1-\delta}O$ and $(Fe_{1-x}Mn_x)_{1-\delta}O$ with the manganese content of x_{Mn} = 0.037, 0.122 and 0.295 mol/mol for the "pseudo-phases" w_1 and w_2 (presented in Figure 17.6). (b) in the oxide $(Fe_{1-x}Mn_x)_{1-\delta}O$ with the manganese content of x_{Mn} = 0.2–1.0 mol/mol at 1273 K (presented in Figure 17.7a). (c) in $Mn_{1-\delta}O$ and $(Fe_{1-x}Mn_x)_{1-\delta}O$ at 1473 K (presented in Figure 17.7c).

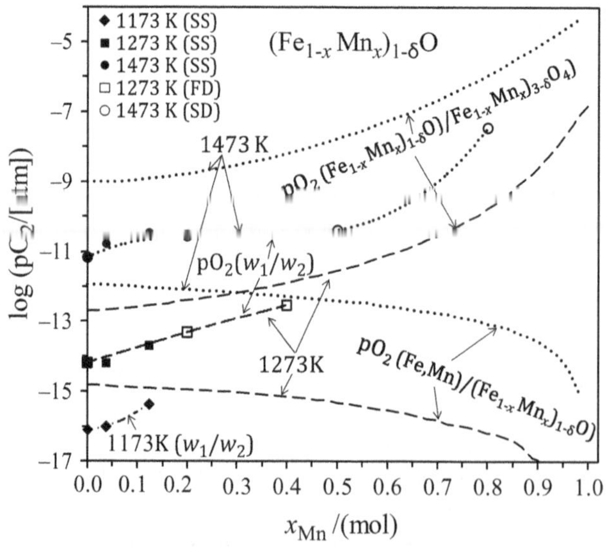

FIGURE 17.9 Dependence of the oxygen pressure at which there occurs the "phase" transition w_1/w_2 on the manganese content in the oxide $(Fe_{1-x}Mn_x)_{1-\delta}O$ at 1273 K ((□) (FD) points) and 1473 K ((○) (SD) points), obtained using the results of the studies on the deviation from the stoichiometry extracted from Franke and Dieckmann (1989), Subramanian and Dieckmann (1993) and Soliman and Stokłosa (1990) at 1173–1473 K ((◆,■,●) (SS) points). Dotted lines and dashed lines denote the equilibrium oxygen pressures at the phase boundary $Fe,Mn/(Fe_{1-x}Mn_x)_{1-\delta}O$ and $(Fe_{1-x}Mn_x)_{1-\delta}O/(Fe_{1-x}Mn_x)_{3\pm\delta}O_4$ at 1273 K and 1473 K (Franke and Dieckmann 1989, Subramanian and Dieckmann 1993).

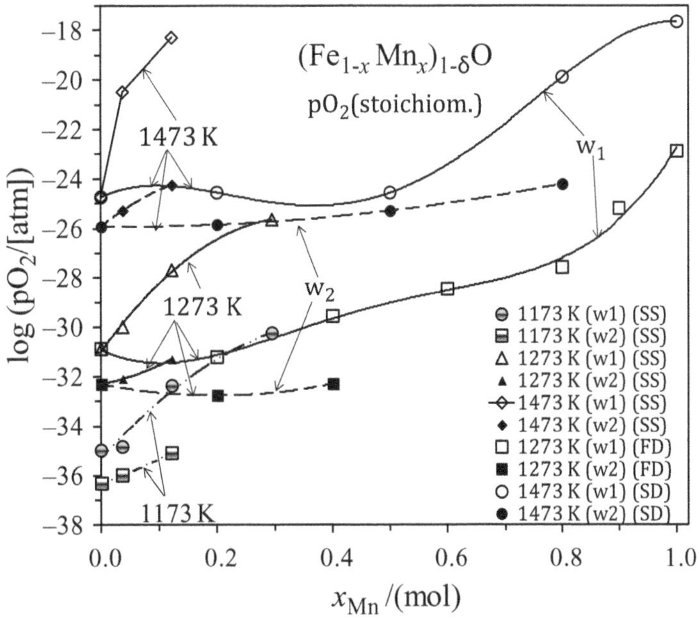

FIGURE 17.10 Dependence of the hypothetical oxygen pressure whereby the oxide $(Fe_{1-x}Mn_x)_{1-\delta}O$ could have reached the stoichiometric composition at 1273 K and 1473 K obtained using the values of the deviation from the stoichiometry reported in Franke and Dieckmann (1989) ((□,■) (FD) points), Subramanian and Dieckmann (1993) ((○,●) (SD) points) and Soliman and Stokłosa (1990) at 1173–1473 K ((⊟,⊖,△,▲,◆,◇) (SS) points).

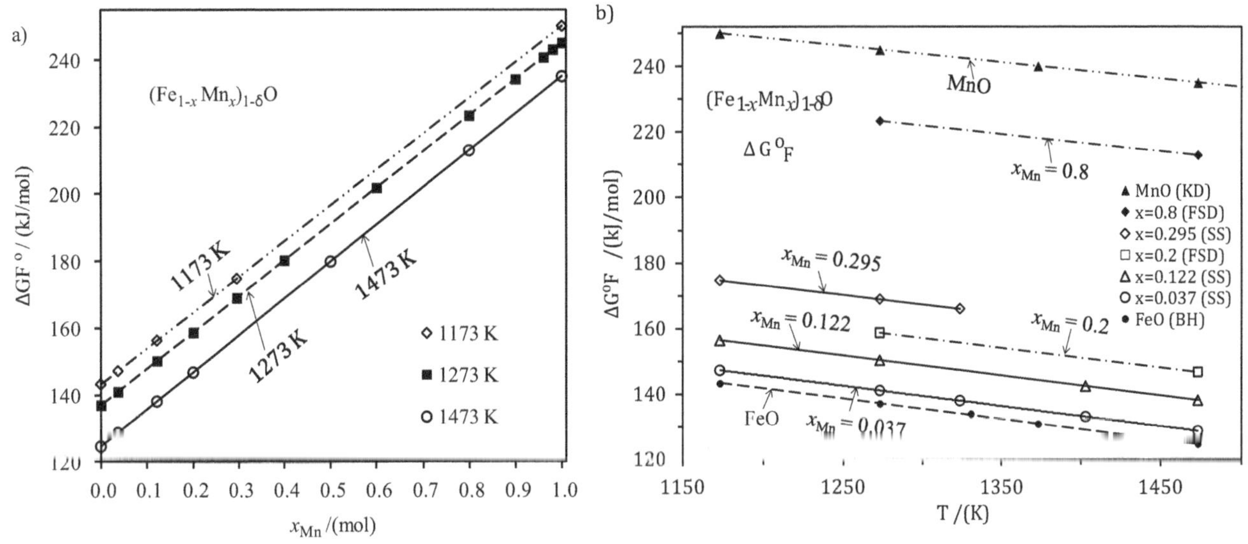

FIGURE 17.11 (a) Dependence of standard Gibbs energy ΔG_F^o of formation of intrinsic Frenkel defects, a) on the manganese content in the oxide $(Fe_{1-x}Mn_x)_{1-\delta}O$ at 1173–1473 K ((■,○,◇) points), (b) on the temperature in the oxide with the manganese content of x_{Mn} = 0.0, (●) (BH) points; 0.037 – (○) (SS); 0.122 – (△) (SS); 0.295 – (◇) (SS); 0.2 – (□) (FSD); 0.8 – (◆) (FSD) and 1.0 mol/mol (▲) (KD), obtained using the values of ΔG_F^o for oxides $Fe_{1-\delta}O$ and $Mn_{1-\delta}O$, determined in the work (Stokłosa 2015).

Figure 17.12a presents the dependence of the fitted values of $\Delta G^o_{V''_M}$ of formation of cation vacancies on the manganese concentration in the oxide $(Fe_{1-x}Mn_x)_{1-\delta}O$, at 1173–1473 K.

As can be seen in Figure 17.12a, when the manganese concentration in the oxide $(Fe_{1-x}Mn_x)_{1-\delta}O$ increases, the values of $\Delta G^o_{V''_M}$ increase monotonically, but there are distinct differences depending on the results used in the calculations, either Franke and Dieckmann (1989), Subramanian and Dieckmann (1993) or Soliman and Stokłosa (1990). In turn, Figure 17.12b presents the temperature dependence of $\Delta G^o_{V''_M}$ for the manganese content of x_{Mn} = 0.0, 0.037, 0.122, 0.2, 0.295 and 0.8 mol/mol. As can be seen in Figure 17.12b, the values of $\Delta G^o_{V''_M}$ increase with the temperature. Table 17.1 compares the values of enthalpy and entropy of formation of cation vacancies V''_M, calculated based on the relations shown in Figure

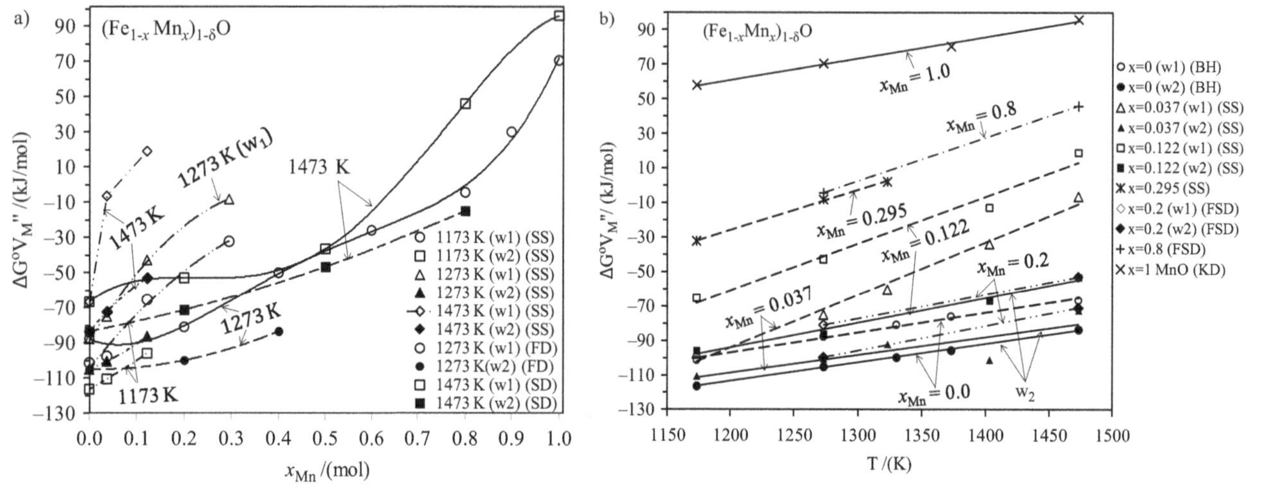

FIGURE 17.12 (a) Dependence of standard Gibbs energy $\Delta G^{o}_{V_M''}$ of formation of cation vacancies V_M'' in the oxide $(Fe_{1-x}Mn_x)_{1-\delta}O$ a) on the manganese content x_{Mn} at 1173 K ((▬,◒) points) (SS), 1273 K – (○,●) (FD), (△,▲) (SS) and 1473 K – (□,■)(SD), (◇,◆) (SS), (b) on the temperature in the oxide with the manganese content of x_{Mn} = 0.0 (●,○) points; 0.037 – (△,▲) (SS); 0.122 – (□,■) (SS), 0.295 – (∗), 0.2 (◇,◆) (FSD), 0.8 – (+) (FSD) and 1.0 mol/mol (×) (KD) (empty points – "pseudo-phase" w_1, solid points – "pseudo-phase" w_2).

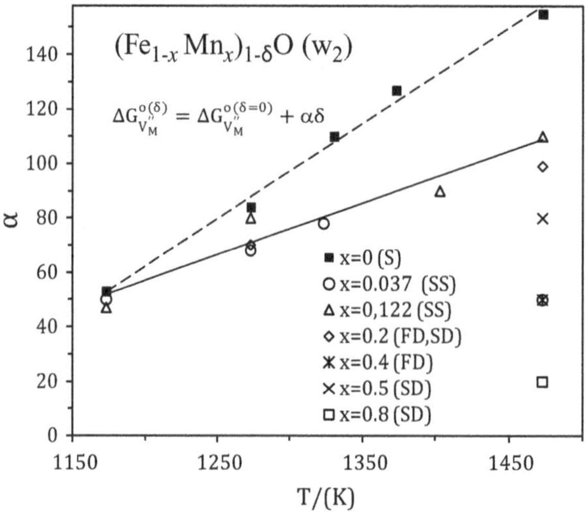

FIGURE 17.13 Temperature dependence of parameter α in Equation 17.4 describing the dependence of $\Delta G^{o}_{V_M''}$ of formation of cation vacancies V_M'' on the deviation from the stoichiometry in the oxide $Fe_{1-\delta}O$ ((■) (S) points) and $(Fe_{1-x}Mn_x)_{1-\delta}O$ with the manganese content of: x_{Mn} = 0.037 mol/mol ((○) (SS) points), 0.122 – (△)(SS), 0.2 – (◇)(FD,SD), 0.4 – (∗)(FD), 0.5 – (×) (SD) and 0.8 – (□)(SD).

17.12b. Figure 17.15a presents the dependence of the enthalpy of formation of cation vacancies $\Delta H^{o}_{V_M''}$ on the manganese content in the oxide $(Fe_{1-x}Mn_x)_{1-\delta}O$, which, as can be seen, quite significantly differ, depending on the results of studies used either Franke and Dieckmann (1989), Subramanian and Dieckmann (1993) or Soliman and Stokłosa (1990).

Figure 17.13 presents the dependence of the parameter α in Equation 17.4 describing the dependence of $\Delta G^{o}_{V_{Fe}''}$ on the deviation from the stoichiometry in the range of existence of w_2 pseudo-phase in wüstite and in the oxide $(Fe_{0.963}Mn_{0.037})_{1-\delta}O$.

In turn, Figure 17.14a presents, calculated according to Equation 17.3, values of standard Gibbs energy $\Delta G^{o}_{M_i^{\bullet\bullet}}$ of formation of interstitial cations depending on the manganese content in the oxide $(Fe_{1-x}Mn_x)_{1-\delta}O$, at 1173–1473 K.

As can be seen in Figure 17.14a, the values of $\Delta G^{o}_{M_i^{\bullet\bullet}}$ are high and positive and when the manganese content in the oxide $(Fe_{1-x}Mn_x)_{1-\delta}O$ increases, they decrease, but the characters of the dependences are different. The existing discrepancies are higher than in the case of $\Delta G^{o}_{V_M''}$ and they are associated with an error in the determination of the deviation from the

TABLE 17.1
Values of Enthalpies ΔH_i^o and Entropies ΔS_i^o of formation of intrinsic Frenkel defects, cation vacancies V_M'' and V_M' and interstitial cations $M_i^{\circ\circ}$ in the oxide $(Fe_{1-x}Mn_x)_{1-\delta}O$, determined based on the dependences shown in Figures 17.11–17.14 when using the results of the studies of the deviation from the stoichiometry reported in (Bransky and Hed 1968) (BH), (Franke and Dieckmann 1989, Subramanian and Dieckmann 1993) (FSD), (Keller and Dieckmann 1985) (KD), (Soliman and Stokłosa 1990) (SS).

Equation	x_{Mn} (mol)	ΔH_i^o (kJ/mol)	ΔS_i^o (J/(mol·K))	ΔT (K)	R^2	Ref
$[V_M''][M_i^{\bullet\bullet}] = K_F$	0	216	62	1173–1473		(BH)
Equation 16.7	0.037	219	62	1173–1473		(SS)
-,,-	0.122	227	61	1173–1473		(SS)
-,,-	0.2	235	60	1273–1473		(FSD)
-,,-	0.295	243	59	1173–1473		(SS)
-,,-	0.8	290	52	1273–1473		(FSD)
-,,-	1	309	50	1173–1473		(KD)
V_M'' (w_1)	0	-236 ± 8	−116 ± 6	1173–1473	0.9913	(BH)
Equation 16.9	0.037	−459 ± 28	−304 ± 20	1173–1473	0.9862	(SS)
-,,-	0.122	−386 ± 40	−271 ± 30	1173–1473	0.9760	(SS)
-,,-	0.2	−259	−140	1173–1473		(FSD)
-,,-	0.295	−302 ± 11	−230 ± 8	1173–1473	0.9987	(SS)
-,,-	0.8	−326	-253	1173–1473		(FSD)
-,,-	1	−88 ± 10	−124 ± 7	1173–1473	0.9930	(KD)
V_M'' (w_2)	0	−243 ± 4	−98 ± 4	1173–1473	0.9979	(BH)
Equation 16.9	0.037	−232 ± 54	−103 ± 34	1173–1473	0.6862	(SS)
-,,-	0.122	−266 ± 17	−143 ± 13	1173–1473	0.9842	(SS)
-,,-	0.2	−283	−143	1273–1473		(FSD)
$M_i^{\circ\circ}$ (w_1)	0	458 ± 10	182 ± 8	1173–1473	0.9948	(BH)
Equation 16.10	0.037	679 ± 28	366±20	1173–1473	0.9904	(SS)
-,,-	0.122	613 ± 40	331 ± 30	1173–1473	0.9838	(SS)
-,,-	0.2	494	200	1273–1473		(FSD)
-,,-	0.295	546 ± 11	289 ± 8	1173–1473	0.9992	(SS)
-,,-	0.8	616	305	1273–1473		(FSD)
-,,-	1	396 ± 10	174 ± 7	1173–1473	0.9964	(KD)
$M_i^{\bullet\bullet}$ (w_2)	0	459 ± 5	170 ± 3	1173–1473	0.9989	(BH)
Equation 16.10	0.037	452 ± 539	165 ± 40	1173–1473	0.8481	(SS)
-,,-	0.122	494 ± 17	204 ± 13	1173–1473	0.9921	(SS)
-,,-	0.2	517	202	1273–1473		(FSD)
V_M' (w_1)	0	−31 ± 49	−14 ± 3	1173–1473	0.9167	(BH)
Equation 16.12	0.037	127 ± 14	122 ± 10	1173–1473	0.9783	(SS)
-,,-	0.122	102 ± 6	108 ± 4	1173–1473	0.9970	(SS)
-,,-	0.2	−25 ± 6	-5	1273–1473		(FSD)
-,,-	0.295	1 ± 6	44 ± 5	1173–1473	0.9877	(SS)
-,,-	0.8	63	80	1273–1473		(FSD)
-,,-	1	54 ± 8	74 ± 6	1173–1473	0.9876	(KD)

stoichiometry and the assumed approximate values of ΔG_F^o of formation of Frenkel defects (see Equation 17.3. In turn, Figure 17.14b presents the temperature dependences $\Delta G^o_{M_i^{\bullet\bullet}}$, which, when the temperature increases, decrease in an approximately linear way. Table 17.1 compares the values of enthalpy and entropy of formation of interstitial cations calculated based on the relations shown in Figure 17.14b. Figure 17.15b presents the enthalpy of formation of interstitial cations $\Delta H^o_{M_i^{\bullet\bullet}}$ depending on the manganese content in the oxide $(Fe_{1-x}Mn_x)_{1-\delta}O$, which, as can be seen, quite significantly differ, depending on the results of studies used either (Franke and Dieckmann 1989, Subramanian and Dieckmann 1993) or (Soliman and Stokłosa 1990).

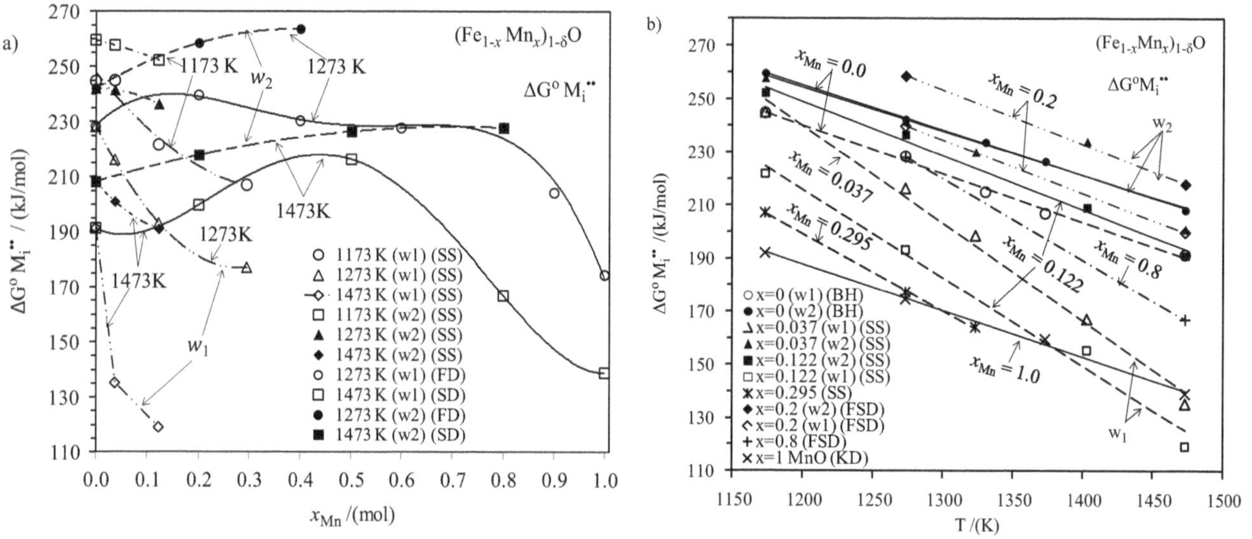

FIGURE 17.14 Dependence of the standard Gibbs energy $\Delta G^{o}_{M_i^{\bullet\bullet}}$ of formation of interstitial cations $\left(M_i^{\bullet\bullet}\right)$, calculated according to Equation 17.3 in the oxide $(Fe_{1-x}Mn_x)_{1-\delta}O$ a) on the manganese content x_{Mn} at temperatures of: 1173 K ((▬,◒ ((SS) points), 1273 K – (○,●) (FD), (△,▲) (SS) and 1473 K – (□,■) (SD), (◇,◆) (SS), (b) on the temperature, for an oxide with the manganese content of x_{Mn} = 0.0 ((●,○) (BH) points); 0.037 – (△,▲) (SS); 0.122 – (□,■) (SS); 0.295 – (✳) (SS), 0.2 – (◇,◆) (FSD), 0.8 – (+) (FSD) and 1.0 mol/mol (×) (KD) (empty points – "pseudo-phase" w_1, solid points – "pseudo-phase" w_2).

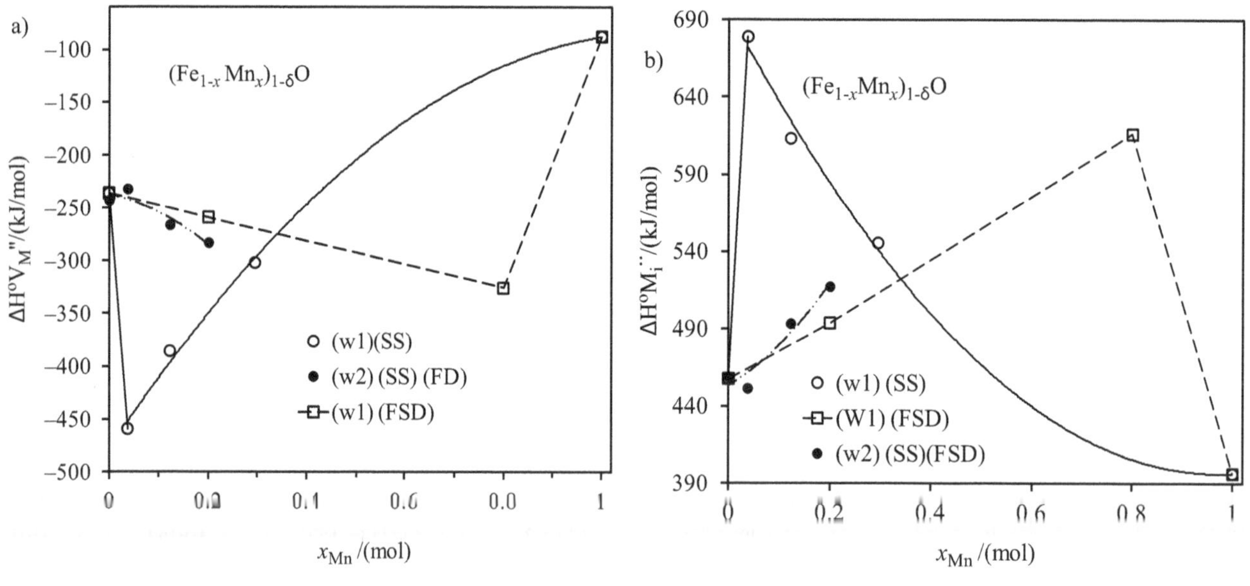

FIGURE 17.15 Dependence of: a) enthalpy $\Delta H^{o}_{V_M''}$ of formation of cation vacancies V_M'', b) enthalpy $\Delta H^{o}_{M_i^{\bullet\bullet}}$ of formation of interstitial cations $\left(M_i^{\bullet\bullet}\right)$, in the oxide $(Fe_{1-x}Mn_x)_{1-\delta}O$ on the manganese content x_{Mn} obtained using the results of the studies on the deviation from the stoichiometry extracted from Franke and Dieckmann (1989), Subramanian and Dieckmann (1993) ((□) (FSD) points) and Soliman and Stokłosa (1990) ((○) (SS) points).

In turn, Figure 17.16a presents the adjusted values of $\Delta G^{o}_{V_M'} = \Delta G^{o}_{V_M^x}$ of formation of cation vacancies V_M' and V_M^x and defect complexes with lower ionisation degrees, depending on the manganese content in the oxide $(Fe_{1-x}Mn_x)_{1-\delta}O$ at 1173–1473 K.

As can be seen in Figure 17.16a, when the manganese content in the oxide $(Fe_{1-x}Mn_x)_{1-\delta}O$ increases, the above values decrease, but the character is different, depending on the temperature and on the results of the deviation that were the basis for the calculations (Franke and Dieckmann 1989, Subramanian and Dieckmann 1993 or Soliman and Stokłosa 1990). The values of $\Delta G^{o}_{V_M'} = \Delta G^{o}_{V_M^x}$ in the range of existence of w_2 pseudo-phase are positive, which results from a low concentration

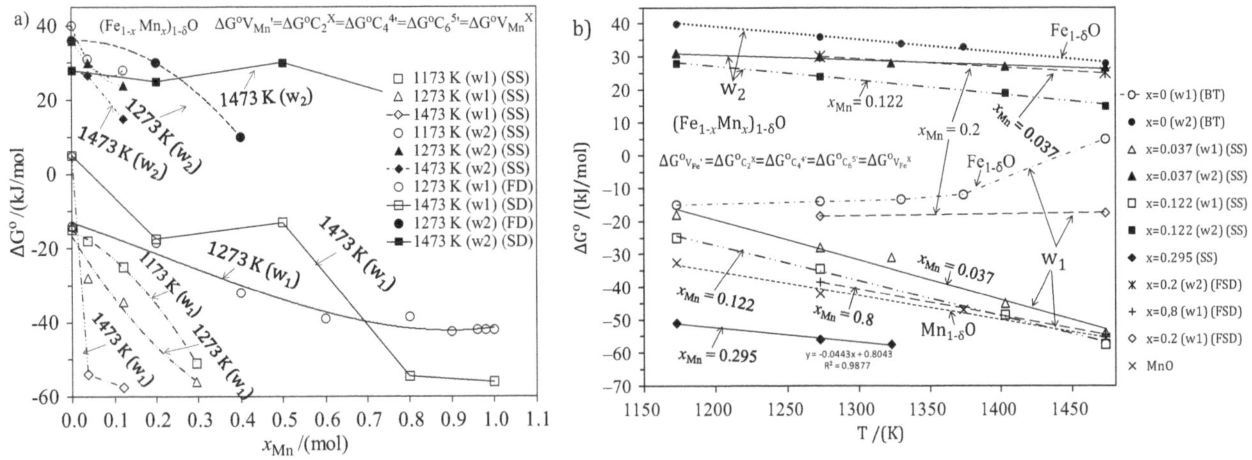

FIGURE 17.16 Dependence $\Delta G^o_{V_M'} = \Delta G^o_{V_M^x}$ of formation of vacancies with lower ionisation degrees and defect complexes in the oxide $(Fe_{1-x}Mn_x)_{1-\delta}O$, a) on the manganese content x_{Mn} at 1173 K ((▬,●) (SS) points), 1273 K – (○,●) (FD), (△,▲) (SS) and 1473 K – (□,■) (SD), (◇,◆) (SS), b) on the temperature in the oxide with the manganese content of x_{Mn} = 0.0 ((●,○) (BT) points) 0.037 – (△,▲) (SS); 0.122 – (□,■) (SS), 0.295 – (◆) (SS), 0.2 (◇,✶) (FSD), 0.8 – (+) (FSD) and 1.0 mol/mol (×) (KD) (empty points – "pseudo-phase" w_1, solid points – "pseudo-phase" w_2).

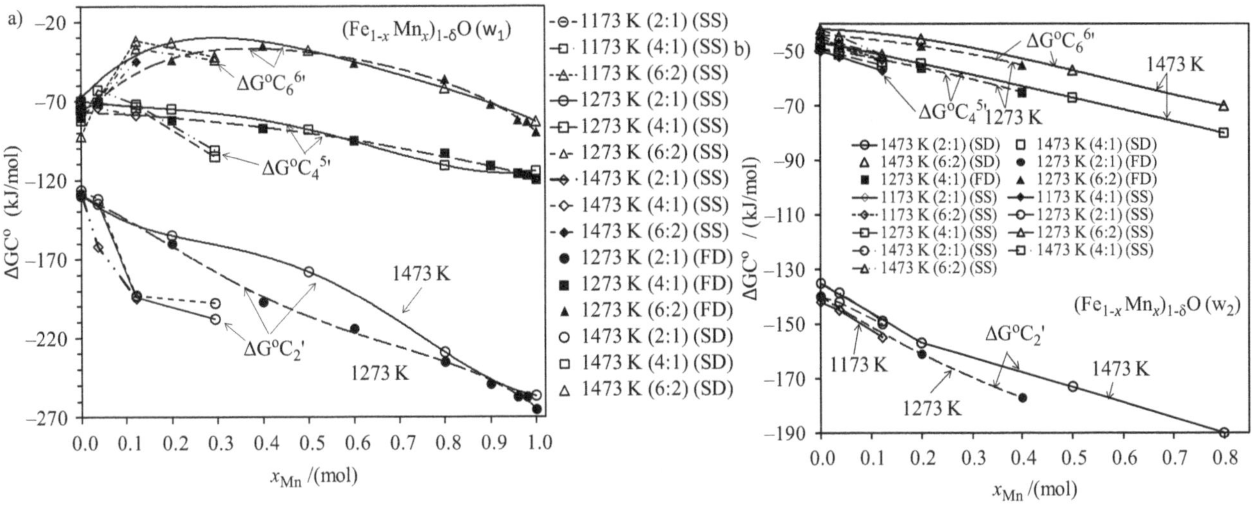

FIGURE 17.17 Dependence of $\Delta G^o_{C_2'}$, $\Delta G^o_{C_4^{5'}}$, $\Delta G^o_{C_6^{6'}}$ of formation of complexes $\{(V_M'')_2 M_i^{3\bullet}\}'$ – (C_2'), $\{(V_M'')_4 M_i^{3\bullet}\}^{5'}$ – ($C_4^{5'}$) and $\{(V_M'')_6 (M_i^{3\bullet})_2\}^{6'}$ – ($C_6^{6'}$) in the oxide $(Fe_{1-x}Mn_x)_{1-\delta}O$ on the manganese content at temperatures: 1173–1473 K, a) for "pseudo-phase" w_1, b) for "pseudo-phase" w_2.

of defects with lower ionisation degree (they do not affect the match between the dependence of δ on p_{O_2} and the experimental results).

In turn, Figure 17.16b presents the temperature dependence of $\Delta G^o_{V_M'} = \Delta G^o_{V_M^x}$ of formation of cations vacancies V_M' and V_M^x. As can be seen in Figure 17.16b, when the temperature increases, the values of $\Delta G^o_{V_M'} = \Delta G^o_{V_M^x}$ decrease practically linearly. Only in the case of pure wüstite and for the manganese content of 0.2 mol/mol these values increase when the temperature increases. Table 17.1 compares the values of enthalpy and entropy of formation of cation vacancies V_M' and V_M^x calculated based on the relations shown in Figure 17.16b.

Figure 17.17 presents the fitted values of standard Gibbs enthalpy of formation of the individual defect complexes ΔG^o_C depending on the manganese content in the oxide $(Fe_{1-x}Mn_x)_{1-\delta}O$ at 1173–1473 K, a) for w_1 pseudo-phase and b) for w_2 pseudo-phase. As can be seen in Figure 17.17, the values of ΔG^o_C for complexes $\{(V_M'')_2 M_i^{3\bullet}\}$ in w_1 pseudo-phase decrease when the manganese content increases, to a smaller extent in w_2 pseudo-phase. For the remaining defect complexes the changes are smaller. These are maximum values; thus, the actual values and the character of the dependence can be different.

Figure 17.18a presents the temperature dependence of ΔG^o_C of formation of the individual complexes in the oxide $(Fe_{1-x}Mn_x)_{1-\delta}O$ with the content of $x_{Mn} = 0.0$, 0.037, 0.2 mol/mol, in w_1 pseudo-phase, and in Figure 17.18b, in w_2 pseudo-phase. Figure 17.18c shows their temperature dependence for the content of $x_{Mn} = 0.122$, 0.295 and 0.8 mol/mol.

The dependences are linear in certain ranges, but in many cases, there are clear deviations from linearity. They arise from the fact that the maximum values of ΔG^o_C were fitted, which, as already mentioned, can differ from the actual values, because the calculated concentration of defects did not affect the degree of match between the dependence of the deviation from the stoichiometry on p_{O_2} and the experimental values of δ.

17.4 DIFFUSION IN MANGANESE-DOPED WÜSTITE

In the case of oxides in which cation vacancies dominate, it is possible to assume that the coefficient of diffusion of metal ions depends on the concentration of cation vacancies and on the coefficient of diffusion of defects, according to the relation:

$$D^*_M \cong D_{V(M)} f^*_{M(V)} \frac{c_V}{c_M} = D^o_{V(M)} \sum_i [V_M] \tag{17.9a}$$

where D^*_M denotes the coefficient of self-diffusion of tracer ions M*, $D_{V(M)}$ – the coefficients of diffusion of defects (metal ions M via cation vacancies), $f^*_{M(V)}$ – the coefficient of correlation for the vacancy mechanism, $D^o_{V(M)}$ is the effective coefficient of diffusion (mobility) of metal ions M via cation vacancies, $[V_M]$ denotes the sum of concentration of quasi-free cation vacancies ($[V_M] = [V''_M] + [V'_M] + [V^x_M]$), $\sum_i [V_M] = [V_M] + n[C_i]$ – concentrations of all cation vacancies, also the vacancies present in defect complexes. Thus, using the concentration of cation vacancies resulting from the diagram of concentrations of defects and the values of coefficients of self-diffusion of tracer (radioisotope), we can determine the coefficients of diffusions of defects, and, more precisely, the mobility of a given type of tracer ions via cation vacancies. Transforming Equation 17.9a we obtain:

$$\sum_i [V_M] = \frac{D^*_M}{D^o_{V(M)}} \tag{17.9b}$$

Therefore, if we assume that the diffusion of metal ions occurs via cation vacancies, then according to Equation 17.9b we can adjust such a value of the coefficient $D^o_{V(M)}$ that the values of the sum of concentrations of vacancies $\sum_i [V_M]$ calculated using the values of coefficients of self-diffusion of tracer M* (D^*_M) are consistent with the dependence of this sum on p_{O_2}, resulting from the diagram of point defects.

Figure 17.19 presents the oxygen pressure dependence of the concentration of quasi-free cation vacancies $[V_M]$ in the oxide $(Fe_{1-x}Mn_x)_{1-\delta}O$ with the manganese content of $x_{Mn} = 0.0$–1.0 mol/mol at 1273 K, resulting from the diagram of point defects (solid lines) (for the content of $x_{Mn} = 0.96$ and 0.98 mol/mol calculated based on extrapolated values of ΔG^o_C of formation of defects). Dashed lines ( ) represent the dependence of the "sum" of concentrations of quasi-free cation vacancies and cation vacancies forming defect complexes ($[V_M]+n[C_i]$) on p_{O_2}. The points denote the values of the concentration of cation vacancies calculated according to Equation 17.9b for the fitted value of $D^o_{V(M)}$ when using the coefficients of self-diffusion of iron ^{59}Fe for pure wüstite, reported in Franke and Dieckmann (1989) ((◇) points) and Chen and Peterson (1975) – (△) for the oxide $(Fe_{1-x}Mn_x)_{1-\delta}O$, extracted from Franke and Dieckmann (1989) ((○) points) and $Mn_{1-\delta}O$ ((◇) points). Also, the values of concentration of vacancies obtained when using the coefficient of diffusion of manganese [54]Mn, taken from Peterson and Chen (1982) at 1273 K are marked ((△) points) and 1473 K – (▲). As can be seen in Figure 17.19b for wüstite and for the oxide $(Fe_{1-x}Mn_x)_{1-\delta}O$ with the manganese content of $x_{Mn} = 0.2$ and 0.4 mol/mol, the dependence of the concentration of cation vacancies on the oxygen pressure significantly deviates from the dependence of the sum of concentrations of all vacancies ($[V_M]+n[C_i]$) on p_{O_2} and the differences decrease when the manganese content increases. The values of the concentration of cation vacancies calculated according to Equation 17.9b at low manganese content are fully consistent with the dependence of the sum of quasi-free vacancies $[V_M]$ on p_{O_2}, resulting from the diagram of concentrations of defects. For higher manganese concentrations in the oxide $(Fe_{1-x}Mn_x)_{1-\delta}O$, owing to a lower concentration of defect complexes and the measurement error, it is difficult to draw conclusions, although there are discrepancies at the highest oxygen pressures. Also, a good consistency, within the error limit, was obtained in the case of manganese oxide for the diffusion of tracers, Fe* as well as Mn*.

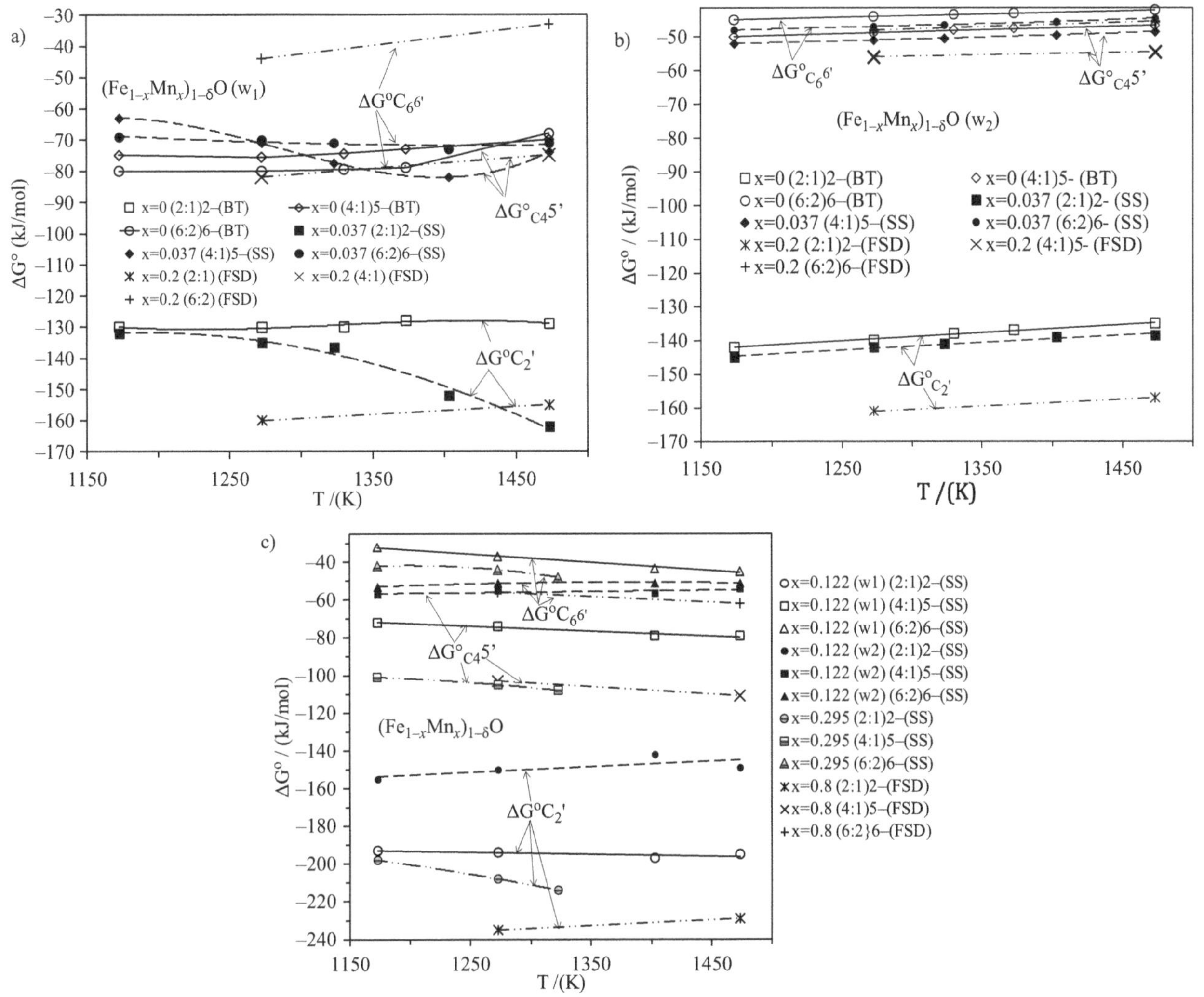

FIGURE 17.18 Temperature dependence of $\Delta G^o_{C_2'}$, $\Delta G^o_{C_4^{5'}}$, $\Delta G^o_{C_6^{6'}}$ of formation of complexes $\{(V''_M)_2 M_i^{3\bullet}\}' \equiv (C_2')$, $\{(V''_M)_4 M_i^{3\bullet}\}^{5'} \equiv (C_4^{5'})$ and $\{(V''_M)_6 (M_i^{3\bullet})_2\}^{6'} \equiv (C_6^{6'})$ in the oxide $(Fe_{1-x}Mn_x)_{1-\delta}O$ with the manganese content of x_{Mn} = 0.0, 0.037, 0.2 mol/mol, a) for "pseudo-phase" w_1, b) for "pseudo-phase" w_2, c) for the content of x_{Mn} = 0.122, 0.295 and 0.8 mol/mol.

Figure 17.20 presents the manganese concentration dependence of the coefficient of diffusion of iron ions via cation vacancies $D_{V(Fe)}$ (their mobility) in the oxide $(Fe_{1-x}Mn_x)_{1-\delta}O$, in w_1 pseudo-phase ((○) points) and in w_2 pseudo phase ((●) points), which, as can be seen, decrease when the manganese content increases. The values of $D^o_{V(Fe)}$ for the manganese content of x_{Mn} = 0.9–0.98 mol/mol deviate, which can indicate, despite a possible measurement error, an increase in the mobility of cations via vacancies in the manganese oxide doped with a small amount of iron. The values of mobility of iron ions Fe* and manganese ions Mn* in manganese oxide, at 1273 K, are $D^o_{V(Fe)} = 7.1 \cdot 10^{-7}$ cm²/s and $D^o_{V(Mn)} = 2.9 \cdot 10^{-7}$ cm²/s, thus, they are significantly different. This could be related to a systematic error in the values of coefficients of self-diffusion reported in Franke and Dieckmann (1989) and Peterson and Chen (1982).

17.5 CONCENTRATION OF ELECTRONIC DEFECTS AND THEIR MOBILITY

From the studies of the electrical conductivity of wüstite it results that it is proportional to the deviation from the stoichiometry and the charge transport occurs via electron jumping between ions Fe^{3+}, according to the hopping mechanism of small polarons (Hodge and Bowen 1981, Molenda et al. 1987). As shown in various works (Bocquet et al. 1967,Duquesnoy 1965, Duquesnoy and Marion 1963, Eror and Wagner Jr. 1971, Franke and Dieckmann 1989, Hed and Tannhauser 1967, Le Brusq and Delmaire 1973, Le Brusq et al. 1968), in manganese oxide, the mobility of electrons is much higher than the mobility of electron holes; due to that, in the middle of the range of the oxide's existence there occurs a *p*/*n*-type transition. Therefore, manganese doping in wüstite should significantly affect the electrical conductivity of oxides $(Fe_{1-x}Mn_x)_{1-\delta}O$.

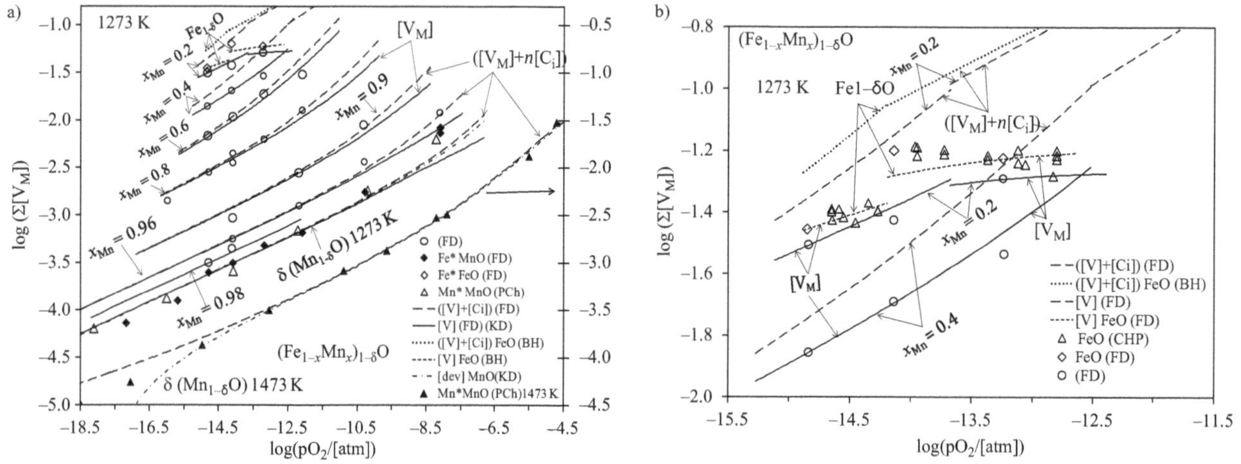

FIGURE 17.19 Oxygen pressure dependence of the sum of concentrations of quasi-free cation vacancies $[V_M]$ (solid lines) in $Fe_{1-\delta}O$ and $Mn_{1-\delta}O$ and in the oxide $(Fe_{1-x}Mn_x)_{1-\delta}O$, with the manganese content of: a) x_{Mn} = 0.2–0.98 mol/mol, b) x_{Mn} = 0.2–0.4 mol/mol obtained at 1273 K using the results of the studies of the deviation from the stoichiometry (Franke and Dieckmann 1989). Dashed lines (---) denote the dependence of the sum of concentrations of all the cation vacancies ($[V_M]+n[C_i]$) on p_{O_2}. The points represent the values of the "sum" of concentrations of cation vacancies calculated according to Equation (17.9b) for the fitted value of $D^o_{V(Fe)}$ when using the values of coefficients of self-diffusion of iron ^{59}Fe for pure wüstite ((▲,◇) points) reported in Chen and Peterson (1975), Franke and Dieckmann (1989) and for manganese-doped oxide (○) and $Mn_{1-\delta}O$ (◆)(FD), taken from Franke and Dieckmann (1989) and self-diffusion of manganese ^{54}Mn extracted from Peterson and Chen (1982) at 1273 K ((△) (PCh) points) and 1473 K – (▲)(PCh).

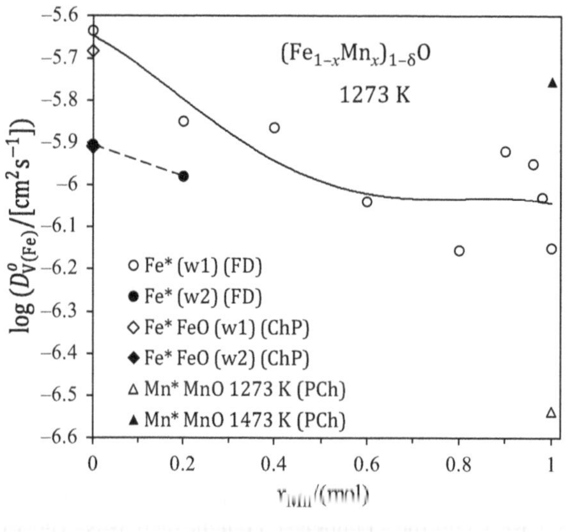

FIGURE 17.20 Dependence of the coefficient of diffusion of iron ions via cation vacancies $D^o_{V(Fe)}$ on the manganese content in the oxide $(Fe_{1-x}Mn_x)_{1-\delta}O$ at 1273 K in "pseudo-phase" w_1 – points (○) (FD), and "pseudo-phase" w_2 – points (●) (FD) and in $Fe_{1-\delta}O$ – (◇,◆) (ChP) and $D^o_{V(Mn)}$ in $Mn_{1-\delta}O$ at 1273 K ((△) (PCh)) and 1473 K (▲) (PCh).

Thus, we assume that in oxides $(Fe_{1-x}Mn_x)_{1-\delta}O$ in the temperature range 1173–1473 K a mixed electrical conductivity occurs, described by a general equation:

$$\sigma = \sigma_p + \sigma_n = \frac{F}{V_{(Fe,Mn)O}}\left(\mu_h\left[h^{\bullet}\right] + \mu_e\left[e'\right]\right) \tag{17.10a}$$

where σ_n and σ_p denote, respectively, electron and hole conductivity, μ_e and μ_h denote mobilities of electrons and holes, F – Faraday constant, $V_{(Fe,Mn)O}$ – molar volume of the oxide $(Fe_{1-x}Mn_x)_{1-\delta}O$, which changes from 12.03 cm³ in wüstite to 13.22 cm³ in the oxide $Mn_{1-\delta}O$. Factoring out the mobility of electron holes we get:

$$\sigma = \frac{F}{V_{(Fe,Mn)_{1-\delta}O}}\mu_h\left(\left[h^{\bullet}\right] + \frac{\mu_e}{\mu_h}\left[e'\right]\right) = \frac{F}{V_{(Fe,Mn)_{1-\delta}O}}\mu_h\left(\left[h^{\bullet}\right] + b_{e/h}\left[e'\right]\right) \tag{17.10b}$$

where $b_{e/h}$ denotes the ratio of the mobility of electrons to the mobility of electron holes ($b_{e/h} = \mu_e/\mu_h$). Transforming Equation 17.10b we get:

$$\left[h^{\bullet}\right] + b_{e/h}\left[e'\right] = \frac{\sigma V_{(Fe,Mn)_{1-\delta}O}}{F\mu_h} \tag{17.10c}$$

If the concentration of electrons is small and the electron conductivity does not contribute to the total electrical conductivity, it will depend on the concentration of electron holes and Equation 17.10c assumes the approximate form of:

$$\left[h^{\bullet}\right] = \frac{\sigma V_{(Fe,Mn)_{1-\delta}O}}{F\mu_h} \tag{17.10d}$$

In turn, if holes are localised on cations, then, in the charge transport, all ions M^{3+} should participate and the "hole" conductivity will depend on their concentration. Therefore, the concentration of charge carriers will depend on the concentration of holes localised on metal ions ($M_M^{\bullet}$), on the concentration of cation vacancies V_M' and V_M^x, which are complexes $\left\{V_M''M_M^{\bullet}\right\}$ and $\left\{M_M^{\bullet}V_M''M_M^{\bullet}\right\}$, and it should also depend on the concentration of complexes containing ions $M_i^{3\bullet}$, which, despite their small concentration, can contribute to the electrical conductivity. Introducing the sum of concentration of ions $[M^{3+}]$ into Equation 17.10d, instead of the concentration of holes, we get:

$$\left[M^{3+}\right] = \frac{\sigma V_{(Fe,Mn)_{1-\delta}O}}{F\mu_h} \tag{17.10e}$$

Therefore, if we assume that the electrical conductivity in wüstite with a low manganese content depends on the concentration of holes or on the concentration of ions $[M^{3+}]$, we can adjust the value of mobility of holes μ_h, in order to obtain the values of concentrations of these ions, calculated according to Equation 17.10e, using the values of the electrical conductivity, consistent with the dependence of the concentration of electron holes or the sum of concentrations of ions $[M^{3+}]$ on the oxygen pressure, resulting from the diagram of concentrations of defects.

For oxides with a higher manganese content, when at low pressures there are discrepancies between the dependence of the concentration of electron holes or concentration of ions $[M^{3+}]$ on p_{O_2} and the values of the concentration of charge carriers calculated according to Equation 17.10c, it is necessary, apart from fitting, for the highest oxygen pressures, the mobility of electron holes μ_h, fit also the parameter $b_{e/h}$, in order for the dependence of the dependence of the sum of concentrations of electronic defects ($[h^{\bullet}] + b_{e/h}[e']$) or ($[M^{3+}] + b_{e/h}[e']$) on p_{O_2} to be consistent with the values of this sum (calculated according to Equation 17.10c). The mobilities of electron holes and electrons are thus determined.

Figure 17.21a–d present the dependence of the concentration of electron holes on the oxygen pressure in the oxide $(Fe_{1-x}Mn_x)_{1-\delta}O$ with the manganese content of x_{Mn} = 0.0, 0.037, 0.122, 0.295 mol/mol at 1173–1473 K (solid lines), which were presented in diagrams of concentrations of defects (Figure 17.3a). Dashed lines (---) represent the dependence of the concentration of ions $[M^{3+}]$ on p_{O_2}. Empty and solid points denote the values of the concentration of charge carriers calculated according to Equation 17.10d, 17.10e for the appropriate fitted values of μ_h and using the values of the electrical conductivity reported in Franke and Dieckmann (1989), Molenda et al. 1987, Molenda et al. 1991).

As can be seen in Figure 17.21a and b, in the range of existence of w_1 pseudo-phase there is a big difference between the concentration of electron holes and the concentration of ions $[M^{3+}]$; also the character (slope) of the above dependences is different. The values of the concentration of charge carriers calculated according to Equation 17.10e in the temperature range of 1173–1323 K are fully consistent with the dependence of the concentration of ions $[M^{3+}]$ on p_{O_2}. This confirms the fact that the electrical conductivity in the oxide $Fe_{1-\delta}O$ and $(Fe_{1-x}Mn_x)_{1-\delta}O$ with a low manganese content depends on the concentration of ions $[M^{3+}]$.

As can be seen in Figure 17.21, at 1403 K and 1473 K, in the oxides $Fe_{1-\delta}O$ and $(Fe_{1-x}Mn_x)_{1-\delta}O$ with the manganese content of x_{Mn} = 0.037 and 0.122 mol/mol there is a discrepancy between the dependence of the concentration of $[M^{3+}]$ on p_{O_2} and the values for charge carriers calculated based on the electrical conductivity (empty points). There is a match with the dependence of the concentration of electron holes on p_{O_2} (solid points). In the case of pure wüstite, the differences in slope are smaller, but there is a better match between the dependence of the concentration of electron holes on p_{O_2} and the calculated values of the concentration of charge carriers, based on the electrical conductivity. This indicates that above 1400 K in the oxide $Fe_{1-\delta}O$ and $(Fe_{1-x}Mn_x)_{1-\delta}O$ with a low manganese content there occurs a change in the mechanism of charge transport. Based on the studies on the electrical conductivity and the thermoelectric power in wüstite, Molenda

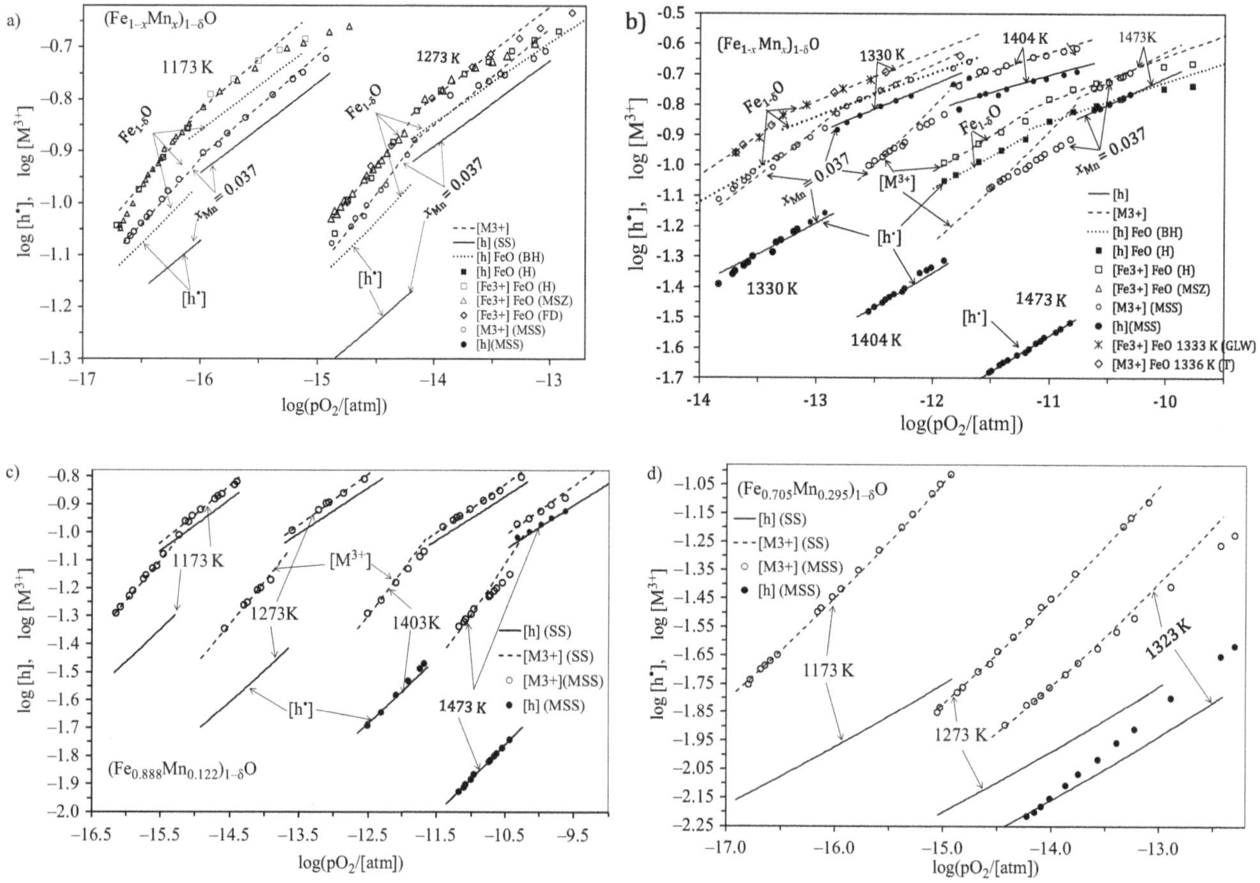

FIGURE 17.21 (a and b) Fragments of diagrams of concentrations of point defects for the oxide $Fe_{1-\delta}O$ and $(Fe_{1-x}Mn_x)_{1-\delta}O$ with the manganese content of 0.0 and 0.037 mol/mol representing the dependence of the concentration of electron holes on p_{O_2} (solid lines) and the dependence of the sum of concentration of ions $[M^{3+}]$ ((- - -) lines) at the temperatures of a) 1173 K and 1273 K, b) 1330 K, 1404 K and1473 K. Empty points represent the values of the concentration of charge carriers determined according to Equation 17.10e using the values of the electrical conductivity reported in Molenda et al. (1991) ((○) (MSS) points), Hillegas (1968) – (□) (H), Molenda et al. (1987) – (△) (MSZ), Tannhauser (1962) – (◇) (T). Solid points represent the values of the concentration of holes $[h^•]$ determined according to Equation 17.10d using the values of the electrical conductivity taken from (Molenda et al. 1991) (points (●) (MSS), (Hillegas 1968) – (■) (H). (c) For the oxide $(Fe_{0.878}Mn_{0.122})_{1-\delta}O$. (d) For the oxide $(Fe_{0.705}Mn_{0.295})_{1-\delta}O$.

et al. (1987) suggest that due to interactions between ions, the acceptor band could overlap the valence band and the metallic band could be created (there occurs a semiconductor/metal transition). Similar effect, even more clear, is present in the case of the oxide $(Fe_{1-x}Mn_x)_{1-\delta}O$ with a low manganese content.

As can be seen in Figure 17.21a–d, in the range of existence of w_1 pseudo-phase, due to a significantly lower concentration of cation vacancies with lower ionisation degree and defect complexes, the differences between the concentration of electron holes and the total concentration of ions $[M^{3+}]$ are small. There is also an agreement between the values of the concentration of ions $[M^{3+}]$ ((○) points) calculated based on the values of the electrical conductivity and the dependence of the concentration of charge carriers on p_{O_2}. Within the error limit, the values of mobility of charge carriers (electron holes) are similar.

In turn, Figure 17.22a and b present the dependence of the concentration of electron holes on the oxygen pressure in the oxide $(Fe_{1-x}Mn_x)_{1-\delta}O$ with the manganese content of $x_{Mn} = 0.0–1.0$ mol/mol, at 1273 K (solid lines), which were presented in Figure 17.1a. Lines (- - -) denote the dependence of the concentration of charge carriers (ions $[M^{3+}]$) on p_{O_2}, and lines (-·-·-) represent the dependence of the sum of ($[M^{3+}] + b_{e/h}[e']$) on p_{O_2} for the fitted values of $b_{e/h}$. Points denote the values of the concentration of ions $[M^{3+}]$ or the values of the sum ($[M^{3+}] + b_{e/h}[e']$) calculated according to Equation 17.10e or 17.10c for the fitted value of μ_h at higher oxygen pressures and when using the values of the electrical conductivity reported in Franke and Dieckmann (1989) ((○,●,◆) points) and for the manganese oxide extracted from Eror and Wagner Jr. (1971) ((■) points).

As can be seen in Figure 17.22a and b, for the manganese content of $x_{Mn} = 0.0$, 0.2 and 0.4 mol/mol there are similar differences between the dependence of the concentration of electron holes and the dependence of the concentration of ions

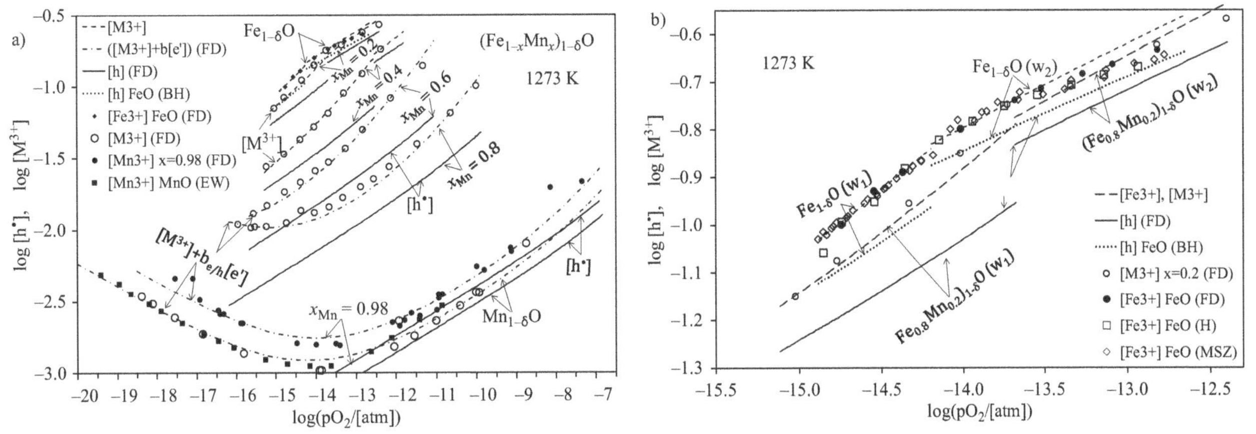

FIGURE 17.22 (a) Fragments of diagrams of concentrations of point defects for the oxide $Fe_{1-\delta}O$ and $(Fe_{1-x}Mn_x)_{1-\delta}O$ with the manganese content of x_{Mn} = 0.2–1.0 mol/mol representing the dependence of the concentration of electron holes (solid lines) and the dependence of the sum of concentration of ions $[M^{3+}]$ (dashed lines (---)) on the oxygen pressure at 1273 K. For the manganese content of x_{Mn} = 0.6, 0.8 and 1.0 mol/mol, the dependence of the sum $([M^{3+}] + b_{e/h}[e'])$ on p_{O_2} was determined ((-·-) line). Points represent the values of the concentration of charge carriers determined according to Equation 17.10c using the values of the electrical conductivity extracted from Franke and Dieckmann (1989) – (○) (FD) points, for $Fe_{1-\delta}O$ (Franke and Dieckmann 1989) – (◆) (FD), (Hillegas 1968) – (□), (Molenda et al. 1987) (H) – (◇) (MSZ), for $Mn_{1-\delta}O$ (Eror and Wagner Jr. 1971) – (■) (EW). (b) For the oxide $(Fe_{1-x}Mn_x)_{1-\delta}O$ with the manganese content of x_{Mn} = 0.0–0.2 mol/mol.

$[M^{3+}]$. The concentration of charge carriers, based on the values of the conductivity, is consistent with the dependence of the concentration of $[M^{3+}]$ on p_{O_2} resulting from the diagram of point defects. As can be seen in Figure 17.22a, in the case of pure manganese oxide and the oxide $(Fe_{1-x}Mn_x)_{1-\delta}O$ with the manganese content of x_{Mn} = 0.6, 0.8 and 0.98 mol/mol, there is a consistency between the dependence of the sum of concentrations of electronic defects $([M^{3+}] + b_{e/h}[e'])$ on p_{O_2} (for the fitted parameter $b_{e/h}$) and the values of this sum calculated according to Equation 17.10c based on the electrical conductivity and the fitted value of the mobility of electron holes μ_h at higher oxygen pressures. The differences between the concentration of ions $[Mn^{3+}]$ and the concentration of electron holes are small, however, in the range of higher oxygen pressures, where there is a significant concentration of cation vacancies with lower ionisation degrees, the concentration of charge carriers calculated according to Equation 17.10c, despite a significant measurement error (scattering of measurement points) shows a higher degree of match with the dependence of $([M^{3+}] + b_{e/h}[e'])$ on p_{O_2}. The fitted values of the parameter $b_{e/h}$ for the manganese content of x_{Mn} = 0.6 mol/mol is $b_{e/h}$ = 20,000, for x_{Mn} = 0.8 $b_{e/h}$ = 19,000, for x_{Mn} = 0.98 $b_{e/h}$ = 1,400, and for the manganese oxide $b_{e/h}$ = 750. The obtained values of the parameter $b_{e/h}$ are quite high, but for the pure manganese oxide they agree with the values obtained using the results of conductivity of several authors (Bocquet et al. 1967, Duquesnoy 1965, Duquesnoy and Marion 1963, Eror and Wagner Jr. 1971, Hed and Tannhauser 1967, Le Brusq and Delmaire 1973, Le Brusq et al. 1968, O'Keeffe and Valigi 1970, Oehlig et al. 1967) (these values depend on impurities and on the content of doping ions Me^{3+} (Stokłosa 2015). The calculated mobility of electrons in the manganese oxide is μ_e = 12 $cm^2\ V^{-1}\ s^{-1}$, while in oxides with the content of x_{Mn} = 0.98 mol/mol it is μ_e = 19 $cm^2\ V^{-1}\ s^{-1}$, for the content of x_{Mn} = 0.8 it is μ_e = 120 $cm^2\ V^{-1}\ s^{-1}$, and for the content of x_{Mn} = 0.6 it is μ_e = 360 $cm^2\ V^{-1}\ s^{-1}$. The character of the dependence of the electrical conductivity on p_{O_2} indicates that in the oxide $(Fe_{1-x}Mn_x)_{1-\delta}O$ with the manganese content above x_{Mn} = 0.6 mol/mol, despite significantly lower concentration of electrons than the concentration of electron holes, the mobility of electrons is high enough to contribute to the total conductivity. From the obtained calculations it results that when the concentration of iron ions in the manganese oxide increases, the mobility of electrons strongly increases.

A different character of the dependence of the concentration of electron holes and the concentration of ions $[M^{3+}]$ on the oxygen pressure is visible in the dependence of the derivative $d\log(p_{O_2})/d\log[h] = n_h$ and $d\log(p_{O_2})/d\log[M^{3+}] = n_M^{3+}$ on the oxygen pressure; these dependences are shown in Figure 17.23.

As can be seen in Figure 17.23a, in the oxide $(Fe_{1-x}Mn_x)_{1-\delta}O$ with a low manganese content, in the range of existence of w_1 pseudo-phase, the exponents n_M^{3+} and n_h differ significantly and they depend on the oxygen pressure. The values and the dependence of n_M^{3+} on p_{O_2} is close to the exponent n_δ of the dependence of the deviation from the stoichiometry on p_{O_2} (see Figure 17.8). The values of n_h are higher and their dependence on the oxygen pressure is weaker. In turn, in the range of existence of w_2 pseudo-phase, the values of the above derivatives are similar, which results from a much lower concentration of vacancies with lower ionisation degrees and defect complexes. Similar differences are present between the oxygen pressure dependence of the derivative $d\log(p_{O_2})/d\log[h^\bullet] = n_h$ and $d\log(p_{O_2})/d\log[M^{3+}] = n_M^{3+}$ for the oxides $(Fe_{1-x}Mn_x)_{1-\delta}O$ with higher manganese content, which are presented in Figure 17.23b. Due to a much wider range of existence of oxides and the change in relative concentrations of defects, the changes in the derivatives are higher. For the

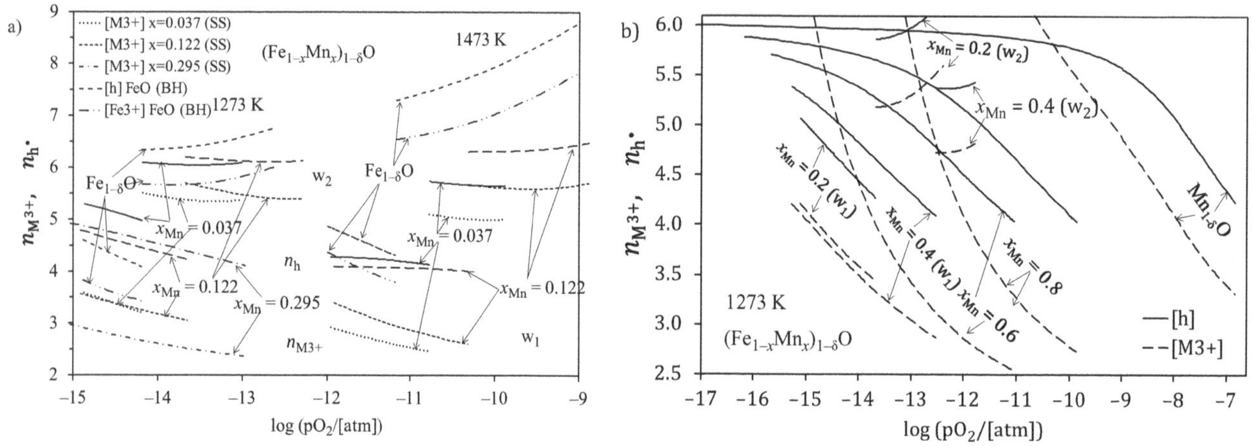

FIGURE 17.23 Oxygen pressure dependence of the derivative $(d\log p_{O_2}/d\log[h^\bullet]) = n_h$ (solid lines) and $(d\log p_{O_2}/d\log[M^{3+}]) = n_{M^{3+}}$ (dashed lines – see legend) a) at temperatures of 1273 and 1473 K, obtained for the dependence of the concentration of holes and the concentration of ions $[M^{3+}]$ on p_{O_2} in the oxide $(Fe_{1-x}Mn_x)_{1-\delta}O$ with the manganese content of x_{Mn} = 0.0, 0.037, 0.122 and 0.295 mol/mol (presented in Figure 17.21b) with the manganese content of x_{Mn} = 0.2–1.0 mol/mol at 1273 K (presented in Figure 17.21).

manganese content of x_{Mn} = 0.6 and 0.8 mol/mol and for the pure manganese oxide, owing to the fact that the dependence of $([M^{3+}] + b_{e/h}[e'])$ on p_{O_2} reaches the maximum, the character of the dependence of the derivative of this function is different (it is increasing more strongly).

Therefore, the match obtained at 1173–1323 K between the calculated values of the concentration of charge carriers (according to Equation 17.10e) and the oxygen pressure dependence on the concentration of ions $[M^{3+}]$ (resulting from the diagram of concentrations of defects) in the oxide $Fe_{1-\delta}O$ and $(Fe_{1-x}Mn_x)_{1-\delta}O$ with the manganese content up to x_{Mn} = 0.4 mol/mol confirms the assumption that the electrical conductivity depends on the concentration of ions $[M^{3+}]$ and the charge transport occurs according to the hopping mechanism, where electrons jump between these ions. Above 1400 K, a change in the mechanism of charge carriers' transport occurs (there occurs a semiconductor/metal-type transition).

In turn, in the oxide $(Fe_{1-x}Mn_x)_{1-\delta}O$ with a higher manganese content, in the range of higher oxygen pressures, hole conductivity dominates and hole transport occurs according to the hopping mechanism of small polarons. At low oxygen pressures, in the manganese oxide and in the oxide with the iron content above 0.4 mol/mol, the contribution of electron

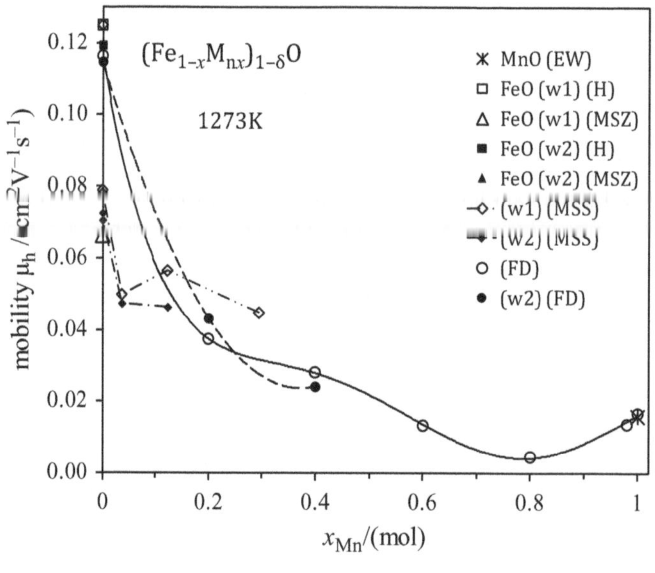

FIGURE 17.24 Dependence of the mobility of charge carriers (electron holes) μ_h on the manganese content in the oxide $Fe_{1-\delta}O$ and $(Fe_{1-x}M_x)_{1-\delta}O$ at 1273 K, obtained using the values of electrical conductivity extracted from (Franke and Dieckmann 1989) ((○,●) (FD) points) and (Molenda et al. 1991) – (◇,◆)(MSS), for wüstite (Hillegas 1968) – (□,■)(H), (Molenda et al. 1987) – (△,▲) (MSZ) and for $Mn_{1-\delta}O$ (Eror and Wagner Jr. 1971) – (*)(EW) and (Franke and Dieckmann 1989) – (○)(FD) (empty points – "pseudo-phase" w_1, solid points – "pseudo-phase" w_2).

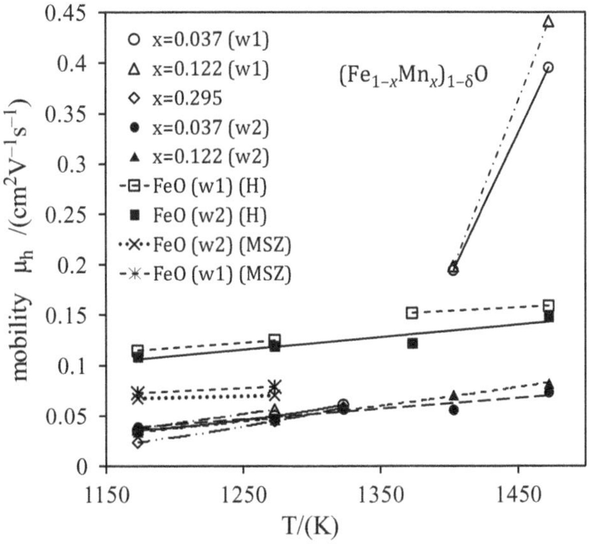

FIGURE 17.25 Dependence of the mobility of charge carriers μ_h (electron holes) on the temperature in the oxide $Fe_{1-\delta}O$ and $(Fe_{1-x}M_x)_{1-\delta}O$ with the manganese content of x_{Mn} = 0.037 ((○,●) points), 0.122 – (△,▲) and 0.295 mol/mol – (◇) obtained using the values of electrical conductivity (Molenda et al. 1991). The mobility of electron holes in wüstite was obtained using the values of electrical conductivity reported in (Hillegas 1968) – (□,■)(H), (Molenda et al. 1987) – (✳,x)(MSZ) (empty points, – "pseudo-phase" w_1; solid points, – "pseudo-phase" w_2).

conductivity is significant. This is an effect of a much higher mobility of electrons than the mobility of electron holes in these oxides.

Figure 17.24 presents the dependence of the mobility of electron holes on the manganese content in the oxide $(Fe_{1-x}Mn_x)_{1-\delta}O$, which, as can be seen, decreases when the manganese content increases.

In turn, Figure 17.25 presents the temperature dependence of the mobility of electron holes in the oxide $(Fe_{1-x}Mn_x)_{1-\delta}O$ with the manganese content of x_{Mn} = 0.037 ((○,●) points), 0.122 – (△,▲) and 0.295 mol/mol – (◇) obtained when using the values of the electrical conductivity (Molenda et al. 1991) (empty points, – pseudo-phase w_1, solid points – pseudo-phase w_2). Also, the dependence of the mobility of electron holes in wüstite is shown, obtained using the values of the electrical conductivity taken from Hillegas (1968) and Molenda et al. 1987).

As can be seen, when the temperature increases, the mobility of electron holes in the oxide $(Fe_{1-x}Mn_x)_{1-\delta}O$ increases, while at 1403 K and 1473 K the increase in mobility in the oxide $(Fe_{1-x}Mn_x)_{1-\delta}O$ with the manganese content x_{Mn} = 0.037, 0.122 mol/mol is significantly stronger. In the case of wüstite, in the temperature range of 1173–1323 K, the mobilities are higher than in the doped wüstite, while above 1400 K the increase in the mobility is smaller than in the doped compound.

For the increase in the mobility of electron holes (charge carriers), when the temperature increases, its temperature dependence should be described with one of the following equations:

$$\mu_h = \mu_h^o T^{-3/2} \exp(-\Delta H_m^h/RT) \tag{17.11a}$$

or

$$\mu_h = \mu_h^o T \exp(-\Delta H_m^h/RT) \tag{17.11b}$$

or

$$\mu_h = \mu_h^o \exp(-\Delta H_m^h/RT) \tag{17.11c}$$

where ΔH_m^h denotes the activation enthalpy of charge carriers' transport (mobility of electron holes).

Figure 17.26 presents the dependence of a) $\log(\mu_h T^{3/2})$, b) $\log(\mu_h T)$, c) $\log(\mu_h)$ on the inverse of the temperature in wüstite and in the oxide $(Fe_{1-x}Mn_x)_{1-\delta}O$. As can be seen in Figure 17.26, the obtained relations are approximately linear. In Table 17.2, the obtained values of the activation enthalpy of the mobility of electron holes, resulting from the relations presented in Figures 17.26, are given. Therefore, the mobility of electron holes occurs according to the mechanism of small polarons and it is activated, however, due to a small number of measurement values, the obtained values do not allow drawing

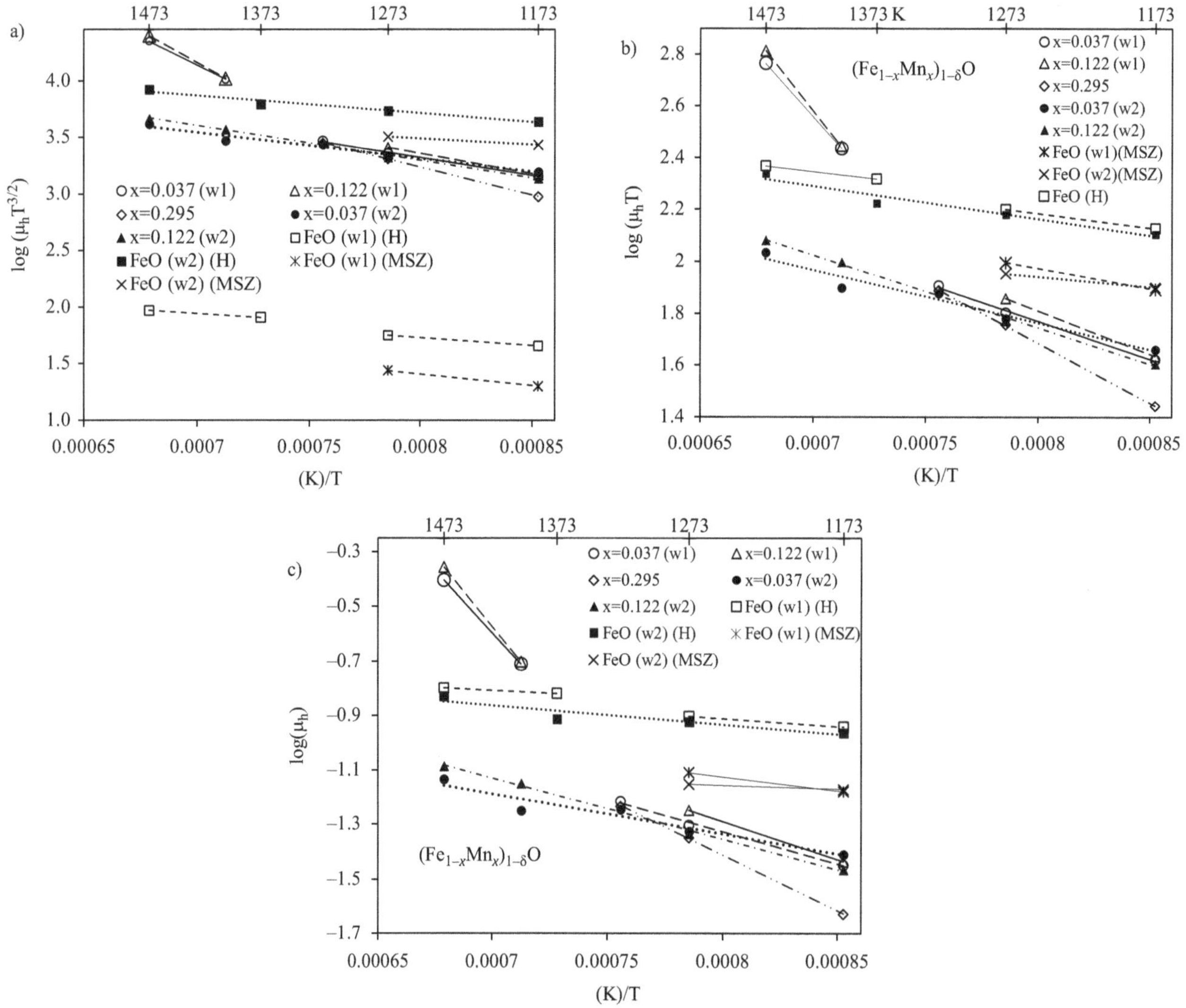

FIGURE 17.26 Temperature dependence of: a) log (μ_c), b) log ($\mu_c T$), c) log ($\mu_c T^{3/2}$) in the oxide $(Fe_{1-x}Mn_x)_{1-\delta}O$ with the manganese content x_{Mn} = 0.037 ((○,●) points), 0.122 – (△,▲) and 0.295 mol/mol – (◇).The mobility of electron holes in wüstite was obtained using the values of electrical conductivity reported in (Hillegas 1968) – (□,■)(H), (Molenda et al. 1987) – (✳,×) (MSZ) (empty points, – "pseudo-phase" w_1; solid points, – "pseudo-phase" w_2) (empty points, – "pseudo-phase" w_1; solid points, – "pseudo-phase" w_2).

Further conclusions on the mechanism of transport of electron holes (the lowest values of the confidence interval and coefficients of the linear regression are present in the case of the relation described with Equation 17.11a).

17.6 CONCLUSIONS

Using the results of the studies of the deviation from the stoichiometry in the oxide $(Fe_{1-x}Mn_x)_{1-\delta}O$ with the manganese content of x_{Mn} = 0.2–1 mol/mol at 1273 K and 1473 K (Franke and Dieckmann 1989, Subramanian and Dieckmann 1993), and, for low contents of x_{Mn} = 0.037–0.295 mol/mol at 1173–1473 K (Soliman and Stokłosa 1990), diagrams of concentrations of defects were determined. In the calculations, simple defects (cation vacancies $[V_M'']$ $[V_M']$, $[V_M^x]$) and defect complexes $\{(V_M'')_2M_i^{3}\}'$, $\{(V_M'')_4M_i^{3}\}^{5'}$ and $\{(V_M'')_6(M_i^{3})_2\}^{6'}$ were taken into account. Similarly, as for pure wüstite, in the oxides $(Fe_{1-x}Mn_x)_{1-\delta}O$ with a low manganese content it was shown that the dependence of the deviation from the stoichiometry on the oxygen pressure can be described with two different curves (with different slopes). For the values of the deviation from the stoichiometry (Soliman and Stokłosa 1990), such a character is present at the manganese content of x_{Mn} = 0.037 and 0.122 mol/mol at 1173–1473 K, while for the results reported elsewhere (Franke and Dieckmann 1989, Subramanian and Dieckmann 1993) at the content of x_{Mn} = 0.2–0.4 mol/mol at 1273 K and x_{Mn} = 0.2–0.8 mol/mol at 1473 K. The above fact indicates that also in wüstite doped with manganese, in the range of low oxygen pressures, the so-called pseudo-phase w_1 exists, and in the range of higher pressures, pseudo-phase w_2. Diagrams of concentrations of defects were determined

TABLE 17.2

Activation Enthalpy of the Mobility of Electron Holes $\left(\Delta H_m^h\right)$ in the Oxide $Fe_{1-\delta}O$ and $(Fe_{1-x}Mn_x)_{1-\delta}O$ with the Manganese Content of x_{Mn} = 0.037–0.295 mol/mol and the Parameter μ_h^o in Equation 17.35 for "Pseudo-phases" w_1 and w_2, Determined based on the Values of Electrical Conductivity Extracted from (Hillegas 1968)(H), (Molenda et al. 1987) (MSZ), and (Molenda et al. 1991) (MSS).

x_{Mn} Equation	ΔH_m^h (kJ/mol)	ΔH_m^h (eV)	$\log \mu_h^o$ (cm^2/Vs)	R^2	ΔT (K)	(Ref)
$\mu_h = \mu_h^o T^{-3/2} \exp\left(-\Delta H_m^h/RT\right)$						
0.0 (w_1)	−26.5	−0.27	2.84		1173–1273	(H)
0.0 (w_2)	−29.7 ± 4	−0.31 ± 0.04	4.66 ± 0.76	0.9635	1173–1473	(H)
0.0 (w_1)	−39.7	−0.41	3.07		1173–1273	(MSZ)
0.0 (w_2)	−20.7	−0.21	4.35		1173–1273	(MSZ)
0.037 (w_1)	−61 ± 4	−0.63 ± 0.04	5.87 ± 0.16	0.9963	1173–1323	(MSS)
0.037 (w_2)	−45 ± 5	−0.46 ± 0.05	5.17 ± 0.19	0.9678	1173–1473	(MSS)
0.122 (w_1)	−69	−0.71	6.23		1173–1323	(MSS)
0.122 (w_2)	−59 ± 2	−0.61 ± 0.02	5.76 ± 0.08	0.9977	1173–1323	(MSS)
0.295 (w_1)	−94 ± 1	−0.97 ± 0.01	7.2 ± 0.01	0.9999	1173–1323	(MSS)
0.0 (w_1)	−23.1	−0.24	2.73		1373–1473	(H)
0.037 (w_1)	−192.0	−1.99	11.2		1403–1473	(MSS)
0.122 (w_1)	−214	−2.22	12.09		1403–1473	(MSS)
$\mu_h = \mu_h^o T^{-1} \exp(-\Delta H_m^h/RT)$						
0.0 (w_1)	−21.0	−0.22	3.06		1173–1273	(H)
0.0 (w_2)	−24.3 ± 4	−0.25 ± 0.04	3.18 ± 0.2	0.9486	1173–1473	(H)
0.0 (w_1)	−29.9	−0.31	3.22		1173–1273	(MSZ)
0.0 (w_2)	−18.2	−0.19	2.6		1173–1273	(MSZ)
0.037 (w_1)	−56 ± 4	−0.58 ± 0.04	4.11 ± 0.15	0.9957	1173–1323	(MSS)
0.037 (w_2)	−39 ± 5	−0.41 ± 0.05	3.40 ± 0.25	0.9592	1173–1323	(MSS)
0.122 (w_1)	−64	−0.66	4.50		1173–1323	(MSS)
0.122 (w_2)	−54 ± 2	−0.5 ± 0.02	4.00 ± 0.08	0.9974	1173–1323	(MSS)
0.295 (w_1)	−89 ± 1	−1.16 ± 0.9	5.39 ± 0.05	0.9998	1173–1323	(MSS)
0.0 (w_1)	−19.3	−0.20	3.05		1373–1473	(H)
0.037 (w_1)	−186	−1.93	9.36		1403–1473	(MSS)
0.122 (w_1)	−208	−2.16	10.19		1403–1473	(MSS)
$\mu_h = \mu_h^o \exp\left(-\Delta H_m^h/RT\right)$						
0.0 (w_1)	−10.9	−0.11	−0.46		1173–1273	(H)
0.0 (w_2)	−13 ± 4	−0.14 ± 0.04	−0.38 ± 0.13	0.8812	1173–1473	(H)
0.0 (w_1)	−19.7	−0.20	−0.2996		1173–1273	(MSZ)
0.0 (w_2)	−8.0	−0.08	−0.8223		1173–1273	(MSZ)
0.037 (w_1)	−46±4	−0.47 ± 0.04	0.57 ± 0.15	0.9941	1173–1323	(MSS)
0.037 (w_2)	−28 ± 5	−0.29 ± 0.058	−0.15 ± 0.1	0.9262	1173–1473	(MSS)
0.122 (w_1)	−53.5	−0.55	0.90		1173–1323	(MSS)
0.122 (w_2)	−43 ± 2	−0.44 ± 0.02	0.45 ± 0.07	0.9964	1173–1473	(MSS)
0.295 (w1)	−78 ± 1	−0.81 ± 0.01	1.86 ± 0.05	0.9997	1173–1323	(MSS)
0.0 (w_1)	−7.5	−0.08	−0.53		1373–1473	(H)
0.037 (w_1)	−174	−1.80	5.77		1403–1473	(MSS)
0.122 (w_1)	−196	−2.03	6.60		1403–1473	(MSS)

independently in the range of existence of pseudo-phases w_1 and w_2. It was found that when the manganese content in the oxide $(Fe_{1-x}Mn_x)_{1-\delta}O$ increases, the range of its existence increases. The range of existence of w_1 pseudo-phase increases, but the range of existence of w_2 pseudo-phase decreases, particularly at 1273 K. At 1473 K, changes of the ranges of existence of w_2 pseudo-phase are smaller. When the manganese content in the oxide $(Fe_{1-x}Mn_x)_{1-\delta}O$ increases, near the decomposition pressure, the deviation from the stoichiometry decreases and the concentrations of simple defects and defect complexes significantly decrease. The oxygen pressure at which w_1/w_2 "phase transition" occurs increases monotonically when the manganese content in the oxide $(Fe_{1-x}Mn_x)_{1-\delta}O$ increases.

From the calculations it results that the hypothetical oxygen activity at which the oxides $(Fe_{1-x}Mn_x)_{1-\delta}O$ with the manganese content up to 0.4 mol/mol could have reached the stoichiometric composition are 10^{-31}–10^{-25} atm at 1273 K and 10^{-23}–10^{-18} atm at 1473 K. In the case of oxides with a low manganese content these oxygen activities are by over 10 orders of magnitude lower than the decomposition pressures of the oxides. At higher manganese contents in the oxide $(Fe_{1-x}Mn_x)_{1-\delta}O$, the difference between the decomposition pressure of the oxide and the pressure at which it could have reached the stoichiometric composition, gets smaller. The reason for the above large differences is a strong interaction between ions, which leads to the decomposition of the oxide (despite a significant concentration of cation vacancies that reduce this interaction). For spinels, which have the same cubic crystallographic structure of NaCl-type but much "looser", the stoichiometric composition is reached in the middle of their range of existence.

From the performed calculations of the diagrams of the concentrations of defects it results that in the range of existence of the w_1 pseudo-phase, cation vacancies $\left[V_M''\right]$ dominate, but the deviation from the stoichiometry is significantly affected by the concentration of vacancies with lower ionisation degrees. During the calculations it was found that when the concentration of complexes increases, the concentration of vacancies $\left[V_M''\right]$ decreases. At the limit concentration of complexes, the concentration of vacancies $\left[V_M''\right]$ becomes independent of the oxygen pressure and further increase in the concentration of complexes makes the calculations impossible. Thus, the determined concentration of defect complexes is the maximum concentration in the oxide $(Fe_{1-x}Mn_x)_{1-\delta}O$ (fulfilling the assumptions for the method of calculations). It is lower than the concentration of cation vacancies $\left[V_M''\right]$ by over one order of magnitude and it does not affect the degree of match between the dependence of the deviation from the stoichiometry on the oxygen pressure and the experimental values of δ. From the performed calculations it results that in the oxide $(Fe_{1-x}Mn_x)_{1-\delta}O$, the highest concentration of complexes $\{(V_M'')_2M_i^{3\bullet}\}'$ and $\{(V_M'')_4M_i^{3\bullet}\}^{5'}$ is present; it is of an order of 0.01–0.001 mol/mol and it decreases when the manganese content increases.

In turn, in the range of existence of w_2 pseudo-phase, the oxygen pressure dependence of the deviation from the stoichiometry is practically equal to the concentration of vacancies $\left[V_M''\right]$. For low manganese content, similarly as for pure wüstite, the concentration of vacancies $\left[V_M''\right]$ was higher than the experimental values of the deviation. The agreement between the dependence of the concentration of vacancies $\left[V_M''\right]$ and the deviation from the stoichiometry was obtained when assuming a linear dependence of the standard Gibbs energy $\Delta G^o_{V_M''}$ of formation of cation vacancies on the deviation from the stoichiometry. In order to obtain a higher concentration of defect complexes in the oxide $(Fe_{1-x}Mn_x)_{1-\delta}O$ with a higher manganese concentration, it was also assumed that $\Delta G^o_{V_M''}$ of formation of cation vacancies depends on the deviation from the stoichiometry and a correction of the value of $\Delta G^o_{V_M''}$ was introduced by fitting the parameter α. Therefore, in the range of existence of w_2 pseudo-phase, the maximum concentration of defect complexes and the maximum concentration of vacancies with lower ionisation degree, which does not deteriorate the match between the dependence of the deviation from the stoichiometry on the oxygen pressure and the experimental values of δ were determined.

It should be noted that the concentration of complexes is high enough for their dense-packing in the oxide structure and the formation of larger defect clusters. As the concentration of defect complexes does not affect the dependence of the deviation from the stoichiometry on p_{O_2}, it is impossible to determine their concentration unambiguously. Similarly, in the range of existence of w_1 pseudo-phase, it is lower than the one calculated, and in the w_2 pseudo-phase it is higher. A limit case of an ordered structure of defect complexes is the spinel structure (see Section 1.1, Figure 1.1), where defect complexes lost the properties of point defects.

When calculating the diagrams of concentrations of defects it was impossible to unambiguously determine the standard Gibbs energy of formation of Frenkel defects ΔG^o_F, because it was assumed that the values of ΔG^o_F change in a linear way when the manganese content increases (for the oxides $Fe_{1-\delta}O$ and $Mn_{1-\delta}O$, the values determined in the work of Stokłosa (2015) were used).

The adjusted standard Gibbs energies $\Delta G^o_{V_M''}$ of formation of cation vacancies in the oxide $(Fe_{1-x}Mn_x)_{1-\delta}O$ change within the range from –90 to 70 kJ/mol at 1273 K, and from –70 to 100 kJ/mol at 1473 K. They increase monotonically when the temperature increases and when the manganese content in the oxide $(Fe_{1-x}Mn_x)_{1-\delta}O$ increases, while there are significant differences when using the results of the studies of the deviation from the stoichiometry either Franke and Dieckmann (1989), Subramanian and Dieckmann (1993) or Soliman and Stokłosa (1990). The differences result from the fact of different values of the deviation from the stoichiometry and from the differences in the character of the dependence of the deviation from the stoichiometry on the oxygen pressure obtained by these authors.

In turn, the calculated values of the standard Gibbs energies of formation of interstitial cations $\Delta G^o_{M_i^{\bullet\bullet}}$ in the oxide $(Fe_{1-x}Mn_x)_{1-\delta}O$ are positive and they decrease when the temperature increases and when the manganese content increases, from 230–180 kJ/mol at 1273 K, and from 190–140 kJ/mol at 1473 K. The dependences of $\Delta G^o_{M_i^{\bullet\bullet}}$ on the manganese content in the oxide $(Fe_{1-x}Mn_x)_{1-\delta}O$ are more complex and they significantly differ from the dependence of $\Delta G^o_{V_M''}$ (the changes are not entirely symmetrical).

The adjusted standard Gibbs energies $\Delta G^{o}_{V'_M} = \Delta G^{o}_{V^x_M}$ of formation of cation vacancies with lower ionisation degree in the oxide $(Fe_{1-x}Mn_x)_{1-\delta}O$, in the range of existence of w_1 pseudo-phase change from −10 to −60 kJ/mol. In the range of existence of w_2 pseudo-phase, the adjusted values of ΔG_i^o of formation of vacancies with lower ionisation degrees and defect complexes will differ from the actual values, as the determined concentration of these defects does not affect the oxygen pressure dependence of the deviation from the stoichiometry.

Using the values of coefficients of iron diffusion in the oxides $(Fe_{1-x}Mn_x)_{1-\delta}O$ at 1273 K, obtained with the tracer method (^{59}Fe) (Franke and Dieckmann 1989), and the concentration of cation vacancies resulting from diagrams of concentrations of defects, the coefficients of diffusion of defects were determined, in particular the mobility of iron ions via cation vacancies, which are of the order of 10^{-6} cm²/s and to a small degree decrease when the manganese content in the oxide $(Fe_{1-x}Mn_x)_{1-\delta}O$ increases.

In turn, using the values of the electrical conductivity in the oxides $(Fe_{1-x}Mn_x)_{1-\delta}O$ reported in Franke and Dieckmann (1989) and Molenda et al. (1991), and the concentration of charge carriers, their mobility was determined. It was shown that the character of the oxygen pressure dependence of electrical conductivity is strictly dependent on the concentration of ions $[M^{3+}]$. Therefore, it is dependent on the concentration of electron holes localised on cations M_M, on the concentration of cation vacancies with a lower ionisation degree containing ions $[M_M]$ and on the concentration of ions $\left[M_i^{3}\right]$ forming defect complexes. The mobility of charge carriers (electron holes) at 1273 K decreases when the manganese content in the oxide $(Fe_{1-x}Mn_x)_{1-\delta}O$ increases, from μ_h = 0.12–0.07 cm²/Vs to 0.01. In turn, for a low manganese content in the oxide $(Fe_{1-x}Mn_x)_{1-\delta}O$ in the temperature range of 1173–1323 K, the mobility increases when the temperature increases, thus, charge transport occurs according to the hopping mechanism of small polarons. Above 1400 K, there occurs a change in the transport mechanism (there occurs a semiconductor/metal transition).

It was shown that similarly as in the manganese oxide, also in the oxides $(Fe_{1-x}Mn_x)_{1-\delta}O$ with the manganese content of x_{Mn} = 0.6–1.0 mol/mol the mobility of electrons is much higher than the mobility of electron holes and despite a low concentration of electrons, they contribute to the total electrical conductivity. Based on the values of the electrical conductivity, and concentrations of holes and electrons resulting from the diagram of point defects, the ratio of mobilities μ_e/μ_h was determined; it decreases when the manganese content in the oxide $(Fe_{1-x}Mn_x)_{1-\delta}O$ increases. In the oxide with the manganese content x_{Mn} = 0.6 mol/mol it is $b_{e/h}$ = 20,000, for x_{Mn} = 0.8 $b_{e/h}$ = 19,000, for x_{Mn} = 0.98 $b_{e/h}$ = 1,400, and for the manganese oxide $b_{e/h}$ = 750. The calculated mobility of electrons in the manganese oxide is μ_e = 12 cm² V⁻¹ s⁻¹, while in oxides with the content of x_{Mn} = 0.98 mol/mol it is μ_e = 19 cm² V^{-1} s^{-1}, for the content of x_{Mn} = 0.8 it is μ_e = 120 cm² V^{-1} s^{-1}, and for the content of x_{Mn} = 0.6 it is μ_e = 360 cm² V^{-1} s^{-1}. From the calculations it results that when the concentration of iron ions in the manganese oxide increases, the mobility of electrons increases.

REFERENCES

Bocquet, J. P., M. Kawahara, and P. Lacombe. 1967. Conductivité électronique, conductibilité ionique et diffusion thermique dans MnO à haute température (de 900 à 1150 °C). *C. R. Acad. Sci. (Paris) Sér. C.* 265: 1318–1321.

Bransky, I., and A. Z. Hed. 1968. Thermogravimetric Determination of the Composition-Oxygen Partial Pressure Diagram of Wüstite $(Fe_{1-y}O)$. *J. Am. Ceram. Soc.* 51: 231–232.

Chen, W. K., and N. L. Peterson. 1975. Effect of the Deviation from Stoichiometry on Cation Self-Diffusion and Isotope Effect in Wüstite, $Fe_{1-x}O$. *J. Phys. Chem. Solids.* 36: 1097–1097.

Duquesnoy, A. 1965. Sur les Variations de la Conductivité Électrique de Quelques Oxydes des Métaux de Transition en Fonction de la Pression d'oxygène d'équilibre à Haute Température. *Rev. Hautes Tempr. Refract.* [illegible]: [illegible]01–[illegible].

Duquesnoy, A., J. Couzin, and P. Gode. 1975. Isothermal Representation of Ternary Phase Diagrams ABO. Study of the System Mn-Fe-O. *C. R. Acad. Sci. (Paris) Sér. C.* 281: 107–109.

Duquesnoy, A., and F. Marion. 1963. Sur les Variations de la Conductivité Électrique des Oxydes CoO, NiO et MnO en Fonction de la Pression Partielle d'oxygène d'équilibre à Haute Température. *C. R. Acad. Sci. (Paris).* 256: 2862–2865.

Engell, H.-J., and H. K. Kohl. 1962. Über die Reduktion von Oxyd-Mischkristallen. *Z. Elektrochem., Ber. Bunsenges. phys. Chem.* 66: 684–689.

Eror, N. G., and J. B. Wagner Jr. 1971. Nonstoichiometric Disorder in Single Crystalline MnO. *J. Electrochem. Soc.* 118: 1665–1670.

Foster, P. K., and A. J. E. Welch. 1956. Metal-Oxide Solid Solutions. Part 1.—Lattice-Constant and Phase Relationships in Ferrous Oxide (Wüstite) and in Solid Solutions of Ferrous Oxide and Manganous Oxide. *Trans. Faraday Soc.* 52: 1636–1642.

Franke, P. 1987. Disorder and Transport in Mixed Oxides of the Types (Fe,Mn) O and $(Fe,Mn)_3O_4$. PhD Thesis, University of Hannover, Germnay.

Franke, P., and R. Dieckmann. 1989. Defect Structure and Transport Properties of Mixed Iron-Manganese Oxides. *Solid State Ionics.* 32–33: 817–823.

Franke, P., and R. Dieckmann. 1990. Thermodynamics of Iron Manganese Mixed Oxides at High Temperatures. *J. Phys. Chem. Solids.* 51: 49–57.

Giddings, R. G., and R. S. Gordon. 1973. Review of Oxygen Activities and Phase Boundaries in Wüstite as Determined by Electromotive Force and Gravimetric Methods. *J. Am. Ceram. Soc.* 56: 111–116.

Hed, A. Z., and D. S. Tannhauser. 1967. Contribution to the Mn-O Phase Diagram at High Temperature. *J. Electrochem. Soc.* 114: 314–318.

Hillegas, J. 1968. Seebeck Coefficient and Electrical Conductivity Measurements on Doped and Undoped Wüstite. PhD Thesis. Northwestern University, Evanston, Ill.

Hodge, J. D., and H. K. Bowen. 1981. High-Temperature Thermoelectric Power Measurements in Wüstite. *J. Am. Ceram. Soc.* 64: 431–436.

Keller, M., and R. Dieckmann. 1985. Defect Structure and Transport Properties of Manganese Oxides: (I) The Nonstoichiometry of Manganosite ($Mn_{1-\Delta}O$). *Ber. Bunsenges. Phys. Chem.* 89: 883–893.

Kjellqvist, L., and M. Selleby. 2010. Thermodynamic Assessment of the Fe-Mn-O System. *J. Phase Equilib. Diff.* 31: 113–134.

Le Brusq, H., and J. P. Delmaire. 1973. Sur les Propriétiés Électriques des Oxides Non Stochiométriques Nb_2O_5, MnO et CoO a Haute Température. *Rev. Int. Hautes Tempér. Réfract.* 10: 15–26.

Le Brusq, H., J. J. Oehlig, and F. Marion. 1968. Sur l'évolution de la Nature des Défaults des Oxides MnO et CoO en Function de la Pression Partielle d'oxygéne à Haute Température. *C. R. Acad. Sci. (Paris) Sér. C.* 266: 965–968.

Libanati, N., J. Philibert, and J. Manenc. 1965. Étude au Moyen de la Microsonde de la Diffusion du Fer et du Manganèse dans les Solutions Solides FeO-MnO. *C. R. Acad. Sci. (Paris).* 260: 1156–1159.

Molenda, J., A. Stokłosa, and H. M. Soliman. 1991. Transport Properties of Manganese-Doped Ferrous Oxide at High Temperature. *Solid State Ionics.* 45: 109–121.

Molenda, J., A. Stokłosa, and W. Znamirowski. 1987. Transport Properties of Ferrous Oxide $Fe_{1-y}O$ at High Temperature. *Phys. Status Solidi (b).* 142: 517–529.

O'Keeffe, M., and M. Valigi. 1970. The Electrical Properties and Defect Structure of Pure and Chromium-Doped MnO. *J. Phys. Chem. Solids.* 31: 947–962.

Oehlig, J. J., H. Le Brusq, A. Duquesnoy, and F. Marion. 1967. Sur les Variations du Pouvoir Thermoélectrique des Oxides CoO, Co_3O_4, MnO, Mn_3O_4 à Haute Température. *C. R. Acad. Sci. (Paris) Sér. C.* 265: 421–424.

Okinaka, H., K. Kosuge, and S. Kachi. 1968. Phase Diagram of (Fe, Mn) O at 1260°K. *J. Jap. Soc. Powder Metall.* 15: 295–301.

Pelton, A. D., and W. T. Thompson. 1975. Phase diagrams. *Prog. Solid State Chem.* 10: 119–155.

Peterson, N. L., and W. K. Chen. 1982. Cation Self-Diffusion and the Isotope Effect in $Mn_{1-\Delta}O$. *J. Phys. Chem. Solids.* 43: 29–38.

Picard, C., and M. Dodé. 1970. Ferrous Oxide at 900°C. *Bull. Soc. Chim. Fr.* 7: 2486.

Picard, C., and P. Gerdanian. 1974. High Temperature Study Of Manganese Monoxide. *J. Solid State Chem.* 11: 190–202.

Schwerdtfeger, K., and A. Muan. 1967. Equilibria in System Fe-Mn-O Involving (Fe,Mn) O and (Fe, Mn)$_3O_4$ Solid Solutions. *Trans. Metall. Soc. AIME.* 239: 1114–1119.

Sockel, H. G., and H. Schmalzried. 1968. Coulometrische Titration an Übergangsmetalloxiden. *Ber. Bunsenges. Phys. Chem.* 72: 745–754.

Soliman, H. M., and A. Stokłosa. 1990. Deviation from Stoichiometry of Mn-Doped Ferrous Oxid. *Solid State Ionics.* 42: 85–91.

Stokłosa, A. 2015. *Non-Stoichiometric Oxides of 3d Metals*. Pfäffikon: Trans Tech Publications Ltd.

Subramanian, R., and R. Dieckmann. 1993. Nonstoichiometry and Thermodynamics of the Solid Solution $(Fe,Mn)_{1-\Delta}O$ at 1200°C *J. Phys. Chem. Solids.* 54: 991–1000.

Takayama, E., and N. Kimizuka. 1980. Thermodynamic Properties and Subphases of Wüstite Field Determined by Means of Thermogravimetric Method in the Temperature Range of 1100–1300ºC. *J. Electrochem. Soc.* 127: 970–976.

Takeuchi, S., K. Furukawa, and K. Gunji. 1960. Statistico-Thermodynamical Studies on Oxidation and Reduction Equilibriums of Mn-Wüstite with Gas Phases. *J. Japan Inst. Metals.* 24: 187–191.

Tannhauser, D. S. 1962. Conductivity in Iron Oxides. *J. Phys. Chem. Solids.* 23: 25–34.

Vallet, P., and P. Raccah. 1965. Contribution to the Study of the Thermodynamic Properties of Solid Iron Protoxide. *Mém. Sci. Rev. Métall.* 62: 1–29.

18 Cobalt-Wüstite – Halite $(Fe_{1-x}Co_x)_{1-\delta}O$

18.1 INTRODUCTION

Cobalt-wüstite is analogous to manganese-wüstite system, it forms a solid solution in the whole range of concentrations and it shows a real deficiency of metal, $(Fe_{1-x}Co_x)_{1-\delta}O$. The range of existence of oxide phases $(Fe_{1-x}Co_x)_{1-\delta}O$ was determined by Aukrust and Muan (1964) and for single compositions, by other authors (Jung et al. 2004), and they are in agreement with the thermodynamic calculations of Subramanian and Dieckmann (1994), Subramanian et al. (1994a) and Zhang and Chen (2013). From the studies of the deviation from the stoichiometry (Aukrust and Muan 1964, Lykasov et al. 1976, Maksutov 1974, Reader et al. 1984, Subramanian et al. 1994b), it results that iron doping of cobalt oxide causes a significant increase in the deviation from the stoichiometry (concentration of defects). It was shown that that the transport of ions and electron holes in the oxide $(Fe_{1-x}Co_x)_{1-\delta}O$ with the cobalt content of $0.6 > x_{Co} > 1.0$ mol/mol at 1473 K depends on the deviation from the stoichiometry (Subramanian et al. 1994b, Tinkler et al. 1994). Studies on the diffusion of tracer ions Co* and Fe* in the cobalt oxide and in the oxide doped with a small amount of iron (Hoshino and Peterson 1985) are consistent with the results obtained by Dieckmann group (Tinkler et al. 1994). Using the results of deviation from the stoichiometry studies (Subramanian et al. 1994b), it was decided to determine the diagrams of defects' concentrations in the oxides $(Fe_{1-x}Co_x)_{1-\delta}O$.

18.2 CALCULATIONS OF DEFECT CONCENTRATIONS DIAGRAMS

For the determination of the diagrams of concentrations of point defects in the oxide $(Fe_{1-x}Co_x)_{1-\delta}O$, an analogous method was used as in the case of oxides $(Fe_{1-x}Mn_x)_{1-\delta}O$ (see Section 17.2). The oxygen pressure dependences of the deviation from the stoichiometry (Subramanian et al. 1994b), plotted in a log–log system, similarly as for oxides $(Fe_{1-x}Mn_x)_{1-\delta}O$, show a change in character and consist of two curves with different slopes. This indicates that also in oxides $(Fe_{1-x}Co_x)_{1-\delta}O$, there are two different pseudo-phases, w_1 and w_2, and this occurs up to the cobalt content $x_{Co} > 0.97$ mol/mol. The observed change in the character of the oxygen pressure dependence of the deviation from the stoichiometry is not an effect of doping the cobalt oxide with Fe^{3+} iron ions. As has been shown (Stokłosa 2015), doping with M^{3+} ions (Cr, Al, Ti) causes a different change in the character of the oxygen pressure dependence of the deviation from the stoichiometry. This is due to an increase in the concentration of cation vacancies $\left[V_M''\right]$ and the deviation from the stoichiometry increases, which in a wide range is independent from the oxygen pressure (see Figure 18.1a). Due to that, the diagrams of point defects for the oxides $(Fe_{1-x}Co_x)_{1-\delta}O$ were determined independently for two ranges of oxygen pressures (in the range of existence of pseudo-phases w_1 and w_2). In the calculations it was assumed that the standard Gibbs energy of formation of Frenkel defects ΔG_F^o and electronic defects ΔG_{eh}^o vary linearly when the cobalt content in the oxide $(Fe_{1-x}Co_x)_{1-\delta}O$ increases. For pure oxides $Fe_{1-\delta}O$ and $Co_{1-\delta}O$, the values extracted from (Stokłosa 2015) were used; they are, respectively: $\Delta G_F^o = 124.7$ and 206 kJ/mol and $\Delta G_{eh}^o = 220$ and 259 kJ/mol. A change in the value of ΔG_F^o and ΔG_{eh}^o within the range of 20 kJ/mol does not cause a change in the degree of match between the dependence of the deviation from the stoichiometry and the experimental values of δ, but only a change in the value of ΔG_i^o of formation of individual defects. Due to that, the same values of ΔG_F^o were assumed for w_1 and w_2 pseudo-phase. The performed initial calculations have shown that in the range of existence of w_1 pseudo-phase, in order to obtain a match between the oxygen pressure dependence of the deviation from the stoichiometry and the experimental results, it is necessary to consider defects with lower ionisation degrees. The determined concentration of defect complexes is the maximum concentration which does not deteriorate the match between the dependence of δ on p_{O_2} and the experimental values of δ.

In turn, in the case of w_2 pseudo-phase, it was revealed that the calculated values of the deviation from the stoichiometry, when assuming that only cation vacancies V_{Fe}'' dominate, are higher than the experimental values (similarly as for pure wüstite). The match between the calculated dependence of the deviation from the stoichiometry on p_{O_2} and the experimental values of δ was obtained when assuming that $\Delta G_{V_{Fe}''}^o$ is a linear function of the deviation δ (see Equation 17.4). Due to that, such maximum concentrations of defect complexes and cation vacancies with lower ionisation degrees were determined that do not deteriorate the degree of match between the dependence of δ on p_{O_2} and the experimental results of the deviation δ.

A detailed method for calculations of diagram of the concentrations of defects is presented in Section 17.2.

18.3 POINT DEFECTS CONCENTRATIONS DIAGRAMS – RESULTS

In Figure 18.1a–e, diagrams of concentrations of point defects (ionic and electronic) are presented, for pure oxide $Co_{1-\delta}O$ at 1473 K, using the results of the deviation from the stoichiometry reported in (Aggarwal and Dieckmann 1991, Bransky and Wimmer 1972, Eror and Wagner Jr. 1968, Fisher and Tannhauser 1966, Sockel and Schmalzried 1968), and for the oxide $(Fe_{1-x}Co_x)_{1-\delta}O$ with the cobalt content of x_{Co} = 0.6, 0.8, 0.925 and 0.97 mol/mol, using the results taken from (Subramanian et al. 1994b) ((○) points). In the case of cobalt oxide (Figure 18.1a), in the range of low oxygen pressures, there are discrepancies between the results reported in (Sockel and Schmalzried 1968) ((□) points) and in Aggarwal and Dieckmann (1991) ((◇) points); due to that, such values of $\Delta G^o_{V_{Co}''}$ and $\Delta G^o_{V_{Co}'} = \Delta G^o_{V_{Co}^x}$ were fitted in order to achieve a match with the experimental results reported in Aggarwal and Dieckmann (1991) ((---) line).

As can be seen in Figure 18.1a–e, a good match was obtained between the dependence of the deviation from the stoichiometry and the experimental values of δ. Similarly, as for pure wüstite, also in oxides $(Fe_{1-x}Co_x)_{1-\delta}O$ up to the cobalt content of x_{Co} = 0.97 mol/mol, a change occurs in the character of the dependence of δ on p_{O_2}, which indicates the existence of two pseudo-phases, w_1 and w_2. In Figure 18.1a the dashed line (–––) and in Figure 18.1b, the dotted line (···) represent the dependence of the deviation from the stoichiometry for the case of doping the cobalt oxide with M^{3+} ions in the amount of $y_{M_{Co}^\bullet}$ = 0.003 mol/mol (see doping of spinels, Section 3.2 and Section 8.2 in the work of Stokłosa (2015)). As can be seen in Figure 18.1b, this is the concentration of dopant $y_{M_{Co}^\bullet}$ which causes an increase in the deviation from the stoichiometry to the average value present in the oxide $(Fe_{0.03}Co_{0.97})_{1-\delta}O$, but in a wide range it is independent of the oxygen pressure. The effect of a significant increase in the deviation from the stoichiometry in the oxide $(Fe_{1-x}Co_x)_{1-\delta}O$ is therefore related to the incorporation of iron ions Fe^{2+} in the sublattice of cobalt ions, which cause very significant changes in interactions between ions. As a result, the standard Gibbs energies of the formation of individual point defects change, their concentration increases and the character of the dependence on the oxygen pressure changes.

In turn, for cobalt oxide (Figure 18.1a), in order to obtain, in the range of low oxygen pressures, a consistency with the experimental results (Aggarwal and Dieckmann 1991) ((◇) points), the values of $\Delta G^o_{V_{Co}''}$ were increased (173 kJ/mol) and $\Delta G^o_{V_{Co}'}$ were decreased (−40 kJ/mol). As a result, as can be seen in Figure 18.1a, the concentration of vacancies $[V_{Co}'']$ and electron holes increased and the concentration of interstitial ions $[Co_i^{\bullet\bullet}]$ and vacancies with lower ionisation degrees decreased ((---) lines). Higher values of the deviation from the stoichiometry at low oxygen pressures, reported in Aggarwal and Dieckmann (1991), could be also a result of contamination with M^{3+} ions. As it results from the performed calculations, the concentration of M^{3+} impurities in the amount of $y_{M_{Co}^\bullet} = 1\cdot10^{-4}$ mol/mol causes an increase in the deviation from the stoichiometry. However, as can be seen in Figure 18.1a dotted line (···)), there are discrepancies between the dependence of δ on p_{O_2} and the experimental values of δ. The above concentration of impurities practically does not cause a change in the concentration of cation vacancies.

As can be seen in Figure 18.1, in the oxide $(Fe_{1-x}Co_x)_{1-\delta}O$ in w_1 pseudo-phase, cation vacancies $[V_M'']$ dominate, but there is also a significant concentration of vacancies with lower ionisation degrees, $[V_M']$ and $[V_M^x]$. The concentration of defect complexes in the oxides $(Fe_{1-x}Co_x)_{1-\delta}O$ with the cobalt content of x_{Co} = 0.6 mol/mol reaches 10^{-2} mol/mol and it is close to the value that is present in pure wüstite. When the cobalt content in the oxide increases, the concentration of defect complexes gradually decreases, to about 10^{-3} mol/mol. As it was mentioned, the calculated concentration of defect complexes is the maximum concentration that does not deteriorate the match of the dependence of δ on p_{O_2}. At the above concentrations of defect complexes, there occurs a decrease in the concentration of cation vacancies $[V_M'']$, and, as a result, it is practically independent of the oxygen pressure.

As it was mentioned, in the range of existence of w_2 pseudo-phase, in order to obtain a match between the dependence of the deviation from the stoichiometry p_{O_2} and the experimental values of δ, it was assumed that $\Delta G^o_{V_{Fe}''}$ is a linear function of the deviation δ (see Equation 17.4). Due to that, the deviation from the stoichiometry in the oxides $(Fe_{1-x}Co_x)_{1-\delta}O$ practically equals the concentration of cation vacancies $[V_M'']$. Similarly, as in w_1 pseudo-phase, as a result of an increase in the concentration of defect complexes, the concentration of vacancies $[V_M'']$ becomes independent of the oxygen pressure, which prevents further increase in the concentration of defect complexes. At the highest concentration of defect complexes, it was necessary to correct the parameter α (in Equation 17.4). Thus, the calculated concentration of defect complexes and concentration of cation vacancies with lower ionisation degrees is the maximum concentration that does not deteriorate the match of the dependence of δ and p_{O_2} and the experimental values of δ. As can be seen in Figure 18.1a–e, the obtained distribution of the concentration of defects in w_1 and w_2 pseudo-phases are different. However, as shown in the work of Stokłosa (2015), the concentration of defect complexes is high enough to allow their dense-packing (it is close to the concentration in wüstite).

Figure 18.2 presents the dependence of the parameter α in Equation 17.4 describing the dependence of $\Delta G^o_{V_M''}$ on the deviation from the stoichiometry in the oxide $(Fe_{1-x}Co_x)_{1-\delta}O$ in the range of existence of w_2 pseudo-phase.

As can be seen, they are multiple times higher than in the case of wüstite, which indicates that cobalt doping of wüstite, or, inversely, iron doping of cobalt oxide, causes a significant increase in repulsive interactions between ions. This might

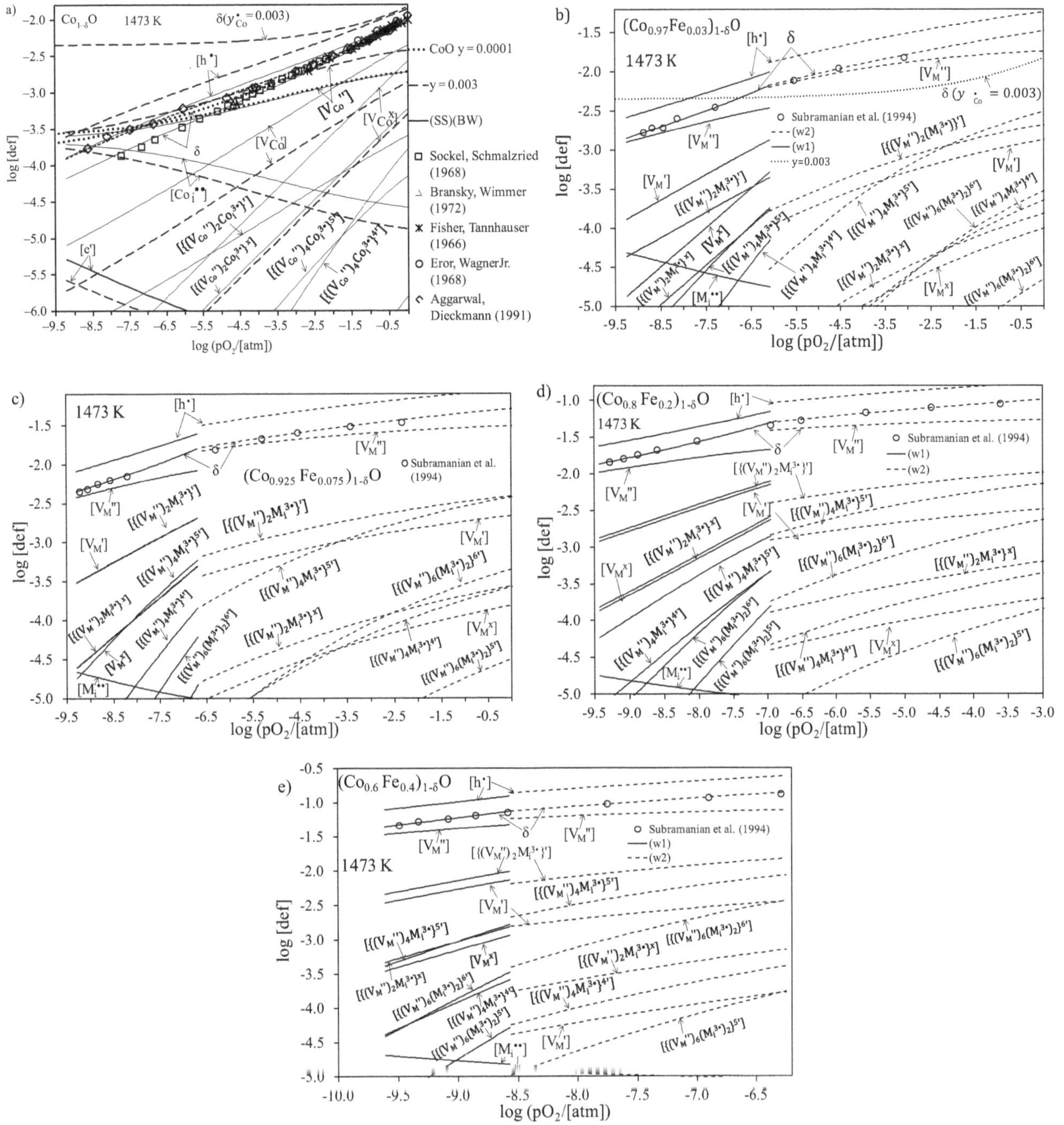

FIGURE 18.1 (a) Diagram of concentration of point defects for the oxide $Co_{1-\delta}O$ at 1473 K, taking into account simple defects and defect complexes $\{(V''_{Co})_2 Co_i^{3\bullet}\}'$, $\{(V''_{Co})_4 Co_i^{3\bullet}\}^{5'}$ and clusters $\{(V''_{Co})_6 (Co_i^{3\bullet})_2\}^{6'}$, as well as defects with lower ionisation degrees (solid lines), obtained using the results of the studies on the deviation from the stoichiometry extracted from (Sockel and Schmalzried 1968) – (□) points, (Fisher and Tannhauser 1966) – (✳), (Eror and Wagner Jr. 1968) – (○), (Bransky and Wimmer 1972) – (△), (Aggarwal and Dieckmann 1991) – (◇). Dashed lines (---) denote the concentrations of defects determined based on the data from (Aggarwal and Dieckmann 1991). The deviation from the stoichiometry in the case of doping the oxide with M^{3+} ions in the amount of $y_M^{\bullet} = 0.0001$ mol/mol is denoted by dotted lines (···), and in the case of $y_M^{\bullet} = 0.003$ mol/mol by (——).lines. (b) for the oxide $(Co_{0.97}Fe_{0.03})_{1-\delta}O$ obtained using the results of the studies on the deviation from the stoichiometry (Subramanian et al. 1994b) – (○). The solid line denotes w_1 "pseudo-phase", the dashed line – w_2 "pseudo-phase". The dotted line denotes the deviation from the stoichiometry in the case of doping the cobalt oxide with M^{3+} ions in the amount of $y_M^{\bullet} = 0.003$ mol/mol. (c) for the oxide $(Co_{0.925}Fe_{0.075})_{1-\delta}O$. (d) for the oxide $(Co_{0.8}Fe_{0.2})_{1-\delta}O$. (e) for the oxide $(Co_{0.6}Fe_{0.4})_{1-\delta}O$.

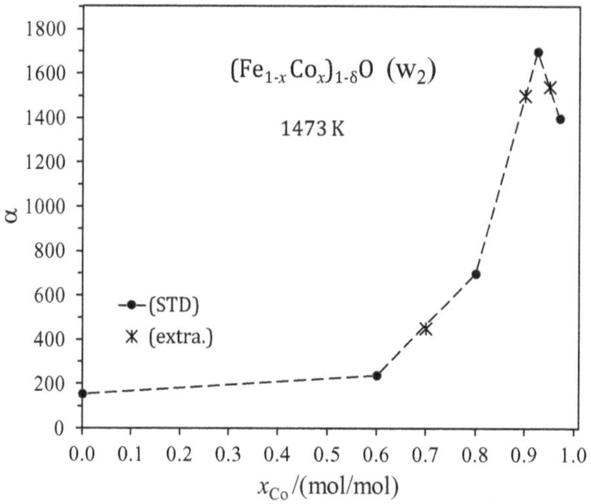

FIGURE 18.2 Dependence of the parameter α of the dependence of $\Delta G^{o}_{V_M''}$ of formation of cation vacancies V_M'' on the deviation from the stoichiometry (in Equation 17.4) on the cobalt content in the oxide $(Fe_{1-x}Co_x)_{1-\delta}O$

be the reason for the formation of w_2 pseudo-phase and for the increase in the concentration of defect complexes and their ordering.

Figure 18.3 compares the dependence of deviation from the stoichiometry on the oxygen pressure (fragments of defects' diagrams) in the oxide $(Fe_{1-x}Co_x)_{1-\delta}O$ at 1473 K (presented in Figure 18.1a–e). As can be seen in Figure 18.3, there occurs a clear change in the character of dependence of the deviation from the stoichiometry on p_{O_2}, which, as it was mentioned, similarly as for pure wüstite, also for cobalt-doped wüstite, indicates the presence of w_1 and w_2 pseudo-phases, differing by the concentration and ordering of defects and defect complexes. The range of existence of w_2 pseudo-phase is much wider than in the oxides $Fe_{1-\delta}O$ and $(Fe_{1-x}Co_x)_{1-\delta}O$, especially above the cobalt content of x_{Co} = 0.8 mol/mol. At the cobalt content above 0.9 mol/mol, as it results from the phase system (Zhang and Chen 2013), the decay pressure is higher than 1 atm of oxygen (see Figure 18.5). Figure 18.3 shows also the results of the studies of deviation from the stoichiometry (Aukrust and Muan 1964) for the cobalt content of x_{Co} = 0.6, 0.8 and 0.925 mol/mol ((▲) points) and (Reader et al. 1984) ((◆,□) points) for the cobalt content of x_{Co} = 0.2, 0.81 and 0.94 mol/mol, which remain in a good agreement with the results reported in (Subramanian et al. 1994b). The discrepancies are present only at the cobalt content of x_{Co} = 0.81 mol/mol in the range of existence of w_2 pseudo-phase, which might be related to a lower concentration of iron and its influence on the deviation from the stoichiometry, especially in the w_2 pseudo-phase.

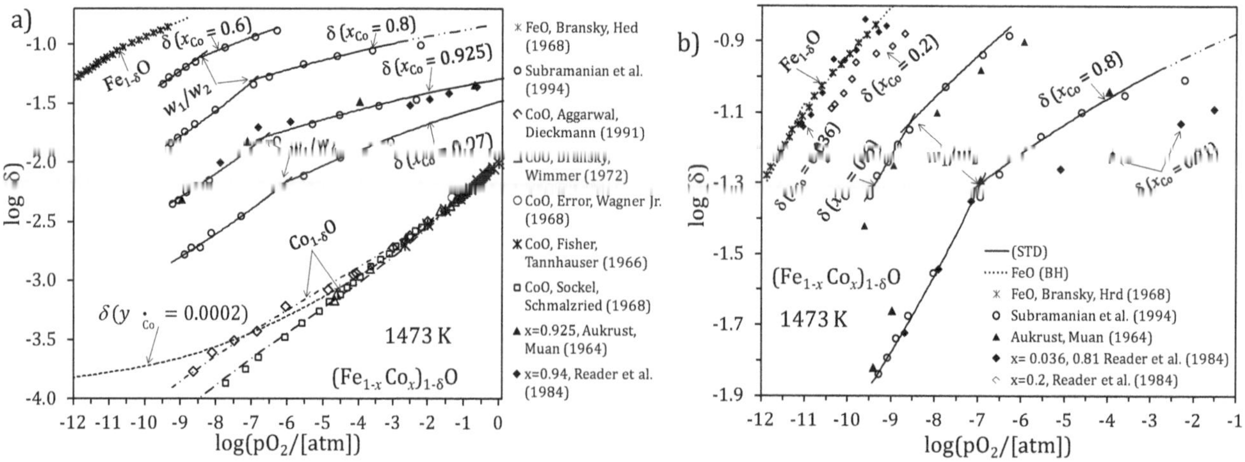

FIGURE 18.3 Fragments of diagrams of concentrations of point defects presenting the dependences of the deviation from the stoichiometry on the oxygen pressure at 1473 K a) in the oxide $Fe_{1-\delta}O$, $Co_{1-\delta}O$ and $(Fe_{1-x}Co_x)_{1-\delta}O$ with the cobalt content of x_{Co} = 0.6–0.97 mol/mol, b) in the oxide $Fe_{1-\delta}O$ and $(Fe_{1-x}Co_x)_{1-\delta}O$ with the manganese content of x_{Mn} = 0.03–0.8 mol/mol, obtained using the results of the studies on the deviation from the stoichiometry reported in (Subramanian et al. 1994b) – (○), (Aukrust and Muan 1964) for x_{Co} = 0.925, 0.6, 0.8 – (▲), (Reader et al. 1984) for x_{Co} = 0.94, 0.81, 0.2 – (◆,◇) and 0.03 – (□), for $Fe_{1-\delta}O$ (Bransky and Hed 1968) – (∗) and $Co_{1-\delta}O$ taken from (Sockel and Schmalzried 1968) – (□) points, (Fisher and Tannhauser 1966) – (×), (Eror and Wagner Jr. 1968) – (∗), (Eror and Wagner Jr. 1968) – (◓), (Bransky and Wimmer 1972) – (△), (Aggarwal and Dieckmann 1991) – (◇).

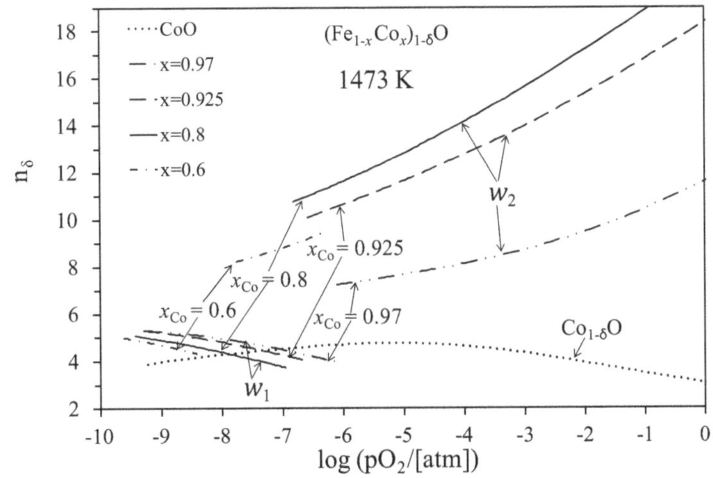

FIGURE 18.4 Oxygen pressure dependence of the derivative: $(d\log p_{O_2}/d\log\delta) = n_\delta$ at 1473 K, obtained for the dependence of the deviation from the stoichiometry on the oxygen pressure in the oxide $Co_{1-\delta}O$ and $(Fe_{1-x}Co_x)_{1-\delta}O$ with the manganese content of x_{Co} = 0.6–0.97 mol/mol for the "pseudo-phases" w_1 and w_2 (presented in Figure 18.3).

In order to compare the character of the dependence from the stoichiometry on the oxygen pressure in the oxides $(Fe_{1-x}Co_x)_{1-\delta}O$, Figure 18.4 presents the oxygen pressure dependence of the derivative $(d\log p_{O_2}/d\log\delta) = n_\delta$. As can be seen in Figure 18.4, in the range of existence of w_1 pseudo-phase, changes in the value of the derivative n_δ are small and when the oxygen pressure decreases, they increase from n_δ = 4–5. This results from the decrease in the concentration of cation vacancies with lower ionisation degree and defect complexes when the oxygen pressure decreases. The value of the derivative n_δ changes in a similar range for cobalt oxide, which, as can be seen, exists in a much wider range of oxygen pressures. When the oxygen pressure decreases, the derivative increases from n_δ = 3, it reaches the maximum value of 4.8, and then it decreases, which is related to the increase in the concentration of interstitial cobalt ions. In the case of w_2 pseudo-phase, the values are large, and when the oxygen pressure increases, they increase from n_δ = 7–18. Thus, the exponents $1/n_\delta$ in the dependence of δ on p_{O_2} are very small, but similar as in pure wüstite (see Figure 17.8). As it was mentioned, such a large change in n_δ results mainly from the change in concentration of cation vacancies when the deviation from the stoichiometry increases, due to the dependence of $\Delta G^o_{V''_M}$ on δ and from quite a wide range of existence of w_2 pseudo-phase.

Figure 18.5 presents the range of oxygen pressures where the oxides $(Fe_{1-x}Co_x)_{1-\delta}O$ exist, at 1473 K, as well as the oxygen pressure at which the phase transition w_1/w_2 occurs.

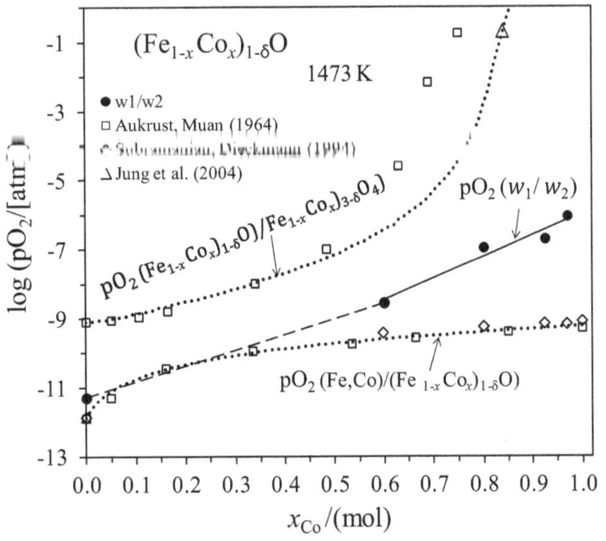

FIGURE 18.5 Dependence of the oxygen pressure at which there occurs the "phase" transition w_1/w_2 on the cobalt content in the oxide $(Fe_{1-x}Co_x)_{1-\delta}O$ at 1473 K ((●) points). Dotted lines denote the equilibrium oxygen pressures at the phase boundary (Fe,Co)/$(Fe_{1-x}Co_x)_{1-\delta}O$ and $(Fe_{1-x}Co_x)_{1-\delta}O/(Fe_{1-x}Co_x)_{3+\delta}O_4$ suggested in (Zhang and Chen 2013). The points denote the equilibrium oxygen pressures extracted from: (Aukrust and Muan 1964) (□) points, (Jung et al. 2004) – (△), (Subramanian and Dieckmann 1994) – (◇).

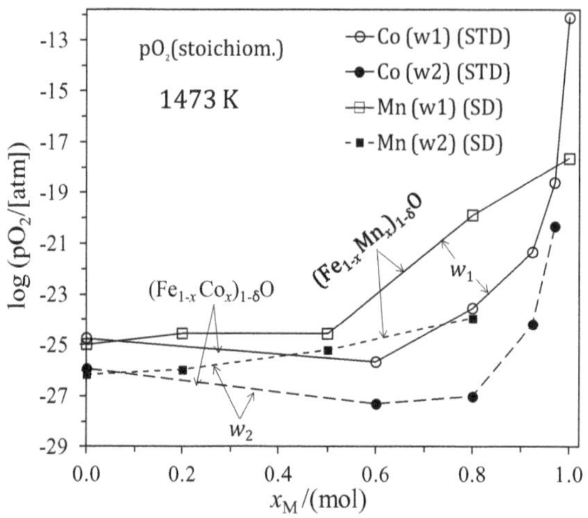

FIGURE 18.6 Dependence of the hypothetical oxygen pressure whereby the oxide $(Fe_{1-x}Co_x)_{1-\delta}O$ and $(Fe_{1-x}Mn_x)_{1-\delta}O$ could have reached the stoichiometric composition at 1473 K obtained using the values of the deviation from the stoichiometry reported in: (Subramanian et al. 1994b) ((○,●) (STD) points), (Subramanian and Dieckmann 1993) ((□,■) (SD) points) (empty points – w_1 "pseudo-phase", solid points – w_2 "pseudo-phase").

As can be seen in Figure 18.5, the oxygen pressure at which the "phase" transition w_1/w_2 occurs, increases when the cobalt content increases. When the cobalt content increases above $x_{Co} = 0.6$ mol/mol, the range of oxygen pressures where the oxides $(Fe_{1-x}Co_x)_{1-\delta}O$ exist gets wider; especially the equilibrium pressures at the oxide/spinel phase boundary increases; it reaches 100 atm (Zhang and Chen 2013).

Figure 18.6 shows the dependence of the hypothetical oxygen activity ($p_{O_2}^{(s)}$), at which the oxides $(Fe_{1-x}Co_x)_{1-\delta}O$ should reach the stoichiometric composition.

As can be seen, these pressures are by over 10 orders of magnitude lower than the decomposition pressure of oxides and when the cobalt content increases above $x_{Co} = 0.6$ mol/mol in the oxide $(Fe_{1-x}Co_x)_{1-\delta}O$, they increase (the difference between the pressure $p_{O_2}^{(s)}$ and the decomposition pressure of the oxide decreases). Higher decomposition pressures than the pressure at which the oxide would reach the stoichiometric composition result from strong repulsive interactions between ions in densely-packed cubic structure of NaCl type. It clearly decreases, but not earlier than above the cobalt content of $x_{Co} = 0.6$ mol/mol. It should be noted that the calculated pressures ($p_{O_2}^{(s)}$) depend on the assumed value of ΔG_{eh}^o of formation of electronic defects and on ΔG_F^o of formation of Frenkel defects. Moreover, the calculated equilibrium constants based on the fitted values of ΔG_i^o of formation of defects are the effective constants, being a product of their value and the mean activity coefficient of components in the range of higher oxygen pressures. Thus, the determined values of $p_{O_2}^{(s)}$ are approximative, however, as it results from the performed calculations, the differences should not be large, in the order of one order of magnitude.

Figure 18.7 presents the dependence of standard Gibbs energy ΔG_F^o of the formation of Frenkel defects on the cobalt content in the oxide $(Fe_{1-x}Co_x)_{1-\delta}O$ ((○,●) (STD) points). As previously mentioned, it was assumed that these values change in a linear way with the cobalt concentration in wüstite. For pure oxides $Fe_{1-\delta}O$ and $Co_{1-\delta}O$ the values determined in the work of Stokłosa (2015) were used. For w_1 and w_2 pseudo-phases, the same values of ΔG_F^o were assumed.

Figure 18.8 presents the dependence of the fitted values $\Delta G_{V_M''}^o$ of formation of cation vacancies on the cobalt concentration in the oxide $(Fe_{1-x}Co_x)_{1-\delta}O$ at 1473 K. As can be seen in Figure 18.8, when the cobalt concentration in the oxide $(Fe_{1-x}Co_x)_{1-\delta}O$ increases, the values $\Delta G_{V_M''}^o$ increase monotonically.

In turn, Figure 18.9 presents, calculated according to Equation 16.17, values of the standard Gibbs energy $\Delta G_{M_i^{\bullet\bullet}}^o$ of formation of interstitial cations depending on the cobalt content in the oxide $(Fe_{1-x}Co_x)_{1-\delta}O$, at 1473 K. As can be seen in Figure 18.9, the values $\Delta G_{M_i^{\bullet\bullet}}^o$ are high and positive and when the cobalt content in the oxide $(Fe_{1-x}Co_x)_{1-\delta}O$ increases, they decrease; the changes are approximately symmetrical related to the changes of the value of $\Delta G_{V_M''}^o$.

In turn, Figure 18.10 presents the adjusted values of $\Delta G_{V_M'}^o = \Delta G_{V_M^x}^o$ of formation of cation vacancies V_M' and V_M^x and defect complexes with lower ionisation degrees, depending on the cobalt content in the oxide $(Fe_{1-x}Co_x)_{1-\delta}O$ at 1473 K.

As can be seen in Figure 18.10, when the cobalt content in the oxide $(Fe_{1-x}Co_x)_{1-\delta}O$ increases, the above values decrease. The values $\Delta G_{V_M'}^o = \Delta G_{V_M^x}^o$ in the range of existence of w_2 pseudo-phase are positive. This results from a low concentration of vacancies with lower ionisation degree (this is the maximum concentration that does not affect the degree of match between the dependence of δ on p_{O_2} and the experimental results at the maximum concentration of defect complexes).

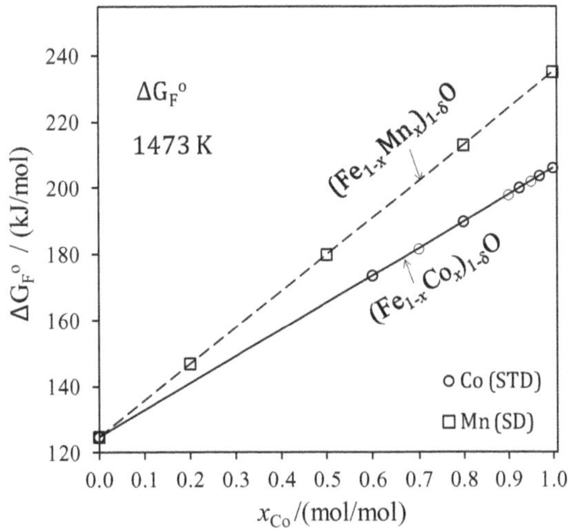

FIGURE 18.7 Dependence of standard Gibbs energy ΔG_F^o of formation of intrinsic Frenkel defects in the oxide $(Fe_{1-x}Co_x)_{1-\delta}O$ ((○) (STD) points) and $(Fe_{1-x}Mn_x)_{1-\delta}O$ ((□) (SD) points), at 1473 K, on the content of cobalt or manganese, obtained using the values of ΔG_F^o for the oxides $Fe_{1-\delta}O$, $Co_{1-\delta}O$ and $Mn_{1-\delta}O$, determined in the work of Stokłosa (2015).

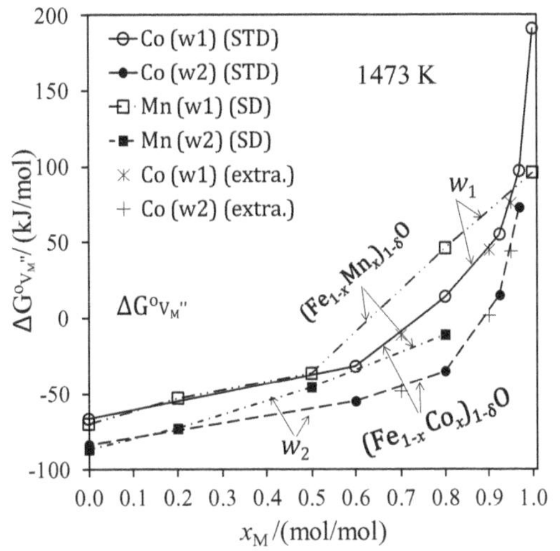

FIGURE 18.8 Dependence of standard Gibbs energy $\Delta G_{V_M''}^o$ of formation of cation vacancies V_M'' in the oxide $(Fe_{1-x}Co_x)_{1-\delta}O$ and $(Fe_{1-x}Mn_x)_{1-\delta}O$ on the content of cobalt or manganese, at 1473 K, obtained using the values of the deviation from the stoichiometry extracted from Subramanian et al. (1994b) ((○,●) (STD) points), Subramanian and Dieckmann (1993) – (□,■) (SD) (empty points – w_1 "pseudo-phase", solid points – w_2 "pseudo-phase")

Figure 18.11 presents the fitted values of standard Gibbs energy of formation of the several defect complexes ΔG_C^o depending on the cobalt content in the oxide $(Fe_{1-x}Co_x)_{1-\delta}O$ at 1473 K a) for w_1 pseudo-phase and b) for w_2 pseudo-phase.

As can be seen in Figure 18.11, the values of ΔG_C^o of formation of complexes $\left\{\left(V_M''\right)_2 M_i^{3\bullet}\right\}$ in w_1 pseudo-phase decrease when the cobalt content increases. On the other hand, the values of ΔG_C^o in w_2 pseudo–phase are lower, but the character of the dependence on the cobalt content is similar. For the remaining defect complexes the changes are smaller. These are maximum values; thus, the actual values and the character of the dependence can be different.

Figures 18.7–18.11 present, for comparison, the values of the standard Gibbs energies of formation of individual defects depending on the manganese content in the oxide $(Fe_{1-x}Mn_x)_{1-\delta}O$ at 1473 K. As can be seen, the influence of cobalt and manganese ions on the standard Gibbs energies of formation of defects in doped wüstite is similar, but the character of the above dependence on the dopant content is significantly different, especially in the range of higher concentrations of cobalt. It could therefore be concluded that the influence of iron ions on defects in the cobalt oxide is stronger than the influence on defects in the manganese oxide.

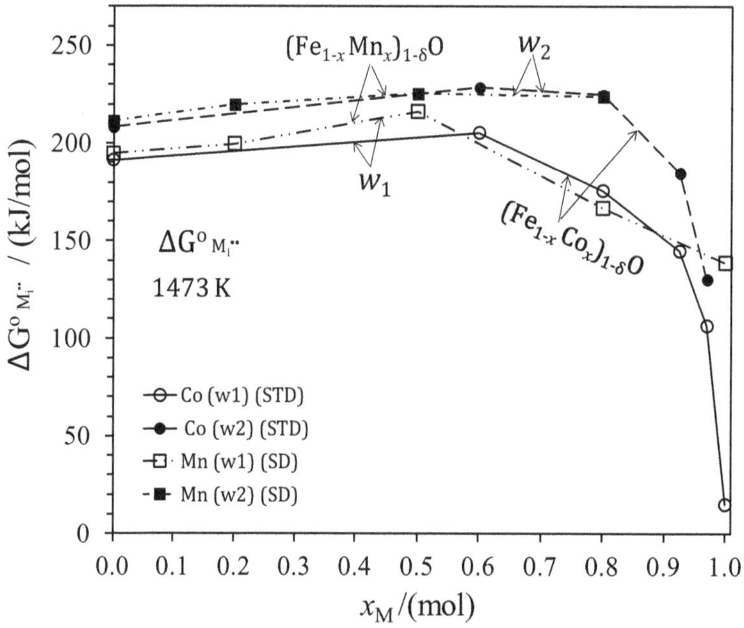

FIGURE 18.9 Dependence of the standard Gibbs energy $\Delta G^{o}_{M_i^{\bullet\bullet}}$ of formation of interstitial cations $M_i^{\bullet\bullet}$, calculated according to Equation 17.3 in the oxide $(Fe_{1-x}Co_x)_{1-\delta}O$ and $(Fe_{1-x}Mn_x)_{1-\delta}O$ on the content of cobalt or manganese, at 1473 K, obtained using the values of the deviation from the stoichiometry reported in Subramanian et al. (1994b) – (○,●) (STD) points, (Subramanian and Dieckmann (1993) – (□,■) (SD)) (empty points – w_1 "pseudo-phase", solid points – w_2 "pseudo-phase").

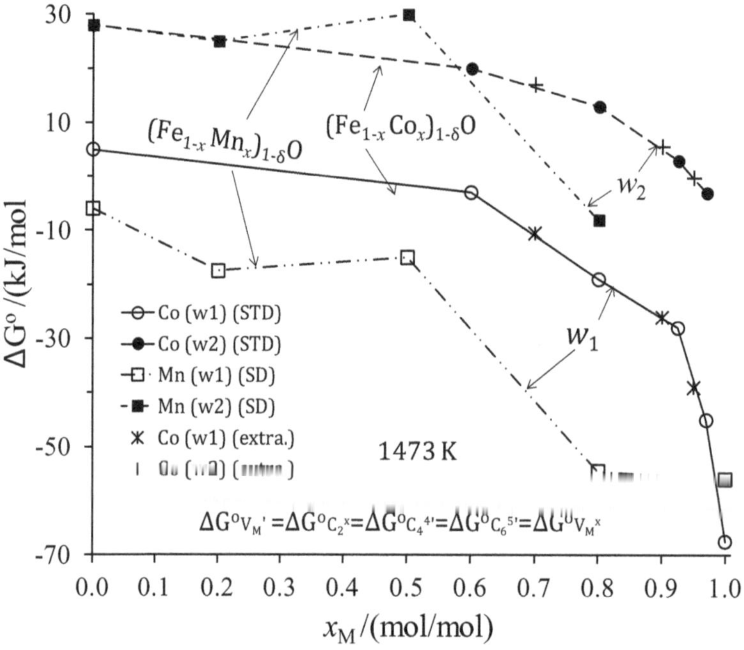

FIGURE 18.10 Dependence of $\Delta G^{o}_{V_M'} = \Delta G^{o}_{V_M^x}$ of formation of cation vacancies with lower ionisation degrees and of defect complexes in the oxide $(Fe_{1-x}Co_x)_{1-\delta}O$ and $(Fe_{1-x}Mn_x)_{1-\delta}O$ on the content of cobalt or manganese x_{Mn}, at 1473 K, obtained using the values of the deviation from the stoichiometry taken from: (Subramanian et al. 1994b) – (○,●) (STD) points, (Subramanian and Dieckmann 1993) – (□,■) (SD)) (empty points – w_1 "pseudo-phase", solid points – w_2 "pseudo-phase").

18.4 DIFFUSION IN COBALT-DOPED WÜSTITE

If we assume that the diffusion of metal ions in the oxide $(Fe_{1-x}Co_x)_{1-\delta}O$ occurs via cation vacancies, then, using their concentration resulting from the diagram of concentrations of defects and the values of coefficients of self-diffusion of a tracer (radioisotope), we can determine the coefficients of diffusions of defects $D^{o}_{V(M)}$, or, more precisely, the mobility of a

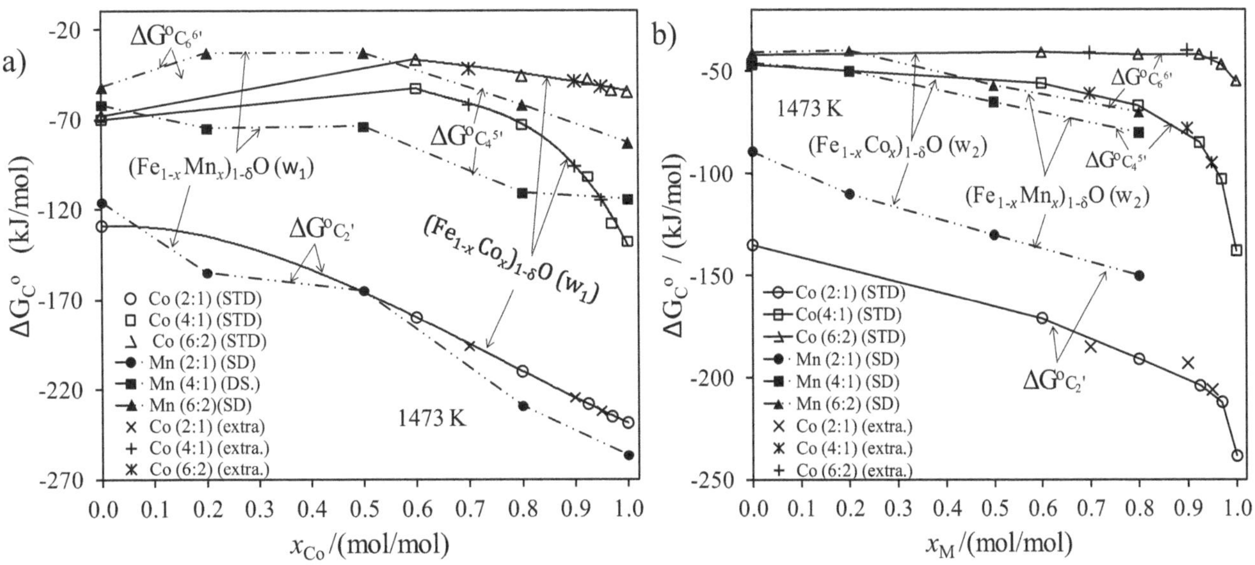

FIGURE 18.11 Dependence of $\Delta G^o_{C_2'}$, $\Delta G^o_{C_4^{5'}}$, $\Delta G^o_{C_6^{6'}}$ of formation of complexes $\{(V''_M)_2 M_i^{3\bullet}\}' - (C_2')$ ((○,●) points), $\{(V''_M)_4 M_i^{3\bullet}\}^{5'} - (C_4^{5'})$ – (□,■) and $\{(V''_M)_6 (M_i^{3\bullet})_2\}^{6'} - (C_6^{6'})$ – (▲,△) in the oxide $(Fe_{1-x}Co_x)_{1-\delta}O$ and $(Fe_{1-x}Mn_x)_{1-\delta}O$ on the content of cobalt or manganese, at 1473 K, obtained using the values of the deviation from the stoichiometry extracted from Subramanian et al. (1994b), Subramanian and Dieckmann (1993); a) for "pseudo-phase" w_1, b) for "pseudo-phase" w_2.

given type of tracer ion via cation vacancies. This is because according to Equation 17.9b we can adjust such a value of the coefficient $D^o_{V(M)}$ that the values of the sum of concentrations of vacancies $\sum_i [V_M]$ calculated using the values of coefficients of self-diffusion of tracer M* (D^*_M) are consistent with the dependence of this sum on p_{O_2}, resulting from the diagram of point defects (see Section 17.2).

Figure 18.12 shows the oxygen pressure dependence of the sum of quasi-free vacancies ($[V_M] = ([V''_M] + [V'_M] + [V^x_M])$ (solid lines) and concentrations of a sum of all the cation vacancies, also these forming defect complexes ($[V_M] + n[C_i]$) (dashed lines) at 1473 K, (a) in the oxide $Fe_{1-\delta}O$ and $(Fe_{1-x}Co_x)_{1-\delta}O$ with the cobalt content of x_{Co} = 0.7 and 0.8 mol/mol, (b) in the oxide $Co_{1-\delta}O$ and $(Fe_{1-x}Co_x)_{1-\delta}O$ with the cobalt content of x_{Co} = 0.9–1.0 mol/mol (resulting from the diagram of defects' concentrations), (c) in $Co_{1-\delta}O$, at the temperature of 1273–1673 K (resulting from diagrams published in Stokłosa (2015), based on the results reported in Bransky and Wimmer (1972), Eror and Wagner Jr. (1968), Fisher and Tannhauser (1966) and Sockel and Schmalzried (1968). For the cobalt content of x_{Co} = 0.7, 0.9 and 0.95 mol/mol, the concentration of cation vacancies was based on the extrapolated values of ΔG^o_i of formation of defects (see Figures 18.7–18.11 ((∗,×,+) points). The points in Figure 18.12 denote the values of the concentration of cation vacancies calculated according to Equation 17.9b for the fitted value of $D^o_{V(M)}$, when using the values of the coefficients of self-diffusion of tracers: iron ^{59}Fe, for pure wüstite (Chen and Peterson 1975) ((◇) (ChP) points), for the oxide $(Fe_{1-x}Co_x)_{1-\delta}O$ (Tinkler et al. 1994), for ions: ^{59}Fe ((●,■,◆) points), ^{54}Mn – (∗,×,+) and ^{60}Co – (○,□,◇) and for $Co_{1-\delta}O$, extracted from (Dieckmann 1977, Tinkler et al. 1994) for ions: ^{59}Fe ((▲) points), ^{54}Mn – (∗) and ^{60}Co – (△,■), (Hoshino and Peterson 1985) for ions: ^{59}Fe ((●) points) and ^{60}Co – (▲) and (Martin and Dorris 1987) for ^{59}Fe ions ((◆) points). As can be seen in Figure 18.12 for wüstite and for oxides $(Fe_{1-x}Co_x)_{1-\delta}O$ with the cobalt content of x_{Mn} = 0.6–1.0 mol/mol, the oxygen pressure dependence of the concentration of cation vacancies $[V_M]$ (solid lines) significantly deviates from the dependence of the sum of concentrations of all cation vacancies ($[V_M]+n[C_i]$) (dashed lines) on p_{O_2}. When the cobalt content in the oxide $(Fe_{1-x}Co_x)_{1-\delta}O$ increases, the differences between the above dependences decrease, and in the case of cobalt oxide they are already small, which results from a decrease in the concentration of defect complexes. In the range of existence of w_1 pseudo-phase, the concentrations of cation vacancies, calculated according to Equation 17.9b, when using the values of coefficients of self-diffusion of tracers and the fitted value of $D^o_{V(M)}$, are consistent with the dependence of the sum of concentrations of quasi-free cation vacancies $[V_M]$ on p_{O_2}. In several oxides, in the range of existence of w_1 phase there is only one measurement point and it is difficult to draw conclusions.

As can be seen in Figure 18.12a, for pure wüstite, the calculated values of concentration of vacancies, based on the coefficients of diffusion of Fe* tracer (Chen and Peterson 1975), are consistent with the dependence of the concentration of quasi-free cation vacancies which in the range of existence of w_2 pseudo-phase are practically independent of the oxygen pressure. In the oxide $(Fe_{1-x}Co_x)_{1-\delta}O$ with the cobalt content of x_{Co} = 0.6–0.8 mol/mol in the range of w_2 pseudo-phase, the

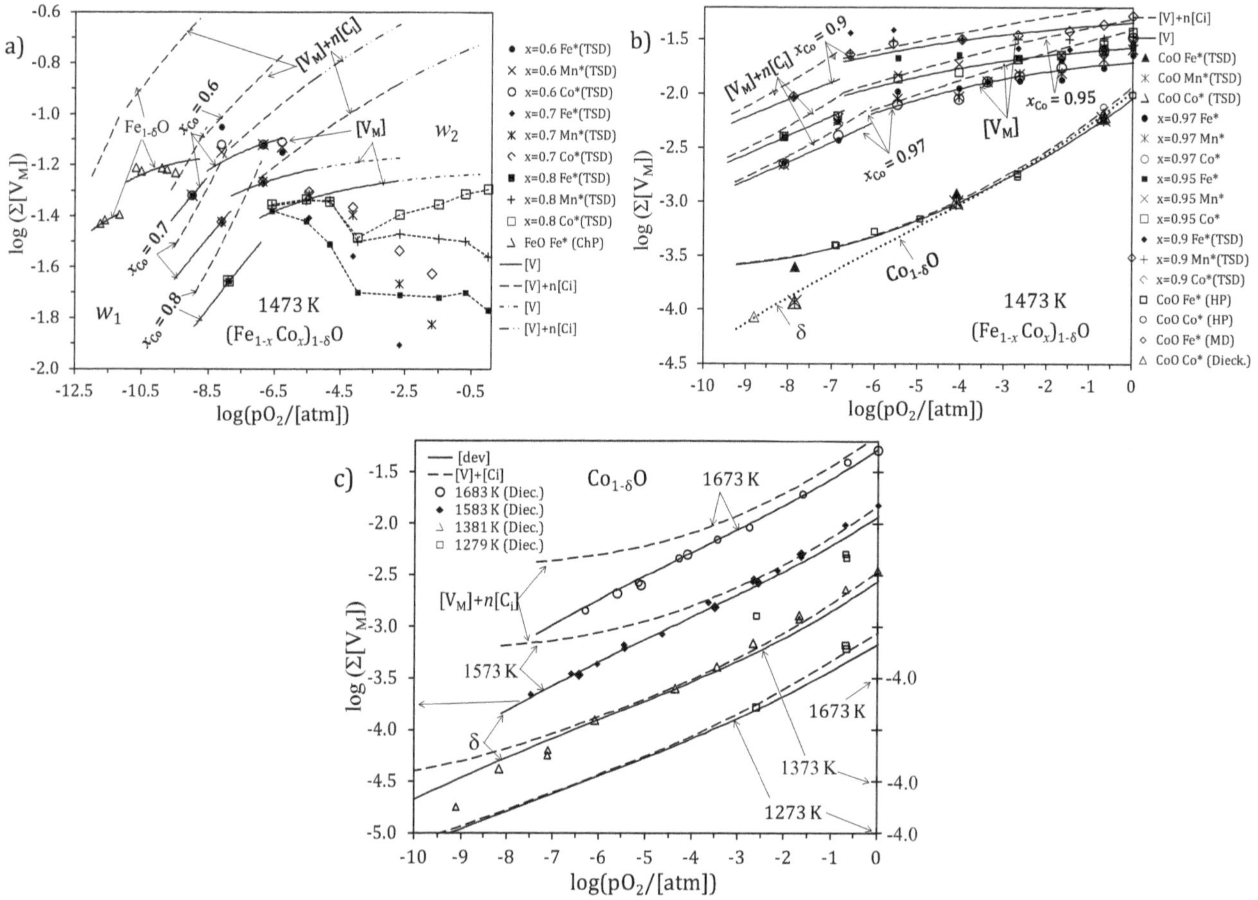

FIGURE 18.12 Oxygen pressure dependence of concentration of quasi-free cation vacancies $[V_M]$ (solid lines) and of the sum of concentrations of quasi-free vacancies and vacancies forming defect complexes ($[V_M]+n[C_i]$) (dashed line) a) in $Fe_{1-\delta}O$ and in the oxide $(Fe_{1-x}Co_x)_{1-\delta}O$ with the cobalt content of $x_{Co} = 0.6$–0.8 mol/mol, b) in $Co_{1-\delta}O$ and in the oxide with the cobalt content of $x_{Co} = 0.9$–0.97 mol/mol, at 1473 K (resulting from the diagram of concentrations of defects), c) in $Co_{1-\delta}O$ at 1273–1673 K, resulting from the diagrams of concentrations of defects presented in (Stokłosa 2015), based on the results reported in (Bransky and Wimmer 1972, Eror and Wagner Jr. 1968, Fisher and Tannhauser 1966, Sockel and Schmalzried 1968). The points denote the values of the concentrations of cation vacancies calculated according to Equation (17.9b) for the fitted value of $D^o_{V(M)}$, using the values of coefficients of self-diffusion of tracers M*, extracted from (Chen and Peterson 1975) for pure iron wüstite (^{59}Fe ((◇) (ChP) points), (Tinkler et al. 1994) for the oxide $(Fe_{1-x}Co_x)_{1-\delta}O$ (^{59}Fe ((●,■,◆) points), ^{54}Mn – (✳,×,+) and ^{60}Co – (O,□,◇) and for $Co_{1-\delta}O$ (Tinkler et al. 1994) (TSD) (^{59}Fe ((▲) points), ^{54}Mn – (✳) and ^{60}Co – (△), (Dieckmann 1977) (▣) (D), (Hoshino and Peterson 1985) (^{59}Fe ((◓) (HP) and ^{60}Co – (▲) (HP)), (Martin and Dorris 1987) –^{59}Fe (◈) (MD).

values of the concentration of cation vacancies, based on the coefficients of self-diffusion of tracers, show large discrepancies. For the oxide with the cobalt content of $x_{Co} = 0.6$ mol/mol, in a relatively narrow range of oxygen pressures, the calculated values of concentration of vacancies are within the error limit close to the dependence of $[V_M]$ on p_{O_2}. For the oxide with the cobalt content of $x_{Co} = 0.7$ and 0.8 mol/mol, the calculated values of the concentration of cation vacancies drastically deviate from the oxygen pressure dependence of their concentration, resulting from the diagram of defects' concentrations. The discrepancies might be related to exceeding the equilibrium oxygen pressure at the $(Fe_{1-x}Co_x)_{1-\delta}O$/spinel phase boundary. For the oxide with the cobalt content of $x_{Co} = 0.7$ mol/mol this pressure could be $4.5 \cdot 10^{-6}$ atm ($\log p_{O_2} = -5.35$) (Zhang and Chen 2013). Therefore, a decrease in the value of the calculated concentration of vacancies in the range of w_2 pseudo-phase (Figure 18.12a) might result from non-homogeneity of the sample due to its continuous transition into spinel. For the cobalt content of $x_{Co} = 0.8$ mol/mol, the range of existence is wider, up to $1.6 \cdot 10^{-3}$ atm ($\log p_{O_2} = -2.8$) and it is difficult to explain the occurring discrepancies. Thus, the problem of the equilibrium oxygen pressures at the $(Fe_{1-x}Co_x)_{1-\delta}O$/spinel phase boundary requires verification.

In turn, as can be seen in Figure 18.12b, in the oxide $(Fe_{1-x}Co_x)_{1-\delta}O$ with the cobalt content of $x_{Co} = 0.9$–1.0 mol/mol, in the range of existence of w_2 pseudo-phase, in the case of diffusion of ions Co* and Mn* within the error limit there is a good consistency between the values of concentration of cation vacancies based on the coefficient of diffusion of tracer and the dependence of the sum of concentrations of vacancies $[V_M]$ on p_{O_2}. The points are more scattered for the diffusion of Fe* ((◆) points) and the calculated concentration of cation vacancies based on the coefficients of diffusion of iron Fe* practically decreases when the oxygen pressure increases. The lowest discrepancies are present at the cobalt content of 0.97 mol/mol. However, a clear increase in the calculated values of the concentration of vacancies cam be seen near the oxygen pressure of 1 atm, which might be a result of an increase in the mobility of ions when the concentration of cation vacancies increases.

Despite the present discrepancies, the obtained results indicate that in the oxide $(Fe_{1-x}Co_x)_{1-\delta}O$, the transport of tracer ions occurs via quasi-free cation vacancies, not bound with the complexes.

As can be seen in Figure 18.12b and c, for the case of cobalt oxide, due to possible relatively low concentration of defect complexes, the difference between quasi-free cation vacancies and the total concentration of vacancies is small. At low oxygen pressures there is quite a big difference between the concentration of cation vacancies $[V_M]$ and the deviation from the stoichiometry, which is related to a significant concentration of interstitial cobalt ions $\left[Co_i^{\bullet\bullet}\right]$ (see Figure 18.1a). In turn, the calculated values of the concentration of cation vacancies based on the coefficients of diffusion of iron Fe* (Chen and Peterson 1980, Hoshino and Peterson 1985, Tinkler et al. 1994) are, within the error limit, consistent with the oxygen pressure dependence of the concentration $[V_M]$. The values of the concentration of cation vacancies determined based on the coefficients of diffusion of cobalt Co* and manganese Mn* (Dieckmann 1977, Tinkler et al. 1994) in the range of low oxygen pressures are consistent with the dependence of the deviation from the stoichiometry. The agreement is present especially at 1583 and 1683 K (see Equation 18.12c). Due to a very weak temperature dependence of deviation from the stoichiometry in cobalt oxide, no correction resulting from the difference between the temperatures of measurement of the deviation and the diffusion coefficients was introduced. This would indicate that at low oxygen pressures, due to interactions between interstitial cobalt cations and cobalt vacancies and the formation of, for example, complexes $\left[V_{Co}'',Co_i^{\bullet\bullet}\right]$, the effective concentration of cation vacancies via which there occurs ion diffusion, decreases.

Figure 18.13 presents the dependence of the coefficient of diffusion of tracer ions Co*, Mn* and Fe* via cation vacancies $D^o_{V(M)}$ (their mobility) on the cobalt concentration in the oxide $(Fe_{1-x}Co_x)_{1-\delta}O$. For the w_1 pseudo-phase, the values $D^o_{V(M)}$ determined for the lowest oxygen pressure are shown (empty points). In the range of existence of w_2 pseudo-phase (solid points) for the oxide $(Fe_{1-x}Co_x)_{1-\delta}O$ with the cobalt content of $x_{Co} = 0.9–1.0$ mol/mol, the values $D^o_{V(M)}$ were determined for the oxygen pressures in the middle of the range, and for the cobalt content of $x_{Co} = 0.6–0.8$ mol/mol for the lowest oxygen pressures.

As can be seen in Figure 18.13, due to a single measurement point in the range of existence of w_1 pseudo-phase, it is difficult to conclude about the extent of error of the determined mobility of ions $D^o_{V(M)}$. However, the mobilities of individual tracer ions in the oxide $(Fe_{1-x}Co_x)_{1-\delta}O$ with the cobalt content of $x_{Co} = 0.6–0.9$ mol/mol clearly differ; the highest mobility is shown by iron ions, the lowest – by cobalt ions. For Co* and Mn* ions, their mobilities are practically

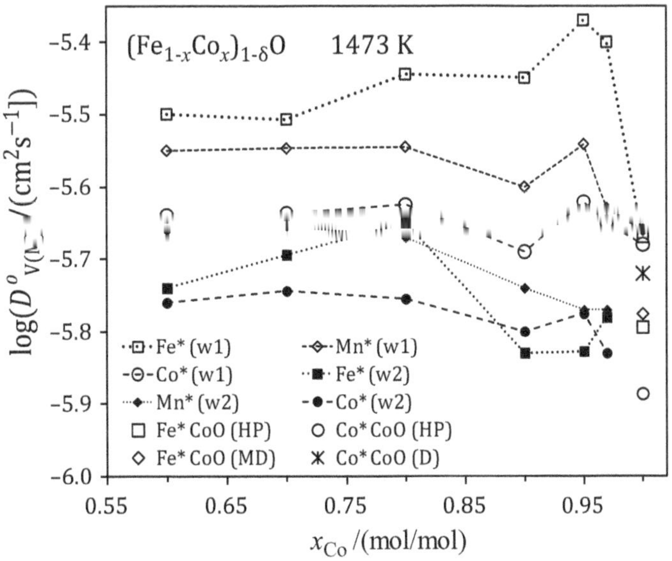

FIGURE 18.13 Dependence of the coefficient of diffusion $D^o_{V(M)}$ of tracer ions ^{59}Fe ((□,■) points), ^{54}Mn – (◇,◆) and ^{60}Co – (○,●)) via cation vacancies on the cobalt concentration in the oxide $(Fe_{1-x}Co_x)_{1-\delta}O$, in w_1 "pseudo-phase" (empty points) and w_2 "pseudo-phase" (solid points), obtained using the values of coefficients of self-diffusion (Tinkler et al. 1994) and in $Co_{1-\delta}O$, reported in: (Tinkler et al. 1994) ((□,◇,○) (TSD) points), (Chen and Peterson 1975) (^{59}Fe – (□) (HP) and ^{60}Co – (○) (HP)), (Dieckmann 1977) (^{60}Co – (✳) (D)), (Chen and Peterson 1980) (^{59}Fe – (◇) (MD)).

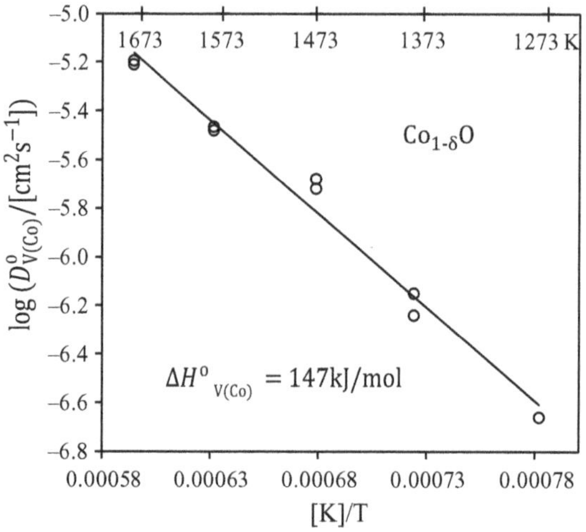

FIGURE 18.14 Temperature dependence of the coefficient of diffusion $D^{o}_{V(M)}$ of tracer ions ^{60}Co (points – (O)) via cation vacancies in the oxide $Co_{1-\delta}O$ in the temperature range of 1273–1673 K, obtained using the values of the coefficients of self-diffusion (Dieckmann 1977).

independent of the concentration of cobalt in the oxide, while the mobility of Fe* ions slightly increases. At higher cobalt contents, the mobilities of individual tracer ions decrease. The mobilities of tracer ions Fe*, Co* and Mn* in cobalt oxide, are, within the error limit, similar. The determined mobilities of tracer ions in the range of existence of w_2 pseudo-phase are lower, but the character of the dependence on the cobalt content in the oxide is similar. In the range of cobalt content of x_{Co} = 0.9–0.97 mol/mol, the mobilities of ions, within the error limit, are similar and equal $1.6 \cdot 10^{-6}$ cm²/s. Taking into account the discrepancies between the calculated values of the concentration of vacancies when using the coefficients of diffusion of tracers and the dependence of their concentration resulting from the diagram of concentrations of defects, the determined mobilities of tracer ions might be burdened with a significant error. Figure 18.13 presents also mobilities of Fe* and Co* ions in cobalt oxide, determined based on the studies on diffusion reported in (Chen and Peterson 1975, Chen and Peterson 1980, Dieckmann 1977), which, as can be seen, quite significantly differ.

Figure 18.14 presents the temperature dependence of the coefficient of diffusion $D^{o}_{V(Co)}$ of cobalt ions via cation vacancies in the cobalt oxide. From the obtained linear dependence, it results that the activation energy of the mobility of cobalt ions is 147 kJ/mol.

18.5 CONCENTRATION OF ELECTRONIC DEFECTS AND THEIR MOBILITY

As mentioned in Chapter 16, in iron oxide and cobalt oxide the dominating electronic defects are electron holes and the electrical conductivity depends on their concentration. Similarly, in the case of oxides $(Fe_{1-x}Co_x)_{1-\delta}O$, the electrical conductivity should be determined by the concentration of electron holes or it should depend on the concentration of $[M^{3+}]$ ions, if charge carriers are localised on these ions. Therefore, according to Equation 17.10c, we can adjust such a value of the mobility of holes μ_h that the values of the concentration of charge carriers (holes or $[M^{3+}]$ ions), based on the values of the electrical conductivity, are consistent with the dependence of the concentration of electron holes on the oxygen pressure dependence of the sum of concentration of ions $[M^{3+}]$, resulting from the diagram of concentrations of defects.

Figure 18.15 presents the dependence of the concentration of electron holes on the oxygen pressure in the oxide $(Fe_{1-x}Co_x)_{1-\delta}O$ with the cobalt content of x_{Co} = 0.0–1.0 mol/mol at 1473 K (solid lines), which were presented in the diagrams of concentrations of defects (Figure 18.1). Dashed lines (---) represent the dependence of the concentration of ions $[M^{3+}]$ on p_{O_2}. The points denote the values of the concentration of charge carriers calculated according to Equation 17.10e for the fitted value of μ_h when using the values of the electrical conductivity (Tinkler et al. 1994).

As can be seen in Figure 18.15, in the range of existence of w_1 pseudo-phase there is a significant difference between the concentration of electron holes and the concentration of ions $[M^{3+}]$; also the character (slope) of the above dependences is different. The values of the concentration of charge carriers calculated according to Equation (17.10e), using the values of the electrical conductivity, are consistent with the dependence of the concentration of electron holes on p_{O_2}. This indicates that the electrical conductivity in the oxide $Fe_{1-\delta}O$ and $(Fe_{1-x}Co_x)_{1-\delta}O$ at 1473 K in the range of existence of w_1 pseudo-phase depends on the concentration of electron holes and their transport occurs in the band, analogously to the case of wüstite and oxide $(Fe_{1-x}Mn_x)_{1-\delta}O$ with a low manganese content.

As can be seen in Figure 18.15a, in the range of existence of w_2 pseudo-phase in the oxide $(Fe_{1-x}Co_x)_{1-\delta}O$ with the cobalt content of x_{Co} = 0.7 and 0.8 mol/mol, the concentration of charge carriers, based on the conductivity, behaves in a not

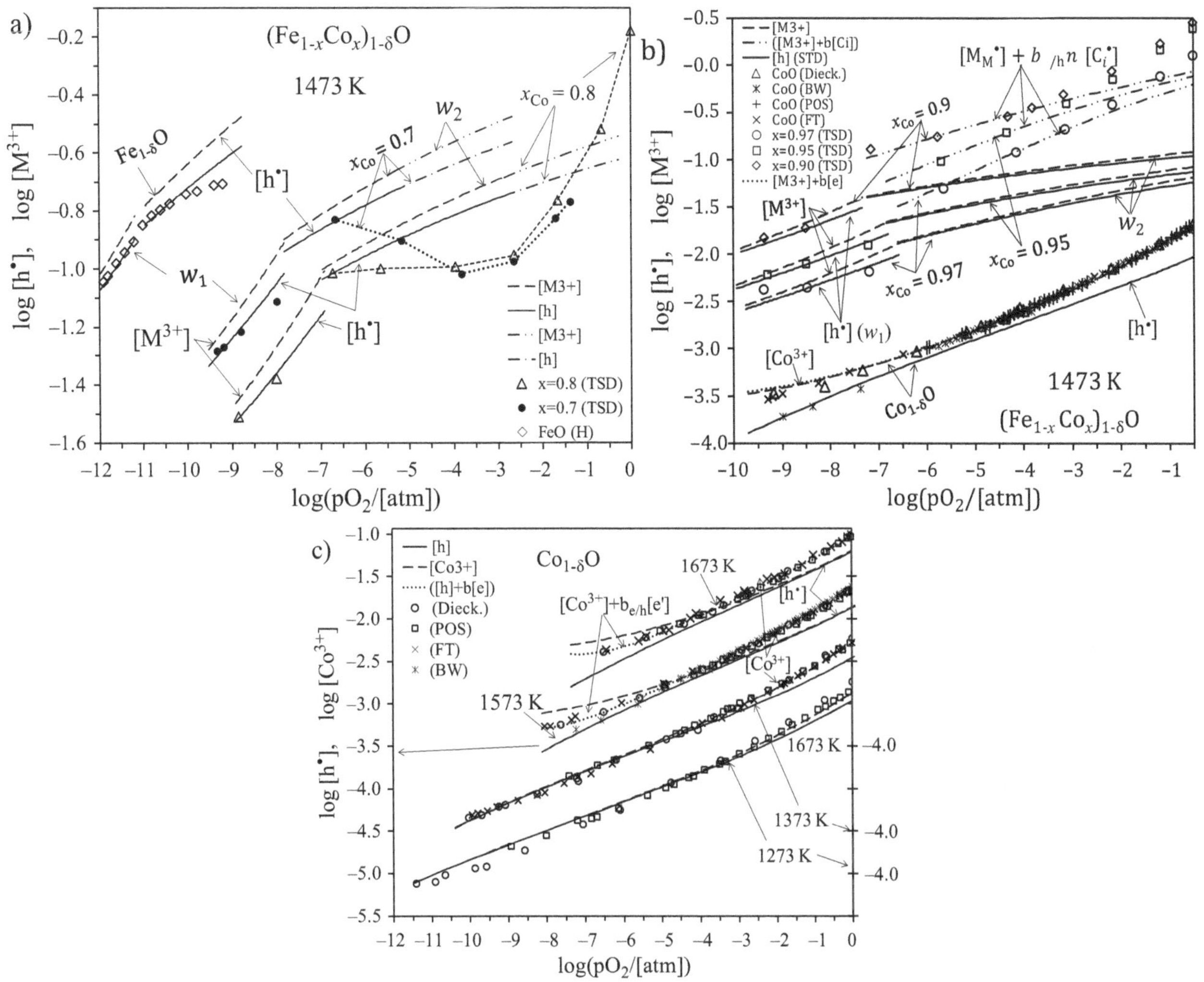

FIGURE 18.15 Dependence of the concentration of electron holes on p_{O_2} (solid lines) and the sum of concentration of ions $[M^{3+}]$ (dashed lines) at 1473 K a) in the oxide $Fe_{1-\delta}O$ and $(Fe_{1-x}Co_x)_{1-\delta}O$ with the cobalt content of $x_{Co} = 0.7$ and 0.8 mol/mol, b) in the oxide $Co_{1-\delta}O$ and $(Fe_{1-x}Co_x)_{1-\delta}O$ with the cobalt content of $x_{Co} = 0.9–0.97$ mol/mol, c) in $Co_{1-\delta}O$, in the temperature range of 1273–1673 K (resulting from diagrams of concentrations of defects (Stokłosa 2015), based on the results extracted from (Bransky and Wimmer 1972, Eror and Wagner Jr. 1968, Fisher and Tannhauser 1966, Sockel and Schmalzried 1968)). Dashed lines represent the oxygen pressure dependence of the sum $[M^{3+}]$, the (-···) lines the dependence of the sum $\left(\left[M_M^\bullet\right] + b_{C/h} n\left[C_i^\bullet\right]\right)$ for the fitted values of the parameter $b_{C/h}$, the dotted lines the dependence of the sum $\left(\left[M_M^\bullet\right] + b_{e/h}\left[e'\right]\right)$ for the fitted values of the parameter $b_{e/h}$. Points represent the values of the concentrations of charge carriers determined according to Equation (18.1b or 17.10c 17.10e) using the values of the electrical conductivity reported in (Tinkler et al. 1994) ((△) (TSD) points) and for $Fe_{1-\delta}O$ (Hillegas 1968) – (◇) (H), for $(Fe_{1-x}Co_x)_{1-\delta}O$ (Tinkler et al. 1994) ((○, □, △, ◇, ●) (TSD) points) and $Co_{1-\delta}O$ reported in (Bransky and Wimmer 1972) – (✳) (BW), (Fisher and Tannhauser 1966) – (×) (FT), (Dieckmann 1984) – (△) (D), (Petot-Ervas et al. 1984) – (+) (POS).

typical way when the oxygen pressure increases. For the oxide $(Fe_{0.3}Co_{0.7})_{1-\delta}O$, the decrease in the calculated concentration of charge carriers and then its increase can be related to exceeding the equilibrium pressure at the phase boundary $(Fe_{1-x}Co_x)_{1-\delta}O$/spinel and a transition of oxide into spinel. For the oxide $(Fe_{0.2}Co_{0.8})_{1-\delta}O$, the independence of the concentration of charge carriers of the oxygen pressure could be related to the increase in the concentration of defect complexes and the decrease in the concentration of electron holes, and, as a result, their independence of the oxygen pressure. However, it is not possible to adjust the values of ΔG_C° of formation of defects in such a way to obtain an independence of the concentration of electron holes of the oxygen pressure. Thus, it can be assumed that at 1473 K, in the oxide $(Fe_{1-x}Co_x)_{1-\delta}O$ with the cobalt content up to $x_{Co} = 0.8$ mol/mol, the charge transport would occur in a narrow band, similarly as in the case of wüstite.

As can be seen in Figure 18.15b, in the oxide $(Fe_{1-x}Co_x)_{1-\delta}O$ with the cobalt content of $x_{Co} = 0.9–0.97$ mol/mol, in the range of existence of w_2 pseudo-phase, the problem is very complex. Due to a significantly lower concentration of cation vacancies with lower ionisation degree and defect complexes, the differences between the concentration of electron holes and the total concentration of ions [M^{3+}] are small. As can be seen in Figure 18.15b, the concentrations of charge carriers, calculated according to Equation (17.10e), are much higher and the oxygen pressure dependence is much stronger ((○,□,◇) points). The match was obtained after assuming that the mobility of charge carriers via interstitial cations $M_i^{3\bullet}$ forming the defects complexes is higher than the mobility via cations $\left[M_M^{\bullet}\right]$ with octahedral coordination. Thus, the electrical conductivity can be presented with the equation:

$$\sigma = \frac{F}{V_{(\mathrm{Fe,Co})_{1-\delta}\mathrm{O}}}\left(\mu_h\left[M_M^{\bullet}\right] + \mu_C n\left[C_i^{\bullet}\right]\right) = \frac{F}{V_{(\mathrm{Fe,Co})_{1-\delta}\mathrm{O}}}\mu_h\left(\left[M_M^{\bullet}\right] + b_{C/h} n\left[C_i^{\bullet}\right]\right) \tag{18.1a}$$

where μ_h denotes the mobility of charge carriers via ions $M_M^{\bullet}$, (holes localised on these ions), with the concentration: $\left[M_M^{\bullet}\right] = \left(\left[h^{\bullet}\right] + \left[V_M'\right] + 2\left[V_M^x\right]\right)$; $n\left[C_i^{\bullet}\right]$ is the sum of concentrations of defect complexes containing n interstitial cations, $b_{C/h}$ denotes the ratio of the mobility of charge carriers via interstitial ions forming complexes to the mobility via ions $M_M^{\bullet}$ ($b_{C/h} = \mu_C/\mu_h$). Transforming Equation (18.1a) we obtain:

$$\left[M_M^{\bullet}\right] + b_{C/h} n\left[C_i^{\bullet}\right] = \frac{\sigma V_{(\mathrm{Fe,Co})_{1-\delta}\mathrm{O}}}{F\mu_h} \tag{18.1b}$$

The match between the calculated concentration of charge carriers and the dependence of $\left(\left[M_M^{\bullet}\right] + b_{C/h} n\left[C_i\right]\right)$ on the oxygen pressure was obtained at the values of $b_{C/h}$ about 100 (see Figure 18.16a). Therefore, in the oxide $(Fe_{1-x}Co_x)_{1-\delta}O$ with the cobalt content above $x_{Co} = 0.9$ mol/mol, in the range of existence of w_2 pseudo-phase, the mobility of charge carriers via interstitial ions $M_i^{3\bullet}$ with tetrahedral coordination would be 100 times higher than via M^{3+} ions with octahedral coordination. This effect was not found for pure wüstite, where an inverse dependence is present. At higher oxygen pressures, the electrical conductivity increases with the oxygen pressure, but weaker than the concentration of electron holes (see Figure 18.15a), which could be related to a decrease in the mobility when the deviation from the stoichiometry increases (Hillegas 1968, Molenda et al. 1987).

In turn, as can be seen in Figure 18.15b, the concentration of charge carriers in the oxide $Co_{1-\delta}O$, calculated based on the values of the electrical conductivity (Chen and Peterson 1975, Chen and Peterson 1980, Dieckmann 1977), are consistent with the dependence of the concentration of ions [Co^{3+}]. Such a large difference between the concentration of electron holes and the sum of concentration of ions [Co^{3+}] unambiguously indicates that the transport of charge carriers occurs according to the mechanism of small polarons and it depends on the concentration of cobalt ions [Co^{3+}]. Thus, it depends on the concentration of holes localised on ions $\left[Co_{Co}^{\bullet}\right]$ and also on the concentration of ions $\left[Co_{Co}^{\bullet}\right]$ forming cation vacancies with a lower ionisation degree and on the concentration of interstitial cobalt ions $Co_i^{3\bullet}$ forming defect complexes, and on the concentration of quasi-free interstitial ions $Co_i^{\bullet\bullet}$; their concentration is significant at low oxygen pressures (see Figure 18.1a). An increase in the concentration of charge carriers (also of the electrical conductivity) at low oxygen pressures does not exclude a contribution of the concentration of electrons to the total electrical conductivity, despite their low concentration (see Figure 18.1a). In Figure 18.15b, the dotted line presents the dependence of the sum of concentration of ions [Co^{3+}] and the concentration of electrons ([Co^{3+}] $+b_{e/h}$[e']). At the value of $b_{e/h} = 4$, thus, when the mobility of electrons is four times higher than the mobility of small polarons, the dotted curve starts to deviate from the dashed curve that does not take into account the concentration of electrons. The contribution of electron conductivity to the total electrical conductivity is visible at higher temperatures.

Figure 18.15c presents the oxygen pressure dependence of the concentration of electron holes and the concentration of ions [Co^{3+}], in $Co_{1-\delta}O$, at 1273–1673 K (resulting from diagrams of concentrations of defects (Stokłosa 2015) and based on the results of the electrical conductivity reported in: (Fisher and Tannhauser 1966) – (×) (FT), (Bransky and Wimmer 1972) – (✳) (BW), (Dieckmann 1984) – (△) (D), (Petot-Ervas et al. 1984) – (□,+) (POS). As can be seen in Figure 18.15c, a very good agreement was obtained between the concentration of charge carriers calculated based on the values of electrical conductivity (Equation 17.10c) and the concentration of ions [Co^{3+}] (the agreement is slightly worse at 1273 K). At the temperature of 1573 K and 1673 K, a clear contribution of electron conductivity is visible. An agreement of the oxygen pressure dependence of ([Co^{3+}] $+b_{e/h}$[e']) with the values calculated according to Equation (17.10c) was obtained for the values $b_{e/h} = 20$ and 30, respectively. In such a case the calculated mobility of electrons at 1573 K is $\mu_e = 4.5$ cm^2 V^{-1} s^{-1}, and at 1673 K it is about $\mu_e = 2$ cm^2 V^{-1} s^{-1}.

Figure 18.16 shows the dependence of the parameter $b_{C/h} = \mu_C/\mu_h$ on the cobalt content in the oxide $(Fe_{1-x}Co_x)_{1-\delta}O$ at 1473 K (see Equation (18.1b)). Figure 18.16b shows the temperature dependence of the values of $b_{e/h}$ for cobalt oxide (see Equation (17.10c)).

In turn, Figure 18.17 presents the dependence of the mobility of charge carriers via ions $M_M^{\bullet}$ on the cobalt content in the oxide $(Fe_{1-x}Co_x)_{1-\delta}O$, which decreases when the cobalt content increases, from $\mu_h = 0.12$ cm^2 V^{-1} s^{-1} in pure wüstite to

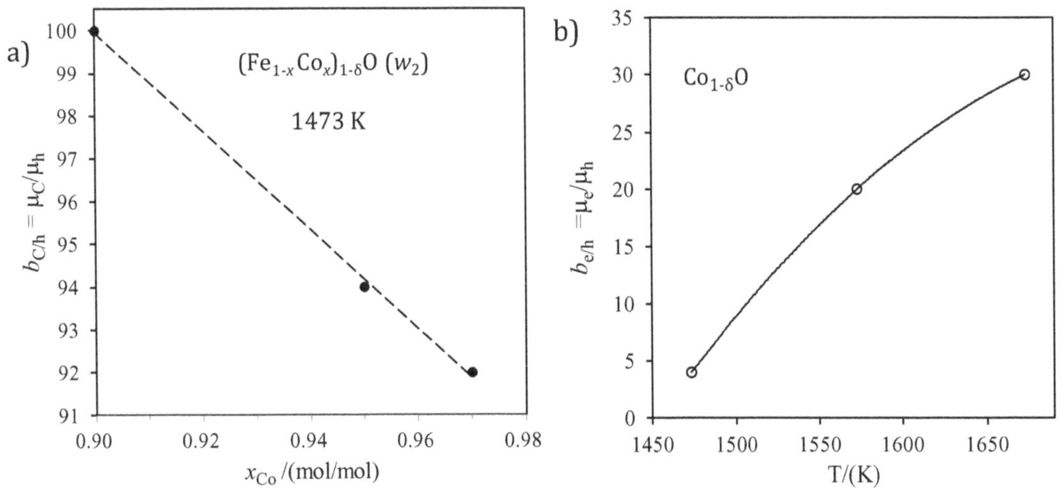

FIGURE 18.16 Dependence of the parameter $b_{C/h}$ being the ratio of the mobility of charge carriers via interstitial ions $M_i^{3\bullet}$ forming complexes to the mobility via ions $M_M^{\bullet}$ ($b_{C/h} = \mu_C/\mu_h$) on the cobalt content in the oxide $(Fe_{1-x}Co_x)_{1-\delta}O$ (see Equation (18.1b)), b) temperature dependence of the parameter $b_{e/h} = \mu_e/\mu_h$ being the ratio of the mobility of electrons to the mobility of electron holes for $Co_{1-\delta}O$ (see Equation (17.10c)).

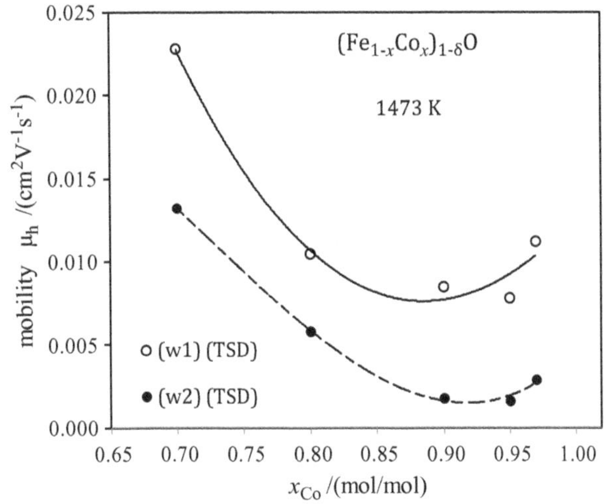

FIGURE 18.17 Dependence of the mobility of charge carriers (electron holes) μ_h on the cobalt content in the oxide $(Fe_{1-x}Co_x)_{1-\delta}O$ at 1473 K, obtained using the values of electrical conductivity (Tinkler et al. 1994) (○) points – w_1 "pseudo-phase", (●) points – w_2 "pseudo-phase") and for $Co_{1-\delta}O$ reported in (Bransky and Wimmer 1972) – (□), (Fisher and Tannhauser 1966) – (△).

0.01 cm^2 V^{-1} s^{-1} at the cobalt content of x_{Co} = 0.97 mol/mol. In the oxide $(Fe_{1-x}Co_x)_{1-\delta}O$ with the cobalt content above x_{Co} = 0.9 mol/mol in the range of existence of w_2 pseudo-phase, as was shown above, the mobility of charge carriers via interstitial cations $M_i^{3\bullet}$ (with tetrahedral coordination) is 100 times higher.

Figure 18.18a presents the dependence of the mobility of electron holes on the temperature in cobalt oxide; as can be seen, it decreases when the temperature increases, which indicates that during their transport, the effect of electron scattering on lattice ions dominates.

The mobility of electron holes, similarly as the electrical conductivity, should be therefore dependent on the temperature, according to the relation:

$$\mu = \frac{\alpha}{T^n} + \beta \qquad (18.2a)$$

After applying logarithm for Equation 18.2a we obtain:

$$\log(\mu - \beta) = \log\alpha + n\log T \qquad (18.2b)$$

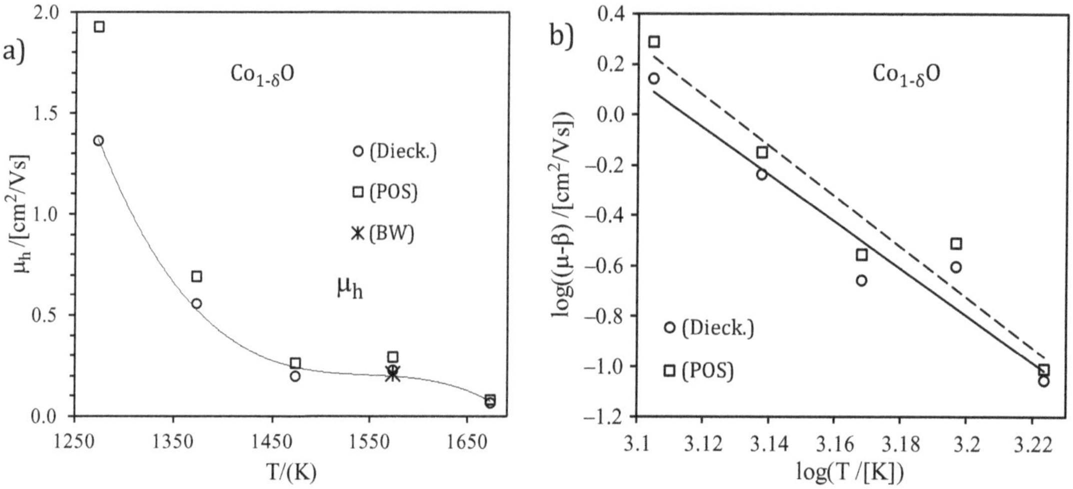

FIGURE 18.18 Dependence of the mobility of charge carriers (electron holes) μ_h in the cobalt oxide $Co_{1-\delta}O$, a) on the temperature b) in a log–log system, obtained using the values of electrical conductivity (Dieckmann 1984) ((O) (D) points), (Petot-Ervas et al. 1984) – (□) (POS).

Figure 18.18b presents the dependence of the mobility of electronic defects on the temperature in a log–log system. As can be seen, the points form a straight line quite well. If the parameter β had a small value, it would be possible to determine the parameter n characterising the temperature dependence of the mobility of holes associated with a scattering effect. The value of mobility depends on the measured values of the electrical conductivity, which are most often burdened with a systematic error resulting for example from the measurement of the sample geometry. Thus, apart from the constant β, there is also the value of the systematic error ($\Delta\mu$) of the determined mobility. In calculations, the value of the sum $(\beta + \Delta\mu) = \beta'$ (which is subtracted from the value of mobility $(\log[\mu_h - (\beta + \Delta\mu_h)])$) was chosen in such a way as to obtain the highest possible value of the linear regression coefficient R^2 (close to 1). The optimum value of the parameter β was 0.023 when using the results reported in Dieckmann (1984) ((O) (D) points), and $\beta = 0.0175$ (Petot-Ervas et al. 1984) – (□) (POS), and the determined values of the exponent n are, respectively, $n \cong 10\pm2$ and 9 ± 1.

18.6 CONCLUSIONS

Using the results of the studies of deviation from the stoichiometry for the oxide $(Fe_{1-x}Co_x)_{1-\delta}O$ with the cobalt content of $x_{Co} = 0.6$–1 mol/mol, at 1473 K, (Tinkler et al. 1994), diagrams of concentrations of defects were determined. In the calculations, simple defects (cation vacancies $\left[V_M''\right]\left[V_M'\right],\left[V_M^x\right]$ and $\left[M_i^{3\bullet}\right]$) and defect complexes $\left\{\left(V_M''\right)_2 M_i^{3\bullet}\right\}',\left\{\left(V_M''\right)_4 M_i^{3\bullet}\right\}^{5'}$ and $\left\{\left(V_M''\right)_6\left(M_i^{3\bullet}\right)_2\right\}^{6'}$ were taken into account. Similarly, as for pure wüstite, in the oxides $(Fe_{1-x}Co_x)_{1-\delta}O$ with the cobalt content of $x_{Co} = 0.6$–0.97 mol/mol it was shown that the oxygen pressure dependence of the deviation from the stoichiometry can be described with two different curves (with different slopes). The above fact indicates that also in cobalt-doped wüstite in the range of low oxygen pressures, the so-called w_1 pseudo-phase exists, and in the range of higher oxygen pressures, the w_2 pseudo-phase. Diagrams of concentrations of defects were determined independently in the range of existence of pseudo-phases w_1 and w_2. It was found that in the oxide $(Fe_{1-x}Co_x)_{1-\delta}O$ above the cobalt content of $x_{Co} = 0.6$ mol/mol, the range of its existence broadens, especially the range of existence of w_2 pseudo-phase enlarges. When the cobalt content in the oxide $(Fe_{1-x}Co_x)_{1-\delta}O$ increases, near the decomposition pressure, the deviation from the stoichiometry decreases and the concentrations of simple defects and defect complexes significantly decrease. The oxygen pressure at which the "phase transition" w_1/w_2 occurs in the oxide $(Fe_{1-x}Co_x)_{1-\delta}O$, up to the cobalt content of $x_{Co} = 0.97$ mol/mol increases monotonically.

From the calculations it results that the hypothetical activity of oxygen, at which the oxides $Fe_{1-x}Co_xO$ could have reached the stoichiometric composition is about 10^{-27} atm at the content of $x_{Co} = 0.6$ mol/mol (it is close to the value for pure wüstite) and it increases monotonically up to 10^{-12} atm in the case of CoO. At the cobalt content of $x_{Co} = 0.6$ mol/mol, these oxygen activities are by over 10 orders of magnitude lower than the decomposition pressure of the oxide. When the cobalt content in the oxide $(Fe_{1-x}Co_x)_{1-\delta}O$ increases further, the difference between the decomposition pressure of the oxide and the pressure at which it could have reached the stoichiometric composition, decreases by two orders of magnitude. Similarly, as for the oxide $(Fe_{1-x}Mn_x)_{1-\delta}O$, the reason for the above large differences is a strong repulsive interaction between ions, which leads to the decomposition of the oxide despite a significant concentration of cation vacancies that reduce this interaction. For spinels, which have the same cubic crystallographic structure of NaCl type but much "looser", the stoichiometric composition is reached in the middle of their range of existence.

From the performed calculations of the diagrams of the concentrations of defects it results that in the range of existence of the w_1 pseudo–phase, cation vacancies $\left[V_M''\right]$ dominate, but the deviation from the stoichiometry is significantly affected by the concentration of vacancies with lower ionisation degrees. The determined concentration of defect complexes in the oxide $(Fe_{1-x}Co_x)_{1-\delta}O$ is the maximum concentration that does not affect the degree of match between the dependence of the deviation from the stoichiometry on the oxygen pressure and the experimental values of δ. It is lower than the concentration of cation vacancies $\left[V_M''\right]$ by over one order of magnitude. From the performed calculations it results that in the oxide $(Fe_{1-x}Co_x)_{1-\delta}O$, the highest concentration of complexes $\left\{\left(V_M''\right)_2 M_i^{3\bullet}\right\}'$, $\left\{\left(V_M''\right)_4 M_i^{3\bullet}\right\}^{5'}$ is present; it is about 0.01 mol/mol and it decreases down to 0.001 when the cobalt content increases.

Similarly, as for pure wüstite, in the range of existence of w_2 pseudo-phase, the calculated oxygen pressure dependence of the deviation from the stoichiometry was higher than the experimental values of the deviation δ. The agreement was obtained when assuming a linear dependence of the standard Gibbs energy $\Delta G^o_{V_M''}$ of formation of cation vacancies on the deviation from the stoichiometry. In association with the above character of the oxygen pressure dependence of the deviation from the stoichiometry in the range of existence of w_2 pseudo-phase, the maximum concentration of defect complexes and the maximum concentration of vacancies with lower ionisation degree, which do not deteriorate the match between the oxygen pressure dependence of the deviation from the stoichiometry and the experimental values of δ were determined. Similarly, as in the case of w_1 pseudo-phase, the calculated concentration of cation vacancies $\left[V_M''\right]$ is only slightly lower than the deviation from the stoichiometry. The concentrations of vacancies with lower ionisation degrees, compared to the concentrations of these defects in the w_1 pseudo-phase, are lower. The concentration of defect complexes is similar to that in w_1 pseudo-phase (below 0.01 mol/mol), but the dependence on the oxygen pressure is much weaker. It should be noted that the concentration of complexes is high enough for their dense-packing in the oxide structure and the formation of larger defect clusters. As the concentration of defect complexes does not affect the dependence of the deviation from the stoichiometry on p_{O_2}, it is impossible to determine their concentration unambiguously. A limit case of an ordered structure of defect complexes is the spinel structure (see Section 1.1, Figure 1.1), where defect complexes lost the properties of point defects.

When calculating the diagrams of concentrations of defects it was impossible to unambiguously determine the standard Gibbs energy of formation of Frenkel defects ΔG^o_F; due to that it was assumed that the values of ΔG^o_F vary linearly when the cobalt content increases (for the oxides $Fe_{1-\delta}O$ and $Co_{1-\delta}O$, the values were determined in the work of Stokłosa (2015)).

The fitted standard Gibbs energies $\Delta G^o_{V_M''}$ of formation of cation vacancies in the oxide $(Fe_{1-x}Co_x)_{1-\delta}O$ monotonically increase when the cobalt content increases and they change from −70 to 190 kJ/mol.

In turn, the calculated values of standard Gibbs energies of formation of interstitial cations $\Delta G^o_{M_i^{\bullet\bullet}}$ in the oxide $(Fe_{1-x}Co_x)_{1-\delta}O$ are positive and they decrease from 190 to 10 kJ/mol when the cobalt content increases. The dependence of $\Delta G^o_{M_i^{\bullet\bullet}}$ on the cobalt content in the oxide $(Fe_{1-x}Co_x)_{1-\delta}O$ differs from the dependence of $\Delta G^o_{V_M''}$, but its character is approximately symmetrical. The changes in the value of ΔG^o_i of formation of defects in the oxide $(Fe_{1-x}Co_x)_{1-\delta}O$ are similar to those of the oxide $(Fe_{1-x}Mn_x)_{1-\delta}O$, but the character of the above dependences on the dopant content is quite significantly different, especially in the range of higher concentrations of cobalt. It could therefore be concluded that the influence of iron ions on defects in the cobalt oxide is stronger than the influence on defects in the manganese oxide.

The adjusted standard Gibbs energies $\Delta G^o_{V_M'} = \Delta G^o_{V_M^x}$ of formation of cation vacancies with lower ionisation degree in the oxide $(Fe_{1-x}Co_x)_{1-\delta}O$, in the range of existence of w_1 pseudo-phase change from −10 to 70 kJ/mol. In the range of existence of w_2 pseudo-phase, a similar character of the change in the value of ΔG^o_i of formation of vacancies with lower ionisation degrees and defect complexes was obtained, which will are negative and will differ from the actual values, as the determined concentration of these defects does not affect the oxygen pressure dependence of the deviation from the stoichiometry.

Using the values of coefficients of diffusion of tracers (^{59}Fe, ^{54}Mn and ^{60}Co) in oxides $(Fe_{1-x}Co_x)_{1-\delta}O$ at 1473 K (Tinkler et al. 1994), and the concentration of cation vacancies resulting from diagrams of concentrations of defects, the coefficients of diffusion of defects, and specifically the mobility of tracer ions via cation vacancies, were determined. From the performed calculations it results that the ion transport occurs via quasi-free cation vacancies (not bound with defect complexes). In the range of cobalt content of $x_{Co} = 0.6–0.8$ mol/mol in the oxide $(Fe_{1-x}Co_x)_{1-\delta}O$, the values $D^o_{V(M)}$ differ, but, within the error limit, they are independent of the cobalt content. The highest mobility is that of iron ions, the lowest is that of cobalt ions. When the cobalt content increases further, they decrease down to the value of mobility in the cobalt oxide (about $D^o_{V(M)} = 2 \cdot 10^{-6}$ cm²/s). It was also shown that the coefficients of diffusion of tracers in pure wüstite are dependent on the concentration of quasi-free cation vacancies; their concentration is practically independent of the oxygen pressure, due to a significant concentration of defect complexes. In the cobalt oxide, due to small concentration of defect complexes, it is difficult to unambiguously determine whether the cation vacancies bound with defect complexes participate in ion transport. However, at oxygen pressures near 1 atm, higher concentrations of vacancies were obtained, which might indicate a dependence (increase) of the coefficient of diffusion $D^{o}_{V(M)}$ on the concentration of defects. At low oxygen pressures, a lower concentration of cation vacancies was found (equal the deviation from the stoichiometry), based on the coefficients of diffusion of tracer ions Co* and Mn*. A decrease in the effective concentration of cation vacancies indicates an interaction between cobalt vacancies and interstitial cations and formation of, for example, complexes $\left[V''_{Co}, Co_i^{\bullet\bullet}\right]$.

In turn, using the values of the electrical conductivity in the oxides $(Fe_{1-x}Co_x)_{1-\delta}O$ with the cobalt content of x_{Co} = 0.6–0.97 mol/mol (Tinkler et al. 1994) and the concentration of charge carriers, their mobility was determined. The obtained result indicate that at 1473 K in the oxides $(Fe_{1-x}Co_x)_{1-\delta}O$ with the cobalt content of x_{Co} = 0.6–0.97 mol/mol, in the range of existence of w_1 pseudo-phase, charge transport occurs in the band, analogously to the case of the oxide $Fe_{1-\delta}O$ and $(Fe_{1-x}Mn_x)_{1-\delta}O$ with a low manganese content.

In the range of existence of w_2 pseudo-phase, the problem is very complex. In the oxide $(Fe_{1-x}Co_x)_{1-\delta}O$ with the cobalt content x_{Co} = 0.9–0.97 mol/mol, charge transport occurs according to the mechanism of small polarons and it depends on the concentration of ions $[M^{3+}]$, but, as it was shown, the mobility of charge carriers via interstitial cations forming defect complexes is about 100 times higher than the mobility via holes localised on ions $M^{\bullet}_{M}$, which is about $\mu_h = 0.01\ cm^2\ V^{-1}\ s^{-1}$. In the case of oxides $(Fe_{1-x}Co_x)_{1-\delta}O$ with the cobalt content of x_{Co} = 0.7 and 0.8 mol/mol, the obtained results do not allow drawing conclusions about the mechanism of charge transport and it is necessary to verify the range of existence of oxide phases and the results of the electrical conductivity.

In the oxide $Co_{1-\delta}O$, the transport of charge carriers depends, similarly, on the concentration of ions $[Co^{3+}]$, thus, on the concentration of cobalt ions $\left[Co^{\bullet}_{Co}\right]$ (holes localised on Co^{3+} ions), also those forming cation vacancies with a lower ionisation degree, and on the concentration of interstitial cobalt ions forming defect complexes and on the concentration of quasi-free interstitial cations, which is significant at low oxygen pressures. The determined mobility of electron holes decreases when the temperature increases, which indicates that in their transport, the effect of electron scattering on lattice ions dominates. At temperatures above 1473 K, at low oxygen pressures, the contribution of the electron conductivity to the total electrical conductivity becomes visible.

REFERENCES

Aggarwal, S., and R. Dieckmann. 1991. Non-Stoichiometry and Point Defect Structure in Cobaltous Oxide. *Ceram. Trans.* 24: 23–30.

Aukrust, E., and A. Muan. 1964. Thermodynamic Properties of Solid Solutions with Spinel-Type Structure. 2. System Co_3O_4-Fe_3O_4 at 1200°C. *Trans. Metall. Soc. AIME.* 230: 1395–1399.

Bransky, I., and A. Z. Hed. 1968. Thermogravimetric Determination of the Composition-Oxygen Partial Pressure Diagram of Wüstite ($Fe_{1-y}O$). *J. Am. Ceram. Soc.* 51: 231.

Bransky, I., and J. M. Wimmer. 1972. The High Temperature Defect Structure of CoO. *J. Phys. Chem. Solids.* 33: 801–812.

Chen, W. K., and N. L. Peterson. 1975. Effect of the Deviation from Stoichiometry on Cation Self-Diffusion and Isotope Effect in Wüstite, $Fe_{1-x}O$. *J. Phys. Chem. Solids.* 36: 1097–1097.

Chen, W. K., and N. L. Peterson. 1980. Isotope Effect for Cation Diffusion in CoO. *J. Phys. Chem. Solids.* 41: 647–652.

Dieckmann, R. 1977. Cobaltous Oxide Point Defect Structure and Non-Stoichiometry, Electrical Conductivity, Cobalt Tracer Diffusion. *Z. Phys. Chem. N. F.* 107: 189–210.

Dieckmann, R. 1984. Point Defects and Transport Properties of Binary and Ternary Oxides. *Solid State Ionics.* 12: 1–22.

Eror, N. G., and J. B. Wagner Jr. 1968. Electrical Conductivity and Thermogravimetric Studies of Single Crystalline Cobaltous Oxide. *J. Phys. Chem. Solids.* 29: 1597–1611.

Fisher, B., and D. S. Tannhauser. 1966. Electrical Properties of Cobalt Monoxide. *J. Chem. Phys.* 44: 1663–1671.

Hillegas, J. 1968. Seebeck Coefficient and Electrical Conductivity Measurements on Doped and Undoped Wustite. PhD Thesis. Northwestern University, Evanston, Ill.

Hoshino, K., and N. L. Peterson. 1985. Diffusion and Correlation Effects in Iron-Doped CoO. *J. Phys. Chem. Solids* 46: 229–240.

Jung, I. H., S. A. Decterov, P. D. Pelton, and H. M. Kim. 2004. Thermodynamic Evaluation and Modeling of the Fe-Co-O System. *Acta Mater.* 52: 507–519.

Lykasov, A. A., I. A. Maksutov, and V. I. Shiskov. 1976. Thermodynamics of Solid Solutions of Cobaltous Oxide and Wustite. *Izv. Vyssh. Uchebn. Zaved. Chern. Metall.* 4: 13–16.

Maksutov, I. A. 1974. Study of Thermodynamic Properties of Solid Wüstite Solutions in System Fe-Co-O. PhD Thesis, University of Chelyabinsk.

Martin, M., and S. Dorris. 1987. Impurity Diffusion in a Chemical Potential Gradient (II): Iron Tracer Diffusion in Cobalt Oxide in an Oxygen Potential Gradient. *Ber. Bunsenges. Phys. Chem.* 91: 779–785.

Molenda, J., A. Stokłosa, and W. Znamirowski. 1987. Transport Properties of Ferrous Oxide $Fe_{1-y}O$ at High Temperature. *Phys. Status Solidi (b).* 142: 517–529.

Morin, F., and R. Dieckmann. 1982. The Determination of Chemical Diffusivity in Cobaltous Oxide by Means of Electrical Conductivity. *Z. Phys. Chem. N. F.* 129: 219–237.

Petot-Ervas, G., P. Ochin, and B. Sossa. 1984. Transport Properties in Pure and Lithium-Doped Cobaltous Oxide. *Solid State Ionics.* 12: 277–293.

Reader, J. H., J. L. Holm, and O. T. Sørensen. 1984. Defects in Metal-Deficient Cobalt-Wüstites, $(Co, Fe)_{1-y}O$. *Solid State Ionics.* 12: 155–159.

Roiter, B. D., and A. E. Paladino. 1962. Phase Equilibria in the Ferrite Region of the System Fe-Co-O. *J. Am. Ceram. Soc.* 45: 128–133.

Sockel, H. G., and H. Schmalzried. 1968. Coulometrische Titration an Übergangsmetalloxiden. *Ber. Bunsenges. Phys. Chem.* 72: 745–754.

Stokłosa, A. 2015. Non-Stoichiometric Oxides of 3d Metals. Pfäffikon: Trans Tech Publications Ltd.

Subramanian, R., and R. Dieckmann. 1993. Nonstoichiometry and Thermodynamics of the Solid Solution $(Fe, Mn)_{1-\Delta}O$ at 1200°C. *J. Phys. Chem. Solids.* 54: 991–1000.

Subramanian, R., and R. Dieckmann. 1994. Thermodynamics of the oxide Solid Solution $(Co_xFe_{1-x})_{1-\Delta}O$ at 1200°C. *J. Phys. Chem. Solids.* 55: 59–67.

Subramanian, R., R. Dieckmann, G. Eriksson, and A. D. Pelton. 1994a. Model Calculations of Phase Stabilities of Oxide Solid Solutions in the Co-Fe-Mn-O System at 1200°C. *J. Phys. Chem. Solids.* 55: 391–404.

Subramanian, R., S. Tinkler, and R. Dieckmann. 1994b. Defects and transport in the solid solution $(Co,Fe)_{1-\delta}O$ at 1200°C I. Nonstoichiometry. *J. Phys. Chem. Solids.* 55: 69–75.

Tinkler, S., R. Subramanian, and R. Dieckmann. 1994. Defects and Transport in the Solid Solution $(Co,Fe)_{1-\delta}O$ at 1200°C - II. Cation Tracer Diffusion and Electrical Conductivity. *J. Phys. Chem. Solids.* 55: 273–286.

Zhang, W. W., and M. Chen. 2013. Thermodynamic Modeling of the Co-Fe-O System. *CALPHAD Computer Coupling Phase Diagrams Thermochem.* 41: 76–88.

Index